T0396138

Polymer Nanocomposites Containing Graphene

Woodhead Publishing Series in Composites Science and Engineering

Polymer Nanocomposites Containing Graphene

Preparation, Properties, and Applications

Edited by

Mostafizur Rahaman

Lalatendu Nayak

Ibnelwaleed A Hussein

Narayan Chandra Das

Editor-in-Chief

Professor Costas Soutis

Series Editor

Professor Suresh G. Advani
Professor Leif Asp
Professor Yuris A. Dzenis
Professor Ing. Habil. Bodo Fiedler
Professor Adrian Mouritz
Professor Chun H. Wang

Woodhead Publishing is an imprint of Elsevier
The Officers' Mess Business Centre, Royston Road, Duxford, CB22 4QH, United Kingdom
50 Hampshire Street, 5th Floor, Cambridge, MA 02139, United States
The Boulevard, Langford Lane, Kidlington, OX5 1GB, United Kingdom

British Library Cataloguing-in-Publication Data
A catalogue record for this book is available from the British Library

Library of Congress Cataloging-in-Publication Data
A catalog record for this book is available from the Library of Congress

ISBN: 978-0-12-821639-2 (print)

ISBN: 978-0-12-821640-8 (online)

For information on all Woodhead Publishing publications
visit our website at https://www.elsevier.com/books-and-journals

Publisher: Matthew Deans
Acquisitions Editor: Gwen Jones
Editorial Project Manager: Megan Healy
Production Project Manager: Surya Narayanan Jayachandran
Cover Designer: Alan Studholme

Working together
to grow libraries in
developing countries
www.elsevier.com • www.bookaid.org

Typeset by MPS Limited, Chennai, India

Contents

List of contributors

Umer Abid Department of Polymer and Process Engineering (PPE), University of Engineering & Technology (UET), Lahore, Pakistan

Ajitha Achuthanunni Department of Chemical Engineering, Indian Institute of Science Education and Research Bhopal, Bhopal, India

Monalisa Adhikari Department of Polymer Science and Technology, University of Calcutta, Kolkata, India

Mir Sahidul Ali Department of Polymer Science and Technology, University of Calcutta, Kolkata, India

Susanta Banerjee School of Nanoscience and Technology, Indian Institute of Technology, Kharagpur, India; Materials Science Centre, Indian Institute of Technology, Kharagpur, India

Biswajit Bera Department of Chemical Engineering, Indian Institute of Technology Kharagpur, Kharagpur, India

Purabi Bhagabati School of Chemistry, The Centre for Research on Adaptive Nanostructures and Nanodevices (CRANN), Trinity College Dublin, Dublin, Ireland

Subhendu Bhandari Department of Plastic and Polymer Engineering, Maharashtra Institute of Technology, Aurangabad, India

Swarup Krishna Bhattacharyya School of Nanoscience and Technology, Indian Institute of Technology, Kharagpur, India

Dipankar Chattopadhyay Department of Polymer Science and Technology, University of Calcutta, Kolkata, India; Center for Research in Nanoscience and Nanotechnology, Acharya Prafulla Chandra Roy Sikhsha Prangan, University of Calcutta, Kolkata, India

Mahuya Das Greater Kolkata College of Engineering and Management, Baruipur, India

Narayan Chandra Das Rubber Technology Centre, Indian Institute of Technology Kharagpur, Kharagpur, India

Paramita Das Department of Chemical Engineering, Indian Institute of Science Education and Research Bhopal, Bhopal, India

Poushali Das School of Nanoscience and Technology, Indian Institute of Technology, Kharagpur, India; Bar-Ilan Institute for Nanotechnology and Advanced Materials and Department of Chemistry, Bar-Ilan University, Ramat-Gan, Israel

Tushar Kanti Das Rubber Technology Centre, Indian Institute of Technology Kharagpur, Kharagpur, India

Ayan Dey Indian Institute of Packaging, Mumbai, India

Sayan Ganguly Department of Chemistry, The Institute of Nanotechnology & Advanced Materials, Bar-Ilan University, Ramat Gan, Israel

Prosenjit Ghosh Center for Carbon Fiber and Prepregs, CSIR-National Aerospace Laboratories, Bangalore, India

Suman Kumar Ghosh Rubber Technology Centre, Indian Institute of Technology Kharagpur, Kharagpur, India

Yasir Qayyum Gill Department of Polymer and Process Engineering (PPE), University of Engineering & Technology (UET), Lahore, Pakistan

Prashant Gupta Department of Plastic and Polymer Engineering, Maharashtra Institute of Technology, Aurangabad, India

Abdulrehman Ishfaq Department of Polymer and Process Engineering (PPE), University of Engineering & Technology (UET), Lahore, Pakistan

Chun-Won Kang Department of Housing Environmental Design and Research Institute of Human Ecology, College of Human Ecology, Jeonbuk National University, Jeonju, Republic of Korea

Haradhan Kolya Department of Housing Environmental Design and Research Institute of Human Ecology, College of Human Ecology, Jeonbuk National University, Jeonju, Republic of Korea

Abhilash J. Kottiyatil Center for Carbon Fiber and Prepregs, CSIR-National Aerospace Laboratories, Bangalore, India

Pankaj Kumar Department of Chemical Engineering, Indian Institute of Science Education and Research Bhopal, Bhopal, India

Roop Singh Lodhi Department of Chemical Engineering, Indian Institute of Science Education and Research Bhopal, Bhopal, India

Umer Mehmood Department of Polymer and Process Engineering (PPE), University of Engineering & Technology (UET), Lahore, Pakistan

Subhadip Mondal Department of Polymer-Nano Science and Technology, Jeonbuk National University, Jeonju, Republic of Korea

Mosongo Moukwa Sushila Goenka R&D Centre, Phillips Carbon Black Ltd, Bharuch, India

Changwoon Nah Department of Polymer-Nano Science and Technology, Jeonbuk National University, Jeonju, Republic of Korea

Muhammad Baqir Naqvi Department of Polymer and Process Engineering (PPE), University of Engineering & Technology (UET), Lahore, Pakistan

Krishnendu Nath Rubber Technology Centre, Indian Institute of Technology Kharagpur, Kharagpur, India

Lalatendu Nayak Sushila Goenka Research & Development Centre, Phillips Carbon Black Limited, Palej, India; Sushila Goenka R&D Centre, Phillips Carbon Black Ltd, Bharuch, India

Suryakanta Nayak Department of Electrical and Computer Engineering, National University of Singapore, Singapore

Jonathan Tersur Orasugh Department of Polymer Science and Technology, University of Calcutta, Kolkata, India; Department of Jute and Fiber Technology, Institute of Jute Technology, University of Calcutta, Kolkata, India; Center for Research in Nanoscience and Nanotechnology, Acharya Prafulla Chandra Roy Sikhsha Prangan, University of Calcutta, Kolkata, India

Srinivas Pagidi Institute of Quantum Systems and Department of Physics, Chungnam National University, Daejeon, Republic of Korea

Chandrika Pal Department of Polymer Science and Technology, University of Calcutta, Kolkata, India

P. Porkodi Center for Carbon Fiber and Prepregs, CSIR-National Aerospace Laboratories, Bangalore, India

Mostafizur Rahaman Department of Chemistry, College of Science, King Saud University, Riyadh, Saudi Arabia

Sanjay Remanan Rubber Technology Centre, Indian Institute of Technology Kharagpur, Kharagpur, India

Banalata Sahoo Department of Chemistry, Binayak Acharya College, Berhampur, Odisha, India

B.L.N. Krishna Sai School of Mechanical Engineering, VIT-AP University, Amravati, India

Sheeja Sunil Center for Carbon Fiber and Prepregs, CSIR-National Aerospace Laboratories, Bangalore, India

Pankaj Tambe School of Mechanical Engineering, VIT-AP University, Amravati, India

Monica Tanniru School of Mechanical Engineering, VIT-AP University, Amravati, India

Rajesh Theravalappil Center for Refining and Petrochemicals, Research Institute, King Fahd University of Petroleum and Minerals, Dhahran, Saudi Arabia

B.G. Toksha Basic Sciences and Humanities Department, Maharashtra Institute of Technology, Aurangabad, India

Preface

Nowadays, graphene-polymer composites are of great interest because of its practical application point of views. Hence, in this book, a discussion has been made on the different properties and applications of polymer-graphene nanocomposites. Initially, we have focused on synthesis techniques of graphenes, its types, and different processes of its modification/functionalization. Among different types of graphenes, neat graphene, graphene oxide, reduced graphene oxide, graphene nanoplatelet, chemically reduced graphene oxide, and thermally reduced graphene oxide are of great interest. A brief concept has been provided on polymer nanocomposites. The nanocomposites had been prepared by physical and covalent mixing methods. Physical mixing method includes melt blending, solution mixing, in-situ polymerization, latex phase blending, and electro-polymerization. Covalent mixing followed the grafting technology. Different properties of polymer-graphene composites are reported in detail, which include mechanical properties, electrical conductivity, dielectric properties, thermal properties, thermal conductivity, rheological properties, and electromagnetic interference shielding effectiveness. Another topic of discussion is the dispersion state of graphene within the polymer matrices govern the physical properties of the composites. The structure of graphenes and polymers also govern the different properties of composites. Graphenes can be used as a reinforcing agent for thermoset resins as discussed herein. In this book, the published patents based on polymer-graphene composites are reported.

The polymer-graphene composites are used in electrical and electronic fields. It is used in transistors, as supercapacitors, and as the radiation protection materials. These composites are also strong enough to be used in structural and load bearing fields. Discussions have been made for its applications as sensors, in fuel cell and solar energy, and in medical fields. As a sensor material, it includes sensing of motions, dopamine, glucose, ammonia, hydrazine, oxygen, nitrogen dioxide, LPG, methanol, VOCs, DNA, temperature, pH, near infra-red light, magnetic field etc. In bio-medical field, it is used in drug delivery, tissue engineering, antibacterial study, as biosensors, flexible supercapacitors, and in body implants. In addition, some other applications of these composites in catalysis hydrogen production, aerospace engineering, as membranes, anticorrosive materials, packaging materials, flame and fire retardant materials have been discussed separately in details.

In this book, discussions have been made on the synthesis of graphene, nanocomposites, preparation methods of polymer-graphene nanocomposites, their properties, related applications, and patented work files up to now. Hence, this book will have great interest to the materials scientists, research communities, industry people, and academicians. In the future, this book can be used as a text book in polymer science and technology departments, materials science department, nano-technology department, etc.

Mostafizur Rahaman[1], Lalatendu Nayak[2], Ibnelwaleed A. Hussein[3] and Narayan Chandra Das[4]

[1]Department of Chemistry, College of Science, King Saud University, Riyadh, Saudi Arabia, [2]Sushila Goenka Research & Development Centre, Phillips Carbon Black Limited, Baroda, India, [3]Gas Processing Center, Qatar University, Doha, Qatar, [4]Rubber Technology Centre, Indian Institute of Technology Kharagpur, Kharagpur, India

Synthesis/preparation and surface modification/functionalization of graphene, and concept of nanocomposites

1

Lalatendu Nayak[1], Mostafizur Rahaman[2] and Mosongo Moukwa[1]
[1]Sushila Goenka R&D Centre, Phillips Carbon Black Ltd, Bharuch, India, [2]Department of Chemistry, College of Science, King Saud University, Riyadh, Saudi Arabia

1.1 Introduction

Recently, academic as well as industrial attention has turned to structural and electronic properties of carbon-based nanomaterials. Among the carbon nanomaterials, now graphene is the hottest topic in condensed-mater physics and material science. Graphene, a two-dimensional (2D) carbon nanomaterial was discovered in 2004 by A.K. Geim and K.S. Novoselov. First graphene synthesis process was published in October 2004 by K.S. Novoselov and A.K. Geim [1]. Graphene was prepared by a sticky-tape method. Since then, graphene research has significantly increased in terms of published research articles. In 2010, Andre Geim and Konstantin Novoselov on Nobel Prize in Physics. Commonly graphene is a single layer of graphite comprising carbon atoms packed in the form of a hexagonal rings in a two-dimensional sheet just one atom thick. This is considered as thinnest material in the universe. This unique structure gives graphene some surprising physical properties—more than 100 times stronger than steel, electrical conductivity higher than silver, lightest known material (1-meter square weighing approximately 0.77 mg), and extremely flexible. Because of excellent electronic, physical, and thermal properties, non-toxicity, highly chemical and thermal tolerant, graphene has been considered as a promising candidate for a wide range of industrial applications like structural and electronic components, batteries/capacitor, adsorbents, catalyst support, thermal transport media, and even application in biotechnology. It is also being a suitable material for energy technologies such as fuel cells, solar cells, hydrogen storage, batteries, and capacitors, filed effect transistor and transparent electrodes. To synthesize graphene, several approaches have been applied and some approaches also have been utilized to produce graphene in commercial scale. Despite strong interest and rapid progress in research on graphene and graphene related materials, there is still a long way for the widespread implementation of graphene. Primary difficulty is quantity and quality imbalance, but there's also issues

© 2022 Elsevier Ltd. All rights reserved.

with large-scale production while maintaining high quality, high yield, and controllable tuning of bandgap of graphene.

This chapter deals with different methods used for graphene synthesis, advantages and disadvantages of different graphene synthesis techniques, market scenario based on commercial graphene production and the methods used for commercial production. This chapter aims to highlight some of the key challenges to produce graphene/graphene oxides (GO) in large scale This chapter also includes different surface functionalization processes of graphene based on different applications [1].

1.2 Scope, challenges, and scale of graphene market

Because of its excellent mechanical, electrical, and thermal properties, graphene has been accepted in broad application areas by the academic, industry as well as others. Large-scale production while maintaining a consistent quality is a great challenge for graphene commercialization and market uptake. The scale of graphene production is very small relative to other carbon nanomaterials like carbon fibers (CF), carbon nanotubes (CNTs) as shown in Fig. 1.1.

To produce graphene in commercial scale with high quality and yield, strong efforts have been given from the academic industry to others. A list of large-scale manufactures, process to manufacture, and targeted industrial applications are presented in Table 1.1.

From the table it is observed that number of manufacturers are available to produce graphene and graphene oxides. Maximum manufacturers are using liquid intercalation

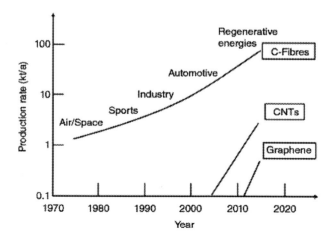

Figure 1.1 Annual production rates of carbon fibers, carbon nanotubes and graphene, 1970–2014. The industries in which carbon fibers found a market and approximate uptake times are also plotted [2].

Table 1.1 List of large-scale manufactures, manufacture processes, and targeted industrial applications [3].

Graphene manufacturers	Synthesis method	Graphene properties	Production capacity	Applications
Angstron Materials (USA)	Liquid exfoliation	Pristine nanographene plateletsThickness: <100 nmCarbon content: ≥95% (for product under three layers in average thickness)	300 tonnes/y	Graphene/silicon anode materials
Thomas Swan (UK)	Liquid exfoliation	Graphene nanoplateletsThickness: 5–7 layers (average)Size: 0.5–2 µmSheet resistance: $10 \pm 5\,\Omega^{-2}$ (25-µm film, equal to 27–80 S/cm) Impurity: ∼8 wt.%	kg/day(September 2014)	Conductive inks and coatings, graphene rubbers, flexible battery straps
Vorbeck Materials (USA)	Oxidation-thermal exfoliation	Functionalized grapheneThickness: mostly 1–3 layersMorphology: wrinkled	40 tonnes/yOctober 2012)	
Sixth Element Materials(China)	Oxidation–exfoliation–reduction	Graphene oxide, reduced graphene oxideThickness: ≤10 nmCarbon content: ≥95%Electrical conductivity: ≥2 S/cm (for electrical conductive-type product)	100 tonnes/y (November 2013)	Mechanical/thermally enhanced composites, anticorrosion coatings
XG Sciences (USA)	Intercalation–exfoliation	Graphene nanoplateletsThickness: 2–15 nm (average) Electrical and thermal conductivity: 3800 S/cm and 500 W/mK, respectively (for 30-µm-thick graphene paper product)	80 tonnes/y(August 2012)	Graphene/silicon anode composite, supercapacitor electrode materials, conductive inks and coatings, graphene paper (for thermal spreading, and electrically conductive applications, and so on)

Table 1.1 (Continued)

Graphene manufacturers	Synthesis method	Graphene properties	Production capacity	Applications
Ningbo Morsh (China)	Intercalation–expansion–exfoliation	Graphene nanoplateletsAverage thickness: 3 nm	300 tonnes/y (December 2013)	Graphene conductive additives and graphene-coated current collector for lithium ion batteries, conductive inks, heat-radiating coatings, anticorrosion coatings, thermally/electrically conductive master batch
DeyangCarboneneTech (China)	Intercalation–expansion–exfoliation	Graphene sheets (few-layer) Thickness: $\leqq 10$ layersElectrical conductivity: ~ 1000 S/cm (for ~ 15-μm-thick membrane)	1.5 tonnes/y (October 2012)300 tonnes/y (2017–19)	Battery materials, thermal management materials, conductive inks, conductive anticorrosion coatings
Bluestone Global Tech (USA)	Chemical vapor deposition (CVD)	Graphene films on Cu, SiO$_2$/Si, polyethylene terephthalate (PET) Film on PET: available maximum size 8×10 inch^2, $<30\,\Omega^{-2}$ at $>85\%$ transmittance (excluding substrate), $<800\,\Omega^{-2}$ at 95% transmittance (excluding substrate)Film on SiO$_2$/Si: available maximum size 4-inch wafer, monolayer $>95\%$, average Hall mobility: $2000{-}4000$ cm^2/V/s		Field-effect transistors, touch panels

2D CarbonTech (China)	CVD	Graphene films on Cu, SiO_2/Si, glass, PETFilm on PET: available maximum size 450 × 550 mm^2, monolayer >90%, 200—400 Ω^{-2} at >85% transmittance (including substrate)	30,000 m^2(May 2013) 200,000 m^2 (December 2014)	Touch panels
WuxiGraphene Film (China)	CVD	Graphene films on Cu, PETFilm on PET: ~600 Ω^{-2} at >97% transmittance (excluding substrate)	80,000 m^2(December 2013)	Touch panels, touch sensors (5 million pieces, December 2013)
Power booster (China)	CVD	Graphene films on Cu, PETFilm on Cu: available maximum size 7.5 m^2 (2013)Film on PET: 50—140 Ω^{-2} at 95.5% transmittance		Touch panels

and exfoliation process for conductive ink and coatings, thermally/mechanical enhanced composites, and for battery applications. Some manufacturers are synthesizing graphene film by chemical vapor deposition (CVD) process for different electronics applications like field effect transistors, touch panels, and touch sensors.

1.3   Synthesis of graphene

Several methods have been reported for the synthesis of graphene. The synthesis methods are divided into two main categories: the top-down approach and the bottom-up approach. Graphite is a stack of graphene layers combined by strong Van der Waals force of attraction. Through top-down processes, graphene layers in graphite are separated individually by applying forces higher than the existing Van der Waal forces. Top-down approaches focus on breaking of graphene precursor (graphite) into atomic layers from stack by mechanical, thermal, chemical, or electrochemical process. Graphenes are synthesized from graphite by applying force higher than the Van der Waal forces between individual sheets of graphene in graphite. Several challenges in top-down approaches are surface defects that occur during sheet separation and re-agglomeration of separated graphene sheets.

In bottom-up approaches carbon molecules from different sources are used as building blocks to build graphene layers. Through this approach, defect free and high pure graphenes are synthesized, but the main challenges of this approach are low yield, very low production, and high setup installation cost.

1.3.1   Top-down approaches

Different top-down approaches are.

1. Mechanical exfoliation
2. Chemical intercalation and exfoliation
3. Electrochemical exfoliation
4. Liquid phase exfoliation
    a. Shear exfoliation in liquid
    b. Microfluidization

1.3.1.1   Mechanical exfoliation

Graphite is composed of number of graphene layers stacked by weak Van der Waals force of interaction with inter layer space 0.34 nm. The bond between layers is broken by mechanical or chemical energy approaches and individual graphene sheets were separated. Chemical molecules having size slightly higher than the graphite interlayer space are interacted in between graphene layers. These interacted molecules create the interlayer space and then weaken the force of interaction, which leads to exfoliate the graphene layers from graphite. This type of work was first started by Viculis et al. [4], who had intercalated potassium ions in graphite by heating potassium hydroxide with graphite in a Pyrex tube at 200°C, then

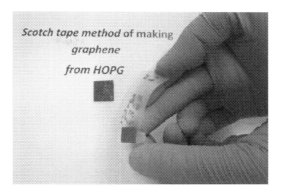

Figure 1.2 Mechanical exfoliation of graphene using scotch tape [6].

intercalated in ethanol by sonication. During sonication carbon nanosheets having 40 ± 15 layers were formed and curled to form nano scrolls. Although prepared nano scrolls were much thicker than few layers graphene (FLG), it gives an idea how to separate graphene layers from graphite.

Mechanical/micromechanical exfoliation is the first technique to produce graphene from graphite precursor. This technique is the key technique for researchers to get high quality thin graphene film from graphite. Through this technique Geim and Konstantin Novoselov produced graphene [1] and were awarded Nobel Prize in Physics in 2010. They peel out graphene layers repeatedly from graphite by using a scotch tape as shown in Fig. 1.2. This tape containing transparent graphene flakes was dissolved in acetone, then after few steps, monolayer and multilayer graphenes were deposited on silicon wafer for microscopic study and obtained free standing graphene plane of one atomic thickness. Prior to this study, the thought was that free standing atomic plane does not exist [1]. Due to irregularity in film thickness, high production cost, and very low yield, this technique is not suitable for commercial production. Jayasena et al. [5] used a different type of mechanical approach to obtain graphene from graphite, where they used an ultra-sharp single crystal diamond wedge to cleave a highly ordered pyrolytic graphite sample to generate the graphene layers. The diamond wedge was mounted on an ultrasonic oscillation system capable of providing tunable frequencies with an amplitude of vibration of a few tens of nanometers. The principle of this process is similar to the process of Novoselvo et al. [1], but due to the ability to control frequency of oscillations and contact pressure, this process allows more control to get consistent properties of graphene. This method has the potential to provide high quality graphene for experimental purposes but does not seem suitable for being scaled up.

1.3.1.2 Chemical intercalation and exfoliation

Chemical intercalation and exfoliation are a widely used process by the academic as well as industries for the production of GO and graphene. Staudenmaier et al. [7] and Hummers et al. [8] processes are basically used for the synthesis of graphene and graphene oxides. To increase quality and quantity of graphene and graphene oxides, Staudenmaier and Hammers processes have been modified. Hummers process is better than Staudenmaier process on safety and yield point of view. Here modified Hummers process has been described for the synthesis of graphene from graphite precursors. This process is accepted by different industries. Different steps to produce graphene by this process are mentioned below.

Step 1: Selection of graphite precursor [Natural graphite (NG), expandable graphite (EG), or Microcrystalline graphite (MG)]

Step 2: Oxidation (reaction media: H_2SO_4, $NaNO_3$, $KMnO_4$)

Step 3: Reaction quenching (H_2O, H_2O_2)

Step 4: Purification of GO (HCL, water or other solvents)

Step 5a: Exfoliation and delamination or called reduction of GO

Step 5b: Wastewater treatment

Step 6: Drying or dispersion of graphene

Step 1: Selection of graphite precursor

Quality and yield of graphene or GO depends on the type of graphite precursors. Different type of graphite precursors like flake natural graphite (NG), EG, and microcrystalline graphite (called amorphous graphite) are used. EG is a pretreated graphite prepared by treating natural graphite flake by an intercalating agent. EG is oxidized easily compared to non-expanded graphite. Graphene prepared from EG reduces synthesis time and increases the yield, but it has limitation like one additional step and cost for graphite expansion. GO can also be synthesized from MG or amorphous graphite (lack of large and stacked lamellar graphene sheets). Due to disordered stacking of graphene layers, MG can be intercalated easily by H_2SO_4. A schematic presentation of intercalation and exfoliation of different types of graphite are given in Fig. 1.3. Dimiev et al. [9] in their study have shown that the difference in graphite flake morphologies causes the difference in intercalant diffusion rate and different in degree of oxidation (Fig. 1.3). From the same source of graphite, small size flakes are oxidized significantly faster than large flakes. In their study, they have also stated that highly crystalline graphite samples are converted to GO at a slower rate when compared to the more disordered samples. The same

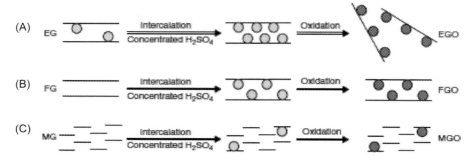

Figure 1.3 Intercalation and oxidation of different form of graphite: (A) Expanded graphite (EG), (B) Flake graphite (FG), and (C) Microcrystalline graphite (MG) [3,10].

crystalline flake graphite is oxidized significantly faster if it is subjected to thermal expansion prior to oxidation.

Step 2: Oxidation and intercalation

NG or pretreated graphites are oxidized by different oxidizing agents and intercalated by intercalating agent to provide GO. Different types of oxidizing agents and intercalating agent used for graphene oxides are mentioned below.

Oxidizing agent:

- Nitric Acid (HNO_3)
- Potassium Nitrate (KNO_3)
- Hydrogen Peroxide (H_2O_2)
- Potassium Permanganate ($KMnO_4$)
- Potassium Dichromate ($K_2Cr_2O_7$)
- Sodium Periodate ($NaIO_4$)
- Sodium Chlorate ($NaClO_3$)

Intercalating agent:

- H_2SO_4—Intercalated in between graphene layers in graphite and adds functional groups in the presence of oxidizer.

In order to intercalate the intercalating agent in between the graphene layers of graphite, it is necessary to oxidize the starting graphite sheets. A small amount of oxidant is used to initiate the electron transfer reaction. By the oxidation, the edges of graphite sheets are opened and the charges (negative or positive depending upon the oxidizing agent) are generated. These charges create repulsion between the graphite layers and increase their relative distance up to around 7.98°A. This increase in interlayer gap allows intercalating agents to enter in between graphene layers of graphite. During sulfuric acid intercalation, graphite bisulfates are formed [9−12]. A schematic presentation (Fig. 1.4) has been given for electron transfer reaction and intercalation of graphite, where electrons from graphite carbon is transferred to the oxidizing agent and charge balance is maintained by proton hopping through the hydrogen bond network of the sulfuric acid.

$$C_X + HSO_4^- \rightarrow C_X^+ HSO_4^- + e^- \tag{1.1}$$

A schematic presentation of oxidation and intercalation of graphite has been shown in Fig. 1.5, where graphite layers are intercalated with both bisulfate anion and sulfuric acid molecules. Intercalated graphite remained in ionic form having carbon as cation and bisulfate as anion.

Depending on the nature of the intercalating agent, the type of oxidant, and the experimental conditions, it is possible to obtain intercalated graphite of different stages as "Stage-I," "Stage-II," "Stage-III," etc. [14]. As shown in Fig. 1.6, Stage-I represents a single layer of graphene is alternated regularly with intercalated species. In a stage—II and stage—III, two graphene layers and three graphene layers are separated by intercalating agent respectively [15].

Due to high oxidizing efficiency in acidic medium, environmentally friendly, lower price, $KMnO_4$ has achieved great importance as an oxidizing agent to

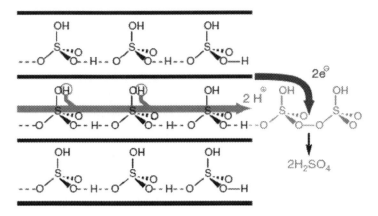

Figure 1.4 Sketch of proposed oxidation mechanism [13].

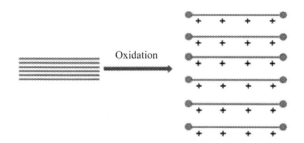

Figure 1.5 Oxidation graphite scheme.

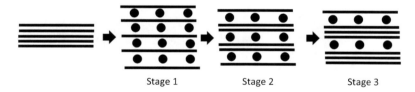

Stage 1   Stage 2   Stage 3

Figure 1.6 Different type of intercalation stage by Rudorff model.

produce graphene oxide both by industries and academics. Lee et al. [16] in their study have proposed an oxidation and intercalation process to prepare graphene in large scale (2 kg/day). In their study, during oxidation process, graphite was mixed with cold sulfuric acid and then KMnO₄ was slowly added to the suspension of graphite for oxidation. KMnO₄ is added in small quantity at different batches. Green color of solution is disappeared after complete reaction of added KMnO₄. Initially, small quantity of KMnO₄ is added to the solution. After complete reaction of added KMnO₄, the green color disappeared. Then another small amount of

$KMnO_4$ is added. In similar way all the calculated amount of $KMnO_4$ is added to the solution. The end of the oxidation is always determined by the disappearance of the color after each $KMnO_4$ addition. The disappearance of green color of $KMnO_4$ in H_2SO_4 indicates additional $KMnO_4$ requirement for graphite intercalation. The reaction initially produces permanganic acid $HOMnO_3$, which is dehydrated by cold sulfuric acid to form its anhydride, Mn_2O_7 and dark green oil type solution is formed.

$$2KMnO_4 + 2H_2SO_4 \rightarrow Mn_2O_7 + H_2O + 2KHSO_4 \tag{1.2}$$

Mn_2O_7 is a highly reactive oxidizing agent and started decomposition at temperature above—$10°C$. During decomposition ozone gas is also produced as a by-product. As ozone is a highly oxidizing agent, both ozone and Mn_2O_7 might be involved in the oxidation of graphite [17]. Graphite oxidation is a highly exothermic and temperature is increased up to $60°C$ when reactor is removed out from ice bath. If reaction temperature will be above $50°C$, Mn_2O_7 will be severely exploded. Hence during oxidation and intercalation temperature of reaction system should be as low as possible ($0°C$—$10°C$).

$$4Mn_2O_7(T > -10°C) \rightarrow 8MnO_2 + 2nO_3 + (6 - 3n)O_2(0 < n < 2) \tag{1.3}$$

Mn_2O_7 can react further with sulfuric acid to give the remarkable manganyl (VII) cation $MnO^+{}_3$.

$$Mn_2O_7 + 2H_2SO_4 \rightarrow 2[MnO_3]^+[HSO_4]^- + H_2O \tag{1.4}$$

During oxidation and intercalation, number of reactions like Eqs. (1.1–1.3) are carried out and in all cases bisulfates are formed for intercalation and the graphite layers are intercalated and oxidized. During oxidation step, both oxidation of peripheral carbon atoms of graphite and intercalation take place. After complete reaction, the solution changed from dark green to brown.

Step 3: Reaction quenching (H_2O, ice, H_2O_2)
A higher degree of oxidation creates flaws in graphene layers. So, to stop excess oxidation, the reaction mixture is quenched by addition of small amount of H_2O, where Mn_2O_7 is quenched in the form of following equation.

$$Mn_2O_7 + H_2O \rightarrow 2HMnO_4 \tag{1.5}$$

Dilution of H_2SO_4 is highly exothermic. To avoid explosion, ice water is added to the solution and solution is externally cooled using ice water.

As Mn_2O_7 decomposes to water insoluble MnO_2, to remove all manganese impurities, MnO_2, and MnO_4^-, some amount of H_2O_2 is added dropwise to the solution until the color of solution changes from brown to brilliant yellow and no more bubbles are formed in the solution. H_2O_2 reduces manganese oxides to water soluble Mn^{2+} as given in following equation described by Hummer et al. [8].

$$MnO_2 + H_2O_2 + 2H^+ \rightarrow Mn^{2+} + O_2 + 2H_2O \tag{1.6}$$

$$2MnO_4^- + 5H_2O_2 + 6H^+2 \rightarrow Mn^{2+} + 5O_2 + 8H_2O \tag{1.7}$$

Oxygen development during the above reactions generates sever foaming in reaction mixture. So slow addition of peroxide with constant stirring and enough free space in reaction vessel are required. The oxidation of graphite with $KMnO_4$ results in the formation of manganate ester which will create a vicinal diol. If left unprotected, the vicinal diol may be oxidized to diketone, which leads to the formation of holes in the graphene basal plane. During oxidation process, nitric acid as oxidant and $NaNO_3$ as an oxidation catalyst has been used by a number of researchers, but due to formation of different toxic environmental polluted gases like NOx, use of this type of oxidant may be avoided.

Step 4: Purification of GO (HCL, water or other solvents)

Purification of GO is important to ensure stability and safety of the final product. In many downstream applications, even very small amounts of impurities are not tolerated. The presence of potassium increases the flammability of GO. When it goes thermal reduction, GO can violently combust in the presence of air and potassium [18]. Therefore GO slurry is thoroughly washed after oxidation using press filter. Initially GO slurry is washed 2−3 times by water, then washed with dilute HCl (1 M) or in a volumetric ratio of 1:10 (HCl:H_2O) 1−2 times to remove metallic impurities. Acid washed GO is again washed with water until pH of solution reaches to 7.

A challenge for GO filtration is that GO tends to gelate in aqueous media. This can greatly complicate certain types of filtration (e.g., gravity and vacuum-assisted filtration), as the filter cake can block the flow of solvent. One method of overcoming this is to use acetone instead of water after washing with HCl, as graphite oxide does not gelate in acetone [18,19].

Removal of non-oxidized graphite:

In order to achieve good quality single to few layer graphene (FLG) or GO, non-oxidized graphite flakes should be removed from GO slurry. On laboratory scale basically centrifugation process is used. For industrial production continuous online centrifugation process can be used. Another alternative to centrifugation is the use of settling tank where denser graphite flakes are settled down to the bottom of tank. Important factor during settling down in tank is to maintain the pH of the solution and concentration of ionic salts. If the pH of the solution is too low and acidic in nature or the salt concentration is too high, GO will precipitate out and settle. Hence, in order to remove the unoxidized graphite from the GO mixture, the ideal stage might be after the ion/salt purification stage [20].

Drying:

In order to get GO powder, it is important to dry the washed GO, and it is a challenging factor in the case of industrial production. In literature it is observed that GO is freeze—dried in order to create a dry powder, but vacuum freeze-drying is expensive on an industrial scale. Peng et al. [21] in their study reported that commercially available freeze-dried GO is very difficult to disperse. The freeze-drying process may lead to partial restacking of the GO sheets and create aggregates. Spray drying process may solve this type of problem observed in freeze drying process [21]. Industrial scale spray dryers are available and can be cost-effective GO drying.

Exfoliation and delamination:

After oxidation of graphite flakes, the product becomes graphite oxides which is the combination of mono to multilayers graphene oxides. Ultrasonication is used to delaminate the graphite oxides to single or very few layers GO. This sonication is a time-consuming process, so it is not suitable for industrial scale production. Microfluidization process may be suitable process for graphite oxide delamination. Thermal exfoliation is also suitable for industrial production of graphene, where graphite oxides are reduced by thermal shock at a temperature of around 1000°C for 30—60 s. During this thermal shock, a violent reaction takes place between carbon of the graphite and the bisulfate ions present in the graphite inter-layers. This reaction occurs with the release of different gases (reaction 1.8) like carbon dioxide (CO_2), sulfur dioxide (SO_2), and water vapor (H_2O). This gas release increases the distance between graphite layers and decreases the Van der Waals force of attraction, which in turn delaminates the graphite layers and produces graphene.

$$C + 2H_2SO_4 \rightarrow CO_2 + 2SO_2 + 2H_2O \tag{1.8}$$

1.3.1.3 Electrochemical exfoliation

Graphene synthesized through chemical or mechanical based routes have several processing limitations like time consuming laborious process, use of environment and health hazardous solvents/reagents, and quality inconsistency. For high quality graphene production some processes like micromechanical cleavage, thermal decomposition of silicon carbide (SIC), and CVD have been approached, but they are impractical for commercial applications due to low production rate and high cost.

Electrochemical approach has been reported as a suitable approach for the mass production of graphene at lesser time and lower cost. This process does not use any harsh chemicals and is a safer process A list of electrochemical approaches to prepare graphene has been shown in Table 1.2. It is a process where graphite electrode is exfoliated in a liquid electrolyte by the application of electrical current as shown in Fig. 1.7. In this process one electrode is considered as working electrode which is typically a graphite rod/film/highly orientated pyrolytic graphite sample. Another counter electrode may be from same graphite/platinum/iron/nickel etc. When potential difference is generated between two electrodes, positively charged ions from electrolyte migrates towards the working graphite electrodes and are intercalated among the graphene layers of graphite. This intercalation process provides driving force to break the van der Waals forces, leading to the structural expansion of graphite. Ion intercalations are also expected to be influenced by the nature of graphite electrodes like graphite particulate sizes, defects, layer arrangement, thickness, and suitable pretreatment. No. of layers, defect density, oxygen content, and lateral size of prepared graphene are controlled by different processing parameters like applied current, electrical potential, type of electrolyte, graphite thickness, electrolyte concentration, reaction time, even also distance between two electrodes, and electrolyte temperature.

Table 1.2 List of electrochemical approaches to prepare grapheme.

Electrolytes	Graphite electrodes	Power supplies	Additional conditions	Properties of preparedgraphene C/O ratio	Size (μm)	No. of layers	I_D/I_G	Yield (%)	References
0.1 M K$_2$SO$_4$	Graphite foil	Anodic, +10 V DC	Graphite powder, HOPG	17.2	0.2–0.6	1–2 nm	0.38		[22]
0.2 M (NH$_4$)$_2$SO$_4$	Graphite rod	Anodic, +10 V, DC	+2 V, DC; sonication	11.7	0.5–1	2	1.76		[23]
10 M H$_2$SO$_4$	Expanded graphite	Anodic, +1 and +2 V, DC		7.5	10–30	2–4	0.3		[24]
0.1 M (NH$_4$)$_2$SO$_4$	Graphite flake	Anodic, +10 V, DC	Packed in dialysis tubing	4.98	>30	2–7 nm	0.9/1.2		[25]
0.5 M H$_2$SO$_4$	Graphite foil	Anodic, +10 V, DC	Pretreatment in 1 M NaOH	11	2.2	2–4	0.29		[26]
0.1 M TBA·HSO$_4$, NaOH	Graphite sheet	Both, 10 V, 0.1 Hz AC		21.2	5	1–3	0.15		[27]
0.1 M C$_6$H$_5$COOH	Graphite rod	+3 and −1.5 V (20 s)	Pulse electrolysis	25	2–5	13	0.28		[28]
0.5 M (NH$_4$)$_2$SO	Graphite foil	Anodic, +30 V, DC	Lifting electrolyte	8	0.3–5	<3	0.8		[29]
0.1 M H$_2$SO$_4$	HOPG	Anodic, +1 to 10 V, DC	Shear field (up to 74,400 s^{-1})	NA	10	1–3	0.21–0.32		[30]

Electrolyte	Electrode	Process	Notes						Ref.
0.01–5 M $(NH_4)_2SO_4$	Graphite flake	Anodic, +10 V, DC	Compare different salts, 3–5 min treatment time	17.2	5	1–3	0.42	>85	[31]
0.1 M $(NH_4)_2SO_4$	Graphite rod	Anodic, +10 V, DC	Inkjet printing ink	NA	0.5–0.65	2 nm	NA		[32]
0.05 M NaCl	Graphite foil	Anodic, +10 V, DC		16.7	0.6	2–3 nm	0.8		[33]
0.1 M Na_2SO_4, 0.1 M NaCl	Graphite foil	Anodic, +10 V, DC		33.3–50	1	<5	0.21		[34]
0.5 M Na_2SO_4	Graphite foil	Anodic, +20 V, DC	0.05 M $CoSO_4$	36	NA	1 and few	0.05		[35]
Sulfonate-/sulfatehydrocarbons (e.g., 0.2 M SNDS)	Graphite foil	Anodic, +10 V, DC		50		2.5 nm	0.2		[36]
1 M $Na_2S_2O_3$, 0.5 M $NaClO_4$,	Graphite sheet	Anodic, +10 V, DC	0.11 A/cm^2. sonication	27.7	4	4–6	0.35		[37]
13.5 wt.% H_2SO_4, 30 wt.% KOH	Graphite plate	Both, 15 V DC		NA	NA	6	0.061		[38]
0.2 M H_2SO_4, 30 wt.% KOH	Graphene flakes	Both, 10–15 V, 1 A, 0.2 Hz AC		NA	3.7	2.5 nm	NA		[39]
0.05 M $(NH_4)_2SO_4$, 50 mg	Graphite foil	Anodic, +10 V, DC		25.3	5–10	1–3	0.1		[40]
37.4 wt.% acetamide, 28 wt.% urea, 34.6 wt.% NH_4NO_3	Graphite foil	CV, −0.6 −1.5 V		7.3	0.2–0.4	1–5	0.6		[41]

(Continued)

Table 1.2 (Continued)

Electrolytes	Graphite electrodes	Power supplies	Additional conditions	Properties of prepared graphene C/O ratio	Size (μm)	No. of layers	I_D/I_G	Yield (%)	References
6.5 g H_2SO_4/100 mL H_2O, melamine (10–200 g/100 mL)	Graphite powder	20 V, AC		26.17	18	<3	0.2–0.45		[42]
8 M $HClO_4$	Graphite flake	LSV, 1.4 V	Thermal reduction at 500°C	19.6	0.15	6–7	0.59		[43]
NaOH/H_2O_2/H_2O	High purity graphite rods	3 V, Anodic	2-Electrode cell, Pt sheet 10 min treatment time			3–6	0.67	95	[44]
Propylene carbonate	Graphite from Li ion battery	Minus 15 plus 5 V	Sonication-assisted			<5		>70	[45]

HOPG, Highly orientated pyrolytic graphite; CV, Cyclic voltammetry; LSV, Linear sweep voltammetry.

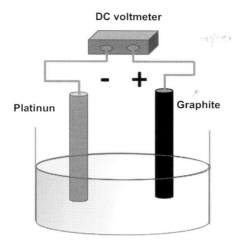

Figure 1.7 Schematic presentation of electrochemical setup.

1.3.1.4 Liquid phase exfoliation

Liquid phase exfoliation (LPE) is a top-down process, where a monolayer or few-layer graphenes are formed by the exfoliation of graphite via high shear force in a liquid medium. Different methodology like sonication [46], jet cavitation [47], high shear mixing [48], and microfluidization [49] have been adopted.

In graphite, graphene layers are stacked by Van der Waal force of attraction at a distance of 0.34 nm. Effectively overcome of Van der Waal force and removal of graphene layers from graphite substrate in liquid phase exfoliation process depends on type of shear force, shear rate, type of solvents, and type of surfactants etc.

In liquid phase exfoliation process, interfacial surface tension plays a vital role. Interfacial tension is reduced by the selection of suitable solvents, surfactants, and type of vehicle mediums. The number. of organic solvents like N-methyl-2-pyrroli-done [46], γ-butyrolactone [50], N, N dimethylacetamide [50], N, N dimethyl formamide [51], Dimethyl sulfoxide [51], and Ortho dichloro benzene [52] have been used to reduce the interfacial tension for better exfoliation of graphene. Suitable surface tension range of organic solvent for graphene exfoliation is $40-50$ mJ/m^2 [50].

One of the most serious problem in the case of pure organic solvents is the very low concentration of graphene formation that is, very low yield. Another problem is these solvents are not volatile and of a high boiling point, which is extremely difficult to remove from graphene.

To increase the exfoliation efficiency, some researchers have used inorganic salts [53] and organic salts [54]. Liu et al. [53] in their study improved exfoliation efficiency 20 times higher by adding NaOH to the different organic solvents as shown in Fig. 1.8.

In a study, Du et al. [54] used different organic salts with different organic solvents to produce graphene and observed that organic solvents can significantly increase the exfoliation efficiency in the presence of organic salts (Fig. 1.9). In only

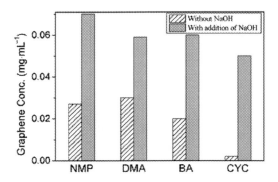

Figure 1.8 Concentrations of graphene sheets dispersed in organic solvents with or without addition of NaOH [53].

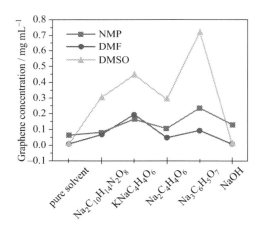

Figure 1.9 Graphene concentration in organic solvents without or with various additives [54].

2 h sonication using dimethyl siloxane (DMSO) with $Na_3C_6H_5O_7$, they achieved monolayer or few layers defect free graphene at 0.72 mg/mL concentration.

In order to enhance the graphene exfoliation, some researchers have used ionic liquid [55–57]. Wang et al. [55] in their study achieved exfoliated graphene concentration up to 0.95 mg/mL via sonication using an ionic liquid 1-butyl-3-methyl-imidazolium bis (trifluoro-methane-sulfonyl) imide. Up to 5.33 mg/mL exfoliated graphene concentration was achieved using commercially available ionic liquid 1-hexyl-3-methylimidazolium hexafluorophosphate [56]. Another study by Ager et al. [57] demonstrated a complete exfoliation of graphene in water at a concentration up to 5 wt.% using an ionic liquid. Although high concentration graphene can be achieved using ionic liquid, but because of high expensive, high viscosity, and difficult to remove from graphene, ionic liquid is not suitable for a potential industrially scalable process.

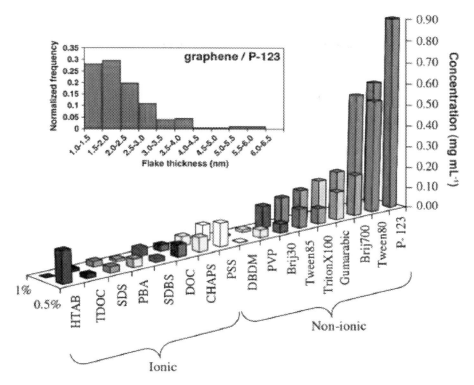

Figure 1.10 Effect of different surfactants on the concentration of graphene in aqueous dispersions. Two surfactant concentrations are shown: 0.5% and 1.0% wt./vol. Inset figure shows the distribution of apparent flake thickness measured by Atomic force microscopy (AFM) on 200 objects from dispersions stabilized by the nonionic triblock copolymer P-123 [58].

There are a number of commercially available surfactants that are suitable for exfoliation of graphene in aqueous medium as shown in Fig. 1.10.

It is observed that nonionic surfactants perform better exfoliation compared to ionic surfactants. A commercially available nonionic surfactant Pluronic P 123 provided highest concentration of graphene dispersions (up to 1 mg/mL). Dispersion of graphene also depends on the sequence of surfactant addition. Notley et al. [59] achieved high concentration of graphene dispersion up to 10.23 mg/mL by the continuous addition of surfactant Pluronic 108. A comparative study on graphene concentration prepared from both batch and continuous surfactant addition process is given in Table 1.3.

Surfactant-assisted liquid phase exfoliation of graphite can offer high-concentration defect-free graphene. But the main disadvantage of using surfactants over organic solvents is the removal of adsorbed surfactants from the surface of graphene. During the exfoliation process, surfactants are adsorbed onto the surface of the graphene and the adsorbed surfactant prevent the re-aggregation of graphene

Table 1.3 Production of aqueous graphene dispersions using different surfactants and addition methods [59].

Surfactant	Type of surfactant	Concentration to achieve = 41 mJ/m^2	Batch concentration (mg/mL)	Continuous concentration (mg/mL)
Pluronic 108	Nonionic	0.10%	0.11	10.23
Pluronic 127	Nonionic	0.10%	0.078	6.55
Sodium dodecyl sulfate	Anionic	7 mM	0.06	4.92
Hexadecyl trimethylammonium bromide	Cationic	0.6 mM	0.05	4.05
Tetradecyl trimethylammonium bromide	Cationic	2.1 mM	0.055	5.01
Dodecyl trimethylammonium bromide	Cationic	10 mM	0.06	5.22

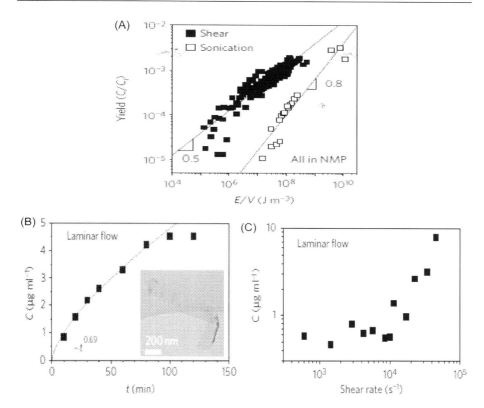

Figure 1.11 (A) Graphene yield versus energy density in sonication and shear process. (B) Concentration of graphene produced in a rotating Couette as a function of mixing time (rotation rate 3000 r.p.m.). Inset: Transmission electron microscopy (TEM) of Couette-produced graphene. (C) Concentration of graphene produced in a rotating Couette as a function of shear rate (mixing time 60 min) [48].

sheets through hydrophilic interactions. An extra processing step is required to remove adhered surfactants before the graphene is used for different applications.

Shear exfoliation in liquid

Shear exfoliation is a physical exfoliation process where graphenes are synthesized from graphite powders by shearing force in a liquid medium. Usage of sonication to separate the graphene layers are replaced by shearing force. As shown in Fig. 1.11A, shear exfoliation is a less energy consuming, and higher efficient process compared to sonication process. In order to separate out the graphene layers from graphite, shearing rate must be higher than 10^4/s [48]. The properties of graphene obtained by this process is similar to the graphene prepared by sonication of graphite in solvent and surfactants. Defect free and un—oxidized graphenes are produced through this process. In 2014, Paton and coworkers developed a shear exfoliation technique for the graphene production using N-methyl-2-pyrrolidone as a dispersing medium and sodium cholate as surfactant [48]. They have used a 300 L high shear mixer (rotor diameter $=$ 11 cm) and yielded 21 g of graphene per batch and production rate as high as 5.3 g/h. Using scaling law, they estimated that a high concentrated dispersion of graphene nanosheets (GNS) can be prepared by high shear mixing of graphite in appropriate solvent and a rate of up to 100 g/h can be achieved when a liquid of 10 m^3 is reached. It is observed that in case of sonication-based exfoliation, increase in graphene concentration is sublinear with sonication time (Graphene yield is directly proportional to the square root of sonication time), but for shear-based exfoliation, graphene concentration is directly proposal to shearing time in surfactant solution Fig. 1.11B and the concentration is dependent on shear rate above 10^4/s (Fig. 1.11C).

Shearing force may be generated using a rotating blade or rotor in a solvent, surfactant, or aqueous medium mixed with graphite. To achieve a higher yield of graphene, the level of shear stress needs to be higher. One way to increase shear rate is by incorporating fluid dynamics phenomenon in the exfoliation process such as turbulence induced shear stress [60]. Different approaches have been applied to prepare graphene through shearing process as mentioned in Table 1.4.

This process is simple, but due to the use of surfactant and huge amount of polar and high-cost solvents, conductive properties of graphene will be reduced, and pure dry graphene production is difficult and the yield is too low. Hence this process has restrictions for the industrial scale production of graphene.

Microfluidization

Microfluidization is a high-pressure technique, where graphenes are prepared by passing graphite dispersed fluid through number of microchannels (diameter $d < 100$ micrometers) at high shear rate $> 10^6 S^{-1}$. This process is more efficient for defect-free high yield graphene compared to other LPE processes like sonication and high shear mixing.

During microfluidization, the graphite exfoliation is mainly controlled by three factors—shear stress, collision, and cavitation. These three factors are also dependent on the design of the microchannels. A schematic presentation of microfluidization process designed by Karagiannidis et al. [66] is shown in Fig. 1.12. Karagiannidis et al. in their study achieved 100% yield by exfoliating graphite in sodium dodecyl sulfate (SDS) aqueous solution under high shear rate ($10^8 S^{-1}$) condition.

When graphite dispersed fluid is passed through microchannels at high pressure ($\approx$ 207 MPa), graphite particles are collided with each other and high shear forces and cavitations are generated. By these high shear stress, collision, and cavitation, monolayer and few layer graphenes are peeled out to solution. Function of cavitation is similar to the sonication-assisted graphene dispersion.

1.3.2   Bottom-up approaches

In bottom-up approach, carbon molecules act as building blocks for the graphene synthesis. These carbon molecules are obtained from the decomposition of different hydrocarbons. This process produces graphene of low yield but produces graphene

Table 1.4 Different approaches to prepare graphene through shearing process.

Sl. no.	Equipment & shearing conditions	Exfoliation medium	Exfoliation time, speed	Graphene quality	References
1	Silverson model L5M mixer, rotor diameter = 32 mm, liquid volume = 4.5 L	N-Methyl 2-pyrrolidone (NMP) and Sodium cholate	20 min, 4500 rpm	Few layers graphene nanosheets, size = 300–800 nm, thicknesses = 4–7, I_D/I_G = 0.17–0.37	[48]
2	High-shear mixer, rotor diameter = 110 mm, liquid volume = 300 L		5 min to 4 h, 3000 rpm	Production rate = 5.3 g/h, conc. = 0.07 mg/m, I_D/I_G = 0.18	[48]
3		Water/SC	2 h	1.1 mg/mL	[61]
4		Water/black liquor	10 h	10 mg/mL	[62]
5		NMP	1 h	1 mg/mL	[63]
6	Two coaxial cylinders Taylor–Couette flow reactor	NMP	1 h, 3000 rpm	Few-layer GFs, lateral size = 500–1500 nm, thickness = <3 nm, I_D/I_G = 0.14, yield = 5%	[64]
7	Pro Scientific PRO250 rotor stator, shear rate = 33 000/s		1 h, 6000 rpm	Size = 0.4–1.5 mm, thickness = 4–6 layers, yield = 16%, I_D/I_G = 0.24	[65]

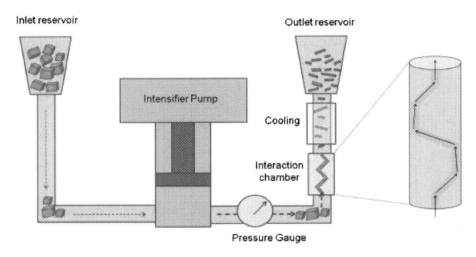

Figure 1.12 A schematic presentation of microfluidization process designed by Karagiannidis et al. [66].

sheets of large surface area. Different bottom-up approaches like graphene synthesis by chemical vapour deposition on the metal surface, synthesis by epitaxial growth on the surface of silicone carbide have been carried out by different researchers [67–73]. CVD process has been widely accepted by industries due to the higher yield, better quality of graphene production compared to epitaxial growth on SiC. Here CVD process has been described because of the widely acceptance of this technology for graphene synthesis.

1.3.2.1 Chemical vapor deposition method

Several unsolved issues like controlling the number of graphene layers, minimizing the folding of graphene, defects in graphene layers, and high quality single crystalline graphene observed in mechanical or chemical exfoliation are solved by chemical vapour deposition process. In this process, high quality, and defect free graphene of desired number of layers are produced. Some of companies those are producing graphene through CVD are shown in Table 1.1

CVD is a process where organic volatile carbon precursor in vaporized form is injected to a reactor where carbon precursor is deposited on the surface of a transition metal plate and then decomposed by thermal or plasma process to produce graphene. Transition metal acts as catalyst for the formation of graphene. Based on the way of hydrocarbon decomposition, CVD are of two types: Thermal CVD and Plasma CVD. When hydrocarbon is decomposed on metal surface by thermal process, it is called thermal CVD, where decomposition takes place by plasma, process is called plasma CVD. Thermal CVD is a high temperature process where hydrocarbon is decomposed at temperature from 800°C–1100°C, but in plasma CVD, graphene is developed at low temperature in plasma environment. A schematic

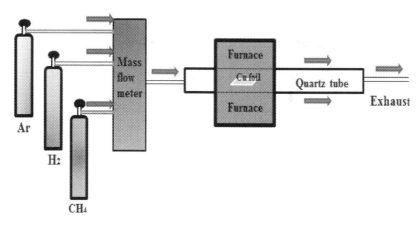

Figure 1.13 The schematic diagram of APCVD [74].

diagram has been shown in Fig. 1.13, where Cu foil has been used as a catalyst and methane gas is used as a carbon precursor for the production of graphene.

1.3.2.2 Factors affecting to the graphene synthesis in chemical vapor deposition process

1. Type of carbon precursor
2. Type of substrate
3. Gas flow rate
4. Temperature
5. Pressure
6. Annealing/cooling rate time
7. Reaction time

Type of carbon precursor

Energy needed to produce graphene depends on the type of precursor [67]. To reduce energy consumption during graphene synthesis, basically low energy C-H bond precursors are considered. Graphene has been produced by different hydrocarbons and polymers (C and H based). Different types of hydrocarbon gases like methane, ethylene, and acetylene having the dehydrogenation energy value of 410, 443, and 506 KJ/mol, respectively are used as gaseous carbon precursors [75]. Liquid like toluene [76], ethanol [77], isopropyl alcohol [78] are also used by different researchers and single to multilayers graphenes are formed. Solid precursor like polymers films are also used to prepared graphene [79]. Among all types of precursors, gaseous hydrocarbon methane is popularly used for the synthesis of graphene because of its low energy C-H bonds and requires less energy consumption.

When methane is used as carbon precursor, hydrogen gas is required during graphene synthesis process to promote carbon deposition on the substrate. The role of methane gas is to provide carbon source, while role of hydrogen gas is to provide H atoms to corrode the amorphous carbon and improve the quality of graphene.

Excessive hydrogen flow can also corrode the synthesized graphene. So, flow rate of methane and hydrogen should be optimized to achieve high quality graphene.

Type of substrate

The growth of graphene by CVD process is a catalyst dependent. The properties of as grown graphene like quality, quantity, number of layers, and layer number distribution are dependent on the type of catalyst used. From a literature survey it is observed that metallic substrates are used as the catalyst for the synthesis of graphene. Transition metals like Ni [80], Cu [81], Pd, Ru [82], and Ir and some of their compounds are used as substrates. The catalytic power of these metals is due to their partially filled d-orbitals and due to the formation of intermediate compounds that favours the reactivity of the precursor gases [79]. Catalytic behavior, atomic packing, and the solubility of different metals are different. Hence different metal substrates show different types of graphene growth mechanism.

In CVD process, graphene films are removed by wet etching of used metal substrates. Due to difficulty in wet etching of metals like Ru, Ir, Pd, and Pt from graphene film, these metals are not favored for graphene synthesis. Cu and Nickel are easily etched out from graphene film, so are the most widely used for graphene synthesis.

The solubility of carbon into the nickel increases with the increase in temperature and it is being difficult to control the precipitation of carbon during cooling. Due to high solubility of nickel at high temperature, carbon precipitation is not controlled and usually graphene with multiple layers is formed. In the case of copper, solubility of carbon atoms in copper is relatively low at high temperature and graphene is formed on the copper substrate by surface adsorption process, where in nickel, graphene is formed by dissolution process. The low solubility of carbon in copper at high temperature is able to control the growth of graphene easily and single crystalline graphene layer of large domain size are formed.

It is very important to grow high-quality, large single-crystalline domains of graphene directly on a dielectric substrate like BN, Si, SiO_2, Al_2O_3, MgO, and Si_3N_4 for applications in electronics and optoelectronics. Some reports are available to replace conventional transparent conductive film like InO_3: Sn (ITO) and SnO_2: F (FTO) on the dielectric substrate by the synthesis of graphene [83–85]. Due to insulating nature of dielectric materials, it is difficult to prepare good quality highly conductive graphene film on the surface.

In a study, Tang et al. [86] synthesized single layer crystalline graphene of domain size 20 μm using silane as gaseous catalyst and h-BN as substrate. This combined advantages of both catalytic CVD, ultra-flat dielectric substrate, and the use of gaseous catalyst will pave the way for synthesizing high quality graphene on the dielectric substrate for different electronics applications.

Gas flow rate

The size and number of graphene layers depend also on the flow rate of carbon precursor gas. The decrease in gas flow rate decreases the density of graphene nuclei and increases the domain size [87]. In a study by Xue et al. [88], bilayer graphene

Table 1.5 Raman characteristics with respect to the processing temperature for graphene synthesis.

Temperature ($^\circ$C)	2D band (cm^{-1})		I$_{2D}$/I$_G$	I$_D$/I$_G$
	Position	FWHM		
850	2687.16	38.88	7.45	0.35
750	2680.77	43.60	4.25	0.51
650	2677.77	51.12	0.68	0.76

was formed instead of multilayer by reducing the methane flow from 300 to 180 sccm using Fe foil as substrate.

The flow rate of gas should be optimized to achieve the graphene of selective layers. Barcenas et al. [89] in their study synthesized graphene on a copper foil at temperature 1000°C varying the flow rate of acetylene gas from 10 sccm to 100 cm and observed that the number. of graphene layers was reduced with the decrease in gas flow, but at every low rate 10 sccm, no graphene layer was formed.

Temperature

Temperature plays a vital role for the graphene synthesis. Controlling the temperature of reaction, the number of graphene layers, size of domain, and yield of graphene can be controlled. Depending on the catalytic activity of metal substrate, the process temperature is basically varied between 800°C−1100°C. When Ni is used as a substrate, at above 800°C the metastable formation of Ni$_3$C phase promotes the precipitation of carbon out of Ni. Graphene synthesis using Co or Fe takes place at a temperature above 850°C. A high temperature above 1035°C yields very low-density graphene nuclei which leads to the formation of a large domain size. Dathbun et al. [77], in their study observed the effect of temperature on the graphene quality. The effect of temperature on the graphene quality has been given through Raman spectra as shown in Table 1.5. I$_D$/I$_G$ ratio decreased with the increase in temperature from 650°C to 850°C. This I$_D$/I$_G$ intensity is calculated to characterize the defect quantity in graphene and to estimate the graphene domain size. Lower value indicates lower defect in graphene. I$_{2D}$/I$_g$ value represents the layer type. This I$_{2D}$/I$_g$ ratio value is increased with the decrease in the number of layers. From Table 1.5, it is observed that the graphene quality is increased with the increase in temperature from 650°C to 850°C. At a temperature below 850°C, a large number edge defects and disorder are observed as evidenced by the presence of D band and strong (G + D) band at about 2940 cm^{-1}. This may be due to the poor graphitization at low temperature.

1.4   Surface functionalization of graphene

The properties of graphene reinforced nanocomposites depend on the extent of graphene dispersion. Pure graphene is hydrophobic and non-polar in nature. Due to high Van der Waals forces, graphenes are agglomerated and not uniformly dispersed inside the matrix. Aggregation of graphenes results in poor physical and mechanical properties in nanocomposites if the aggregates act as stress concentration points. To reduce this agglomeration tendency and to explore their application areas, different surface modification/functionalization processes have been successfully developed by researchers. Surface functionalization reduces the agglomerating tendency of graphene and increases the carbon-polymer interfacial adhesion through covalent or ionic bonds. Based on applications, different types of organic groups are attached to the surface of graphene.

One of the most important application of functionalized graphene is the fabrication of biosensor, where graphene acts as an electron wire and reduces the electron transfer distance between the functional groups of biomolecules. The efficiency of functionalized graphene-based biosensors is much better than that of pure graphene in some cases. Due to the absence of hydrophilic groups on the surface, pristine graphene is not suitable for biosensor applications, but surface functionalized graphenes are very well dispersed in matrix and very easily interacted with biomolecules compared to pristine graphenes. A biosensor is formed by the attachment of fluorescent molecules to the single strained DNA and then attachment of DNA to functionalized graphene. Cost effective and very stable biosensors are formed by the immobilization of different enzymes to the functionalized graphene. Functionalized graphene-based biosensors are also used to detect glucose, NADH, cholesterol, hydrogen peroxide, and other molecules.

In case of supercapacitor, the capacitance performance of pristine graphenes is limited by the aggregation and poor interaction between graphene and electrolyte. To overcome this problems, functionalized graphenes are used for supercapacitor electrode material. Due to the absence of band gap, pristine graphene is not suitable for transistor application. Hence for the use of graphene in transistor, the band gap of graphene is increased by the surface functionalization.

Based on the type of interactions involved between the matrix and carbon atoms on the graphene, there is two categories of functionalization: chemical functionalization (or covalent functionalization) and physical modification (or noncovalent functionalization)

1.4.1 Covalent functionalization

During chemical functionalization, different functional groups are attached to the surface of filler by covalent bonds. By the functionalization, the electronic configuration of graphene is also modified due to the rehybridization of one or more sp2 carbon atoms of the carbon network into the sp3 configuration. The chemically functionalized filler can produce strong interfacial bonds with polar or semi polar mediums/polymers and allow the composites to possess high mechanical strength and other functional properties. Functionalized fillers disperse easily in different polar media and enhance color properties in paint, costing, and ink.

GO formed by the top-down approaches of graphene synthesis is also one kind of functionalized graphene. Covalent attachment of organic functional groups takes

Table 1.6 Modifying agents used for covalent attachment to the surface of grapheme.

Chemicals used for surface modification	Medium for dispersion	Dispersibility (gm/mL)	References
Allylamine	Water, DMF	1.55	[90]
3-aminopropyltriethoxysilane	Water, ethanol, DMF, DMSO	0.5	[91]
Ionic liquid-NH_2 (IL-NH_2)	Water, DMF, DMSO	0.5	[92]
Polyglycerol	Water	3	[93]
6-amino-4-hydroxy-2-naphthalenesulfonic acid	Water	3	[94]
Sulfanilic acid	Water	2	[95]
NMP	Ethanol, DMF, NMP, PC, THF	0.2−1.4	[96]
Organic isocyanate	DMF	1	[97,98]
ODA	THF	0.5	[99]
PEG-NH_2	Water	1	[100]
Chitosan (CS)	Water	2	[101]
a-CD, b-CD, c-CD	Water, ethanol, DMF, DMSO	>2.5	[102]

place by two routes. (1) Organic attachment to the $C = C$ backbone of graphene and (2) Attachment of organic groups to the oxygen groups attached to the surface of GO.

Modifying agents used for covalent attachment to the surface of graphene are shown in Table 1.6. The organic modifiers like allylamine, APTS, and Ionic liquid-NH_2 having amine ($-NH_2$) functionality are reacted with the epoxy groups of GO through nucleophilic substitution mechanism for the dispersion of graphene in water medium [90−92]. Reaction mechanism of graphene functionalization by alkylamine is shown in Fig. 1.14. A significant increase in hydrophilic affinity of graphene is observed by the reaction of graphene with allylamine. The maximum solubilities of GO and allylamine-functionalized GO in water were determined to be 0.69 and 1.55 mg/mL, respectively. Graphene is also surface modified with polyglycerol through nucleophilic substitution mechanism to disperse graphene at high concentration in water and the modified graphene also shows dispersion of around 3 mg/mL graphene in water medium [93].

Si et al. [95] prepared water-soluble graphene by the reaction of graphene oxides with aryl diazonium salt of sulfonic acid as shown in Fig. 1.15. Grafting of graphene by aryl diazonium salt is an electrophilic substitution, where a hydrogen atom from graphene surface is displaced by an electrophile aryl group. The prepared lightly sulfonated graphene was readily dispersed in water at reasonable concentrations (2 mg/mL) in the pH range of 3−10.

Graphenes are also chemically modified with isocyanate and amine compounds by condensation reaction mechanism, where isocyanate, diisocyanate, and amine

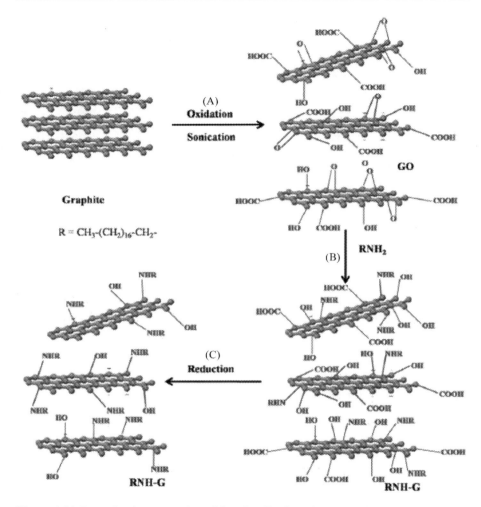

Figure 1.14 Route for the preparation of functionalized graphene: (A) Oxidation of natural flake graphite to graphite oxide, followed by ultrasonication; (B) an aqueous GO dispersion is treated with alkylamine to obtain amine-modified GO (RNH-GO); (C) RNH-GO is reduced using hydrazine to produce amine-modified graphene (RNH-G) [90].

compounds are attached to the graphene surface through the formation of amides and carbonate ester linkages. In a study Stankovich et al. [98] functionalized graphene with isocyanates using DMF as a solvent. The isocyanate modified graphene showed good dispersion in DMF up to 1 mg/mL. A schematic presentation of condensation reaction between isocyanate and graphene is shown in Fig. 1.16.

A biocompatible compound chitosan (CS) was attached to the surface of graphene by a condensation reaction mechanism under microwave irradiation, where carboxylic groups of GO nanosheets were reacted with the amino groups of chitosan. The

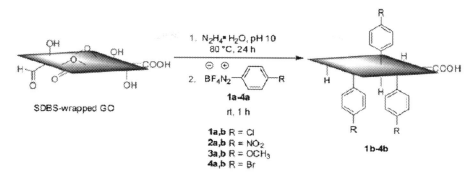

Figure 1.15 Reduction and functionalization of intermediate sodium dodecylbenzene sulfonate (SDBS)-wrapped CCG with diazonium salts, starting with SDBS-wrapped GO [95].

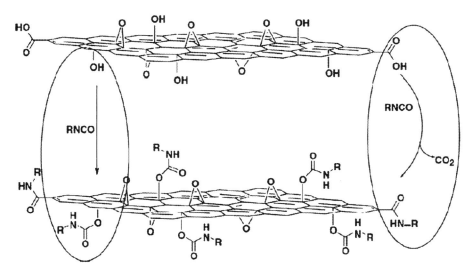

Figure 1.16 Proposed reactions on the functionalization of GO with isocyanate, in which organic isocyanates react with the hydroxyl (*left oval*) and carboxyl groups (*right oval*) of GO sheets to form carbamate and amide functionalities, respectively [98].

prepared intermediate was reduced by hydrazine to produce chitosan grafted graphene nanocomposite. Schematic of the formation procedure for the Chitosan (CS) chains grafting onto the surface of GNS is given Fig. 1.17. After functionalization by chitosan, the dispersibility of graphene was significantly increased up to 2 mg/mL in water.

1.4.2 Noncovalent functionalization

Noncovalent functionalization is a nondestructive technique, where suitable molecules are physically adsorbed on the surface of graphene. In this technique, the

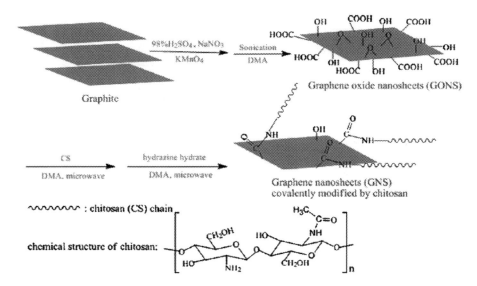

Figure 1.17 Schematic of the formation procedure for the CS chains grafting onto the surface of graphene nanosheets [101].

Table 1.7 Examples of different modifying agents for noncovalent modification of GO and the dispersibility of modified graphene.

Modifying agent	Dispersing medium	Dispersibility (mg/mL)	References
PSS	Water	1	[103]
SPANI	Water	>1	[104]
PSS-g-PPY	Water	3	[105]
SDBS	Water	1	[106,107]
SLS, SCMC, HPC-Py	Water	0.6–2	[108]
PIL	Water	1.5	[109]

HPC-Py, Pyrene-containing hydroxypropyl cellulose; *PSS*, poly (sodium 4-styrene sulfonate); *SCMC*, sodium carboxymethyl cellulose; *SDBS*, sodium dodecylbenzene sulfonate; *SLS*, sodium lignosulfonate; *SPANI*, sulfonated polyaniline, *PSS-g-PPY*, Polystyrene sulfonate grafted polypyrrole.

original properties of the graphene are retained by preventing the damage of π-conjugated structure. The noncovalent functionalization is carried out by basically following two ways.

1. Polymer wrapping
2. Surfactant adsorption

Examples of different modifying agents for noncovalent modification of GO and the dispersibility of modified graphene are shown in Table 1.7.

Highly water dispersible graphene nanoplatelets were prepared by Stankovich et al. [103] using poly (sodium 4-styrene sulfonate) (PSS), where both graphene exfoliation and reduction were carried out in the presence of PSS. Graphene nano-platelets were coated with PSS polymers and the dispersion stability was increased upto 1 mg/mL in water. In a study, Bai et al. [10] prepared a stable dispersion of graphene in water more than 1 mg/mL by adding a conductive polymer sulfonated polyaniline (SPANI) with graphene. They reduced the graphene oxides in the presence of SPANI and the composite film of SPANI-functionalized graphene showed improved electrochemical stability and enhanced electrocatalytic activity. A novel electrolytic biosensing material was developed by the functionalization of reduced graphene oxide with poly (styrene sulfonic acid-g-pyrrole) (PSSA-g-PPY), via—noncovalent interaction [105]. The resulting nanocomposite exhibited well dispersion in water for at least 2 months with a solubility of 3.0 mg/mL.

Apart from polymer wrapping, colloidal stabilization of graphene using different surfactants has received great attention in biomedical and polymer composite filed. Different ionic and nonionic surfactants are used for stable colloidal suspension of graphene. These surfactants are adsorbed on the surface of graphene and keep separate graphene particles from one another by either electrostatic repulsion or steric hindrance. The repulsive force introduced by surfactant molecules overcomes the van der wall force of interaction among the graphene particles.

Anionic surfactant like SDS, lithium dodecyl sulfate, sodium dodecylbenzene sulfonate (SDBS), and nonionic surfactant like Triton X-100 are widely used for colloidal stabilization of carbon materials. Chang et al. [106] and Zeng et al. [107] have prepared SDBS functionalized graphene sheets, where the hydrophilic part (alkyl group along with benzene rings) of SDBS is adsorbed to the graphitic surface of graphenes. A negative charged environment generated by surfactant around the graphene prevents the aggregation and creates a stable colloidal suspension in aqueous phase. The electrochemical stability of SDBS wrapped graphene remains almost unchanged after 1000 cycles. SDBS wrapped graphene also acts as a biosensor for the detection of hydrogen peroxide. Stable aqueous dispersion of graphene is also possible through noncovalent functionalization using lignin and cellulose derivatives. Yang et al., [108] prepared a stable aqueous suspensions of graphene (G) nanosheets with high concentration (0.6−2 mg/mL) through chemical reduction of exfoliated graphite oxide with the aid of sodium lignosulfonate (SLS), sodium carboxymethyl cellulose, and pyrene-containing hydroxypropyl cellulose.

1.5 Graphene-based nanocomposites

Because of extremely high electrical conductivity, chemical resistance, mechanical strength, and reinforcing efficiency, graphene has been considered as a major filling agent for composite applications in many industries, like aerospace, electronics, energy, structural and mechanical, environmental, medicine, and food and beverage. Compared to conventional fillers, graphene is incorporated in different substrates at a very small concentration. This small concentration of graphene not only makes the material lighter, but also makes it stronger for various multifunctional applications.

1.5.1   Graphene-based polymer nanocomposites

Polymers in general are of low modulus and nonconductive material. Thus conductive fillers like carbon black, graphite, carbon fiber, and metal particles, etc. are widely used in polymers to achieve desired conductivity and other properties for different electrical and electronic applications. It has been reported that conductive micro fillers are added to polymers in high volume to improve electrical conductivity. This high filler loading may adversely affect the processability and sometimes mechanical properties of polymer composites. So, the selection of suitable conductive filler is important to design a conductive composite for specific application. To overcome the disadvantages of micro fillers in polymer composites, conductive fillers with nano dimension have attracted recent interest due to their high aspect ratio in combination with excellent intrinsic conductivity. In fact, nanofillers have ultra-high interfacial surface area which permits strong polymer-filler interaction at low filler concentration. Huge interfacial area and large number of nanoparticles per unit volume of composite are the key factors for the improvement of different properties of the composite. Graphene nanocomposites are used as reinforcement for polymer matrix and for different applications like supercapacitor, Li ion battery, sensor, fuel cell, solar cell as presented in Table 1.8.

D.W. Wang et al. [142] in their study prepared flexible graphene−PANI nanocomposite by in-situ electro polymerization for supercapacitor applications, where a good electrochemical capacitance of 233 F/g, and 135 F/cm^3 for gravimetric and volumetric capacitance, respectively were obtained. In another study, Yan et al. [145] prepared electrochemical capacitor by coating GNS with ~2 nm PANI nanoparticles and achieved a high capacitance of 1046 F/g compared to 115 F/g for pure PANI [145]. In addition to PANI nanoparticles, Wang et al. used PANI nanofibers with GO sheets to prepare super capacitor with a specific capacitance of 531 F/g at a mass ratio of 100:1 for PANI: GO. [143]

Santos et al. [114,115] prepared graphene-based nanocomposites using poly-*N* vinyl carbazole (PVK) for antimicrobial application which exhibited more than 80% microbial inhibition and toxicity toward a broad array of bacteria. Antimicrobial activity of PVK-GO nanocomposites was also studied by Carpio et al. [116], where Carpio et al. studied the toxicity effects of PVK-GO nanocomposite on planktonic microbial cells, *Escherichia coli*, *Cupriavidus metallidurans*, *Bacillus subtilis* and *Rhodococcus opacus*, biofilms, and mammalian fibroblast cells (NIH$_3$T$_3$). Their results showed that PVK-GO presented a stronger antimicrobial effect than pristine GO.

1.5.2   Graphene-based ceramic nanocomposites

Because of high reinforcing efficiency and high electrical conductivity, graphene has been a great choice for ceramic-based electronic applications. Graphene acts as a good charge carrier in ceramic nanocomposite and results exceptional electrochemical performance. Graphene in ceramic nanocomposites not only enhances the electrical properties, but also increases thermal conductivity, mechanical,

Table 1.8 Examples of some graphene-based nanocomposites with their specific applications.

Applications	Matrix	References
Reinforcement	PMMA	[110]
Reinforcement	Chitosan	[111]
Reinforcement	PVA	[112]
Reinforcement and conductivity	Cellulose acetate	[113]
Antimicrobial (microbial inhibition)	Poly N-vinyl carbazole	[114,115]
Antimicrobial (microbial inhibition)	Poly N-vinyl carbazole	[116]
Ammonia sensor	PANI	[117]
Benzene	SnO_2	[118]
Leukemia	Poly-L-lysine	[119]
Tryptophan	Silver	[120]
Li ion battery	Carbon coated SnO_2	[121]
Li ion battery	$CoFe_2O_4$	[122]
Li ion battery	Carbon doped Li_2SnO_3	[123]
Li ion battery	$ZnFe_2O_4$ nanoparticles	[124]
Li ion battery	$Li_3V_2(PO_4)3@C$	[125]
Li ion battery	Tin indium oxide	[126]
Li ion battery	Sulfur-coated SnO_2	[127]
Fuel cell	Platinum	[128]
Fuel cell	Xerogel- based nonprecious metal catalyst	[129]
Fuel cell	AuPtPd	[130]
Fuel cell	CeO_2	[131]
Fuel cell	Poly (ethylene oxide)	[132]
Solar cell	PANI	[133]
Solar cell	Poly(3,4-ethylenedioxythiophene) doped with poly(4-styrenesulfonate)	[134]
Solar cell	NiO	[135]
Solar cell	Resorcinol-formaldehyde based aerogel	[136]
Field emission	PANI	[137]
Field emission	Pt	[138]
Supercapacitor	Polyethylene dioxythiophene	[139]
Supercapacitor	Poly (arylene ether nitrile)	[140]
Supercapacitor	Polypyrrole nanofiber	[141]
Supercapacitor	PANI	[142]
Supercapacitor	PANI	[143]
Supercapacitor	V_2O_5	[144]

anticorrosive, and antifriction. Brittleness is reduced and fracture toughness is increased by the addition of graphene to the ceramic matrices. Graphene has been used widely for ceramic-based composite preparation using different types of ceramics like SiC [146], Si_3N_4 [147,148], Al_2O_3 [149], ZrB_2 [150], ZrO-Al_2O_3 [151],

BN [152]. Presently, graphene-based zirconium composite [150] has been developed for aerospace industry as a high temperature barrier for space vehicle during the re-entry event. Because of ultra-high temperature resistance, these materials are consistently used as the primal infrastructure for the nose caps in space shuttles and military ballistic equipment. TiN-graphene composites have shown promising results as a selective permeable membrane for hydrogen. Kim and Hong [153] in their study showed that the hydrogen permeability of TiN graphene composites was better than the Pd-Ag amorphous membrane at 1.67, 2.09, and 2.83 $\times$ 10^{-7} mol/msPa$^{1/2}$ at 673K under 0.3 MPa, respectively [153]. Lee et al. (2013) in their study had also achieved almost similar results (2.62 $\times$ 10^{-7} mol/ms Pa$^{1/2}$ at 673K under 0.3 MPa) of hydrogen permeation with the use of Al_2O_3/CeO_2/graphene composite membranes prepared by hot-press method [154]. By exploiting the pore size distribution, surface area, and elasticity, one can use such kinds of membranes for high purity separation and filtration of chemicals.

1.6 Conclusions

Due to exceptional mechanical, thermal, chemical, and electrical properties, graphene has captured a wide area of applications like supercapacitor, sensor, solar cells, Li ion battery, microbial inhibition, and spacecraft etc. Different approaches have been established for the synthesis of graphene and graphene oxide. A few approaches have been adopted for the large-scale production of graphene, but it is still a challenge to produce high quality single layer graphene in large scale to fulfill worldwide requirements. Based on the requirement for different applications, graphene and graphene oxide have been surface functionalized. Different methods to prepare graphene and graphene oxides, different surface functionalization methods based on applications, and the concept of graphene nanocomposites are well explained in this chapter. This chapter provided knowledge on different aspects of graphene from synthesis, functionalization to application.

References

[1] K.S. Novoselov, A.K. Geim, S.V. Morozov, D. Jiang, Y. Zhang, S.V. Dubonos, et al., Electric field effect in atomically thin carbon films, Science 306 (2004) 666−669.
[2] P. Greil, Perspectives of nano-carbon based engineering materials, Adv. Eng. Mater. 17 (2) (2015) 124−137.
[3] S.E. Lowe, Y.L. Zhong, Challenges of industrial-scale graphene oxide production, Graphene Oxide: Fundamentals and Applications, first ed, John Wiley & Sons, Ltd, 2017, pp. 410−431.
[4] L.M. Viculis, J.J. Mack, R.B. Kaner, A chemical route to carbon nanoscrolls, Science 299 (5611) (2003) 1361.
[5] B. Jayasena, S. Subbiah, A novel mechanical cleavage method for synthesizing few-layer graphenes, Nanoscale Res. Lett. 6 (2011) 95.

[6] V. Singh, D. Joung, L. Zhai, S. Das, S.I. Khondaker, S. Seal, Graphene based materials: Past, present, and future, Prog. Mat. Sci. 56 (8) (2011) 1178–1271.

[7] L. Staudenmaier, Verfahren zur Darstellung der Graphitsäure, Ber. Dtsch. Chem. Ges. 31 (1898) 1481.

[8] W.S. Hummers, R.E. Offeman, Preparation of graphitic oxide, J. Am. Chem. Soc. 80 (6) (1958). 1339–1339.

[9] A.M. Dimiev, J.M. Tour, Mechanism of graphene oxide formation, ACS Nano 8 (3) (2014) 3060–3068.

[10] X. Hu, Y. Yu, J. Zhou, I. Song, Effect of graphite precursor on oxidation degree, hudrophilicity and microstructure of graphene oxide, Nano 9 (3) (2014) 1450037.

[11] Y. Yosida, S. Tanuma, K. Okabe, In situ observation of X-ray diffraction in a synthesis of H_2SO_4-GICs, Synth. Met. 34 (1989) 341–346.

[12] A.M. Dimiev, G. Ceriotti, N. Behabtu, D. Zakhidov, M. Pasquali, R. Saito, et al., Direct real-time monitoring of stage transitions in graphite intercalation compounds, ACS Nano 7 (2013) 2773–2780.

[13] S. Seiler, C.E. Halbig, F. Grote, P. Rietsch, F. Börrnert, U. Kaiser, et al., Effect of friction on oxidative graphite intercalation and high-quality graphene formation, Nat. Commun. 9 (2018) 836.

[14] R. Matsumoto, Y. Hoshina, N. Azukawa, Thermoelectric properties and electrical transport of graphite intercalation compounds, Mater. Trans. 50 (2009) 1607–1611.

[15] H.P. Boehm, R. Setton, E. Stumpp, Nomenclature and terminology of graphite intercalation compounds, Pure Appl. Chem. 66 (1994) 1893–1901.

[16] S. Lee, S.H. Eom, J.S. Chung, S.H. Hur, Large scale production of high quality reduced graphene oxide, Chem. Eng. J. 233 (2013) 297–304.

[17] F.J. Tolle, K. Gamp, R. Mulhaupt, Scale-up and purification of graphite oxide as intermediate for functionalized graphene, Carbon 75 (2014) 432–442.

[18] F. Kim, J. Luo, R. Cruz-Silva, L.J. Cote, K. Sohn, J. Huang, Self-propagating domino-like reactions in oxidized graphite, Adv. Funct. Mater. 20 (17) (2010) 2867–2873.

[19] D. Krishnan, F. Kim, J. Luo, R. Cruz-Silva, L.J. Cote, H.D. Jang, et al., Energetic graphene oxide: challenges and opportunities, Nano Today 7 (2) (2012) 137–152.

[20] W. Ren, H.M. Cheng, The global growth of graphene, Nat. Nanotechnol. 9 (10) (2014) 726–730.

[21] L. Peng, Z. Xu, Z. Liu, Y. Wei, H. Sun, Z. Li, et al., An iron-based green approach to 1-h production of single-layer graphene oxide, Nat. Commun. 6 (2015) 5716.

[22] J.M. Munuera, J.I. Paredes, S.V. Rodil, M.A. Varela, A. Pagán, S.D. Aznar-Cervantes, et al., High quality, low oxygen content and biocompatible graphene nanosheets obtained by anodic exfoliation of different graphite types, Carbon 94 (2015) 729–739.

[23] M. Sevilla, G.A. Ferrero, A.B. Fuertes, Aqueous dispersions of graphene from electrochemically exfoliated graphite, Chem. Eur. J. 22 (2016) 17351–17358.

[24] L. Wu, W. Li, P. Li, S. Liao, S. Qiu, M. Chen, et al., Powder, paper and foam of few-layer graphene prepared in high yield by electrochemical intercalation exfoliation of expanded graphite, Small 10 (2014) 1421–1429.

[25] T.C. Achee, W. Sun, J.T. Hope, S.G. Quitzau, C.B. Sweeney, S.A. Shah, et al., High-yield scalable graphene nanosheet production from compressed graphite using electrochemical exfoliation, Sci. Rep. 8 (2018) 14525.

[26] X. Huang, S. Li, Z. Qi, W. Zhang, W. Ye, Y. Fang, Low defect concentration few-layer graphene using a two-step electrochemical exfoliation, Nanotechnology 26 (2015) 105602.

[27] S. Yang, A.G. Ricciardulli, S. Liu, R. Dong, M.R. Lohe, A. Becker, et al., Ultrafast delamination of graphite into high-quality graphene using alternating currents, Angew. Chem. 56 (2017) 6669–6675.

[28] Y.I. Kurys, O.O. Ustavytska, V.G. Koshechko, V.D. Pokhodenko, Structure and electrochemical properties of multilayer graphene prepared by electrochemical exfoliation of graphite in the presence of benzoate ions, RSC Adv. 6 (2016) 36050–36057.

[29] Q. Zhou, Y. Lu, H. Xu, High-yield production of high-quality graphene by novel electrochemical exfoliation at air-electrolyte interface, Mater. Lett. 235 (2019) 153–156.

[30] D.B. Shinde, J. Brenker, C.D. Easton, R.F. Tabor, A. Neild, M. Majumder, Shear assisted electrochemical exfoliation of graphite to graphene, Langmuir 32 (2016) 3552–3559.

[31] K. Parvez, Z.S. Wu, R. Li, X. Liu, R. Graf, X. Feng, et al., Exfoliation of graphite into graphene in aqueous solutions of inorganic salts, J. Am. Chem. Soc. 136 (2014) 6083–6091.

[32] F. Miao, S. Majee, M. Song, J. Zhao, S.L. Zhang, Z.B. Zhang, Inkjet printing of electrochemically exfoliated graphene nanoplatelets, Synth. Met. 220 (2016) 318–322.

[33] J.M. Munuera, J.I. Paredes, M. Enterría, A. Pagán, S. Villar-Rodil, M.F.R. Pereira, et al., Electrochemical exfoliation of graphite in aqueous sodium halide electrolytes toward low oxygen content graphene for energy and environmental applications, ACS Appl. Mater. Interfaces. 9 (2017) 24085–24099.

[34] J.M. Munuera, J.I. Paredes, S. Villar-Rodil, A. Castro-Muñiz, A. Martínez-Alonso, J.M.D. Tascón, High quality, low-oxidized graphene via anodic exfoliation with table salt as an efficient oxidation-preventing co-electrolyte for water/oil remediation and capacitive energy storage applications, Appl. Mater. Today 11 (2018) 246–254.

[35] A. Ejigu, K. Fujisawa, B.F. Spencer, B. Wang, M. Terrones, I.A. Kinloch, et al., On the role of transition metal salts during electrochemical exfoliation of graphite: antioxidants or metal oxide decorators for energy storage applications, Adv. Funct. Mater. 28 (2018) 1804357.

[36] J.M. Munuera, J.I. Paredes, S. Villar-Rodil, M. Ayán-Varela, A. Martínez-Alonso, J.M.D. Tascón, Electrolytic exfoliation of graphite in water with multifunctional electrolytes: An route towards high quality, oxide-free graphene flakes, Nanoscale 8 (2016) 2982–2998.

[37] N. Parveen, M.O. Ansari, M.H. Cho, Simple route for gram synthesis of less defective few layered graphene and its electrochemical performance, RSC Adv. 5 (2015) 44920–44927.

[38] C.H. Chuang, C.Y. Su, K.T. Hsu, C.H. Chen, C.H. Huang, C.W. Chu, et al., A green, simple and cost-effective approach to synthesize high quality graphene by electrochemical exfoliation via process optimization, RSC Adv. 5 (2015) 54762–54768.

[39] S.M. Jung, D.L. Mafra, C.T. Lin, H.Y. Jung, J. Kong, Controlled porous structures of graphene aerogels and their effect on supercapacitor performance, Nanoscale 7 (2015) 4386–4393.

[40] S. Yang, S. Bruller, Z.S. Wu, Z. Liu, K. Parvez, R. Dong, et al., Organic radical-assisted electrochemical exfoliation for the scalable production of high-quality graphene, J. Am. Chem. Soc. 137 (2015) 13927–13932.

[41] Y. Zhang, Y. Xu, J. Zhu, L. Li, X. Du, X. Sun, Electrochemically exfoliated high-yield graphene in ambient temperature molten salts and its application for flexible solid-state supercapacitors, Carbon 127 (2018) 392–403.

[42] C.H. Chen, S.W. Yang, M.C. Chuang, W.Y. Woon, C.Y. Su, Towards the continuous production of high crystallinity graphene via electrochemical exfoliation with molecular in situ encapsulation, Nanoscale 7 (2015) 15362–15373.

[43] B. Gurzęda, P. Florczak, M. Wiesner, M. Kempinski, S. Jurga, P. Krawczyk, Graphene material prepared by thermal reduction of the electrochemically synthesized graphite oxide, RSC Adv. 6 (2016) 63058–63063.

[44] K.S. Rao, J. Senthilnathan, Y.F. Liu, M. Yoshimura, Role of peroxide ions in formation of graphene nanosheets by electrochemical exfoliation of graphite, Sci. Rep. 4 (1) (2015) 4237.

[45] Y. Yang, F. Lu, Z. Zhou, W. Song, Q. Chen, X. Ji, Electrochemically cathodic exfoliation of graphene sheets in room temperature ionic liquids Nbutyl, methylpyrrolidinium bis (trifluoromethylsulfonyl) imide and their electrochemical properties, Electrochim. Acta. 113 (2013) 9−16.

[46] J.N. Coleman, Liquid-phase exfoliation of nanotubes and graphene, Adv. Funct. Mater. 19 (2009) 3680−3695.

[47] Z. Shen, J. Li, M. Yi, X. Zhang, S. Ma, Preparation of graphene by jet cavitation, Nanotechnology 22 (2011) 365306.

[48] K.R. Paton, E. Varrla, C. Backes, R.J. Smith, U. Khan, A. O'Neill, et al., Scalable production of large quantities of defect-free few-layer graphene by shear exfoliation in liquids, Nat. Mater. 13 (2014) 624−630.

[49] Y. Wang, X. Zhang, H. Liu, X. Zhang, SMA-assisted exfoliation of graphite by microfluidization for efficient and large-scale production of high-quality graphene, Nanomaterials 9 (2019) 1653.

[50] Y. Hernandez, V. Nicolosi, M. Lotya, F.M. Blighe, Z. Sun, S. De, et al., High-yield production of graphene by liquid-phase exfoliation of graphite, Nat. Nanotechnol. 3 (2008) 563−568.

[51] J.N. Coleman, Liquid exfoliation of defect-free graphene, Acc. Chem. Res. 46 (2013) 14−22.

[52] C.E. Hamilton, J.R. Lomeda, Z. Sun, J.M. Tour, A.R. Barron, High-yield organic dispersions of unfunctionalized graphene, Nano Lett. 9 (2009) 3460−3462.

[53] W.W. Liu, J.N. Wang, Direct exfoliation of graphene in organic solvents with addition of NaOH, Chem. Commun. 47 (2011) 6888−6890.

[54] W. Du, J. Lu, P. Sun, Y. Zhu, X. Jiang, Organic salt-assisted liquid-phase exfoliation of graphite to produce high-quality graphene, Chem. Phys. Lett. 568 (2013) 198−201.

[55] X. Wang, P.F. Fulvio, G.A. Baker, G.M. Veith, R.R. Unocic, S.M. Mahurin, et al., Direct exfoliation of natural graphite into micrometre size few layers graphene sheets using ionic liquids, Chem. Commun. 46 (2010) 4487−4489.

[56] D. Nuvoli, L. Valentini, V. Alzari, S. Scognamillo, S.B. Bon, M. Piccinini, et al., High concentration few-layer graphene sheets obtained by liquid phase exfoliation of graphite in ionic liquid, J. Mater. Chem. 21 (2011) 3428−3431.

[57] D. Ager, V.A. Vasantha, R. Crombez, J. Texter, Aqueous graphene dispersions-optical properties and stimuli-responsive phase transfer, ACS Nano 8 (2014) 11191−11205.

[58] L. Guardia, M.J. Fernandez-Merino, J.I. Paredes, P. Solis-Fernandez, S. Villar-Rodil, A. Martinez-Alonso, et al., High-throughput production of pristine graphene in an aqueous dispersion assisted by non-ionic surfactants, Carbon 49 (2011) 1653−1662.

[59] S.M. Notley, Highly concentrated aqueous suspensions of graphene through ultrasonic exfoliation with continuous surfactant addition, Langmuir 28 (2012) 14110−14113.

[60] M. Yi, Z. Shen, A review on mechanical exfoliation for the scalable production of graphene, J. Mater. Chem. A 3 (22) (2015) 11700−11715.

[61] J. Phiri, P. Gane, T.C. Maloney, High-concentration shear-exfoliated colloidal dispersion of surfactant-polymer-stabilized few-layer graphene sheets, J. Mater. Sci. 52 (2017) 8321−8337.

[62] J. Ding, H. Zhao, Q. Wang, W. Peng, H. Yu, Ultrahigh performance heat spreader based on gas-liquid exfoliation boron nitride nanosheets, Nanotechnology 28 (2017) 475602.

[63] E. Varrla, K.R. Paton, C. Backes, A. Harvey, R.J. Smith, J. McCauley, et al., Turbulence-assisted shear exfoliation of graphene using household detergent and a kitchen blender, Nanoscale 6 (2014) 11810−11819.

[64] T.S. Tran, S.J. Park, S.S. Yoo, T.-R. Lee, T. Kim, High shear-induced exfoliation of graphite into high quality graphene by Taylor−Couette flow, RSC Adv. 6 (15) (2016) 12003−12008.

[65] E.T. Bjerglund, M.E.P. Kristensen, S. Stambula, G.A. Botton, S.U. Pedersen, K. Daasbjerg, Efficient graphene production by combined bipolar electrochemical intercalation and high-shear exfoliation, ACS Omega 2 (10) (2017) 6492−6499.

[66] P.G. Karagiannidis, S.A. Hodge, L. Lombardi, F. Tomarchio, N. Decorde, S. Milana, et al., Microfluidization of graphite and formulation of graphene-based conductive inks, ACS Nano 11 (2017) 2742−2755.

[67] W. Zhang, P. Wu, Z. Li, J. Yang, First-principles thermodynamics of graphene growth on Cu surfaces, J. Phys. Chem. C. 115 (36) (2011) 17782−17787.

[68] R.M. Jacobberger, R. Machhi, J. Wroblewski, B. Taylor, A.L. Gillian-Daniel, M.S. Arnold, Simple graphene synthesis via chemical vapor deposition, J. Chem. Educ. 92 (11) (2015) 1903−1907.

[69] P. Sutter, Epitaxial graphene: How silicon leaves the scene, Nat. Mater. 8 (2009) 171.

[70] T. Ohta, A. Bostwick, J. Mcchesney, T. Seyller, K. Horn, E. Rotenberg, Interlayer interaction and electronic screening in multilayer graphene investigated with angle-resolved photoemission spectroscopy, Phys. Rev. Lett. 98 (2007) 206802.

[71] S. Morozov, K. Novoselov, M. Katsnelson, F. Schedin, L. Ponomarenko, D. Jiang, Strong suppression of weak localization in graphene, Phys. Rev. Lett. 97 (2006) 016801.

[72] J. Jobst, D. Waldmann, F. Speck, R. Hirner, D.K. Maude, T. Seyller, et al., arXiv:0908.1900v1 (unpublished).

[73] T. Shen, J. Gu, M. Xu, Y. Wu, M. Bolen, M. Capano, Observation of quantum-Hall effect in gated epitaxial graphene grown on SiC (0001), Appl. Phys. Lett. 95 (2009) 172105.

[74] B.B. Jiang, M. Pan, C. Wang, M.H. Wu, K. Vinodgopal, G.P. Dai, Controllable synthesis of circular graphene domains by atmosphere pressure chemical vapor deposition, J. Phys. Chem. C. 122 (2018) 13572−13578.

[75] Y.R. Luo, Comprehensive Handbook of Chemical Bond Energies, CRC Press, 2007.

[76] Y. Zhang, L. Zhang, P. Kim, M. Ge, Z. Li, C. Zhou, Vapor trapping growth of single-crystalline graphene flowers: synthesis, morphology, and electronic properties, Nano Lett. 12 (6) (2012) 2810−2816.

[77] A. Dathbun, S. Chaisitsak, Effects of three parameters on graphene synthesis by chemical vapor deposition, NEMS, Suzhou, China, April 7−10, 2013.

[78] M. Robert, Jacobberger, R. Machhi, J. Wroblewski, B. Taylor, A. Lynn Gillian-Daniel, et al., Simple graphene synthesis via chemical vapor deposition, J. Chem. Educ. 92 (11) (2015) 1903−1907.

[79] R. Munoz, C. Gómez-Aleixandre, Review of CVD synthesis of graphene, Chem. Vap. Depos. 19 (2013) 297−322 (10-11-12).

[80] A. Cupolillo, N. Ligato, et al., Low energy two-dimensional plasmon in epitaxial graphene on Ni (111), Surf. Sci. 608 (2013) 88−91.

[81] H. Kim, C. Mattevi, et al., Activation energy paths for graphene nucleation and growth on Cu, ACS Nano 6 (4) (2012) 3614−3623.

[82] S. Marchini, S. Gunther, et al., Scanning tunneling microscopy of graphene on Ru (0001), Phys. Rev. B 76 (2007) 075429.

[83] S.K. Jerng, D.S. Yu, Y.S. Kim, J. Ryou, S. Hong, C. Kim, et al., Nanocrystalline graphite growth on sapphire by carbon molecular beam epitaxy, J. Phys. Chem. C. 115 (11) (2011) 4491–4494.

[84] H. Bi, S. Sun, F. Huang, X. Xie, M. Jiang, Direct growth of few-layer graphene films on SiO_2 substrates and their photovoltaic applications, J. Mater. Chem. 22 (2) (2012) 411–416.

[85] A. Ismach, C. Druzgalski, S. Penwell, A. Schwartzberg, M. Zheng, A. Javey, et al., Direct chemical vapor deposition of graphene on dielectric surfaces, Nano Lett. 10 (5) (2010) 1542–1548.

[86] T. Shujie, W. Haomin, W. Huishan, Silane-catalysed fast growth of large single-crystalline graphene on hexagonal boron nitride, Nat. Commun. 6 (2015) 6499. Available from: https://doi.org/10.1038/ncomms7499.

[87] X. Li, C.W. Magnuson, A. Venugopal, J. An, J.W. Suk, B. Han, et al., Graphene films with large domain size by a two-step chemical vapour deposition process, Nano Lett. 10 (11) (2010) 4328–4334.

[88] Y. Xue, B. Wu, Y. Guo, L. Huang, L. Jiang, J. Chen, et al., Synthesis of large-area, few-layer graphene on iron foil by chemical vapour deposition, Nano Res. 4 (12) (2011) 1208–1214.

[89] A. Moreno-Bárcenas, J.F. Perez-Robles, Y.V. Vorobiev, N. Ornelas-Soto, A. Mexicano, A.G. García, Graphene synthesis using a CVD reactor and a discontinuous feed of gas precursor at atmospheric pressure, J. Nanomater. 2018 (2018) 3457263. Available from: https://doi.org/10.1155/2018/3457263. 11 pages.

[90] S. Stankovich, D.A. Dikin, O.C. Compton, G. H. B., R.S. Ruoff, S.T. Nguyen, Systematic post-assembly modification of graphene oxide paper with primary alkyla-mines, Chem. Mater. 22 (2010) 4153–4157.

[91] H. Yang, F. Li, C. Shan, D. Han, Q. Zhang, L. Niu, et al., Covalent functionalization of chemically converted graphene sheets via silane and its reinforcement, J. Mater. Chem. 19 (2009) 4632–4638.

[92] H. Yang, C. Shan, F. Li, D. Han, Q. Zhang, L. Niu, Covalent functionalization of polydisperse chemically converted graphene sheets with amine-terminated ionic liq-uid, Chem. Commun. (2009) 3880–3882.

[93] T.A. Pham, N.A. Kumar, Y.T. Jeong, Covalent functionalization of graphene oxide with polyglycerol and their use as templates for anchoring magnetic nanoparticles, Synth. Met. 160 (2010) 2028–2036.

[94] T. Kuila, P. Khanra, S. Bose, N.H. Kim, B.C. Ku, B. Moon, et al., Preparation of water-dispersible graphene by facile surface modification of graphite oxide, Nanotechnology 22 (2011) 305710.

[95] Y. Si, E.T. Samulski, Synthesis of water-soluble graphene, Nano Lett. 8 (2008) 1679–1682.

[96] V.H. Pham, T.V. Cuong, S.H. Hur, E. Oh, E.J. Kim, E.W. Shin, et al., Chemical func-tionalization of graphene sheets by solvothermal reduction of a graphene oxide sus-pension in N-methyl-2-pyrrolidone, J. Mater. Chem. 21 (2011) 3371–3377.

[97] S. Stankovich, D.A. Dikin, G.H.B. Dommett, et al., Graphene based polymer compo-sites, Nature 442 (2006) 282–286.

[98] S. Stankovich, R.D. Piner, S.T. Nguyen, R.S. Ruoff, Synthesis and exfoliation of isocyanate-treated graphene oxide nanoplatelets, Carbon 44 (2006) 3342–3347.

[99] S. Niyogi, E. Bekyarova, M.E. Itkis, J.L. McWilliams, M.A. Hamon, R.C. Haddon, Solution properties of graphite and graphene, J. Am. Chem. Soc. 128 (2006) 7720–7721.

[100] Z. Liu, J.T. Robinson, X. Sun, H. Dai, PEGylated nanographene oxide for delivery of water-insoluble cancer drugs, J. Am. Chem. Soc. 130 (2008) 10876–10877.

[101] H. Hu, X. Wang, J. Wang, F. Liu, M. Zhang, C. Xu, Microwave-assisted covalent modification of graphene nanosheets with chitosan and its electrorheological characteristics, Appl. Surf. Sci. 257 (2011) 2637—2642.

[102] Y. Guo, S. Guo, J. Ren, Y. Zhai, S. Dong, E. Wang, Cyclodextrin functionalized graphene nanosheets with high supramolecular recognition capability: synthesis and host—guest inclusion for enhanced electrochemical performance, ACS Nano 4 (2010) 4001—4010.

[103] S. Stankovich, R.D. Piner, X. Chen, N. Wu, S.T. Nguyen, R.S. Ruoff, Stable aqueous dispersions of graphitic nanoplatelets via the reduction of exfoliated graphite oxide in the presence of poly(sodium 4-styrenesulfonate), J. Mater. Chem. 16 (2006) 155—158.

[104] H. Bai, Y. Xu, I. Zhao, C. Li, G. Shi, Non-covalent functionalization of graphene sheets by sulfonated polyaniline, Chem. Commun. (2009) 1667—1669.

[105] J. Zhang, J. Lei, R. Pan, Y. Xue, H. Ju, Highly sensitive electrocatalytic biosensing of hypoxanthine based on functionalization of graphene sheets with water-soluble conducting graft copolymer, Biosens. Bioelectron. 26 (2010) 371—376.

[106] H. Chang, G. Wang, A. Yang, X. Tao, X. Liu, Y. Shen, et al., A transparent, flexible, low-temperature, and solution-processible graphene composite electrode, Adv. Funct. Mater. 20 (2010) 2893—2902.

[107] Q. Zeng, J. Cheng, L. Tang, X. Liu, Y. Liu, et al., Self-assembled graphene—enzyme hierarchical nanostructures for electrochemical biosensing, Adv. Funct. Mater. 20 (2010) 3366—3372.

[108] Q. Yang, X. Pan, F. Huang, K. Li, Fabrication of high-concentration and stable aqueous suspensions of graphene nanosheets by noncovalent functionalization with lignin and cellulose derivatives, J. Phys. Chem. C 114 (2010) 3811—3816.

[109] T.Y. Kim, H. Lee, J.E. Kim, K.S. Suh, Synthesis of phase transferable graphene sheets using ionic liquid polymers, ACS Nano 4 (2010) 1612—1618.

[110] T. Ramanathan, A.A. Abdala, S. Stankovich, et al., Functionalized graphene sheets for polymer nanocomposites, Nat. Nanotechnol. 3 (6) (2008) 327—331.

[111] J.H. Lee, J. Marroquin, K.Y. Rhee, S.J. Park, D. Hui, Cryomilling application of graphene to improve material properties of graphene/chitosan nanocomposites, Compos. B 45 (1) (2013) 682—687.

[112] J. Guo, L. Ren, R. Wang, C. Zhang, Y. Yang, T. Liu, Water dispersible graphene noncovalently functionalized with tryptophan and its poly(vinyl alcohol) nanocomposite, Compos. B 42 (8) (2011) 2130—2135.

[113] G.W. Jeon, J. An, Y.G. Jeong, High performance cellulose acetate propionate composites reinforced with exfoliated graphene, Compos. B 43 (8) (2012) 3412—3418.

[114] C.M. Santos, M.C.R. Tria, R.A.M.V. Vergara, F. Ahmed, R.C. Advincula, D.F. Rodrigues, Antimicrobial graphene polymer (PVK-GO) nanocomposite films, Chem. Commun. 47 (31) (2011) 8892—8894.

[115] C.M. Santos, J. Mangadlao, F. Ahmed, A. Leon, R.C. Advincula, D.F. Rodrigues, Graphene nanocomposite for biomedical applications: fabrication, antimicrobial and cytotoxic investigations, Nanotechnology 23 (39) (2012). Article ID 395101.

[116] I.E.M. Carpio, C.M. Santos, X. Wei, D.F. Rodrigues, Toxicity of a polymer-graphene oxide composite against bacterial planktonic cells, biofilms, and mammalian cells, Nanoscale 4 (15) (2012) 4746—4756.

[117] Z. Wu, X.D. Chen, S. Zhu, et al., Enhanced sensitivity of ammonia sensor using graphene/polyaniline nanocomposite, Sens. Actuators B 178 (2013) 485—493.

[118] F.L. Meng, H.H. Li, L.T. Kong, et al., Parts per billion-level detection of benzene using SnO_2/graphene nanocomposite composed of sub-6 nm SnO_2 nanoparticles, Anal. Chim. Acta 736 (2012) 100−107.

[119] L. Zhang, S. Wang, D. Cai, et al., $Li_3V_2(PO_4)3$@C/graphene composite with improved cycling performance as cathode material for lithium-ion batteries, Electrochim. Acta 91 (2013) 108−113.

[120] J. Li, D. Kuang, Y. Feng, et al., Green synthesis of silver nanoparticles-graphene oxide nanocomposite and its application in electrochemical sensing of tryptophan, Biosens. Bioelectron. 42 (2013) 198−206.

[121] J. Cheng, H. Xin, H. Zheng, B. Wang, One-pot synthesis of carbon coated-SnO_2/graphene-sheet nanocomposite with highly reversible lithium storage capability, J. Power Sources 232 (2013) 52−158.

[122] H. Xia, D.D. Zhu, Y. Fu, X. Wang, $CoFe_2O_4$-graphene nanocomposite as a high-capacity anode material for lithium ion batteries, Electrochim. Acta 83 (2012) 166−174.

[123] Y. Zhao, Y. Huang, Q. Wang, X. Wang, M. Zong, Carbondoped Li_2SnO_3/graphene as an anode material for lithium-ion batteries, Ceram. Int. 39 (2) (2013) 1741−1747.

[124] H. Xia, Y. Qian, Y. Fu, X. Wang, Graphene anchored with $ZnFe_2O_4$ nanoparticles as a high-capacity anode material for lithium-ion batteries, Solid. State Sci. 17 (2013) 67−71.

[125] D. Zhang, Y. Zhang, L. Zheng, Y.Z. Zhan, L.C. He, Graphene oxide/poly-L-lysine assembled layer for adhesion and electrochemical impedance detection of leukemia K562 cancer cells, Biosens. Bioelectron. 42 (2013) 112−118.

[126] H. Yang, T. Song, S. Lee, et al., Tin indium oxide/graphene nanosheet nanocomposite as an anode material for lithium ion batteries with enhanced lithium storage capacity and rate capability, Electrochim. Acta 91 (2013) 275−281.

[127] J. Zhu, D. Wang, L. Wang, X. Lang, W. You, Facile synthesis of sulfur coated SnO_2-graphene nanocomposites for enhanced lithium ion storage, Electrochim. Acta 91 (2013) 323−329.

[128] S. Liu, J. Wang, J. Zeng, et al., Green electrochemical synthesis of Pt/graphene sheet nanocomposite film and its electrocatalytic property, J. Power Sources 195 (15) (2010) 4628−4633.

[129] X. Fu, J. Jin, Y. Liu, et al., Graphene- xerogel- based nonprecious metal catalyst for oxygen reduction reaction, Electrochem. Commun. 28 (2013) 5−8.

[130] L. Wang, Y. Zhang, Z. Li, Chemical reduced graphene oxide/AuPtPd nanocomposite for enhanced electrocatalytic ability, Mater. Lett. 94 (2013) 179−182.

[131] S. Yu, Q. Liu, W. Yang, K. Han, Z. Wang, H. Zhu, Graphene- CeO_2 hybrid support for Pt nanoparticles as potential electrocatalyst for direct methanol fuel cells, Electrochim. Acta 94 (2013) 245−251.

[132] Y. Cao, C. Xu, X. Wu, X. Wang, L. Xing, K. Scott, A poly (ethylene oxide)/graphene oxide electrolyte membrane for low temperature polymer fuel cells, J. Power Sources 196 (20) (2011) 8377−8382.

[133] G. Wang, S. Zhuo, W. Xing, Graphene/polyaniline nanocomposite as counter electrode of dye-sensitized solar cells, Mater. Lett. 69 (2012) 27−29.

[134] J.J. Zeng, C.L. Tsai, Y.J. Lin, Hybrid photovoltaic devices based on the reduced graphene oxide-based polymer composite and n-type GaAs, Synth. Metals 162 (15−16) (2012) 1411−1415.

[135] R. Bajpai, S. Roy, N. Koratkar, D.S. Misra, NiO nanoparticles deposited on graphene platelets as a cost-effective counter electrode in a dye sensitized solar cell, Carbon 56 (2013) 56−63.

[136] W.Y. Cheng, C.C. Wang, S.Y. Lu, Graphene aerogels as a highly efficient counter electrode material for dye-sensitized solar cells, Carbon 54 (2013) 291−299.

[137] S. Goswami, U.N. Maiti, S. Maiti, S. Nandy, M.K. Mitra, K.K. Chattopadhyay, Preparation of graphene-polyaniline composites by simple chemical procedure and its improved field emission properties, Carbon 49 (7) (2011) 2245−2252.

[138] Z.J. Li, B.C. Yang, G.Q. Yun, S.R. Zhang, M. Zhang, M.X. Zhao, Synthesis of Sn nanoparticle decorated graphene sheets for enhanced field emission properties, J. Alloy. Compd. 550 (2013) 353−357.

[139] F. Alvi, M.K. Ram, P.A. Basnayaka, E. Stefanakos, Y. Goswami, A. Kumar, Graphene polyethylenedioxythiophene conducting polymer nanocomposite-based supercapacitor, Electrochim. Acta 56 (25) (2011) 9406−9412.

[140] Y. Zhan, X. Yang, H. Guo, J. Yang, F. Meng, X. Liu, Cross linkable nitrile functionalized graphene oxide/poly (arylene ether nitrile) nanocomposite films with high mechanical strength and thermal stability, J. Mater. Chem. 22 (12) (2012) 5602−5608.

[141] S. Sahoo, S. Dhibar, G. Hatui, P. Bhattacharya, C.K. Das, Graphene/polypyrrole nano-fiber nanocomposite as electrode material for electrochemical supercapacitor, Polymer 54 (3) (2013) 1033−1042.

[142] D.W. Wang, F. Li, J.P. Zhao, W.C. Ren, Z.G. Chen, J. Tan, et al., Fabrication of gra-phene/polyaniline composite paper via in situ anodic electro-polymerization for high-performance flexible electrode, ACS Nano 3 (7) (2009) 1745−1752.

[143] H.L. Wang, Q.L. Hao, X.J. Yang, L.D. Lu, X. Wang, Graphene oxide doped polyani-line for supercapacitors, Electrochem. Commun. 11 (2009) 1158.

[144] S.D. Perera, A.D. Liyanage, N. Nijem, J.P. Ferraris, Y.J. Chabal, J.K.J. Balkus, Vanadiumoxide nanowire-Graphene binder free nanocomposite paper electrodes for supercapacitors: a facile green approach, J. Power Sources 230 (2013) 130−137.

[145] J. Yan, T. Wei, B. Shao, Z.J. Fan, W.Z. Qian, M.L. Zhang, et al., Preparation of a graphene nanosheet/polyaniline composite with high specific capacitance, Carbon 48 (2010) 487.

[146] P. Miranzo, C. Ramırez, B.R. Manso, et al., In situ processing of electrically conduct-ing graphene/SiC nanocomposites, J. Eur. Ceram. 33 (10) (2013) 1665−1674.

[147] P. Hvizdos, J. Dusza, C. Balazsi, Tribological properties of Si_3N_4-graphene nanocom-posites, J. Eur. Ceram. 33 (12) (2013) 2359−2364.

[148] P. Kun, O. Tapaszto, F. Weber, C. Balazsi, Determination of structural and mechani-cal properties of multilayer graphene added silicon nitride-based composites, Ceram. Int. 38 (1) (2012) 211−216.

[149] F. Inam, B.R. Bhat, T. Vo, W.M. Daoush, Structural health monitoring capabilities in ceramic-carbon nanocomposites, Ceram. Int. 40 (2) (2013) 3793−3798.

[150] G.B. Yadhukulakrishnan, S. Karumuri, A. Rahman, R.P. Singh, A.K. Kalkan, S.P. Harimkar, Spark plasma sintering of graphene reinforced zirconium diboride ultra-high temperature ceramic composites, Ceram. Int. 39 (6) (2013) 6637−6646.

[151] J. Liu, H. Yan, M.J. Reece, K. Jiang, Toughening of zirconia/alumina composites by the addition of graphene platelets, J. Eur. Ceram. 32 (16) (2012) 4185−4193.

[152] H. Liem, H.S. Choy, Superior thermal conductivity of polymer nanocomposites by using graphene and boron nitride as fillers, Solid. State Commun. 163 (2013) 41−45.

[153] K.I. Kim, T.W. Hong, Hydrogen permeation of TiN—graphene membrane by hot press sintering (HPS) process, Solid. State Ion. 225 (2012) 699−702.

[154] N.R. Lee, S.S. Lee, K.I. Kim, et al., Fabrications and evaluations of hydrogen perme-ation on Al_2O_3/CeO_2/graphene (ACG) composites membrane by Hot Press Sintering (HPS), Int. J. Hydrog. Energy 38 (18) (2013) 7654−7658.

Preparation/processing of polymer-graphene composites by different techniques

2

Sayan Ganguly
Department of Chemistry, The Institute of Nanotechnology & Advanced Materials, Bar-Ilan University, Ramat Gan, Israel

2.1   Introduction

Polymer composite is a special class of material having diversified phases on which reinforcing inclusions are amalgamated with a polymer matrix resulting synergistic improvement in its physicochemical properties [1,2]. In the case of nanocomposites, a similar kind of definition obeys except the filler inclusion is in nanodimension. There are several filler inclusions that are reported by various researchers [3,4]. The inclusions are normally metal oxides, carbon black, nanotubes, layered nanomaterials (phyllosilicates, graphene, and carbides) [5,6]. The incorporation of fillers into polymer matrices could improve the composites' mechanical, thermal, electrical, and gas barrier properties. Among these materials, the most nurtured one in today's research is graphene. Graphene-based polymeric nanocomposites are significantly drawing research attention due to their extraordinary mechanical properties in very low filler loading comparative to other conventional fillers [7,8]. The versatility of the composites with various polymers is depicted in brief in Fig. 2.1.

Before discussing polymer-graphene composites, we will cover graphene. Graphene is a 2D allotrope of carbon having sp^2 hybridized atoms arrayed in a honeycomb structure. It is also a building block of other graphitic carbonaceous materials. As an example, graphite is 3D allotrope of carbon which is made of stacked 2D sheets of graphene anchored by weak Van der Waal forces. In graphene, the repeating unit is benzene ring which is aromatic in nature due to planar π electrons. The floatable electrons present in graphene sheet act like massless relativistic particles leading to anomalous quantum Hall effect [9]. Graphene is unique because of its exceptional properties like high electronic mobility at ambient temperature (6000 S/cm) [10], high thermal conductivity (5000 W/mK) [11], and excellent elasticity (Young modulus is 1 TPa) [12]. The preparation of graphene is categorized into two major strategies; one is bottom-up graphene and another one is top-down graphene. In bottom-up approach graphene is prepared from small molecular states via special adopted methods like chemical vapor deposition [13], arc discharge [14], unzipping carbon nanotubes [15], chemically grown/nucleation [15], epitaxial growth over silicon carbide [16], and

Polymer Nanocomposites Containing Graphene. DOI: https://doi.org/10.1016/B978-0-12-821639-2.00015-X

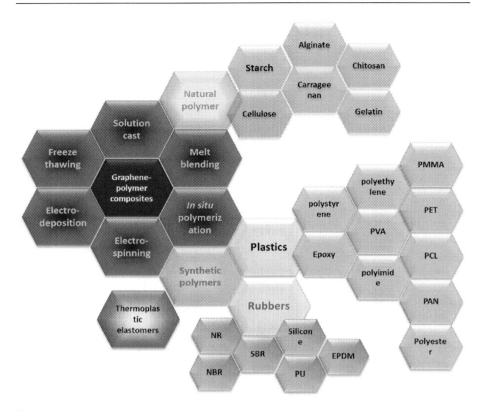

Figure 2.1 Brief outline of graphene-polymer composite preparation strategies and most well-practiced polymers.

self-assembly of surfactants [17]. Bottom-up method is suitable for small scale production and better scientific researches, but not good for large scale synthesis. In that context, "top-down" approach became more popular. Top-down process follows the route of dislodging 2D graphene sheets from the graphitic array by means of forces or chemical methods. Generally this method is applied already industrially for polymer composite manufacturing.

Polymer composites are already a vehemently uttered area of research and commercially successful domain. Various fillers are incorporated into polymer matrices for synergistic improvement in properties. There are several fillers that are already used in preparation of polymer composites to boost their mechanical properties. Among the fillers, graphene has drawn much attraction due to its extraordinary features. Graphene-based polymer composite is superior because of (1) their excellent improvement in Young modulus, (2) high aspect ratio, (3) minimally possible filler loading which leads to small increment in specific gravity, (4) easily achievable homogeneity, (5) very high strength to weight ratio compared to metals. As graphene is not abundant in polar functionalities in their basal planes, the dispersibility is relatively poor rather than other low cost fillers like phyllosilicates. But to achieve better physicomechanical properties, ensuring the proper compatibility of graphene into polymer matrix is important [18,19]. Therefore, functionalization of graphene sheets is a common task opted by various researchers in order to promote better solubility into several solvents and polymer matrices. The functionalization of graphene is also a pretreatment process of preparing graphene-polymer nanocomposites. This functionalization could be categorized into covalent and noncovalent [20−22]. The functionalization is required on graphene basal plane to improve the polymer compatibility of the graphene. As per the various reports covalent functionalization is better compared to noncovalent modification. Most of time graphene acts as a load bearing moiety, so better compatibilization is the most required strategy adopted by researchers. Noncovalent modification is easy and fast but one demerit is their weak van der Waal forces of attraction. Meanwhile, it also should be mentioned that sometimes covalent functionalization also has an adverse effect because of defect generation in graphene basal planes. These defects sometimes deteriorate the other special features of graphene like electrical and thermal conductivities. Covalent functionalization of grafting actually turns the sp^2 carbon atoms of graphene basal planes to sp^3 carbons which restrict the electronic flow in the sheets.

Covalent modification to graphene sheets is much better in terms of their thermal stability. Covalent modification of graphene oxide is a chemical method glass small organic molecules are subjected to attach onto that graphene basal plane by means of chemical bonding. in most of the cases graphene oxide has been selected due to the high abundance of hydroxyl carboxyl and epoxy groups onto its surface. Sometimes covalent attachment of porphyrins, phthalocyanines, and azobenzenes on the graphene surface promotes very significant optoelectronic properties. In the area of energy harvesting systems, the aforementioned covalent modified graphemes are very much not chart YouTube they are hi visible light extension coefficient. In these systems, porphyrins and phthalocyanine behave as antennas far energy harvesting via photons. The most common strategy adopted to prepare organically modified graphene sheets is via the reaction with octadecylamide (ODA) and consecutive reduction by hydroquinone. Such organo-modified graphene sheets are much compatible to the less polar/nonpolar polymer matrices. Polystyrene (PS) and poly(methyl methacrylate) (PMMA) blends were reported earlier where ODA modified graphene oxide was used as filler [23]. Polymer grafting on to graphene basal planes are also another way to promote dissolvability of graphene sheets into polymer matrices. As grafted polymer chains are lengthy resulting diffusive nature into base macromolecular phase which enhances the compatibility and stress transfer character. Polymer modified graphene oxides actually heightens the interfacial interaction among the macromolecular phases and filler particles. More the interfacial interaction more will be their extent of stress transfer. The most significant point of discussion in this chapter is their two large nurtured domains of functionalization; one is "grafting to" and another one "grafting from" techniques. Besides these, some noncovalent approaches and other commonly practiced synthetic methods are also discussed in the further area.

2.2   Preparation techniques of polymer graphene nanocomposites

Synergistic improvement of the desired properties of the composites is a key objective for developing new composites. The fabrication methods are depicted in Fig. 2.2. Dispersion of nanofillers into polymer matrix is the important step to prepare graphene-polymer nanocomposites. For addition of small amount of nanofillers normally improve the properties in several folds. Graphene sheets are anisotropic in nature and thus for more surface contact becomes more feasible for graphene-polymer nanocomposites. But die to their high surface energy it is a quite difficult task to distribute uniformly of the graphene platelets into the whole polymer matrix. Most of the graphene reinforced polymeric nanocomposites are physically prepared in terms of three basic methods: (1) melt blending, (2) solution mixing, and (3) in situ polymerization.

2.3   Physical mixing methods

2.3.1   Melt blending

Melt blending is a practical, versatile, and industrially viable technique for the preparation of graphene/polymer nanocomposites. The melt blending is a less hazardous

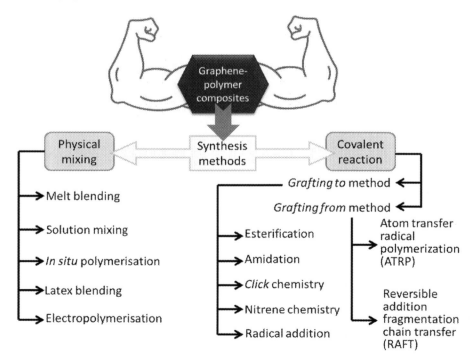

Figure 2.2 Synthesis methods of graphene-polymer composites.

process in terms of usage of organic volatile solvents. This method normally uses shear force and heat to mix the graphene filler into polymer matrix [24]. This process is much more acceptable due to mass scale production of nanocomposites. There are several practical examples of melt mixing based nanocomposite fabrication. The most common polymers used in this method are polyethylene (PE) [25−27], polypropylene (PP) [28−30], poly(ethylene terephthalate) (PET) [31−33], polyurethane (PU) [34], poly(methyl methacrylate) (PMMA) [35], poly(lactic acid) (PLA) [36−38], and polyphenylene Sulfide (PPS) [39]. The fabrication of composites consists of two stages: one is mechanical mixing of graphene sheets into the polymer and the second one is the use of some molding methods viz. injection, extrusion or melt spinning of the composite dough. Polyamide-based graphene nanocomposite had been reported by melt mixing method [40]. The prepared nanocomposites are superior with respect to their virgin polymer state regarding their tensile strength, toughness, and permanent deformation. Besides these, graphene oxide and nylon six-based nanocomposites were reported by Nguyen and his coworkers [41]. When graphene and polymers is melt mixed, the shear force are the main contributing force which made the graphene platelets intercalated followed by increment in gallery gap. The applied heat during the melt mixing method acts as plasticizer in mixing. Due to extensive mixing process the macromolecular chains are going to diffuse in the gallery gap of the graphene. Sometimes polar functional groups present in the basal planes of graphene oxide help to improve the interaction between the polar macromolecular chains and graphene sheets.

Melt blending has another significance of in situ reduction of GO to graphene in the presence of heat and shear forces. Commercially viable melt interaction already has enough impact in low cost fabrication domain. GO is sometimes less interested to be filled into commodity polymers because of its high polarity. In situ reduced graphene-based polymer nanocomposites are relatively new in this area. GO can be reduced inside polymer matrix by means of heat and vigorous shear forces [42]. High shear force dislodges the graphene platelets inside the mixing mill followed by the stabilization of the platelets by polymer matrix [42]. This method is not only fast but also relatively energy saving compared to other methods. Morphological variation is also another crucial feature of all graphene polymer nanocomposites as already reported elsewhere [43]. Melt mixing method showed better orientation of graphene platelets compared to solution and in situ polymerization method as shown in Fig. 2.3. Melt mixing composites showed more thick and clumsy morphology. This typical morphology is better for uniaxial tensile property. Melt mixing method could not exfoliate all platelets. Here the composite contains a mixture of intercalated and exfoliated graphene platelets in all over the polymer matrix. Now melt compounding is comprised of two different type graphenes; one is thermally reduced graphene oxide (TRGO) which is an outcome of thermal reduction of GO and another one is functionalized GO-based melt mixed compounds.

2.3.1.1 Thermally reduced graphene oxide-polymer composite fabrication

Thermally stable polymers are mostly used in this method. Graphene reinforced PET nanocomposite was reported earlier which was prepared by internal mixer [31]. The report revealed that melt mixed graphene-PET nanocomposite showed 10% rising of

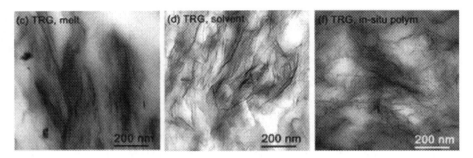

Figure 2.3 Comparative images of the graphene reinforced PU nanocomposites in three different processing methods. First one is the melt mixing, middle one the solution method, and the right one is in situ polymerization method. All composites contain 3 wt.% of graphene filler. The images provide a clear idea of dispersion of graphene platelets in various mixing methods. Melt minx showed most oriented morphology than others.
Source: Reprinted with permission from H. Kim, Y. Miura, C.W. Macosko, Graphene/polyurethane nanocomposites for improved gas barrier and electrical conductivity, Chem. Mater. 22 (2010) 3441−3450. © 2010 American Chemical Society.

their elastic modulus and 40% enhancement of tensile property compared to virgin PET. Later Arby et al. developed graphene reinforced styrene butadiene rubber (SBR) nanocomposites for high mechanical and electrical property [44]. For this they used graphene nanoplatelets (GNPs) which were intercalated by shear force enhanced mixing method. In another work polyamide-12 based blend was melt mixed with graphene for better mechanical property [45]. For this type of blend polyethylene octane rubber phase was mixed with polyamide-12 and twin screw was used at 220°C for better shear action. This blend was reported as electrically conductive with a very low percolation threshold concentration of graphene (0.3 vol.%). Melt blending is not only improving the dispersion of filler into polymer matrix but also it helps to make a better bond between filler particles and macromolecular chains. In order to evaluate the interfacial interactions play in this graphene-polymer nanocomposites TRGO, and polycarbonate (PC) nanocomposite was studied [46]. As per the report, they showed that melt mixing infers two types of interactions acting in between filler particles and polymer chains. The major factor is the chemical bonding between PC and graphene basal polar groups. This leads to formation of ester linkages which are liable for holding the graphene platelets by PC chains. The second type of interaction was purely physical; named as $\pi-\pi$ interaction between PC and graphene sheets. In another work, TRGO was melt mixed with maleated liner low density polyethylene in the presence of pyridine derivatives [47]. They revealed that $\pi-\pi$ interaction also happened in this system.

2.3.1.2 Functionalized graphene oxide-polymer composite fabrication

Though interfacial interaction has been discussed in the melt blending process, there are some problems about interactions and compatibility. That's why scientists proposed some functionalization over graphene sheets which could be a better choice comparative to the pure graphene. In this context, "grafting to" method was adopted by researchers where maleated polypropylene was grafted over graphene sheets [48]. Such typical functionalized graphene oxide (FGOs) are compatible to mix to the other polymer matrices. Grafted reduced graphene oxide (RGO) enhances the interfacial interactions and better dispersion of filler into polymer matrix. In another work functionalized graphene was mixed with PLA in two roll mixer to form master batch [49]. They showed that functionalized graphene improved the composite crystallinity, mechanical properties, electrical conductivity, and fire resistance. These features have obvious connections to the dispersion of fillers. A similar kind of master batch mixing was also reported by another group of researchers where functionalized graphene was melt mixed to PP granules.

2.3.2   Solution mixing

Solution mixing or solution casting technique is a wet mixing method to prepare graphene polymer composites in a homogeneous manner. This is done by dispersing of graphene filler in a solvent followed by sonication and addition of polymer. In liquid medium graphene could be easily exfoliated into separate platelets which could be stabilized by the macromolecular chains. In this method choice of solvent should be done in such way where both the graphene and the polymer both have sufficient affinity to get dispersed as uniform as possible. If the whole procedure of fabrication could be summed up, it can be stated that solvent casting follows three general steps; (1) the first one is the dispersion of filler (here graphene of any functionalized graphene), (2) second one is insertion of polymer into the solvent for proper dissolution and finally (3) removing of the solvent and settling down the whole system into a composite. The most common polymers which are studied in solvent mixing method are mostly used in several literatures are PU [50], poly(vinyl alcohol) (PVA) [51−53], PMMA [54], PS [55,56], epoxy [57,58] etc. Besides these, several thermoplastic elastomers are also handled in wet mixing method in order to prepare graphene polymer nanocomposites. The most common is poly(styrene-b-isoprene-b-styrene) or SIS [59]. Solvent assisted prolonged time sonication normally increase the gallery gap of the graphene sheets and weakens the van der Waal interactions. This leads to the proper exfoliation of graphene sheets in solvent medium. This opening of interlayer spacing in graphene is permanent and already proved [60]. It has been already proved by solid state C^{13} NMR spectra. As per the reports, after solvent exfoliation and drying, a small fraction of liquid solvent has always been entrapped inside graphene lattice/interlayers. In another research, it was reported that solution casted polyamide-graphene nanocomposites are thermally stable [61]. Relative to other common nanocomposite preparation methods, solvent mixing is easy but it has one major demerit of solvent vaporization. Solvent removal step is not environmentally friendly. But the environmentally friendly method was developed in another research where graphene oxide and polyamide were taken for composite preparation. It was gradual grafting of polymer chains onto graphene oxide basal planes which improved the compatibility of graphene sheets into polymer matrix resulting excellent enhancement in mechanical property. Solvent casting also delayed the decomposition of the prepared composites with gaining sufficient heat resistance.

2.3.3 In situ polymerization

In situ polymerization technique is a highly effective method to fabricate graphene polymer nanocomposites with excellent dispersability and homogeneity. It is also true that in situ method is also superior to the aforementioned methods. In this preparation, monomer molecules have been taken in the presence of graphene/functionalized graphene. The polymerization system consists of monomer, filler/ reinforcement particles, initiator(s), and solvent. The demerit of this process is viscosity of the polymer mix. The increment of viscosity is inevitable in this case, so bulk phase polymerization is partially risky and inefficient for this method due to heat generation. Now, this technique could be categorized into various aspects of interactions between fillers and polymers. In this chapter, those typical areas/fabrication procedures are discussed here.

2.3.3.1 Noninteractive filler-polymer nanocomposites

Thermosetting graphene polymer composites are major systems in this category. RGO filled phenol-formaldehyde was reported earlier [62]. In this work researchers took GO as filler which was mixed to phenol and formaldehyde. Here phenol acted as a reducing agent, thus GO was reduced to RGO during in situ reduction process. The also prepared differently loaded RGO thermoset composites. As per their microscopic study of the RGO-based composites it was seen that the miscibility of RGO sheets inside the thermoset matrix was comparatively better than melt and solution mixing processes. That's why it is also a good method to achieve better thermal and electrical conductivity. Moreover, such good dispersion is also effective for enhancing their mechanical properties. In another work, Paszkiewicz and coworkers reported poly(trimethylene terephthalate-blocktetramethylene oxide) copoymeric system reinforced with GO by in situ polymerization strategy [63]. They established a novel finding where oxygenated functionalities present in GO basal planes take part in polymerization centers. In this work they also got exceptionally good mechanical features of the GO reinforced copolymeric matrix. Water soluble graphene-poly(aniline) was also reported with good electrical conductivity in another work [64]. In several reports, some researches showed that surface functionalized graphene is much better than GO respect to their compatibility and dispersion in polymer matrices. Cyclohexyl isocyanate (CI) grafted GO was filed into polyimide via in situ polymerization method [65]. Here, the grafting agent (CI) was actually a precursor material of polyamic acid which is the basic monomer unit of PI. That's why CI grafted GO has an effective chance to get attached and compatible to the PI matrix without any phase separation problem. Graphene loaded in situ polymerized nanocomposite hydrogels are also another domain of interest in biomedical area. There are several reported on graphene reinforced hydrogels which showed robust mechanical features associated to electrical conductivity [66]. Pristine graphene was prepared from expanding graphite via acoustic cavitation which had been used as reinforcing filler for graphene nanocomposite hydrogel [67]. The preparation was ecofriendly and without using any vigorous reaction condition. The fabrication strategy has been given in Fig. 2.4.

In hydrogel systems, graphene colloidal sheets were arrested into gel matrices resulting highly stretchable character also. The gel matrix is actually playing a role of stabilization of the graphene nanosheets inside the matrix. Moreover, graphene also acted as a polymerization center for in situ polymerization process which could directly affect the mechanical quality of the end products. In the case of in situ nanocomposites, polymeric chains are diffused to the inter-sheet area (gallery gaps) of the graphene/GO resulting intercalation and physisorption onto basal planes. Such physisorption enhances the better dispersion into matrix. Better dispersion in in situ polymerized systems also affects the delayed rupturing phenomenon of the nanocomposites via immediate stress transferring from graphene sheets to viscoelastic phases.

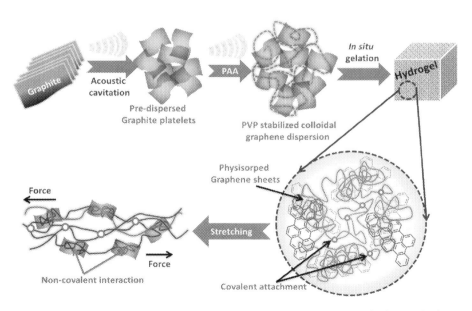

Figure 2.4 Schematic of in situ graphene-based hydrogel via acoustic cavitation method. The image showed the synthesis procedure, their microstructure, and alteration of their microstructure after experiencing external uniaxial load.
Source: Reprinted with permission from S. Ganguly, D. Ray, P. Das, P.P. Maity, S. Mondal, V. Aswal, et al., Mechanically robust dual responsive water dispersible-graphene based conductive elastomeric hydrogel for tunable pulsatile drug release, Ultrason. Sonochem. 42 (2018) 212–227. © 2018 Elsevier.

2.3.3.2 Interactive filler-polymer nanocomposites

The interaction between graphene/GO and polymers are the most crucial for better interactions with respect to the noninteractive graphene-polymer composites. Poly (butylene succinate) and GO-based composite was prepared for establishing this feature [68]. As GO has plenty of polar functionalities in their basal planes, it has a chance to attach covalently to the polar group containing polymers. Here, the author also proposed ester formation between Poly(butylene succinate) polymer and GO polar groups. This covalent interaction causes better adhesion to the filler particles and effective homogenization. In another work, GO filled polyamide six nanocomposite was reported by in situ anionic ring-opening polymerization. In their experiment initially 4,4'-methylenebis(phenylisocyanate) grafted GO was dispersed into monomer solution followed by in situ polymerization technique. As per their report 74 wt.% grafted GO was quite effective for homogenous dispersion of the fillers into polymer matrix. Moreover, in this case, the physicomechanical properties were also enhanced to some extent.

Interaction between filler and polymer is not only restricted into the covalent reactions, but also noncovalent adsorption was another area where physically interactive attachment possibly took place. GO-polyaniline (PANI) composite is a very familiar nanocomposite in this area [69]. The composite was evaluated by various characterizations inferring that GO and PANI had $\pi-\pi$ interaction between the graphene sheets and the benzene rings present in PANI. Moreover, as the PANI was in doped state, there was an electrostatic interaction between the anionic GO sheets and the PANI macrochains. This type of composite showed excellent electrochemical performances. PANI deposited over graphene paper based composite was another work via in situ electropolymerization [59]. Graphene-based PANI composite was also prepared by spin coating method. In situ polymerization of graphene and PANI was also reported via surfactant stabilized system [70]. This composite was evaluated to be 526 F/g of specific capacitance. Layered PANI-graphene composite was fabricated by Tong et al which showed eco-friendly route to prepare such composite [71]. This composite showed binder free fabrication strategy having $\pi-\pi$ interactions, H-bonding, and strong electrostatic interaction between the graphitic plane and PANI benzene rings. In a similar fashion in situ graphene -Poly (3,4-ethyldioxythiophene) (PEDOT) nanocomposite was prepared [72]. Graphene-PEDOT composite was very stable at high temperature and showed excellent electrical conductivity. There are several thermoplastic matrices also showed excellent homogenous in situ graphene-polymer nanocomposites. PS filled with graphene nanosheets was developed by Hu et al. via in situ emulsion polymerization [73]. As PS has benzene rings in their pendent chain structure, the interaction between PS and graphene planes are very much promising as already observed and discussed before. In general virgin PS has comparatively low degradation temperature than other commercially available commodity plastics. But after incorporation of graphene into PS matrix the thermal stability improves drastically. Graphene nanosheets also have severe impact over the electrical conductivity in polymer-graphene composites as reported elsewhere [74]. Water borne polyurethane

(WPU)-based nanocomposite was fabricated via in situ polymerization in presence of functionalized graphene sheets [75]. This work improved the electrical conductivity at least 105 times than their unfilled WPU matrix.

2.3.3.3 In situ Ziegler-Natta polymerization

Graphene reinforced polymer composites have enormous demand not only in academic/research purposes but also in several commercial product manufacturing. But it is quite difficult to disperse graphene into polyolefinic materials due to high polarity differences. But this process normally consists of two steps. One is graphene supported Zeigler-Natta (ZN) catalyst and the second one is the coordination polymerization. Huang et al. prepared GO reinforced PP nanocomposites via ZN polymerization [76]. They initially developed ZN catalyst which was based on ZN catalyst. In that case they followed the Grignard reaction over graphene sheets. The whole reaction procedure and composite preparation has been shown in Fig. 2.5.

Zhao et al. reported isotactic polypropylene (iPP) and graphene nanocomposite prepared via in situ ZN polymerization [76]. They took GO as the raw materials which was reduced in presence of Grignard reagent (n-BuMgCl) forming RGO supported ZN catalyst. In a similar fashion GNPs reinforced in situ ZN polymerized nanocomposite was also reported in another work [77]. The most promising result was that this composite showed excellent homogeneity in matrix phase which was confirmed from transmission electron microscopic images. In this case the electrical

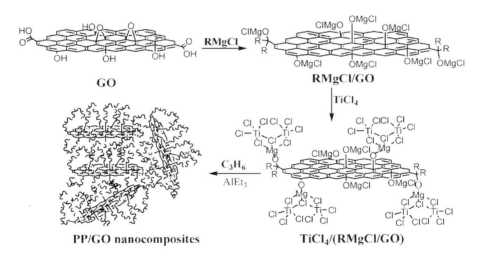

Figure 2.5 Fabrication of GO filled polypropylene nanocomposite via in situ Ziegler-Natta polymerization.
Source: Reprinted with permission from Y. Huang, Y. Qin, Y. Zhou, H. Niu, Z.-Z. Yu, J.-Y. Dong, Polypropylene/graphene oxide nanocomposites prepared by in situ Ziegler − Natta polymerization, Chem. Mater. 22 (2010) 4096−4102. © 2010 American Chemical Society.

conductivity was increased up to 10^8 times than unfilled iPP. Stürzel et al. proposed a new method of fabrication where they used a dual-site catalyst system [78]. In this case they used ultra-high molecular weight polyethylene (UHMWPE) based nanocomposites master batches.

2.3.4 Latex phase blending

Graphene dispersions are not easy to process due to their high surface energy and prone to agglomeration. That's why several researchers tried to make nanocomposites via high temperature and high shear force mixing techniques. Besides these, graphene incorporation also increases the medium viscosity of the whole system which difficult to flow. Thus to low down the medium viscosity was also a major part of research for better surface finish of the end products. Several researchers studied the latex phase blending of the graphene fillers into polymer matrices. This mixing method was done into two different domains; one is plastic based system and another one is the rubber based systems. This in this portion we will discuss about graphene-plastic nanocomposites and graphene-rubber nanocomposites where both were prepared in latex blending procedure.

2.3.4.1 Graphene-plastic latex blending

UHMWPE and RGO-based nanocomposites were developed in water/ethanol mixture which was termed as latex mixing [79]. In this work researchers first dispersed the Go into UHMWPE system where the GO sheets coated the polymer globules. After that the whole system was reduced by hydrazine followed by compression molding. In another report by Pang et al. also reported RGO-UHMWPE nanocomposites fabricated by similar fashion [80]. For achieving the high conductivity in electronic applications PMMA and RGO was mixed in latex phase [81]. The composite was self-assembled and surfactant free fabrication method as shown in Fig. 2.6. This method was established effective for homogeneous and high conductive nanocomposites. The mechanism behind this self-assembly was pure electrostatic interaction between PMMA particles and GO particles. PMMA was cationic here and GO colloidal system was negative in nature. After anchoring together, they were reduced by hydrazine where RGO-based nanocomposite was developed.

2.3.4.2 Graphene-rubber latex blending

Mixing of graphene into rubber latex mixture is a highly practiced field. Zhan et al. reported simple preparation of graphene-natural rubber (NR) nanocomposite preparation via latex stage mixing followed by hot pressing and vulcanization [82]. The mixing of GO into latex was relatively easy with respect to other polymer matrices because of its polar character. Here, the author initially mixed the GO by ultrasonication and recued it by hydrazine. The recued GO was assembled inside the latex which acted as stabilizing agent. Graphene rubber composite was also fabricated by Ting et al. by similar kind of ultrasonication assisted mixing method [83]. The fabrication

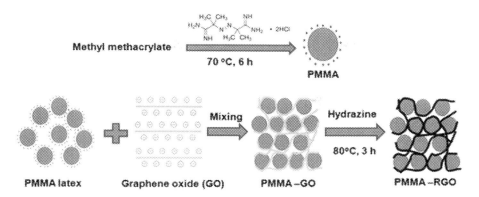

Figure 2.6 Fabrication of PMMA based self-assembled graphene nanocomposites.
Source: Reprinted with permission from V.H. Pham, T.T. Dang, S.H. Hur, E.J. Kim, J.S. Chung, Highly conductive poly (methyl methacrylate)(PMMA)-reduced graphene oxide composite prepared by self-assembly of PMMA latex and graphene oxide through electrostatic interaction, ACS Appl. Mater. Interfaces 4 (2012) 2630–2636. © 2012 American Chemical Society.

strategy has been given in Fig. 2.7. In this work they took carboxylated nitrile rubber (XNBR) latex as the matrix which has excellent compatibility to GO by their polar carboxylic acid groups. They proposed spin flash drying process followed by in situ thermal reduction of GO. Surfacetant and ultrasonic field free latex mixing was also adopted elsewhere which showed mixing of graphene into SBR latex having excellent gas barrier property, high heat resistance, low heat build-up and good abrasion resistance [84]. Ammonia stabilized NR latex (NRL) was reinforced by RGO nanosheets by latex mixing process which was designated to be a green process [85]. NRL and graphene nanocomposite could be arranged as ordered at microscopic range by a method called vacuum assisted self-assembly or VASA. In this procedure very high amount of RGO could be incorporated into NRL matrix which any phase separation [86]. For electrically conductive nanocomposites segregated microstructure is very important for low percolation threshold. In this context, NRL latex was mixed with Poly(diallyldimethylammonium chloride) modified graphene [87]. Such modified graphene has cationic in nature which was anchored to negatively charged NRL particles and vulcanized. This type of nanocomposite showed excellent electrical conductivity and very good mechanical properties with respect to other latex based nanocomposites. Suriani et al. reported electrochemically exfoliated graphene as conductive filler into NRL for supercapacitor applications [88]. Beside these works several researchers also reported lots of NRL and graphene nanocomposites for different applications [89–91]. Emulsion polymerized PS latex was used with RGO in another work [92]. The researchers modified graphene by PS sulfonate (PSS) macrochains for increasing the compatibility to the PS latex phase. After that they mixed modified graphene with the PS latex by simple single stage blending followed by lyophilization and compression melding. The uniqueness of this process was that it helped to reduce the

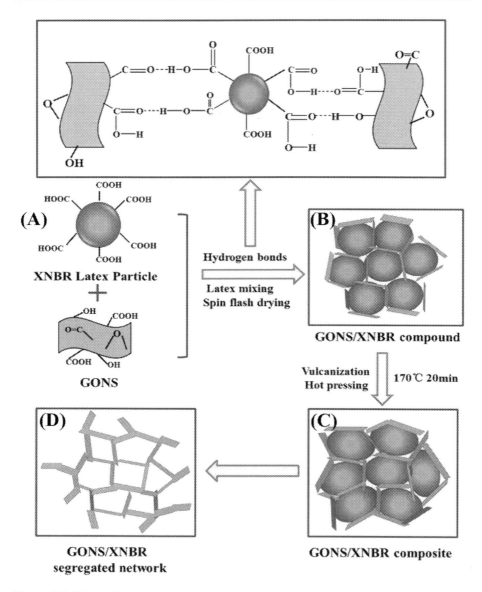

Figure 2.7 Schematic representation of preparation of graphene-XNBR nanocomposites by latex mixing procedure.
Source: Reprinted with permission from M. Tian, J. Zhang, L. Zhang, S. Liu, X. Zan, T. Nishi, et al., Graphene encapsulated rubber latex composites with high dielectric constant, low dielectric loss and low percolation threshold, J. Colloid Interface Sci. 430 (2014) 249–256. © 2014 Elsevier.

filler volume percentage to attain electrical percolation. Heterocoagulation process is another method of latex blending which had been adopted by Kim et al. cationic stabilizer was initially used to stabilize multilayer graphene sheets in order to improve its cationic charge on the basal planes. After that scientists mixed that with SBR latex followed by curing. They reported at least 6 fold increments in electrical conductivity after addition of 0.5 wt.% filler which was relatively lower amount observed in other graphene-latex nanocomposites.

2.3.5 Electropolymerization

Electropolymerization is relatively new and efficient method of fabrication of graphene-polymer composites for conductive applications. This method is easy to operate, time saving, and ecofriendly in nature. There are two typical electrode systems applied for this process; one is "two electrodes" and another one "three electrode." The end products obtained from this method are generally applied for the electrochemical sensing and energy storage like supercapacitors and batteries.

The most common electropolymerized composite is PANI and graphene based. Hu et al. prepared a graphene and PANI based nanocomposite via in situ electropolymerization in reverse micelle electrolyte [93]. Reverse micelle is a water-oil mixture where indium tin oxide (ITO) electrode was taken as working electrode, graphene normally taken as counter electrode. Electrophoresis was taken place after external pulsed current in the cell resulting pulling of water body toward working electrode. After the collision of water body to electrode, the monomers (aniline) molecules are getting oxidized and deposited over the electrode. In the case of graphene dispersion in aniline medium, the monomers are physically adsorbed over the graphene sheets. These sheets were then pulled toward working electrode forming graphene-PANI interconnects. Aniline monomer is very much susceptible to adhere with the graphene sheets by means of hydrogel bonding, thus graphene and aniline both traversed to electrode and as a result a homogeneous conducting composite might be obtained. Wang et al. fabricated free standing graphene PANI composite membrane for supercapacitor application [94]. They took three electrode system for in situ anodic electropolymerization where Pt was taken as counter electrode and SCE as reference electrode. The working electrode was specially designed by graphene paper via reduction of GO thick film. The monomer aniline was dissolved into sulfuric acid followed by electropolymerization. The whole process has been depicted in Fig. 2.8.

Liu et al. reported polypyrrole (ppy) and graphene nanocomposite for dye sensitized solar cell application [95]. In this work they used 0.5 M Na_2SO_4 aqueous solution in nitrogen medium at a scan rate of 50 mV/s. In this preparation method they used fluorine doped tin oxide (FTO) or ITO as working electrode, Ag/AgCl as reference electrode and Pt wire as counter electrode. In this case they got ppy wrapped GO which showed excellent electrical conductivity. Ppy-graphene composite was tested by electrochemical impedance spectroscopy and showed excellent catalytic activity against triiodide reduction. In the similar fashion as discussed above, graphene sheets were wrapped by PANI in several reported literatures [93,96–99]. Damlin reported

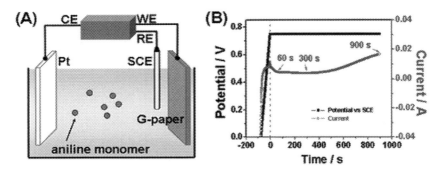

Figure 2.8 (A) Schematic illustrating the anodic electropolymerization of aniline monomer on G-paper (B) the potential and current response recorded during GO-polyaniline electropolymerization on G-paper. *CE*, counter electrode (Pt plate); *RE*, reference electrode (SCE); *WE*, working electrode (G-paper).
Source: Reprinted with permission from D.-W. Wang, F. Li, J. Zhao, W. Ren, Z.-G. Chen, J. Tan, et al., Fabrication of graphene/polyaniline composite paper via in situ anodic electropolymerization for high-performance flexible electrode, ACS Nano 3 (2009) 1745−1752. © 2009 American Chemical Society.

PEDOT and GO-based nanocomposite in presence of ionic liquid [100]. In this work they used polished Pt electrode and FTO electrode system. Li et al. fabricated PEDOT-graphene composite via single step cyclic voltammetry without using any surfactant [101]. They developed Electrochemiluminescence Immunoassay for quantum dots. Graphene quantum dots (GQDs) are relatively new comer in polymer composite domain which has exceptionally good optical features, noncytotoxicity and reinforcing character in polymer matrices [102−104]. PEDOT-GQDs composites had been prepared via electropolymerization of 3,4-ethylene dioxythiophene (EDOT) by Abidin et al. which was applied for high performance supercapacitor [105].

2.4 Covalent mixing method

Graphene to graphene oxide synthesis was first reported by Hummer which could produce bulk amount of GO. Reduction of GO sheets produces RGO where GO sheets are partially reduced. Both the GO and RGO are consists of polar functional groups. These functionalities are very important for a reaction to further with other functional groups of materials. Modification of GO/RGO nanosheets are carried out by covalent attachment of molecules in two ways; one is "*grafting to*" method and another one is "*grafting from*" method.

2.4.1 "Grafting to" method

This method is a two-step process where the initial step is the synthesis of polymer followed by attachment reaction to GO/RGO basal plane. At the first step a polymer has been made which was chosen in such a way that the functionalities of graphene surface could attach covalently with the polymer pendent groups. There are several commonly practiced reactions by which *"grafting to"* method could be inferred. The significant reactions are esterification, amidation, click chemistry, nitrene, and radical addition types. The reactions are discussed here in brief.

2.4.1.1   Esterification chemistry

Easterificaiton in between the carboxylic acid group of GO and hydroxyl group of PVA was carried out by Salavagione et al. in presence of N,N0-dicyclohexylcarbo-diimide (DCC) and 4-dimethylaminopyridine (DMAP). DCC and DMAP are acted as zero chain length crosslinker and catalyst for grafting of PVA onto the basal planes of GO via esterification reaction [106]. After grafting the whole system was reduced by hydrazine which was depicted in Fig. 2.9.

Grafting of PVA onto GO surface affects the glass transition temperature, crystallinity, and thermal stability of the PVA matrix. They showed that 10 wt.% GO restricted the movement of PVA macrochains and increased was evaluated of 35°C. An important finding was reported here that the crystallinity of PVA was totally diminished after incorporation of GO via esterification. In another experiment, GO was treated with $SOCl_2$ to convert GO to acid chloride. Similar kind of experiment was also reported with poly(vinyl chloride) which showed $\sim 30°C$ increment of T_g after addition of 1.2 wt.% GO as filler. The mechanical property was also drastically changed after this modification which reflected in the 70% increment in storage modulus [107]. Moreover, as PVA is always suffering from premature degradation by heat, small amount of grafted GO improved the thermal stability. Some biopolyemrs

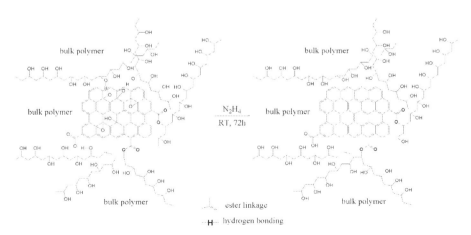

Figure 2.9 Schematic illustration of the reduction of GO-es-PVA with hydrazine.
Source: Reprinted with permission from Reprinted with permission from H.J. Salavagione, M.A. Gomez, G. Martinez, Polymeric modification of graphene through esterification of graphite oxide and poly (vinyl alcohol), Macromolecules 42 (2009) 6331–6334. © 2009 American Chemical Society.

are also in this grafting list especially the water soluble polysaccharide community. Hydroxypropyl cellulose (HPC) and chitosan are two widely used biopolymers which were reported in several works showed esterification onto GO surface. In one article it was seen that HPC and chitosan were grafted around 30% and 20% respectively [108]. poly(piperazine spirocyclic pentaerythritol bisphosphonate) was grafted to GO surface followed by blending with elthylene vinyl acetate copolymer rubber [109]. This nanocomposite showed good fire retardant property due to excellent dispersed graphene system. Graphene sheets are thermal conductor and they heat distribution was fast because of their heat diffusion throughout the composite. The esterification grafting reaction was also performed by conducting polymer poly(3-hexyl thiophene) by donor-acceptor adduct formation.

2.4.1.2 Amidation reaction

Amide linkage formation to graphene plane is another way of covalent modification of GO. Liu et al. reported PEG grafted GO via DCC chemistry where PEG and GO were attached by amide bond [110]. This PEG grafted GO was biologically compatible and applied for insoluble cancer drug delivery via $\pi-\pi$ interaction. In another work, GO was coupled with various molecular weight PEG by amidation reaction [111]. Poly(ethylene imine) (PEI) modified GO was prepared by using DCC as coupling agent [112]. They used this material to in situ reduction to prepare silver nanoparticles for bactericidal applications. Due to grafting of PEI on the plane of GO improve the water tolerance of the sample. Covalent addition of poly(L-lysine) to GO was another work which was carried out in alkaline medium [113]. This sample was conjugated to the horseradish peroxidase. This material was used to fabricate gold electrode for hydrogen peroxide sensing (Fig. 2.10). Triphenylamine and polyazimethine conjugated GO was prepared for fabrication of nonvolatile rewritable memory device with electric switching [114]. Epoxy based graphene composite is a common nanocomposite developed by various scientists. Wang and his coworkers developed epoxy polymer grafted GO followed by azo coupling reaction in mild condition [115]. They established that one azobenzene moiety was grafted to the GO plane where 30 was the carbon number per moiety as per the XPS study.

Ramezanzadeh et al. reported GO reinforced PU nanocomposite preparation via precursor modification strategy where they initially graft polyisocyanate (PI) to the GO surface followed by using that as functionalized reinforcing filler [116]. polyisocyanate modified GO is highly compatible in PU matrix. They showed that 0.1 wt.% PI grafted GO enhanced the Yung modulus, fracture toughness, elongation at break and ultimate tensile strength. In a similar fashion they also proved that this composite was efficient for gas barrier even in saline environment [117]. PANI grafted GO in also reported ion another article where p-phenylenediamine was taken as precursor monomer to graft onto GO. Then the p-phenylenediamine was covalently attached to the GO sheets which was polymerized in presence of aniline monomer via free radical addition [118]. All the aim of the reports by several researchers throughout the world is to compatibilize GO as a pretreatment followed by polymerization.

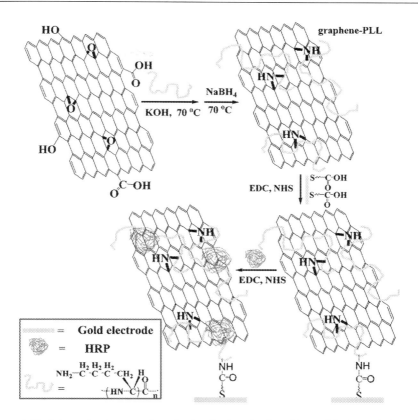

Figure 2.10 Schematic siagram of synthesis of poly(L-lactic acid) modified GO their activity towards horseradish peroxidase and gold electrode synthesis.
Source: Reprinted with permission from C. Shan, H. Yang, D. Han, Q. Zhang, A. Ivaska, L. Niu, Water-soluble graphene covalently functionalized by biocompatible poly-L-lysine, Langmuir, 25 (2009) 12030–12033. © 2009 American Chemical Society.

2.4.1.3 Click chemistry

Water dispersible grafted graphene oxide was reported by Pan et al. where azido termi-nated polymer chains were grafted to GO via click chemistry [119]. That material was again grafted to poly(*n*-isopropylacrylamide) (PNIPAM) which was an excellent vector for drug delivery applications. PNIPAM is an excellent materials showing sol-gel trans-formation upon temperature variation. Similar kind of work was also accounted by PEG for anticancer drug delivery application [120]. There are several polymers which were prepared by using this method due to low amount of toxicity and easy solvation. The azide termination click chemistry was also applied for PS, PMMA, poly(methyl acrylic acid), poly(dimethyl aminoethyl methacrylate), poly(4-vinyl pyridine) [121,122]. When solubility of functionalized graphene is a major issue, that time click chemistry is a good choice due to fine tunablity in solvent tolerance. Click chemistry based synthesized products could be soluble in oil, acidic, alkaline, and a wide range of polar to nonpolar solvents. Another important merit of this method is relatively mild reaction condition with respect to the other chemical treatment methods.

2.4.1.4 Nitrene chemistry

It is also another approach of grafting which also showed excellent dispersion of functionalized GO in solvents. PEG and PS were grafted to graphene surface by azido termination process [123]. PEG modified graphene showed excellent dispersion due to high amount of exfoliation in aqueous environment and PS grafted graphene showed effective dispersibility in THF, DMF and other less protic solvents. Conducting polymer, polyacetylene was also grafted on graphene surface resulting excellent solvent dispersibility [124–126]. The work also suggested that after grafting of polyacetylene onto the graphene surface did not affect their electronic and fluorescent behavior.

2.4.1.5 Radical addition reaction

Radical addition is a fast process of grafting polymer chains onto graphene plane via free radical generation. Vuluga prepared PMMA grafted GO with in situ reduction of GO [127]. They showed high electrical conductivity could be achieved by this process. Acrylic acid and acrylamide copolymeric nanocomposite was prepared by the similar process of radical generation in presence of GO dispersion [128]. Amphiphilic GO was prepared by radical addition reaction in between GO and PS/ polyacrylamide block copolymer [129]. Fig. 2.11 shows the basic graphical illustration of the nanocomposite formation.

Block copolymer nanocomposite has a wide range of solubility in different solvents. In a similar fashion glycidyl methacrylate (GMA) was free radically polymerized by thermal initiator system, azobisisobutyronitrile onto GO surface [130]. GMA grafting minimizes the intrinsic viscosity of the dispersion compared to the pristine GO dispersion. The lowering of intrinsic viscosity resembles the GMA-g-GO as linear polymer in solution behavior. This infers GMA grafted GO could be a useful filler as well as process aid for polymers. In a similar fashion other vinylic polymers were also used as hairy brush like polymeric assembly when grafted to GO plane [131]. Poly(vinyl acetate) (PVAc) grafted GO was reported by γ-ray irradiation method which was less chemically hazardous [79]. PVAc-g-GO was highly effective for preparing solution based graphene reinforced nanocomposites because of their excellent miscibility in various solvents. Successive intercalation and dislodging of graphene platelets from bulk GO in monomer solution was another approach adopted by Chen et al. for preparing radical addition polymer nanocomposites [132]. They also did γ-ray grafting method for fabrication of the nanocomposite.

2.4.2 "Grafting from" method

"Grafting from" method is initiated from the surface of the graphene via generation of macro-radical. The initiators used here are covalently attached to the graphene

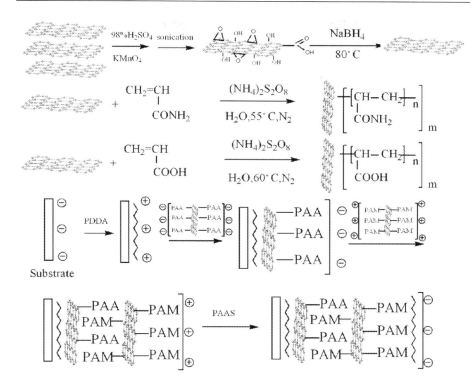

Figure 2.11 Synthesis of poly(acrylic acid-co-acrylamide) copolymer based graphene nanocomposite.
Source: Reprinted with permission from J. Shen, Y. Hu, C. Li, C. Qin, M. Shi, M. Ye, Layer-by-layer self-assembly of graphene nanoplatelets, Langmuir 25 (2009) 6122–6128.
© 2009 American Chemical Society.

surface from which polymer macrochains are subjected to grow. But before the initiator the graphene surface treatment is a must process. The main significance of this process is polymer chain growth never faces steric hindrance. That's why chain growth is never restricted during polymerization. Liu et al. fabricated tough nanocomposite hydrogel by using graphene peroxide (GPO) [133]. GPO is a special class of material which is manufactured after treatment the GO in γ-ray and excess oxygen atmosphere. GPO has a dual activity of initiation as well crosslinking. They reported highly stretchable hydrogel with 5300% elongation, low hysteresis and high resilience properties. In another work, Bergman cyclization was also adopted to prepare conjugated polymer based graphene nanocomposite where "enediyne" containing groups were used [134]. They revealed that such functionalized graphenes were easily soluble in several solvents and show good electrical conductivity. PANI grafted GO showed an anomalous feature in microstructure [135]. Here PANI changed the morphology from nanotube to flat rectangular nanopipe by free radical polymerization of aniline. Acyl group functionalized GO was grafted by

3-(2-hydroxyethyl)-2,5-thienylene by $FeCl_3$ [136]. This material was used as solar cell sensitizer having power conversion efficiency of 3.06%. Besides these another two major techniques are also there under the subgroup of *"grafting from"* method; those are atom transfer radical polymerization and reversible addition-fragmentation chain transfer. But those are beyond the scope of this chapter to discuss.

2.5 Summary and future perspective

The chapter summarizes the methods of the fabrication of graphene-based nano-composites. The methods were justified into two separate broad categories; one is physical mixing and another one is chemical/covalent grafting. Physical blending is fast and a less expensive compared to covalent functionalization. In physical mixing several methods were applied where in situ fabrication is the most effective in terms of compatibility between graphene and polymer matrices. Industrial production is always given preference in solvent-free approach due less possible pollution. That's why melt mixing was the mostly used industrial synthesis or fabricate mode. The fabrication strategy has a direct impact on the quality of the end product especially when the requirement of tunable electrical conductivity, thermal conductivity, and mechanical properties. Covalent grafting is much better than physical mixing because of their excellent compatibility toward polymer matrix. Actually graphene was grafted in such a way that the grafted moiety is normally a precursor material of the polymer used. The grafting methods are categorized into two subgroups; one is *"grafting to"* and another one *"grafting from."* Both have been discussed here which inferred that *"grafting from"* is qualitatively superior than *"grafting to"* method because of less steric effect. The functionalization of graphene basal planes also affects the electrical conductivity and extent of increment in tensile property of the nanocomposites. The versatile aspect of graphene reinforced polymer composites is primarily depending upon the mode of fabrication which actually enhances their significance in technological and engineering applications.

References

[1] B.T. Astrom, Manufacturing of Polymer Composites, CRC Press, 1997.
[2] F. Gao, Clay/polymer composites: the story, Mater. Today 7 (2004) 50–55.
[3] A.C. Balazs, T. Emrick, T.P. Russell, Nanoparticle polymer composites: where two small worlds meet, Science 314 (2006) 1107–1110.
[4] J. Holbery, D. Houston, Natural-fiber-reinforced polymer composites in automotive applications, Jom 58 (2006) 80–86.
[5] S. Ganguly, N.C. Das, Synthesis of a novel pH responsive phyllosilicate loaded poly-meric hydrogel based on poly (acrylic acid-co-N-vinylpyrrolidone) and polyethylene gly-col for drug delivery: modelling and kinetics study for the sustained release of an antibiotic drug, RSC Adv. 5 (2015) 18312–18327.

[6] S. Ganguly, S. Mondal, P. Das, P. Bhawal, P.P. Maity, S. Ghosh, et al., Design of psyllium-g-poly (acrylic acid-co-sodium acrylate)/cloisite 10A semi-IPN nanocomposite hydrogel and its mechanical, rheological and controlled drug release behaviour, Int. J. Biol. Macromol. 111 (2018) 983—998.

[7] P. Bhawal, S. Ganguly, T. Chaki, N. Das, Synthesis and characterization of graphene oxide filled ethylene methyl acrylate hybrid nanocomposites, RSC Adv. 6 (2016) 20781—20790.

[8] S. Mondal, S. Ganguly, P. Das, P. Bhawal, T.K. Das, L. Nayak, et al., High-performance carbon nanofiber coated cellulose filter paper for electromagnetic interference shielding, Cellulose 24 (2017) 5117—5131.

[9] Y. Zhang, Y.-W. Tan, H.L. Stormer, P. Kim, Experimental observation of the quantum Hall effect and Berry's phase in graphene, Nature 438 (2005) 201.

[10] X. Du, I. Skachko, A. Barker, E.Y. Andrei, Approaching ballistic transport in suspended graphene, Nat. Nanotechnol. 3 (2008) 491.

[11] A. Balandin, S. Ghosh, W. Bao, I. Calizo, D. Teweldebrhan, F. Miao, et al., Extremely high thermal conductivity of graphene: experimental study, arXiv preprint arXiv:0802. 1367, (2008).

[12] C. Lee, X. Wei, J.W. Kysar, J. Hone, Measurement of the elastic properties and intrinsic strength of monolayer graphene, Science 321 (2008) 385—388.

[13] X. Wang, H. You, F. Liu, M. Li, L. Wan, S. Li, et al., Large-scale synthesis of few-layered graphene using CVD, Chem. Vap. Depos. 15 (2009) 53—56.

[14] N. Li, Z. Wang, K. Zhao, Z. Shi, Z. Gu, S. Xu, Large scale synthesis of N-doped multi-layered graphene sheets by simple arc-discharge method, Carbon 48 (2010) 255—259.

[15] A. Hirsch, Unzipping carbon nanotubes: a peeling method for the formation of graphene nanoribbons, Angew. Chem. Int. Ed. 48 (2009) 6594—6596.

[16] M.S.A. Bhuyan, M.N. Uddin, M.M. Islam, F.A. Bipasha, S.S. Hossain, Synthesis of graphene, Int. Nano Lett. 6 (2016) 65—83.

[17] W. Zhang, J. Cui, Ca Tao, Y. Wu, Z. Li, L. Ma, et al., A strategy for producing pure single-layer graphene sheets based on a confined self-assembly approach, Angew. Chem. Int. Ed. 48 (2009) 5864—5868.

[18] S. Ganguly, S. Mondal, P. Das, P. Bhawal, T.K. Das, S. Ghosh, et al., An insight into the physico-mechanical signatures of silylated graphene oxide in poly (ethylene methyl acrylate) copolymeric thermoplastic matrix, Macromol. Res. 27 (2019) 268—281.

[19] D. Mondal, S. Ghorai, D. Rana, D. De, D. Chattopadhyay, The rubber—filler interaction and reinforcement in styrene butadiene rubber/devulcanize natural rubber composites with silica—graphene oxide, Polym. Compos. 40 (2019) E1559—E1572.

[20] J.R. Lomeda, C.D. Doyle, D.V. Kosynkin, W.-F. Hwang, J.M. Tour, Diazonium functionalization of surfactant-wrapped chemically converted graphene sheets, J. Am. Chem. Soc. 130 (2008) 16201—16206.

[21] S. Mallakpour, A. Abdolmaleki, S. Borandeh, Covalently functionalized graphene sheets with biocompatible natural amino acids, Appl. Surf. Sci. 307 (2014) 533—542.

[22] W.S. Sarsam, A. Amiri, S. Kazi, A. Badarudin, Stability and thermophysical properties of non-covalently functionalized graphene nanoplatelets nanofluids, Energy Convers. Manage. 116 (2016) 101—111.

[23] C. Mao, Y. Zhu, W. Jiang, Design of electrical conductive composites: tuning the morphology to improve the electrical properties of graphene filled immiscible polymer blends, ACS Appl. Mater. Interfaces 4 (2012) 5281—5286.

[24] R. Verdejo, M.M. Bernal, L.J. Romasanta, M.A. Lopez-Manchado, Graphene filled polymer nanocomposites, J. Mater. Chem. 21 (2011) 3301—3310.

[25] Z. Han, Y. Wang, W. Dong, P. Wang, Enhanced fire retardancy of polyethylene/alumina trihy-drate composites by graphene nanoplatelets, Mater. Lett. 128 (2014) 275−278.

[26] V. Mittal, G.E. Luckachan, N.B. Matsko, PE/chlorinated-PE blends and PE/chlori-nated-PE/graphene oxide nanocomposites: morphology, phase miscibility, and interfa-cial interactions, Macromol. Chem. Phys. 215 (2014) 255−268.

[27] V. Mittal, G.E. Luckachan, N.B. Matsko, Polyolefin-graphene oxide nanocomposites: interfacial interactions and low temperature brittleness reduction, Macromolecular Symposia, Wiley Online Library, 2014, pp. 37−43.

[28] F. You, D. Wang, X. Li, M. Liu, G.-H. Hu, Z.-M. Dang, Interfacial engineering of polypropylene/graphene nanocomposites: improvement of graphene dispersion by using tryptophan as a stabilizer, RSC Adv. 4 (2014) 8799−8807.

[29] S.H. Ryu, A. Shanmugharaj, Influence of hexamethylene diamine functionalized gra-phene oxide on the melt crystallization and properties of polypropylene nanocompo-sites, Mater. Chem. Phys. 146 (2014) 478−486.

[30] M. El Achaby, F.E. Arrakhiz, S. Vaudreuil, A. el Kacem Qaiss, M. Bousmina, O. Fassi-Fehri, Mechanical, thermal, and rheological properties of graphene-based poly-propylene nanocomposites prepared by melt mixing, Polym. Compos. 33 (2012) 733−744.

[31] O.M. Istrate, K.R. Paton, U. Khan, A. O'Neill, A.P. Bell, J.N. Coleman, Reinforcement in melt-processed polymer−graphene composites at extremely low graphene loading level, Carbon 78 (2014) 243−249.

[32] I.M. Inuwa, A. Hassan, S.A. Samsudin, M.H. Mohamad Kassim, M. Jawaid, Mechanical and thermal properties of exfoliated graphite nanoplatelets reinforced polyethylene terephthalate/polypropylene composites, Polym. Compos. 35 (2014) 2029−2035.

[33] I. Inuwa, A. Hassan, D.-Y. Wang, S. Samsudin, M.M. Haafiz, S. Wong, et al., Influence of exfoliated graphite nanoplatelets on the flammability and thermal proper-ties of polyethylene terephthalate/polypropylene nanocomposites, Polym. Degrad. Stab. 110 (2014) 137−148.

[34] M. Raja, S.H. Ryu, A. Shanmugharaj, Influence of surface modified multiwalled carbon nanotubes on the mechanical and electroactive shape memory properties of polyure-thane (PU)/poly (vinylidene diflouride)(PVDF) composites, Colloids Surf. A: Physicochem. Eng. Asp. 450 (2014) 59−66.

[35] L. Xie, G. Duan, W. Wang, M. Wang, Q. Wu, X. Zhou, et al., Effect of γ-ray-radia-tion-modified graphene oxide on the integrated mechanical properties of PET blends, Ind. Eng. Chem. Res. 55 (2016) 8123−8132.

[36] H. Norazlina, Y. Kamal, Graphene modifications in polylactic acid nanocomposites: a review, Polym. Bull. 72 (2015) 931−961.

[37] C. Sriprachuabwong, S. Duangsripat, K. Sajjaanantakul, A. Wisitsoraat, A. Tuantranont, Electrolytically exfoliated graphene−polylactide-based bioplastic with high elastic performance, J. Appl. Polym. Sci. 132 (2015).

[38] V. Mittal, A.U. Chaudhry, G.E. Luckachan, Biopolymer−thermally reduced graphene nanocomposites: structural characterization and properties, Mater. Chem. Phys. 147 (2014) 319−332.

[39] S. Deng, Z. Lin, B. Xu, W. Qiu, K. Liang, W. Li, Isothermal crystallization kinetics, morphology, and thermal conductivity of graphene nanoplatelets/polyphenylene sulfide composites, J. Therm. Anal. Calorim. 118 (2014) 197−203.

[40] T.D. Thanh, L. Kaprálková, J. Hromádková, I. Kelnar, Effect of graphite nanoplatelets on the structure and properties of PA6-elastomer nanocomposites, Eur. Polym. J. 50 (2014) 39−45.

[41] L. Nguyễn, S.-M. Choi, D.-H. Kim, N.-K. Kong, P.-J. Jung, S.-Y. Park, Preparation and characterization of nylon 6 compounds using the nylon 6-grafted GO, Macromol. Res. 22 (2014) 257−263.

[42] P. Bhawal, S. Ganguly, T.K. Das, S. Mondal, S. Choudhury, N. Das, Superior electromagnetic interference shielding effectiveness and electro-mechanical properties of EMA-IRGO nanocomposites through the in-situ reduction of GO from melt blended EMA-GO composites, Compos. Part. B Eng. 134 (2018) 46−60.

[43] H. Kim, Y. Miura, C.W. Macosko, Graphene/polyurethane nanocomposites for improved gas barrier and electrical conductivity, Chem. Mater. 22 (2010) 3441−3450.

[44] S. Araby, L. Zhang, H.-C. Kuan, J.-B. Dai, P. Majewski, J. Ma, A novel approach to electrically and thermally conductive elastomers using graphene, Polymer 54 (2013) 3663−3670.

[45] D. Yan, H.-B. Zhang, Y. Jia, J. Hu, X.-Y. Qi, Z. Zhang, et al., Improved electrical conductivity of polyamide 12/graphene nanocomposites with maleated polyethylene-octene rubber prepared by melt compounding, ACS Appl. Mater. Interfaces 4 (2012) 4740−4745.

[46] B. Shen, W. Zhai, M. Tao, D. Lu, W. Zheng, Enhanced interfacial interaction between polycarbonate and thermally reduced graphene induced by melt blending, Compos. Sci. Technol. 86 (2013) 109−116.

[47] A.A. Vasileiou, M. Kontopoulou, A. Docoslis, A noncovalent compatibilization approach to improve the filler dispersion and properties of polyethylene/graphene composites, ACS Appl. Mater. Interfaces 6 (2014) 1916−1925.

[48] B. Yuan, C. Bao, L. Song, N. Hong, K.M. Liew, Y. Hu, Preparation of functionalized graphene oxide/polypropylene nanocomposite with significantly improved thermal stability and studies on the crystallization behavior and mechanical properties, Chem. Eng. J. 237 (2014) 411−420.

[49] C. Bao, L. Song, W. Xing, B. Yuan, C.A. Wilkie, J. Huang, et al., Preparation of graphene by pressurized oxidation and multiplex reduction and its polymer nanocomposites by masterbatch-based melt blending, J. Mater. Chem. 22 (2012) 6088−6096.

[50] S.K. Khanna, H.T. Phan, High strain rate behavior of graphene reinforced polyurethane composites, J. Eng. Mater. Technol. 137 (2015).

[51] J. Jose, M.A. Al-Harthi, M.A.A. AlMa'adeed, J. Bhadra Dakua, S.K. De, Effect of graphene loading on thermomechanical properties of poly (vinyl alcohol)/starch blend, J. Appl. Polym. Sci. 132 (2015).

[52] D.S. Yu, T. Kuila, N.H. Kim, J.H. Lee, Enhanced properties of aryl diazonium salt-functionalized graphene/poly (vinyl alcohol) composites, Chem. Eng. J. 245 (2014) 311−322.

[53] R.K. Layek, A.K. Das, M.U. Park, N.H. Kim, J.H. Lee, Layer-structured graphene oxide/polyvinyl alcohol nanocomposites: dramatic enhancement of hydrogen gas barrier properties, J. Mater. Chem. A 2 (2014) 12158−12161.

[54] E. Arda, Ö.B. Mergen, Ö. Pekcan, Electrical and optical percolations in PMMA/GNP composite films, Phase Transit. 91 (2018) 546−557.

[55] P. Ding, J. Zhang, N. Song, S. Tang, Y. Liu, L. Shi, Anisotropic thermal conductive properties of hot-pressed polystyrene/graphene composites in the through-plane and in-plane directions, Compos. Sci. Technol. 109 (2015) 25−31.

[56] L. Zhang, T. Shi, D. Tan, H. Zhou, X. Zhou, Pickering emulsion polymerization of styrene stabilized by the mixed particles of graphene oxide and NaCl, Fuller. Nanotub. Carbon Nanostructures 22 (2014) 726−737.

[57] B. Ahmadi-Moghadam, F. Taheri, Effect of processing parameters on the structure and multi-functional performance of epoxy/GNP-nanocomposites, J. Mater. Sci. 49 (2014) 6180−6190.

[58] D. Galpaya, M. Wang, G. George, N. Motta, E. Waclawik, C. Yan, Preparation of graphene oxide/epoxy nanocomposites with significantly improved mechanical properties, J. Appl. Phys. 116 (2014) 053518.

[59] L. Peponi, A. Tercjak, R. Verdejo, M.A. Lopez-Manchado, I. Mondragon, J.M. Kenny, Confinement of functionalized graphene sheets by triblock copolymers, J. Phys. Chem. C. 113 (2009) 17973–17978.

[60] F. Barroso-Bujans, S. Cerveny, R. Verdejo, Jd del Val, J.M. Alberdi, A. Alegría, et al., Permanent adsorption of organic solvents in graphite oxide and its effect on the thermal exfoliation, Carbon 48 (2010) 1079–1087.

[61] K. Scully, R. Bissessur, Decomposition kinetics of nylon-6/graphite and nylon-6/graphite oxide composites, Thermochim. Acta 490 (2009) 32–36.

[62] X. Zhao, Y. Li, J. Wang, Z. Ouyang, J. Li, G. Wei, et al., Interactive oxidation−reduction reaction for the in situ synthesis of graphene−phenol formaldehyde composites with enhanced properties, ACS Appl. Mater. Interfaces 6 (2014) 4254–4263.

[63] S. Paszkiewicz, A. Szymczyk, Z. Špitalský, J. Mosnáček, K. Kwiatkowski, Z. Rosłaniec, Structure and properties of nanocomposites based on PTT-block-PTMO copolymer and graphene oxide prepared by in situ polymerization, Eur. Polym. J. 50 (2014) 69–77.

[64] L. Wan, B. Wang, S. Wang, X. Wang, Z. Guo, H. Xiong, et al., Water-soluble polyaniline/graphene prepared by in situ polymerization in graphene dispersions and use as counter-electrode materials for dye-sensitized solar cells, React. Funct. Polym. 79 (2014) 47–53.

[65] Y. Qian, Y. Lan, J. Xu, F. Ye, S. Dai, Fabrication of polyimide-based nanocomposites containing functionalized graphene oxide nanosheets by in-situ polymerization and their properties, Appl. Surf. Sci. 314 (2014) 991–999.

[66] S. Ganguly, P. Das, P.P. Maity, S. Mondal, S. Ghosh, S. Dhara, et al., Green reduced graphene oxide toughened semi-ipn monolith hydrogel as dual responsive drug release system: rheological, physicomechanical, and electrical evaluations, J. Phys. Chem. B 122 (2018) 7201–7218.

[67] S. Ganguly, D. Ray, P. Das, P.P. Maity, S. Mondal, V. Aswal, et al., Mechanically robust dual responsive water dispersible-graphene based conductive elastomeric hydrogel for tunable pulsatile drug release, Ultrason. Sonochem. 42 (2018) 212–227.

[68] X. Wang, C.-A. Zhang, P. Wang, J. Zhao, W. Zhang, J. Ji, et al., Enhanced performance of biodegradable poly (butylene succinate)/graphene oxide nanocomposites via in situ polymerization, Langmuir 28 (2012) 7091–7095.

[69] H. Wang, Q. Hao, X. Yang, L. Lu, X. Wang, Effect of graphene oxide on the properties of its composite with polyaniline, ACS Appl. Mater. Interfaces 2 (2010) 821–828.

[70] L. Mao, K. Zhang, H.S.O. Chan, J. Wu, Surfactant-stabilized graphene/polyaniline nanofiber composites for high performance supercapacitor electrode, J. Mater. Chem. 22 (2012) 80–85.

[71] Z. Tong, Y. Yang, J. Wang, J. Zhao, B.-L. Su, Y. Li, Layered polyaniline/graphene film from sandwich-structured polyaniline/graphene/polyaniline nanosheets for high-performance pseudosupercapacitors, J. Mater. Chem. A 2 (2014) 4642–4651.

[72] Y. Xu, Y. Wang, J. Liang, Y. Huang, Y. Ma, X. Wan, et al., A hybrid material of graphene and poly (3, 4-ethyldioxythiophene) with high conductivity, flexibility, and transparency, Nano Res. 2 (2009) 343–348.

[73] H. Hu, X. Wang, J. Wang, L. Wan, F. Liu, H. Zheng, et al., Preparation and properties of graphene nanosheets−polystyrene nanocomposites via in situ emulsion polymerization, Chem. Phys. Lett. 484 (2010) 247–253.

[74] X. Zhao, Q. Zhang, D. Chen, P. Lu, Enhanced mechanical properties of graphene-based poly (vinyl alcohol) composites, Macromolecules 43 (2010) 2357–2363.
[75] Y.R. Lee, A.V. Raghu, H.M. Jeong, B.K. Kim, Properties of waterborne polyurethane/functionalized graphene sheet nanocomposites prepared by an in situ method, Macromol. Chem. Phys. 210 (2009) 1247–1254.
[76] Y. Huang, Y. Qin, Y. Zhou, H. Niu, Z.-Z. Yu, J.-Y. Dong, Polypropylene/graphene oxide nanocomposites prepared by in situ Ziegler – Natta polymerization, Chem. Mater. 22 (2010) 4096–4102.
[77] S. Zhao, F. Chen, C. Zhao, Y. Huang, J.-Y. Dong, C.C. Han, Interpenetrating network formation in isotactic polypropylene/graphene composites, Polymer 54 (2013) 3680–3690.
[78] M. Stürzel, Y. Thomann, M. Enders, R. Mülhaupt, Graphene-supported dual-site catalysts for preparing self-reinforcing polyethylene reactor blends containing UHMWPE nanoplatelets and in situ UHMWPE shish-kebab nanofibers, Macromolecules 47 (2014) 4979–4986.
[79] B. Zhang, Y. Zhang, C. Peng, M. Yu, L. Li, B. Deng, et al., Preparation of polymer decorated graphene oxide by γ-ray induced graft polymerization, Nanoscale 4 (2012) 1742–1748.
[80] H. Pang, T. Chen, G. Zhang, B. Zeng, Z.-M. Li, An electrically conducting polymer/graphene composite with a very low percolation threshold, Mater. Lett. 64 (2010) 2226–2229.
[81] V.H. Pham, T.T. Dang, S.H. Hur, E.J. Kim, J.S. Chung, Highly conductive poly (methyl methacrylate)(PMMA)-reduced graphene oxide composite prepared by self-assembly of PMMA latex and graphene oxide through electrostatic interaction, ACS Appl. Mater. Interfaces 4 (2012) 2630–2636.
[82] Y. Zhan, M. Lavorgna, G. Buonocore, H. Xia, Enhancing electrical conductivity of rubber composites by constructing interconnected network of self-assembled graphene with latex mixing, J. Mater. Chem. 22 (2012) 10464–10468.
[83] M. Tian, J. Zhang, L. Zhang, S. Liu, X. Zan, T. Nishi, et al., Graphene encapsulated rubber latex composites with high dielectric constant, low dielectric loss and low percolation threshold, J. Colloid Interface Sci. 430 (2014) 249–256.
[84] W. Xing, M. Tang, J. Wu, G. Huang, H. Li, Z. Lei, et al., Multifunctional properties of graphene/rubber nanocomposites fabricated by a modified latex compounding method, Compos. Sci. Technol. 99 (2014) 67–74.
[85] C. Li, C. Feng, Z. Peng, W. Gong, L. Kong, Ammonium-assisted green fabrication of graphene/natural rubber latex composite, Polym. Compos. 34 (2013) 88–95.
[86] H. Yang, P. Liu, T. Zhang, Y. Duan, J. Zhang, Fabrication of natural rubber nanocomposites with high graphene contents via vacuum-assisted self-assembly, RSC Adv. 4 (2014) 27687–27690.
[87] Y. Luo, P. Zhao, Q. Yang, D. He, L. Kong, Z. Peng, Fabrication of conductive elastic nanocomposites via framing intact interconnected graphene networks, Compos. Sci. Technol. 100 (2014) 143–151.
[88] A. Suriani, M. Nurhafizah, A. Mohamed, I. Zainol, A. Masrom, A facile one-step method for graphene oxide/natural rubber latex nanocomposite production for supercapacitor applications, Mater. Lett. 161 (2015) 665–668.
[89] Y. Zhan, J. Wang, K. Zhang, Y. Li, Y. Meng, N. Yan, et al., Fabrication of a flexible electromagnetic interference shielding Fe3O4@ reduced graphene oxide/natural rubber composite with segregated network, Chem. Eng. J. 344 (2018) 184–193.
[90] X. Zhang, J. Wang, H. Jia, S. You, X. Xiong, L. Ding, et al., Multifunctional nanocomposites between natural rubber and polyvinyl pyrrolidone modified graphene, Compos. Part. B: Eng. 84 (2016) 121–129.

[91]   A. Suriani, M. Nurhafizah, A. Mohamed, A. Masrom, V. Sahajwalla, R. Joshi, Highly
        conductive electrodes of graphene oxide/natural rubber latex-based electrodes by
        using a hyper-branched surfactant, Mater. Des. 99 (2016) 174−181.
[92]   E. Tkalya, M. Ghislandi, A. Alekseev, C. Koning, J. Loos, Latex-based concept for
        the preparation of graphene-based polymer nanocomposites, J. Mater. Chem. 20
        (2010) 3035−3039.
[93]   L. Hu, J. Tu, S. Jiao, J. Hou, H. Zhu, D.J. Fray, In situ electrochemical polymerization
        of a nanorod-PANI−Graphene composite in a reverse micelle electrolyte and its appli-
        cation in a supercapacitor, Phys. Chem. Chem. Phys. 14 (2012) 15652−15656.
[94]   D.-W. Wang, F. Li, J. Zhao, W. Ren, Z.-G. Chen, J. Tan, et al., Fabrication of gra-
        phene/polyaniline composite paper via in situ anodic electropolymerization for high-
        performance flexible electrode, ACS Nano 3 (2009) 1745−1752.
[95]   W. Liu, Y. Fang, P. Xu, Y. Lin, X. Yin, G. Tang, et al., Two-step electrochemical
        synthesis of polypyrrole/reduced graphene oxide composites as efficient Pt-free
        counter electrode for plastic dye-sensitized solar cells, ACS Appl. Mater. Interfaces 6
        (2014) 16249−16256.
[96]   D. Li, Y. Li, Y. Feng, W. Hu, W. Feng, Hierarchical graphene oxide/polyaniline nano-
        composites prepared by interfacial electrochemical polymerization for flexible solid-
        state supercapacitors, J. Mater. Chem. A 3 (2015) 2135−2143.
[97]   T. Lindfors, R.-M. Latonen, Improved charging/discharging behavior of electropoly-
        merized nanostructured composite films of polyaniline and electrochemically reduced
        graphene oxide, Carbon 69 (2014) 122−131.
[98]   H.-P. Cong, X.-C. Ren, P. Wang, S.-H. Yu, Flexible graphene−polyaniline composite
        paper for high-performance supercapacitor, Energy Environ. Sci. 6 (2013)
        1185−1191.
[99]   Y. Jafari, S. Ghoreishi, M. Shabani-Nooshabadi, Polyaniline/graphene nanocomposite
        coatings on copper: electropolymerization, characterization, and evaluation of corro-
        sion protection performance, Synth. Met. 217 (2016) 220−230.
[100] P. Damlin, M. Suominen, M. Heinonen, C. Kvarnström, Non-covalent modification of gra-
        phene sheets in PEDOT composite materials by ionic liquids, Carbon 93 (2015) 533−543.
[101] W. Li, W. Dai, L. Ge, S. Ge, M. Yan, J. Yu, Electropolymerized poly (3, 4-ethylendioxythio-
        phene)/graphene composite film and its application in quantum dots electrochemilumines-
        cence immunoassay, J. Inorg. Organomet. Polym. Mater. 23 (2013) 719−725.
[102] P. Das, S. Ganguly, S. Banerjee, N.C. Das, Graphene based emergent nanolights: a
        short review on the synthesis, properties and application, Res. Chem. Intermed. 45
        (2019) 3823−3853.
[103] S. Ganguly, P. Das, S. Banerjee, N.C. Das, Advancement in science and technology
        of carbon dot-polymer hybrid composites: a review, Funct. Compos. Struct. 1 (2019)
        022001.
[104] P. Das, S. Ganguly, S. Mondal, M. Bose, A.K. Das, S. Banerjee, et al., Heteroatom
        doped photoluminescent carbon dots for sensitive detection of acetone in human
        fluids, Sens. Actuators B Chem. 266 (2018) 583−593.
[105] S.N.J.S.Z. Abidin, M.S. Mamat, S.A. Rasyid, Z. Zainal, Y. Sulaiman,
        Electropolymerization of poly (3, 4-ethylenedioxythiophene) onto polyvinyl alcohol-
        graphene quantum dot-cobalt oxide nanofiber composite for high-performance super-
        capacitor, Electrochim. Acta 261 (2018) 548−556.
[106] H.J. Salavagione, M.A. Gomez, G. Martinez, Polymeric modification of graphene
        through esterification of graphite oxide and poly (vinyl alcohol), Macromolecules 42
        (2009) 6331−6334.

[107] L.M. Veca, F. Lu, M.J. Meziani, L. Cao, P. Zhang, G. Qi, et al., Polymer functionalization and solubilization of carbon nanosheets, Chem. Commun. (2009) 2565–2567.

[108] Q. Yang, X. Pan, K. Clarke, K. Li, Covalent functionalization of graphene with polysaccharides, Ind. Eng. Chem. Res. 51 (2012) 310–317.

[109] G. Huang, S. Chen, S. Tang, J. Gao, A novel intumescent flame retardant-functionalized graphene: nanocomposite synthesis, characterization, and flammability properties, Mater. Chem. Phys. 135 (2012) 938–947.

[110] Z. Liu, J.T. Robinson, X. Sun, H. Dai, PEGylated nanographene oxide for delivery of water-insoluble cancer drugs, J. Am. Chem. Soc. 130 (2008) 10876–10877.

[111] L. Jin, K. Yang, K. Yao, S. Zhang, H. Tao, S.-T. Lee, et al., Functionalized graphene oxide in enzyme engineering: a selective modulator for enzyme activity and thermostability, ACS Nano 6 (2012) 4864–4875.

[112] X. Cai, M. Lin, S. Tan, W. Mai, Y. Zhang, Z. Liang, et al., The use of polyethyleneimine-modified reduced graphene oxide as a substrate for silver nanoparticles to produce a material with lower cytotoxicity and long-term antibacterial activity, Carbon 50 (2012) 3407–3415.

[113] C. Shan, H. Yang, D. Han, Q. Zhang, A. Ivaska, L. Niu, Water-soluble graphene covalently functionalized by biocompatible poly-L-lysine, Langmuir 25 (2009) 12030–12033.

[114] X.D. Zhuang, Y. Chen, G. Liu, P.P. Li, C.X. Zhu, E.T. Kang, et al., Conjugated-polymer-functionalized graphene oxide: synthesis and nonvolatile rewritable memory effect, Adv. Mater. 22 (2010) 1731–1735.

[115] W. Yong, C. Fen-Xiong, G. Hong-Xiang, Kernel-based target tracking with spatial histogram and template drift correction, Acta Automatica Sin. 38 (2012) 430–435.

[116] B. Ramezanzadeh, E. Ghasemi, M. Mahdavian, E. Changizi, M.M. Moghadam, Characterization of covalently-grafted polyisocyanate chains onto graphene oxide for polyurethane composites with improved mechanical properties, Chem. Eng. J. 281 (2015) 869–883.

[117] B. Ramezanzadeh, E. Ghasemi, M. Mahdavian, E. Changizi, M.M. Moghadam, Covalently-grafted graphene oxide nanosheets to improve barrier and corrosion protection properties of polyurethane coatings, Carbon 93 (2015) 555–573.

[118] T. Remyamol, H. John, P. Gopinath, Synthesis and nonlinear optical properties of reduced graphene oxide covalently functionalized with polyaniline, Carbon 59 (2013) 308–314.

[119] Y. Pan, H. Bao, N.G. Sahoo, T. Wu, L. Li, Water-soluble poly (N-isopropylacrylamide)–graphene sheets synthesized via click chemistry for drug delivery, Adv. Funct. Mater. 21 (2011) 2754–2763.

[120] Z. Jin, T.P. McNicholas, C.-J. Shih, Q.H. Wang, G.L. Paulus, A.J. Hilmer, et al., Click chemistry on solution-dispersed graphene and monolayer CVD graphene, Chem. Mater. 23 (2011) 3362–3370.

[121] Y.-S. Ye, Y.-N. Chen, J.-S. Wang, J. Rick, Y.-J. Huang, F.-C. Chang, et al., Versatile grafting approaches to functionalizing individually dispersed graphene nanosheets using RAFT polymerization and click chemistry, Chem. Mater. 24 (2012) 2987–2997.

[122] L. Kou, H. He, C. Gao, Click chemistry approach to functionalize two-dimensional macromolecules of graphene oxide nanosheets, Nano-Micro Lett. 2 (2010) 177–183.

[123] H. He, C. Gao, General approach to individually dispersed, highly soluble, and conductive graphene nanosheets functionalized by nitrene chemistry, Chem. Mater. 22 (2010) 5054–5064.

[124] X. Xu, Q. Luo, W. Lv, Y. Dong, Y. Lin, Q. Yang, et al., Functionalization of graphene sheets by polyacetylene: convenient synthesis and enhanced emission, Macromol. Chem. Phys. 212 (2011) 768–773.

[125] S. Ganguly, P. Bhawal, R. Ravindren, N.C. Das, Polymer nanocomposites for electro-magnetic interference shielding: a review, J. Nanosci. Nanotechnol. 18 (2018) 7641−7669.

[126] S. Ghosh, S. Ganguly, P. Das, T.K. Das, M. Bose, N.K. Singha, et al., Fabrication of reduced graphene oxide/silver nanoparticles decorated conductive cotton fabric for high performing electromagnetic interference shielding and antibacterial application, Fibers Polym. 20 (2019) 1161−1171.

[127] D. Vuluga, J.-M. Thomassin, I. Molenberg, I. Huynen, B. Gilbert, C. Jérôme, et al., Straightforward synthesis of conductive graphene/polymer nanocomposites from graphite oxide, Chem. Commun. 47 (2011) 2544−2546.

[128] J. Shen, Y. Hu, C. Li, C. Qin, M. Shi, M. Ye, Layer-by-layer self-assembly of gra-phene nanoplatelets, Langmuir 25 (2009) 6122−6128.

[129] J. Shen, Y. Hu, C. Li, C. Qin, M. Ye, Synthesis of amphiphilic graphene nanoplatelets, Small 5 (2009) 82−85.

[130] L. Kan, Z. Xu, C. Gao, General avenue to individually dispersed graphene oxide-based two-dimensional molecular brushes by free radical polymerization, Macromolecules 44 (2011) 444−452.

[131] B. Wang, D. Yang, J.Z. Zhang, C. Xi, J. Hu, Stimuli-responsive polymer covalent functionalization of graphene oxide by Ce (IV)-induced redox polymerization, J. Phys. Chem. C. 115 (2011) 24636−24641.

[132] L. Chen, Z. Xu, J. Li, Y. Li, M. Shan, C. Wang, et al., A facile strategy to prepare functionalized graphene via intercalation, grafting and self-exfoliation of graphite oxide, J. Mater. Chem. 22 (2012) 13460−13463.

[133] J. Liu, C. Chen, C. He, J. Zhao, X. Yang, H. Wang, Synthesis of graphene peroxide and its application in fabricating super extensible and highly resilient nanocomposite hydrogels, ACS Nano 6 (2012) 8194−8202.

[134] X. Ma, F. Li, Y. Wang, A. Hu, Functionalization of pristine graphene with conjugated polymers through diradical addition and propagation, Chem. Asian J. 7 (2012) 2547−2550.

[135] S. Chatterjee, R.K. Layek, A.K. Nandi, Changing the morphology of polyaniline from a nanotube to a flat rectangular nanopipe by polymerizing in the presence of amino-functionalized reduced graphene oxide and its resulting increase in photocurrent, Carbon 52 (2013) 509−519.

[136] S. Chatterjee, A.K. Patra, A. Bhaumik, A.K. Nandi, Poly [3-(2-hydroxyethyl)-2, 5-thienylene] grafted reduced graphene oxide: an efficient alternate material of TiO 2 in dye sensitized solar cells, Chem. Commun. 49 (2013) 4646−4648.

Mechanical properties of polymer/graphene composites

3

Roop Singh Lodhi[1,]*, Pankaj Kumar[1,]*, Ajitha Achuthanunni[1], Mostafizur Rahaman[2] and Paramita Das[1]

[1]Department of Chemical Engineering, Indian Institute of Science Education and Research Bhopal, Bhopal, India, [2]Department of Chemistry, College of Science, King Saud University, Riyadh, Saudi Arabia

3.1 Introduction

A composite material is formed by combining two or more materials with the intention to enhance the physical and chemical properties, which will be higher than the individual components. In this approach, polymer composites can be prepared in a multiphase manner as reinforcement filler will integrate with the polymer matrix. The synergistic mechanical properties will be higher as compared to the individual component. A polymer/graphene composite is a material in which graphene is incorporated into the polymer matrix to obtain the combined properties of both the components. In this current scenario, graphene has received tremendous attention for scientific research and industrial purposes. Graphene was first discovered by Novoselov et al. in 2004 [1]. Graphene is composed of monolayer sp^2-hybridized carbon atoms that are grouped in a hexagonal lattice. It was found to be the strongest material having high thermal stability, exceptional electron mobility, large surface area, and high aspect ratio. Thermal stability is the ability of a material to resist the action of heat energy by maintaining its mechanical properties like strength, toughness, or elasticity at a given temperature. Graphene oxide (GO) exhibits good thermal stability but it starts to decompose at about $200°C$ due to the decomposition of carboxylic groups leading to release of CO_2 [2]. The defect-free monolayer graphene has the intrinsic tensile strength of 130 GPa (42 N/m), which is the strongest material ever tested [3].

Electron mobility is the measure of how fast an atom can move through a metal or semiconductor under the influence of the external electric field. The electrical conductivities of graphene and graphene nanoplatelets (GNPs) are 6000 and 2420 S/m, respectively; whereas, the electrical conductivity of GO is only 0.02 S/m [4,5]. Graphene has a large theoretical surface area of about 2675 m^2/g [6]. These remarkable properties of graphene make their potential use as a filler in high-performance polymer composites. Graphene can drastically enhance the properties of polymer composites even at extremely small loadings of these atomically thin carbon sheets in

---

* Roop Singh Lodhi and Pankaj Kumar contributed equally to this chapter.

Polymer Nanocomposites Containing Graphene. DOI: https://doi.org/10.1016/B978-0-12-821639-2.00019-7

the host polymer matrix. Graphene and its other forms are potential nanofillers that are widely used with various polymers to produce polymer/graphene composites. Graphene and its derivatives like GO and reduced graphene oxide (RGO) both in the form of flakes/sheets as well as fibers have been used as nanofillers along with various polymers. The incorporation of these fillers leads to the improvement of various mechanical and other physical properties of the composites.

In this chapter, we have discussed the mechanical properties such as tensile properties, compressive properties, dynamic mechanical properties, and flexural properties of polymer/graphene composites (see Fig. 3.1 and Table 3.1).

3.2 Tensile properties of polymer/graphene composites

Tensile properties of the composites are expressed by the action of the composites to resist forces when the forces are applied in tension. Determination of the tensile properties is vital because it provides details about the strength, stiffness, toughness, elasticity, and percent elongation [35]. The incorporation of graphene and its different derivatives within various polymeric matrices enhances various tensile properties like strength, stiffness, toughness, fatigue, and elastic modulus of the composites compared to the pure polymer matrix. A chart, having comparison of graphene properties with more traditional materials in terms of Young's modulus as a function of density, is represented in Fig. 3.2A. The chart shows that the defect-free graphene is the stiffest and strongest material reported in the literature and its magnificent intrinsic properties are because of its larger surface areas sustained within the polymer composites. Therefore, polymer/graphene composites are gaining attention from researchers and are one of the most prospected areas in the field of polymeric composite materials [36].

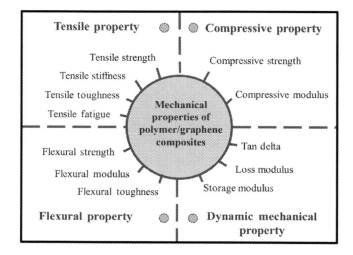

Figure 3.1 Mechanical properties of polymer/graphene composites.

Table 3.1 Mechanical properties of polymer/graphene composites materials.

Tensile mechanical properties							
Graphene type	Matrix	Preparation method	Filler content (wt.%)	Tensile modulus of matrix, E_m (GPa)	Increase in tensile modulus (%)	Increase in tensile strength (%)	Reference
GO	TPU/PLA	Blending	0.5	–	75.5	–	[7]
GNPs	Epoxy/CF	Hot press molding and postcuring	0.25	55	–	20	[8]
GN	Epoxy	Solution blending	1	1.48	11.5	11	[9]
f-GNR	Epoxy	Ball milling	0.15	1.1	14	7	[10]
GNS	PVC	Solution blending	2	0.85	135	71	[11]
f-GO	PI	Solution blending	0.75	1.8	61	59	[12]
GO-epoxy	PC	Solution blending	1	1.1	72	20	[13]
GO	PMMA	Melt mixing	1	2.2	35	21	[14]
GO	NR	Latex mixing	1	0.0015	47	34	[15]
RGO	PVA	Wet spinning	2	5.4	294	244	[16]
RGO	PC	Solution blending	0.25	1.8	10	2	[17]

Compressive mechanical properties

Graphene type	Matrix	Preparation method	Filler content (wt.%)	Increase in compressive modulus (%)	Compressive strength of the matrix (MPa)	Increase in compressive strength (%)	References
f-GS	Silicon	Foaming by condensation	0.25	136	–	–	[18]
GO	TPU/PLA	Blending	5	167	–	–	[7]
GO	Geopolymers	Extrusion	2	50	–	43	[19]
GO	Epoxy resin	Two-phase extraction	0.0375		–	48.3 at failure	[20]
GO	PVA	Selective laser sintering	2.5	152	–	60	[21]
GO	Glutaraldehyde	One pot synthesis	5	195	0.146	288	[22]
GO	PAM/CMC	Free radical polymerization	0.8		0.759	37	[23]
GO	PAM/CMC	Free radical polymerization	1.6		2.87	172	[23]
RGO	Poly (AMPS-co-AAm)	Thermal-induced polymerization	0.5		0.4	250	[24]
RGO	PVA	Solution casting method	0.1	77	–	25.7	[25]

Dynamic mechanical properties

Graphene type	Matrix	Preparation method	Filler content	The storage modulus of the matrix	Increase in storage modulus (%)	References
GO	BFDE	Dispersing and curing	0.5wt.%	1.1 GPa	48	[26]
f-GO	Polystyrene	Solution casting followed by compression molding	0.5wt.%	–	41	[27]
GO	Gelatin	One-pot syntheses	0.1wt.%	10.4 kPa	11.53	[22]
GO	Gelatin	One-pot synthesis	1wt.%	10.4 kPa	93.26	[22]

(Continued)

Table 3.1 (Continued)

Dynamic mechanical properties

Graphene type	Matrix	Preparation method	Filler content	The storage modulus of the matrix	Increase in storage modulus (%)	References
GO	Gelatin	One-pot syntheses	3wt.%	10.4 kPa	159.6	[22]
GO	PVA	Freeze-thaw	2wt.%	~0.14 MPa	50	[28]
GNS	PHBV	Solvent casting	0.1wt.%	1.395 GPa	25	[29]
GNS	PHBV	Solvent casting	0.5wt.%	1.395 GPa	87	[29]
GNS	PHBV	Solvent casting	1wt.%	1.395 GPa	91	[29]
RGO	Gelatin	One-pot synthesis	10 mg/mL	64.4 kPa at 2 mg/mL gelatin	–	[30]
RGO	Gelatin	One-pot synthesis	10 mg/mL	91.1 kPa at 5 mg/mL gelatin		[30]
RGO	Gelatin	One-pot synthesis	10 mg/mL	172.3 kPa at 10 mg/mL gelatin		[30]
GO	woven CF/Epoxy resin	Vacuum-assisted resin transfer molding	0.2wt.%	16 GPa	56	[31]

Flexural mechanical properties

Graphene type	Matrix	Preparation method	Filler (wt.%)	Flexural strength of the matrix (MPa)	Flexural strength increases (%)	Flexural modulus of the matrix (GPa)	Flexural modulus increases (%)	References
GNPs	Epoxy/CF	Hot press molding and post curing	0.25	106	9	2.22	12	[8]

(Continued)

Table 3.1 (Continued)

Flexural mechanical properties

Graphene type	Matrix	Preparation method	Filler (wt.%)	Flexural strength of the matrix (MPa)	Flexural strength increases (%)	Flexural modulus of the matrix (GPa)	Flexural modulus increases (%)	References
GNPs	Epoxy/CF	Hot press molding and post curing	1.5	106	9	2.22	19	[8]
f-GO	BFDE	Dispersing and curing	0.5	71.8 ± 3.2	49	1,525 ± 21	21	[26]
GO	woven CF/Epoxy resin	Vacuum-assisted resin transfer molding	0.2	337	55	22	31	[31]
GO coated SGS	PES	Extrusion compounding and injection molding	0.5	160	9.3	–	15.9 at 1wt.%	[32]
RGO	Geopolymer	In situ mixing and curing	0.5	3.2	134	–	–	[33]
GNPs	PEN with 20wt.% CF	Melt-mixing	5	PEN/CF:193	47	PEN/CF: 11.0	54.5	[34]
GNPs	PEN with 20wt.% CF	Melt-mixing	10	PEN/CF:193	30.5	PEN/CF: 11.0	69	[34]

Here increase in mechanical properties of the composites with respect to the matrix is mentioned. BFDE, biobased bis-furan di-epoxide; CF, carbon fiber; CMC, carboxymethyl cellulose; GNPs, graphite nanoplatelets; GO, graphene oxide; GNR, graphene nanoribbons; GNS, graphene nanosheet; GN, Graphene nanoparticles; f-GS, functionalized graphene sheet; TPU, thermoplastic polyurethane; PI, polyimide; PAM, polyacrylamide; PC, polycarbonate; PEN, poly(arylene ether nitrile); PES, polyethersulphone; SGS, short glass fiber; PHBV, poly(3-hydroxybutyrate-co-3-hydroxy valerate); PLA, polylactic acid; PMMA, polymethyl methacrylate); PVA, Poly (vinyl alcohol); PVC, polyvinyl chloride; RGO, reduced graphene oxides.

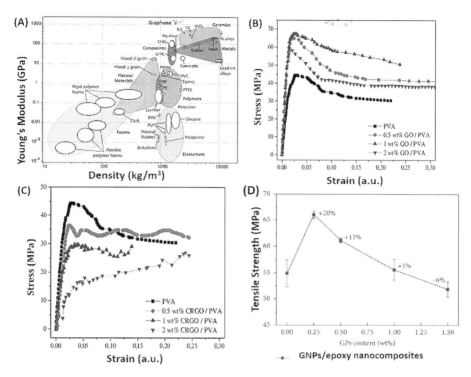

Figure 3.2 Tensile mechanical properties of polymer/graphene composites. (A) Chart of Young's modulus as a function of density comparing graphene properties to more traditional materials. (The axes are in logarithmic scale. Graphene density was taken as 2200 kg/m³), reproduced with permission from Verdejo R., Bernal M.M., Romasanta L.J., Lopez-Manchado M.A., Graphene filled polymer nanocomposites, J. Mater. Chem. 2011 3301−3310 [36]. Typical stress−strain curves of (B) GO/PVA composites with varying GO loadings and (C) CRGO/PVA composites with varying CRGO loadings, reproduced with permission from Goumri M., Poilâne C., Ruterana P., Ben Doudou B., Wéry J., Bakour A., Synthesis and characterization of nanocomposites films with graphene oxide and reduced graphene oxide nanosheets, Chin. J. Phys. 2017 412−422 [37], (D) Tensile strength of graphene nanoplatelets (GNPs)/epoxy composites, reproduced with permission from Shen M.Y., Chang T.Y., Hsieh T.H., Li Y.L., Chiang C.L., Yang H., Mechanical properties and tensile fatigue of graphene nanoplatelets reinforced polymer nanocomposites, J. Nanomater. 2013; Creative Commons Attribution License [8]. *CRGO, Chemically reduced graphene oxide; GO, graphene oxide; PVA, poly (vinyl alcohol).*

3.2.1 Tensile strength

The major intention of the incorporation of nanofillers like graphene and its other forms with various polymers is to enhance the properties of the resulting composites. Graphene and its derivatives like GO and RGO are used along with various polymers like epoxy, Poly (vinyl alcohol) (PVA), Poly (ethylene oxide), polystyrene (PS), etc., are used to prepare different composites. The dispersion of fillers within the polymer matrix is an important factor for the improvement of mechanical properties of the polymer composites. The low cost and magnificent properties of polymer/graphene composites have enabled them for various wide-range applications. The increase in loading level of graphene enhances the strength and elastic modulus of the polymer composites. Other than the loading of graphene, some important factors like aspect ratio, surface area, degree of dispersion etc., are also responsible for the properties of composites. Moreover, the functional groups, present on the surface of graphene sheets, play important roles to disperse the graphene sheets easily within the polymer matrix [38].

The graphene and its derivatives are mixed or dispersed in the polymer matrix using different processing techniques [38,39]. It is reported that GO nanofillers were distributed within the polymer matrix like PVA to make PVA/GO composites by simple water-based solution processing methods [40]. In the polymer composites, the formation of H-bonding between the PVA and GO have enhanced its strength [41]. It was reported that 76% increase in tensile strength has been achieved by the addition of only 0.7wt.% of GO. This increase in the tensile strength indicated that a small quantity of uniformly dispersed graphene was sufficient to get the desired strength of the composite [42]. Besides the excellent mechanical properties, it has also improved electrical and thermal conductivity. As a graphene derivative, RGO is a more advantageous nanofiller because it has high mechanical properties similar to that of GO. RGO is produced by reducing GO via removal of the functional groups and restoring the carbon structure; whereas, GO is an oxidized form of graphene laced with oxygen-containing groups. The GO can be reduced either through chemical or thermal methods to produce either chemically RGO (CRGO) or thermally RGO (TRGO). RGO shows improvement of many properties when incorporated within the polymer matrices and it is also cost-effective and abundant in nature. The main reasons for the differences are the surface chemistry, the dispersion state of fillers in matrices, interfacial interactions between fillers and matrices, and the crystallinity of polymer matrices.

Fig. 3.2B shows the stress—strain curve of polymer/graphene composites and it is believed that the strong interaction between GO and the PVA matrix is because of the homogenous dispersion of graphene sheets in PVA matrix. The graphene sheets started staking when the content loading reached to 2wt.% because of Van der Waals forces. Fig. 3.2C shows that the maximum stress was reduced with the increase in CRGO loadings in PVA matrix. It was also reported that better dispersion of GO inside the PVA matrix occurs because of strong hydrogen bond between the components, which leads to higher mechanical strength [37].

The wrinkled topology of graphene as well as crystallinity of polymer matrix are the key factors for the load transfer in composites. The wrinkled topology of the graphene sheets depends on the small thickness with distortion and oxygen functionalization at the nanoscale. The mechanical interlocking of polymer chains and load transfer between graphene and polymer matrix are the functions of the nanoscale surfaces [43].

While comparing the properties of composites reinforced with graphene as well as with other nanofillers like carbon nanotubes (CNT), graphene reinforced composites show an enhancement in the tensile properties. The results show that when CNT and graphene nanosheets (GNS) were reinforced with poly(methyl methacrylate) (PMMA) to prepare CNT/PMMA, and GNS/PMMA composites, respectively, the tensile strength of the CNT/PMMA and GNS/PMMA composites are reported as 103 and 112 MPa, respectively. The tensile strength of GNS incorporated composites showed 8.7% higher value as compared to the CNTs loaded composites [44].

Like graphene flakes/sheets, graphene nanoparticles (GN) also exhibit better reinforcement in polymer composites due to their high aspect ratio (600−10,000) and higher surface area. The loading amount of graphene nanofiller largely affects the tensile properties of the composites [44]. Shen et al. prepared GNPs/epoxy composite and GNPs/epoxy/CF composite laminates and studied the effect of GNPs loading amount on the tensile properties of the composites. Various amounts of GNPs (0, 0.25, 0.5, 1, and 1.5wt.%) were uniformly dispersed in epoxy resin for the preparation of GNPs/epoxy composites. The ultimate tensile strength of the composites is shown in Table 3.1. The experimental results shown in Fig. 3.2D indicates that the tensile strength of the composites gets enhanced by the reinforcement of graphene. The tensile strength of the composites with 0.25wt.% loading of GNPs shows the best enhancement (20%) as compared to the composite without any GNPs (see Fig. 3.2D). The strength begins to degrade at 0.5wt.% GNPs loading [8]. The decrease in tensile strength with high nanofiller loading can be attributed to the following two effects: (1) nonuniform dispersion of the nanofillers in higher loading systems, and (2) the presence of voids which was produced during the fabrication process and was increased with higher nano-particle content [8].

3.2.2   Tensile stiffness

Stiffness is one of the key tensile properties of composite that is, the extent to which a material can resist deformation in response to an applied force. It is the mechanical property of graphene, which makes it to stand both as an individual material and as a reinforcing agent in composites. Tensile strength influences the stiffness of the material. Young's modulus is a mechanical property that measures the stiffness of a composite. Higher value of the modulus means more stiff material. The reinforcement of graphene and its derivatives can enhance the stiffness of the composites to a greater extent. The research revealed that GO with good dispersion in the polymer matrix significantly enhances the stiffness of the composites. Higher loading of functional GO exhibited increased stiffness [45]. A comparison of the stiffness performance of different types of functionalized GO (f-GO) and TRGO have shown a nearly comparable increase in tensile modulus, however, TRGO improved the stiffness for rigid glassy polymers like poly(ethylene naphthalene) and polycarbonate (PC), while the incorporation of RGO with PVA increased the Young's modulus by ten times at 1.8vol.% loading of RGO [46]. When the RGO sheets are dispersed in a polymer matrix, the effective stiffness may be limited because of issues with its dispersion and exfoliation in the polymer matrix. There should be complete exfoliation of sheets so that dispersion in the polymer matrix is better. A lower aspect ratio leads to lower stiffness and the agglomeration of graphene. The strong bond between RGO sheets and polymer matrices contribute mostly to the mechanical property of composites. Studies also reveal that the functionalized graphene (FG) showed better reinforcement than the non FG. The PS/ functional graphene composites show increase in Young's modulus with the increase in filler content. The maximum improvement was found at 0.5wt.% loading [47]. However, the stiffness decreases at higher filler loadings. The reduction in stiffness at higher graphene content is associated with filler aggregation [48].

3.2.3   Tensile toughness

One of the most important mechanical properties of graphene is the toughness since it is a property very relevant to composite applications. Toughness is a quantitative way of expressing a resistance of the material to crack propagation. The toughness is the amount of absorbed energy per unit volume of a material before its rupture. It is defined as the area under the stress—strain curve of a material. Besides graphene and its derivatives, FG shows a major significance in the toughness of the composites. FG nanoplatelets reinforced epoxy composites show that at 4wt.% loading of filler, the fracture toughness of composite is increased by 85% compared to the base matrix. The toughness of the graphene/epoxy composites was studied by Zaman et al. where they incorporated TRGO and GNPs within the epoxy matrix [49].

It is shown in Fig. 3.3A that fracture toughness has increased for low filler content (0—0.5wt.%). Fracture toughness starts to decrease when the GNPs weight fraction is increased more than 1.0wt.%. This indicates that the fracture toughness value decreases beyond a certain limit of weight fraction of GNPs [50]. A smooth mirror-like fracture surface showing brittle failure of unfilled epoxy composite and the corrugated crevices in the matrix for GNPs/epoxy composites can be seen from the SEM images shown in Fig. 3.3B and C, respectively. It is reported that GNPs cross-linked in the crevices of the corrugation area and hence restrained creviced growth. Therefore, the corrugation and GNPs improves the interfacial fraction between GNPs and epoxy leading to enhancement of static mechanical properties. It can also be interpreted that the performance of TRGO was outstanding compared to the other fillers. The improvement in fracture toughness for GNP/epoxy composite is 24% while that of for TRGO/epoxy composite is 40%. Both TRGO/epoxy and GNP/epoxy composites show higher toughness but TRGO shows a two-fold increase in toughness as compared to GNP toughened epoxy, which is due to the partial functionalization and fewer numbers of layer stacked in TRGO. The covalent bonding increases because of the presence of functional groups on the edges and surface of the graphene sheets and within the epoxy resin. Graphene and its derivatives reinforced epoxy composites show more significant increase in toughness but only in low filler loading. An increase in the filler content within the polymer matrix revealed saturation point in the toughness value [50].

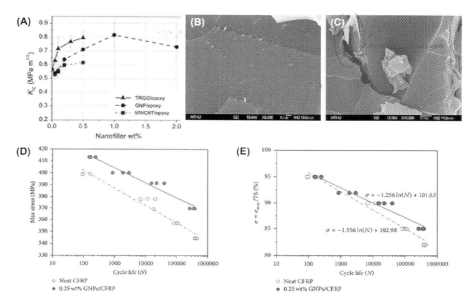

Figure 3.3 (A) The fracture toughness of epoxy composites in terms of weight percent of fillers, reproduced with permission from Chandrasekaran S., Sato N., Tölle F., Mülhaupt R., Fiedler B., Schulte K., Fracture toughness and failure mechanism of graphene based epoxy composites, Compos. Sci. Technol. 2014 90–99 [50], (B) Fracture surface of pure epoxy composite, (C) Fracture surface of 0.25wt.% GNPs/epoxy composite, (D) S-N curve of the CFRP composite laminates with and without 0.25wt.% GNPs content, (E) Graph showing normalized ($\sigma = \frac{\sigma \max}{TS}$, %) S-N curve of the CFRP composite laminates with and without 0.25wt.% GNPs, reproduced with permission from Shen M.Y., Chang T.Y., Hsieh T.H., Li Y.L., Chiang C.L., Yang H., Mechanical properties and tensile fatigue of graphene nanoplatelets reinforced polymer nanocomposites, J. Nanomater. 2013; Creative Commons Attribution License [8]. *GNPs, Graphene nanoplatelets; CFRP, carbon fiber reinforced plastic.*

3.2.4 Tensile fatigue

Tensile fatigue is also a relevant tensile property of composites. Fatigue strength is the stress level below which an infinite number of loading cycles can be applied to a material without causing fatigue failure. The behavior of materials under a much lower stress level than ultimate tensile strength and to examine long-term dynamic reliability, materials can suffer mechanical fatigue. The fatigue life of the composites was increased by the incorporation of nanofillers like graphene and its derivatives. It was reported in the macroscopic studies that a small amount of graphene can improve fatigue life. While investigating the fatigue properties, it was observed that 0.25wt.% of GNP loading improve the fatigue properties at all levels of cyclic stress for epoxy/carbon fiber (CF) composite [8].

Table 3.1 demonstrates the experimental results of the fatigue test for the pure and GNPs reinforced epoxy/CF composites. S-N curve is the plot of the magnitude of alternating stress versus the number of cycles of failure for a given material. In Fig. 3.3D, it is shown that at σ_{max} = 370 MPa and 400 MPa, the durability of composite significantly increased to more than 15.30 to 37.07 times with the addition of 0.25wt.% GNPs. Moreover, pure composite laminate shows a deeper slope in the S-N curves compared to GNPs-added composites. This implies that the fatigue life of the pure composite laminate was more sensitive to applied stress levels compared to the GNPs-added composite laminate and the shift in S-N slope may result because of the earlier fracture occurred in pure composite laminate at all cycle stress. Fig. 3.3E, represents the normalized ($\sigma = \frac{\sigma max}{TS}$, %) S-N curve for the pure and GNPs reinforced epoxy/CF composite. The results demonstrate that the fatigue life of GNPs reinforced epoxy/CF composites extended from 1.21 to 5.39 times as compared to pure composite at 95% and 85% normalized ($\sigma = \frac{\sigma max}{TS}$%) cyclic stress. Thus it is found that embedding GNPs in CFRP composites can restrain the creviced growth as well as restrict the expansion of these cracks in the nanocomposites. As a result, this further lowers the effects of delamination and minimizes fractures, while improves the overall lifespan of CFRP composite laminates [8].

Zakaria et al. have studied GNPs and multiwalled CNT (MWCNT)-based epoxy polymer composite. They have prepared GNPs/epoxy composite that resulted an improved tensile strength, thermal conductivity, flexural strength, and dielectric constant up to 11%, 126%, 17%, and 171%, respectively [9]. Similarly, Nadiv et al. have developed graphene nanoribbon-polymer composites by reinforcing edge-FG nanoribbons (EF-GNRs) and polyvinylamine (PVAM) into the brittle epoxy polymer matrix (PVAM EF-GNRs/) up to 0.15wt.% of FG loading. This composite material shows very good flexural, shear, and fracture properties at a very low loading level. The EF-GNR loaded nanocomposite materials have enhanced the flexural and elastic modulus up to 7% and 14%, respectively, while preserved its ductility [10]. Vadukumpully et al. have developed a composite film of polyvinyl chloride (PVC) and surfactant-wrapped graphene nanoflakes. The mechanical properties for graphene/PVC composite films were increased as compared to the pure PVC film as shown in Fig. 3.4A and B. The PVC film obtained with 2wt.% graphene reinforcement exhibits an increment in Young's modulus and tensile strength up to 58% and 130%, respectively.

The composite of isocyanate functionalized GO (FGS) with polyimide exhibited significant increment in tensile strength and modulus by 60% at 0.75wt.% FGS loading [12]. The composites, prepared by adding GO sheets and FGS sheets with PMMA via Hummers' method exhibited an enhanced Young's modulus of 2.9 and 2.7 GPa, respectively, compared to pure PMMA (2.1 GPa) [14]. In another study, epoxy (Anhydride-cured diglycidyl ether of bisphenol A (DGEBA) reinforced) composite reinforced with GO platelets have been prepared by Yasmin and Daniel. Mechanical analysis of this composite shows an increment of 10% in elastic modulus at 5% GO reinforcement and an increment of 21% in tensile strength at 2.5% GO reinforcement in comparison with pure epoxy, which is having modulus and strength of 3.0 $\pm$ 0.1 and 34 $\pm$ 3 GPa, respectively [15]. The composite developed by reinforcement of GO into polyvinyl alcohol fiber (PVA/RGO) exhibited a

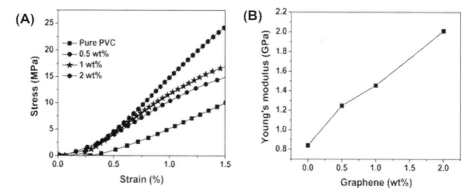

Figure 3.4 (A) Stress–strain curves and (B) Young's modulus of the graphene/polyvinyl chloride composite films having various weight fractions of graphene, reproduced with permission from Vadukumpully S., Paul J., Mahanta N., Valiyaveettil S., Flexible conductive graphene/poly(vinyl chloride) composite thin films with high mechanical strength and thermal stability Carbon198-2052011Vadukumpully S., Paul J., Mahanta N., Valiyaveettil S., Flexible conductive graphene/poly(vinyl chloride) composite thin films with high mechanical strength and thermal stability, Carbon 2011 198–205 [11].

significant improvement in tensile strength and Young's modulus which are 244% and 294% for 2.0wt.% RGO reinforcement [16].

3.3 Compressive properties

In this section, we are discussing the compressive mechanical properties of GO and RGO-based polymer composites. Here compressive property describes the behavior of a material under compressive load where loading rate is uniform and relatively low. Compressive strength and modulus are two common values generated by the test. This discussion includes the change in the compressive behavior of GO and/or RGO reinforcement within the polymer composites.

3.3.1 Compressive property of polymer/graphene oxide composites

Graphene has attracted much attention in recent years due to its extraordinary mechanical properties. In this section we are discussing the compressive properties of some GO reinforced polymer composite materials prepared in different forms such as bead, porous foam, film, hydrogel, resin, disc, scaffold, and some specially designed 3D printed material. These composite materials are prepared by reinforcing exfoliated GO sheets into the polymer matrix where the reinforcement exhibits the enhancement of mechanical properties of the composites. Further improvements in their mechanical properties were reported by surface functionalization of rein-forced GO. GO is commonly used as a reinforcing material because of its high sur-face area and high surface functionality, which results in the formation of facilitating strong interfacial interaction (mostly hydrogen bonding) and produced a path for the transfer of load from the polymer matrix to the reinforcing material and this is why it is known as a good mechanical property modifier. In literature, a dis-cussion on GO reinforced inorganic polymer composite was made. The composite was prepared in the form of a porous foam structure using thermally FGS and sili-cones. In this work, FGS is produced by oxidation of raw graphene at 300°C by the Brodie method. This functionalization of GO sheets results in the formation of hydrogen bond between the polymer matrix and graphene sheets, which is responsi-ble for the enhanced compressive strength of the composite material. Compression testing of commercially available silicon foam reinforced with 0.25wt.% FGS shows more than 200% increment in normalized modulus. This mechanical property improvement shows good interactions between FGS and polymer [18]. The GO is also used to prepare 3D printed biocompatible polymer composite materials. For example, Chen et al. have prepared a biocompatible thermoplastic polyurethane/poly(lactic acid)/GO composite by blending. The composite TPU/PLA/GO is pre-pared in two different orientations: standing specimen (denoted as S) and lying specimen (denoted as L). The compressive properties of composite material shows an enhancement in compression modulus for both S and L specimens. The incre-ment in compressive modulus was 167% in S and 56% in L, respectively at 5wt.% GO loading (see Table 3.1) [7]. Similarly, another 3D printed GO grafted polymer composite material was prepared by Zhong et al. via extrusion-based 3D printing by reinforcing GO into the aqueous mixture geopolymers (aluminosilicate and alkaline-source particles). The composite showed enhanced compressive strength up to 36 MPa after reinforcing at 10wt.% of GO loading. In this composite, the mechanical reinforcement by GO is mainly resulting from the glue effects of the GO layers. Further increase in GO reinforcement up to 20wt.% resulted in the decrease in compressive strength due to the accumulation of GO layers between the interstitials of GO coated HGPPs (hydrated geopolymer particles) when GO layers are more than enough to coat HGPPs and hence initiate the cracks by forming weak zone [19].

The compressive property of some organic polymer/graphene-based films, beads and hydrogels are discussed herein. For example, Tai et al. have synthesized GO−polyacrylic acid composite (GO−PAA) hydrogel by crosslinking PAA incor-porated with GO sheets. The GO-PAA hydrogel composite at 3wt.% GO reinforce-ment has enhanced the compressive modulus three times as compared to neat PAA hydrogel due to efficient load transfer between GO sheets and PAA polymer matrix, which provided better mechanical stability in PAA-GO composite. In Fig. 3.5A, it is shown that the compressive modulus of GO−PAA nanocomposite hydrogel is three fold higher than that of PAA hydrogel [51]. In the energy sector, GO plays an important role as the energy storage devices. For example, boron cross-linked GO/PVA composite gel is used as a flexible solid-state electric double layer electrolyte for a capacitor. The compressive properties for these boron cross-linked (B-PVA)

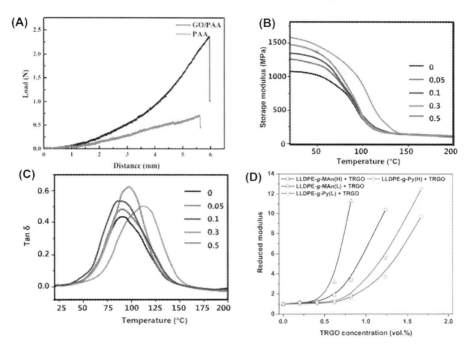

Figure 3.5 (A) Compressive load—distance curves of the PAA hydrogel and the GO—PAA composite hydrogel at room temperature, reproduced with permission from Tai Z., Yang J., Qi Y., Yan X., Xue Q., Synthesis of a graphene oxide-polyacrylic acid nanocomposite hydrogel and its swelling and electroresponsive properties, RSC Adv. 2013 12751—12757 [51], Dynamic mechanical properties of bGO/BFDE composites showing (B) Variation of storage modulus with temperature change and (C) Change in tan δ with temperature, reproduced with permission from Zhao H., Ding J., Yu H., Variation of mechanical and thermal properties in sustainable graphene oxide/epoxy composites, Sci. Rep. 2018 1—8; Creative Commons CC BY License [26], (D) Reduced storage modulus G_r', (where, $G_r' = G'/G_0'$, G_0' is the storage modulus of the matrix and G' is the modulus of the composite) versus TRGO concentration at 0.1 rad/s and 190°C, reprinted with permission from Vasileiou A.A., Kontopoulou M., Docoslis A., A noncovalent compatibilization approach to improve the filler dispersion and properties of polyethylene/graphene composites, ACS Appl. Mater. Interfaces 2014 1916—1925. Copyright (2014) American Chemical Society [52]. *BFDE, biobased bis-furan di-epoxide; PAA, polyacrylic acid; TRGO, thermally reduced graphene oxide.*

and GO modified (GO-B-PVA/KOH) composites were measured. The result showed that compression ratio of ∼ 63% and compression strength of 0.085 MPa were obtained for B-PVA composite; whereas, the compression ratio of ∼ 37% and compression strength of 0.056 MPa were obtained for 2wt.% loaded GO-B-PVA composite [28]. Another GO grafted polymeric composite hydrogel was developed by Piao and Chen using gelatine, glutaraldehyde (GTA), and GTA-grafted GO double chemical cross-linkers. These gelatin-GO hydrogels showed very good mechanical property enhancement by GO reinforcement, for example, the highest compressive strength value of 566 kPa was obtained at 5 mg/mL of GO reinforcement; whereas, the compressive strength for neat gelatine hydrogel was 146 kPa. The enhanced mechanical properties were because of the formation of double cross-linked structure by GTA-grafted GO sheets reinforcement [22].

Hydrogels now-a-days play an important role in biomedical and agricultural applications due to its stimuli responsive behavior. Here, we discussed about pH-responsive hydrogels developed by Peng et al. In this work, biomimetic jellyfish-like PVA/GO composite was prepared using a convenient, effective and directional freezing-thawing technique in which hydrogels mimic the hierarchical structures. These biomimetic jellyfish-like PVA/GO nanocomposite hydrogels are anisotropic and pH-responsive, which contained 97−99wt.% water. The composite exhibited high compressive strengths (greater than 2 MPa) but this mechanical property was weakened when hydrogels were swollen in acidic and basic solutions. These hydrogels could be used in some future applications in the field of biomedical, industrial, and soft robotics as bio-inspired hydrogels [53]. Another pH-responsive hydrogel, GO/sodium alginate (SA)/polyacrylamide (PAM) composite semi-IPN hydrogel, having enhanced mechanical strength is prepared by reinforcing GO by dipping into the sodium alginate and PAM with a cross-linking agent (N, N′-methylenebisacrylamide or BIS). The obtained GO/SA/PAM/BIS nanocomposite hydrogels have both stimuli responsive behavior as well as superior compressive strength of 65.6 MPa at 0.018 g of GO reinforcement where the quantities of other components were 0.10 g of SA (sodium alginate), 0.010 g of BIS, and 3.6 g of AM (acrylamide). Further increase of GO reinforcements leads to decrease in compressive strength. In this study, it is shown that the compressive properties depend on the GO dispersion and their interfacial interaction with polymer chains [54]. Similarly, when GO was incorporated with PAM and carboxymethyl cellulose sodium (CMC), it formed GO/PAM/CMC composite hydrogels. Blend of PAM/CMC hydrogels were very brittle, so the compressive test was unable to perform. However, the addition of GO sheets up to 1.6wt.% enhanced the compressive strength of the GO/PAM/CMC composite hydrogel to 2.87 MPa [23].

A stretchable, tough, and compressive novel polymer/GO nanocomposite hydrogels have been developed and tested for cyclic compressive test by Pan et al. that exhibited excellent self-healing performance. This composite hydrogel was prepared by free radical polymerization of 2-(dimethylamino)ethylacrylatemethochloride (DAC) and AM in the presence of GO as a reinforcing agent to improve the mechanical properties. It was shown that the P(AM-co-DAC)/GO composite hydrogel can recover completely at 80% strain rate up to 30 cycles due to the formation of the hydrogen bonds between GO and AM and ionic bonds between GO and DAC, which are responsible for the efficient dissipation of energy and rebuilding of the network. The resulting composite hydrogels also possess high stiffness (Young's modulus: ∼ 1.1 MPa), high fatigue resistance, high toughness (∼ 9.3 MJ/m^3), and high self-healing efficiency (> 99% of tensile strain, >92% of tensile strength, and >93% of toughness) [55].

Yang et al. tunably loaded chemically converted GO into epoxy resin and prepared the epoxy/GO nanocomposites through two-phase extraction. It was shown that 0.0375wt.% incorporation of chemically converted GO sheets into the epoxy resin could increase both the compressive failure strength by 48.3% and toughness by 1185.2% due to strong interfacial linkages [20]. A GO reinforced PVA composite scaffold was developed by Shuai et al. This polymer composite had an interconnected porous structure and was prepared by selective laser sintering technique for tissue engineering applications. In this composite, the GO reinforcement improves the compressive strength by 60% at 2.5wt.% GO loading within the PVA polymer matrix. The factor responsible for the mechanical properties enhancement is the strong hydrogen bonding interactions between PVA matrix and GO sheets and the homogenous dispersion of the GO sheets within the polymer matrix but a higher amount of reinforcement led to the agglomeration, which resulted in the decrease of mechanical properties [21].

3.3.2 Compressive property of polymer/reduced graphene oxide composites

In this section, we have discussed the compressive property of some reduced graphene-based polymer composite materials and the effect of RGO reinforcement on its compressive properties. We have already discussed the compressive property of composite hydrogels reinforced with GO sheets in the previous section. Herein, the discussion has been made on the compressive property of some composite hydrogels reinforced with RGO. A novel RGO/poly (2-acrylamide-2-methyl propane sulfonic acid-co-acrylamide) [RGO/poly (AMPS-co-AAm)] composite hydrogel was developed by Yang et al. through thermal-induced polymerization technique. This composite material exhibited excellent mechanical properties because of the hydrogen bonding type interactions generated in between RGO sheets and the polymer chains. This could dissipate the strain energy in the polymer network and enhance the mechanical compressive strength of the composite hydrogel. The compressive strength analysis shows that the RGO/poly (AMPS-co-AAm) composite hydrogels reinforced at 0.5wt.% RGO loading are much tougher than the blank hydrogel under compressive stress and can be quickly recovered to their original shapes on releasing the compressive stress. Initially, the blank hydrogel breaks at the compressive stress of 0.4 MPa but on reinforcing at 0.5wt.% RGO loading, it enhanced the compressive stress at break to 1.4 MPa [24]. Like GO, the RGO has also reinforced the PVA polymer to obtain mechanically improved composite materials. Pourjavadi et al. prepared composite of PVA reinforced with functionalized RGOs (f-RGO), which shows excellent mechanical properties. The composite material was prepared using f-RGO as a reinforcing agent, citric acid as a cross-linker, and glycerol as a plasticizing agent. GO was reduced and stabilized by hydrolyzed Salep, which was known as a biodegradable polysaccharide. These f-RGO/PVA hydrogels were also analyzed to find their compression strengths. The objective of the mechanical analysis of this composite material was to confirm whether its mechanical features were for tissue scaffold mimicking or not. It was shown that on adding 0.1g f-RGO, the young's modulus and compressive strength were increased to 77% and 25.7%, respectively, when calculated at 65% deformation. This study showed that the compressive strength was improved by reinforcing RGO with the PVA composite hydrogels, which made appropriate mechanical characteristics for soft tissue scaffolds [25].

3.4 Dynamic mechanical properties

Dynamic mechanical analysis (DMA) of composite materials show their viscoelastic behavior results in various engineering, biomedical, and industrial applications. In this section, we are discussing the dynamic mechanical analysis of some GO and/or RGO reinforced polymer composite materials. In DMA, the input is given as sinusoidal stress and the output strain is measured to determine the complex modulus. In this measurement, the temperature and frequency are varied, which leads to the change in the complex modulus; this approach is also used to find out the glass transition temperature of the material. Moreover, the variation of different parameters such as storage modulus, complex modulus, loss factor, and the phase change behavior of composite materials are also studied.

3.4.1 Dynamic mechanical properties of polymer/graphene oxide composites

The dynamic mechanical properties of GO reinforced polymer composites have been briefly discussed in this section. These composite materials were prepared by reinforcing the polymer matrix with exfoliated GO sheets and this reinforcement enhances the dynamic mechanical properties of composites. Further improvements in their dynamic mechanical properties are done by surface functionalization of reinforcing GO sheets. Zhao et al. had synthesized sustainable GO/epoxy composites (bGO/BFDE) by grafting FGS into a biobased bis-furan di-epoxide (BFDE) polymer matrix. The functionalization of GO was done by simply adding BFDE to the GO dispersion followed by sonication in the presence of NaOH (act as a catalyst) and then the mixture was placed at 75°C for 8 h. The mixture was then stirred under an inert atmospheric condition. The obtained bGO was incorporated into the BFDE polymer to enhance the dynamic mechanical properties of the composite. The bGO/BFDE composite, containing 0.05−0.5wt.% of bGO exhibits high thermal and mechanical properties. The bGO/BFDE composite was analyzed for DMA within the temperature range 25°C−130°C at the constant heating rate 3°C/min and frequency 1 Hz. The result showed a decrease of storage modulus value with the increase in temperature. Fig. 3.5B shows that how the storage modulus value increases with the increase in reinforcement concentrations and the highest storage modulus value is observed for the composite having 0.5wt.% GO loading. The storage modulus value at 25°C for neat polymer is 1100 MPa, which is increased to

1645 MPa for 0.5wt.% GO reinforced composite; this meant about 48% increment compared to the neat polymer. The tan δ value of cured BFDE composites with varying amount of bGO incorporation are shown in Fig. 3.5C. It is observed that the T_g value increases from 90.3°C to 116.8°C with the increase in amount of bGO incorporation. Here the wrinkled morphology of bGO sheets with high specific surface area constrains the segmental movement of polymer chains to a certain degree and hence are responsible for the change in the T_g value [26].

The composite based on natural rubber (NR) also shows a great mechanical property enhancement by reinforcing NR with GO. For example, graphene reinforced and vinyltriethoxysilane (VTES) grafted NR composite was developed by Yin et al. in the presence of potassium persulfate (KPS) as an initiator. This (NR-g-VTES) was reinforced with GO at different loadings by mechanical mixing method. The specimens of size 15 × 6 × 0.5 mm were taken in tension mode for dynamic mechanical characterization, which was measured within the temperature range of −40°C to +110°C at the constant heating rate 5°C/min, frequency 1 Hz, stress 0.1 N, and strain 0.1%. The analysis was performed to determine the loss factor and storage modulus of the NR/GO and NR-g-VTES/GO composites. The DMA analysis showed higher storage modulus value for graphene reinforced composite material, which indicated the formation of chemical bonding between filler and rubber and also the transfer of stress from rubber to filler. Moreover, NR-g-VTES/GO showed higher tan δ value than the NR/GO composite at 0°C that indicated better wet-skid resistance property of NR-g-VTES/GO, and high tan δ value of NR-g-VTES/GO composite at 60°C meant high rolling resistance than NR/GO composite. The glass transition temperature for NR-g-VTES/GO composite was higher compared to NR/GO composite. This is because of additional crosslinking among rubber chains, which is formed due to hydrolysis and condensation of $-Si(OC_2H_5)_3$ groups present in NR-g-VTES. This reduces the mobility of molecular chains and increases the T_g [56]. Another NR/GO composites was developed via the latex mixing process by Zhao et al. [57]. The composite was reinforced by GO for mechanical property enhancement and tested for DMA. The result showed an increase in storage modulus value of the composite with the increase in graphene loading. This is because, the molecular mobility depressed, and hence the glass transition value and tan δ value increases with the increase in reinforcement of graphene within the NR. Improvement in storage modulus was observed at −10°C for NR/GO composites [57].

The dynamic mechanical properties are also enhanced in double cross-linked gelatine/GO hydrogels. The gelatine/GO hydrogels were tested for DMA with respect to varying angular frequency measured at room temperature and 0.1% constant strain using a parallel plate geometry (diameter 25 mm) with the gap of 2 mm. The storage modulus of the pure gelatine hydrogel was 10.4 kPa, which was enhanced up to 11.6, 20.1, and 27 kPa on reinforcement with 0.1, 1, and 3wt.% of GO, respectively, when measured at 10 rad/s. Further increase in the reinforcement declines the storage modulus value to 23.6 kPa for 5wt.% GO reinforcement due to the GO aggregation and lower cross-linking density, which weakened the structure of hydrogel [22]. It is observed that the storage modulus has decreased at high temperature for NR-g-VTES/GO, gelatine/GO, and bGO/BFDE composites, PVA has also shown the same behavior. In a recent report, boron cross-linked GO/polyvinyl alcohol composite gel electrolyte (GO-B-PVA) was tested for DMA within the temperature range 30°C to 100°C or 120°C when measured at 1 Hz frequency and 5°C/min heating rate. It has shown the same behavioral decrement in storage modulus with the increase in temperature but the storage modulus was drastically increased with the incorporation of GO. In this study, 2wt.% of GO incorporation resulted in shear storage modulus value G' $\sim$0.21 MPa, which was higher compared to neat B-PVA [(G') $\sim$0.14 MPa] [28].

These studies showed that the storage modulus for the polymer composites was increased by incorporating GO up to a certain level, above which the GO resulted in agglomeration within the polymer matrix and hence decreased the storage modulus value on further reinforcement.

3.4.2 Dynamic mechanical properties of polymer/reduced graphene oxide composites

GOs can be reduced by various methods such as sonication, chemical, and thermal treatment. In this section, we have discussed about the dynamic mechanical properties of some RGO reinforced polymer composites in brief. For example, graphene/carbon black/natural rubber (RGO/CB/NR) composites were developed by Wang et al. via WCL method (wet compounding and latex mixing) using RGO and CB as reinforcing materials, and NR as a polymer matrix. The RGO/CB/NR composites showed enhanced mechanical properties than the composites prepared by conventional latex mixing. The DMA of the composite showed an increase in storage modulus (G') and decrease in tan δ value of the composite when CB was substituted with RGO incorporation. Here, the elastic modulus increased with RGO incorporation because of the decrease of chain mobility during dynamic mechanical deformation due to the adsorption of the rubber chain on to the surface of RGO. Incorporation of RGO up to 1wt.% resulted a significant decrease in rolling resistance. On the contrary, the addition of RGO >2wt.% increased the rolling resistance value due to the agglomeration of RGO within the polymer matrix [58]. Vasileiou et al. have prepared linear low-density polyethylene (LLDPE)/TRGO composite where RGO was exfoliated and reduced by thermally induced expansion method under high vacuum atmosphere. The resulting TRGO consisted of single as well as multilayer graphene, which had 586 m^2/g of specific surface area. TRGO was added in different amount to the LLDPE-g-Man (LLDPE grafted maleic anhydride) and LLDPE-g-Py (LLDPE-graft-aminomethylpyridine) matrices to get different composites ranging from 0.5 to 4.0wt.% loading. It was reported that the enhanced noncovalent interactions between the fillers and the aromatic moieties grafted onto the polymer matrix was responsible for the enhancement of mechanical and thermal properties of the polymer composite. Viscoelastic behavior of the composites was analyzed at the frequency 0.1 Hz and temperature 190°C within the linear viscoelastic region. The variations of reduced modulus with respect to TRGO loading are presented in Fig. 3.5D, where H and L denoted in the figure represented for high and low viscosity, respectively [52]. It was reported that the dynamic mechanical property was enhanced for RGO reinforced gelatine composite hydrogel compared to neat gelatine hydrogel [30]. The composite was prepared by the one-pot synthesis method mainly for drug delivery applications. The rheological measurement was conducted at 20°C to determine the storage moduli (G') and loss moduli (G") over 0.1−100 rad/s frequency range at 0.1% strain. The results exhibited the enhancement of the storage modulus from 64.4 to 91.1 kPa, and then become 172.3 kPa after reinforcing with RGO at 2, 5, and 10 mg/mL, respectively [30]. Hence, like GO, the RGO reinforcement also greatly enhanced the dynamic mechanical properties of the polymer composites. The enhancement in mechanical properties make these composites suitable for various future applications in the biomedical, agricultural, and industrial fields.

While comparing the properties of composites reinforced with GO or RGO, functionalized graphene (FG) reinforced composites also show an enhancement in the dynamic mechanical properties. For example, functionalized or CTAB stabilized graphene reinforced PVC composite was developed by Vadukumpully et al. [11]. The functionalization of graphene with CTAB (cetyltrimethylammonium bromide) was carried out by sonication of the graphite powder with 0.5 M CTAB solution for several hours in glacial acetic acid and then heated at 100°C for 48 h under nitrogen atmosphere. The incorporation of these surfactant-wrapped graphene nanoflakes enhanced the storage modulus from 0.5 GPa for pure PVC film to almost 3 GPa for graphene/PVA composite film measured at 20°C for 2wt.% incorporation of GO, whereas, the highest peak for tan δ was obtained for 0.5wt.% incorporation over all temperature range from 25°C−150°C [11].

Similarly, Wang et al. had prepared PS/FGs composite using raw PS and PS-FG by a sonochemical method in which graphite flakes are directly exfoliated in styrene. This PS-FG/PS composite was tested for DMA and it was investigated that the uniform distribution of PS-FG sheets into the PS matrix was responsible for the increment in storage modulus up to 41% for the sample corresponding to 0.5wt.% FGs loading [27].

Wang et al. developed poly(3-hydroxybutyrate-co-3-hydroxy valerate) (PHBV)/GNS composites, which showed a great improvement in mechanical property by reinforcing with GNSs [29]. In this work, GNSs were obtained by thermal exploitation of GO. Dynamic mechanical analysis of polymer composite showed that the storage modulus (E') was generally increased by reinforcing with GNSs, when tested within the temperature range −50°C to 80°C. For example, the values of E' at −50°C increased from 1395 to 2670 MPa on reinforcing with 1wt.% GNSs. Similarly, in comparison with neat PHBV, the storage modulus had increased by 25%, 87%, and 91% after incorporating 0.1, 0.5 and 1.0wt.% GNSs within the polymer matrix, respectively [29]. The surface properties are responsible for well dispersion of graphene into the polymer matrix, which results in the enhancement of mechanical characteristics.

3.5 Flexural properties of polymer/graphene composites

The flexural properties, mainly flexural strength, flexural modulus, and flexural toughness of different polymer composites reinforced with GO and RGO, have been briefly discussed in this section. Flexural strength represents the maximum stress when a material breaks under specified deflection. Flexural modulus represents the tendency for a material to resist bending when a force is applied perpendicular to the long edge of the sample. Flexural modulus is defined as the ratio of stress and strain of the material within its linear elastic region, whereas, flexural toughness is calculated by considering the required area under the load-deflection curve in flexure.

Tang et al. have studied the properties of GO/epoxy resin composites and reported that pure epoxy resin has impact strength of 29 kJ/m^2 and with increasing the GO content from 0.1 to 1wt.%, its impact strength has increased from 97% to 176% and then reduced to 77% due to agglomeration of graphene within the epoxy matrix at higher loading [59]. Graphene had a large specific surface area with high mechanical strength, which enhanced the toughness of composite material. However, the decrement in impact strength at high concentration of graphene was due to its agglomeration and stress concentration within the polymer matrix [60]. The effects of different weight ratios of GNPs with epoxy resin were investigated and flexural strength and flexural modulus were discussed [8]. It was reported that with an increase in the incorporation of GNP from 0.25 to 1.5wt.%, there was an increment in flexural modulus from 12% to ∼19%, respectively, as compared to pure epoxy. The highest improvement in flexural strength (115.5 MPa) was 9% at 0.25wt.% loading of GNP compared to pure epoxy (106 MPa). Flexural strength reduces with increase in the GNP because of the following possible reasons: (1) GNP makes aggregates and generates low surface area with millimeter size fillers, (2) GNP agglomerates and creates a steric obstacle so that polymer gets restricted to flow by resulting noncontacting area between GNP and epoxy. Corrugation and interfacial fraction were the reason for increased static mechanical properties. Aggregation of particles increased the stiffness of material, which increased the modulus of the composite [8].

Functionalized GO (f-GO) was also synthesized using GO and BFDE. The composite exhibited an increment of 108 ± 2 MPa flexural strength and 1840 ± 50 MPa flexural modulus because f-GO had better dispersibility within the solvent and polymer matrix [26]. FG was also used to prepare PS-based polymer composite and its fractural property was investigated. Single and multiple layers of the FG were present in the composite. These were of hundreds of nanometers to several microns in sizes. Woven CF reinforced epoxy composite showed an increase in flexural strength with the increase in GO concentration. At 0.05, 0.1, 0.2, and 0.4wt.% loading of GO, the increment in flexural strength was 34%, 43%, 56%, and 51%, respectively, compared to pure epoxy. It was found that there was greater adhesion between GO and epoxy because of the higher surface area of GO particles. When the load was applied to the composite, the two-dimensional GO sheets get tilted, absorb higher energy, and then lead to crack propagation. At a certain point, strength reaches its maximum point and then decreases due to failure in all the stresses (tension, compression, and bending) [31]. CF/epoxy composites with GO and f-GO [functionalized with ethylenediamine (EDA), 4, 40-diamino diphenyl sulfone (DDS), and p-phenylenediamine (PPD)] were investigated to test their bond strengths. It was found that the flexural strength was increased with an increase in the GO concentration and it was maximum at 0.3wt.% concentration. The flexural strengths were 900 and 1050 MPa at 0.3wt.% loading of GO when it was functionalized with EDA and DDS, respectively. On the contrary, it was 1200 and 1480 MPa at 0.5wt.% of GO loading and when functionalized with PPD, respectively. Flexural modulus was also increased with an increase in the GO and f-GO. Flexural modulus was 64, 64, and 62 GPa for EDA, DDS, and PPD f-GO at 0.3% loading, respectively. Tensile fracture morphology was also investigated for these composites using SEM. It was found that there was a gap between fibers and epoxy materials that represented the weak binding in the case of the controlled composite. The gap was reduced in GO and DDS-f-GO composites, which showed better adhesion between fibers and epoxy. The composites, having f-GO functionalized with EDA and PPD, showed that high amount of epoxy was attached with CF surfaces and a gap was remained in between fibers and epoxy that resulted in less adhesion [60]. The mechanical properties of short glass fiber (SGF) reinforced polyethersulphone (PES) composites were increased with GO coating on SGF [32]. Flexural properties of the composites were examined and found that initially it was increased with the increase in GO loading. The strength was increased from 160 MPa (at 0% of GO) to 175 MPa (at 0.5% of GO) and then decreased with further increase in the GO content. The surface morphology of different composites was examined using SEM. Smooth surface was observed for raw SGF but it became rough after coated with GO indicating that the SGF was successfully coated with GO. It was reported that higher amount of polymer matrix was attached with short fibers that created strong interfacial bonding. Agglomeration of the GO content at higher loading reduced the interfacial adhesion, which decreased the strength of composite after a particular concentration of GO [32].

In a report, RGO was used with fly ash-based geopolymers for the increment of mechanical strength of the composite material. There was an increment in flexural strength of 49%, 130%, and 134% for 0.10, 0.35, and 0.50wt.% loaded RGO composites compared to neat polymer. The flexural toughness of the composites was increased by 12%, 56%, and 48% with the addition of 0.10, 0.35, and 0.50wt.% RGO, respectively. These results showed higher mechanical properties compared to CNTs based composites [33]. GNP, as a reinforcing agent, was used to investigate its effect on the composite of wood-flour/ polypropylene (PP). In this composite material, maleic anhydride grafted PP (MAPP) was added for the better interphase adhesion. The flexural strength at 0.2 and 5wt.% GNP loading was reported as 35 and 80 MPa, respectively. The modulus of the composites was not decreased though the graphene nanoplatelets were agglomerated at higher concentrations [61]. Two different kinds of carbons namely graphene nanoplatelet and CNTs were used to improve the mechanical properties of pure epoxy. All composites showed higher mechanical properties compared to pure epoxy where larger size of the graphene flakes supported crack deflection. It was reported that both the flexural modulus and the fracture toughness were higher for 25 μm flake-based composite compared to 5 μm based composite one when measured for all loading level (0.1 to 2wt.%). Fracture toughness for the pure epoxy was 0.50 MPa.m$^{1/2}$, which was continuously increased with an increase in the content of graphene. Its maximum value was 0. 91 and 0. 8 MPa.m$^{1/2}$ for 25 and 5 μm flaked based composites, respectively. The highest improvement in the fracture toughness was 77% for the composite having 9:1 ratio of CNT: GNP. It was reported that the flexural modulus value of the composite, having CNT:GNP ratio of 9:1 had increased by 17%, which was higher compared to individual CNT and GNP based composites (9% and 5%) [62].

In a very interesting recent study, it was reported that the multi-scaled CF and graphene nanoplatelet as reinforcements had increased the mechanical and thermal properties of the poly (arylene ether nitrile) (PEN) based composites [34]. The flexural modulus of CF-based PEN composites was increased with the increase in CF loading. It was reported that at 10wt.% CF loading, the flexural modulus was increased by 179.3% compared to pure PEN (2.9 GPa). At 30wt.% CF loading, the flexural modulus value of the composite was increased five times compared to its initial value. However, the flexural strength of the composite increased with increasing CF loading but then became saturated at 20wt.% CF loading. It was mentioned that the increase in viscosity at higher loading of CF which led to a heterogeneous dispersion state could be the probable reason behind this behavior. Further enhancement of mechanical properties were shown by PEN/CF/GN composites which were prepared by melt-mixing process. The flexural modulus of PEN/CF/GN composites with 10wt.% GN loading was 18.6 GPa, which is 1.7, 4.5 and 6.4 times larger than those of PEN/CF composites, PEN/GN composites and PEN host, respectively Maximum flexural strength of 283.7 MPa in PEN/CF/GN composites was acheived with 5wt.% GN loading (see Table 3.1). Such enhancement in mechanical properties were due to the synergetic effect of micro-scale CF and nano-scale GN in the PEN matrix and enhanced interfacial interactions [34]. In another study, a small amount of GO sheets changed the shear failure in the composites. There was a needle-like and leaf-like structure that appeared in the fracture surface of 5wt.% GO loaded short carbon fiber/graphene oxide/epoxy resin (SCF-GO5-EP) composite. This facilitated the increase of the fractured surface area and degree of microcracking, hence increased the strength and toughness of the composites [64].

In another study, silane functionalized graphene sheet (sGO) was fabricated with four different self-assembly monolayer (SAM) to increase the strength of the composite. The sGO was functionalized with (3-aminopropyl) trimethoxysilane (APTMS), (3-aminopropyl) triethoxysilane (APTES), triethoxymethylsilane (MTES), and (3-glysidyloxypropyl)trimethoxysilane (GPTMS). It was reported that the bond strength was increased from 20.87 MPa (for GO) to 31.97, 27.72, 24.52, and 26.91 MPa for the APTMS-sGO, APTES-sGO, MTES-sGO, and GPTMS treated composites. In this study, the functionalized group plays an important role in enhancing the interaction energy between fiber and the adhesive. It was reported that the amine functional group is more active than alkyl and epoxy in the bonding process. The reason behind the activeness of the amine group was that it has an unpaired electron, which was responsible for the hydrogen bonding. There was 17.5% to 53.2% increment in the bonding strength by GO and sGO incorporation, which was comparable to the conventional methods like plasma treatment and acid treatment [63]. Composites of exfoliated graphite nanoplatelets (xGNP) reinforced PP was prepared by melt mixing and polymer solution coating. It was reported that the coating method was more effective compared to the melt mixing method when considered in terms of flexural properties. Flexural strength and modulus were investigated for the composites having filler loadings 0, 1, 3 and 10vol.%. It was analyzed that the composites prepared by coating method had higher strength (48 and 51 MPa at 3 and 10vol.%) compared to the composites prepared by melt mixing method (46 and 47 MPa). Both composites were analyzed through SEM. Agglomerates of xGNP-15 were reported when the composite was prepared by melt mixing; whereas, the graphene particles were well-dispersed within the PP matrix when it was prepared by coated method. In this study, it was also reported that there were two different areas of the graphene dispersion; One area showed larger flat graphite platelets and the other showed very small graphite particles, which represented that melt mixing had not provided enough dispersion of graphite platelets within the composites [65]. Epoxy and graphite based composites were prepared for the analysis of different mechanical properties. Elastic modulus was increased with an increase in the graphite content up to 10wt.%. The maximum value of elastic modulus was 58 GPa, which was 26% higher compared to the neat epoxy. Pure epoxy had 45 GPa elastic modulus. Flexural modulus and flexural shear modulus were also increased with an increase in graphite content within the composite. When graphite concentration was increased from 0 to 11.5wt.%, there was 68% increment in flexural modulus. The reported flexural modulus for neat epoxy and 11.5wt.% loaded epoxy/graphite composites was 25 and 42 GPa, respectively [66].

Three different kinds of nanofillers were used with epoxy to study the toughness behavior of the composites. In this study, the fillers used were MWCNT, TRGO, and GNP. It was reported that at lower concentration of filler, the fracture toughness was increased and after a certain loading, the value was declined. The increment in fracture toughness values of 0.5wt.% loaded MWCNT/epoxy, GNP/epoxy, and TRGO/epoxy composites were 8%, 24%, and 40%, respectively, compared to the pure epoxy. It was reported that the toughness value was 3 times higher for GNP/epoxy composite compared to MWCNT/epoxy composite. The presence of the different number of layers and the functional group was responsible for the variation in fracture toughness value for GNP/epoxy and TRGO/epoxy composites [50]. In another study, aluminum oxide (Al_2O_3) was reinforced with RGO (5vol.%) and the composite was characterized along different orientations. The fracture toughness value for the Al_2O_3/RGO composite along cross-sectional orientation was increased by 200%. On the contrary, there was no toughness observed along in-plane orientation. However, moderate toughness was observed for random orientation. The fracture toughness values for pure Al_2O_3, composites with cross-sectional, in-plane, and random orientations were 2.3, 6.7, 2.1, and 4.7 MPa, respectively [67].

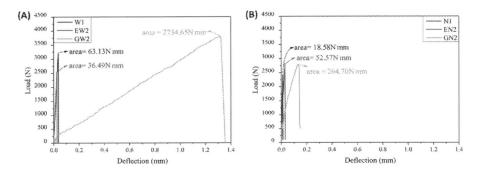

Figure 3.6 Fracture toughness of EP and RGO/EP alkali-activated slag. (A) Water-glass activated slag mortar, (B) NaOH-activated slag mortar, reproduced with permission from Guo S., Lu Y., Wan X., Wu F., Zhao T., Shen C., Preparation, characterization of highly dispersed reduced graphene oxide/epoxy resin and its application in alkali-activated slag composites, Cem. Concr. Compos. 2020 103424 [68]. *RGO, Reduced graphene oxide, EP, epoxy resin.*

The load-deflection curves in the four-point bending flexural tests of an alkali-activated slag (AAS) composite mortar with 0.3% EP and 3% RGO/epoxy resin (EP) with water glass and NaOH as activator are shown in the Fig. 3.6. It is observed that the addition of RGO/EP significantly improves the fracture toughness of alkali-activated slag mortar, especially for water activated slag mortar, the toughness increased more than seventy times compared the one blend EP [68].

3.6   Conclusions

Graphene-based polymer composites are the most developed materials used for mechanically strengthening the polymer matrices. Graphene is exfoliated into its GO and RGO states via thermal, chemical, and sonication techniques in the form of either sheets or nanoplatelets. Various reports have been discussed in this chapter about the effects of loadings of graphene and its derivatives. The result showed that the polymer composite reinforced with GO and RGO exhibited significant improvement in mechanical properties such as tensile, fatigue, fractural, compressive, and dynamic mechanical properties even at their very low loading for example, the reinforcement showed an enhancement up to 244% in tensile strength at very low loading (2wt.%) for PVA/RGO composite, enhancement in compressive strengths up to 288% at loading of 5 mg/mL of GO for gelatin/GO composite hydrogel, enhancement in dynamic storage modulus up to 160% at loading 3wt.% for GO/gelatin composite, and enhancement in flexural strength about 134% at 0.5wt.% loading for RGO/geopolymer composite. Surface fucntionalization of graphene sheets further improves the mechanical properties of the composites due to better interfacial interactions and dispersion in the polymer matrix. However, higher loading of graphene leads to deterioration in mechanical proeprties due to agglomeration in the polymer matrix.

3.7 Future outlook

In the present market, there are many products that are available on large and commercial scales that are prepared by reinforcing GO and RGO into the polymer matrix. These polymeric composites have been found beneficial in the fields of chemical, petrochemical, biological, electrical, energy, industrial, household design, 3D printing material, and biomedical applications mostly as structural materials [69,70]. These high-strength multifunctional graphene-based polymer composites with improved mechanical strength offer new opportunities for the development of a large range of cost-effective, environmentally friendly, and abundantly available high strength structural materials. Surface functionalization and surface engineering of the graphene nanoparticles will lead to further enhancement of mechanical properties in the polymer/graphene composites and also tailor its applications.

Acknowledgment

Paramita Das gratefully acknowledges the infrastructural support from Indian Institute of Science Education and Research (IISER) Bhopal.

References

[1] K.S. Novoselov, A.K. Geim, S.V. Morozov, D. Jiang, Y. Zhang, S.V. Dubonos, et al., Electric field in atomically thin carbon films, Science 306 (2004) 666−669.
[2] F. Najafi, M. Rajabi, Thermal gravity analysis for the study of stability of graphene oxide−glycine nanocomposites, Int. Nano Lett. 5 (4) (2015).
[3] C. Lee, X. Wei, J.W. Kysar, J. Hone, Measurement of the elastic properties and intrinsic strength of monolayer graphene, Science 321 (2008) 385−388.
[4] X. Du, I. Skachko, A. Barker, E.Y. Andrei, Approaching ballistic transport in suspended graphene, Nat. Nanotechnol. 3 (8) (2008) 491−495.
[5] S. Bai, X. Shen, Graphene−inorganic nanocomposites, RSC Adv. 2 (1) (2012) 64−98.
[6] Y. Huang, J. Liang, Y. Chen, An overview of the applications of graphene-based materials in supercapacitors, Small 8 (12) (2012) 1805−1834.
[7] Q. Chen, J.D. Mangadlao, J. Wallat, A. De Leon, J.K. Pokorski, R.C. Advincula, 3D printing biocompatible polyurethane/poly(lactic acid)/graphene oxide nanocomposites: anisotropic properties, ACS Appl. Mater. Interfaces 9 (4) (2017) 4015−4023.
[8] M.Y. Shen, T.Y. Chang, T.H. Hsieh, Y.L. Li, C.L. Chiang, H. Yang, et al., Mechanical properties and tensile fatigue of graphene nanoplatelets reinforced polymer nanocomposites, J. Nanomater. 2013 (2013).

[9] M.R. Zakaria, M.H. Abdul Kudus, H. Md Akil, M.Z. Mohd Thirmizir, Comparative study of graphene nanoparticle and multiwall carbon nanotube filled epoxy nanocomposites based on mechanical, thermal and dielectric properties, Compos. B Eng. 119 (2017) 57–66.

[10] R. Nadiv, M. Shtein, M. Buzaglo, S. Peretz-Damari, A. Kovalchuk, T. Wang, et al., Graphene nanoribbon-polymer composites: the critical role of edge functionalization, Carbon 99 (2016) 444–450.

[11] S. Vadukumpully, J. Paul, N. Mahanta, S. Valiyaveettil, Flexible conductive graphene/poly(vinyl chloride) composite thin films with high mechanical strength and thermal stability, Carbon 49 (1) (2011) 198–205.

[12] L. Bin Zhang, J.Q. Wang, H.G. Wang, Y. Xu, Z.F. Wang, Z.P. Li, et al., Preparation, mechanical and thermal properties of functionalized graphene/polyimide nanocomposites, Compos. A Appl. Sci. Manuf. 43 (9) (2012) 1537–1545.

[13] B. Shen, W. Zhai, M. Tao, D. Lu, W. Zheng, Chemical functionalization of graphene oxide toward the tailoring of the interface in polymer composites, Compos. Sci. Technol. 77 (2013) 87–94.

[14] C. Vallés, I.A. Kinloch, R.J. Young, N.R. Wilson, J.P. Rourke, Graphene oxide and base-washed graphene oxide as reinforcements in PMMA nanocomposites, Compos. Sci. Technol. 88 (2013) 158–164.

[15] A. Yasmin, I.M. Daniel, Mechanical and thermal properties of graphite platelet/epoxy composites, Polymer 45 (24) (2004) 8211–8219.

[16] Y. Li, J. Sun, J. Wang, C. Qin, L. Dai, Preparation of well-dispersed reduced graphene oxide and its mechanical reinforcement in polyvinyl alcohol fibre, Polym. Int. 65 (9) (2016) 1054–1062.

[17] H.K. Jang, H.I. Kim, T. Dodge, P. Sun, H. Zhu, J. Do Nam, et al., Interfacial shear strength of reduced graphene oxide polymer composites, Carbon 77 (2014) 390–397.

[18] R. Verdejo, F. Barroso-Bujans, M.A. Rodriguez-Perez, J.A. De Saja, M.A. Lopez-Manchado, Functionalized graphene sheet filled silicone foam nanocomposites, J. Mater. Chem. 18 (19) (2008) 2221–2226.

[19] J. Zhong, G.X. Zhou, P.G. He, Z.H. Yang, D.C. Jia, 3D printing strong and conductive geo-polymer nanocomposite structures modified by graphene oxide, Carbon 117 (2017) 421–426.

[20] H. Yang, C. Shan, F. Li, Q. Zhang, D. Han, L. Niu, Convenient preparation of tunably loaded chemically converted graphene oxide/epoxy resin nanocomposites from graphene oxide sheets through two-phase extraction, J. Mater. Chem. 19 (46) (2009) 8856–8860.

[21] C. Shuai, P. Feng, C. Gao, X. Shuai, T. Xiao, S. Peng, Graphene oxide reinforced poly (vinyl alcohol): nanocomposite scaffolds for tissue engineering applications, RSC Adv. 5 (32) (2015) 25416–25423.

[22] Y. Piao, B. Chen, Synthesis and mechanical properties of double cross-linked gelatin-graphene oxide hydrogels, Int. J. Biol. Macromol. 101 (2017) 791–798.

[23] H. Zhang, D. Zhai, Y. He, Graphene oxide/polyacrylamide/carboxymethyl cellulose sodium nanocomposite hydrogel with enhanced mechanical strength: preparation, characterization and the swelling behavior, RSC Adv. 4 (84) (2014) 44600–44609.

[24] C. Yang, Z. Liu, C. Chen, K. Shi, L. Zhang, X.J. Ju, et al., Reduced graphene oxide-containing smart hydrogels with excellent electro-response and mechanical properties for soft actuators, ACS Appl. Mater. Interfaces 9 (18) (2017) 15758–15767.

[25] A. Pourjavadi, B. Pourbadiei, M. Doroudian, S. Azari, Preparation of PVA nanocomposites using salep-reduced graphene oxide with enhanced mechanical and biological properties, RSC Adv. 5 (112) (2015) 92428–92437.

[26] H. Zhao, J. Ding, H. Yu, Variation of mechanical and thermal properties in sustainable graphene oxide/epoxy composites, Sci. Rep. 8 (1) (2018) 1−8.

[27] X. Wang, D. Tan, Z. Chu, L. Chen, X. Chen, J. Zhao, et al., Mechanical properties of polymer composites reinforced by functionalized graphene prepared via direct exfoliation of graphite flakes in styrene, RSC Adv. 6 (113) (2016) 112486−112492.

[28] Y.F. Huang, P.F. Wu, M.Q. Zhang, W.H. Ruan, E.P. Giannelis, Boron cross-linked graphene oxide/polyvinyl alcohol nanocomposite gel electrolyte for flexible solid-state electric double layer capacitor with high performance, Electrochim. Acta 132 (2014) 103−111.

[29] B.J. Wang, Y.J. Zhang, J.Q. Zhang, Q.T. Gou, Z.B. Wang, P. Chen, et al., Crystallization behavior, thermal and mechanical properties of PHBV/graphene nanosheet composites, Chin. J. Polym. Sci. 31 (4) (2013) 670−678.

[30] Y. Piao, B. Chen, One-pot synthesis and characterization of reduced graphene oxide-gelatin nanocomposite hydrogels, RSC Adv. 6 (8) (2016) 6171−6181.

[31] N.C. Adak, S. Chhetri, N.H. Kim, N.C. Murmu, P. Samanta, T. Kuila, Static and dynamic mechanical properties of graphene oxide-incorporated woven carbon fiber/epoxy composite, J. Mater. Eng. Perform. 27 (3) (2018) 1138−1147.

[32] S. Sen, Du, F. Li, H.M. Xiao, Y.Q. Li, N. Hu, S.Y. Fu, Tensile and flexural properties of graphene oxide coated-short glass fiber reinforced polyethersulfone composites, Compos. B Eng. 99 (2016) 407−415.

[33] M. Saafi, L. Tang, J. Fung, M. Rahman, J. Liggat, Enhanced properties of graphene/fly ash geopolymeric composite cement, Cem. Concr. Res. 67 (2015) 292−299.

[34] X. Yang, Z. Wang, M. Xu, R. Zhao, X. Liu, Dramatic mechanical and thermal increments of thermoplastic composites by multi-scale synergetic reinforcement: carbon fiber and graphene nanoplatelet, Mater. Des. 44 (2013) 74−80.

[35] J.V. Koleske, L.W. Hill, Dynamic mechanical and tensile properties, Paint. Coat. Test. Man. 14 (2012) 534.

[36] R. Verdejo, M.M. Bernal, L.J. Romasanta, M.A. Lopez-Manchado, Graphene filled polymer nanocomposites, J. Mater. Chem. 21 (2011) 3301−3310.

[37] M. Goumri, C. Poilâne, P. Ruterana, B. Ben Doudou, J. Wéry, A. Bakour, et al., Synthesis and characterization of nanocomposites films with graphene oxide and reduced graphene oxide nanosheets, Chin. J. Phys. 55 (2) (2017) 412−422.

[38] R.K. Layek, A.K. Nandi, A review on synthesis and properties of polymer functionalized graphene, Polymer 54 (19) (2013) 5087−5103.

[39] N. Kumari, K. Kumar, Mechanical behaviour of graphene and carbon fibre reinforced epoxy based hybrid nanocomposites for orthotic callipers, J. Exp. Nanosci. 13 (Suppl. 1) (2018) S14−S23.

[40] P. Das, V.C. Mai, H. Duan, Flexible bioinspired ternary nanocomposites based on carboxymethyl cellulose/nanoclay/graphene oxide, ACS Appl. Polym. Mater. 1 (6) (2019) 1505−1513.

[41] S.K. Sharma, J. Prakash, P.K. Pujari, Effects of the molecular level dispersion of graphene oxide on the free volume characteristics of poly(vinyl alcohol) and its impact on the thermal and mechanical properties of their nanocomposites, Phys. Chem. Chem. Phys. 17 (43) (2015) 29201−29209.

[42] J. Liang, Y. Huang, L. Zhang, Y. Wang, Y. Ma, T. Cuo, et al., Molecular-level dispersion of graphene into poly(vinyl alcohol) and effective reinforcement of their nanocomposites, Adv. Funct. Mater. 19 (14) (2009) 2297−2302.

[43] M. Bothe, F. Emmerling, T. Pretsch, Poly(ester urethane) with varying polyester chain length: polymorphism and, Macromol. Chem. Phys. 214 (2013) 2683−2693.

[44] Y. Li, S. Wang, Q. Wang, M. Xing, A comparison study on mechanical properties of polymer composites reinforced by carbon nanotubes and graphene sheet, Compos. B Eng. 133 (2018) 35−41.

[45] M.M. Gudarzi, F. Sharif, Enhancement of dispersion and bonding of graphene-polymer through wet transfer of functionalized graphene oxide, Express Polym. Lett. 6 (12) (2012).

[46] X. Zhao, Q. Zhang, D. Chen, P. Lu, Enhanced mechanical properties of graphene-based polyvinyl alcohol composites, Macromolecules 43 (5) (2010) 2357−2363.

[47] A. Kausar, Wajid-Ullah, B. Muhammadb, Processing and characterization of fire-retardant modified polystyrene/functional graphite composites, Compos. Interfaces 22 (6) (2015) 517−530.

[48] J. Chu, R.J. Young, T.J.A. Slater, T.L. Burnett, B. Coburn, L. Chichignoud, et al., Realizing the theoretical stiffness of graphene in composites through confinement between carbon fibers, Compos. A Appl. Sci. Manuf. 113 (2018) 311−317.

[49] I. Zaman, T.T. Phan, H.C. Kuan, Q. Meng, L.T. Bao La, L. Luong, et al., Epoxy/graphene platelets nanocomposites with two levels of interface strength, Polymer 52 (7) (2011) 1603−1611.

[50] S. Chandrasekaran, N. Sato, F. Tölle, R. Mülhaupt, B. Fiedler, K. Schulte, Fracture toughness and failure mechanism of graphene based epoxy composites, Compos. Sci. Technol. 97 (2014) 90−99.

[51] Z. Tai, J. Yang, Y. Qi, X. Yan, Q. Xue, Synthesis of a graphene oxide-polyacrylic acid nanocomposite hydrogel and its swelling and electroresponsive properties, RSC Adv. 3 (31) (2013) 12751−12757.

[52] A.A. Vasileiou, M. Kontopoulou, A. Docoslis, A noncovalent compatibilization approach to improve the filler dispersion and properties of polyethylene/graphene composites, ACS Appl. Mater. Interfaces 6 (3) (2014) 1916−1925.

[53] X. Peng, C. He, J. Liu, H. Wang, Biomimetic jellyfish-like PVA/graphene oxide nanocomposite hydrogels with anisotropic and pH-responsive mechanical properties, J. Mater. Sci. 51 (12) (2016) 5901−5911.

[54] H. Zhang, X. Pang, Y. Qi, pH-sensitive graphene oxide/sodium alginate/polyacrylamide nanocomposite semi-IPN hydrogel with improved mechanical strength, RSC Adv. 5 (108) (2015) 89083−89091.

[55] C. Pan, L. Liu, Q. Chen, Q. Zhang, G. Guo, Tough, stretchable, compressive novel polymer/graphene oxide nanocomposite hydrogels with excellent self-healing performance, ACS Appl. Mater. Interfaces 9 (43) (2017) 38052−38061.

[56] C. Yin, Q. Zhang, J. Liu, Y. Gao, Y. Sun, Q. Zhang, Preparation and characterization of grafted natural rubber/graphene oxide nanocomposites, J. Macromol. Sci. B 58 (7) (2019) 645−658.

[57] L. Zhao, X. Sun, Q. Liu, J. Zhao, W. Xing, Natural rubber/graphene oxide nanocomposites prepared by latex mixing, J. Macromol. Sci. B 54 (5) (2015) 581−592.

[58] J. Wang, K. Zhang, Z. Cheng, M. Lavorgna, H. Xia, Graphene/carbon black/natural rubber composites prepared by a wet compounding and latex mixing process, Plast. Rubber Compos. 47 (9) (2018) 398−412.

[59] J. Tang, H. Zhou, Y. Liang, X. Shi, X. Yang, J. Zhang, Properties of graphene oxide/epoxy resin composites, J. Nanomat. 2014 (2014).

[60] A. Ashori, H. Rahmani, R. Bahrami, Preparation and characterization of functionalized graphene oxide/carbon fiber/epoxy nanocomposites, Polym. Test. 48 (2015) 82−88.

[61] S. Sheshmani, A. Ashori, M. Arab, Fashapoyeh, Wood plastic composite using graphene nanoplatelets, Int. J. Biol. Macromol. 58 (2013) 1−6.

[62] S. Chatterjee, F. Nafezarefi, N.H. Tai, L. Schlagenhauf, F.A. Nu, Size and synergy effects of nanofiller hybrids including graphene nanoplatelets and carbon nanotubes in mechanical properties of epoxy composites, Carbon 50 (15) (2012) 5380−5386.

[63] C. Yeong, J. Bae, T. Kim, S. Chang, S. Young, Using silane-functionalized graphene oxides for enhancing the interfacial bonding strength of carbon/epoxy composites, Compos. A Appl. Sci. Manuf. 75 (2015) 11−17.

[64] X. Zhu, J.A. Joyce, Review of fracture toughness (G, K, J, CTOD, CTOA) testing and standardization, Eng. Fract. Mech. 85 (2012) 1−46.

[65] K. Kalaitzidou, H. Fukushima, L.T. Drzal, A new compounding method for exfoliated graphite—polypropylene nanocomposites with enhanced flexural properties and lower percolation threshold, Compos. Sci. Technol. 67 (10) (2007) 2045−2051.

[66] R. Baptista, A. Mendão, M. Guedes, R. Marat-mendes, An experimental study on mechanical properties of epoxy-matrix composites containing graphite filler, Proc. Struct. Integr. 1 (2016) 074−081.

[67] Q. Wang, C. Ramírez, C.S. Watts, O. Borrero-López, A.L. Ortiz, B.W. Sheldon, et al., Fracture, fatigue, and sliding-wear behavior of nanocomposites of alumina and reduced graphene-oxide, Acta Mater. 186 (2020) 29−39.

[68] S. Guo, Y. Lu, X. Wan, F. Wu, T. Zhao, C. Shen, Preparation, characterization of highly dispersed reduced graphene oxide/epoxy resin and its application in alkali-activated slag composites, Cem. Concr. Compos. 105 (2020) 103424.

[69] P. Avouris, C. Dimitrakopoulos, Graphene: synthesis and applications, Mater. Today 15 (3) (2012) 86−97.

[70] X. Huang, Z. Yin, S. Wu, X. Qi, Q. He, Q. Zhang, et al., Graphene-based materials: synthesis, characterization, properties, and applications, Small 7 (14) (2011) 1876−1902.

Electrical conductivity of polymer-graphene composites

Mostafizur Rahaman[1], Rajesh Theravalappil[2], Subhendu Bhandari[3], Lalatendu Nayak[4] and Purabi Bhagabati[5]

[1]Department of Chemistry, College of Science, King Saud University, Riyadh, Saudi Arabia, [2]Center for Refining and Petrochemicals, Research Institute, King Fahd University of Petroleum and Minerals, Dhahran, Saudi Arabia, [3]Department of Plastic and Polymer Engineering, Maharashtra Institute of Technology, Aurangabad, India, [4]Sushila Goenka Research & Development Centre, Phillips Carbon Black Limited, Palej, India, [5]School of Chemistry, The Centre for Research on Adaptive Nanostructures and Nanodevices (CRANN), Trinity College Dublin, Dublin, Ireland

4.1 Introduction

Most of the polymers are insulating in nature. Though some conductive polymers have been synthesized in the recent years they still suffer from mechanical strength and are not usable alone. Hence, these polymers cannot be suitable for some particular applications in electrical and electronic sectors. However, by adding conductive additive with insulating polymers, conductive composites can be made that can serve for many applications like as a conductive adhesive [1], electromagnetic interference (EMI) shielding [2−4], antistatic materials [5], sensors [6−10], biomedical fields [11], etc. Electrically conductive polymer composites are made by adding metallic fillers, traditional carbon fillers, and conductive polymers within the insulating polymer matrix [8,12−14]. Metallic particles are heavy in weight and hence are avoided in lightweight related applications. Traditional carboneous fillers graphite, expanded graphite (EG), carbon black (CB), carbon fiber (CF), carbon nanofiber (CNF), and carbon nanotubes (CNTs) are proven good ingredients as conductive additives for insulating polymers [15−20]. These materials provide good electrical conductivity and mechanical strength to the polymer composites. CNTs have a high aspect ratio and hence, low electrical percolation threshold is observed in polymer composites [21]. But when there is the priority of getting high ultimate electrical conductivity, then there needs to be a very high filler loading. This makes the composite brittle in nature [22]. Hence, we need to search a filler that will provide both low electrical percolation threshold and high ultimate electrical conductivity at low filler loading.

Graphene is a promising carboneous material with superb electrical conductivity of 6×10^5 S/m in a suitably isolated environment [23]. The graphene is a two-dimensional material with hexagonal arrangement of carbon with sp^2 hybridization

Polymer Nanocomposites Containing Graphene. DOI: https://doi.org/10.1016/B978-0-12-821639-2.00025-2

[24]. There is many graphene related materials (GRM) like pristine graphene, graphene oxide (GO), reduced GO (rGO), graphene nanoplatelets (GNPs), etc. These GRM differ in their structure and are of great interest in the present time for composite science and technology. The electrical conductivity of composites varies when there is the variation of graphene types within the polymer matrix. Moreover, the ultimate conductivity of composite also depends on different processing methods and post-processing treatment.

In this chapter, we have focused on the electrical conductivity of different graphenes obtained from several routes. It has been highlighted that which production route gives better quality graphene. The fundamental aspect of electrical percolation theory has been discussed. The conductive composites based on different graphenes are presented with respect to their different processing methods and effectiveness. The percolation threshold value and processing method of some more composites are reported separately within the table. Different factors that govern the electrical conductivity are discussed herein and finally some applications of these composites are presented.

4.2 Electrical conductivity of different graphenes

The graphenes are synthesized from different sources and in different ways. Accordingly, the number of graphene layers varies. Moreover, the GO is reduced and functionalized. Hence, its conductivity will be varied as per their origin. The electrical conductivity of different types of graphenes are discussed one by one here under as follows:

4.2.1 Pristine graphene

The monolayer graphene is known as the pristine graphene. This type of graphene has very high electrical conductivity. This is because, this type of graphene possesses very high charge concentration and the mobility of charge carriers is also very high. Electrical conductivity of mechanically exfoliated graphene has been reported as 6×10^5 S/m when measured on a suspended sheet [23]. However, this mechanically exfoliated method is not suitable for large scale of production. Hence, some alternative processes have been developed through which we can get scalable production but these processes suffer from the deterioration of its properties and getting pure monolayer flake. The broad categories of GRM are given below.

4.2.2 Chemically vapor deposited graphene

Graphene with a large area of monolayer or multilayer can be produced by chemical vapor deposited (CVD) technique, which depend on the condition of growth and the use of substrate [25,26]. The electrical conductivity of graphene, produced by CVD technique is lower compared to graphene produced by mechanical exfoliation method. This can be attributed to the fact that the produced graphene by CVD technique is polycrystalline in nature that means there are boundaries between the grain scattering charges [27]. Hence, there is a reduction in electrical conductivity. However, single crystal graphene has been developed by various authors [28,29]. There is a report of the preparation of single crystal graphene on copper substrate that showed good electrical conductivity without any backscattering [30]. This is because there was the suppression of forming any defect and grain boundaries [30]. However, this CVD process suffers from limitations. Still the graphene is grown on metal substrate when the process through CVD and transferring graphene from this substrate becomes a challenging work [31].

4.2.3 Graphene oxide

GO is synthesized by the chemical oxidation of graphite through different techniques [32−35]. It is a highly exfoliated and functionalized form of graphene. The oxygenated functionalized groups like diols, hydroxyl, epoxide, ketone, and carboxyl are present within the graphene as shown in Fig. 4.1 [36]. Because of functionalization, the conjugation within the chemical structure of graphene breaks down in many places, as a result, the density of charge carriers as well as their mobility is decreased. Hence, deterioration of electrical conductivity is observed. In a study, the electrical conductivity of GO was reported as 2×10^{-2} S/m, substantially less compared to pristine graphene [37]. This is why, GO is not used in many conductive applications.

4.2.4 Reduced graphene oxide

The electrical conductivity of GO can be improved by the process of its reduction. There are some reports where the increment of electrical conductivity has been mentioned after its reduction [38,39]. The reduction process reduces the number of oxygenated group within graphene by converting sp^3 into sp^2 network. Hence, the number of conjugation increases, which facilitates the mobility of charge carriers and thereby increases the electrical conductivity. The reduction of GO is mainly carried out by the thermal treatment or chemical treatment or by both of them. The structural changes of GO after reduction are shown in Fig. 4.1 [36]. The chemical reduction of GO is carried out by treating it with hydrazine, sodium borohydride, hydroquinone, various alcohols, sulfur containing compounds, etc. [40,41]. Stankovich et al. have reported the conductivity recovery of GO at the level of 200 S/m by treating with hydrazine [37]. The conductivity recovery for monolayer graphene was reported as 50−200 S/m [42]. In another study, chemical treatment of GO was carried out by immersing in FeI_2 at 95°C where conductivity recovery was achieved at 6×10^4 S/m [43].

The GO can also be reduced by electrochemical reduction process. In this reduction process, hydrogen ions are generated by the oxidation of water present on the surface of GO. In this way conductivity was recovered up to 1.0 S/m, which makes using this rGO in micro devices and circuits possible [44]. Mohan et al. synthesized

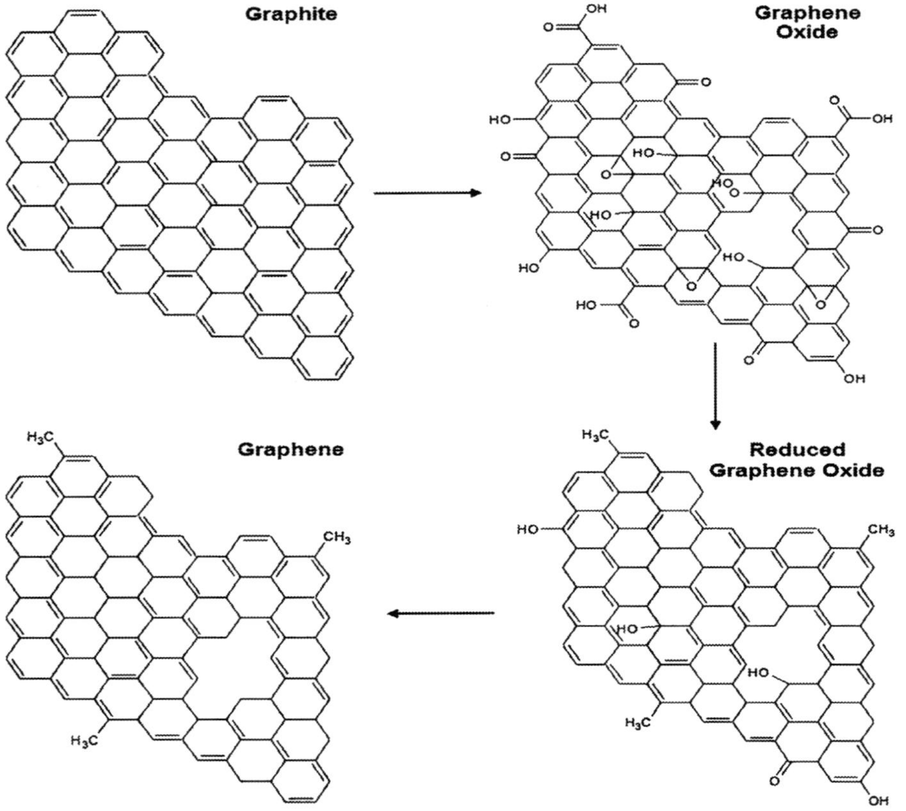

Figure 4.1 Chemical structures of graphite, GO, rGO, and grapheme.
Source: Reproduced after permission from V.B. Mohan, R. Brown, K. Jayaraman, D. Bhattacharyya, Characterisation of reduced graphene oxide: effects of reduction variables on electrical conductivity, Mater. Sci. Eng. B 193 (2015) 49−60.

the GO and then reduced it by hydroiodic acid (HI) to increase its conductivity [36]. As reported, they achieved the electrical conductivity 10^3 S/m for rGO film. Moreover, in another study, the reduction of GO has also been carried out by using $NaBH_4$ as a reducing agent [45]. Boron oxide was formed during the reaction process, and the electrical conductivity and C/O ratios were increased with the increase in $NaBH_4$ concentration. They have investigated that the use of $NaBH_4$ as a reducing agent is more advantageous compared to hydrazine because the generation of nitrogen containing functional group results in the deterioration of electrical conductivity [45].

The thermal treatment process to reduce GO is more beneficial to recover its electrical conductivity compared to chemical treatment process. It has been shown that when a thin film of rGO is annealed from temperature 200°C−940°C, its electrical conductivity is improved from 3 to 10^5 S/m [46]. A conductivity of 5×10^4 S/m as achieved for rGO by heating GO at temperature 1100°C [47]. However, the same study reported the conductivity 5×10^3 S/m after being treated with hydrazine and subsequent annealed at 400°C. The electrical conductivity of rGO reached up to 9×10^4 S/m when annealed at 1000°C under Ar/H$_2$ environment [48]. Moreover, when annealing was carried out above 2000°C, the electrical conductivity was recovered up to 2×10^5 S/m [49]. Whatever may be, though the complete removal of oxygen group is possible from GO it cannot achieve the electrical conductivity equal to pristine graphene because of some remaining defects in the lattice. This is why, the covalent functionalized rGO may not be suitable for high conductivity applications and the unmodified graphene flakes can be more advantageous [50].

4.2.5   Liquid exfoliated graphene

The graphite can also be exfoliated in a liquid medium and there is no need of heavy functionalization of graphene sheet. It has been reported that the exfoliated graphene was produced in liquid medium using surfactant as a stabilizer [51]. Hernandez et al. investigated the production of graphene flakes containing 1−10 number of layers in thickness where the lateral dimension was 500−1000 nm [52]. This solution processable method yielded very high quantity of graphene containing few defects. A large volume of graphene can also be obtained through this method. The use of surfactant may decrease the electrical conductivity of graphene as the former one is nonconducting in nature [53]. Nowadays, there is also a report of making inkjet printed thin film of liquid exfoliated graphene [54]. In this method, the graphene was printed several times to make different thicknesses for thin films. As investigated the repeated printing of film increased the electrical conductivity of the graphene film. The conductivity was reported as 3×10^3 S/m, which was relatively higher compared to the other solution processable methods. This avoids the use of harmful chemicals and also the power loss due to preparation at high temperature.

4.2.6   Graphene nanoplatelets

GNPs are cheaper graphene compared to its other forms and are commercially easily available at large scale. There are different production methods of GNPs but the most important methods are by thermal expansion and mechanical agitation or by both of them. The electrical conductivity of GNPs depend on its number of graphene layers that is its thickness; the thickness is lesser, and its electrical conductivity is higher. It has been reported that the electrical conductivity of GNPs, having thickness 50, 5, and 3 nm is 7×10^4, 1×10^4, and 1.5×10^5 S/m, respectively [55]. This increment in electrical conductivity of GNPs with the decrease in layer thickness can be attributed to the poor c-place conductivity of graphene. Hence, for the preparation of polymer/GNPs composites, the priority is the maintaining of GNP thickness having less than 10 layers of graphene thin flakes.

From the above ongoing discussion, it is revealed that there is approximately 10^8 order of difference in electrical conductivity when compared to highly conductive pristine graphene (10^6 S/m) with the highly functionalized GO (10^{-2} S/m). However, from the application point of view with polymer composites, one needs to compromise the quality with the quantity of large scale production. The electrical conductivity of mechanically exfoliated pristine graphene are CVD graphene are quite high but suffer from large scale of production to be required for composites. Moreover, though the highly functionalized GO can be produced in large scale but it lacks electrical conductivity value because of its oxygenated functional group, which reduces its conductivity. This problem can be overcome up to a certain extent using rGO but still requires to recover conductivity of pristine graphene. Hence, the better choice is to use liquid exfoliated graphene and GNPs. These give scalable production as well as better electrical conductivity, and consequently can be effectively used as a conductive filler within polymer matrix to prepare polymer conductive composites.

4.3   Electrical percolation theory

The electrical percolation threshold is a parameter that tells us about the effectiveness of a composite material with regards to its electrical conductivity. For electrically conductive composites, a low value of percolation threshold is the most desirable. The electrical percolation threshold is defined as the concentration of filler at which the maximum change in electrical conductivity is observed [56]. This happened when there was a formation of continuous conductive network of filler particles within the polymer matrix. At this stage, a drastic change in electrical conductivity is observed. When a conductive filler is added gradually within an insulating polymer matrix, the variation in electrical conductivity of the composite system is observed according to the plot shown in Fig. 4.2 [56]. Initially, the increase in electrical conductivity with respect to filler loading is marginal. This is because, at this stage, the conductive additives are separated from each other within the polymer matrix. Hence, the increase in conductivity is marginal. The composite in this region is insulating in nature. After a certain filler loading, there is a drastic change in conductivity. At this stage, a continuous conductive network of filler particles within the polymer matrix has taken place and hence the conductivity has increased suddenly. The filler loading, at which this drastic change in conductivity is observed is known as the percolation threshold for that composite. After further loading of filler, the increase in conductivity again becomes marginal. This is because, when a continuous conductive network is formed, the addition of more filler only increases the number of such conductive paths and hence contributes less to the increase in conductivity. The electrical percolation threshold of a conductive system can be determined by the classical power law Eq. (4.1) [57,58] and Sigmoidal-Boltzmann model Eq. (4.2) [59] as given below;

$$\sigma_c = \sigma_0 (Wf - Wfc)^t; \text{ for } Wf > Wfc \qquad (4.1)$$

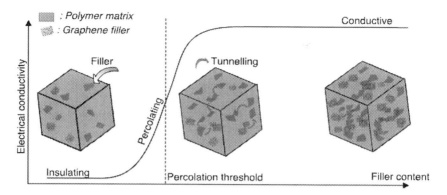

Figure 4.2 Variation of electrical conductivity of polymer/graphene composites with respect to graphene loading.
Source: Reproduced after permission from A. Marsden, D. Papageorgiou, C. Vallés, A. Liscio, V. Palermo, M. Bissett, et al., Electrical percolation in graphene—polymer composites, 2D Mater. 5 (3) (2018) 032003.

$$\sigma_c = \sigma_1 + \frac{\sigma_2 - \sigma_1}{1 + e^{\left(\frac{Wf - Wfc}{\Delta Wf}\right)}} \qquad (4.2)$$

where, σ_c and σ_0 are the electrical conductivities of composite and filler, respectively; Wf and Wfc are the filler loading of composite material and the composite at percolation threshold, respectively; t is the critical exponent whose value should be within $1.65-2.0$ for three-dimensional composites [58]; σ_1 and σ_2 are the initial and final value of electrical conductivity, respectively; and ΔWf is a fitting parameter. The electrical percolation threshold and electrical conductivity of a composite system depend on many factors. These will be discussed point-by-point later on.

4.4 Electrical conductivity of polymer/graphene composites

In this section, we have covered the electrical conductivity of and percolation threshold of polymer-based graphene composites. The composites of different polymers with CVD graphene, pristine graphene, GO, rGO, GNPs, and solution processed graphene are discussed with special focus on their percolation threshold value, ultimate conductivity, and process methodologies. To get a clear picture for comparison, it has also been shown in tabular form for some important composites.

4.4.1 Polymer/chemical vapor deposited graphene composites

Polymer/CVD graphene-based composites are considered as ideal composites because the graphene prepared by this CVD method has a large lateral size, which

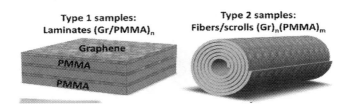

Figure 4.3 Ideal composite prepared from PMMA/CVD-graphene layers that can either be flat or fibers/rolled into scrolls.
Source: Reproduced after permission from I. Vlassiouk, G. Polizos, R. Cooper, I. Ivanov, J. K. Keum, F. Paulauskas, et al., Strong and electrically conductive graphene-based composite fibers and laminates, ACS Appl. Mater. Interfaces 7(20) (2015) 10702−10709.

helps to disperse well within the polymer matrix. Hence, a high quality polymer/ graphene composite is formed. The electrical conductivity of polymethyl methacrylate (PMMA) and CVD-based graphene nanocomposite, prepared by layer-by-layer assembly method, has been investigated and its results are shown in Fig. 4.3 [60]. They prepared the composites in two different forms: a flat laminated form and a rolled scrolled form for making a fiber. The electrical conductivity of flat laminated formed composites was reported 810 S/m at the graphene loading 0.13 vol.%. This value of electric conductivity at this loading is the highest value compared to other graphene-based polymer nanocomposites. This is a type of ideal composite with perfect orientation of graphene layers where no crumpling/rolling of graphene is present. These composites possess sufficient conductivity and strength to support and provide electric power supply through this as shown in Fig. 4.4 [60]. In another study, Liu et al. prepared polycarbonate (PC)/CVD graphene nanocomposite, where the electrical conductivity was achieved 420 S/m at the graphene loading 0.19 vol. % and the percolation threshold value was 0.003 vol.% [61]. This procedure is of great importance in the preparation of advance multifunctional graphene-based polymer composites. However, it is still challenging work to prepare the composite in bulk quantities. Table 4.1 represents the summary of this type of more composites and the composites of those will be discussed later on.

4.4.2 Polymer/graphene nanoplatelets nanocomposites

GNPs are very important carbon materials. This is used as a conductive additive in polymer matrices because of its heavy mechanical strength, electrical and thermal conductivities, and more importantly its abundance in large quantity. However, the processing methods of GNPs with polymer is very important because of its crumpling, wrinkling, and rolling nature during high shear processing specially in melt mixing process. On the contrary, solution mixing technique is more beneficial compared to melt mixing process. This is because, in solution mixing process, there is less shearing force exerted on GNPs and hence its structure is almost retained. In addition, due to lower viscosity of mixture, the polymer macromolecular chain can intercalate between the graphite layers at the time of mixing process [94]. For

Figure 4.4 LED lamp weighting 100 g suspended on the transparent and electrically conductive (PMMA/Graphene)$_{16}$ laminate.
Source: Reproduced after permission from I. Vlassiouk, G. Polizos, R. Cooper, I. Ivanov, J. K. Keum, F. Paulauskas, et al., Strong and electrically conductive graphene-based composite fibers and laminates, ACS Appl. Mater. Interfaces 7(20) (2015) 10702−10709.

example, ultra-high molecular weight polyethylene (UHMWPE)/GNP nanocomposite was prepared by solvent casting method and then followed by compression molding at high temperature [95]. It was reported that they achieved the percolation threshold value at 0.07 vol.% of GNP loading.

In situ polymerization was also found to be advantageous with regards to the uniform dispersion of GNP flakes throughout the polymer matrix. Kim et al. have studied the effect of different processing methods like in situ polymerization, solution mixing, and melt mixing on the electrical conductivity of polyurethane (PU)/graphene composites [96]. It was reported that the highest conductivity was achieved for solution mixed composites at their similar filler loading. The composites, prepared by melt mixing technique, lead to aggregation of graphene within the polymer matrix and reduction of its lateral dimension. There was the formation of covalent bonding between the matrix polymer and the filler in case of composites prepared by in situ polymerization method. This leads to hindrance in the formation of direct contact among graphene fillers and hence the effective aspect ratio was reduced. Similar observation was also reported by Kim et al. [97] and Xu et al. [98]. However, the post processing methods like hot compressing, injection molding, etc. also affect the dispersion of graphene within polymer matrix and consequently affect the electrical conductivity.

The electrical conductivity of polymer composites depends on the orientation of filler within the polymer matrix. The orientation of filler is affected by different processing parameters and methods of preparations. The most usable industrial method for the preparation of composite is extrusion. However, different types of extrusion methods lead to different extent and types of orientation of filler within the polymer matrix. It has been investigated that the small scale extrusion [99,100]

Table 4.1 Electrical conductivity of some polymer/graphene composites.

Neat polymers	Conductive additives	Processing methods	Percolation threshold	Ultimate DC conductivity (S/m)	References
PC	rGO	Solution blending/in situ thermal reduction	0.21	0.1	[62]
PC	f-GNP	Emulsion mixing	0.14 vol.%	51	[63]
POSS-PCL	rGO	Solution blending	2.5 vol.%	0.1	[64]
PE	f-GNP	Melt mixing	0.83 vol.%	0.01	[65]
LLDPE	TRGO	Melt mixing	0.5 vol.%	10^{-4}	[66]
HDPE	f-GNP	Solution blending	0.37–0.74 vol.%	27	[67]
UHMWPE	rGO	Solution blending and hydrazine reduction	0.028 vol.%	5	[68]
sPS	GNP	Solution blending	0.46 vol.%	470	[69]
Epoxy	f-rGO	Solution blending	0.71 vol.%	10^{-6}	[70]
Epoxy	GNP	Solution blending	0.5–1 vol.%	10^{-2}	[71]
Epoxy	GNP	Layer by Layer	0.6 vol.%	10^{-4}	[72]
Epoxy	CRGO	In situ polymerization	0.12 vol.%	0.1	[73]
Epoxy	CRGO aerogel	Vacuum-assisted infiltration method	<0.25 wt.%	20	[74]
Epoxy	f-GNP	Solution blending	0.16 vol.%	10	[75]
Epoxy	GNP	Three roll mill	0.22 vol.%	10^{-5}	[76]
Epoxy	f-GO	Solution blending	0.78 vol.%	1	[77]
PI	f-GNP	In situ polymerization	0.5 vol.%	0.1	[78]
PI	f-GO	In situ polymerization	0.25 vol.%	0.092	[79]
NR	GNP	Latex self-assembly	0.62 vol.%	0.03	[80]
NR	CRGO	Coagulation method	3 wt.%	10^{-4}	[81]
NR	f-rGO	Electrostatic self-assembly	0.21	7.3	[82]
NR	rGO	Vacuum-assisted self-assembly	10 phr	100	[83]
NR	TRGO	Latex technologically	3 phr	10^{-4}	[84]

NR	rGO	Electrostatic self-assembly	0.23	1	[85]
SBR	f-3D-GO	Latex coagulation	0.39 vol.%	10^{-2}	[86]
XNBR	TRGO	Electrostatic LbL self-assembly	–	0.82	[87]
ABS	GO	Coagulation blending	0.13 vol.%	0.1	[88]
PLA	TRGO	In situ polymerization	0.5–0.75 vol.%	0.01	[89]
P(MMA-co-BA)	GNP	Latex blending	0.1 vol.%	217	[90]
PMMA	f-GNP	Solution blending	0.3	10^{-3}	[91]
PMMA	f-TRG	Self-assembly	0.06 vol.%	1.2	[92]
PMMA	f-GNP	Solution blending	0.8 vol.%	20	[93]

and mono extrusion processes [101] result in randomly oriented GNPs within the polymer matrix. In another study, it has been reported that the multilayer coextrusion [101,102] and blown film extrusion [103,104] methods can produce composite with linearly oriented GNPs. There are also some post processing methods after extrusion process: these are injection molding, melt mixing, and hot pressing, which give the material a shape for their testing. As a result, the orientation of GNPs can also be changed during these post processing methods and will affect their morphology [94,105−107]. Hence, for the proper alignment of filler to a particular direction, different strategies can be adopted.

Selective localization is another method to increase the electrical conductivity of polymer-based conductive composites. In this method, the conductive filler is localized within a selective matrix. This is done by adding other polymeric material within the selective polymer and the filler is localized only/mostly within the selective polymer and can form the conductive network at low concentration of filler particles. This is because, if the filler is distributed uniformly within the both polymer phase, then it requires more amount of filler to form conductive networks within the polymer matrix. Qi et al. have studied the electrical property of polystyrene (PS)/graphene composite prepared by solution mixing technique [108]. They have achieved the electrical percolation threshold at 0.33 vol.% (shown in Fig. 4.5) and the ultimate electrical conductivity was higher by two to four order in magnitude compared to CNT-based composite studied at the same condition. Thereafter they prepared PS/polylactic acid (PLA)/graphene composite at the same condition and observed that the graphene was mostly localized within the PS phase region and resulted in the electrical percolation threshold value 0.075 vol.% of graphene loading. Fig. 4.6A explain how the selective localization facilitates to form the conductive networks within the polymer matrices and reduce the value of percolation threshold. If the graphene would distribute uniformly within the both polymer matrix phase, then it would not have formed the conductive network (Fig. 4.6B);

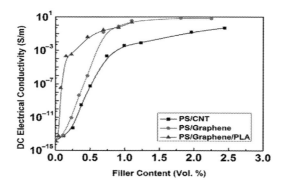

Figure 4.5 Electrical conductivity versus filler content for neat PS and its nanocomposites. *Source:* Reproduced after permission from X.-Y. Qi, D. Yan, Z. Jiang, Y.-K. Cao, Z.-Z. Yu, F. Yavari, et al., Enhanced electrical conductivity in polystyrene nanocomposites at ultra-low graphene content, ACS Appl. Mater. Interfaces 3 (8) (2011) 3130−3133.

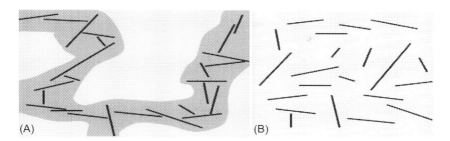

Figure 4.6 (A) Selective localization of graphene facilitating the formation conductive network at lower filler contents compared to non-selective one (B).
Source: Reproduced after permission from H. Kim, S. Kobayashi, M.A. AbdurRahim, M.J. Zhang, A. Khusainova, M.A. Hillmyer, et al., Graphene/polyethylene nanocomposites: effect of polyethylene functionalization and blending methods, Polymer 52 (8) (2011) 1837−1846.

whereas, if the same amount of filler is localized within the small volume then it can form network very easily. For obtaining high value of electrical conductivity, this unfunctional graphene is beneficial. Moreover, functionalization of graphene helps in better dispersion of it within the polymer matrix because of the increase in polymer-filler interaction. On the contrary, the covalent functionalization disrupts the sp^2 hybridization of conductive path. As a result, electrical properties will be reduced [109,110].

4.4.3 Polymer/reduced graphene oxide conductive composites

GO is the highly functionalized graphene and hence possesses low electrical conductivity. But the advantage of using GO as a filler within polymer matrices is that it can make Van der Waal force of interaction with the polymers and hence can disperse uniformly within the polymer matrices. Hence, the composite can result good mechanical property. Due to insulating in nature, it is not used in most electrical applications. But it has been mentioned earlier that its conductivity can be recovered by reduction it by thermal or chemical processes. The product is known as thermally RGO (TRGO) or chemically RGO (CRGO). A lot of studies on GO and rGO-based composite have been made with regards to their processing techniques, polymer type used, filler types, etc. Hence, the electrical conductivity of TRGO and CRGO based polymer composites are discussed here.

4.4.4 Polymer/thermally reduced graphene oxide based conductive composites

TRGO is obtained by oxidation and then thermal expansion of graphite. The thermal reduction increases the carbon content in rGO by 97% and increases its electrical conductivity up to 10^5 S/m. Though the oxygen content is reduced in TRGO, it is still sufficient to disperse well within the polymer matrices by making interfacial adhesion. The dispersion is more facilitated within the polar polymers [66,96,97,111,112]. There are many reports of forming conductive composites with thermoplastics like PE [97,113], maleic anhydride grafted polypropylene (MAPP) [114], PS [115] ethylene/methyl acrylate/acrylic and copolymer [116], PMMA [93], and TPU [96]. Steurer and his coworkers prepared the composites of PP, poly (styrene-co-acrylonitrile) (SAN), polyamide-6 (PA6), and PC with TRGO by melt mixing technique through a twin screw extruder [111]. It is reported that the electrical percolations are 4.0, 2.5, 4.5 and 8.6 wt.% for SAN, PC, PP, and PA6-based graphene composites, respectively. They also have reported that the TRGO, having aspect ratio >200 are dispersed well within the polymer matrix during melt mixing.

The electrical conductivity of elastomer-based TRGO composites has also been studied in the recent years. Natural rubber (NR)/TRGO composite showed electrical percolation threshold at 3 phr loading of graphene and the electrical conductivity was reported as 10^{-4} S/m [84]. The electrical AC conductivity with respect to frequency has been studied for styrene butadiene rubber (SBR) composite separately made with TRGO and CRGO. It is investigated that TRGO-based composite showed good electrical performance [112]. The SBR/TRGO-based composites also showed a better electrical performance compared to other nanomaterials like CB, EG and CNTs [112], when mixed with SBR. There are also reports of improvement in electrical conductivity in SBR/TRGO [117] and NR composites [81,118].

In the all above cases, the TRGO effectively formed percolative networks within the polymer matrices. The TRGO-based composites showed lower percolation threshold compared to other CNT or nanoparticle-based composites. This is because of the higher aspect ratio TRGO compared to 0D or 1D nanoparticles [66,111]. This low percolation threshold value is helpful because it required less filler to make nanocomposites. However, in terms of maximum achievement of conductivity, the TRGO-based composites show about 10 times lower conductivity compared to CNT-based composites. This is because of the presence of oxygenated group and defects within the TRGO that adversely affect the electrical conductivity. This is one of its limiting factor for using it as a conductive filler within the polymer matrices.

The use of compatibilizer improve the adhesion between the filler and polymer matrix. This results in the better dispersion and preferential distribution of filler particles within the polymer matrix. Hence, low electrical percolation threshold of electrical conductivity is obtained. Shim and his coworkers prepared the SBR composites with neat TRGO, carboxylated TRGO, and cetyl trimethyl ammonium bromide TRGO [117,119]. The obtained improved electrical conductivity and percolation threshold was at 0.5 wt.%. In another study, the linear low density polyethylene (LLDPE) was compatibilized/grafted with maleic anhydride (MA) and pyridine derivatives and prepared composites with TRGO by melt mixing technique [66]. The electrical conductivity of these composites are shown in Fig. 4.7. The percolation threshold was achieved in between 0.5 and 0.9 vol.%. However, preparing melt mixed composite is quite difficult because of low density and high surface area of TRGO.

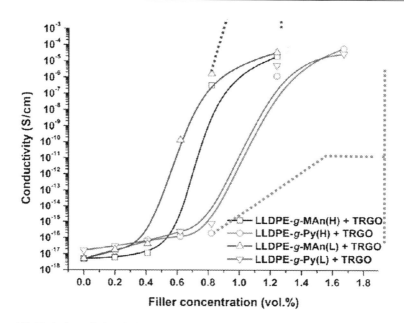

Figure 4.7 Variation of electrical conductivity after chemical compatibilization of the composites.

Source: Reproduced after permission from A.A. Vasileiou, M. Kontopoulou, A. Docoslis, A noncovalent compatibilization approach to improve the filler dispersion and properties of polyethylene/graphene composites, ACS Appl. Mater. Interfaces 6 (3) (2014) 1916−1925.

4.4.5 Polymer/Chemically Reduced Graphene Oxide (CRGO) Composites

The chemical reduction of GO can be performed in aqueous medium by the strong reducing agent hydrazine. This reduction process cannot remove oxygen completely from the graphene sheet. Approximately 15% of oxygen still remains to be removed from the backbone of graphene sheet. After reduction, the CRGO can agglomerate and form restack. The formation of this agglomerate can be prevented by adding surfactant or polymer during the reduction process [120−125]. The researchers have made a number of studies regarding CRG-based polymer composites and reported in literatures [73,81,112,126−128]. The electrical properties of CRGO-based latex polymer composite was studied where electrical conductivity was achieved 15 S/m at 1.8 wt.% and percolation threshold value was 0.9 wt.% of CRGO loading [125,126]. In another study, Potts and his coworkers reported the electrical property of NR-based composites filled with both CRGO and TRGO, where both composites showed better electrical properties compared CB filled composites [81,118]. Schopp and his team compared the electrical conductivity of different nanofillers in their study [112]. They prepared different composites of these nanofillers with SBR at identical processing conditions. It was observed that the electrical conductivity of SBR/EG composite, SBR/CB composite, and SBR/CRGO composite was approximately four to five order of magnitude less compared to the composites made with SBR/TRGO, SBR/MLG and SBR/CNTs, when measured at the same filler loading [112].

4.4.6   Polymer/graphene/carbon nanotubes hybrid composites

Nowadays, the interest has increased for the preparation of hybrid composites. This is because, in many cases, the hybrid fillers show better electrical conductivity or synergistic effect compared to individual filler. The same result is also obtained for composites. The composite prepared by hybrid filler shows a better performance compared to the individual filler-based composites. Hence, the hybrid filler based has been successfully prepared to improve their mechanical strength and electrical conductivity for application in EMI shielding [129]. The electrical conductivity of graphene paper was reported as 1.8×10^5 S/m [130]. To improve its electrical conductivity, it was mixed with CNT and was formed a hybrid paper. The measured electrical conductivity of hybrid paper was 2.7×10^5 S/m, higher value compared to individual graphene paper. The synergistic effect was also noticed for PU/GNP/CNT composite, which showed higher electrical conductivity (10 S/m) compared to individual PU/GNP and PU/CNT composites [129]. This increment in electrical conductivity for hybrid composite can be explained to the fact that the added CNTs acted as a conductive bridge between the graphene flakes and helped in the formation of conductive networks. Moreover, CNTs also prevent restacking of the graphene flakes and thus helped to disperse well within the polymer matrix [131]. Another beneficial fact is that the rheological property of composite in improved because of the presence of graphene that resulted in good dispersion [132] and the brittleness of the composite was reduced [133].

4.4.7   Polymer/functionalized graphene composites

It has been discussed earlier that TRGO or CRGO provide good electrical conductivity to the composite but these cannot be added in more quantity within the polymer matrix. Moreover, there is the lack of good dispersion and hence satisfactory mechanical strength cannot be obtained. On the contrary, functionalization of graphene, though suffer from less electrical conductivity, provide interfacial coupling between polymer and filler, and hence reinforce the polymer composite. Another advantage is that functionalized graphene provide better dispersion and more amount of filler can be added within the polymer matrix. The effect of functionalization has been investigated by incorporating nitrogen containing functional group in polymer matrix prepared by dry grinding method in nitrogen atmosphere [134]. It was observed that the functionalization had imparted interfacial coupling at the filler/epoxy interface, provided good dispersion but inferior electrical conductivity was observed. The dry milled graphene (DMG) composite showed higher percolation threshold compared to TRGO/EP and N-TRGO composites because of its low aspect ratio and lower viscosity. However, we cannot ignore DMGs because of it is very cheat compared to TRGO and higher amount of filler can be loaded without hampering the processibility.

Another method to increase the electrical conductivity of polymer composite is the doping of graphene by heteroatoms like nitrogen [135]. There are the reports of doping TRGO with nitrogen (N-TRGO) where the electrical conductivity of graphene as well as composite were improved [136]. In this study, the increment in nitrogen content from 0 to 12 wt.% resulted the improvement in electrical conductivity from 4800 to 47000 S/m [136]. There are many literatures regarding the doping of nitrogen to improve the electrical conductivity of PA6 [136], PE [137], PP [138], PU [139], rubber [112], PS [140,141], and PA [142]-based composites made with TRGO. The electrical percolation threshold of TRGO and N-TRGO-based composites was quite low which is in between 1 and 2.5 wt.% because of their high aspect ratio and good dispersion.

4.5   Factors affecting electrical conductivity

The electrical conductivity of a material depends on many matters. From the very basics, it can be said that the electrical conductivity primarily depends on the nature of materials, dimension of materials, temperature of environment, pressure exerted on materials during testing, frequency of electric filed, etc. For any composite system, except the above mentioned field, the electrical conductivity also depends on the proportion of ingredients, and their inherent conductivity value. The electrical conductivity of polymer/graphene composites will also depend on the above mentioned facts and the details of these are given below;

4.5.1   Effect of graphene loading

It is a well-known fact that the addition of any conductive additive within the insulating polymer matrix increases the electrical conductivity of the composite system. Same phenomena also happen in case of graphene loading within any insulating polymer matrix. The electrical percolation threshold value and the trend of increase in conductivity depend on the nature of graphene used. The nature of conductivity plot with respect to graphene loading is S-shaped or distorted S-shaped. Pang et al have studied the electrical properties of UHMWPE/graphene sheets composites [95]. The electrical percolation threshold was observed at 0.07 vol.% loading of graphene. The electrical conductivity of neat polymer was 10^{-13} S/m and was increased up to 10^{-4} S/m after 0.076 vol.% loading of graphene. Yousefi and his coworkers investigated the morphology and electrical percolation phenomena of PU based graphene composites prepared by an environment friendly benign technique [143]. The conductivity of neat PU was 10^{-11} S/cm. They reported the electrical percolation threshold value 0.078 vol.% of graphene loading and the ultimate increment in conductivity was 10^8 folds at 2 vol.% loading. A large number of study have been made by several authors regarding the effect of graphene loading within the polymer matrices in the past [108,144−146].

4.5.2   Nature of polymer and graphene

Each polymer is having different characteristics. They differ in their melting behavior, solubility, molecular weight, viscosity, crystallinity, polarity, inherent conductivity etc. Hence, if a graphene composite is made with different polymer at the identical processing condition, then it may or may not be possible because maintaining identical processing condition in quite difficult. For making a composite we have to use either solvent or high temperature. Hence, we need to choose the respective solvent for polymer or the required temperature above its melting point. GP nanocomposites were prepared with PLA, PS, PVC, and epoxy by solution blending technique [147−150]. The electrical percolation threshold was obtained for PLA/GNP, PS/GNP, PVC/GNP, and epoxy/GNP composites at 0.004, 0.1, 0.1, and 0.52 vol.% of GNP loading, respectively. Moreover, the ultimate electrical conductivity was 0.1, 13.8, 5.8, and 0.05 S/m for PLA, PS, PVC, and epoxy based GNP composites, respectively. This is not in order with their percolation threshold value. This is because the level of dispersion of GNP within the polymer matrices are different. The polymer composites based on rGO have also been prepared by various authors [143,151,152]. These composites were prepared by solution blending technique. The percolation threshold value for PMMA/rGO, PU/rGO, and NR/rGO composites were achieved at 0.25, 0.078, and 0.21 vol.% of rGO loading; whereas, the ultimate conductivity was 0.01, 0.001, and 0.23 S/m, respectively. Herein, the lowest percolation threshold value and the ultimate electrical conductivity was observed for PU/rGO composite though it should show the highest ultimate conductivity. Hence, the conductive network formation and the ultimate conductivity depend on the nature of polymer that how the conductive additive will distribute and disperse within it at lower as well as higher loading.

The nature of graphene also influences the electrical percolation threshold and ultimate electrical conductivity of polymer composites. Beckert et al have studied the electrical conductivity of PA12 based composites filled with TRGO and N-TRGO [136]. The composites were prepared by melt compounding technique. The PA12/N-TRGO composite showed low electrical percolation threshold value (at 1 wt.% loading) and high ultimate conductivity (10^{-4} S/m) compared to PA12/TRGO composite where electrical percolation threshold was reported in between 1 and 2.5 wt.% and ultimate conductivity 5.2×10^{-6} S/m. Chen and his coworkers have investigated the electrical property of polydimetyl siloxane (PDMS)/CRGO, PDMS/TRGO, and PDMS/CVD graphene nanocomposites [153]. The PDMS/CVDG composite exhibited the electrical conductivity 10^3 S/m at the graphene loading 0.5 wt.%. This conductivity value is quite high compared to the PDMS/CRGO and PDMS/TRGO composites at their similar graphene loading. This can be attributed to the three-dimensional connected conductive networks of CVDG sheets with one another without any break/defect within the polymer matrix. However, in another study, the electrical percolation of epoxy filled TGO and

N-TRGO composites was showed at 1 wt.% loading and for epoxy/DMG composite at 5 wt.% loading [134]. The composites were prepared by solvent free mixing method. The ultimate electrical conductivity was reported as 2×10^{-6}, 10^{-5}, and 8×10^{-5} S/m for epoxy/TRGO, epoxy/N-TRGO, and epoxy/DMG composites. There are also some studies about the effect of graphene types and other carbon types on the electrical conductivity of their polymer based composites as presented in Table 4.2.

4.5.3 Processing methods of composites

The electrical conductivity of polymer/graphene composites is greatly depending on how the composite is going to make. The composite, prepared by in situ polymerization method and solvent casting technique show higher conductivity compared to composite prepared by melt mixing technique when consider at the same filler loading [154]. This is because, the conductive particle breakage is high for melt blending technique and good dispersion is also not completely achieved. PS/GNP composites were prepared separately by solution blending [148], electrostatic self-assembly [155], and electrostatic assembly [156] methods. It is observed that the composite prepared by electrostatic assembly method showed the lowest percolation threshold value (at 0.054 vol.%) and the highest ultimate conductivity (46 S/m) compared to the composites prepared by solution blending (at 0.1 vol.% and 13.8 S/m) and electrostatic self-assembly (at 0.09 vol.% and 25 S/m) techniques. Epoxy/f-GNP composites were fabricated by solution blending and sonication assisted calendaring processes [157,158], where the composite prepared by solution blending method showed lower percolation threshold value (at 1.3 vol.%) compared to other one (at vol.%). Hence, the electrical conductivity is greatly dependent on processing condition of polymer composites.

4.5.4 Some other factors

It has been mentioned earlier that the electrical conductivity of polymer/graphenes varies with the functionalization/modification of graphene either by chemical or thermal treatment. In addition, it was reported that the blending of one polymer with another polymer leads to the increase in electrical properties because of preferential distribution of graphene within any one polymer phase [108].

The electrical conductivity of polymer/graphene composites are also affected by the alignment of graphene within the polymer matrix [159]. Kim and his coworkers prepared PC/graphene composites by melt mixing technique and prepared the samples for testing by three different methodologies [105]. The have shown that compression molded samples showed more aligned graphene composite but suffer from deterioration of electrical conductivity. In another study, epoxy/GNP composites were prepared where the graphene flakes were aligned by applying external electric field [76]. It was observed that the conductivity towards the aligned direction was two to three order of magnitude higher compared to its transverse direction.

Table 4.2 Effect of graphene types and other carbon types on the electrical conductivity of their polymer based composites.

Neat polymers	Conductive additives	Processing methods	Percolation threshold	Ultimate DC conductivity (S/m)	References
PA6	TRGO	Melt mixing	7.5 wt.%	7.1×10^{-3}	[111]
	CB		7.5 wt.%	2.2×10^{-8}	
	CNT		–	–	
PC	TRGO	Melt mixing	2.5 wt.%	0.1	[111]
	CB		2.5 wt.%	9.1	
	CNT		2.5 wt.%	0.56	
iPP	TRGO	Melt mixing	5 wt.%	5.3×10^{-2}	[111]
	CB		5 wt.%	3.3	
	CNT		5 wt.%	3.1	
SAN	TRGO	Melt mixing	4 wt.%	0.123	[111]
	CB		4 wt.%	9	
	CNT		12 wt.%	7.4×10^{-4}	
PP	TRGO	Melt mixing	<5 wt.%	10^{-4}	[138]
	MLG		5 wt.%	3×10^{-3}	
	CB		7.5 wt.%	3×10^{-5}	
	CNT		7.5 wt.%	4×10^{6}	
	EG		–	–	
PU	f-TRGO	In situ polymerization	0.5–2 wt.%	1.4×10^{-7}	[139]
	CB		2 wt.%	1.3×10^{-7}	
	CNT		2 wt.%	1.9×10^{-7}	

Layer-by-layer (LbL) assembly is another method to prepare polymer/graphene ultra-thin composites where contact between fillers increases [160−163]. This LbL process helps to assemble the fillers and polymers as layer form in nanometer level to prepare ideal polymer composites. Elastomer based LbL composite multilayer films were fabricated using polyether imide (PEI), carboxylated acrylonitrile butyl rubber (XNBR) and GO by electrostatic self-assembly process as shown in Fig. 4.8A [87]. The electrical conductivity of multilayer film was increased with the increase in the number of GO layers within the composite as shown in Fig. 4.8B. The conductivity was increased from 6.5×10^{-2} to 0.82 S/m when the number of deposition layers of GO was increased from 2 to 30. In another study, PS/LbL composites were prepared at three different layers [127]. The conductivity for single layer was 7×10^{-7} S/m, which was increased up to 0.02 and 0.05 S/m for bilayer and trilayer composites, respectively.

The electrical conductivity of polymer composites is also affected by applied strain, pressure, and temperature on it. These causes changes in the microstructure of composites and hence depending on that, the conductivity is either increased or decreased. The applied strain results in the breakage of conductive network and hence the electrical conductivity is reduced [10,164]. On the contrary, the applied pressure on composite bring closer of the conductive networks and thus facilitate the flow of charge carriers, which results in the increase in electrical conductivity [7,8]. However, there is a critical pressure above which the damaging of test sample take place and lead to decrease in conductivity. Upon raise in temperature, a conductive composite may exhibit either positive temperature coefficient (PTC) or negative temperature coefficient (NTC) behavior of resistivity. Mostly PTC behavior of resistivity is observed up to the melting temperature of polymer; whereas, after melting temperature, NTC behavior is noticed [165−167].

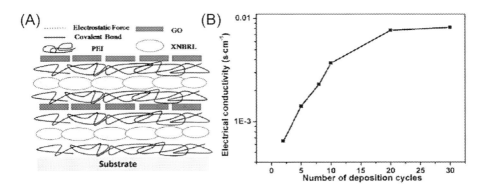

Figure 4.8 (A) A diagram of a layer-by-layer assembly and (B) the electrical conductivity of (PEI/XNBR/PEI/GO)n films with respect to the number of deposition cycles.
Source: Reproduced after permission from L. Wang, W. Wang, Y. Fu, J. Wang, Y. Lvov, J. Liu, et al., Enhanced electrical and mechanical properties of rubber/graphene film through layer-by-layer electrostatic assembly, Compos. Part B Eng. 90 (2016) 457−464.

4.6 Applications

Electrically conductive composites have versatility on their application point of view. It can be used as an EMI shielding material to protect the electronic equipment from harmful radiation [168,169]. Moreover, the electrical conductivity varied when the composite is stretched, pressed or heated at high temperature. Hence, it can serve as pressure sensor, strain sensor or temperature sensor materials [8−10,170,171]. Low conductive composites can be used as antistatic coating materials [172−174]. It is also used in memory and energy storage devices. Biopolymer-based graphene composites are used in biomedical fields [175,176]. Finally, elastomer based composites are used as flexible electrode in light emitting diode and transparent conductive coating in solar cell [177−179].

4.7 Conclusions

From this chapter, it is revealed that the electrical conductivity of graphenes differ from each other based on their method of synthesis/manufacturing. Mechanically exfoliated graphene showed the highest electrical conductivity (10^6 S/m) and the highly functionalized GO showed the lowest conductivity (10^{-2} S/m). However, though the mechanically exfoliated graphene and CVD graphene show high conductivity, these cannot be produced in large scale to meet the present days' requirements. Highly functionalized GO is scalable but lacks in conductivity. Hence, GNP and liquid exfoliated graphene are more suitable because of its high conductivity, low cost, and large scale availability. Electrically conductive composites of polymer/graphenes were prepared successfully by different authors. The electrical percolation threshold and ultimate conductivity were reported and discussed. The electrical conductivity is dependent on may factors like the nature of polymer and graphenes, graphene variation, processing conditions, dispersion of graphene, applied strain, pressure, etc. These conductive composites can be used for EMI shielding effectiveness, conductive strain sensor, pressure sensitive sensor, antistatic coating, etc.

References

[1] J. Kim, B.-S. Yim, J.-M. Kim, J. Kim, The effects of functionalized graphene nanosheets on the thermal and mechanical properties of epoxy composites for anisotropic conductive adhesives (ACAs), Microelectron. Reliab. 52 (3) (2012) 595−602.
[2] R. Jan, A. Habib, M.A. Akram, I. Ahmad, A. Shah, M. Sadiq, et al., Flexible, thin films of graphene−polymer composites for EMI shielding, Mater. Res. Express 4 (3) (2017) 035605.
[3] M. Rahaman, T. Chaki, D. Khastgir, Development of high performance EMI shielding material from EVA, NBR, and their blends: effect of carbon black structure, J. Mater. Sci. 46 (11) (2011) 3989−3999.

[4] M. Rahaman, T.K. Chaki, D. Khastgir, High-performance EMI shielding materials based on short carbon fiber-filled ethylene vinyl acetate copolymer, acrylonitrile butadiene copolymer, and their blends, Polym. Compos. 32 (11) (2011) 1790−1805.

[5] H. Wang, G. Xie, M. Fang, Z. Ying, Y. Tong, Y. Zeng, Electrical and mechanical properties of antistatic PVC films containing multi-layer graphene, Compos. Part B Eng. 79 (2015) 444−450.

[6] C.S. Boland, U. Khan, C. Backes, A. O'Neill, J. McCauley, S. Duane, et al., Sensitive, high-strain, high-rate bodily motion sensors based on graphene−rubber composites, ACS Nano 8 (9) (2014) 8819−8830.

[7] M. Rahaman, T. Chaki, D. Khastgir, Polyaniline, ethylene vinyl acetate semiconductive composites as pressure sensitive sensor, J. Appl. Polym. Sci. 128 (1) (2013) 161−168.

[8] H.B. Yao, J. Ge, C.F. Wang, X. Wang, W. Hu, Z.J. Zheng, et al., A flexible and highly pressure-sensitive graphene−polyurethane sponge based on fractured microstructure design, Adv. Mater. 25 (46) (2013) 6692−6698.

[9] J. Yang, Y. Ye, X. Li, X. Lü, R. Chen, Flexible, conductive, and highly pressure-sensitive graphene-polyimide foam for pressure sensor application, Compos. Sci. Technol. 164 (2018) 187−194.

[10] Y. Tang, Z. Zhao, H. Hu, Y. Liu, X. Wang, S. Zhou, et al., Highly stretchable and ultrasensitive strain sensor based on reduced graphene oxide microtubes−elastomer composite, ACS Appl. Mater. Interfaces 7 (49) (2015) 27432−27439.

[11] J.K. Wang, G.M. Xiong, M. Zhu, B. Özyilmaz, A.H. Castro Neto, N.S. Tan, et al., Polymer-enriched 3D graphene foams for biomedical applications, ACS Appl. Mater. Interfaces 7 (15) (2015) 8275−8283.

[12] M. Rahaman, L. Nayak, T. Chaki, D. Khastgir, Conductive composites made from in-situ polymerized polyaniline in ethylene vinyl acetate copolymer (EVA): mechanical properties and electromagnetic interference shielding effectiveness, Adv. Sci. Lett. 18 (1) (2012) 54−61.

[13] R. Ram, M. Rahaman, D. Khastgir, Mechanical, electrical, and dielectric properties of polyvinylidene fluoride/short carbon fiber composites with low-electrical percolation threshold, J. Appl. Polym. Sci. 131 (3) (2014).

[14] M. Iqbal, G. Mamoor, T. Bashir, M. Irfan, M. Manzoor, A study of polystyrene-metal powder conductive composites, J. Chem. Eng. (2010) 61−64.

[15] W. Bauhofer, J.Z. Kovacs, A review and analysis of electrical percolation in carbon nanotube polymer composites, Compos. Sci. Technol. 69 (10) (2009) 1486−1498.

[16] I. Balberg, A comprehensive picture of the electrical phenomena in carbon black−polymer composites, Carbon 40 (2) (2002) 139−143.

[17] N. Sohi, M. Rahaman, D. Khastgir, Dielectric property and electromagnetic interference shielding effectiveness of ethylene vinyl acetate-based conductive composites: effect of different type of carbon fillers, Polym. Compos. 32 (7) (2011) 11 bg 48−1154.

[18] L. Nayak, M. Rahaman, D. Khastgir, T.K. Chaki, Thermal degradation kinetics of polyimide nanocomposites from different carbon nanofillers: applicability of different theoretical models, J. Appl. Polym. Sci. 135 (7) (2018) 45862.

[19] N.M. Abdullah, M.S. Kamarudin, A.Z.M. Rus, M. Abdullah, Preparation of conductive polymer graphite (PG) composites, IOP Conference Series: Materials Science and Engineering, IOP Publishing, 2017, p. 012181.

[20] W. Zheng, S.-C. Wong, Electrical conductivity and dielectric properties of PMMA/expanded graphite composites, Compos. Sci. Technol. 63 (2) (2003) 225−235.

[21] F.H. Gojny, M.H. Wichmann, B. Fiedler, I.A. Kinloch, W. Bauhofer, A.H. Windle, et al., Evaluation and identification of electrical and thermal conduction mechanisms in carbon nanotube/epoxy composites, Polymer 47 (6) (2006) 2036−2045.

[22] F. Li, L. Qi, J. Yang, M. Xu, X. Luo, D. Ma, Polyurethane/conducting carbon black composites: structure, electric conductivity, strain recovery behavior, and their relationships, J. Appl. Polym. Sci. 75 (1) (2000) 68−77.

[23] X. Du, I. Skachko, A. Barker, E.Y. Andrei, Approaching ballistic transport in suspended graphene, Nat. Nanotechnol. 3 (8) (2008) 491.

[24] I. Suarez-Martinez, N. Grobert, C. Ewels, Nomenclature of sp2 carbon nanoforms, Carbon 50 (2011) 741−747.

[25] X. Li, Y. Zhu, W. Cai, M. Borysiak, B. Han, D. Chen, et al., Transfer of large-area graphene films for high-performance transparent conductive electrodes, Nano Lett. 9 (12) (2009) 4359−4363.

[26] A. Reina, X. Jia, J. Ho, D. Nezich, H. Son, V. Bulovic, et al., Large area, few-layer graphene films on arbitrary substrates by chemical vapor deposition, Nano Lett. 9 (1) (2009) 30−35.

[27] F. Banhart, J. Kotakoski, A.V. Krasheninnikov, Structural defects in graphene, ACS Nano 5 (1) (2011) 26−41.

[28] Y. Hao, M. Bharathi, L. Wang, Y. Liu, H. Chen, S. Nie, et al., The role of surface oxygen in the growth of large single-crystal graphene on copper, Science 342 (6159) (2013) 720−723.

[29] G.-D. Lee, E. Yoon, K. He, A.W. Robertson, J.H. Warner, Detailed formation processes of stable dislocations in graphene, Nanoscale 6 (24) (2014) 14836−14844.

[30] J.D. Buron, F. Pizzocchero, B.S. Jessen, T.J. Booth, P.F. Nielsen, O. Hansen, et al., Electrically continuous graphene from single crystal copper verified by terahertz conductance spectroscopy and micro four-point probe, Nano Lett. 14 (11) (2014) 6348−6355.

[31] Y. Chen, X.L. Gong, J.G. Gai, Progress and challenges in transfer of large-area graphene films, Adv. Sci. 3 (8) (2016) 1500343.

[32] W.S. Hummers Jr, R.E. Offeman, Preparation of graphitic oxide, J. Am. Chem. Soc. 80 (6) (1958) 1339. 1339.

[33] S.H. Huh, Thermal reduction of graphene oxide, Physics and Applications of Graphene—Experiments, 2011, pp. 73−90.

[34] L. Staudenmaier, Method for the preparation of graphitic acid, Ber. Dtsch. Chem. Ges. 31 (1898) 1481−1487.

[35] D. Marcano, D. Kosynkin, J. Berlin, A. Sinitskii, Z. Sun, A. Slesarev, et al., ACS Nano 4 (2010) 4806−4814.

[36] V.B. Mohan, R. Brown, K. Jayaraman, D. Bhattacharyya, Characterisation of reduced graphene oxide: effects of reduction variables on electrical conductivity, Mater. Sci. Eng.: B 193 (2015) 49−60.

[37] S. Stankovich, D.A. Dikin, R.D. Piner, K.A. Kohlhaas, A. Kleinhammes, Y. Jia, et al., Synthesis of graphene-based nanosheets via chemical reduction of exfoliated graphite oxide, Carbon 45 (7) (2007) 1558−1565.

[38] C. Mattevi, G. Eda, S. Agnoli, S. Miller, K.A. Mkhoyan, O. Celik, et al., Evolution of electrical, chemical, and structural properties of transparent and conducting chemically derived graphene thin films, Adv. Funct. Mater. 19 (16) (2009) 2577−2583.

[39] V.B. Mohan, L. Jakisch, K. Jayaraman, D. Bhattacharyya, Role of chemical functional groups on thermal and electrical properties of various graphene oxide derivatives: a comparative x-ray photoelectron spectroscopy analysis, Mater. Res. Express 5 (3) (2018) 035604.

[40] D.R. Dreyer, S. Park, C.W. Bielawski, R.S. Ruoff, The chemistry of graphene oxide, Chem. Soc. Rev. 39 (1) (2010) 228−240.

[41] W. Chen, L. Yan, P. Bangal, Chemical reduction of graphene oxide to graphene by sulfur-containing compounds, J. Phys. Chem. C 114 (47) (2010) 19885−19890.

[42] C. Gómez-Navarro, R.T. Weitz, A.M. Bittner, M. Scolari, A. Mews, M. Burghard, et al., Electronic transport properties of individual chemically reduced graphene oxide sheets, Nano Lett. 7 (11) (2007) 3499−3503.

[43] C. Liu, F. Hao, X. Zhao, Q. Zhao, S. Luo, H. Lin, Low temperature reduction of free-standing graphene oxide papers with metal iodides for ultrahigh bulk conductivity, Sci. Rep. 4 (1) (2014) 1−6.

[44] J.M. Mativetsky, A. Liscio, E. Treossi, E. Orgiu, A. Zanelli, P. Samorì, et al., Graphene transistors via in situ voltage-induced reduction of graphene-oxide under ambient conditions, J. Am. Chem. Soc. 133 (36) (2011) 14320−14326.

[45] H.-J. Shin, K.K. Kim, A. Benayad, S.-M. Yoon, H.K. Park, I.-S. Jung, et al., Efficient reduction of graphite oxide by sodium borohydride and its effect on electrical conductance, Adv. Funct. Mater. 19 (12) (2009) 1987−1992.

[46] A. Vianelli, A. Candini, E. Treossi, V. Palermo, M. Affronte, Observation of different charge transport regimes and large magnetoresistance in graphene oxide layers, Carbon 89 (2015) 188−196.

[47] H.A. Becerril, J. Mao, Z. Liu, R.M. Stoltenberg, Z. Bao, Y. Chen, Evaluation of solution-processed reduced graphene oxide films as transparent conductors, ACS Nano 2 (3) (2008) 463−470.

[48] H. Yang, Y. Cao, J. He, Y. Zhang, B. Jin, J.-L. Sun, et al., Highly conductive free-standing reduced graphene oxide thin films for fast photoelectric devices, Carbon 115 (2017) 561−570.

[49] Z.-S. Wu, W. Ren, L. Gao, J. Zhao, Z. Chen, B. Liu, et al., Synthesis of graphene sheets with high electrical conductivity and good thermal stability by hydrogen arc discharge exfoliation, ACS Nano 3 (2) (2009) 411−417.

[50] S. De, P.J. King, M. Lotya, A. O'Neill, E.M. Doherty, Y. Hernandez, et al., Flexible, transparent, conducting films of randomly stacked graphene from surfactant-stabilized, oxide-free graphene dispersions, Small 6 (3) (2010) 458−464.

[51] K.R. Paton, E. Varrla, C. Backes, R.J. Smith, U. Khan, A. O'Neill, et al., Scalable production of large quantities of defect-free few-layer graphene by shear exfoliation in liquids, Nat. Mater. 13 (6) (2014) 624.

[52] Y. Hernandez, V. Nicolosi, M. Lotya, F.M. Blighe, Z. Sun, S. De, et al., High-yield production of graphene by liquid-phase exfoliation of graphite, Nat. Nanotechnol. 3 (9) (2008) 563.

[53] M. Lotya, Y. Hernandez, P.J. King, R.J. Smith, V. Nicolosi, L.S. Karlsson, et al., Liquid phase production of graphene by exfoliation of graphite in surfactant/water solutions, J. Am. Chem. Soc. 131 (10) (2009) 3611−3620.

[54] D.J. Finn, M. Lotya, G. Cunningham, R.J. Smith, D. McCloskey, J.F. Donegan, et al., Inkjet deposition of liquid-exfoliated graphene and MoS 2 nanosheets for printed device applications, J. Mater. Chem. C. 2 (5) (2014) 925−932.

[55] Q. Meng, J. Jin, R. Wang, H.-C. Kuan, J. Ma, N. Kawashima, et al., Processable 3-nm thick graphene platelets of high electrical conductivity and their epoxy composites, Nanotechnology 25 (12) (2014) 125707.

[56] A. Marsden, D. Papageorgiou, C. Vallés, A. Liscio, V. Palermo, M. Bissett, et al., Electrical percolation in graphene−polymer composites, 2D Mater. 5 (3) (2018) 032003.

[57] R. Ram, M. Rahaman, A. Aldalbahi, D. Khastgir, Determination of percolation thresh-
      old and electrical conductivity of polyvinylidene fluoride (PVDF)/short carbon fiber
      (SCF) composites: effect of SCF aspect ratio, Polym. Int. 66 (4) (2017) 573−582.
[58] M. Rahaman, T. Chaki, D. Khastgir, Modeling of DC conductivity for ethylene vinyl
      acetate (EVA)/polyaniline conductive composites prepared through insitu polymeriza-
      tion of aniline in EVA matrix, Compos. Sci. Technol. 72 (13) (2012) 1575−1580.
[59] M. Rahaman, A. Aldalbahi, P. Govindasami, N. Khanam, S. Bhandari, P. Feng, et al.,
      A new insight in determining the percolation threshold of electrical conductivity for
      extrinsically conducting polymer composites through different sigmoidal models,
      Polymers 9 (10) (2017) 527.
[60] I. Vlassiouk, G. Polizos, R. Cooper, I. Ivanov, J.K. Keum, F. Paulauskas, et al., Strong
      and electrically conductive graphene-based composite fibers and laminates, ACS Appl.
      Mater. Interfaces 7 (20) (2015) 10702−10709.
[61] P. Liu, Z. Jin, G. Katsukis, L.W. Drahushuk, S. Shimizu, C.-J. Shih, et al., Layered and
      scrolled nanocomposites with aligned semi-infinite graphene inclusions at the platelet
      limit, Science 353 (6297) (2016) 364−367.
[62] C. Xu, J. Gao, H. Xiu, X. Li, J. Zhang, F. Luo, et al., Can in situ thermal reduction be
      a green and efficient way in the fabrication of electrically conductive polymer/reduced
      graphene oxide nanocomposites? Compos. Part A Appl. Sci. Manuf. 53 (2013) 24−33.
[63] M. Yoonessi, J.R. Gaier, Highly conductive multifunctional graphene polycarbonate
      nanocomposites, Acs Nano 4 (12) (2010) 7211−7220.
[64] T. Nezakati, A. Tan, A.M. Seifalian, Enhancing the electrical conductivity of a hybrid
      POSS−PCL/graphene nanocomposite polymer, J. Colloid Interface Sci. 435 (2014)
      145−155.
[65] C. Tu, K. Nagata, S. Yan, Influence of melt-mixing processing sequence on electrical
      conductivity of polyethylene/polypropylene blends filled with graphene, Polym. Bull.
      74 (4) (2017) 1237−1252.
[66] A.A. Vasileiou, M. Kontopoulou, A. Docoslis, A noncovalent compatibilization
      approach to improve the filler dispersion and properties of polyethylene/graphene com-
      posites, ACS Appl. Mater. Interfaces 6 (3) (2014) 1916−1925.
[67] M. Castelaín, G. Martínez, C. Marco, G. Ellis, H.J. Salavagione, Effect of click-
      chemistry approaches for graphene modification on the electrical, thermal, and mechan-
      ical properties of polyethylene/graphene nanocomposites, Macromolecules 46 (22)
      (2013) 8980−8987.
[68] H. Hu, G. Zhang, L. Xiao, H. Wang, Q. Zhang, Z. Zhao, Preparation and electrical con-
      ductivity of graphene/ultrahigh molecular weight polyethylene composites with a segre-
      gated structure, Carbon 50 (12) (2012) 4596−4599.
[69] Y.-C. Chiu, C.-L. Huang, C. Wang, Rheological and conductivity percolations of syn-
      diotactic polystyrene composites filled with graphene nanosheets and carbon nanotubes:
      a comparative study, Compos. Sci. Technol. 134 (2016) 153−160.
[70] Y. Li, J. Tang, L. Huang, Y. Wang, J. Liu, X. Ge, et al., Facile prep2aration, characteriza-
      tion and performance of noncovalently functionalized graphene/epoxy nanocomposites
      with poly (sodium 4-styrenesulfonate), Compos. Part A Appl. Sci. Manuf. 68 (2015) 1−9.
[71] M. Monti, M. Rallini, D. Puglia, L. Peponi, L. Torre, J. Kenny, Morphology and elec-
      trical properties of graphene−epoxy nanocomposites obtained by different solvent
      assisted processing methods, Compos. Part A Appl. Sci. Manuf. 46 (2013) 166−172.
[72] Q. Meng, H. Wu, Z. Zhao, S. Araby, S. Lu, J. Ma, Free-standing, flexible, electrically
      conductive epoxy/graphene composite films, Compos. Part A Appl. Sci. Manuf. 92
      (2017) 42−50.

[73] N. Yousefi, X. Lin, Q. Zheng, X. Shen, J.R. Pothnis, J. Jia, et al., Simultaneous in situ reduction, self-alignment and covalent bonding in graphene oxide/epoxy composites, Carbon 59 (2013) 406−417.

[74] Z. Wang, X. Shen, M. Akbari Garakani, X. Lin, Y. Wu, X. Liu, et al., Graphene aerogel/epoxy composites with exceptional anisotropic structure and properties, ACS Appl. Mater. Interfaces 7 (9) (2015) 5538−5549.

[75] S. Zhao, H. Chang, S. Chen, J. Cui, Y. Yan, High-performance and multifunctional epoxy composites filled with epoxide-functionalized graphene, Eur. Polym. J. 84 (2016) 300−312.

[76] S. Wu, R.B. Ladani, J. Zhang, E. Bafekrpour, K. Ghorbani, A.P. Mouritz, et al., Aligning multilayer graphene flakes with an external electric field to improve multifunctional properties of epoxy nanocomposites, Carbon 94 (2015) 607−618.

[77] G. Tang, Z.-G. Jiang, X. Li, H.-B. Zhang, S. Hong, Z.-Z. Yu, Electrically conductive rubbery epoxy/diamine-functionalized graphene nanocomposites with improved mechanical properties, Compos. Part B Eng. 67 (2014) 564−570.

[78] O.-K. Park, J.-Y. Hwang, M. Goh, J.H. Lee, B.-C. Ku, N.-H. You, Mechanically strong and multifunctional polyimide nanocomposites using amimophenyl functionalized graphene nanosheets, Macromolecules 46 (9) (2013) 3505−3511.

[79] O.-K. Park, S.-G. Kim, N.-H. You, B.-C. Ku, D. Hui, J.H. Lee, Synthesis and properties of iodo functionalized graphene oxide/polyimide nanocomposites, Compos. Part B Eng. 56 (2014) 365−371.

[80] Y. Zhan, M. Lavorgna, G. Buonocore, H. Xia, Enhancing electrical conductivity of rubber composites by constructing interconnected network of self-assembled graphene with latex mixing, J. Mater. Chem. 22 (21) (2012) 10464−10468.

[81] J.R. Potts, O. Shankar, L. Du, R.S. Ruoff, Processing−morphology−property relationships and composite theory analysis of reduced graphene oxide/natural rubber nanocomposites, Macromolecules 45 (15) (2012) 6045−6055.

[82] Y. Luo, P. Zhao, Q. Yang, D. He, L. Kong, Z. Peng, Fabrication of conductive elastic nanocomposites via framing intact interconnected graphene networks, Compos. Sci. Technol. 100 (2014) 143−151.

[83] H. Yang, P. Liu, T. Zhang, Y. Duan, J. Zhang, Fabrication of natural rubber nanocomposites with high graphene contents via vacuum-assisted self-assembly, RSC Adv. 4 (53) (2014) 27687−27690.

[84] H. Aguilar-Bolados, J. Brasero, M.A. López-Manchado, M. Yazdani-Pedram, High performance natural rubber/thermally reduced graphite oxide nanocomposites by latex technology, Compos. Part B Eng. 67 (2014) 449−454.

[85] C. He, X. She, Z. Peng, J. Zhong, S. Liao, W. Gong, et al., Graphene networks and their influence on free-volume properties of graphene−epoxidized natural rubber composites with a segregated structure: rheological and positron annihilation studies, Phys. Chem. Chem. Phys. 17 (18) (2015) 12175−12184.

[86] Y. Lin, S. Liu, J. Peng, L. Liu, Constructing a segregated graphene network in rubber composites towards improved electrically conductive and barrier properties, Compos. Sci. Technol. 131 (2016) 40−47.

[87] L. Wang, W. Wang, Y. Fu, J. Wang, Y. Lvov, J. Liu, et al., Enhanced electrical and mechanical properties of rubber/graphene film through layer-by-layer electrostatic assembly, Compos. Part B Eng. 90 (2016) 457−464.

[88] C. Gao, S. Zhang, F. Wang, B. Wen, C. Han, Y. Ding, et al., Graphene networks with low percolation threshold in ABS nanocomposites: selective localization and electrical and rheological properties, ACS Appl. Mater. Interfaces 6 (15) (2014) 12252−12260.

[89] J.-H. Yang, S.-H. Lin, Y.-D. Lee, Preparation and characterization of poly (l-lactide)—graphene composites using the in situ ring-opening polymerization of PLLA with graphene as the initiator, J. Mater. Chem. 22 (21) (2012) 10805—10815.

[90] A. Noël, J. Faucheu, M. Rieu, J.-P. Viricelle, E. Bourgeat-Lami, Tunable architecture for flexible and highly conductive graphene—polymer composites, Compos. Sci. Technol. 95 (2014) 82—88.

[91] İ. Mutlay, L.B. Tudoran, Percolation behavior of electrically conductive graphene nanoplatelets/polymer nanocomposites: theory and experiment, Fuller. Nanotub. Car. N. 22 (5) (2014) 413—433.

[92] N.H. Vo, T.D. Dao, H.M. Jeong, Electrically conductive graphene/poly (methyl methacrylate) composites with ultra-low percolation threshold by electrostatic self-assembly in aqueous medium, Macromol. Chem. Phys. 216 (7) (2015) 770—782.

[93] H.-B. Zhang, W.-G. Zheng, Q. Yan, Z.-G. Jiang, Z.-Z. Yu, The effect of surface chemistry of graphene on rheological and electrical properties of polymethylmethacrylate composites, Carbon 50 (14) (2012) 5117—5125.

[94] H. Wu, B. Rook, L.T. Drzal, Dispersion optimization of exfoliated graphene nanoplatelet in polyetherimide nanocomposites: extrusion, precoating, and solid state ball milling, Polym. Compos. 34 (3) (2013) 426—432.

[95] H. Pang, T. Chen, G. Zhang, B. Zeng, Z.-M. Li, An electrically conducting polymer/graphene composite with a very low percolation threshold, Mater. Lett. 64 (20) (2010) 2226—2229.

[96] H. Kim, Y. Miura, C.W. Macosko, Graphene/polyurethane nanocomposites for improved gas barrier and electrical conductivity, Chem. Mater. 22 (11) (2010) 3441—3450.

[97] H. Kim, S. Kobayashi, M.A. AbdurRahim, M.J. Zhang, A. Khusainova, M.A. Hillmyer, et al., Graphene/polyethylene nanocomposites: effect of polyethylene functionalization and blending methods, Polymer 52 (8) (2011) 1837—1846.

[98] H. Xu, L.-X. Gong, X. Wang, L. Zhao, Y.-B. Pei, G. Wang, et al., Influence of processing conditions on dispersion, electrical and mechanical properties of graphene-filled-silicone rubber composites, Compos. Part A Appl. Sci. Manuf. 91 (2016) 53—64.

[99] C. Vallés, A.M. Abdelkader, R.J. Young, I.A. Kinloch, Few layer graphene—polypropylene nanocomposites: the role of flake diameter, Faraday Discuss. 173 (2014) 379—390.

[100] D. Yan, H.-B. Zhang, Y. Jia, J. Hu, X.-Y. Qi, Z. Zhang, et al., Improved electrical conductivity of polyamide 12/graphene nanocomposites with maleated polyethylene-octene rubber prepared by melt compounding, ACS Appl. Mater. Interfaces 4 (9) (2012) 4740—4745.

[101] Y. Gao, O.T. Picot, W. Tu, E. Bilotti, T. Peijs, Multilayer coextrusion of graphene polymer nanocomposites with enhanced structural organization and properties, J. Appl. Polym. Sci. 135 (13) (2018) 46041.

[102] X. Li, G.B. McKenna, G. Miquelard-Garnier, A. Guinault, C. Sollogoub, G. Regnier, et al., Forced assembly by multilayer coextrusion to create oriented graphene reinforced polymer nanocomposites, Polymer 55 (1) (2014) 248—257.

[103] G. Carotenuto, S. De Nicola, M. Palomba, D. Pullini, A. Horsewell, T.W. Hansen, et al., Mechanical properties of low-density polyethylene filled by graphite nanoplatelets, Nanotechnology 23 (48) (2012) 485705.

[104] K. Gaska, X. Xu, S. Gubanski, R. Kádár, Electrical, mechanical, and thermal proper-
       ties of LDPE graphene nanoplatelets composites produced by means of melt extrusion
       process, Polymers 9 (1) (2017) 11.
[105] H. Kim, C.W. Macosko, Processing-property relationships of polycarbonate/graphene
       composites, Polymer 50 (15) (2009) 3797–3809.
[106] H. Wu, L.T. Drzal, Graphene nanoplatelet-polyetherimide composites: revealed
       morphology and relation to properties, J. Appl. Polym. Sci. 130 (6) (2013)
       4081–4089.
[107] H. Wu, L.T. Drzal, Multifunctional highly aligned graphite nanoplatelet-polyether
       imide composite in film form, Mater. Chem. Phys. 182 (2016) 110–118.
[108] X.-Y. Qi, D. Yan, Z. Jiang, Y.-K. Cao, Z.-Z. Yu, F. Yavari, et al., Enhanced electrical
       conductivity in polystyrene nanocomposites at ultra-low graphene content, ACS Appl.
       Mater. Interfaces 3 (8) (2011) 3130–3133.
[109] A. Arzac, G.P. Leal, R. Fajgar, R. Tomovska, Comparison of the emulsion mixing
       and in situ polymerization techniques for synthesis of water-borne reduced graphene
       oxide/polymer composites: advantages and drawbacks, Part. Part. Syst. Charact. 31 (1)
       (2014) 143–151.
[110] A. Arzac, G.P. Leal, J.C. de la Cal, R. Tomovska, Water-borne polymer/graphene
       nanocomposites, Macromol. Mater. Eng. 302 (1) (2017) 1600315.
[111] P. Steurer, R. Wissert, R. Thomann, R. Mülhaupt, Functionalized graphenes and ther-
       moplastic nanocomposites based upon expanded graphite oxide, Macromol. Rapid
       Commun. 30 (4-5) (2009) 316–327.
[112] S. Schopp, R. Thomann, K.F. Ratzsch, S. Kerling, V. Altstädt, R. Mülhaupt,
       Functionalized graphene and carbon materials as components of styrene-butadiene
       rubber nanocomposites prepared by aqueous dispersion blending, Macromol. Mater.
       Eng. 299 (3) (2014) 319–329.
[113] W. Zheng, X. Lu, S.C. Wong, Electrical and mechanical properties of expanded
       graphite-reinforced high-density polyethylene, J. Appl. Polym. Sci. 91 (5) (2004)
       2781–2788.
[114] F.T. Cerezo, C.M. Preston, R.A. Shanks, Morphology, thermal stability, and mechani-
       cal behavior of [poly (propylene)-grafted maleic anhydride]-layered expanded graphite
       oxide composites, Macromol. Mater. Eng. 292 (2) (2007) 155–168.
[115] G. Chen, W. Weng, D. Wu, C. Wu, PMMA/graphite nanosheets composite and its
       conducting properties, Eur. Polym. J. 39 (12) (2003) 2329–2335.
[116] F.T. Cerezo, C. Preston, R. Shanks, Structural, mechanical and dielectric properties of
       poly (ethylene-co-methyl acrylate-co-acrylic acid) graphite oxide nanocomposites,
       Compos. Sci. Technol. 67 (1) (2007) 79–91.
[117] J.S. Kim, S. Hong, D.W. Park, S.E. Shim, Water-borne graphene-derived
       conductive SBR prepared by latex heterocoagulation, Macromol. Res. 18 (6) (2010)
       558–565.
[118] J.R. Potts, O. Shankar, S. Murali, L. Du, R.S. Ruoff, Latex and two-roll mill proces-
       sing of thermally-exfoliated graphite oxide/natural rubber nanocomposites, Compos.
       Sci. Technol. 74 (2013) 166–172.
[119] J.S. Kim, J.H. Yun, I. Kim, S.E. Shim, Electrical properties of graphene/SBR nano-
       composite prepared by latex heterocoagulation process at room temperature, J. Ind.
       Eng. Chem. 17 (2) (2011) 325–330.
[120] D. Li, M.B. Müller, S. Gilje, R.B. Kaner, G.G. Wallace, Processable aqueous disper-
       sions of graphene nanosheets, Nat. Nanotechnol. 3 (2) (2008) 101.

[121] M. Fang, K. Wang, H. Lu, Y. Yang, S. Nutt, Single-layer graphene nanosheets with controlled grafting of polymer chains, J. Mater. Chem. 20 (10) (2010) 1982–1992.

[122] Y. Liang, D. Wu, X. Feng, K. Müllen, Dispersion of graphene sheets in organic solvent supported by ionic interactions, Adv. Mater. 21 (17) (2009) 1679–1683.

[123] D.Y. Lee, S. Yoon, Y.J. Oh, S.Y. Park, I. In, Thermo-responsive assembly of chemically reduced graphene and poly (N-isopropylacrylamide), Macromol. Chem. Phys. 212 (4) (2011) 336–341.

[124] S. Stankovich, R.D. Piner, X. Chen, N. Wu, S.T. Nguyen, R.S. Ruoff, Stable aqueous dispersions of graphitic nanoplatelets via the reduction of exfoliated graphite oxide in the presence of poly (sodium 4-styrenesulfonate), J. Mater. Chem. 16 (2) (2006) 155–158.

[125] R. Wissert, P. Steurer, S. Schopp, R. Thomann, R. Mülhaupt, Graphene nanocomposites prepared from blends of polymer latex with chemically reduced graphite oxide dispersions, Macromol. Mater. Eng. 295 (12) (2010) 1107–1115.

[126] E. Tkalya, M. Ghislandi, A. Alekseev, C. Koning, J. Loos, Latex-based concept for the preparation of graphene-based polymer nanocomposites, J. Mater. Chem. 20 (15) (2010) 3035–3039.

[127] W. Fan, C. Zhang, W.W. Tjiu, T. Liu, Fabrication of electrically conductive graphene/polystyrene composites via a combination of latex and layer-by-layer assembly approaches, J. Mater. Res. 28 (4) (2013) 611–619.

[128] H.W. Ha, A. Choudhury, T. Kamal, D.-H. Kim, S.-Y. Park, Effect of chemical modification of graphene on mechanical, electrical, and thermal properties of polyimide/graphene nanocomposites, ACS Appl. Mater. Interfaces 4 (9) (2012) 4623–4630.

[129] M. Verma, S.S. Chauhan, S. Dhawan, V. Choudhary, Graphene nanoplatelets/carbon nanotubes/polyurethane composites as efficient shield against electromagnetic polluting radiations, Compos. Part B Eng. 120 (2017) 118–127.

[130] E. Zhou, J. Xi, Y. Guo, Y. Liu, Z. Xu, L. Peng, et al., Synergistic effect of graphene and carbon nanotube for high-performance electromagnetic interference shielding films, Carbon 133 (2018) 316–322.

[131] J.Y. Oh, G.H. Jun, S. Jin, H.J. Ryu, S.H. Hong, Enhanced electrical networks of stretchable conductors with small fraction of carbon nanotube/graphene hybrid fillers, ACS Appl. Mater. Interfaces 8 (5) (2016) 3319–3325.

[132] S. Kuester, N.R. Demarquette, J.C. Ferreira Jr, B.G. Soares, G.M. Barra, Hybrid nanocomposites of thermoplastic elastomer and carbon nanoadditives for electromagnetic shielding, Eur. Polym. J. 88 (2017) 328–339.

[133] S. Kugler, K. Kowalczyk, T. Spychaj, Hybrid carbon nanotubes/graphene modified acrylic coats, Prog. Org. Coat. 85 (2015) 1–7.

[134] K. Tschoppe, F. Beckert, M. Beckert, R. Mülhaupt, Thermally reduced graphite oxide and mechanochemically functionalized graphene as functional fillers for epoxy nanocomposites, Macromol. Mater. Eng. 300 (2) (2015) 140–152.

[135] T. Van Khai, H.G. Na, D.S. Kwak, Y.J. Kwon, H. Ham, K.B. Shim, et al., Significant enhancement of blue emission and electrical conductivity of N-doped graphene, J. Mater. Chem. 22 (34) (2012) 17992–18003.

[136] M. Beckert, F.J. Tölle, B. Bruchmann, R. Mülhaupt, Nitrogen-doped multilayer graphene as functional filler for carbon/polyamide 12 nanocomposites, Macromol. Mater. Eng. 300 (8) (2015) 785–792.

[137] M. Stürzel, Y. Thomann, M. Enders, R. Mülhaupt, Graphene-supported dual-site catalysts for preparing self-reinforcing polyethylene reactor blends containing UHMWPE

nanoplatelets and in situ UHMWPE shish-kebab nanofibers, Macromolecules 47 (15) (2014) 4979−4986.

[138] D. Hofmann, K.A. Wartig, R. Thomann, B. Dittrich, B. Schartel, R. Mülhaupt, Functionalized graphene and carbon materials as additives for melt-extruded flame retardant polypropylene, Macromol. Mater. Eng. 298 (12) (2013) 1322−1334.

[139] A.-K. Appel, R. Thomann, R. Mülhaupt, Polyurethane nanocomposites prepared from solvent-free stable dispersions of functionalized graphene nanosheets in polyols, Polymer 53 (22) (2012) 4931−4939.

[140] F. Beckert, C. Friedrich, R. Thomann, R. Mülhaupt, Sulfur-functionalized graphenes as macro-chain-transfer and RAFT agents for producing graphene polymer brushes and polystyrene nanocomposites, Macromolecules 45 (17) (2012) 7083−7090.

[141] F. Beckert, A.M. Rostas, R. Thomann, S. Weber, E. Schleicher, C. Friedrich, et al., Self-initiated free radical grafting of styrene homo-and copolymers onto functionalized graphene, Macromolecules 46 (14) (2013) 5488−5496.

[142] D. Hofmann, M. Keinath, R. Thomann, R. Mülhaupt, Thermoplastic carbon/polyamide 12 composites containing functionalized graphene, expanded graphite, and carbon nanofillers, Macromol. Mater. Eng. 299 (11) (2014) 1329−1342.

[143] N. Yousefi, M.M. Gudarzi, Q. Zheng, S.H. Aboutalebi, F. Sharif, J.-K. Kim, Self-alignment and high electrical conductivity of ultralarge graphene oxide−polyurethane nanocomposites, J. Mater. Chem. 22 (25) (2012) 12709−12717.

[144] J. Ding, Y. Fan, C. Zhao, Y. Liu, C. Yu, N. Yuan, Electrical conductivity of waterborne polyurethane/graphene composites prepared by solution mixing, J. Compos. Mater. 46 (6) (2012) 747−752.

[145] M. Li, C. Gao, H. Hu, Z. Zhao, Electrical conductivity of thermally reduced graphene oxide/polymer composites with a segregated structure, Carbon 65 (2013) 371−373.

[146] V.H. Pham, T.V. Cuong, T.T. Dang, S.H. Hur, B.-S. Kong, E.J. Kim, et al., Superior conductive polystyrene−chemically converted graphene nanocomposite, J. Mater. Chem. 21 (30) (2011) 11312−11316.

[147] M. Sabzi, L. Jiang, F. Liu, I. Ghasemi, M. Atai, Graphene nanoplatelets as poly (lactic acid) modifier: linear rheological behavior and electrical conductivity, J. Mater. Chem. A 1 (28) (2013) 8253−8261.

[148] N. Liu, F. Luo, H. Wu, Y. Liu, C. Zhang, J. Chen, One-step ionic-liquid-assisted electrochemical synthesis of ionic-liquid-functionalized graphene sheets directly from graphite, Adv. Funct. Mater. 18 (10) (2008) 1518−1525.

[149] S. Vadukumpully, J. Paul, N. Mahanta, S. Valiyaveettil, Flexible conductive graphene/poly (vinyl chloride) composite thin films with high mechanical strength and thermal stability, Carbon 49 (1) (2011) 198−205.

[150] Y. Li, H. Zhang, H. Porwal, Z. Huang, E. Bilotti, T. Peijs, Mechanical, electrical and thermal properties of in-situ exfoliated graphene/epoxy nanocomposites, Compos. Part A Appl. Sci. Manuf. 95 (2017) 229−236.

[151] X. Zeng, J. Yang, W. Yuan, Preparation of a poly (methyl methacrylate)-reduced graphene oxide composite with enhanced properties by a solution blending method, Eur. Polym. J. 48 (10) (2012) 1674−1682.

[152] B. Dong, S. Wu, L. Zhang, Y. Wu, High performance natural rubber composites with well-organized interconnected graphene networks for strain-sensing application, Ind. Eng. Chem. Res. 55 (17) (2016) 4919−4929.

[153] Z. Chen, W. Ren, L. Gao, B. Liu, S. Pei, H.-M. Cheng, Three-dimensional flexible and conductive interconnected graphene networks grown by chemical vapour deposition, Nat. Mater. 10 (6) (2011) 424.

[154] T. Kuilla, S. Bhadra, D. Yao, N.H. Kim, S. Bose, J.H. Lee, Recent advances in graphene based polymer composites, Prog. Polym. Sci. 35 (11) (2010) 1350−1375.

[155] P. Zhao, Y. Luo, J. Yang, D. He, L. Kong, P. Zheng, et al., Electrically conductive graphene-filled polymer composites with well organized three-dimensional microstructure, Mater. Lett. 121 (2014) 74−77.

[156] Z. Tu, J. Wang, C. Yu, H. Xiao, T. Jiang, Y. Yang, et al., A facile approach for preparation of polystyrene/graphene nanocomposites with ultra-low percolation threshold through an electrostatic assembly process, Compos. Sci. Technol. 134 (2016) 49−56.

[157] J.-W. Zha, B. Zhang, R.K. Li, Z.-M. Dang, High-performance strain sensors based on functionalized graphene nanoplates for damage monitoring, Compos. Sci. Technol. 123 (2016) 32−38.

[158] R. Moriche, M. Sánchez, A. Jiménez-Suárez, S. Prolongo, A. Ureña, Electrically conductive functionalized-GNP/epoxy based composites: from nanocomposite to multiscale glass fibre composite material, Compos. Part B Eng. 98 (2016) 49−55.

[159] F. Du, J.E. Fischer, K.I. Winey, Effect of nanotube alignment on percolation conductivity in carbon nanotube/polymer composites, Phys. Rev. B 72 (12) (2005) 121404.

[160] S. Ansari, A. Kelarakis, L. Estevez, E.P. Giannelis, Oriented arrays of graphene in a polymer matrix by in situ reduction of graphite oxide nanosheets, Small 6 (2) (2010) 205−209.

[161] D.D. Kulkarni, I. Choi, S.S. Singamaneni, V.V. Tsukruk, Graphene oxide − polyelectrolyte nanomembranes, ACS Nano 4 (8) (2010) 4667−4676.

[162] D. Wang, X. Wang, Self-assembled graphene/azo polyelectrolyte multilayer film and its application in electrochemical energy storage device, Langmuir 27 (5) (2011) 2007−2013.

[163] X. Zhao, Q. Zhang, Y. Hao, Y. Li, Y. Fang, D. Chen, Alternate multilayer films of poly (vinyl alcohol) and exfoliated graphene oxide fabricated via a facial layer-by-layer assembly, Macromolecules 43 (22) (2010) 9411−9416.

[164] M. Amjadi, K.U. Kyung, I. Park, M. Sitti, Stretchable, skin-mountable, and wearable strain sensors and their potential applications: a review, Adv. Funct. Mater. 26 (11) (2016) 1678−1698.

[165] L. He, S.-C. Tjong, Electrical behavior and positive temperature coefficient effect of graphene/polyvinylidene fluoride composites containing silver nanowires, Nanoscale Res. Lett. 9 (1) (2014) 1−8.

[166] P. Zhang, D. Cao, S. Cui, Resistivity-temperature behavior and morphology of low density polyethylene/graphite powder/graphene composites, Polym. Compos. 35 (8) (2014) 1453−1459.

[167] M. Rahaman, T.K. Chaki, D. Khastgir, Control of the temperature coefficient of the DC resistivity in polymer-based composites, J. Mater. Sci. 48 (21) (2013) 7466−7475.

[168] D.X. Yan, H. Pang, B. Li, R. Vajtai, L. Xu, P.G. Ren, et al., Structured reduced graphene oxide/polymer composites for ultra-efficient electromagnetic interference shielding, Adv. Funct. Mater. 25 (4) (2015) 559−566.

[169] N. Yousefi, X. Sun, X. Lin, X. Shen, J. Jia, B. Zhang, et al., Highly aligned graphene/polymer nanocomposites with excellent dielectric properties for high-performance electromagnetic interference shielding, Adv. Mater. 26 (31) (2014) 5480−5487.

[170] R. Hodlur, M. Rabinal, Self assembled graphene layers on polyurethane foam as a highly pressure sensitive conducting composite, Compos. Sci. Technol. 90 (2014) 160−165.

[171] H. Pang, Y.-C. Zhang, T. Chen, B.-Q. Zeng, Z.-M. Li, Tunable positive temperature coefficient of resistivity in an electrically conducting polymer/graphene composite, Appl. Phys. Lett. 96 (25) (2010) 251907.

[172] C.-H. Chang, T.-C. Huang, C.-W. Peng, T.-C. Yeh, H.-I. Lu, W.-I. Hung, et al., Novel anticorrosion coatings prepared from polyaniline/graphene composites, Carbon 50 (14) (2012) 5044−5051.

[173] Y. Tong, S. Bohm, M. Song, Graphene based materials and their composites as coatings, Austin, J. Nanomed. Nanotechnol. 1 (1) (2013) 1003.

[174] D. Pierleoni, Z.Y. Xia, M. Christian, S. Ligi, M. Minelli, V. Morandi, et al., Graphene-based coatings on polymer films for gas barrier applications, Carbon 96 (2016) 503−512.

[175] M. Silva, N.M. Alves, M.C. Paiva, Graphene-polymer nanocomposites for biomedical applications, Polym. Adv. Technol. 29 (2) (2018) 687−700.

[176] C.M. Santos, J. Mangadlao, F. Ahmed, A. Leon, R.C. Advincula, D.F. Rodrigues, Graphene nanocomposite for biomedical applications: fabrication, antimicrobial and cytotoxic investigations, Nanotechnology 23 (39) (2012) 395101.

[177] K. Deshmukh, M.B. Ahamed, K.K. Sadasivuni, D. Ponnamma, M.A.-A. AlMaadeed, S.K. Pasha, et al., Graphene oxide reinforced poly (4-styrenesulfonic acid)/polyvinyl alcohol blend composites with enhanced dielectric properties for portable and flexible electronics, Mater. Chem. Phys. 186 (2017) 188−201.

[178] R.-H. Kim, M.-H. Bae, D.G. Kim, H. Cheng, B.H. Kim, D.-H. Kim, et al., Stretchable, transparent graphene interconnects for arrays of microscale inorganic light emitting diodes on rubber substrates, Nano Lett. 11 (9) (2011) 3881−3886.

[179] K.S. Lee, Y. Lee, J.Y. Lee, J.H. Ahn, J.H. Park, Flexible and platinum-free dye-sensitized solar cells with conducting-polymer-coated graphene counter electrodes, ChemSusChem 5 (2) (2012) 379−382.

Dielectric properties of polymer-graphene composites

Banalata Sahoo[1] and Suryakanta Nayak[2]
[1]Department of Chemistry, Binayak Acharya College, Berhampur, Odisha, India,
[2]Department of Electrical and Computer Engineering, National University of Singapore, Singapore

5.1   Introduction

Composites are materials that are prepared from two or more constituent materials and these constituents have different physical and chemical properties. However, composites have properties that are different from the individual constituents [1−5]. Moreover, nanocomposites contain at least one dispersed phase in nanometer range and these composites are better than the normal composites due to the large surface to volume ratio of nanomaterials present [6−10]. Polymer composites (polymers reinforced with filler particles) have attracted great attention to many researchers due to their improved properties (mechanical, electrical, thermal; and other barrier properties) by the addition of different filler particles. There are added advantages to these composites if the composites are reinforced with nanoparticles instead to conventional microparticles. Nanoparticles have higher surface to volume ratio which helps to improve certain properties like thermomechanical properties and the degradation stability of polymer composites/nanocomposites without affecting the dielectric strength of the composites [11−15]. There are different kinds of nanoparticles/nanomaterials [inorganic oxide nanoparticles, nanoclay, carbon fillers, nanofiber, carbon nanotubes (CNTs), and graphene] used in the preparation of various polymer composites/nanocomposites with suitable applications [2,10,16−19]. Polymer composites with the above types of filler particles have various applications depending on the nature of polymer and the filler particles present. There are anew possible applications using polymer composite materials which are listed as follows: electromagnetic interference (EMI) shielding, photovoltaic devices, capacitors, actuators, and static-charge dissipation. Polymer/polymer composite materials are easily processable and can be molded into any complex shape/geometries. Hence, these composite materials can also be used in compact electronics [14,20].

There are different kinds of composite materials which play an important role in making our lives comfortable. Conducting polymers prepared from different conducting fillers can be used in flexible interconnects and flexible/stretchable electrodes/electronics. Moreover, conducting polymer (polyacetylene, polyaniline,

Polymer Nanocomposites Containing Graphene. DOI: https://doi.org/10.1016/B978-0-12-821639-2.00008-2

polypyrrole, polythiophene, polyindoles, and polyphenylene vinylene) can be used as filler/dispersed phase in polymer matrices to improve the permittivity, conductivity, and other properties of the end composites [21−26]. In the current chapter, we focus on the electrical properties (permittivity and conductivity) of polymer-graphene based composites and also discuss the effect of various factors on the conductivity and dielectric properties of polymer-graphene composites.

5.2 Graphene and its application in polymer composites

There are various carbon fillers available on the market, which are used to reinforce the polymer matrices and improve the mechanical, thermal, electrical properties. However, graphene-based polymer-composites/nanocomposites have attracted great attention from scientists because of their potential applications in various fields. In the recent decade, the usage of graphene has increased significantly due to its unique properties and potential application in different fields like supercapacitors, lithium ion batteries, solar cell energy-conversion, microwave absorbing properties, and in catalysis [27−30]. The usage of graphene as filler in the polymer matrices, can enhance various properties like mechanical, electrical/dielectric, and thermal conductivity of the composites.

5.3 Factors affecting the dielectric properties of polymer−graphene composites

In this section, we focus on the effect of various factors, which affects the dielectric/electrical properties of polymer-graphene composites. The different factors are listed as follows: filler dispersion/aggregation, processing condition, surface modification of graphene, filler-filler interactions, interfacial interactions between graphene and the polymer-matrix, concentration, frequency, temperature, shear effects, mixing method, composite morphology, pressure, and electric field. Graphene plays an important role due to their superior functional and structural properties such as high aspect ratio, high mechanical strength, and high electrical properties [2,31,32]. Here, we discuss a few important factors in detail which affect the dielectric/electrical properties of polymer-graphene composites.

5.3.1 Effect of processing condition

Different processing parameters (mixing speed, mixing method, filler concentration, mixing temperature, and shear force) play an important role in modifying the properties of polymer-based composites [33−44]. Here, we discuss the effect of processing conditions on dielectric properties of various polymer−graphene composites/ nanocomposites. M.J. Martínez-Morlanes et al. have studied the effect of processing conditions on mechanical, microstructural, and tribological properties of ultra-high molecular weight polyethylene (PE)—graphene nanoplatelets (UHMWPE-GNPs) composites. There is change in microhardness and average coefficient of friction (COF) by the addition of different concentration of GNPs in PE matrix, mechanical mixing methods [ball milling (BM) and blade mixing (BL)], and by the increase in temperature (Figs. 5.1 and 5.2). The hardness of as-supplied neat PE powder (hot pressed) was 4.15 HV, which was increased to 4.5 HV and 4.75 HV after the BM and BL processes, respectively. There is also an increase in hardness of the composites after increasing the GNPs concentration in the PE matrix. However, the increase is monotonous in case of -BL composite, and the value is 5.2 HV at the highest GNPs loading (10% increase compared to pristine PE). Moreover, -BM composites with more than 0.5 wt.% GNP content have similar nature., The effect of GNPs on microhardness is too high compared to the other PE-graphene oxide (PE-GO) composites. Fig. 5.1B shows the effect of temperature on the microhardness of neat PE, PE-GNP (with 1 and 5 wt.% GNP) composites obtained by the BM process. It is observed that there is an increase in the microhardness of the composites with the molding temperature and the microhardness value at highest molding temperature (240°C) is positive for all the studied materials compared to the values at 175°C, in particular, in pristine PE.

It is observed that the GNPs reinforced -BM composites have a positive effect on the COF (Fig. 5.2A)., The lubrication effect (decrease in friction) of graphene is seen above 0.5 wt.% of GNPs and shows a minimum value of COF of 0.073 ± 0.004 at 3 wt.% GNP contents, representing a reduction of 25%. Moreover,

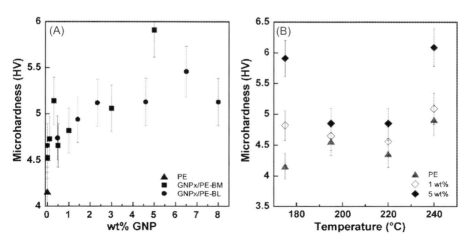

Figure 5.1 (A) Reinforcement effect of graphene nanoplatelets (GNPs) on microhardness of composites processed at 175°C for BL and BM mixing methods, (B) Effect of molding temperature on microhardness for GNPx/PE-BM composite (x = 0, 1 and 5 wt.% GNP content). *Source:* Reprinted with permission from Journal of the Mechanical Behavior of Biomedical Materials 115 (2021): 104248. Copyright (2021) Elsevier.

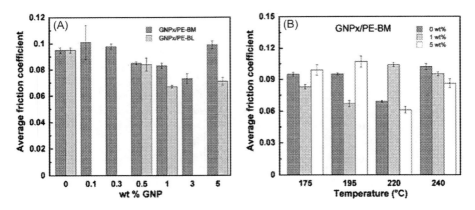

Figure 5.2 Variation of average coefficient of friction (COF) obtained in the range 3000−400 m as a function of (A) GNP loading and (B) temperature.
Source: Reprinted with permission from Journal of the Mechanical Behavior of Biomedical Materials 115 (2021): 104248. Copyright (2021) Elsevier.

similar trend is observed in the case of -BL composites, where the COF value is reaching to a minimum value of 0.067 ± 0.01 at 1.4 wt.%. The above 33% reduction in COF value is due to the lubrication effect, obtained from the interlayer shearing of the graphene nanaoplatelet layers. Fig. 5.2B shows the temperature effect on COF of PE-BM composites containing 1 and 5 wt.% of GNPs. The pristine PE shows a significant decrease (nearly 25%) in COF at 220°C and the trend again changes to the initial value at 240°C. This type of behavior is difficult to explain as a function of microstructural changes, introduced by the high hot-pressed temperatures in the UHMWPE matrix. At high temperature polymer chain scission happens due to the oxidative process, followed by increase in chain mobility. At 5 wt.% of GNPs, the influence is irregular with a decrease in COF at 220°C, but there is an increase at 240°C. This type of complex behavior could be due the presence of GNPs aggregates and free radical scavenger effect of the graphene nanoplatelets [45]. Hui Xu et al. studied the effect of processing conditions on the mechanical, dispersion, and electrical properties of silicone rubber (SR)-graphene composites. In this work, the authors have prepared the composite in three different processes/methods (mechanical mixing, ball milling, and solution blending) (Fig. 5.3). It is observed that the graphene sheets dispersion in the SR matrix highly depends on the processing conditions as evidenced from scanning electron microscopy, transmission electron microscopy, and transmission optical microscopy (TOM). The graphene distribution at both macro and micro scale in SR matrix was best in the case of ball milling process compared to the solution blending process and mechanical mixing. Similarly, the mechanical property enhancement is good in the case of ball milling process (Fig. 5.4). Better dispersion of graphene in the SR matrix enhances the mechanical properties as evidenced from the morphology and the microstructure analysis. However, high electrical conductivity was obtained in

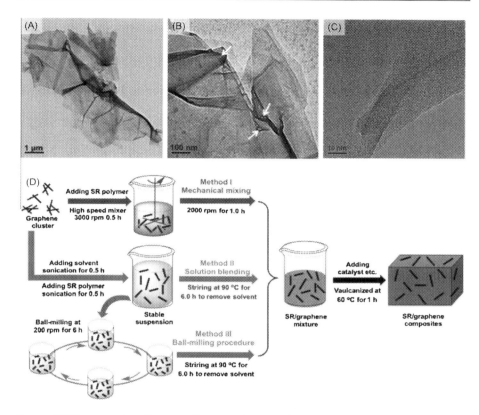

Figure 5.3 Transmission electron microscopy images of (A) graphene sheets agglomerate and (B, C) graphene sheets after the sonication in solvent (the arrows indicate the defects on the graphene sheets); and (D) schematic representation of the preparation of SR-graphene composites via three processing technologies (solution blending, mechanical mixing, and ball-milling process).
Source: Reprinted with permission from Composites Part A: Applied Science and Manufacturing 91 (2016): 53−64. Copyright (2016) Elsevier.

the case of composites prepared through the solution blending process (Fig. 5.5), which is due to the better dispersion and/or exfoliation of graphene in the composites. It is interesting to note that among the three composite systems, best dispersion levels of the graphene sheets for the ball-milling process/method did not produce the highest conductivity value. At a fixed concentration of graphene loading, the inter-sheet contact in solution blending treated sample is significantly greater to provide more conductive pathways than that of a good filler dispersion. However, ball milling process produces a high shear force which improves the dispersion of graphene sheets in the SR matrix and produces SR sheath coated on the graphene sheets, and this would restrict the formation of conductive network to a

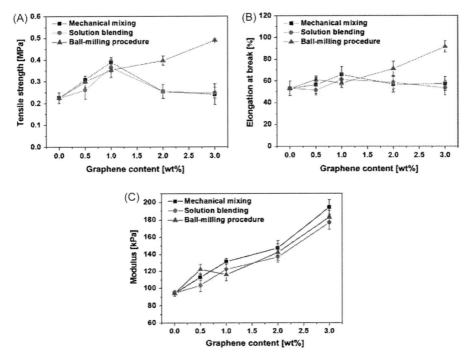

Figure 5.4 Tensile properties of pure SR and SR-graphene composites prepared by different processing techniques: (A) tensile strength and (B) elongation at break.
Source: Reprinted with permission from Composites Part A: Applied Science and Manufacturing 91 (2016): 53−64. Copyright (2016) Elsevier.

certain degree, thus leading to lower values of electrical conductivity above the percolation threshold [46].

M. Hamidinejad et al. have studied the effect of processing conditions (supercritical-fluid treatment and physical foaming) on EMI shielding and electrical properties of polymer-graphene nanoplatelet composites. The composites were prepared by the GNPs incorporation in lightweight high-density PE (HDPE) using two methods namely physical foaming and supercritical-fluid (SCF) treatment in an injection molding process. It is observed that the nanocomposites prepared through foaming process have the higher dielectric permittivity, broadband dielectric permittivity, and electrical conductivity compared to the regular injection molded counterparts (Figs. 5.6, 5.7 and 5.8). Both processes exfoliated the GnPs and changed the flow-induced arrangement by reducing the cellular growth and melt viscosity. In addition, the generated cellular structures rearranged the GnPs to be perpendicular to the radial direction of the bubble growth. This rearrangement enhanced the interconnectivity of GnPs and produced a unique arrangement of GnP around the cells. Hence, the conductivity through the plane increased up to a maximum of nine

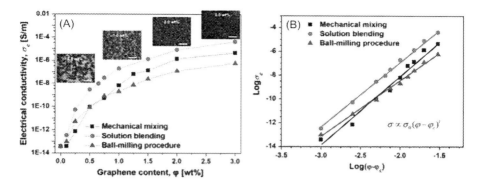

Figure 5.5 (A) Electrical conductivities (σ_c) of pure SR and SR-graphene composites prepared by different processing methods (inset: TOM images of the SR composites at different graphene loadings prepared by solution blending); and (B) $\log\sigma_c$ plotted against $\log(\phi-\phi_c)$, where ϕ_c is the percolation threshold. (Solid lines in the graphs are calculated based on the fitting of experimental data for different processing techniques).
Source: Reprinted with permission from Composites Part A: Applied Science and Manufacturing 91 (2016): 53−64. Copyright (2016) Elsevier.

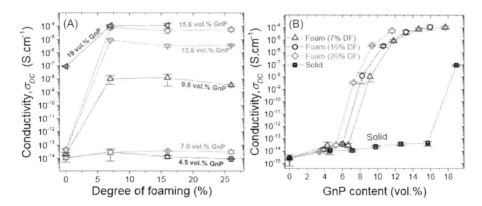

Figure 5.6 (A) Variations of the foaming degree on the electrical conductivity of HDPE-GnP composites; (B) evolution of percolation threshold with the foaming degree.
Source: Reprinted with permission from ACS Appl. Mater. Interfaces 2018, 10, 36, 30752−30761. Copyright (2018) American Chemical Society.

orders of magnitude. The permittivity of nanocomposites (9.8 vol.% GnP) prepared through foaming process have higher real permittivity of $\varepsilon' = 106.4$ compared to that of conventional injection-molded samples ($\varepsilon' = 6.2$). Moreover, composites prepared through the foaming process have reduced density, which helps to produce lightweight composites [47].

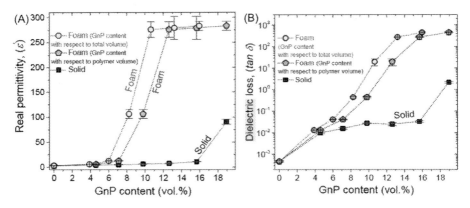

Figure 5.7 Variation of (A) real dielectric permittivity (ε'); and (B) dielectric loss (tan δ) of the solid and foamed (16% degree of foaming) nanocomposites with GNP loading measured at 1×10^3 Hz. (GNP vol.% is reported with respect to the polymer volume).
Source: Reprinted with permission from ACS Appl. Mater. Interfaces 2018, 10, 36, 30752−30761. Copyright (2018) American Chemical Society.

5.3.2 Effect of morphology

The effect of composite morphology on the dielectric and electrical properties of polymer-graphene composites have been discussed here. H. Zhang et al. used two different dimensional nanofillers of GNPs and MWCNTs as bifillers in PMMA matrix to explore their synergistic effect on the electrical and mechanical properties. The well dispersion of MWCNTs-GNPs fillers in PMMA matrix helps in stress transfer when exposed to tensile stress. However, the critical distance between the nanofillers is not enough for effective electron movements. It is observed that there is improvement in electrical/dielectric properties of composites when the composites foams were prepared by supercritical fluid assisted processing. GNPs and MWCNTs reveal prominent synergistic effect in influencing the electrical and tensile properties of ternary composite foam systems. Fig. 5.9 shows the schematic representation for the preparation of the PMMA/MWCNTs-GNPs nanocomposite foams *via* supercritical CO_2 process. First, the specimens were first placed in the high-pressure vessel under an 8.5 MPa pressure at 40°C After that a full sorption of 16 h, the CO_2 pressure was released with a depressurization rate of 0.6 MPa/s. In second step, the gas-saturated specimens were moved into hot-oil at various temperatures (90°C−110°C) for various time (20−60 s) to induce cell growth. Then, the foamed specimens were immersed into ice-water mixture for 5 min (Fig. 5.9). Fig. 5.10 and 5.11 shows the electrical conductivity of ternary PMMA nanocomposites: PMMA/GNPs-MWCNTs composites (solid and foam respectively) with (A) 2 wt.% and (B) 3 wt.% nanofillers. It is observed that foamed composites (Fig. 5.11) have higher electrical conductivity compared to the solid composites (Fig. 5.10) at the same loading which is due the increased inter-connectivity and enhanced conductive network between nanofillers after volume expansion [48,49].

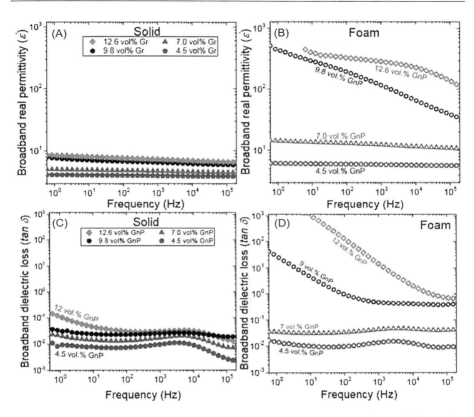

Figure 5.8 Broadband dielectric permittivity of (A) the solid samples, and (B) the foamed 9.8 vol.% HDPE-GnP composites. Broadband dielectric loss of (C) the solid samples, and the foamed 9.8 vol.% HDPE-GnP composites.
Source: Reprinted with permission from ACS Appl. Mater. Interfaces 2018, 10, 36, 30752–30761. Copyright (2018) American Chemical Society.

5.3.3 Effect of frequency and filler concentration

Frequency plays an important role in the dielectric/electrical properties of polymer composites. The electrical, mechanical, and other properties of polymer composites/nanocomposites can be controlled by the different concentration of filler particles. C. Yang et al. have studied the frequency and concentration dependent dielectric properties of poly(vinylidene fluoride)—controllable hydroxylated/carboxylated graphene nanocomposites. The composites were prepared by simple solution cast and hot-pressing methods. In the modified graphene, hydroxy groups act as an electron donor and improve the dielectric properties compared to the carboxyl groups with electron withdrawing nature. So, PVDF composites prepared from GROH have high dielectric constant and low loss compared to GRCOOH. These kind of low loss composites/materials can be used to store electrical charge/energy and play a vital role in modern electronics and electrical power systems. Fig. 5.12A and B

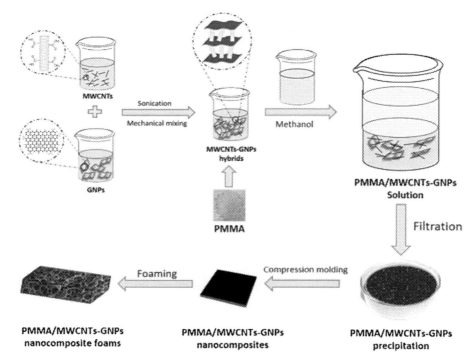

Figure 5.9 Schematic representation of the preparation process of PMMA/MWCNTs-GNPs nanocomposite foams *via* supercritical CO_2 process.
Source: Reprinted with permission from Chemical Engineering Journal 353 (2018): 381−393. Copyright (2018) Elsevier.

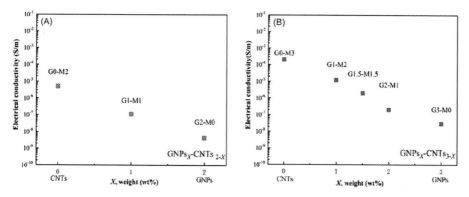

Figure 5.10 Electrical conductivity of ternary PMMA nanocomposites: PMMA/GNPs-MWCNTs composites with (A) 2 wt.% and (B) 3 wt.% nanofillers.
Source: Reprinted with permission from Chemical Engineering Journal 353 (2018): 381−393. Copyright (2018) Elsevier.

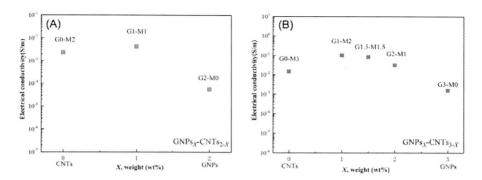

Figure 5.11 Electrical conductivity of ternary PMMA nanocomposite foams: PMMA/GNPs-MWCNTs composite foams with (A) 2 wt.% and (B) 3 wt.% nanofillers.
Source: Reprinted with permission from Chemical Engineering Journal 353 (2018): 381−393. Copyright (2018) Elsevier.

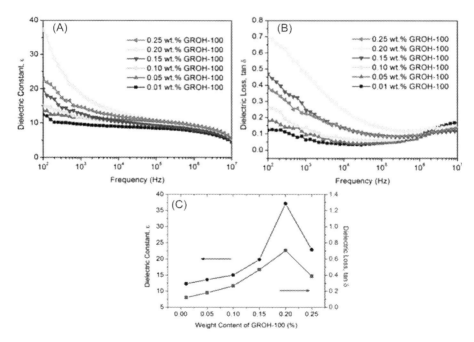

Figure 5.12 Variation of (A) dielectric constant, (B) dielectric loss with frequency for a series of GROH-100/PVDF composites with various weight percent of GROH-100 at room temperature, and (C) shows the dependence of the dielectric constant and dielectric loss of the GROH-100/PVDF composites on the GROH-100 mass fraction, measured at room temperature and 100 Hz.
Source: Reprinted with permission from Carbon 117 (2017): 301−312. Copyright (2017) Elsevier.

shows the dielectric constant and dielectric loss respectively for GROH-100/
PVDF (GROH-100 = molar ratio between successfully functionalized OH group
and benzene ring on the GR sheet is 1) composites with increasing GROH-100
content from 0.01 to 0.25 wt.% at room temperature, over a frequency range of
10^2-e10^7 Hz. Fig. 5.12C is given for better understanding the comparison of permittivity and dielectric loss of all composites obtained at frequency of 100 Hz. It
is observed from Fig. 5.12A that the permittivity of GROH-100/PVDF composite
decreases with the increase in frequency, which is due to the polarization at the
inner structure of the composites [50,51]. As shown in the Fig. 5.13, there will be
hydrogen bonds formation between the hydroxyl groups of GROH-100 and -F of
PVDF after the formation of the composites [50,52−56]. So, there will be strong
interfacial interaction between GROH-100 nanosheets and the PVDF matrix.
Moreover, the non-polar nature of PVDF polymer and the polymer molecular
chains hinder the contribution of electrical polarization. As shown in Fig. 5.13B,
the dielectric loss shows two different dielectric behaviors: at low frequency
dielectric loss decreases while there is sudden increase at high frequency region.
The dielectric behavior at low frequency is due to the interfacial polarization.
However, the dielectric response at high frequency is due to the Debye relaxation

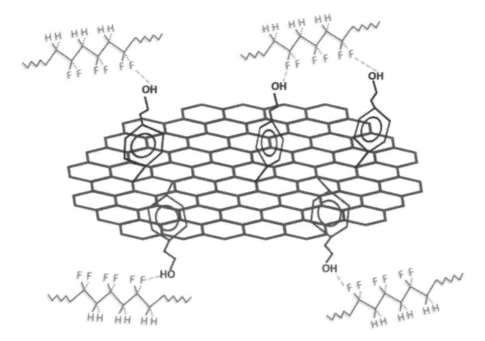

Figure 5.13 Schematic representation of the formation of hydrogen bond in GROH/PVDF
composites.
Source: Reprinted with permission from Carbon 117 (2017): 301−312. Copyright (2017)
Elsevier.

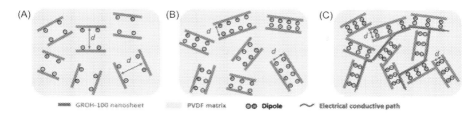

Figure 5.14 Schematic representations of micro-structures of GROH-100/PVDF composites with different mass fraction of GROH-100, (A) below 0.10 wt.%, (B) between 0.10 wt.% and 0.20 wt.%, and (C) above 0.20 wt.%.
Source: Reprinted with permission from Carbon 117 (2017): 301−312. Copyright (2017) Elsevier.

resulting from the C-F dipole orientation polarization of PVDF matrix [50,56]. In Fig. 5.13C, it is observed that there is increase in both permittivity and loss up to a loading of 0.2 wt.% but again slightly decreases at the highest loading. This phenomenon could be understood by the percolation theory where the dielectric properties of the conductive polymer-nanofiller composites depend on the capacitance of the micro-capacitors formed by GROH-100 nanosheets as plates and PVDF as the base matrix and the microstructures of different GROH-100/PVDF composites with increased GROH-100 mass fractions are schematically represented in Fig. 5.14 [50,57−59]. In case of unmodified graphene there will be high chance of particle aggregation but in case of modified graphene (GROH), the particles will be separated from each other due to equal charges available on each graphene sheets. So, initially it will increase the number capacitors with the increase in the graphene loading. But, at high loading there is chance of agglomeration due to reduced space on the polymer matric. So, at higher loading due to graphene sheets agglomeration the number of capacitors will be reduced thus reducing the dielectric permittivity of the composites at the highest loading [50,60−62].

Fig. 5.15 shows the variation of the dielectric constant (permittivity) and dielectric loss GRCOOH-100/PVDF composite with frequency ranging from 10^2 to 10^7 Hz. The percolation threshold is also found to be 0.2 wt.%. However, it is observed that the dielectric constant and dielectric loss of GROH-100/PVDF is larger/lower than that of GRCOOH-100/PVDF with identical weight content, and it agrees with the fact that electron donating groups (−OH) contribute more than the electron withdrawing groups (−COOH) to the improvement of dielectric properties. Therefore, GROH shows superior properties to GRCOOH from the perspective of dielectric materials [50,63−65].

5.3.4 Effect of temperature

The material's permittivity depends on the orientation of the permanent dipoles. These dipoles can be orientated by the change in temperature leading to a change in

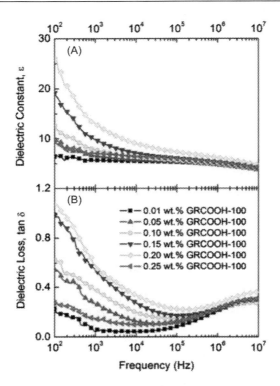

Figure 5.15 Dielectric constant (A) and dielectric loss (Bb) as a function of frequency for a series of GRCOOH-100/PVDF composites with various weight content of GRCOOH100 at room temperature.
Source: Reprinted with permission from Carbon 117 (2017): 301−312. Copyright (2017) Elsevier.

permittivity of the material. However, the change in permittivity with the increase in temperature does not follow the same path. The increase in permittivity with an increase in temperature is due to the increase in the mobility of bound charges such as interfacial space charges and induced dipoles. The thermal excitation of these charges happens at higher temperature which is reason for the increase in permittivity. The electrical conductivity of polymer composites also increases with the increase in temperature but up to the certain limit as reported in our earlier publication [2]. Kai Ke et al. have studied the effect of temperature on dielectric properties of thermoplastic polyurethane (TPU) composite with branched CNTs and graphene nanoplatelets (GNPs) (Fig. 5.16). It is observed from Fig. 5.16 that the TPU composites with carbon nanostructures (CNS) have higher dielectric constant at both frequencies (1.26 and 1187 Hz) compared to both neat TPU and the TPU-0.5GNP composite. The increased permittivity/dielectric constant is due to the increase in interfacial polarization, which is more in case of the CNS filler compared to GNP

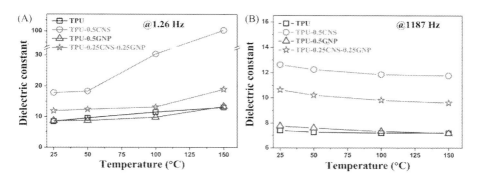

Figure 5.16 Dielectric constant at (A) 1.26 and (B) 1187 Hz for TPU and its composites at various temperatures.
Source: Reprinted with permission from Composites Part B: Engineering 166 (2019): 673–680. Copyright (2019) Elsevier.

as a result of its branched structure. The TPU composites shows temperature dependent dielectric properties, which indicate their potential application temperature sensing [66].

5.3.5 Effect of external applied pressure

The dielectric properties of polymer composites are affected by the external applied pressure. The change in resistance and capacitance (dielectric property) may depend on the type and nature of polymer and filler type. In case of conducting polymer composites, the electrical conduction mainly depends on the electrons/holes, where conductivity increases with the increase in external applied stress/pressure. However, the electrical conductivity of insulating polymer composites increases with the external stress/pressure, where the electrical conduction is due to the ionic charge carriers [2]. C. Feng et al. have prepared solvent crystallization-induced porous polyurethane/graphene (PU/G) composite foams for pressure sensing application. The prepared composite foam can be attached to detect body motions such as walking, finger bending, and jumping, which renders its applications in various areas for pressure sensing. As developed PU/G pressure sensor has demonstrated sensing capabilities in monitoring body motions from subtle actions to vigorous activities. Fig. 5.17A shows the compressibility and recoverability of PU/G foam which is later places in between two aluminum electrodes with the support of polyethylene terephthalate (PET) sheets to make the complete sensor device (Fig. 5.17B). A finger press test was conducted on the as-prepared pressure sensor to investigate the real-time current changes ($\Delta I/I_0$) and response time (Fig. 5.17C). There is change in current once the finger touched on the sensor and decreases upon release. Fig. 5.17D shows the schematic and mechanism illustration of the structural changes of PU/G foam with the applied stress. The PU/G sensors can also be used to detect the finger joint bending (finger force)

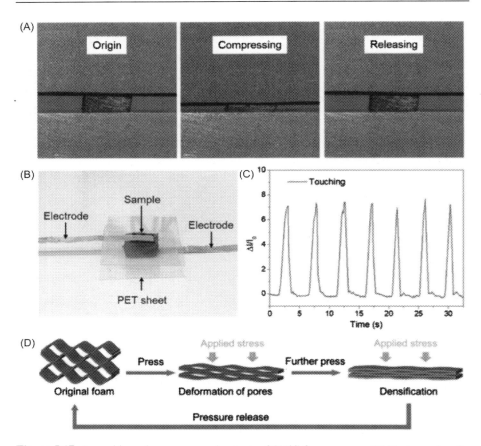

Figure 5.17 Assembly and pressure sensing tests of PU/G foam sensor. (A) Photographs of PU/G foams during compressing and releasing process. (B) Assembly of pressure sensor with aluminum electrodes and supporting PET sheets. (C) Finger touching test of the fabricated PU/G pressure sensor (30 wt.% graphene). The current changes maintain at around 0 without touching and increases quickly upon touching. (D) Schematic and mechanism illustration of the structural changes of PU/G foam with applied stress.
Source: Reprinted with permission from Composites Part B: Engineering 194 (2020): 108065. Copyright (2020) Elsevier.

(Fig. 5.18A), walking and jumping (Fig. 5.8B and C) by real-time monitoring current signals. Moreover, jumping can achieve higher current signal (~ 45 mA) compared to walking (~ 30 mA) because the momentum of falling down human body has increased the pressure applied on the PU/G sensor. These results suggest that PU/G foam sensor can be used in various areas such as helping athletes to improve their exercise efficiency and assist patients to monitor their fitness, health and rehabilitation [67].

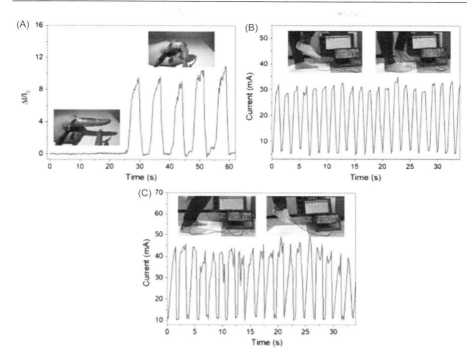

Figure 5.18 Applications of PU/G foam pressure sensor (30 wt.% graphene) for body motion recognition. Current signal changes when detecting finger bending (A), walking (B) and jumping (C). The insets in a are photographs of the assembly of PU/G sensor on a finger with straight and 90° bending. The insets in b and c are the assembly of PU/G sensor at the bottom of foot and two states of walking and jumping. The person for walking and jumping test has a weight of 45 kg.
Source: Reprinted with permission from Composites Part B: Engineering 194 (2020): 108065. Copyright (2020) Elsevier.

5.4 Summary and conclusion

This chapter focused on graphene-based polymer composites for dielectric applications. Here, we discussed the effect of different factors on the dielectric properties of polymer-graphene composites. Graphene based polymer composites/nanocomposites are promising materials which have various applications in the electronics field. These materials can be used in temperature and pressure sensors. Depending upon the application, suitable polymer matrix and preparation methods/process need to be selected. Processing conditions also plays an important role in making the right end materials for any application. The composite property can be improved by the increase in graphene loading and the proper dispersion/distribution of graphene in the polymer matrix.

References

[1] M. Fazeli, P.F. Jennifer, A.S. Renata, Improvement in adhesion of cellulose fibers to the thermoplastic starch matrix by plasma treatment modification, Compos. Part. B: Eng. 163 (2019) 207−216.

[2] S. Nayak, Dielectric Properties of Polymer−Carbon Composites, Carbon-Containing Polymer Composites, Springer, Singapore, 2019, pp. 211−234.

[3] R. Elhajjar, L.S. Valeria, M. Anastasia (Eds.), Smart Composites: Mechanics and Design, CRC Press, 2013.

[4] S. Nayak, B. Sahoo, D. Khastgir, Flexible nanocomposites comprised of poly (dimethylsiloxane) and high-permittivity TiO_2 nanoparticles doped with La^{3+}/Cu^+ for dielectric applications, ACS Appl. Nano Mater. 2 (7) (2019) 4211−4221.

[5] S. Nayak, Y. Li, W. Tay, E. Zamburg, D. Singh, C. Lee, et al., Liquid-metal-elastomer foam for moldable multi-functional triboelectric energy harvesting and force sensing, Nano Energy 64 (2019) 103912.

[6] R. Manna, S. Nayak, M. Rahaman, D. Khastgir, Effect of annealed titania on dielectric and mechanical properties of ethylene propylene diene monomer-titania nanocomposites, e-Polymers 14 (4) (2014) 267−275.

[7] S. Nayak, T.K. Chaki, D. Khastgir, Dielectric relaxation and viscoelastic behavior of polyurethane−titania composites: dielectric mixing models to explain experimental results, Polym. Bull. 74 (2) (2017) 369−392.

[8] E. Manias, Polymer nanocomposite technology, fundamentals of barrier, J. Nat. Mater. 6 (2007) 9−11.

[9] E. Manias, Stiffer by design, Nat. Mater. 6 (1) (2007) 9−11.

[10] S. Nayak, T.K. Chaki, D. Khastgir, Development of flexible piezoelectric poly (dimethylsiloxane)−$BaTiO_3$ nanocomposites for electrical energy harvesting, Ind. Eng. Chem. Res. 53 (39) (2014) 14982−14992.

[11] L. Ramajo, S.C. Miriam, M.R. Maria, Dielectric response of Ag/$BaTiO_3$/epoxy nanocomposites, J. Mater. Sci. 45 (1) (2010) 106−111.

[12] D.P. Dharaiya, S.C. Jana, Nanoclay-induced morphology development in chaotic mixing of immiscible polymers, J. Polym. Sci. Part. B: Polym. Phys. 43 (24) (2005) 3638−3651.

[13] Y. Rao, C.P. Wong, J. Qu, T. Marinis, Effective dielectric constant prediction of polymer-ceramic composite based on self-consistent theory, in: Proceedings of the 50th Electronic Components and Technology Conference (Cat. No. 00CH37070). IEEE, 2000.

[14] S. Nayak, M. Rahaman, A.K. Pandey, D.K. Setua, T.K. Chaki, D. Khastgir, Development of poly (dimethylsiloxane)−titania nanocomposites with controlled dielectric properties: effect of heat treatment of titania on electrical properties, J. Appl. Polym. Sci. 127 (1) (2013) 784−796.

[15] N. Setter, R. Waser, Electroceramic materials, Acta Mater. 48 (1) (2000) 151−178.

[16] Q.T. Nguyen, D.G. Baird, Preparation of polymer−clay nanocomposites and their properties, Adv. Polym. Technol. J. Polym. Process. Inst. 25 (4) (2006) 270−285.

[17] M. Moniruzzaman, K.I. Winey, Polymer nanocomposites containing carbon nanotubes, Macromolecules 39 (16) (2006) 5194−5205.

[18] F.L. Beyer, N.C.B. Tan, A. Dasgupta, M.E. Galvin, Polymer − layered silicate nanocomposites from model surfactants, Chem. Mater. 14 (7) (2002) 2983−2988.

[19] N.G. Sahoo, S. Rana, J.W. Cho, L. Li, S.H. Chan, Polymer nanocomposites based on functionalized carbon nanotubes, Prog. Polym. Sci. 35 (7) (2010) 837–867.

[20] P. Chahal, R.R. Tummala, M.G. Allen, M. Swaminathan, A novel integrated decoupling capacitor for MCM-L technology, IEEE Trans. Compon. Packag. Manuf. Technol. Part. B 21 (2) (1998) 184–193.

[21] Y. Zhang, T. Pan, Z. Yang, Flexible polyethylene terephthalate/polyaniline composite paper with bending durability and effective electromagnetic shielding performance, Chem. Eng. J. 389 (2020) 124433.

[22] J. Sun, L. Wang, Q. Yang, Y. Shen, X. Zhang, Preparation of copper-cobalt-nickel ferrite/graphene oxide/polyaniline composite and its applications in microwave absorption coating, Prog. Org. Coat. 141 (2020) 105552.

[23] S. Iqbal, H. Khatoon, R.K. Kotnala, S. Ahmad, Mesoporous strontium ferrite/polythiophene composite: influence of enwrappment on structural, thermal, and electromagnetic interference shielding, Compos. Part. B: Eng. 175 (2019) 107143.

[24] J. Thekkedath, P.K. Bipinbal, T. Thomas, S.K. Narayanankutty, Polythiophene coated cellulosic fibers from banana stem for improved electrical, mechanical, thermal and dielectric properties of polypropylene composites, J. Sci. Res. 12 (4) (2020) 687–699.

[25] S.P. Raghunathan, S. Narayanan, A.C. Poulosed, R. Joseph, Flexible regenerated cellulose/polypyrrole composite films with enhanced dielectric properties, Carbohydr. Polym. 157 (2017) 1024–1032.

[26] R. Chaturvedi, R.K. Gupta, N.R. Gorhe, P. Tyagi, Percolative polyurethane-polypyrrole-straw composites with enhanced dielectric constant and mechanical strength, Compos. Part. A: Appl. Sci. Manuf. 131 (2020) 105810.

[27] X.-J. Zhang, G.-S. Wang, Y.-Z. Wei, L. Guo, M.-S. Cao, Polymer-composite with high dielectric constant and enhanced absorption properties based on graphene–CuS nanocomposites and polyvinylidene fluoride, J. Mater. Chem. A 1 (39) (2013) 12115–12122.

[28] J. Luo, H.D. Jang, J. Huang, Effect of sheet morphology on the scalability of graphene-based ultracapacitors, ACS Nano 7 (2) (2013) 1464–1471.

[29] K.S. Kim, Y. Zhao, H. Jang, S.Y. Lee, J.M. Kim, K.S. Kim, et al., Large-scale pattern growth of graphene films for stretchable transparent electrodes, Nature 457 (7230) (2009) 706–710.

[30] F. Schedin, A.K. Geim, S.V. Morozov, E.W. Hill, P. Blake, M.I. Katsnelson, et al., Detection of individual gas molecules adsorbed on graphene, Nat. Mater. 6 (9) (2007) 652–655.

[31] J.L. Suter, R.C. Sinclair, P.V. Coveney, Principles governing control of aggregation and dispersion of graphene and graphene oxide in polymer melts, Adv. Mater. 32 (36) (2020) 2003213.

[32] C. Soldano, A. Mahmood, E. Dujardin, Production, properties and potential of graphene, Carbon 48 (8) (2010) 2127–2150.

[33] M.F. Mina, S. Seema, R. Matin, M.J. Rahaman, R.B. Sarker, M.A. Gafur, et al., Improved performance of isotactic polypropylene/titanium dioxide composites: effect of processing conditions and filler content, Polym. Degrad. Stab. 94 (2) (2009) 183–188.

[34] S. Radhakrishnan, B.T.S. Ramanujam, A. Adhikari, S. Sivaram, High-temperature, polymer–graphite hybrid composites for bipolar plates: effect of processing conditions on electrical properties, J. Power Sources 163 (2) (2007) 702–707.

[35] S. Iannace, R. Ali, L. Nicolais, Effect of processing conditions on dimensions of sisal fibers in thermoplastic biodegradable composites, J. Appl. Polym. Sci. 79 (6) (2001) 1084−1091.

[36] B. Alcock, N.O. Cabrera, N.M. Barkoula, T. Peijs, The effect of processing conditions on the mechanical properties and thermal stability of highly oriented PP tapes, Eur. Polym. J. 45 (10) (2009) 2878−2894.

[37] R. Ali, S. Iannace, L. Nicolais, Effect of processing conditions on mechanical and viscoelastic properties of biocomposites, J. Appl. Polym. Sci. 88 (7) (2003) 1637−1642.

[38] W. Yang, E. Fortunati, F. Dominici, J.M. Kenny, D. Puglia, Effect of processing conditions and lignin content on thermal, mechanical and degradative behavior of lignin nanoparticles/polylactic (acid) bionanocomposites prepared by melt extrusion and solvent casting, Eur. Polym. J. 71 (2015) 126−139.

[39] P.V. Joseph, K. Joseph, S. Thomas, Effect of processing variables on the mechanical properties of sisal-fiber-reinforced polypropylene composites, Compos. Sci. Technol. 59 (11) (1999) 1625−1640.

[40] S. Solarski, M. Ferreira, E. Devaux, G. Fontaine, P. Bachelet, S. Bourbigot, et al., Designing polylactide/clay nanocomposites for textile applications: effect of processing conditions, spinning, and characterization, J. Appl. Polym. Sci. 109 (2) (2008) 841−851.

[41] R.G. Raj, B.V. Kokta, The effect of processing conditions and binding material on the mechanical properties of bagasse fibre composites, Eur. Polym. J. 27 (10) (1991) 1121−1123.

[42] N.M. Stark, L.M. Matuana, C.M. Clemons, Effect of processing method on surface and weathering characteristics of wood−flour/HDPE composites, J. Appl. Polym. Sci. 93 (3) (2004) 1021−1030.

[43] H. Takagi, A. Asano, Effects of processing conditions on flexural properties of cellulose nanofiber reinforced "green" composites, Compos. Part. A: Appl. Sci. Manuf. 39 (4) (2008) 685−689.

[44] H. Tang, X. Chen, Y. Luo, Electrical and dynamic mechanical behavior of carbon black filled polymer composites, Eur. Polym. J. 32 (8) (1996) 963−966.

[45] M.J. Martínez-Morlanes, F.J. Pascual, G. Guerin, Influence of processing conditions on microstructural, mechanical and tribological properties of graphene nanoplatelet reinforced UHMWPE, J. Mech. Behav. Biomed. Mater. 115 (2020) 104248.

[46] Hui Xu, et al., Influence of processing conditions on dispersion, electrical and mechanical properties of graphene-filled-silicone rubber composites, Compos. Part. A: Appl. Sci. Manuf. 91 (2016) 53−64.

[47] H. Xu, L.X. Gong, X. Wang, L. Zhao, Y.B. Pei, Enhanced electrical and electromagnetic interference shielding properties of polymer−graphene nanoplatelet composites fabricated via supercritical-fluid treatment and physical foaming, ACS Appl. Mater. Interfaces 10 (36) (2018) 30752−30761.

[48] H. Zhang, G. Zhang, M. Tang, L. Zhou, J. Li, X. Fan, Synergistic effect of carbon nanotube and graphene nanoplates on the mechanical, electrical and electromagnetic interference shielding properties of polymer composites and polymer composite foams, Chem. Eng. J. 353 (2018) 381−393.

[49] H. Zhang, G. Zhang, J. Li, X. Fan, Z. Jing, J. Li, Lightweight, multifunctional microcellular PMMA/Fe$_3$O$_4$@MWCNTs nanocomposite foams with efficient electromagnetic interference shielding, Compos. Part. A: Appl. Sci. Manuf. 100 (2017) 128−138.

[50] C. Yang, S.J. Hao, S.L. Dai, X.Y. Zhang, Nanocomposites of poly (vinylidene fluoride)-Controllable hydroxylated/carboxylated graphene with enhanced dielectric performance for large energy density capacitor, Carbon 117 (2017) 301−312.

[51] Z.M. Dang, Y.H. Lin, C.W. Nan, Novel ferroelectric polymer composites with high dielectric constants, Adv. Mater. 15 (19) (2003) 1625—1629.

[52] T. Zhou, J.W. Zha, R.Y. Cui, B.H. Fan, J.-K. Yuan, Z.-M. Dang, Improving dielectric properties of $BaTiO_3$/ferroelectric polymer composites by employing surface hydroxylated $BaTiO_3$ nanoparticles, ACS Appl. Mater. Interfaces 3 (7) (2011) 2184—2188.

[53] L. Shaohui, Z. Jiwei, W. Jinwen, X. Shuangxi, Z. Wenqin, Enhanced energy storage density in poly (vinylidene fluoride) nanocomposites by a small loading of suface-hydroxylated $Ba_{0.6}Sr_{0.4}TiO_3$ nanofibers, ACS Appl. Mater. Interfaces 6 (3) (2014) 1533—1540.

[54] X. Li, Y.F. Lim, K. Yao, F.E.H. Tay, K.H. Seah, Ferroelectric poly (vinylidene fluoride) homopolymer nanotubes derived from solution in anodic alumina membrane template, Chem. Mater. 25 (4) (2013) 524—529.

[55] G. Hu, F. Gao, J. Kong, S. Yang, Q. Zhang, Z. Liu, et al., Preparation and dielectric properties of poly (vinylidene fluoride)/$Ba_{0.6}Sr_{0.4}TiO_3$ composites, J. Alloy. Compd. 619 (2015) 686—692.

[56] C.V. Chanmal, J.P. Jog, Dielectric relaxations in PVDF/$BaTiO_3$ nanocomposites, Express Polym. Lett. 2 (4) (2008) 294—301.

[57] M. Li, X. Huang, C. Wu, H. Xu, P. Jiang, T. Tanaka, Fabrication of two-dimensional hybrid sheets by decorating insulating PANI on reduced graphene oxide for polymer nanocomposites with low dielectric loss and high dielectric constant, J. Mater. Chem. 22 (44) (2012) 23477—23484.

[58] Z.M. Dang, L. Wang, Y.I. Yin, Q. Zhang, Q.-Q. Lei, Giant dielectric permittivities in functionalized carbon-nanotube/electroactive-polymer nanocomposites, Adv. Mater. 19 (6) (2007) 852—857.

[59] X. Huang, P. Jiang, L. Xie, Ferroelectric polymer/silver nanocomposites with high dielectric constant and high thermal conductivity, Appl. Phys. Lett. 95 (24) (2009) 242901.

[60] L. Chu, Q. Xue, J. Sun, F. Xia, W. Xing, D. Xia, et al., Porous graphene sandwich/poly (vinylidene fluoride) composites with high dielectric properties, Compos. Sci. Technol. 86 (2013) 70—75.

[61] Y. Bai, Z.Y. Cheng, V. Bharti, H.S. Xu, High-dielectric-constant ceramic-powder polymer composites, Appl. Phys. Lett. 76 (25) (2000) 3804—3806.

[62] X. Xu, C. Yang, J. Yang, T. Huang, N. Zhang, Y. Wang, et al., Excellent dielectric properties of poly (vinylidene fluoride) composites based on partially reduced graphene oxide, Compos. Part. B: Eng. 109 (2017) 91—100.

[63] M. Tian, J. Zhang, L. Zhang, S. Liu, X. Zan, T. Nishi, Graphene encapsulated rubber latex composites with high dielectric constant, low dielectric loss and low percolation threshold, J. Colloid Interface Sci. 430 (2014) 249—256.

[64] S.K. Hong, S.M. Song, O. Sul, B.J. Cho, Carboxylic group as the origin of electrical performance degradation during the transfer process of CVD growth graphene, J. Electrochem. Soc. 159 (4) (2012) K107.

[65] J.S. Meena, M.C. Chu, R. Singh, C.S. Wu, U. Chand, H.-C. You, et al., Polystyrene-block-poly (methylmethacrylate) composite material film as a gate dielectric for plastic thin-film transistor applications, RSC Adv. 4 (36) (2014) 18493—18502.

[66] K. Ke, M. McMaster, W. Christopherson, K.D. Singer, I. Manas-Zloczower, Effects of branched carbon nanotubes and graphene nanoplatelets on dielectric properties of thermoplastic polyurethane at different temperatures, Compos. Part. B: Eng. 166 (2019) 673—680.

[67] C. Feng, Z. Yi, X. Jin, S.M. Seraji, Y. Dong, L. Kong, et al., Solvent crystallization-induced porous polyurethane/graphene composite foams for pressure sensing, Compos. Part. B: Eng. 194 (2020) 108065.

Thermal properties of polymer-graphene composites

6

Subhendu Bhandari[1] and Mostafizur Rahaman[2]
[1]Department of Plastic and Polymer Engineering, Maharashtra Institute of Technology, Aurangabad, India, [2]Department of Chemistry, College of Science, King Saud University, Riyadh, Saudi Arabia

6.1   Introduction

Graphene is widely used in polymer composites primarily for different applications depending on its ability to improve mechanical strength [1,2], gas barrier property [3,4], electrical conductivity [5,6], electrochemical behavior [7−9], etc. However, the thermal properties of such composites may influence its stability to perform in service temperature environment. Therefore the thermal degradation characteristics are considered an inevitable aspect of related research.

Thermogravimetry analysis (TGA) reveals the information of thermal stability as well as the percent compositions of the constituents of polymer composites. The differential of TGA curve reveals the change of degradation rate with clearer information about temperatures at the start and end of degradation steps in case of possible apparent overlays of TGA curves. TGA may be carried out in inert (N_2) or oxidizing (O_2) atmosphere. Degradation in oxidizing environment is initiated at a lower temperature, and involves degradation of both carbonaceous filler viz. graphene or its derivatives and the polymer matrix. Glass transition and melting temperatures may be identified from differential scanning calorimetry (DSC) analysis (Fig. 6.1) [10].

Use of graphene in composites may contain single polymer or the blends based on polyester [11], poly(allylamine) [12], polyvinylidene fluoride (PVDF) [13], polyimide [14], poly(vinyl chloride) [15], polyaniline [16], polybenzimidazole [17], phenolic resin [18], ethylene-propylene-diene rubber [19], natural rubber [20], cyanate ester resin [21], poly(butylene terephthalate) [22], polysulfone [23], styrene methylmethacrylate [24], etc. Nevertheless, graphene may also be used in the presence of other filler systems like carbon nanotube [25], zinc ferrite [26], magnetite, [27] etc. For improvement of important composite-forming parameters viz. polymer-filler interaction, filler dispersion, etc., graphene may also be functionalized prior to or during preparation of the composites. Different approaches of modifications of graphene may include the use of octadecyl amine [28], *p*-phenylenediamine [29], polyethyleneimine [30], etc. Moreover, thermal degradation characteristics may differ with the changes of composite preparation techniques

Polymer Nanocomposites Containing Graphene. DOI: https://doi.org/10.1016/B978-0-12-821639-2.00014-8

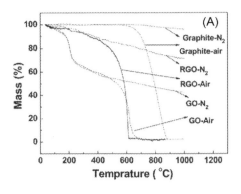
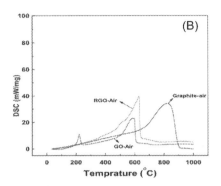

Figure 6.1 Comparative (A) TGA and curves of graphene oxide, reduced graphene oxide, and graphite in air and inert (N_2) environment, and (B) differential scanning calorimetry curves of respective samples. *TGA, thermogravimetry analysis.*
Source: Reproduced (adapted) with permission from Y. He, N. Zhang, F. Wu, F. Xu, Y. Liu, J. Gao, Graphene oxide foams and their excellent adsorption ability for acetone gas, Mater. Res. Bull. 48 (2013) 3553−3558.

viz. melt processing [31], in-situ polymerization [32], solution-phase dispersion [33], etc.

6.2   Thermal properties of unmodified graphene-based composite with single polymer

The unmodified graphene-based nanocomposites using single polymer have been extensively investigated by researchers. Graphene-based elastomeric composites may be prepared via latex stage dispersion [34]. Dong et al. [35] dispersed graphene (GE) in natural rubber (NR) latex (60wt.% dry rubber content) at different loading up to 1wt.%. The degradation temperatures of 10% and 90% weight loss as well as the melting temperatures increased with the increase of graphene loading (Table 6.1).

Moreover, the kinetic analysis of degradation by using Friedman [36,37] and Kissinger-Akahira-Sunose (KAS) model [38,39] revealed about higher activation energy values of degradation (E_a) for all graphene-filled composites in comparison to the unfilled natural rubber. The barrier effect of graphene nanosheets and its free radical scavenging ability may be attributed to the improvement of activation energy of degradation. Migration of polymer in the composites to the surface is delayed or restricted because of its confinement into lamellar structures of graphene with high specific surface area as well as some amount of polymers also get adsorbed on the graphene layers. Moreover, during thermal chain scission of polymer, free radicals are generated which are then removed while migrating to surface and also greatly affected with restricted polymer chain mobility with the increase of

Table 6.1 Degradation temperatures of natural rubber/ graphene composites at different loading.

Graphene loading in NR latex (wt.%)	Temperature of 10% weight loss, $T_{10\%}$ (°C)	Temperature of 10% weight loss, $T_{90\%}$ (°C)	T_m (°C)
0	343.3	441.8	370.3
0.1	346.5	446.0	373.0
0.3	350.5	449.0	376.0
0.5	351.0	454.0	376.5
1.0	352.5	460.5	379.0

Source: Reproduced (adapted) with permission from Z. Dong, F. Cai, Z. Jiang, W. Xu, Thermal property studies of in situ blended graphene/nature rubber nanocomposites, Int. J. Polym. Sci. 2020 (2020).

graphene loading. At lower loading of graphene radical scavenging and gas inhibition predominates which leads to improvement of activation energy. However, beyond certain loading the trend of change of activation energy might be reversed because of faster degradation of polymer confined between graphene layers at increased heat transfer for higher graphene loading [40]. Wang et al. [41] reported on similar findings for poly(butylene succinate) matrix where the increase of loading of graphene nanosheet beyond 2wt.% resulted in lower temperature of maximum weight loss in contrast to the pristine polymer.

Tarani et al. [42] compared the effect of graphene nanoplatelet diameter on the degradation characteristics of its composites with high density polyethylene (HDPE). At the constant loading of 2.5wt.%, the graphene nanoplatelets with a diameter of 5, 15, and 25 μm exhibited the activation energy values of 205.3, 210.8, and 217.8 kJ/mol respectively, whereas the value was found to be 200.2 kJ/mol for the pristine polymer considering autocatalysis mechanism of degradation of n-th order. The improvement of activation energy may be attributed to the nanoconfinement effect of HDPE in between graphene layers, as suggested by Chen et al. [43,44]. With the addition of graphene nanoplatelets, conformational entropy decreases on account of stretching of the chains of HDPE matrix. Therefore the nanolayers are repelled by the stretched polymer chains which create locally confined zones where the regular coil conformation gets interrupted [45,46]. As a consequence, chain mobility of locally confined polymer gets restricted and its chemical activities as well as thermal degradation are retarded.

Bhawal et al. [47] also reported on the competing factors for graphene loading controlling thermal stability (Table 6.2), that is, (1) graphene layers act as a physical barrier and reduce degradation, and (2) the presence of oxygen-containing functional groups affect thermal stability adversely. Both of these two factors apparently influenced thermal degradation exhibiting improvement of onset and maximum temperature of degradation for up to 5wt.% of filler loading, followed by a reverse trend at higher extent of incorporation of graphene in EMA matrix. Initial improvement of glass transition temperature for the composite with 1wt.% filler loading in contrast to the pristine polymer may correspond to the restricted chain mobility of polymer matrix owing to the hydrogen bonding and polar-polar interaction between

Table 6.2 Effect of graphene loading on thermal degradation as well as T_g and T_m of polyethylene methacrylate.

Filler loading (wt. %)	Onset temperature of degradation (°C)	Maximum temperature of degradation (°C)	T_g (°C)	T_m (°C)
0	385	506	−34	83
1	362	505	−31	84
3	383	508	−30	85
5	389	511	−28	85
7	385	509	−30	84

Source: Reproduced (adapted) with permission from P. Bhawal, S. Ganguly, T.K. Chaki, N.C. Das, Synthesis and characterization of graphene oxide filled ethylene methyl acrylate hybrid nanocomposites, RSC Adv. 6 (2016) 20781−20790.

oxygenated functional groups existing on graphene layers and the acrylate functional group of the matrix. However, no significant change in the composites was found for higher filler loading. Likewise, the same phenomena were expected to be responsible for initial slight improvement of the melting point also.

A similar observation was observed by Wang et al. [41] for graphene-reinforced poly(butylene succinate) composite, prepared via solvent casting using chloroform medium. An increase of 16°C for 5% degradation was observed for 2wt.% loading. Derivative of degradation revealed faster degradation rate for 2wt.% filler loading which presumably may be assigned to high heat conductivity of graphene nanosheets. However, melting temperature was found to increase marginally for 2wt.% loading (109.2°C) in contrast to the pristine polymer matrix (108.1°C).

Pham et al. [48] studied the effect of reduced graphene oxide loading in poly-methyl methacrylate prepared via dispersion in latex stage. Glass transition temperature, as determined from dynamic mechanical analysis, increased about 15°C for 0.5wt.% filler loading compared to pristine PMMA (Fig. 6.2). However, significant change of T_g was not found for further addition of reduced graphene oxide (Fig. 6.2A). An increase of onset temperature of degradation about 10°C was observed for 1wt.% filler loading followed by slight decrease. Initial improvement of thermal stability may be attributed to the network formation of reduced graphene oxide of high aspect ratio which acts as a physical barrier, and thus thermal degradation is inhibited [3,49]. Formation of network may also be triggered by temperature [50].

Eswaraiah et al. [51] prepared a composite in a single step via in situ reduction of graphite oxide, mixed with PVDF in solid state. The mixture was then exposed to focus on solar radiation using converging lens which led to simultaneous exfoliation and photoreduction of filler to form graphene oxide as well as melting of PVDF matrix. The composite, containing 2.2wt.% of reduced graphite oxide exhibited an increase of onset temperature of degradation by ca. 35°C, however, the degradation behavior of the composite and the bare PVDF were found quite similar beyond 500°C.

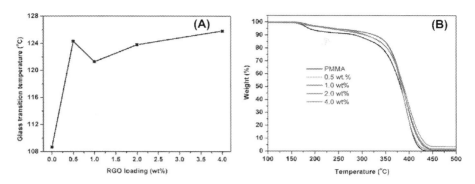

Figure 6.2 Influence of reduced graphene oxide loading on (A) glass transition temperature and (B) thermal degradation behavior.
Source: Reproduced (adapted) with permission from V.H. Pham, T.T. Dang, S.H. Hur, E.J. Kim, J.S. Chung, Highly conductive poly(methyl methacrylate) (PMMA)-reduced graphene oxide composite prepared by self-assembly of PMMA latex and graphene oxide through electrostatic interaction, ACS Appl. Mater. Interf. 4 (2012) 2630−2636.

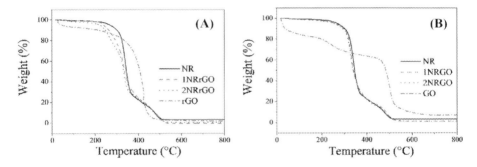

Figure 6.3 Thermogravimetry analysis curves of pristine natural rubber and its composites with (A) reduced graphene oxide and (B) graphene oxide at 1 and 2wt.% of filler loading, carried out in synthetic air atmosphere.
Source: Reproduced (adapted) with permission from C.F. Matos, F. Galembeck, A.J.G. Zarbin, Multifunctional and environmentally friendly nanocomposites between natural rubber and graphene or graphene oxide, Carbon N.Y. 78 (2014) 469−479.

The work of Matos et al. [52] using unmodified graphene oxide and reduced graphene oxide, dispersed in natural rubber latex using surfactant cetyltrimethylammonium bromide (CTAB), has given an insight in comparison to thermal stability in oxidizing atmosphere by using graphene oxide and reduced graphene oxide, that is, graphene. Degradation started for graphene-filled composites earlier than that of pristine natural rubber (less than 300°C) which might correspond to decomposition of CTAB in the range of 260°C−310°C (Fig. 6.3). However, the onset temperatures of the graphene-oxide-filled composites are quite similar to pristine natural rubber signifying no substantial improvement of thermal stability due to the presence of

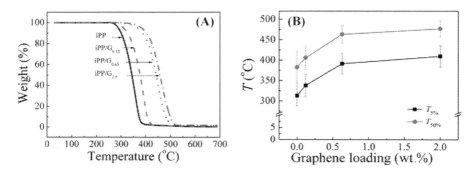

Figure 6.4 (A) Thermogravimetry analysis curves of in-situ synthesized isotactic polypropylene/ graphene composites, and (B) variation of the temperatures of 5% and 50% weight loss of the composites. The error bars signify standard errors.
Source: Reproduced (adapted) with permission from S. Zhao, F. Chen, Y. Huang, J.Y. Dong, C.C. Han, Crystallization behaviors in the isotactic polypropylene/graphene composites, Polymer. 55 (2014) 4125−4135.

graphene oxide in the rubber matrices. The general trend of reduced thermal stability, influenced by initial decomposition of carbonaceous filler-based polymer composites [53,54], was not observed in the investigation of Gkourmpis et al. [55]. In that work, commercially available reduced graphene oxide was mixed with polypropylene of >90% tacticity (iPP) using Brabender internal mixer at 20 rpm. for 15 min at 210°C. The increasing trend of thermal stability even at very low loading was evident from the investigation. Initiation of thermal decomposition prevails on the surface of unfilled polymer, whereas, the presence of carbonaceous fillers facilitates transferring heat from the surface to the bulk more efficiently, thus leading to degradation occurring both on the surface and in the bulk of the samples.

A similar investigation on isotactic polypropylene (iPP)/ graphene composite was reported by Zhao et al. [56] The composites were prepared Ziegler-Natta polymerization on to graphene-supported catalyst, followed by in-situ polymerization for synthesis of iPP. The temperatures of 5% and 50% weight reduction, that is, $T_{5\%}$ and $T_{50\%}$ increased noticeably than the pristine matrix by 97°C and 94°C for 2wt.% of filler loading in contrast to 25°C and 23°C for 0.12wt.% of graphene content (Fig. 6.4). The improvement of thermal stability was found to be approaching plateau near 2% graphene content.

6.3 Thermal properties of modified graphene-based composite with single polymer

Ma et al. [57] used vinyl trimethoxy silane for modification of graphene oxide to anchor terminal vinyl groups on to its surfaces. The functionalized graphene oxide (FGO) was further mixed with two viscous components of silicone. The ultrasonic dispersion of FGO in tetrahydrofuran (THF) was mixed with Si-H compound (Mw = 73267) at high shearing followed by evaporation of THF in vacuum. Then it was mixed with the mixture of Si—CH = CH$_2$ compound (Mw = 73803) and chlor-oplatinic acid (5 ppm) at equimolar ratio. The final mixture was molded at 160°C for 2 h. The FGO loading of 0.5% in the silicone matrix resulted in the increase of the temperature of 5% weight loss by 26.1°C which indicates significant improvement of thermal stability at low loading of FGO. The improvement of thermal stability may occur because the uniformly dispersed FGO nanoplatelets (1) restrict the movement of polymer chains [58], as well as (2) act as physical barriers for degradation of volatile components [48,59]. Yu et al. [60] grafted octaaminophenyl polyhedral oligomeric silsesquioxanes (OapPOSS) with graphene and used in epoxy matrix. Thermogravimetric analysis exhibits synergistic improvement of initial decomposition temperature (394.5°C) in contrast to the pristine graphene oxide (75.2°C) and OapPOSS (176.2°C) counterparts. Fang et al. [61] carried out noncovalent functionalization of graphene foam (GF) via π-π stacking while coating it with polydopamine (PDA), followed by its covalent functionalization with 3-aminopropyltriethoxysilane (APTS). Infiltration of the resulting modified compressed GF yielded the composite c-GF/PDA/APTS/PDMS. Coating with PDA in the modified filler system revealed slower degradation as well as enhanced thermal stability compared to the unmodified graphene composite.

Sharmila et. al. [62] compared silane-modified graphene (SMG) with graphite, commercially available graphene (CG) and reduced graphene (CRG) using the matrix of epoxy resin Araldite GY 250 based on triethylenetetramine hardener (HY 951) and diglycidyl ether of bisphenol A (DGEBA) (epoxy equivalent 187 g/eq). Aqueous colloidal dispersion of graphene oxide was treated with ethylene diamine and 3-amino propyltriethoxysilane to prepare SMG. Composites were prepared ultrasonically in acetone medium followed by addition of 10 phr of hardener. Graphene oxide was found to be thermally unstable and experiences drastic weight loss at 190°C due to rapid thermal expansion, caused by pyrolysis of oxygen containing labile functional groups (Fig. 6.5). The presence of small extent of oxygen in CRG and SMG even after reduction and exfoliation was confirmed from a slight reduction of weight signifying the presence of functional groups in small extent. Silane modification was found to reduce thermal stability in comparison with commercially available graphene and graphene oxide. However, thermal stability of SMG was very similar to the use of CG. According to the investigation by Tung et al. [63], the composites of polyaniline and poly(sodium 4-styrenesulfonate) (PSS)-modified graphene exhibited significant improvement of degradation temperature from 360°C in the pristine polymer to 460°C. The investigation of Xu et al. [64] revealed a comparison between the thermal degradation of two different types of surface modification of graphene oxide in phenyl silicone rubber α,ω-dihydroxy poly dimethyl diphenyl siloxane) matrix. 4,4'-diphenylmethane diisocyanate (MDI) and silane coupling agent (KH550) were used for modification of filler by using ethanol and N,N-dimethylacetamide (DMAc) medium by ultrasonication. Dibutyltin dilaurate was used to catalyze the modification with MDI. Modification of graphene oxide with MDI resulted in shifting of onset of degradation toward higher

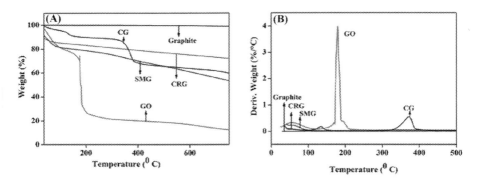

Figure 6.5 (A) Thermogravimetric analysis and (B) differential thermogravimetry graphs of epoxy composite with graphite, graphene oxide, silane-modified graphene, commercially available graphene, and reduced graphene.
Source: Reproduced (adapted) with permission from T.K. Bindu Sharmila, S. Sasi, N.R. Suja, P.M.S. Beegum, E.T. Thachil, A comparative investigation of aminosilane/ethylene diamine−functionalized graphene epoxy nanocomposites with commercial and chemically reduced graphene: static and dynamic mechanical properties, Emergent Mater. 2 (2019) 371−386.

temperature; however, the TGA graph follows approximately the similar trend of pristine filler. The weight loss of the phenyl silicone composite with the unmodified, silane coupling agent-modified and MDI-modified graphene oxide within the temperature range of 120°C−240°C were found to be 25%, 12% and 14% respectively, signifying the pyrolysis of surface oxygen functional groups. Treatment of filler with silane coupling agent resulted in significant reduction of weight loss compared to the unmodified as well as the MDI-modified filler which might correspond to the introduction of Si-O group on the surface of filler. Unlike single-step degradation in the other samples, modification of filler with MDI an additional step of degradation with the temperature range of 240°C−360°C ascribing to the degradation of the grafted functional groups. Wang et al. [65] prepared graphene-based organic-inorganic hybrid flame retardant via sol-gel reaction with phenyl-bis-(triethoxysilylpropyl) phosphamide (PBTP). The composites of the hybrid dispersed in epoxy resin exhibited improvement of thermal stability because of the presence of inorganic content compared to unmodified graphene oxide-based composite. However, both the composites based on unmodified graphene and the hybrid, even though having the presence of inorganic content in the hybrid filler, the onset temperatures of degradation were found to be significantly lower (290°C and 303°C respectively) in contrast to pristine epoxy resin (315°C) which may be ascribed to the pyrolysis of PBTP and oxygen-containing groups on unmodified graphene oxide and its hybrid. Wan et al. [66] used polyoxyethylene octyl phenyl ether-based nonionic surfactant Triton X-100 for preparation of modified graphene for improvement of its dispersion in epoxy resin. Epoxy resin needs hardener for curing [67], and for that purpose 4-methylhexahydrophthalic anhydride was used for diglycidyl ether of bisphenol-A at the resin to hardener ratio of 185:170 (w/w). Even though the degradation characteristics apparently appeared to be nearly identical, a closer observation revealed incremental improvement of thermal stability owing to the sheet barrier effect and higher heat capacity of graphene compared to the polymeric matrix [41]. It is noteworthy that, unlike previous many other cases, here decrease of onset temperature for degradation was not observed. Wang et al. [68] used PSS as surface modifying agent for graphene to prevent agglomeration during dispersion in poly(vinyl alcohol) matrix. A comparative study confirmed about improvement of thermal stability after incorporation of graphene as well as slight improvement of onset temperature of degradation for 1wt.% of reduced graphene oxide loading, whereas no such improvement was observed for incorporation of graphene oxide at the same loading. The restoration of the conjugated structure and thermal conductivity might cause such improvement of thermal stability for reduced graphene oxide-based composite. Chhetri et al. [69] used aqueous solution of L-glutathione as modifier of graphene which was reacted with exfoliated graphene oxide during its reduction. The modified filler was ultrasonically dispersed in the diglycidyl ether of bisphenol-A based epoxy (Lapox-B-11) and polyamide hardener composed of tall-oil fatty acids and triethylenetetramine (Trade name Lapox AH-713). Pristine epoxy exhibited 10% and 50% weight loss at 336°C and 386°C respectively in presence of air atmosphere, whereas, the corresponding degradation temperatures of the composites were found to be 338°C and 399°C respectively for both 0.25% and 0.5% filler loading. The improvement of stability of the composites toward oxidative degradation may be attributed to barrier property of the modified graphene nanosheets which possibly restricted the flow of oxygen into the composite, and thus impeded further oxidative degradation of the epoxy matrix.

6.4   Thermal properties of graphene-based composites with polymer blend

Xue et al. [70] followed dispersion of graphene oxide in latex stage using styrene butadiene rubber (SBR 1502) latex of with 23wt.% of styrene content and carboxylated acrylonitrile butadiene rubber (XNBR 6721) latex of 43wt.% of solid content. The latex was coagulated using NaCl and the solid compound was masticated in two roll mill. The composite, thus formed with incorporation of 15wt.% of graphene oxide exhibited increase of temperature of 50% and maximum weight loss by 11.41°C and 18.34°C respectively. In addition to the barrier effect of graphene oxide, the improvement of thermal stability was also explained by formation of interfacial adhesion between the nanosheets and rubber chains after exfoliation of graphene oxide. Bera et al. [71] used reduced graphene oxide as compatibilizing agent in the blend of PVDF and polyether polyol-based thermoplastic polyurethane (TPU), where thermal properties corroborated the compatibilizing characteristics of the filler for the incompatible PVDF/TPU blend. The TPU was synthesized by using poly(tetramethylene ether)glycol, 4,4'-methylene diphenyl diisocyanate (with [NCO]:[OH] ratio 1:1) and 1,4-butanediol (1,4-BD) at the proportion of 1:2:1. Reduced graphene oxide was dispersed in TPU ultrasonically by using N,N-dimethylformamide (DMF) solvent medium. The master batch was melt blended with PVDF at 70:30 (w/w) ratio of PVDF:TPU by using twin-screw extruder. Gradual improvement of compatibility was evident from reduction of melting temperature of with increasing loading of reduced graphene oxide, and the maximum improvement was observed for 0.3wt.% loading which exhibited a reduction of melting temperature by 27°C. The optimum loading (3wt.%) was also corroborated from the maximum reduction of crystallization temperature owing to the nucleation effect of the filler. With the addition of 0.5wt.% of filler loading the temperatures for 5% and 50% weight loss increased by 10.5°C and 16°C respectively. Several physicochemical interactions [72] as well as enhanced dissipation of thermal energy owing to the high thermal conductivity of the filler system [73] were expected to improve thermal stability of the composite.

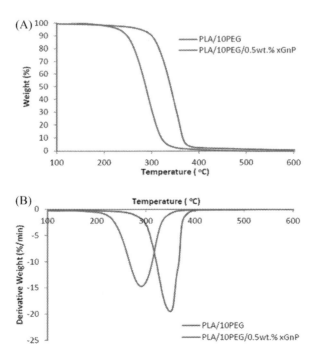

Figure 6.6 Effect of graphene nanoplatelets on thermal stability of poly(lactic acid)/ poly (ethylene glycol) blend: (A) Thermogravimetry analysis and (B) differential thermogravimetry curves.
Source: Reproduced (adapted) with permission from B.W. Chieng, N.A. Ibrahim, W.M.Z.W. Yunus, M.Z. Hussein, Poly(lactic acid)/poly(ethylene glycol) polymer nanocomposites: effects of graphene nanoplatelets, Polymers (Basel). 6 (2014) 93–104.

Table 6.3 Effects of loading of graphene oxide in polyvinyl alcohol/ starch composite on the degradation characteristics at different heating rates.

Loading of graphene oxide in polyvinyl alcohol/ starch blend	Heating rate (°C/ min)	Initial temperature (°C)	Temperature of maximum degradation (°C)	Weight loss (%)
0	5	173	410	4.12
	10	183	441	4.69
	15	187	448	4.80
	20	189	472	4.90
1	5	168	450	7.20
	10	175	470	7.68
	15	183	475	7.73
	20	186	491	7.85
2.5	5	165	471	14.15
	10	171	492	11.70
	15	177	496	11.09
	20	182	502	9.30
5	5	167	465	8.17
	10	173	480	8.43
	15	179	487	8.73
	20	184	495	8.82

Source: Reproduced (adapted) with permission from E. Sedaghat, A.A. Rostami, M. Ghaemy, A. Rostami, Characterization, thermal degradation kinetics, and morphological properties of a graphene oxide/poly (vinyl alcohol)/starch nanocomposite, J. Therm. Anal. Calorim. 136 (2019) 759–769.

Chieng et al. [74] dispersed graphene nanoplatelets in poly(lactic acid)/ poly(ethylene glycol) via melt processing at loading up to 1wt.% using Brabender internal mixer (temperature: 160°C, rotor rpm: 25, duration: 10 min). TGA and DTG curves of 0.5wt.% graphene-loaded composite has been illustrated in Fig. 6.6. Significant improvement of thermal stability was evident from TGA graph. The onset and maximum temperature of degradation was found to be 194.5°C and 291°C respectively for the unfilled polymer blend; whereas, the corresponding temperatures for the 0.5wt.% graphene-filled composite was elevated to 250.4°C and 344°C respectively. Considerable improvement of thermal stability with incorporation of graphene nanoplatelets could be attributed to the thermal stability of filler itself.

The investigation of Sedaghat et al. [75] for the composite of graphene oxide in polyvinyl alcohol/ starch blend revealed the increasing trend of the initial temperature and the temperature of maximum degradation with the increase of heating rate; whereas, the initial temperatures were found to follow a decreasing trend with the increase of filler loading (Table 6.3). In addition to the fact of transport barrier behavior of graphene oxide nanosheets, the formation of hydrogen bending between the blended matrix and the filler decreases the quality of fre -OH groups, and thus the matrix is stabilized.

6.5 Thermal properties of graphene-containing mixed filler-based polymer composite

The investigation of Azizi et al. [76] on the composites of graphene/ inorganic filler combinations using polyethylene propylene diene rubber (EPDM) matrix revealed that the presence of 2wt.% of graphene results in improvement of thermal stability in contrast to the use of individual inorganic fillers. The composites were prepared by using Haake Rheomix OS (internal mixer) at 100 rpm screw speed for 10 min at 60°C. The temperature of 5% degradation in inert condition increases by 24°C and 13°C, whereas in the presence of air it increases by 5°C and 3°C for the use of graphene in combination with modified fumed silica (MFS) (10wt.%) and titania (20wt.%) respectively. Similarly, the increase of the temperature for 50% degradation was also found to be improved by 7°C and 3°C at inert condition, whereas 2°C and 3°C at air atmosphere respectively was for the same loading of MFS and titania (TiO_2). The overall improvement of thermal degradation was ascribed to the formation of inorganic filler network, followed by its integration with graphene which enhanced a physical interaction between the additives and the polymeric matrix [77−81]. The use of the combination of 20wt.% of TiO_2 and 2wt.% of graphene in 50/50 (w/w) blend of EPDM/ Silicone rubber, as investigated by Azizi et al. [76], resulted in an increase of the temperature of 5% degradation by 22°C at inert condition and 28°C at air atmosphere respectively while compared to the use of only TiO_2. For the same compositions, the temperature of 50% weight loss also exhibited an increase by 9°C in air atmosphere, but met an unexpected decrease by 40°C in inert condition.

The investigation of Han et al. [82] revealed the effect of CeO_2 in graphene-filled phenyl methyl vinyl silicone (PMVQ) and methyl vinyl silicone (MVQ) rubber. Interfacial bond formation between nanofiller and polymer matrix can slow down thermal degradation by limiting the movement of polymer chain backbone. The presence of CeO_2 can restrict the movement of polymer chains with graphene. With addition of 2 phr CeO_2 and 0.8 phr graphene, onset temperature of thermal degradation was found to improve by 109°C and 78°C respectively for PMVQ and MVQ, whereas, the corresponding temperatures while using 1.5 phr graphene were found to be elevated by 108°C and 86°C respectively. Nevertheless, the formation of π-π stacking between phenyl methyl vinyl silicone rubber may also facilitate in improvement of thermal stability by reducing the exposure of matrix polymer chain tail, thus hindering the formation of intramolecular cyclic transition state.

Wang et al. [83] prepared attapulgite-graphene hybrid by in-situ addition of attapulgite solution in acetone medium to the graphene oxide dispersion. Epoxy resin (Araldite resin LY5052) was used as the matrix while added to the attapulgite-graphene hybrid dispersion, whereas Aradur hardener HY5052 was used as the curing agent. Major weight loss of graphene oxide was observed at ca. 200°C owing to pyrolysis of oxygen-containing functional groups [84,85]. The residue of graphene oxide at 250°C was found to be 70%, whereas attapulgite exhibited a much higher extent of residue (94%) even at 800°C (Fig. 6.7). Thermal stability of the

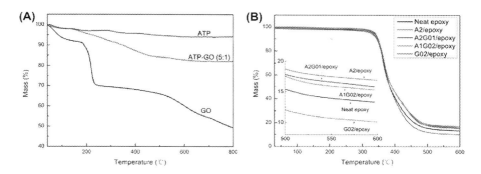

Figure 6.7 Thermogravimetric analysis curves of (A) the fillers: attapulgite, graphene oxide, and attapulgite-graphene oxide hybrid, and (B) epoxy resin-based composites with different fillers (inset: magnified curve within temperature range from 500°C to 600°C).
Source: Reproduced (adapted) with permission from R. Wang, Z. Li, W. Liu, W. Jiao, L. Hao, F. Yang, Attapulgite-graphene oxide hybrids as thermal and mechanical reinforcements for epoxy composites, Compos. Sci. Technol. 87 (2013) 29–35.

attapulgite-graphene oxide hybrid was enhanced in contrast to pristine graphene oxide leaving 82% residue at 800°C. Weight loss increased with introduction of 0.2wt.% graphene oxide [86,87]. Influence of filler on curing system was revealed to be of importance since the presence of functional groups on 0.2wt.% graphene oxide would be theoretically insufficient for a difference of 5% in residual weight at 600°C [88]. The presence of oxygen-containing groups, viz. carboxyl, epoxy etc., might react with the amine-based hardener to form a network structure of epoxy which may exhibit decrease in thermal degradation. However, the graphene nanoribbon-based hybrid with polyaniline in epoxy matrix, as studied by Joshi et al. [89], was not found to exhibit much change in degradation characteristics.

6.6 Conclusion

Thermal properties of graphene-filled polymer composites may be influenced by different mechanisms and preparation strategies. Dispersion of graphene in polymer matrix is influenced by different composite preparation methods and techniques; whereas, thermal degradation of composite is also dependent on dispersion of filler. The interaction of graphene with polymer matrix and possible π-π stacking may facilitate in improvement of thermal stability. Presence of oxygen-containing functional groups may lead to faster degradation. Moreover, different competing effects may improve or reduce thermal stability of the composites. Graphene may act as a physical barrier to gas which may hinder the thermally induced oxidation and enhance thermal stability. However, the high aspect ratio may also influence thermal stability by scavenging free radicals. High heat conductivity of graphene may also facilitate in transferring heat from the surface to the bulk which may lead to higher extent of degradation. To the contrary, high heat capacity may also lead to thermal energy dissipation leading to less degradation. Graphene may restrict movement of polymer chains owing to polymer-filler interactions. Beyond a certain level of graphene loading polymer chains confined between layers of graphene may degrade faster than the lower level of loading. Modification of graphene as well as the presence of other fillers or a number of matrix polymers may also be used to tune thermal degradation characteristics.

References

[1] B. Hui, L. Ye, X. Zhao, In situ preparation of polyurethane-imide/graphene oxide nano-composite foam: intercalation structure and thermal mechanical stability, J. Polym. Res. 25 (2018) 267.

[2] S. Rehman, S. Akram, A. Kanellopoulos, A. Elmarakbi, P.G. Karagiannidis, Development of new graphene/epoxy nanocomposites and study of cure kinetics, thermal and mechanical properties, Thermochim. Acta 694 (2020) 178785.

[3] H. Kim, A.A. Abdala, C.W. MacOsko, Graphene/polymer nanocomposites, Macromolecules 43 (2010) 6515−6530.

[4] M. Raef, M. Razzaghi-Kashani, The role of interface in gas barrier properties of styrene butadiene rubber-reduced graphene oxide composites, Polymer 182 (2019) 121816.

[5] R.M. Hodlur, M.K. Rabinal, Self assembled graphene layers on polyurethane foam as a highly pressure sensitive conducting composite, Compos. Sci. Technol. 90 (2014) 160−165.

[6] P. Kumar, S. Yu, F. Shahzad, S.M. Hong, Y.H. Kim, C.M. Koo, Ultrahigh electrically and thermally conductive self-aligned graphene/polymer composites using large-area reduced graphene oxides, Carbon 101 (2016) 120−128.

[7] S. Sahoo, G. Karthikeyan, G.C. Nayak, C.K. Das, Modified graphene/polyaniline nano-composites for supercapacitor application, Macromol. Res. 20 (2012) 415−421.

[8] X. Feng, N. Chen, J. Zhou, Y. Li, Z. Huang, L. Zhang, et al., Facile synthesis of shape-controlled graphene/polyaniline composites for high performance supercapacitor electrode materials, N. J. Chem. 39 (2015) 2261−2268.

[9] D. Ponnamma, Q. Guo, I. Krupa, M.A.S.A. Al-Maadeed, K.T. Varughese, S. Thomas, et al., Graphene and graphitic derivative filled polymer composites as potential sensors, Phys. Chem. Chem. Phys. 17 (2015) 3954−3981.

[10] Y. He, N. Zhang, F. Wu, F. Xu, Y. Liu, J. Gao, Graphene oxide foams and their excellent adsorption ability for acetone gas, Mater. Res. Bull. 48 (2013) 3553−3558.

[11] K. Liu, L. Chen, Y. Chen, J. Wu, W. Zhang, F. Chen, et al., Preparation of polyester/reduced graphene oxide composites via in situ melt polycondensation and simultaneous thermo-reduction of graphene oxide, J. Mater. Chem. 21 (2011) 8612−8617.

[12] A. Satti, P. Larpent, Y. Gun'Ko, Improvement of mechanical properties of graphene oxide/poly(allylamine) composites by chemical crosslinking, Carbon 48 (2010) 3376−3381.

[13] V. Eswaraiah, V. Sankaranarayanan, S. Ramaprabhu, Functionalized graphene-PVDF foam composites for EMI shielding, Macromol. Mater. Eng. 296 (2011) 894−898.

[14] J.Y. Wang, S.Y. Yang, Y.L. Huang, H.W. Tien, W.K. Chin, C.C.M. Ma, Preparation and properties of graphene oxide/polyimide composite films with low dielectric constant and ultrahigh strength via in situ polymerization, J. Mater. Chem. 21 (2011) 13569−13575.

[15] M.F. Shakir, A.N. Khan, R. Khan, S. Javed, A. Tariq, M. Azeem, et al., EMI shielding properties of polymer blends with inclusion of graphene nano platelets, Results Phys. 14 (2019) 102365.

[16] Z. Gao, F. Wang, J. Chang, D. Wu, X. Wang, X. Wang, et al., Chemically grafted graphene-polyaniline composite for application in supercapacitor, Electrochim. Acta 133 (2014) 325−334.

[17] N. Üregen, K. Pehlivanoğlu, Y. Özdemir, Y. Devrim, Development of polybenzimidazole/graphene oxide composite membranes for high temperature PEM fuel cells, Int. J. Hydrog. Energy 42 (2017) 2636−2647.

[18] A.P. Singh, P. Garg, F. Alam, K. Singh, R.B. Mathur, R.P. Tandon, et al., Phenolic resin-based composite sheets filled with mixtures of reduced graphene oxide, γ-Fe$_2$O$_3$ and carbon fibers for excellent electromagnetic interference shielding in the X-band, Carbon 50 (2012) 3868−3875.

[19] B. Chen, N. Ma, X. Bai, H. Zhang, Y. Zhang, Effects of graphene oxide on surface energy, mechanical, damping and thermal properties of ethylene-propylene-diene rubber/petroleum resin blends, RSC Adv. 2 (2012) 4683−4689.

[20] S. Zhao, S. Xie, P. Sun, Z. Zhao, L. Li, X. Shao, et al., Synergistic effect of graphene and silicon dioxide hybrids through hydrogen bonding self-assembly in elastomer composites, RSC Adv. 8 (2018) 17813−17825.

[21] Q. Lin, L. Qu, Q. Lü, C. Fang, Preparation and properties of graphene oxide nanosheets/cyanate ester resin composites, Polym. Test 32 (2013) 330−337.

[22] P. Fabbri, E. Bassoli, S.B. Bon, L. Valentini, Preparation and characterization of poly (butylene terephthalate)/grapheme composites by in-situ polymerization of cyclic butylene terephthalate, Polymer 53 (2012) 897−902.

[23] S.I. Voicu, M.A. Pandele, E. Vasile, R. Rughinis, L. Crica, L. Pilan, et al., The impact of sonication time through polysulfone-graphene oxide composite films properties, Dig. J. Nanomater. Biostruct. 8 (2013) 1389−1394.

[24] F. Torrisi, D. Popa, S. Milana, Z. Jiang, T. Hasan, E. Lidorikis, et al., Stable, surfactant-free graphene−styrene methylmethacrylate composite for ultrafast lasers, Adv. Opt. Mater. 4 (2016) 1088−1097.

[25] B. Li, S. Dong, X. Wu, C. Wang, X. Wang, J. Fang, Anisotropic thermal property of magnetically oriented carbon nanotube/graphene polymer composites, Compos. Sci. Technol. 147 (2017) 52−61.

[26] G. Ma, Y. Chen, L. Li, D. Jiang, R. Qiao, Y. Zhu, An attractive photocatalytic inorganic antibacterial agent: preparation and property of graphene/zinc ferrite/polyaniline composites, Mater. Lett. 131 (2014) 38−41.

[27] D.K.L. Harijan, V. Chandra, Magnetite/graphene/polyaniline composite for removal of aqueous hexavalent chromium, J. Appl. Polym. Sci. 133 (2016) 44002.

[28] T. Kuila, P. Khanra, A.K. Mishra, N.H. Kim, J.H. Lee, Functionalized-graphene/ethylene vinyl acetate co-polymer composites for improved mechanical and thermal properties, Polym. Test 31 (2012) 282−289.

[29] M. Bera, P. Gupta, P.K. Maji, Efficacy of ultra-low loading of amine functionalized graphene oxide into glycidol-terminated polyurethane for high-performance composite material, React. Funct. Polym. 139 (2019) 60−74.

[30] L. Shao, J. Li, Y. Guang, Y. Zhang, H. Zhang, X. Che, et al., PVA/polyethyleneimine-functionalized graphene composites with optimized properties, Mater. Des. 99 (2016) 235−242.

[31] V. Phetarporn, S. Loykulnant, C. Kongkaew, A. Seubsai, P. Prapainainar, Composite properties of graphene-based materials/natural rubber vulcanized using electron beam irradiation, Mater. Today Commun. 19 (2019) 413−424.

[32] H. Kim, Y. Miura, C.W. MacOsko, Graphene/polyurethane nanocomposites for improved gas barrier and electrical conductivity, Chem. Mater. 22 (2010) 3441−3450.

[33] M.E. Uddin, R.K. Layek, N.H. Kim, D. Hui, J.H. Lee, Preparation and properties of reduced graphene oxide/polyacrylonitrile nanocomposites using polyvinyl phenol, Compos. B Eng. 80 (2015) 238−245.

[34] C. Li, C. Feng, Z. Peng, W. Gong, L. Kong, Ammonium-assisted green fabrication of graphene/natural rubber latex composite, Polym. Compos. 34 (2013) 88−95.

[35] Z. Dong, F. Cai, Z. Jiang, W. Xu, Thermal property studies of in situ blended graphene/nature rubber nanocomposites, Int. J. Polym. Sci. 2020 ()) (2020).

[36] P. Paik, K.K. Kar, Thermal degradation kinetics and estimation of lifetime of polyethylene particles: effects of particle size, Mater. Chem. Phys. 113 (2009) 953−961.

[37] X.-G. Li, M.-R. Huang, G.-H. Guan, T. Sun, Kinetics of thermal degradation of thermotropic poly (p-oxybenzoate-co-ethylene 2,6-naphthalate) by single heating rate methods, Polym. Int. 46 (1998) 289−297.

[38] S. Vyazovkin, A.K. Burnham, J.M. Criado, L.A. Pérez-Maqueda, C. Popescu, N. Sbirrazzuoli, ICTAC kinetics committee recommendations for performing kinetic computations on thermal analysis data, Thermochim. Acta 520 (2011) 1−19.

[39] M.L. Saladino, T.E. Motaung, A.S. Luyt, A. Spinella, G. Nasillo, E. Caponetti, The effect of silica nanoparticles on the morphology, mechanical properties and thermal degradation kinetics of PMMA, Polym. Degrad. Stab. 97 (2012) 452−459.

[40] Y.R. Huang, P.H. Chuang, C.L. Chen, Molecular-dynamics calculation of the thermal conduction in phase change materials of graphene paraffin nanocomposites, Int. J. Heat. Mass. Transf. 91 (2015) 45−51.

[41] X. Wang, H. Yang, L. Song, Y. Hu, W. Xing, H. Lu, Morphology, mechanical and thermal properties of graphene-reinforced poly(butylene succinate) nanocomposites, Compos. Sci. Technol. 72 (2011) 1−6.

[42] E. Tarani, Z. Terzopoulou, D.N. Bikiaris, T. Kyratsi, K. Chrissafis, G. Vourlias, Thermal conductivity and degradation behavior of HDPE/graphene nanocomposites: pyrolysis, kinetics and mechanism, J. Therm. Anal. Calorim. 129 (2017) 1715−1726.

[43] K. Chen, C.A. Wilkie, S. Vyazovkin, Nanoconfinement revealed in degradation and relaxation studies of two structurally different polystyrene-clay systems, J. Phys. Chem. B 111 (2007) 12685−12692.

[44] K. Chen, M.A. Susner, S. Vyazovkin, Effect of the brush structure on the degradation mechanism of polystyrene-clay nanocomposites, Macromol. Rapid Commun. 26 (2005) 690−695.

[45] E. Roumeli, A. Markoulis, K. Chrissafis, A. Avgeropoulos, D. Bikiaris, Substantial enhancement of PP random copolymer's thermal stability due to the addition of MWCNTs and nanodiamonds: decomposition kinetics and mechanism study, J. Anal. Appl. Pyrolysis 106 (2014) 71−80.

[46] Y. Zhan, Y. Lei, F. Meng, J. Zhong, R. Zhao, X. Liu, Electrical, thermal, and mechanical properties of polyarylene ether nitriles/graphite nanosheets nanocomposites prepared by masterbatch route, J. Mater. Sci. 46 (2011) 824−831.

[47] P. Bhawal, S. Ganguly, T.K. Chaki, N.C. Das, Synthesis and characterization of graphene oxide filled ethylene methyl acrylate hybrid nanocomposites, RSC Adv. 6 (2016) 20781−20790.

[48] V.H. Pham, T.T. Dang, S.H. Hur, E.J. Kim, J.S. Chung, Highly conductive poly(methyl methacrylate) (PMMA)-reduced graphene oxide composite prepared by self-assembly of PMMA latex and graphene oxide through electrostatic interaction, ACS Appl. Mater. Interfaces 4 (2012) 2630−2636.

[49] T. Kuila, S. Bose, C.E. Hong, M.E. Uddin, P. Khanra, N.H. Kim, et al., Preparation of functionalized graphene/linear low density polyethylene composites by a solution mixing method, Carbon 49 (2011) 1033–1037.

[50] S. Pal, T. Chatterjee, K. Naskar, Temperature-triggered three-dimensional network formation in graphene–polybutadiene nanocomposite, J. Appl. Polym. Sci. 136 (2019) 48209.

[51] V. Eswaraiah, K. Balasubramaniam, S. Ramaprabhu, One-pot synthesis of conducting graphene-polymer composites and their strain sensing application, Nanoscale 4 (2012) 1258–1262.

[52] C.F. Matos, F. Galembeck, A.J.G. Zarbin, Multifunctional and environmentally friendly nanocomposites between natural rubber and graphene or graphene oxide, Carbon 78 (2014) 469–479.

[53] A.V. Raghu, Y.R. Lee, H.M. Jeong, C.M. Shin, Preparation and physical properties of waterborne polyurethane/ functionalized graphene sheet nanocomposites, Macromol. Chem. Phys. 209 (2008) 2487–2493.

[54] Y.S. Jun, J.G. Um, G. Jiang, A. Yu, A study on the effects of graphene nano-platelets (GnPs) sheet sizes from a few to hundred microns on the thermal, mechanical, and electrical properties of polypropylene (PP)/GnPs composites, Express Polym. Lett. 12 (2018) 885–897.

[55] T. Gkourmpis, K. Gaska, D. Tranchida, A. Gitsas, C. Müller, A. Matic, et al., Melt-mixed 3D hierarchical graphene/polypropylene nanocomposites with low electrical percolation threshold, Nanomaterials 9 (2019) 1766.

[56] S. Zhao, F. Chen, Y. Huang, J.Y. Dong, C.C. Han, Crystallization behaviors in the isotactic polypropylene/graphene composites, Polymer 55 (2014) 4125–4135.

[57] W.S. Ma, J. Li, X.S. Zhao, Improving the thermal and mechanical properties of silicone polymer by incorporating functionalized graphene oxide, J. Mater. Sci. 48 (2013) 5287–5294.

[58] J. Zhang, S. Feng, Q. Ma, Kinetics of the thermal degradation and thermal stability of conductive silicone rubber filled with conductive carbon black, J. Appl. Polym. Sci. 89 (2003) 1548–1554.

[59] X. Wang, W. Xing, P. Zhang, L. Song, H. Yang, Y. Hu, Covalent functionalization of graphene with organosilane and its use as a reinforcement in epoxy composites, Compos. Sci. Technol. 72 (2012) 737–743.

[60] W. Yu, J. Fu, X. Dong, L. Chen, L. Shi, A graphene hybrid material functionalized with POSS: synthesis and applications in low-dielectric epoxy composites, Compos. Sci. Technol. 92 (2014) 112–119.

[61] H. Fang, Y. Zhao, Y. Zhang, Y. Ren, S.L. Bai, Three-dimensional graphene foam-filled elastomer composites with high thermal and mechanical properties, ACS Appl. Mater. Interfaces 9 (2017) 26447–26459.

[62] T.K. Bindu Sharmila, S. Sasi, N.R. Suja, P.M.S. Beegum, E.T. Thachil, A comparative investigation of aminosilane/ethylene diamine–functionalized graphene epoxy nanocomposites with commercial and chemically reduced graphene: static and dynamic mechanical properties, Emergent Mater. 2 (2019) 371–386.

[63] N.T. Tung, T. Van Khai, M. Jeon, Y.J. Lee, H. Chung, J.H. Bang, et al., Preparation and characterization of nanocomposite based on polyaniline and graphene nanosheets, Macromol. Res. 19 (2011) 203–208.

[64] Y. Xu, Q. Gao, H. Liang, K. Zheng, Effects of functional graphene oxide on the properties of phenyl silicone rubber composites, Polym. Test 54 (2016) 168–175.

[65] Z. Wang, P. Wei, Y. Qian, J. Liu, The synthesis of a novel graphene-based inorganic-organic hybrid flame retardant and its application in epoxy resin, Compos. B Eng. 60 (2014) 341–349.

[66] Y.J. Wan, L.C. Tang, D. Yan, L. Zhao, Y.B. Li, L. Bin, et al., Improved dispersion and interface in the graphene/epoxy composites via a facile surfactant-assisted process, Compos. Sci. Technol. 82 (2013) 60−68.

[67] P.A. Gupta, H. Bhayani, S.K. Pramanik, A.C. Rao, S.P. Deshmukh, Cost effective approach of acrylic resin based flooring applications, Constr. Build. Mater. 79 (2015) 48−55.

[68] X. Wang, X. Liu, H. Yuan, H. Liu, C. Liu, T. Li, et al., Non-covalently functionalized graphene strengthened poly(vinyl alcohol), Mater. Des. 139 (2018) 372−379.

[69] S. Chhetri, N.C. Adak, P. Samanta, N.C. Murmu, D. Hui, T. Kuila, et al., Investigation of the mechanical and thermal properties of L-glutathione modified graphene/epoxy composites, Compos. B Eng. 143 (2018) 105−112.

[70] X. Xue, Q. Yin, H. Jia, X. Zhang, Y. Wen, Q. Ji, et al., Enhancing mechanical and thermal properties of styrene-butadiene rubber/carboxylated acrylonitrile butadiene rubber blend by the usage of graphene oxide with diverse oxidation degrees, Appl. Surf. Sci. 423 (2017) 584−591.

[71] M. Bera, U. Saha, A. Bhardwaj, P.K. Maji, Reduced graphene oxide (RGO)-induced compatibilization and reinforcement of poly(vinylidene fluoride) (PVDF)−thermoplastic polyurethane (TPU) binary polymer blend, J. Appl. Polym. Sci. 136 (2019) 47010.

[72] M. Bera, P.K. Maji, Effect of structural disparity of graphene-based materials on thermo-mechanical and surface properties of thermoplastic polyurethane nanocomposites, Polymer 119 (2017) 118−133.

[73] M. Strankowski, P. Korzeniewski, J. Strankowska, A.S. Anu, S. Thomas, Morphology, mechanical and thermal properties of thermoplastic polyurethane containing reduced graphene oxide and graphene nanoplatelets, Materials (Basel), 11, 2018, p. 82.

[74] B.W. Chieng, N.A. Ibrahim, W.M.Z.W. Yunus, M.Z. Hussein, Poly(lactic acid)/poly (ethylene glycol) polymer nanocomposites: effects of graphene nanoplatelets, Polymers (Basel), 6, 2014, pp. 93−104.

[75] E. Sedaghat, A.A. Rostami, M. Ghaemy, A. Rostami, Characterization, thermal degradation kinetics, and morphological properties of a graphene oxide/poly (vinyl alcohol)/ starch nanocomposite, J. Therm. Anal. Calorim. 136 (2019) 759−769.

[76] S. Azizi, G. Momen, C. Ouellet-Plamondon, E. David, Performance improvement of EPDM and EPDM/Silicone rubber composites using modified fumed silica, titanium dioxide and graphene additives, Polym. Test 84 (2020) 106281.

[77] L. Gan, S. Shang, S.X. Jiang, Impact of vinyl concentration of a silicone rubber on the properties of the graphene oxide filled silicone rubber composites, Compos. B Eng. 84 (2016) 294−300.

[78] T.H. Mokhothu, A.S. Luyt, M. Messori, Preparation and characterization of EPDM/silica nanocomposites prepared through non-hydrolytic sol-gel method in the absence and presence of a coupling agent, Express Polym. Lett. 8 (2014) 809−822.

[79] S. Akhlaghi, A.M. Pourrahimi, C. Sjöstedt, M. Bellander, M.S. Hedenqvist, U.W. Gedde, Degradation of fluoroelastomers in rapeseed biodiesel at different oxygen concentrations, Polym. Degrad. Stab. 136 (2017) 10−19.

[80] M. Ali, M.A. Choudhry, Preparation and characterization of EPDM-silica nano/micro composites for high voltage insulation applications, Mater. Sci. Pol. 33 (2015) 213−219.

[81] F. Yan, X. Zhang, F. Liu, X. Li, Z. Zhang, Adjusting the properties of silicone rubber filled with nanosilica by changing the surface organic groups of nanosilica, Compos. B Eng. 75 (2015) 47−52.

[82] R. Han, Z. Wang, Y. Zhang, K. Niu, Thermal stability of CeO_2/graphene/phenyl silicone rubber composites, Polym. Test 75 (2019) 277−283.

[83] R. Wang, Z. Li, W. Liu, W. Jiao, L. Hao, F. Yang, Attapulgite-graphene oxide hybrids as thermal and mechanical reinforcements for epoxy composites, Compos. Sci. Technol. 87 (2013) 29—35.

[84] J. Fan, Z. Shi, L. Zhang, J. Wang, J. Yin, Aramid nanofiber-functionalized graphene nanosheets for polymer reinforcement, Nanoscale 4 (2012) 7046—7055.

[85] M.E. Uddin, T. Kuila, G.C. Nayak, N.H. Kim, B.C. Ku, J.H. Lee, Effects of various surfactants on the dispersion stability and electrical conductivity of surface modified graphene, J. Alloy Compd. 562 (2013) 134—142.

[86] S.L. Qiu, C.S. Wang, Y.T. Wang, C.G. Liu, X.Y. Chen, H.F. Xie, et al., Effects of graphene oxides on the cure behaviors of a tetrafunctional epoxy resin, Express Polym. Lett. 5 (2011) 809—818.

[87] J.H. Park, A. Choudhury, B.L. Farmer, T.D. Dang, S.Y. Park, Chemically modified graphene oxide/polybenzimidazobenzophenanthroline nanocomposites with improved electrical conductivity, Polymer 53 (2012) 3937—3945.

[88] L. Ci, J.B. Bai, The reinforcement role of carbon nanotubes in epoxy composites with different matrix stiffness, Compos. Sci. Technol. 66 (2006) 599—603.

[89] A. Joshi, A. Bajaj, R. Singh, A. Anand, P.S. Alegaonkar, S. Datar, Processing of graphene nanoribbon based hybrid composite for electromagnetic shielding, Compos. B Eng. 69 (2015) 472—477.

Rheological properties of polymer-graphene composites

7

Mahuya Das[1] and Ayan Dey[2]
[1]Greater Kolkata College of Engineering and Management, Baruipur, India,
[2]Indian Institute of Packaging, Mumbai, India

7.1 Introduction

Graphene has attracted significant scientific importance in recent years as a potential nanofiller because of its remarkable properties that include high electrical conductivity, superior thermal conductivity, exceptionally high mechanical flexibility, extraordinary physical properties, and high surface area [1−3]. Graphene is a 2D flat sheet of carbon atoms with the sp^2 hybridized network. It is electrically more conductive than any other substance known, its mechanical strength is 200 times greater than that of steel and it has a thermal conductivity greater than that of diamond [4−6]. Elastic modulus of 1 TPa and thermal conductivity of 5.1 kW/m/K, the intrinsic electrical conductivity of 6×10^5 S/m makes it a promising material for applications in the area of energy conversion, energy storage, conductive ink, transistors, Electromagnetic Interference (EMI) shield coating, thermal interface materials, sensors, and low-density structure materials [7]. Pyrolytic graphite is mechanically cleaved or exfoliated to obtain pristine graphenes. The chemical vapor deposition technique is also popular to obtain graphene where carbon atoms are recombined on the cupper or nickel surface at $1000°C$ in the inter atmosphere. Methane is used as a precursor material in this process [8]. Graphene oxide (GO), an oxidized derivative of graphene has emerged as amphiphilic material that is found to possess compatibility with both the polar and nonpolar polymers. Reduced GO is further developed through the restoration of unsaturation using reducing agents like hydrazine, hydriodic acid, electron complexes in liquid ammonia, metal particles, sodium hydroxide, etc.

The present-day involves the use of different polymer composites in aerospace, automobile, defense industries, etc., with exceptional specific modulus, specific strength. In order to achieve such characteristics graphene sheets have been incorporated both into thermosetting [9−13] and thermoplastic polymers [13−16]. To utilize the full advantage of graphene's wide range of properties for applications in the diversified field the most promising route is the integration of individual graphene in polymer matrices to form advanced multifunctional composites. Fabrication and processing of polymer composites into intricately shaped components with excellent preservation of the structure and properties of graphene can be

Polymer Nanocomposites Containing Graphene. DOI: https://doi.org/10.1016/B978-0-12-821639-2.00021-5

easily carried out using conventional economic processing methods. Works reported by different researchers reveal that graphene can be incorporated in different polymeric matrixes, such as olefinic, acrylic, styrenic, thermoset polymers. Results from an extensive investigation of different properties of graphene/polymer composites (GPC) are promising multifunctional materials with significantly improved tensile strength and elastic modulus, electrical and thermal conductivity, etc. The type of graphene used and its intrinsic properties, the dispersion state of graphene in the polymer matrix and its interfacial interaction, the amount of wrinkling in the graphene, and its network structure in the matrix are the key factors that affect the properties and application of GPC [17].

Compared with carbon nanotubes (CNTs), a promising filler for composites before graphene was isolated, graphene has a higher surface-to-volume ratio because of the inaccessibility of the inner nanotube surface to polymer molecules [16]. This makes graphene potentially more favorable for improving the properties of polymer matrices, such as the mechanical, thermal, microwave absorption properties, electrical and its orientation flexibility (morphology). More importantly, the production of graphene from the graphite precursor is much cheaper than CNTs making graphene-based polymer composites the focus of both academic and industrial interest.

Optimal enhancement of the properties in GPC can only be achieved by resolving the issue of improved dispersion and proper alignment to have good interfacial adhesion in the GPC [18,19] but the interlayer van der Waals force in graphene renders cluster formation throwing a challenge to dispersion and exfoliation during the processing step. Incorporation of filler in the polymer matrix may influence different properties of polymer positively but reduces the ease of processing of polymer. The flow properties can be measured through rheological properties analysis which could indirectly be a measure of the nanofiller dispersion in the polymer [20−22]. The rheology or flow properties of polymer nanocomposites are much diverged and are in between those of the pure polymer and the colloidal suspensions. The dispersion gradient that affects the viscosity and viscoelastic properties of the composite during processing plays a significant role in industrial applications and the development of the production technology for the modern industry related to graphene polymer nanocomposites (GPNs). The rheological analysis [23] is a highly efficient tool in the fabrication of nanocomposites with optimized properties and is a controlling parameter of nanocomposite production technology and processing operation [24−26]. The detailed study of rheological properties during processing can tentatively verify the nanocomposite structure which should subsequently be confirmed by other techniques like microscopy. Among the rheological properties, the platelet dispersion can be predicted by the low-frequency moduli. Linear viscoelastic rheology may provide information on the platelet dispersion. Therefore the evaluation of the dispersion quality following the rheological property measurement is deemed necessary. Change in rheology can be expected mainly due to interaction between graphene and polymers which are found to be affected by the shape, aspect ratio, and functionalization of graphene particles. Several models are noteworthy to mention here that are used to analyze the data obtained through several rheological studies. Hence, this chapter briefs about the rheological characteristics of various GPNs and models used to explain the rheology of such composites.

7.2 Rheological properties of graphene polymer nanocomposites

Rheological characterization of any composite material is composed of the study of the variation in rheological parameters like (1) complex viscosity, (2) storage modulus, (3) loss modulus, (4) tanδ, (5) compliance, (6) normal force against shear stress/strain, frequency/angular frequency, temperature, etc. Besides this, another important study is performed for only graphene-based composites which are the effect of an electric force applied on the rheological parameters. In many published research articles, the Cole-Cole plot and van Gurp analyses are used for understanding the phase miscibility of the blends [27]. Two types of rheological studies are used to analyse the flow behaviour of GPNs. These are (1) melt rheology and (2) solution rheology. The last decade has evidenced the popularity of melt rheology as the majority of the studies involve the development of thermoplastic-based polymer composite. Polyethylene, polypropylene (PP), nylon, acrylonitrile-butadiene-styrene, polyether ether ketone, etc. are few examples of thermoplastics that melt at a specific temperature which is known as the melting point of the polymers. Flow behavior at molten state widely varies with the structural characteristics of the polymers. Such as PP exhibits shear thinning characteristics in the molten state, that is, the complex viscosity decreases with the increase in shear rate or angular frequency [28]. The incorporation of GPNs is found to exhibit a direct influence on melt rheological characteristics of PP-based composites. Li et al. [28] reported a decrease in storage modulus with the incorporation of 5% and 10% graphene nanoparticles (GN) but increases with a further increase in the percentage of GN (Above 10%). Hence, a critical concentration of GNs is essential to experience the reinforcement for any thermoplastic polymer based nanocomposites [28]. Li et al. [28] related the lowering of storage modulus with "the low surface friction and agglomeration" of nanoparticles. Moreover, the polymer-based GPNs exhibit three distinct stages: rubbery, viscoelastic, and glassy state [29,30]. Crosslinked polymer matrix such as gelatin hydrogel exhibits a different observation regarding the influence of graphene incorporation on storage modulus. Here, storage modulus is found to increase due to 0.1vol.% graphene (M grade) incorporation at a frequency of 100 rad/s and 0.1% strain [31]. However, such a phenomenon is not observed in the study reported by Zhang et al. [32]. In this study, storage modulus is found to increase due to GO nanoparticle incorporation in the poly(N-isopropyl acrylamide) based hydrogel. The influence of GO or functionalized GO on various properties of GPNs is discussed in the review of Rohini et al. [33]. The extent of reinforcement achieved by incorporating GO or functionalized GO is shown in Table 7.1.

Han and Chuang described a method for evaluating the compatibility of phases that are obtained by the plot of storage modulus and loss modulus in a frequency

Table 7.1 Extent of increase in storage modulus or elastic modulus reported for various graphene polymer nanocomposites.

Particle	Matrix	Mixing procedure	Increase in storage/elastic modulus (%)	Reference
0.5% GO	Epoxy	Mechanical mixing	35	[33]
0.1wt.% Functionalized Graphene modified using 1,8-diazabicycloundecene/diethyl bromo malonate	Epoxy	Mechanical mixing	18	[33]
0.2wt.% Graphene Oxide modified using DER332, diaminodiphenyl methane	Epoxy	Mechanical mixing	62	[33]
2wt.% Graphene oxide modified with Hexachloro cyclotriphosph azene and glycidol	Epoxy	Mechanical mixing	113	[33]
3wt.% Al-GO	Epoxy	Ball milling	36.4	[33]
0.5wt.% D230-f-GO	Polyether amine	Mechanical mixing	12	[33]
3wt.% Aqueous reduced graphene	Thermoplastic polyurethan	Mechanical mixing	568	[33]
5wt.% Reduced GO	polyurethan	Mechanical mixing	129	[33]
2wt.% Graphene nanoparticles	Polyvinilidine fluoride	Solution mixing	Negligible	[34]
5wt.% Graphene nanoparticles	Polyether ether ketone	Melt mixing	~900	[35]
0.5/1% Graphene	Polyethylene terepththalate	Melt mixing	Lower than the original polymer	[36]
0.5/1% Graphene	poly(ethylene-2,6-naphthalene)	Melt mixing	Lower than the original polymer	[36]

sweep experiment [37,38]. Kashi et al. [39] stated that Dynamic viscoelastic properties obtained at low frequencies in oscillatory rheometry reveal information about the microstructure and the interactions within the system, for instance, the formation of a rigid network of the nanoparticles. At high frequencies, "the rheological properties reflect the motions and mechanical resistance of short polymer segments being less affected by the presence of nanoparticles" [39]. 7% graphene incorporation results in significant improvement in storage modulus (68.83 kPa for virgin PE to 234.1 kPa for 7% graphene based composite at 10 rad/s) which is reported by Mittal and Chaudhry [27]. Similarly, viscosity is also evidenced to increase by ~ 1.75 fold due to only 7% graphene incorporation in PE matrix [27]. But at the same time, 7% incorporation of graphene is found to exhibit maximum deviation from standard semicircular plot, that is, phase incompatibility thus indicating the "higher extent of filler stacks and aggregates" as described by Mittal and Chaudhry [27]. According to Mittal and Chaudhry [27], the 0.5% and 1% dose of graphene results in no phase incompatibility when blended with PE but it shows poor enhancement in storage modulus. Besides this, the curves of the PE and 0.5% graphene incorporated PE is found to overlap at a higher frequency region (100 rad/s), that is, polymer-polymer interaction predominates over polymer-particle interaction at a higher frequency region.

7.3   Factors affecting rheological properties of graphene polymer nanocomposites

Rheological properties are of polymer composites are very much related to the rheological percolation threshold that is nothing but the long-range connectivity of the reinforcement aggregates dispersed in the polymer matrix leading to a supramolecular three-dimensional network. It is important to determine the percolation threshold because the rheological behavior of such composites is generally very different before and after the percolation threshold. Due to this long-range connectivity, there is property enhancement of the composite that is very much different than the previous stage, that is, before the network formation. The rheological percolation indicates a rheological transition; This occurs due to a phase transition from liquid or gel to a solid one. Different rheological properties like complex viscosity, storage modulus, loss modulus are the properties that are to be considered to determine the percolation threshold. Hence, factors affecting the percolation threshold that are discussed below are also responsible parameters to affect the rheological properties of GPN. Graphene-polymer interaction is a predominant factor that controls the rheological behavior of GPNs and this interaction is found to be affected by many factors like physical and chemical characteristics of graphene particles, their distribution as well as the fabrication techniques of the GPNs. Moreover, temperature plays a crucial role in the performance of the GPNs and the phenomenon can be characterized in terms of rheology. These are discussed in brief in the subsequent sections.

7.4   Effect of characteristic of graphenes and modified graphenes on rheology

7.4.1   Shape

It is evidenced that a highly ordered state of anisotropic particles or liquid crystal-line phase exists at mesoscopic scales [40]. Free rotation is getting restricted with the increasing proportion of the anisotropic particles which in turn results in the for-mation of ordered structure. High restriction in free rotation at a higher concentra-tion of anisotropic particle results in the formation of nematic phases or lyotropic LC phases. The aspect ratio is very high for GO (Reported value is 10,000) which makes GO an ideal candidate to form nematic phases [41]. Negative charges exist on GO which provides good dispersion stability of the particles in water. This avoids the aggregation among the particles due to strong π-π interaction and Vander Wall forces of attraction which are found to exist for graphene particles. GO behaves as a "reversible concentrated flocculated networks showing an elastic behavior" [40]. The concentration of GO when varied from 0.1% to 8% storage modulus is found to reach as high as 10^6 Pa at a frequency of 1 rad/s. Moreover, a critical strain value is observed up to which linear viscoelastic region is maintained and after this value, the GO dispersion loses its stability as the network of GO breaks. However, at higher concentrations (i.e., more than 1.23vol.%) both the stor-age and loss modulus exhibits frequency independent behavior. The flow behavior of the poly(methyl methacrylate) (PMMA) is observed to be greatly influenced by the incorporation of multilayered graphenes. Classic linear behavior of PMMA is altered by the incorporation of graphene and the graphene-filled composites show a plateau region at low frequency and then storage modulus increases when frequency increases [42]. The network of graphene is reported to be responsible for achieving higher storage modulus even compared to CNT. Both the polymer-filler and filler-filler interactions are responsible for the enhancement of storage modulus as well as the complex viscosity in the presence of the filler network. Moreover, Graphene has a stable two-dimensional lamellar structure which has been found to offer an increase in complex viscosity up to 22.4% increase from 263.81 Pa.s at 10 rad/s and in the case of graphene/tourmaline composite modified asphalt than that of base asphalt [43]. Similarly, GO nanoparticle is used as modifiers to rubber asphalt (GORA) and the rheological characterization of such materials is found to reveal that the GORA has better high-temperature rutting resistance and lower temperature sensitivity. The reason is the high concentration of oxygen containing group on the base surface and edge of the GO which helped in the formation of dense and stable structure [44]. At low frequency, the complex modulus is found to increase if graphene is incorporated in the composite formulation. However, the complex mod-ulus values for the graphene based composites and graphene free composite approaches closure to each other when frequency increases to 1000 Hz [44]. Jojibabu et al. [45]. studied the rheological properties of epoxy adhesive reinforced with different carbon nanofillers such as multiwalled carbon nanotubes (MWCNT), graphene nanoplatelets (GNP), and single-walled carbon nanohorns. They observed shear thinning behavior with a significant increase in the viscosity of epoxy with the increase in nanoparticle content but a comparatively small increase in viscosity is observed with GNP addition [45]. This fact is attributed to the higher density of GNP and it might align in the direction of flow with shear due to their two dimensional plate like structure. The increase in yield stress is also lower in the case of the GNP filled epoxy compared to the other two nanoparticles because due to their flat morphology and lower volume fraction GNP can not form network structure.

In another study, the percolation network of MWCNT and Functionalized Graphene Sheets (FGS) in epoxy nanocomposites have been studied using rheology. It has been reported that the rod like shape and high aspect ratio of MWCNT enable them to form readily a filler network exhibiting a lower percolation threshold value. But FGS has a planar structure which is not suitable for the formation of network structure. In addition, the functional groups on the surface of FGS did not raise the viscosity even at a higher concentration which benefits the processing [46]. On the other hand, Keramati et al. reported the stiffening effect of graphene nanolayers which in turn causes a slight increase in the storage modulus in the glassy region [47].

7.4.2 Aspect ratio/surface area

The surface area of graphene largely influences the storage modulus of the graphene/gelatin hydrogel. Tungkavet et al. [31] used three different grades of graphene whose surface areas are 40.58, 114.55, and 388.48 m^2/g that are designated as C, H, and M grades respectively. Density is also reported to have a proportionate relationship with the average surface area of the graphene particles. Only 0.01vol.% of graphene results in an increase of storage modulus from 2.56 to 6.36 MPa [31]. Another literature reveals an increase in storage modulus, loss modulus, and complex viscosity as the loading of graphene is varied by the step of 0.25%. The reason is attributed to the large surface area of graphene which is full of mobile electrons. The very percolation threshold occurs at a graphene loading below 0.25% due to the large surface area, high aspect ratio, and proper dispersion of GN in the polymer matrix [48]. Epoxidized natural rubbers (ENR)−graphene (ENR−GE) composites with segregated GE networks were successfully fabricated by He et al. [49]. The rheological percolation threshold was obtained at 0.23vol.%. The average aspect ratio of 255 and a length range of 0.5−4 mm which are much larger than the average diameter of the entanglement distance of ENR molecular chains are responsible for the reduced value of long range segmental motion of the ENR chains. The formation of GE networks in the composites can be achieved by strong interconnection among GE sheets, which restricted the free rotation of GE and influenced strongly the viscoelastic property. Therefore in the vicinity of' GE nanoparticles high density interfacial region has been built up which restrained the motion of resin more with higher loading of GE and as a result, the storage modulus has increased [49]. Two types of PLA/graphene nanocomposites containing different types of graphene nanoplatelets (xGn and NO_2) by solvent casting technique were fabricated by Sabzi et al. [50]. Rheological analysis revealed that NO_2 showed much a lower percolation threshold than xGn because of the former's larger aspect ratio, better dispersion, favorable surface structure, and chemistry. They further used steady state shear, stepwise small-amplitude oscillatory shear (SAOS), large amplitude oscillatory shear (LAOS), and frequency sweep tests immediately after the stepwise shear to determine the disruption and recovery of the filler network structure. The two graphene samples (xGn-6 and N02-1) though exhibited similar linear viscoelastic behavior but showed different recovery processes because the former has a local filler cluster structure while the latter possesses a space-filling percolated network structure [51].

Another investigation was reported by Yao et al. [52] where solution rheology of GO reinforced cellulose carbamate (CC) reveals the fact that the high aspect ratio of nanofiller and the strong polarity of CC along with the preferred orientation and interactions are the reasons behind the exhibited property [52]. As The GO-CC interaction predominated over GO-GO interaction a slight increase in viscosity was observed at 1wt.% loading. With the increase in wt.% of graphene (up to 2), the long range interaction among the graphene sheets prevailed resulting a decrease in viscosity, whereas the formation of the three dimensional network between GO−CC and GO-GO lead to an increase in viscosity when the concentration of graphene exceeds the dose of 3 wt.%.

7.4.3 Particle dispersion/distribution

Graphene, GO, and rGO particles were observed to have different distribution profiles in the case of different composites. It is reported that in the concentrated regime, dispersion of dispersed GO acts as a reversible flocculated network. The GO flakes not only act as a space filling additive but also exhibit electrostatic interaction with the neighboring polymer chains. Such electrostatic interaction results in an improvement of elastic modulus. Shear forces break the flocculated networks of GOs and on the other hand, Brownian motion tries to build up the flocks. Hence there is always a competition that exists which is described by Peclet number (P_e). It is "the relative time scales for Brownian motion of, and the hydrodynamic forces on, the particles" [53]. It is expressed by Eq. (7.1).

$$P_e = \frac{\text{Shear Rate}(\dot{\gamma})}{\text{Rotary brownian diffusion coefficient}(D_y)} = \dot{\gamma}\left(\frac{32\eta_s b^3}{3kT}\right) \tag{7.1}$$

Here, η_s, b, k, T are solvent viscosity, radius of the disk of GO deposits, Boltzmann's constant and temperature in K respectively [54]. The parameter "b" can be measured using atomic force microscopy on the deposits of GO on silicon oxide. Vallés et al. reported a value of 0.6 μm for the parameter b [53]. However, loss modulus was reported to attain plateau region at higher frequencies.

Graphene network was reported to be gradually developed as graphene concentration in GPN increases. It was revealed from steady shear experiments of PMMA based GPNs that the viscosity values increase with the increase in graphene concentration but decrease with an increase in shear rate. In addition, a plateau region in storage modulus at low frequency is also can be attributed to the formation of a "percolating particulate network" [53].

Breaking of the GO network is said to be attributed to the shear thinning region after the plateau region. Zhang et al. pointed out that the increase in C/O ratio of GO dispersion improves in PMMA which is responsible for not only lower rheological percolation threshold but also higher storage modulus and melt viscosity [55]. In the case of reduced GO, the storage modulus is found to achieve a plateau at higher frequencies. However, frequencies at which it attains plateau decrease with the increase in dosage of reduced GO (rGO). At a particular frequency, storage modulus was found to decrease up to a concentration of 0.35% and increased to a very high extent on further increase in the concentration of reduced GOs to 1.79 vol.% ($\sim$3 fold increase in case of ultra high molecular weight polyethylene or UHMWPE) [56]. Hence it can be said that aggregation of rGO is developed at 0.88vol.% which is responsible for drastic increase in storage modulus in the case of UHMWPE. The morphology of GPN showing the distribution profile of graphene was reported by Potts et al. [57] and is shown in Fig. 7.1. The figure clearly indicates the difference in the distribution of thermally expanded GO and

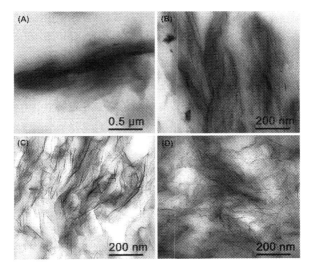

Figure 7.1 Transmission electron microscopy images illustrating the morphological differences in composites with a thermoplastic poly(urethane) matrix filled with (A) unexfoliated graphite in a stacked morphology, and (B) thermally expanded graphite oxide (TEGO), processed by melt mixing. Images (C) and (D) show TEGO/polyurethane composites produced by solution blending and in situ polymerization respectively, illustrating a more exfoliated state of dispersion.
Source: Reprinted with the permission from Elsevier, J.R. Potts, D.R. Dreyer, C.W. Bielawski, R.S. Ruoff, Graphene-based polymer nanocomposites. Polymer 52 (1) (2011) 5−25.

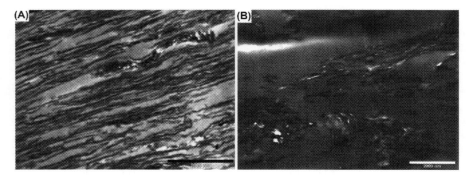

Figure 7.2 Transmission electron microscopy images contrasting (A) the preferential orientation of RGO platelets parallel to the surface of a solution-cast RGO/Nafion and (B) the randomly oriented dispersion of TEGO platelets in Nafion (scale bars ¼ 2 mm) [57]. *RGO, reduced graphene oxide.*

unexfoliated graphite in thermoplastic polyurethane matrix (Fig. 7.1) [57]. This figure also shows the difference in the distribution of graphene particles in GPNs produced by melt-mixing, solution blending, and in situ polymerization.

TEM images of reduced GO/Nafion composite are shown in Fig. 7.2 [57]. It shows a randomly oriented distribution. Storage modulus was observed to increase with the increase in frequency for such composites after attaining an initial plateau in the frequency region from 10^{-3} to 10^{-2} rad/s.

7.4.4 Influence of functionalization of graphenes on rheology of graphene polymer nanocomposites

Introducing functional groups to graphene for improving the dispersion of nanoparticles was reported as an effective solution to increase the polymer-graphene interaction which reduces the difficulties in translating unique properties of graphene to its nanocomposites [58].

Functionalization of graphene can be achieved in two ways: (1) noncovalent interaction and (2) covalent reaction. Noncovalent interaction between organic molecules and graphene occurs mainly through π-π stacking and electrostatic or hydrophobic interactions. Small molecules such as sodium dodecylbenzene sulfonate, 1-pyrenecarboxylic acid, 1-pyrenebutyrate, 3, 4, 9, 10-perylenetetracarboxylic diimide bisbenzenesulfonic acid, Pyrene-1-sulfonic acid sodium salt, and others are used as an agent for functionalizing graphenes. Moreover, polymers like polystyrene, polyaniline, amphiphilic triblock copolymers are used to prepare GPNs through noncovalent interactions [59]. Various biomolecules like diphenylalanine (aromatic peptide), BSA protein are also found to be associated with graphene structures. The peptide can self-assemble with graphene to produce core/shell nanowire [60] and the BSA/graphene act as a natural glue to metal/polymer nanoparticles [61]. The covalent reaction based functionalization process was reported to be carried out via oxidation, amide bonding, diazonium salt, reduction and atom transfer radical polymerization. Isocyanates, amine-functionalized porphyrin were evidenced to act as a modifying agent, which was attached via amide bonding on the graphene surfaces. Compton attempted alkylation via amide bonding to produce uniformly conductive alkylated graphene papers [62]. The reduction process was carried out by sodium borohydride to obtain partially reduced GO. This was further sulfonated using aryl diazonium salt to induce -SO3H group on graphene sheet [59].

Bai et al. reported a significant increase in storage modulus when toluene-2,4-diisocyanate (TDI) functionalized graphenes were incorporated to prepare polyurethane nanocomposites [63]. An increase in TDI functionalized GO concentration results in shifting the point of transition from liquid-like to solid-like toward the lower frequency region [63].

Oh et al. synthesized poly(ethyl methacrylate) (PEMA)/functionalized graphene sheet nanocomposites (FGS) by two different methods; a physical mixing method and an in situ method [64]. An increase in storage modulus and viscosity was observed with a pseudo-solid behavior for in situ composites whereas the physical mixing method established grafting of PEMA chains on FGS. The rheological behavior and viscosities study of epoxy/ graphene decorated with SDS (Grs) composite revealed that functionalization of graphene with SDS renders lower viscosities compared to the composite reinforced with pristine graphene. The dilatancy of epoxy resin was transformed into shear thinning behavior with the loading of Grs of 0.5% [65]. Han and Chun reported a synthesis of graphene functionalized by 2-(4-aminoethyl) ethanol which provides a remarkable improvement for tensile storage modulus from 53 to 150 MPa (for 2% loading of functionalized graphene) when incorporated in polyurethane [60].

Maleic anhydride-grafted polyethylene [66], and chlorinated-polyethylene [67] were reported as compatible for the use in processing of GPNs.

7.4.5   Effect of compatibilizer on rheology of graphene polymer nanocomposites

Rheological analysis of polystyrene-grafted graphene as compatible nanofillers was reported by Beckert et al. [68]. A high degree of compatibility between graphene and polymer matrix was achieved through hydrophobically modified graphite oxide grafted with polystyrene chains (PS-g-FG). The evolution of the solid-like network structure during temperature and shear exposure was well monitored by melt rheological characterization and simultaneously conducted dielectric spectroscopy. This study addressed a nice correlation between rheology, electrical conductivity, and statistical morphology analysis, taking into account the effect of graphene/ polymer compatibilization by grafting polymers onto graphene. Botlhoko et al. used thermally expanded reduced GO (TERGO) as nanofiller in a blend of polylactic acid (PLA) and poly (ε-caprolactone) (PCL) [69] and investigated the influence of TERGO on the rheological parameters of the blend [69]. It was observed in terms of morphology that the filler has reduced the droplet size of the PCL phase as in the case neat blend. The melt viscosity was reported to increase with filler incorporation and hence the rheological properties were improved. The dispersion is uniform with low filler loading and the rheological percolation threshold was obtained at low TERGO loading due to high surface area and better property transfer between filler and matrix at lower loading [69]. Melt rheological analysis of thermally reduced GO (TRG) reinforced PP/maleic anhydride-grafted-ethylene vinyl acetate (PP/EVA-g-MA) blends exhibited a viscous behavior of up to 3wt.% but showed a solid-like behavior at 5wt.% [70]. A composite hydrogel was developed containing thermoresponsive PU and graphene with appropriate rheological properties suitable for the application as a potential bioink for 3D bioprinting and differentiation of neural stem cells. Graphene was coated with a commercial triblock copolymer Pluronic P123 to facilitate the homogeneous dispersion in PU hydrogel while no such coating is required for GO as the surface contains higher oxygen containing group. The storage modulus was slightly decreased with the addition of G-P and GO may be due to interference of the gelation of PU with nanoparticles. Both the PU hydrogels also exhibited shear thinning properties [71].

7.4.6 Fabrication techniques of graphene polymer nanocomposite and its effect on rheology

Ivanov et al. (reported that the rheological percolation threshold of graphene dispersed epoxy resin composite is dependent on processing conditions at low concentration probably due to the dispersion problem of a higher amount of graphene under high speed mechanical stirring [72]. A very low rheological percolation threshold of 0.05wt.% indicated very good dispersion of nanofiller at very low concentration by the high speed mechanical stirring. A low percolation threshold was reported for a graphene concentration of 0.05wt.% which were said to be strong evidence for homogenous dispersion under mechanical agitation of GNP in the polymer matrix. The percolation threshold was observed at 0.5wt.% when ultrasonication was used [72]. The comparison between the effects of solvent casting and melt intercalation mixing processes on different characteristics of polylactide-nanographite platelets composites was evaluated by Narimissa et al. [73]. There are three processing techniques which are: (1) dry mixing and melt intercalation; (2) solvent casting using dichloromethane organic solvent; and (3) a combination of solvent casting and melt intercalation techniques. These processes were performed to compare the effectiveness of fabrication method rheological characteristics of nanocomposites [73]. Morphological analysis of the samples revealed that the solvent casting technique can independently improve the dispersion of NGP (Nanographene platelet) nanofillers within the PLA matrix. However, when this mixing technique is followed by melt intercalation, the optimum level of nanofiller dispersion (i.e. exfoliation) can be achieved. The shear rheological analysis of the samples revealed that although both mixing techniques could individually damage the rheological properties of neat PLA and PLA/ NGP composites and the detrimental effects of solvent casting on samples are extended when it is combined with extrusion [73]. In another work, the effect of graphene nanosheet loading in polyamide 6 (PA6) and acrylonitrile butadiene styrene blend was evaluated. The complex viscosity was found to be dependent directly on the graphene loading due to a parallel increase in loss and storage modulus [74]. Graphene based polyurethane composite was prepared by Canales et al., for the application as hot melt adhesive [75]. They studied the rheological properties in order to evaluate the filler matrix interaction by analyzing the dynamic viscoelastic response at the oscillation frequency tends to zero. In particular, increasing graphene concentration turns the curve for the elastic modulus with frequency from convex to concave curves at $100°C$ that indicates that the elastic modulus tends to level off as the frequency tends to zero. This fact is attributed to the formation of a percolation network with the graphene and polymer chain interaction suppressing the flow. It was reported by the researchers that "for the first time, electrical and rheological percolation thresholds of a polymer/graphene composite in the molten state are matched" [75].

In an effort, acrylonitrile-butadiene-styrene resin (ABS) based nanocomposite was fabricated through a facile coagulation method in DMF (Dimethylformamide) with chemically reduced with graphene nanosheet made with in situ chemical reduction technique in the presence of ABS at the dispersion stage [76]. This technique prevents the agglomeration of graphene. Transmission electron microscopy indicated uniform dispersion of graphene in the styrene-acrylonitrile (SAN) phase of ABS as only SAN is soluble in the DMF but not the polybutadiene (PB) phase. The strong $\pi - \pi$ stacking interaction between aromatic organic molecules and the basal plane of graphite is responsible for this preferential dispersion. Similarly, the linear rheological response of the polymer is influenced by the graphene incorporation and the nanocomposites also showed a transition to solid-like behavior. Moreover, an increase in graphene content leads to the enhancement of the melt mechanical modulus and viscosity.

7.4.7 Interaction between graphene and polymer

A stable hybrid organogel was developed incorporating unfunctionalized and non-oxidized graphene in a fluorescent organogel in o-dichlorobenzene. The dispersion of graphene in the organogel is facilitated by the pyrene-conjugated gelator through noncovalent p−p stacking interactions with peptide. The rheological property measurement revealed that the flow of the hybrid organogel becomes more resistant toward the applied angular frequency due to the presence of graphene imparting about seven times higher rigidity than that of the native gel [77]. Compton et al. used four different grades of graphene depending on their structure polarity flake to reinforce epoxy resin and evaluated the rheological properties [78]. The results showed that the polar graphene flakes have stronger interaction with polar epoxy resin and the grade N006-P graphene, with the highest aspect ratio had the greatest effect on the rheology. The addition of graphene increased the apparent viscosity at all shear rates [78]. Zhang et al. reported the use of GO with different oxygen content in polymethyl methacrylate matrix [55]. The storage modulus showed frequency-independent behavior at the low frequency indicated a solid-like visco-elastic behavior of the composite due to formation. The best interfacial interaction hence the better dispersion was observed for the graphene with a C/O ratio of 13.2 exhibiting a lower rheological percolation threshold. A low cost blended film of chitosan and GO was also reported to exhibit strong H-bondings between hydroxyl groups of the chitosan and hydroxyl groups of the GO. It is a green material suitable in the biomaterial or packaging field. Graphene was homogeneously dispersed in chitosan and storage modulus was observed to be increased with the incorporation of graphene in chitosan [79]. Kotal et al. [80] developed functionalized graphene/bromobutyl rubber nanocomposites using a "grafting to" approach. An excellent enhancement in storage modulus was observed. Grafting with rubber not only improves the dispersion but also facilitates exfoliation [80]. Rheological properties of graphene (G), an extended graphene (EG), and reinforced cycloolefin (COC) composite were studied by Kasgoz et al. [81]. It was reported that the EG formed a honeycomb network structure whereas graphene flakes showed poor dispersion [81]. The rheological percolation threshold value for the COC/G composite is 17.7wt.% while for COC/EG the value is 3.67%. The percolation threshold values are comparatively lower in both cases than those reported in previous reports. They used the reduced modulus as a parameter that is the ratio of the plateau modulus of composite to that of polymer. The reduced modulus was reported to increase slightly up to a particular value of filler loading that was designated as a critical filler amount. Rheological properties study of GO reinforced polydimethylsiloxane (PDMS) composites with systematic change in the molecular weight of PDMS and concentration of GO showed negative normal stress differences during shear experiments in low molecular weight PDMS composites (Mw < Mc) at low shear rates. However, a positive normal stress difference was reported for high molecular weight PDMS composites [82]. Further, Niu et al studied the effect of oxygen concentration on the GN surface on the rheological properties of (GP-x)/polydimethylsiloxane composite [82]. The strong inter-particle attractive interaction between GP and PDMS leads to the vorticity alignment of aggregates under weak shear. The inclusion of GO in glass fiber reinforced epoxy resin showed an increase in viscosity as it affected the resin cure reaction by reducing the resin gel time [83]. GN reinforced polyvinylidene fluoride composite also exhibited shear thinning property with a converge nature of complex viscosity curve at high frequency. This is due to the restricted movement of the polymer chain by the aggregated state of graphene. A deviation from the viscoelastic linearity was reported for the composite [84].

7.4.8 Temperature

Rheological properties are found to be greatly influenced by an increase in temperature. The elastic modulus of nanocomposite is significantly influenced by both the temperature and strain rate. Mahieux and Reifsnider defined the elasticity modulus expression that was modified further by Richeton et al. [85]. This theory is successfully used by Acar et al. for GPN [86]. Graphene particles are mainly used as reinforcing filler which binds the polymer molecules in an array. Such characteristic has a direct impact on chain mobility as well as segmental mobility. With an immediate effect of such phenomenon, both glass transition and melting temperature increase for polymer materials. This in turn affects the flow behavior of the graphene based composites at a particular temperature. Thus both the rate of change in storage modulus and loss modulus with the increase in temperatures is affected due to the incorporation of graphenes. The last decade has evidenced many rheological studies on GPN. The change in modulus is widely varies based on polymers, functionalization of graphenes, state of dispersion, and the graphene polymer interactions. Tungkavet et al. reported a significant increase in storage modulus due to an increase in temperature from 40°C to 50°C for gelatin based composites as a result of graphenes (GO) incorporation [31]. They have explained it as a result of an increase in retractive force and physical entanglement density due to an increase in entropy at a higher temperature for gelatin based GPNs. However, a decrease in G' is observed when the temperature goes above 60°C. Besides this, Kashi et al. reported an anomalous rheological behavior observed for graphene/polylactic acid composites [87]. Generally, for any polymer composite, storage modulus is observed to be reduced due to the increase in temperature for a particular composition of graphene in the polymer matrix. But at low frequency region (0.1 rad/s) the storage modulus is found to increase significantly at 220°C. The magnitude of storage modulus is observed from $\sim 10^2$ to $\sim 10^4$ Pa for the composite containing 6%, 9%, and 12% of graphene respectively [39]. Moreover, the rheological percolation threshold is found to appear at a concentration of 8.5wt.% of graphene in PLA/graphene composite at a temperature of 180°C. The same is found to appear at 5.2wt.% concentration when the temperature reaches 220°C. At higher temperatures, graphene particles are found to exhibit physical gelation with polymer matrix when sheared at a low frequency region (around 0.1 rad/s). In another report, Kashi et al. reported a frequency sweep experiment in linear viscoelastic regions at four different temperatures of 160°C, 180°C, 200°C, and 220°C for PBAT/GNP nanocomposites [87]. According to this article, the complex modulus increases with the increase in graphene loading up to 9 wt.%. The slope of log G'- log ω plot in low frequency region ($<$1 rad/s) was found to increase with the increase in temperature but decreased with an increase in graphene loading at the temperatures 160°C−220°C. Sun et al. (2017) reported a study based on the shear rate-shear stress curve [88]. Here, four different models were used to fit the curve, which are the Bingham-Plastic model, Herschel-Bulkley model, Vocadlo model, and Vom Berg model [88]. These models are explained in the subsequent section. A nonlinear relationship between shear stress and shear rate and a shear thinning behavior of graphene loaded polymer matrix was reported by Sun et al. [88].

According to Kelarakis et al., higher temperature facilitates the adsorption of the molten polymer chains on the nanofiller surface increasing their apparent volume fraction in the interfacial region [89]. Hence the decreasing trend of percolation threshold in PBAT/GNP composite can be said to be attributed to the overall effect of polymer-polymer, polymer-GNP, and GNP-GNP interactions at a higher temperature. Polypyrrole/thermally reduced rGO (Ppy/TRGO) nanocomposites were successfully synthesized by in situ polymerization and the effect of shear stress along with temperature was studied by Manivel et al. [90]. It was reported that shear stress was increased continuously with increasing shear forces for both the neat resin and composite at 25°C but behave like a Bingham fluid when at 150°C. However, the shear stress and viscosity of the nanocomposites are higher than that of Ppy. High shear thinning behavior of Ppy/TRGO nanocomposite exhibited high shear thinning behavior at any temperature due to strong interaction of Ppy with TRGO through both H-bonding and pi-pi interaction preventing agglomeration of polymer chains and hence led to improvement in rheological properties [90].

7.5 Different models to analyze graphene polymer interaction

Viscoelastic models are generally based on continuum theory. There are two classes of it. One of them is classical plasticity theory where time effects (rate, creep, relaxation, and recovery of strain) are excluded. The second class of the continuum theory is based on viscoelasticity where time effects are included. Here, inelasticity is assumed as rate dependent. Cooperative viscoelastic theory based on overstress (VBO) developed for polymers (VBOP) only, was modified by Acar et al. to make it applicable for GPN [86]. The classical form of VBO for amorphous polymers was first proposed by Colak et al. where only "additive form of the rate of deformation tensor" was used [86]. The equation is given below.

$$d = d^e + d^{vp} = \frac{(1+\nu)}{CE}\dot{S} + \frac{3}{2}\dot{\gamma}^{vp}\frac{(s-g)}{\Gamma} \qquad (7.2)$$

Here, d, E, ν, c, and $\dot{\gamma}^{vp}$ are deviator of the arte of total deformation tensor, elastic modulus, Poisson's ratio, parameter corresponds to nonlinear viscoelastic behavior and viscoelastic shear strain rate respectively. $S = dS/dt$ where s is the deviatoric part of the Cauchy stress tensor. "g" is the deviatoric part of equivalent stress tensor. It is also termed as "rate-independent contribution to hardening" [86]. $(s-g)$ denotes overstress tensor. Γ, the "oversress invariant" and "C" are defined by the Eqs. (7.3) and (7.4) respectively.

$$\Gamma^2 = \frac{3}{2}(s-g):(s-g) \qquad (7.3)$$

$$C = 1 - \lambda\left(\frac{|g-k|}{A}\right)^{\phi} \qquad (7.4)$$

The terms are explained by Acer et al. in their published article [86]. "g" is related to kinematic stress tensor (k) by the Eq. (7.5).

$$\dot{g} = \frac{\psi}{E}\dot{s} + \dot{\gamma}^p \left[\frac{(s-g)}{\Gamma} - \frac{(g-k)}{A} \right] + \left(1 - \frac{\psi}{E}\right)\dot{k} \tag{7.5}$$

Here, ψ is scalar shape function ant it should lies between the values of tangent modulus (E_t) and E. In VBOP approach, k covers the Bauschinger effect. The tensor "k" is defined by the Eq. (7.6).

$$\dot{k} = \frac{E_t}{\left(1 - \frac{E_t}{E}\right)} \dot{Y}^p \frac{(S-g)}{\Gamma} \tag{7.6}$$

E_t is a function of viscoelastic strain and the corresponding relation is described by the Eq. (7.7).

$$E_t = \frac{E_{t_0}}{2}\left[1 + \exp(\alpha|\varepsilon^{vp}|)\right]^\beta \tag{7.7}$$

E_{t_0} is the tangent elastic modulus at yield point for plastic and α, β are constant for material under testing. For the GPN, the stiffening was found to depend on the extent of agglomeration of graphene particles in the polymer matrix. Thus Acar et al. made one attempt to include the stiffening effect due to graphene incorporation in the VBOP theory [86]. Here, few assumptions were made that are: (1) graphene sheets behave as transversely isotropic elastic material, (2) behavior of polymer matrix is an isotropic elastic material. Further detail in predicting storage modulus is given in the paper published by Acar et al. [86]. Moreover, many models were established by several research groups. These are used for describing various influencing factors on modulus like crystallinity, aspect ratio, etc. Besides this, many models are derived from the previously established models. These models are found to be effective for determining the relaxation modulus, flow characteristics, etc. All of them are briefed here.

1. *Bingham-Plastic model*: It is a two parameter model. This model is useful for fitting the shear stress-shear rate curve. It is found that fitting is improved with the increase in temperature as R^2 values increases for 0.957–0.982 [88]. The expression for this model is given in Eq. (7.12). Alike this there are many two parameter models, for example, Power Law, Casson, Eyring, etc. [91]. These are also used to describe the rheological characteristics of various polymer systems.
2. The power law was proposed by Ostwald which is based on "the variation of viscosity with velocity gradient of the fluid" [Eqs. (7.8) and (7.9)]

$$\sigma = k\dot{\gamma}^n \tag{7.8}$$

$$\eta = k\dot{\gamma}^{n-1} \tag{7.9}$$

Here k indicates the consistency of liquid and n is an exponent which is equal to unity for Newtonian fluid and deviates from unity for non-Newtonian fluid. However, in reality, the above expression is valid only at a certain range of shear rate. The model of the cross and Carreaue-Yasuda model is an extension of the Power Law model which is expressed by the Eqs. (7.10) and (7.11).

$$\frac{\eta - \eta_\infty}{\eta_0 - \eta_\infty} = \frac{1}{(1 + \lambda.\dot{\gamma}^m)} \tag{7.10}$$

$$\frac{\eta - \eta_\infty}{\eta_0 - \eta_\infty} = (1 + \lambda.\dot{\gamma}^\alpha)^{\frac{(m-1)}{m}} \tag{7.11}$$

The details were given by Oarhim et al. [92]. Both the models have incorporated zero shear viscosity and the shear viscosity at an infinite shear rate. For filler based composites, Quemeda proposed a model to account for the variation of viscosity due to the incorporation of rigid, spherical and noninteractive aggregates which is expressed by Eq. (7.12) [92].

$$\eta = \eta_\infty \left[\frac{\left(1 + \frac{\sigma}{\sigma_c}\right)^p}{\left(\chi + \frac{\sigma}{\sigma_c}\right)^p} \right]^2 \tag{7.12}$$

Here, χ depends on the volume fraction of the polymer. σ, σ_c, and p are the applied stress, characteristic stress and slope of the curve at the inflection point respectively.

3. A *two-parameter model* is also found to be in use which is expressed by Eq. (7.13).

$$\tau = \tau_0 + \mu_p \dot{\gamma} \tag{7.13}$$

4. *Herschel−Bulkley model*: This model fits well with observed values from the rheological characterization of GPNs. Moyano et al. fitted the flow curve to the Herschel−Bulkley model for the GO and PEI-PEG based printable ink formulation [93]. Generally, the model is applied to colloidal type gels. The equation corresponding to this model is given by Eq. (7.14) [93].

$$\tau = \tau_0 + k\dot{\gamma}^n \tag{7.14}$$

Here, τ_0, k, $\dot{\gamma}$, and n are equivalent yield point, consistency parameter, shear rate, and flow index respectively. The flow index denotes the degree of the non-Newtonian characteristic of the fluid material. Flow index is generally below unity for the shear thinning fluid. τ is the yield stress. For a shear thinning material, there is a rapid decline in storage modulus with the increase in loss modulus at the same time. The crossover point indicates the equivalent yield point. It is found to show fairly good presentation of the τ values in the range 01−20 S.

5. *Vocadlo model*: Both the shear softening and shear hardening behavior can be predicted using the Vocaldo model at two different temperatures where maximum shear stress approaches infinity when shear rate approaches to infinite value [94]. The equation for this model is given below.

$$\tau = \left[\sqrt[n]{\tau_0} + \sqrt[n]{k}\sqrt{\dot{\gamma}} \right]^{1/n} \tag{7.15}$$

6. *Vipulanandan rheological model*: This model is used to describe the shear rate-shear stress relationship for the slurry, drilling mud, etc. The relationship is defined by Eq. (7.16).

$$\tau - \tau_y = \frac{\dot{\gamma}}{C + D\dot{\gamma}} \tag{7.16}$$

Here τ, τ_y are the shear stress and yield stress respectively. C and D are model parameters. Graphenes are used in preparing new cement formulation where this model is also used by Akkhamis and Imquam [95]. The tangential viscosity is denoted by $d\tau/d\dot{\gamma}$ which can be obtained by Eq. (7.17).

$$\frac{d\tau}{d\dot{\gamma}} = \frac{(C + D\dot{\gamma}) - \dot{\gamma}D}{(C+D\dot{\gamma})^2} = \frac{C}{(C+D\dot{\gamma})^2} \tag{7.17}$$

The rate of change in tangential viscosity is obtained by differentiating Eq. (7.17) with respect to shear rate ($\dot{\gamma}$) and given by Eq. (7.18).

$$\frac{d^2\tau}{d\dot{\gamma}^2} = \frac{-2CD}{(C+D\dot{\gamma})^3} \tag{7.18}$$

It is stated that if $C > 0$ then the tangential viscosity > 0 and $D < 0$. When shear rate or strain rate approaches to infinite value then the ultimate shear stress (τ_{max}) is defined by Eq. (7.19).

$$\tau_{max} = \frac{1}{D} + \tau_y \tag{7.19}$$

Maxiumum shear stress is the only limit for this model [94].

7. *Vom Berg model*: This model is found to be more consistent with slurry rheology [91]. This model is represented by the equation given below. It is also found to best fit the experimental values obtained for GPNs [88].

$$\tau = \tau_0 + b \sinh^{-1} \frac{\dot{\gamma}}{c} \tag{7.20}$$

8. *Cho-Choi-Jhon model:* Lu et al. found that Cho-Choi-Jhon model fitted the observed value with more accuracy for the GO grafted polyaniline (GO-g-PANI) composites [96]. It is a six parameter model and is expressed by Eq. (7.21).

$$\tau = \frac{\tau_y}{1 + (t_1\dot{\gamma})^\alpha} + \eta_\alpha \left(1 + \frac{1}{(t_2\dot{\gamma})^\beta}\right)\dot{\gamma} \tag{7.21}$$

First term on the right hand side of Eq. (7.21) describes the variation of shear stress with the increase in shear rate and the second term describes the yielding characteristics of τ when critical shear rate exceeds. The parameters t_1, t_2 are the time constants and η_∞ is the shear viscosity at infinite strain rate. The parameter β lies between 0 and 1 [96].

9. *Rheological percolation threshold:* Winter-Chambon criteria are found to be used to determine the percolation threshold for a filler containing composites. As graphene act as a filler then these criteria are also helpful to determine the rheological percolation threshold. Here, it is assumed that a filled polymer system at the percolation threshold behaves as physical gels. The percolation threshold can be determined from frequency independency of loss tangent (tan δ = G''/G) in the low frequency region [39].

10. Relationship between crystallinity and the storage modulus: The increase in storage modulus is assumed to be associated with the "progressive transformation of the initial liquid state of the polymer in a suspension of crystallites" and the "square of the solid volume fraction" as stated by Canales et al. [97]. The relationship can be expressed by Eq. (7.22).

$$G'(t) = G'_0 + G'_\infty (1 - \exp(-Kt^n))^2 \tag{7.22}$$

Here, G'_0 and G'_∞ are elastic modulus at liquid state and at $\varphi = 1$, that is, for the polymer matrix without filler. K and n are adjustable parameters. Here, φ can be obtained by Eqs. (7.23) and (7.24).

$$\varphi = 1 - \exp(-Kt^n) \tag{7.23}$$

$$G'(t) = G'_0 + G'_\infty \varphi^2 \tag{7.24}$$

Half crystallization time can be obtained by the equation which indicate he direct effect of graphene incorporation in to polymer matrix [97].

11. Relationship between entanglement density and shear modulus: Storage modulus increases with the increase in frequency and the curve attains linearity when frequency or angular frequency reaches a threshold value. In the case of UHMWPE, the angular frequency is 10 rad/s above which the plateau region starts [56]. The average storage modulus in this region is termed as plateau modulus which is designated by G_N^0. The plateau modulus was found to be directly related to the average molar mass between entanglements ($<M_c>$) by Eq. (7.25) [56]. As the interaction between the GO and the polymer chains increases, the chain entanglements of the polymer molecules are suppressed resulting in an increase in molar mass between crosslinks. This further causes a reduction in plateau modulus.

$$G_N^0 = \frac{g_N \rho RT}{M_c} \tag{7.25}$$

Weir et al. used (4/5) value in place of the term g_N for the rubbery network in Eq. (7.9) [98]. Doi and Edward proposed the reputation model that is used to replace the g_N with the factor (4/5) [99]. Pandey et al. used a simplified equation to calculate the degree of entanglement which is directly related to tensile storage modulus (E') by Eq. (7.26) [100].

$$\text{Degree of entanglement} = \frac{E'}{6RT} \tag{7.26}$$

Here, R is the universal gas constant. The E' is determined by the DMTA instrument. The entanglement density can be calculated by the ratio of polymer density and M_c [99].

12. Reinforcing factor: Reinforcing efficiency factor can be obtained using DMTA which provides an idea about the filler-polymer bonding. The reinforcing factor is determined using the Einstein equation [Eq. (7.27)] [100].

$$E_c = E_m(1 + r.V_f)$$ (7.27)

Here, E_c, E_m, and V_f are storage modulus of composite, matrix storage modulus, and filler volume respectively.

13. Performance of graphene composites: Performance of graphene composites is evaluated by "C factor". It provides an idea about the effectiveness of graphene particles as filler in the composites. C factor is determined by Eq. (7.27) [100].

$$C = \frac{\left(\frac{E'_g}{E'_r}\right)_{composite}}{\left(\frac{E'_g}{E'_r}\right)_{polymer\ matrix}}$$ (7.28)

Here, E'_g and E'_r are tensile storage modulus in grassy and rubbery resin respectively. These moduli are determined by the study of variation in log(storage modulus) with respect to temperature in tensile mode using DMTA/DMA. Pothan et al. experimented with a constant frequency of 0.1 Hz and the temperatures taken for the analysis are 45°C and 130°C for glassy and rubbery region respectively (Plateau region of the curve) for polyester matrix [101].

14. Aspect ratio of graphene fiber: A model was proposed by Guth to determine the aspect ratio of fibers from the storage modulus data against the variation of frequency (0.01−100 Hz) [75]. It is valid for both the plate-like and rod-like fillers. Canales et al. used the equation derived from Guth's model to determine the aspect ratio of the graphene fibers. The equation is given by Eq. (7.29).

$$G'_{comp} = G'_m \left(1 + 0.67f_s\varphi + 1.62f_s^2\varphi^2\right)$$ (7.29)

where, G'_{comp} and G'_m are the storage modulus of composite and matrix respectively at a frequency of 0.12 Hz [75]. φ and f_s are the volume fraction and the aspect ratio of graphene respectively.

15. *Relaxation modulus:* Relaxation modulus of a system at a gel point can be obtained by Eq. (7.30).

$$G(t) = St^{-n}$$ (7.30)

Here, $G(t)$ and n are the relaxation modulus at gel point and relaxation component respectively. The value of n is zero for a perfectly elastic body and 1 for a perfect viscous body. The S value can be determined using Eq. (7.31).

$$G'(\omega) = \frac{G''(\omega)}{\tan\left(\frac{n\pi}{2}\right)} = S\omega^n\Gamma(1-n)\cos\left(\frac{n\pi}{2}\right)$$ (7.31)

Here, the relaxation exponent n can be calculated from the curves of G' and G'' with respect to the angular frequency at different temperatures and is used to determine S by the above equation [102].

7.6 Conclusion

Interest in GPNS is immensely increasing in recent days due to its unique properties. Besides this, many challenges in processing and optimization of its flow behavior have been faced for these types of composites. Rheology is an important tool to determine the behavioral change of GPNs due to the effect of increasing concentration of graphenes, its size, and shape, distribution in the polymer matrix, functionalization, or structural alteration of the graphenes that are briefed in this chapter. Moreover, several models used to analyze the influence and the graphene-polymer interaction are also discussed which plays a crucial role in determining the characteristics of the GPNs. This chapter helps in finding an effective solution to overcome the challenges in developing GPNS and in optimizing its flow behavior using rheological characterization.

References

[1] L. Jiang, Z. Fan, Design of advanced porous graphene materials: from graphene nanomesh to 3D architectures, Nanoscale 6 (4) (2014) 1922–1945.
[2] Z. Zheng, X. Zheng, H. Wang, Q. Du, Macroporous graphene oxide—polymer composite prepared through pickering high internal phase emulsions, ACS Appl. Mater. Interface 5 (16) (2013) 7974–7982.
[3] C. Wu, X. Huang, X. Wu, R. Qian, P. Jiang, Mechanically flexible and multifunctional polymer-based graphene foams for elastic conductors and oil-water separators, Adv. Mater. 25 (39) (2013) 5658–5662.
[4] C. Lee, X. Wei, J.W. Kysar, J. Hone, Measurement of the elastic properties and intrinsic strength of monolayer graphene, Science 321 (5887) (2008) 385–388.
[5] X. Du, I. Skachko, A. Barker, E.Y. Andrei, Approaching ballistic transport in suspended graphene, Nat. Nanotechnol. 3 (8) (2008) 491–495.
[6] J.H. Seol, I. Jo, A.L. Moore, L. Lindsay, Z.H. Aitken, M.T. Pettes, et al., Two-dimensional phonon transport in supported graphene, Science 328 (5975) (2010) 213–216.
[7] K. Hu, D.D. Kulkarni, I. Choi, V.V. Tsukruk, Graphene-polymer nanocomposites for structural and functional applications, Prog. Polym. Sci. 39 (11) (2014) 1934–1972.
[8] H. Kim, A.A. Abdala, C.W. Macosko, Graphene/polymer nanocomposites, Macromolecules 43 (16) (2010) 6515–6530.
[9] K.S. Novoselov, A.K. Geim, S.V. Morozov, D. Jiang, Y. Zhang, S.V. Dubonos, et al., Electric field effect in atomically thin carbon films, Science 306 (5696) (2004) 666–669.
[10] Y. Hernandez, V. Nicolosi, M. Lotya, F.M. Blighe, Z. Sun, S. De, et al., High-yield production of graphene by liquid-phase exfoliation of graphite, Nat. Nanotechnol. 3 (9) (2008) 563–568.
[11] C.J. Shih, A. Vijayaraghavan, R. Krishnan, R. Sharma, J.H. Han, M.H. Ham, et al., Bi-and trilayer graphene solutions, Nat. Nanotechnol. 6 (7) (2011) 439.
[12] H. Kim, Y. Miura, C.W. Macosko, Graphene/polyurethane nanocomposites for improved gas barrier and electrical conductivity, Chem. Mater. 22 (11) (2010) 3441–3450.

[13] X. Li, G. Zhang, X. Bai, X. Sun, X. Wang, E. Wang, et al., Highly conducting graphene sheets and Langmuir–Blodgett films, Nat. Nanotechnol. 3 (9) (2008) 538–542.

[14] M.D. Stoller, S. Park, Y. Zhu, J. An, R.S. Ruoff, Graphene-based ultracapacitors, Nano Lett. 8 (10) (2008) 3498–3502.

[15] W. Gao, N. Singh, L. Song, Z. Liu, A.L.M. Reddy, L. Ci, et al., Direct laser writing of micro-supercapacitors on hydrated graphite oxide films, Nat. Nanotechnol. 6 (8) (2011) 496–500.

[16] S. Stankovich, D.A. Dikin, G.H. Dommett, K.M. Kohlhaas, E.J. Zimney, E.A. Stach, et al., Graphene-based composite materials, Nature 442 (7100) (2006) 282–286.

[17] J. Du, H.M. Cheng, The fabrication, properties, and uses of graphene/polymer composites, Macromol. Chem. Phys. 213 (10–11) (2012) 1060–1077.

[18] N. Yousefi, M.M. Gudarzi, Q. Zheng, S.H. Aboutalebi, F. Sharif, J.K. Kim, Self-alignment and high electrical conductivity of ultralarge graphene oxide–polyurethane nanocomposites, J. Mater. Chem. 22 (25) (2012) 12709–12717.

[19] N. Yousefi, X. Lin, Q. Zheng, X. Shen, J.R. Pothnis, J. Jia, et al., Simultaneous in situ reduction, self-alignment and covalent bonding in graphene oxide/epoxy composites, Carbon 59 (2013) 406–417.

[20] S.S. Rahatekar, K.K.K. Koziol, S.A. Butler, J.A. Elliott, M.S.P. Shaffer, M.R. Mackley, et al., Optical microstructure and viscosity enhancement for an epoxy resin matrix containing multiwall carbon nanotubes, J. Rheol. 50 (5) (2006) 599–610.

[21] F.J. Galindo-Rosales, P. Moldenaers, J. Vermant, Assessment of the dispersion quality in polymer nanocomposites by rheological methods, Macromol. Mater. Eng. 296 (3–4) (2011) 331–340.

[22] E. Ivanov, R. Kotsilkova, E. Krusteva, Effect of processing on rheological properties and structure development of EPOXY/MWCNT nanocomposites, J. Nanopart. Res. 13 (8) (2011) 3393–3403.

[23] R. Kotsilkova, Thermosetting Nanocomposites for Engineering Application, Rapra Smiths Group, London, 2007, p. 325.

[24] M.J. Solomon, A.S. Almusallam, K.F. Seefeldt, A. Somwangthanaroj, P. Varadan, Rheology of polypropylene/clay hybrid materials, Macromolecules 34 (6) (2001) 1864–1872.

[25] R. Wagener, T.J. Reisinger, A rheological method to compare the degree of exfoliation of nanocomposites, Polymer 44 (24) (2003) 7513–7518.

[26] Q. Zhang, F. Fang, X. Zhao, Y. Li, M. Zhu, D. Chen, Use of dynamic rheological behavior to estimate the dispersion of carbon nanotubes in carbon nanotube/polymer composites, J. Phys. Chem. B 112 (40) (2008) 12606–12611.

[27] V. Mittal, A.U. Chaudhry, Polymer—graphene nanocomposites: effect of polymer matrix and filler amount on properties, Macromol. Mater. Eng. 300 (5) (2015) 510–521.

[28] Y. Li, J. Zhu, S. Wei, J. Ryu, L. Sun, Z. Guo, Poly (propylene)/graphene nanoplatelet nanocomposites: melt rheological behavior and thermal, electrical, and electronic properties, Macromol. Chem. Phys. 212 (18) (2011) 1951–1959.

[29] X. Chen, S. Wei, A. Yadav, R. Patil, J. Zhu, R. Ximenes, et al., Poly (propylene)/carbon nanofiber nanocomposites: ex situ solvent—assisted preparation and analysis of electrical and electronic properties, Macromol. Mater. Eng. 296 (5) (2011) 434–443.

[30] S. Wei, R. Patil, L. Sun, N. Haldolaarachchige, X. Chen, D.P. Young, et al., Ex situ solvent—assisted preparation of magnetic poly (propylene) 8nocomposites filled with Fe@ FeO nanoparticles, Macromol. Mater. Eng. 296 (9) (2011) 850–857.

[31] T. Tungkavet, N. Seetapan, D. Pattavarakorn, A. Sirivat, Graphene/gelatin hydrogel composites with high storage modulus sensitivity for using as electroactive actuator: effects of surface area and electric field strength, Polymer 70 (2015) 242−251.

[32] E. Zhang, T. Wang, C. Lian, W. Sun, X. Liu, Z. Tong, Robust and thermo-response graphene−PNIPAm hybrid hydrogels reinforced by hectorite clay, Carbon 62 (2013) 117−126.

[33] R. Rohini, P. Katti, S. Bose, Tailoring the interface in graphene/thermoset polymer composites: a critical review, Polymer 70 (2015) A17−A34.

[34] Y.J. Xiao, W.Y. Wang, X.J. Chen, T. Lin, Y.T. Zhang, J.H. Yang, et al., Hybrid network structure and thermal conductive properties in poly (vinylidene fluoride) composites based on carbon nanotubes and graphene nanoplatelets, Compos. A Appl. Sci. Manuf. 90 (2016) 614−625.

[35] Á. Alvaredo, M.I. Martín, P. Castell, R. Guzmán de Villoria, J.P. Fernández-Blázquez, Non-isothermal crystallization behavior of PEEK/Graphene nanoplatelets composites from melt and glass states, Polymers 11 (1) (2019) 124.

[36] A. Ghadami, M. Ehsani, H.A. Khonakdar, A comprehensive study on morphological and rheological behavior of poly (ethylene terephthalate) and poly (ethylene-2, 6-naphthalene) nanocomposite blends in presence of graphene, J. Vinyl Addit. Technol. 23 (2017) E160−E169.

[37] H.K. Chuang, C.D. Han, Rheological behavior of polymer blends, J. Appl. Polym. Sci. 29 (6) (1984) 2205−2229.

[38] C.D. Han, H.K. Chuang, Criteria for rheological compatibility of polymer blends, J. Appl. Polym. Sci. 30 (11) (1985) 4431−4454.

[39] S. Kashi, R.K. Gupta, T. Baum, N. Kao, S.N. Bhattacharya, Phase transition and anomalous rheological behaviour of polylactide/graphene nanocomposites, Compos. B Eng. 135 (2018) 25−34.

[40] C. Vallés, Rheology of graphene oxide dispersions, Graphene Oxide: Fundamentals Applications, Wiley, 2017, pp. 121−146.

[41] B. Dan, N. Behabtu, A. Martinez, J.S. Evans, D.V. Kosynkin, J.M. Tour, et al., Liquid crystals of aqueous, giant graphene oxide flakes, Soft Matter 7 (23) (2011) 11154−11159.

[42] S.E. Zakiyan, H. Azizi, I. Ghasemi, Influence of chain mobility on rheological, dielectric and electromagnetic interference shielding properties of poly methyl-methacrylate composites filled with graphene and carbon nanotube, Compos. Sci. Technol. 142 (2017) 10−19.

[43] T. Guo, C. Wang, H. Chen, Z. Li, Q. Chen, A. Han, et al., Rheological properties of graphene/tourmaline composite modified asphalt, Pet. Sci. Technol. 37 (21) (2019) 2190−2198.

[44] M. Lin, Z.L. Wang, P.W. Yang, P. Li, Micro-structure and rheological properties of graphene oxide rubber asphalt, Nanotechnol. Rev. 8 (1) (2019) 227−235.

[45] P. Jojibabu, M. Jagannatham, P. Haridoss, G.J. Ram, A.P. Deshpande, S.R. Bakshi, Effect of different carbon nano-fillers on rheological properties and lap shear strength of epoxy adhesive joints, Compos. A Appl. Sci. Manuf. 82 (2016) 53−64.

[46] M. Martin-Gallego, M.M. Bernal, M. Hernandez, R. Verdejo, M.A. López-Manchado, Comparison of filler percolation and mechanical properties in graphene and carbon nanotubes filled epoxy nanocomposites, Eur. Polym. J. 49 (6) (2013) 1347−1353.

[47] M. Keramati, I. Ghasemi, M. Karrabi, H. Azizi, M. Sabzi, Dispersion of graphene nanoplatelets in polylactic acid with the aid of a zwitterionic surfactant: evaluation of the shape memory behavior, Polym. Technol. Eng. 55 (10) (2016) 1039−1047.

[48] S. Basu, M. Singhi, B.K. Satapathy, M. Fahim, Dielectric, electrical, and rheological characterization of graphene—filled polystyrene nanocomposites, Polym. Compos. 34 (12) (2013) 2082−2093.

[49] C. He, X. She, Z. Peng, J. Zhong, S. Liao, W. Gong, et al., Graphene networks and their influence on free-volume properties of graphene−epoxidized natural rubber composites with a segregated structure: rheological and positron annihilation studies, Phys. Chem. Chem. Phys. 17 (18) (2015) 12175−12184.

[50] M. Sabzi, L. Jiang, F. Liu, I. Ghasemi, M. Atai, Graphene nanoplatelets as poly (lactic acid) modifier: linear rheological behavior and electrical conductivity, J. Mater. Chem. A 1 (28) (2013) 8253−8261.

[51] M. Sabzi, L. Jiang, N. Nikfarjam, Graphene nanoplatelets as rheology modifiers for polylactic acid: graphene aspect-ratio-dependent nonlinear rheological behavior, Ind. Eng. Chem. Res. 54 (33) (2015) 8175−8182.

[52] L. Yao, Y. Lu, Y. Wang, L. Hu, Effect of graphene oxide on the solution rheology and the film structure and properties of cellulose carbamate, Carbon 69 (2014) 552−562.

[53] C. Vallés, R.J. Young, D.J. Lomax, I.A. Kinloch, The rheological behaviour of concentrated dispersions of graphene oxide, J. Mater. Sci. 49 (18) (2014) 6311−6320.

[54] C.W. Macosko, Rheology: Principles, Measurements, and Applications, Wiley-VCH, New York, 1994.

[55] H.B. Zhang, W.G. Zheng, Q. Yan, Z.G. Jiang, Z.Z. Yu, The effect of surface chemistry of graphene on rheological and electrical properties of polymethylmethacrylate composites, Carbon 50 (14) (2012) 5117−5125.

[56] K. Liu, S. Ronca, E. Andablo-Reyes, G. Forte, S. Rastogi, Unique rheological response of ultrahigh molecular weight polyethylenes in the presence of reduced graphene oxide, Macromolecules 48 (1) (2015) 131−139.

[57] J.R. Potts, D.R. Dreyer, C.W. Bielawski, R.S. Ruoff, Graphene-based polymer nanocomposites, Polymer 52 (1) (2011) 5−25.

[58] M.Z. Iqbal, A.A. Abdala, V. Mittal, S. Seifert, A.M. Herring, M.W. Liberatore, Processable conductive graphene/polyethylene nanocomposites: effects of graphene dispersion and polyethylene blending with oxidized polyethylene on rheology and microstructure, Polymer 98 (2016) 143−155.

[59] H. Chang, H. Wu, Graphene-based nanocomposites: preparation, functionalization, and energy and environmental applications, Energy Environ. Sci. 6 (12) (2013) 3483.

[60] T.H. Han, W.J. Lee, D.H. Lee, J.E. Kim, E.Y. Choi, S.O. Kim, Peptide/graphene hybrid assembly into core/shell nanowires, Adv. Mater. 22 (18) (2010) 2060−2064.

[61] J. Liu, S. Fu, B. Yuan, Y. Li, Z. Deng, Toward a universal "adhesive nanosheet" for the assembly of multiple nanoparticles based on a protein-induced reduction/decoration of graphene oxide, J. Am. Chem. Soc. 132 (21) (2010) 7279−7281.

[62] O.C. Compton, D.A. Dikin, K.W. Putz, L.C. Brinson, S.T. Nguyen, Electrically conductive "alkylated" graphene paper via chemical reduction of amine—functionalized graphene oxide paper, Adv. Mater. 22 (8) (2010) 892−896.

[63] J.J. Bai, G.S. Hu, J.T. Zhang, B.X. Liu, J.J. Cui, X.R. Hou, et al., Preparation and rheology of isocyanate functionalized graphene oxide/thermoplastic polyurethane elastomer nanocomposites, J. Macromol. Sci. B 58 (3) (2019) 425−441.

[64] S.M. Oh, H.I. Lee, H.M. Jeong, B.K. Kim, The properties of functionalized graphene sheet/poly (ethyl methacrylate) nanocomposites: the effects of preparation method, Macromol. Res. 19 (4) (2011) 379.

[65] J. Trinidad, B.M. Amoli, W. Zhang, R. Pal, B. Zhao, Effect of SDS decoration of graphene on the rheological and electrical properties of graphene-filled epoxy/Ag composites, J. Mater. Sci. Mater. Electron. 27 (12) (2016) 12955−12963.

[66] J.W. Shen, W.Y. Huang, S.W. Zuo, J. Hou, Polyethylene/grafted polyethylene/graphite nanocomposites: preparation, structure, and electrical properties, J. Appl. Polym. Sci. 97 (1) (2005) 51−59.

[67] A. Chaudhry, V. Mittal, High-density polyethylene nanocomposites using masterbatches of chlorinated polyethylene/graphene oxide, Polym. Eng. Sci. 53 (1) (2013) 78−88.

[68] F. Beckert, A. Held, J. Meier, R. Mülhaupt, C. Friedrich, Shear-and temperature-induced graphene network evolution in graphene/polystyrene nanocomposites and its influence on rheological, electrical, and morphological properties, Macromolecules 47 (24) (2014) 8784−8794.

[69] O.J. Botlhoko, J. Ramontja, S.S. Ray, Morphological development and enhancement of thermal, mechanical, and electronic properties of thermally exfoliated graphene oxide-filled biodegradable polylactide/poly (ε-caprolactone) blend composites, Polymer 139 (2018) 188−200.

[70] A.M. Varghese, V.M. Rangaraj, S.C. Mun, C.W. Macosko, V. Mittal, Effect of Graphene on polypropylene/maleic Anhydride-graft-ethylene−vinyl acetate (PP/EVA-g-MA) blend: mechanical, thermal, morphological, and rheological properties, Ind. Eng. Chem. Res. 57 (23) (2018) 7834−7845.

[71] C.T. Huang, L.K. Shrestha, K. Ariga, S.H. Hsu, A graphene−polyurethane composite hydrogel as a potential bioink for 3D bioprinting and differentiation of neural stem cells, J. Mater. Chem. B 5 (44) (2017) 8854−8864.

[72] E. Ivanov, H. Velichkova, R. Kotsilkova, S. Bistarelli, A. Cataldo, F. Micciulla, et al., Rheological behavior of graphene/epoxy nanodispersions, Appl. Rheol. 27 (2) (2017) 1−9.

[73] E. Narimissa, R.K. Gupta, N. Kao, H.J. Choi, S.N. Bhattacharya, The comparison between the effects of solvent casting and melt intercalation mixing processes on different characteristics of polylactide-nanographite platelets composites, Polym. Eng. Sci. 55 (7) (2015) 1560−1570.

[74] R. Bouhfid, F.Z. Arrakhiz, A. Qaiss, Effect of graphene nanosheets on the mechanical, electrical, and rheological properties of polyamide 6/acrylonitrile−butadiene−styrene blends, Polym. Compos. 37 (4) (2016) 998−1006.

[75] J. Canales, M.E. Muñoz, M. Fernández, A. Santamaría, Rheology, electrical conductivity and crystallinity of a polyurethane/graphene composite: implications for its use as a hot-melt adhesive, Compos. A Appl. Sci. Manuf. 84 (2016) 9−16.

[76] C. Gao, S. Zhang, F. Wang, B. Wen, C. Han, Y. Ding, et al., Graphene networks with low percolation threshold in ABS nanocomposites: selective localization and electrical and rheological properties, ACS Appl. Mater. Interface 6 (15) (2014) 12252−12260.

[77] B. Adhikari, J. Nanda, A. Banerjee, Pyrene-containing peptide-based fluorescent organogels: inclusion of graphene into the organogel, Chem. Eur. J. 17 (41) (2011) 11488−11496.

[78] B.G. Compton, N.S. Hmeidat, R.C. Pack, M.F. Heres, J.R. Sangoro, Electrical and mechanical properties of 3D-printed graphene-reinforced epoxy, JOM 70 (3) (2018) 292−297.

[79] S. Han, B.C. Chun, Preparation of polyurethane nanocomposites via covalent incorporation of functionalized graphene and its shape memory effect, Compos. A Appl. Sci. Manuf. 58 (2014) 65−72.

[80] M. Kotal, S.S. Banerjee, A.K. Bhowmick, Functionalized graphene with polymer as unique strategy in tailoring the properties of bromobutyl rubber nanocomposites, Polymer 82 (2016) 121−132.

[81] A. Kasgoz, D. Akın, A. Durmus, Rheological behavior of cycloolefin copolymer/graphite composites, Polym. Eng. Sci. 52 (12) (2012) 2645−2653.

[82] R. Niu, J. Gong, D. Xu, T. Tang, Z.Y. Sun, Influence of molecular weight of polymer matrix on the structure and rheological properties of graphene oxide/polydimethylsiloxane composites, Polymer 55 (21) (2014) 5445−5453.

[83] R. Umer, Y. Li, Y. Dong, H.J. Haroosh, K. Liao, The effect of graphene oxide (GO) nanoparticles on the processing of epoxy/glass fiber composites using resin infusion, Int. J. Adv. Manuf. Technol. 81 (9−12) (2015) 2183−2192.

[84] B. Yang, Y. Shi, J.B. Miao, R. Xia, L.F. Su, J.S. Qian, et al., Evaluation of rheological and thermal properties of polyvinylidene fluoride (PVDF)/graphene nanoplatelets (GNP) composites, Polym. Test. 67 (2018) 122−135.

[85] J. Richeton, G. Schlatter, K.S. Vecchio, Y. Rémond, S. Ahzi, A unified model for stiffness modulus of amorphous polymers across transition temperatures and strain rates, Polymer 46 (19) (2005) 8194−8201.

[86] A. Acar, O. Colak, J.P.M. Correia, S. Ahzi, Cooperative-VBO model for polymer/graphene nanocomposites, Mech. Mater. 125 (2018) 1−13.

[87] S. Kashi, R.K. Gupta, N. Kao, S.N. Bhattacharya, Viscoelastic properties and physical gelation of poly (butylene adipate-co-terephthalate)/graphene nanoplatelet nanocomposites at elevated temperatures, Polymer 101 (2016) 347−357.

[88] X. Sun, Q. Wu, J. Zhang, Y. Qing, Y. Wu, S. Lee, Rheology, curing temperature and mechanical performance of oil well cement: combined effect of cellulose nanofibers and graphene nano-platelets, Mater. Des. 114 (2017) 92−101.

[89] A. Kelarakis, K. Yoon, R.H. Somani, X. Chen, B.S. Hsiao, B. Chu, Rheological study of carbon nanofiber induced physical gelation in polyolefin nanocomposite melt, Polymer 46 (25) (2005) 11591−11599.

[90] P. Manivel, S. Kanagaraj, A. Balamurugan, N. Ponpandian, D. Mangalaraj, C. Viswanathan, Rheological behavior and electrical properties of polypyrrole/thermally reduced graphene oxide nanocomposite, Colloids Surf. A Physicochem. Eng. Asp. 441 (2014) 614−622.

[91] Z. Tang, R. Huang, C. Mei, X. Sun, D. Zhou, X. Zhang, et al., Influence of cellulose nanoparticles on rheological behavior of oil well cement-water slurries, Materials 12 (2) (2019) 291.

[92] W. Ouarhim, F.Z.S.A. Hassani, R. Bouhfid, Rheology of polymer nanocomposites, Rheology of Polymer Blends and Nanocomposites, Elsevier, 2020, pp. 73−96.

[93] J.J. Moyano, J. Mosa, M. Aparicio, D. Pérez-Coll, M. Belmonte, P. Miranzo, et al., Strong and light cellular silicon carbonitride—reduced graphene oxide material with enhanced electrical conductivity and capacitive response, Addit. Manuf. 30 (2019) 100849. Available from: https://doi.org/10.1016/j.addma.2019.100849.

[94] W. Sarwar, K. Ghafor, A. Mohammed, Modeling the rheological properties with shear stress limit and compressive strength of ordinary Portland cement modified with polymers, J. Build. Pathol. Rehab. 4 (1) (2019) 25.

[95] M. Alkhamis, A. Imqam, New cement formulations utilizing graphene nano platelets to improve cement properties and long-term reliability in oil wells, in: Proceedings of the SPE Kingdom of Saudi Arabia Annual Technical Symposium and Exhibition. Society of Petroleum Engineers, 2018.

[96] Q. Lu, H.S. Jang, W.J. Han, J.H. Lee, H.J. Choi, Stimuli-responsive graphene oxide-polymer nanocomposites, Macromol. Res. 27 (11) (2019) 1061−1070.

[97] J. Canales, M. Fernández, J.J. Peña, M.E. Muñoz, A. Santamaría, Rheological methods to investigate graphene/amorphous polyamide nanocomposites: aspect ratio, processing, and crystallization, Polym. Eng. Sci. 55 (5) (2014) 1142—1151.

[98] M.P. Weir, D.W. Johnson, S.C. Boothroyd, R.C. Savage, R.L. Thompson, S.M. King, et al., Distortion of chain conformation and reduced entanglement in polymer—graphene oxide nanocomposites, ACS Macro. Lett. 5 (4) (2016) 430—434.

[99] M. Xiang, C. Li, L. Ye, Polyamide 6/reduced graphene oxide nano-composites prepared via reactive melt processing: formation of crystalline/network structure and electrically conductive properties, J. Polym. Res. 26 (5) (2019) 104.

[100] A.K. Pandey, R. Kumar, V.S. Kachhavah, K.K. Kar, Mechanical and thermal behaviours of graphite flake-reinforced acrylonitrile—butadiene—styrene composites and their correlation with entanglement density, adhesion, reinforcement and C factor, RSC Adv. 6 (56) (2016) 50559—50571.

[101] L.A. Pothan, Z. Oommen, S. Thomas, Dynamic mechanical analysis of banana fiber reinforced polyester composites, Compos. Sci. Technol. 63 (2) (2003) 283—293.

[102] C. Liu, J. Zhang, J. He, G. Hu, Gelation in carbon nanotube/polymer composites, Polymer 44 (24) (2003) 7529—7532.

Electromagnetic interference shielding property of polymer-graphene composites

8

Jonathan Tersur Orasugh[1,2,3], Chandrika Pal[1], Mir Sahidul Ali[1] and Dipankar Chattopadhyay[1,3]
[1]Department of Polymer Science and Technology, University of Calcutta, Kolkata, India, [2]Department of Jute and Fiber Technology, Institute of Jute Technology, University of Calcutta, Kolkata, India, [3]Center for Research in Nanoscience and Nanotechnology, Acharya Prafulla Chandra Roy Sikhsha Prangan, University of Calcutta, Kolkata, India

8.1 Introduction to electromagnetic radiation

The facile synthesis of novel electromagnetic interference (EMI) shielding materials based on polymer nanocomposites has gained increased responsiveness from the academia industry in comparison to conventional metal-based EMI shielding materials due to their ease of processing, being lightweight, and low cost.

Electromagnetic (EM) radiation is generated whenever charges accelerate. Electromagnetic radiation refers to the waves of the electromagnetic field (one electrical field and the other is magnetic field orthogonal to each other as depicted in (Fig. 8.1) transporting energy and momentum, self-propagating through a medium). EM waves (Fig. 8.2) require no medium, they can travel through empty space and

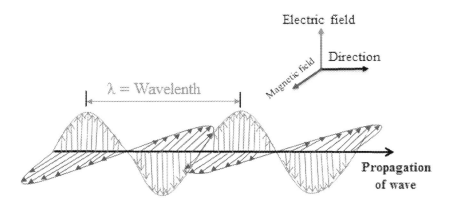

Figure 8.1 Graphical representation of electromagnetic wave.

Polymer Nanocomposites Containing Graphene. DOI: https://doi.org/10.1016/B978-0-12-821639-2.00006-9

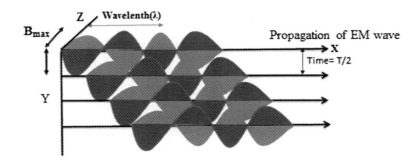

Figure 8.2 A segment of traveling electromagnetic wave with in-phase mutually orthogonal E and B.

also a medium, depending upon its frequency EM waves include radio waves, microwaves (MW), infrared radiation, ultraviolet-visible (UV−Vis) light, X-rays, and gamma (γ) rays, etc. [1−8]. Electromagnetic waves can have any wavelength λ, frequency f, then we can write the formula that relates the frequency (f) and wavelength (λ) of an EM wave is given by $v = f\lambda = \lambda/v$, where v is the velocity and υ is the wavenumber of EM wave. The EM wave traveling in free space has v equivalent to that of the speed of light (c); when electromagnetic waves propagate through a medium, the speed of the EM waves in the medium is $v = c/n(\lambda_{\text{free}})$, where $n(\lambda_{\text{free}})$ is the index of refraction of the medium. The index of refraction n is a property of the medium, and it depends on the wavelength (λ_{free}) of the EM wave. The velocity of EM waves also depends on traveling medium permeability and medium permittivity, $v = \frac{1}{\sqrt{\mu_0 \varepsilon_0}}$ where μ_0 and ε_0 represent the permeability and permittivity of the space medium respectively.

The speed of any electromagnetic waves in free space is the speed of light $v = c = 2.998 \times 10^8$ m/s, where $\mu_0 \simeq 4\pi \times 10^{-7} \frac{H}{m}, \varepsilon_0 \simeq 8.85 \times 10^{-7} \text{F/m}$

The formula that relates electric permittivity (ε) and magnetic permeability (μ) of an EM radiation in a particular medium, can be written *as* $\varepsilon = \varepsilon' - j\varepsilon'' = \varepsilon_0 \varepsilon_r$ and $\mu = \mu_0 \mu_r$ respectively (where ε' represents the real part of permittivity or $\varepsilon = \varepsilon' - j\varepsilon'' = \varepsilon_0 \varepsilon_r$ and $\mu = \mu_0 \mu_r$ dielectric constant ε'' represents the imaginary part of permittivity or dielectric loss and ε_r and μ_r represents relative electric permittivity and relative magnetic permeability of the particular medium respectively).

- The mathematical equation of EM waves

A plane EM wave traveling in the x-direction is the form of

$$E(x, t) = E_{\max} \sin(kx - \omega t), \tag{8.1}$$

$$B(x, t) = B_{\max} \sin(kx - \omega t), \tag{8.2}$$

where E is the electric field vector, and B is the magnetic field vector of the EM radiation, $k = 2\pi/\lambda$, and $\omega = 2\pi/T$. The direction of propagation is the direction of $E \times B$.

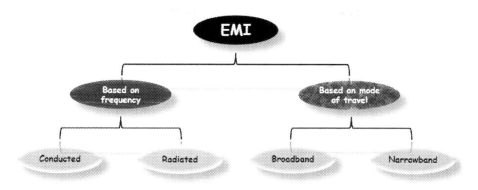

Figure 8.3 Categories of electromagnetic interference.

EMI is generally categorized based on its characteristics of frequency (υ) and mode of travel as shown in Fig. 8.3 [7].

8.2 Effects of electromagnetic radiation on the human body

The area of EM radiation and its interference came into focus in 1933 by CISPR (International Special Committee on Radio Interference), a subcommittee of International notable oil Commission (IEC) in Paris, for analyzing the long-term difficulties arising out from radio frequency uses. The popularity of radio, in 1820, as a primary home appliance, led to the rapid increase of RF radiations that served deleterious to other electronic devices, and thus, the awareness of EMI started to unleash among the electronic industries. CISPR, in 1934, started to build awareness on the approved RF emissions and the compatibility levels for electronic systems, which created a huge impact on the world's electromagnetic compatibility regulations. In the 1960s–1980s, researchers and scientific groups became alarmed about the interference of EM radiations in electronic systems, living beings, and the environment as a whole. The United States military, in 1967, revealed Mil standard 461 An in order to test and verify the radiation limits of electronic systems utilized in the military applications. In 1979, the Federal Communications Commission in the United States also imposed legal limits on EM emissions for all electronic devices, which is still effective today [8]. Knowingly or unknowingly, every human being is exposed to dangerous EM waves at different localities. The EM waves damage human cells and the extent of damage depends on the intensity of E and H, f, polarization, and their direction of propagation. The body temperature raises when the EM waves get in contact with the body tissues. The EM waves enter on the human tissues absorb part of them and get transmitted into the internal organs [9,10]. The effect of EM field strength, as well as specific absorption rate in areas near the cellular phone base stations, can affect human health critically. It is entirely unsafe to reside in these areas and it is advisable to live at least at a distance of 4 m away from these transmission sources [11]. Constant exposure to EM radiations may increase the risks of one's health such as skin problems, cancer, heart problems, headache, and several other minors and acute diseases [12,13].

Several studies reported that the constant use of cordless and cellular phones results in brain cancer and other dangerous diseases [8]. Several standards are developed in order to check and gauge the risk-free use of MW or RF radiations namely Australian Radiation Protection and Nuclear Safety Agency Standard (ARPANSA, 2002), the American Society for Testing and Materials (ASTM D4935-99), National Council on Radiation Protection and Measurement (NCRP, 1986) [8]. The safety limit for a specific absorption rate as set by most of the protection standards is up to 1.6 W/kg [9,11,14].

8.3 Electromagnetic interference shielding property

EMI is an unwanted electromagnetic induction triggered by extensive use of alternating current or voltage which creates corresponding induced signals in the nearby electronic device and tries to interfere with its performance [1]. EMI radiation is blocked which means the generated noises are stopped by using shielding materials (see Fig. 8.4). The conducted EMI uses filters to impede these noise formations in electronic circuits [2].

Various terminologies related to EM shielding and EM waves are very important in order to discuss and elaborate on the EMI shielding of materials [3].

8.3.1 Shielding theory

EMI shielding process is schematically shown in Fig. 8.5. It is basically a barrier to control the transmission of the EM wave through its bulk. In power electronics,

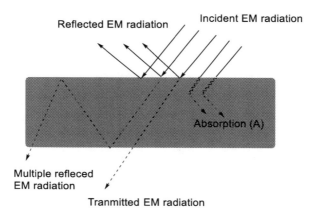

Figure 8.4 Schematic illustration of electromagnetic interference shielding mechanism.

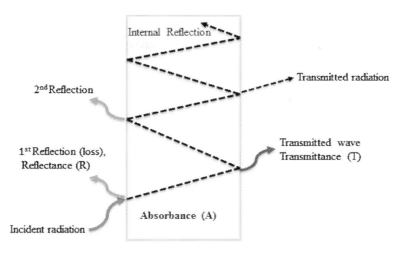

Figure 8.5 Schematic representation of electromagnetic interference shielding process of a material.

term shield usually refers to an enclosure that completely encloses an electronic device or a part of that product and prevents the EM emission from an outside source to deteriorate its electronic performance. The reflection loss or absorption loss or losses associated with multiple reflections are usually measured in decibels (dB). The shielding efficiency is generally measured in terms of reduction in the magnitude of incident power upon moving across the shield. According to the Schelkunoffs theory, mathematically shielding effectiveness or total EMI SE (SE_T) can be expressed in logarithmic function as per expressions below [4,5].

$$SE_T = SE_R + SE_A + SE_M = 10\log_{10}\frac{P_T}{P_I} = 20\log_{10}\frac{E_T}{E_I} = 20\log_{10}\frac{H_T}{H_I} \tag{8.3}$$

Where P_I, E_I, and H_I represent the power intensities, electric and magnetic field intensity of incident radiation while and P_T, E_T, and H_T represents the power intensities, electric and magnetic field intensity respectively of transmitted radiation. As shown in Fig. 8.5, three different mechanisms namely reflection (R), absorption (A), and multiple internal reflections (MIRs) contribute toward overall attenuation with reflection loss(SE_R), absorption loss (SE_A), and multiple internal reflection loss (SE_M)as corresponding shielding effectiveness components.

The reflection loss (SE_R) is associated with the relative impedance mismatch between the shield's surface and propagating wave. The generalized form of the reflection loss equation is given as [15].

$$SE_R = -10\log_{10}\left(\frac{\sigma_T}{16f\varepsilon_r\mu_r}\right) \tag{8.4}$$

Where σ_T is the total conductivity (in S/cm).

From Eq. (8.4) above, it can be observed that SE_R decreases with an increase in frequency for a given material (constant σ_T and μ_r).

Absorption loss (SE_A) is a physical characteristic of the shield that does not depend on the type of source field used [8]. When the EM wave passes via a medium, the amplitude of the EM wave is found to reduce exponentially. This type of decay loss EM radiation happens due to currents induced in the medium produce ohmic losses and heating of the material. E_T and H_T can be expressed as $E_T = E_I e^{-t/\partial}$ and $H_T = H_I e^{-t/\partial}$ [3], where ∂ is skin depth ($\partial = \frac{1}{\sqrt{\pi f \mu \sigma}}$) of the and t is the shield thickness of the shielding material. SE_A can be expressed by the following equation;

$$SE_A = 3.34t\sqrt{f\mu_r\sigma_r} \tag{8.5}$$

Eq. (8.5) revealed that SE_A is proportional to the square root of the product of μ_r and σ_r, By rearranging the above equation we get the following expression for EMI SE_A in decibel;

$$SE_A = -20\frac{t}{\partial}\log_{10}e = -8.68\frac{t}{\partial} = -8.68t\frac{\sqrt{f\sigma_T\mu_r}}{2} \tag{8.6}$$

From Eq. (8.6), it can be said that with an increase in f value, the SE_A also increases.

For a thin layered shielding material, the reflected EM wave reflected again and again in between the first and second boundary is shown in Fig. 8.5. The reduction due to these multiple internal reflections (SE_M) can be mathematically expressed as:

$$SE_M = 20\log_{10}(1 - e^{-2t/\partial}) = 20\log_{10}\left(1 - 10^{-\frac{SE_A}{10}}\right) \tag{8.7}$$

Therefore, it can be understood from the above mathematical expression that SE_M is closely related to absorption loss (SE_A). SE_M plays a crucial role in porous structures, in certain composites materials and also, in certain design morphologies (or geometries). SE_M can be neglected for thick shields wall, which may be due to the fact that the amplitude of the absorbed waves becomes negligible by the time it reaches the second boundary due to high SE_A value.

The shielding basics are based on the transmission line theory and the plane wave shielding theory [6]. By assuming a uniform plane wave characteristic by E and H that varies within a plane only with x-direction as shown in Fig. 8.6, Maxwell's curl equation is given as;

$$\frac{dE}{dx} = -j\omega\mu H \text{ and } \frac{dH}{dx} = -(\sigma + j\omega\varepsilon)E \tag{8.8}$$

where μ is the permeability of the material and $\mu = \mu_0\mu_r$, μ_0 and μ_r are the permeability of air (or free space medium) and shield material respectively, σ is the conductivity of the material in S/m. Where ε is the permittivity of the material, $\varepsilon = \varepsilon_0\varepsilon_r$. ε_0 and ε_r

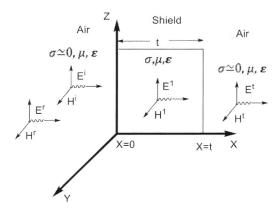

Figure 8.6 Propagation of electromagnetic waves and interaction with shielding material.

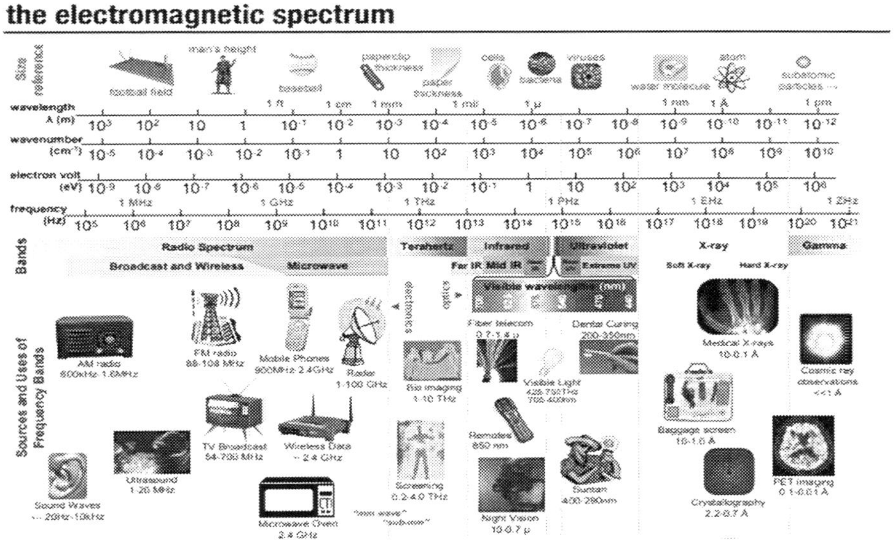

the electromagnetic spectrum

Figure 8.7 Electromagnetic spectrum.

are the permittivity of air (or free space) and shield material respectively, $\omega = 2\pi f$:
where ω is the angular frequency and $f\,(=1/T)$ is the linear frequency in hertz (Hz).

A typical electromagnetic spectrum is represented in Fig. 8.7.

8.4 Polymer composites/nanocomposites

A composite is said to be a mixture of at least two or more immiscible materials
forming a novel material with properties that are quite distinct from those of the

individual material. There are numerous reports on the fabrication of novel graphene polymer composites and their hybrids for advanced applications especially in EMI shielding recently due to the attractive properties of these composites [16−91].

8.5   Graphene-based polymer nanocomposites

This portion of the chapter focuses on using graphene as a nanofiller in diverse polymeric nanocomposite materials such as epoxy, polyaniline (PANI), polystyrene (PS), poly (vinylidene fluoride) (PVDF), polyurethane (PU), Nafion, polycarbonate (PC), Poly(ethylene terephthalate) (PET), poly(3,4-ethyldioxythiophene) (PEDOT), etc. as shown in Table 8.1.

8.5.1   Preparation techniques

The synthesis technique for graphene polymer nanocomposites depends on the molecular weight, polarity, hydrophilicity, hydrophobicity, thermal properties, reactive groups, etc., graphene type, and solvent. There are three main approaches for incorporating graphene nanomaterials into the polymer for the development of diverse polymer-graphene nanocomposites as represented below:

8.5.1.1   In situ intercalative polymerization

In this approach, graphene or its modified form is dispersed and swollen in a liquid monomer. Thereafter, a preselected initiator is added to initiate polymerization by heat or light/radiation [29]. Several polymer-graphene nanocomposites via this method have been reported for the preparation of graphene/PS [29−31], expanded graphite/polymethylmethacrylate (PMMA) [32], etc.

8.5.1.2   Solution intercalation

This method employs a polymer or prepolymer that is soluble in a specific solvent (like water, acetone, chloroform, tetrahydrofuran (THF), dimethylformamide (DMF) or toluene) system and graphene or it is modified is swollen in it and then the polymer is added into the preswollen mixture and afterward the solvent is evaporated [33]. This method has been reportedly employed for the preparation of graphene-polymer nanocomposites [34−41].

8.5.1.3   Melt intercalation

In this system, graphene or its modified form is mixed mechanically with the thermoplastic polymer matrix in the molten form (leading to intercalated or exfoliated of the graphene in the matrix) using conventional polymer processing methods such as extrusion and injection molding [38,39,42]. An extensive variety of graphene-polymer nanocomposites have been reportedly prepared by this approach [39,43,44].

Table 8.1 Some literature on graphene based polymer nanocomposites.

Polymer	% Graphene/Graphite in Polymer	Nanocomposite Nature	EMI Shielding Effectiveness	References
Polycarbonate	**0.5 wt.%**	Foam Slab	$\sim$39 dB $\cdot$ cm^3/g $\sim$1.1 dB $\cdot$ cm^3/g	**[16]**
Epoxy	15 wt.%	Slab	21 dB	[17]
Epoxy	0.1 wt.% rGO and 20.4 wt.% GNPs	Foam	51 dB	[18]
Epoxy	0.5 wt.%	Slab	5.93–38.8	[19]
Epoxy	0.1 wt.%	Laminate slab	10 dB	[20]
Epoxy	0.5 wt.%		20–26 dB	[21]
Epoxy	5 wt.% rGO	Slab	25.748 dB	[22]
poly methyl methacrylate	1.8 vol.%	Foam	13–19 dB	[23]
Polyaniline nanorods embedded in a paraffin matrix	20 wt.%	Foam/nanorods	20–45 dB	[24]
Epoxy-GraftedGraphene	0.5 wt.%	Fibers	70 dB	[25]
PVA	50 wt.%	Slab	17–32 dB	[26]
Polyetherimide (PEI)	0.18 vol.%	Foam	44 dB	[18]
polyvinylidene fluoride (PVDF)	5 wt.%	Foam	20 dB	[27]
Polyurethane (PU)	0.3 wt.%	Foam	16 dB	[28]

8.5.2 Factors affecting electromagnetic interference shielding effectiveness properties of polymer graphene composites

There is a pronounced deal of research focused to the design and engineering of graphene polymer composite materials for EMI SE due to the orientation of graphene nanosheets in a spatially segregated 3D structural design which forms a continuous nanofiller architecture achieved via confinement of the nanosheets into a predetermined volume of the polymer matrix. Undeniably, the actualization of 3D graphene ghettoized structural design permits proper development of the polymer composites performance leading to substantial enhancement with respect to its structural properties such as barrier, dynamic mechanical properties, mechanical properties, etc. along with its functional properties like EMI SE, sensing ability, electrical properties, thermal properties, and adsorption/desorption propensities, etc. Table 8.2 shows the EMI SE property of graphene polymer composites, processed from various polymers and graphenes of varying nature and the effect of their thickness, processing methods, etc. on the EMI SE as found in the literature.

Also, graphene polymer composites are categorized into two different groups; namely bulk and porous (foam or aerogels) possessing the peculiar orientation of graphene nanoplates in the segregated and continuous 3D assemblage. In graphene polymer composites, the synthesis techniques have a great influence on the resulting architecture which correlates their properties of the composite as aforementioned. EMI SE of polymer graphene composite is reliant on the percolation threshold of graphene in the polymer composites along with the properties of the conducting fillers, their dispersion, and interfacial interactions between the filler particles and the polymer matrix [82].

8.5.2.1 Nature of polymers

The nature of the polymeric material used for the preparation of any nanocomposites or composites especially in EMI SE has a great influence on the resultant properties because of the difference in the physicochemical properties of these polymers. Insulating/nonconductive polymers such PVC have been reported to hardly exhibit any EMI SE property [48]. Conductive polymer composites (CPCs) containing conductive graphene nanofillers have been widely considered as an alternative candidate for EMI shielding and electrostatic discharge (ESD) protection applications based on their attractive properties like lightweight, low cost, good processability, and wide range adjustable conductivities. Table 8.3 shows the EMI SE properties of graphene polymer composites within the X-band frequency region as obtained from the literature.

Literature reports have proven that the use of conducting polymers for EMI shielding for the fabricating of conducting polymer composites presents numerous beneficial properties (including high conductivity, environmental stability, lightweight, low cost, microwave absorption, good processability, high resistance to corrosion, etc.) for EMI shielding applications [49]. Graphene-based CPCs shielding behavior is realized through three mechanisms namely: reflection, absorption, and

Table 8.2 EMI SE properties of graphene polymer composites within the X-band frequency region.

Nature of polymers	Composite thickness	Nature of graphene	Concentration/loading of graphene	Preparation technique	EMI SE (dB)	References
PVC	1.80 mm	Graphene	5 wt.%	Solution intercalation-solvent casting	13.0	[48]
PVDF	–	Graphene	7 wt.%	In situ polymerization	28.0	[26]
PEI	2.30 mm	Graphene	10 wt.%	In situ intercalation-solvent casting	11.0	[17]
PU	20.00 mm	rGO	3 mg/mL	dip-coating	12.4	[58]
PU	2.00 mm	Graphene	7.7 wt.%	Solution mixing-solvent casting	35.0	[51]
PEDOT-PSS	1.15 mm	Graphene	58% weight fraction	CVD	91.9	[45]
PMMA	0.79 mm	Graphene	5.0 wt.%	Solution blending-melt compounding-batch foaming process	19.0	[22]
PMMA	3.40 mm	Graphene	4.23 vol.%	Batch foaming process	30.0	[50]
PS	2.80 mm	Graphene	10.0 wt.%	Compression moulding	18.0	[52]
PS	2.50 mm	Graphene	7.0 wt.%	High-pressure solid-phase compression moulding	45.1	[53]
Epoxy	–	rGO	15.0 wt.%	In situ intercalation-solvent casting	21.0	[16]

(Continued)

Table 8.2 (Continued)

Nature of polymers	Composite thickness	Nature of graphene	Concentration/loading of graphene	Preparation technique	EMI SE (dB)	References
Epoxy	3 mm	GNPs/rGO	20.0 wt.%	In situ intercalation-solvent casting-templating method	51.0	[15]
PDMS	1.00 mm	Graphene	0.8 wt.%	Template-directed CVD	30.0	[74]
PI	0.80 mm	rGO	16.0 wt.%	Solvent casting via in situ polymerization	17.0–21.0	[75]
UHMWPE	2.50 mm	rGO	0.66 vol.%	Melt compounding/hot press	28.3–32.4	[76]
PVDF	2.00 mm	Graphene	15.0 wt.%	Water vapour induced phase separation	45.6	[77]
PS	2.50 mm	Graphene	30.0 wt.%	High-pressure compression moulding plus salt leaching	64.4	[78]
PMMA	2.00 mm	Graphene	20.0 wt.%	Solvent casting	21.0	[79]
PVC	2.00 mm	Graphene	20.0 wt.%	Solvent casting	31.0	[79]
PPY	–	Graphene	5.00 wt.%	In situ polymerization	53.0	[49]
PPY	2.50 mm	rGO	–	In situ polymerization	48.0	[81]
PANI	5.00 μm	Graphene	1.00 wt.%	In situ polymerization-solution intercalation-casting	32.0–42.0	[47]

PANI	2.00–7.00 mm	GO	25 wt.%	In situ polymerization-conventional painting technique	20.0–33.0	[85]
PU	2.44 mm	Graphene	20 wt.%	Solution intercalation-solvent casting	17.0–20.0	[86]
PS	–	TrGO	2.24 wt.%	Solution mixing-solvent casting-compression moulding	30.0	[87]
Epoxy	3.00 mm	3D grapheme Nanoplatelets (GNPs)/rGO		Template method	30.0	[88]
PDMS	0.1 mm	Graphene	10.00 wt.%	Solution mixing-solvent casting- microwave irradiation	130–134	[89]

Table 8.3 EMI SE properties of graphene polymer composites within the X-band frequency region.

Nature of polymers	Composite thickness (mm)	Nature of graphene	Concentration/ loading of graphene	Preparation technique	EMI SE (dB)	References
PVC	1.80	Graphene	5 wt.%	Solution intercalation-solvent casting	13.0	[48]
PVDF	–	Graphene	7 wt.%	In situpolymerization	28.0	[26]
PEI	2.30	Graphene	10 wt.%	In situ intercalation-solvent casting	11.0	[17]
PU	20.00	rGO	3 mg/mL	Dip-coating	12.4	[58]
PU	2.00	Graphene	7.7 wt.%	Solution mixing-solvent casting	35.0	[51]
PEDOT-PSS	1.15	Graphene	58% weight fraction	Chemical vapor deposition (CVD)	91.9	[45]
PMMA	0.79	Graphene	5.0 wt.%	Solution blending-melt compounding-batch foaming process	19.0	[22]
PMMA	3.40	Graphene	4.23 vol.%	Batch foaming process	30.0	[50]
PS	2.80	Graphene	10.0 wt.%	Compression molding	18.0	[52]
PS	2.50	Graphene	7.0 wt.%	High-pressure solid-phase compression molding	45.1	[53]
Epoxy	–	rGO	15.0 wt.%	In situ intercalation-solvent casting	21.0	[16]
Epoxy	3	GNPs/rGO	20.0 wt.%	In situ intercalation-solvent casting-templating method	51.0	[15]
PDMS	1.00	Graphene	0.8 wt.%	Template-directed chemical vapor deposition (CVD)	30.0	[74]

Polymer		Filler	Content	Method	Value	Ref.
Polyimide (PI)	0.80	rGO	16.0 wt.%	Solvent casting via in situ polymerization	17.0–21.0	[75]
UHMWPE	2.50	rGO	0.66 vol.%	Melt compounding/hot press	28.3–32.4	[76]
PVDF	2.00	Graphene	15.0 wt.%	Water vapor induced phase separation	45.6	[77]
PS	2.50	Graphene	30.0 wt.%	High-pressure compression molding plus salt leaching	64.4	[78]
PMMA	2.00	Graphene	20.0 wt.%	Solvent casting	21.0	[79]
PVC	2.00	Graphene	20.0 wt.%	Solvent casting	31.0	[79]
PPY	–	Graphene	5.00 wt.%	In situ polymerization	53.0	[49]
PPY	2.50	rGO	–	In situ polymerization	48.0	[81]
PANI	5.00 μm	Graphene	1.00 wt.%	In situ polymerization-solution intercalation-casting	32.0–42.0	[47]
PANI	2.00–7.00	GO	25 wt.%	In situ polymerization—conventional painting technique	20.0–33.0	[85]
PU	2.44	Graphene	20 wt.%	Solution intercalation-solvent casting	17.0–20.0	[86]
PS	–	TrGO	2.24 wt.%	Solution mixing-solvent casting-compression molding	30.0	[87]
Epoxy	3.00	3D graphene Nanoplatelets (GNPs)/rGO		Template method	30.0	[88]
PDMS	0.1 mm	Graphene	10.00 wt.%	Solution mixing-solvent casting—microwave irradiation	130–134	[89]

multiple reflections of ER by the composites depending on the overall composition, design, and fabrication process employed. However, adoption of graphene-based segregated polymer composites (SPCs) processed mostly by the hot compacting method can be a worthy way out for attaining high EMI SE even at low filler content due to the formation of the ordered architecture of the conductive filler within the polymer matrix, as the conductive graphene filler particles are located on boundaries between the polymer matrix chains. Here, since the conductive filler particles such as graphene are packed closely in the partition of the structure at the filler content greater than the percolation threshold, the conductivity is in effect high even at low filler content. This kind of architecture can provide multiple reflections within graphene based SPCs resulting in increased overall absorption loss. Also, Mamunya et al., have shown from work that lower percolation thresholds "ϕc" = 0.21 and 0.55 vol.% were observed for the SPC composites of Gr/PE and TEG/PE in comparison to thermally treated anthracite (A)/PE composite having $\phi c = \sim 3$ vol.% for the SPC and $\phi_c = \sim 25$ vol.% for composite with random filler distribution. They reported that the ϕ_c for Gr and A filler is 100 and 10 times lower in the segregated system was 100 and 10 times lower than that for A filler with its random distribution in polymer matrix along with noticeably enhanced EMI SE value in the SPC system due to multiple internal reflections of EM wave within the structure of the segregated architecture [91]. Consequently, SPCs present itself potential indispensable advantages in EMI SE in comparison to CPCs which possesses random distribution of the conductive filler material.

8.5.2.2 Nature of graphenes

The EMI SE of graphene polymer nanocomposite is greatly affected by the nature of the graphene whether it be single, double, or multilayered; GO, reduced graphene oxide (TrGO), chemically reduced graphene (CrGO), their inorganic composite/hybrid form, modified form, etc. In comparison to the traditional approach of adding 0D, 1D, and 2D conductive graphene nanofillers, it has been shown that the EMI SE of GPC can be significantly enhanced using 3D graphene framework/polymer nanocomposites at a relatively lower filler percentage [15,89].

The used Multilayer Graphene sheets (MLG) has been reported to enhance EMI SE to a large extent [80], in comparison with single/few-layer graphene. In a study by Chen et al., the authors studied two polymer PS composites reinforced with thermally TrGO and CrGO [87]. Their report revealed that much higher electrical conductivity and EMI shielding effectiveness were achieved for TGO containing composites in comparison to those of RGO composites [87]. It has been shown that graphene polymer composites containing only a single-phase 3D graphene network hardly possessed admirable EMI SE due to defects on the surface of the reduced graphene and the low electrical conductivity [15,89], and lack of seamless conducting networks along with the existence of interface incompatibility [15]. However, Liang et al. recommended in their study that these challenges faced with single-phase 3D graphene can to a large extent be addressed by using a hybrid of 3D graphene nanoplatelets/reduced graphene oxide foam/epoxy (GNPs/rGO/EP) composites where 3D GNPs/rGO polymer composites presented better EMI SE compared to GNPs/EP [15]. It has been shown from previous reports that graphene nanoplatelets (Gr), thermally exfoliated graphite (TEG) possess better distribution within the polymer matrix and lower percolation thresholds "ϕc" = 0.21 and 0.55 vol.% were observed for the composites Gr/PE and TEG/PE in comparison to thermally treated anthracite (A)/PE composite having ϕc = ∼3 vol.% for the segregated composite and ϕc = ∼25 vol.% for composite with random filler distribution [91].

8.5.2.3 Concentration/loading of graphene

For the synthesis/preparation of graphene polymer-based s composites (solids and foams) for EMI SE application, a high content of graphene as shielding agents (∼5−30 wt.%) is largely needed for the attainment of the required EMI SE [50,75].

Li et al. found in their study that the EMI SE values revealed an increasing tendency with an increase in graphene content (4−20 wt.%) [86], with the graphene content at 20 wt.%, the EMI SE reached 17−21 dB. Table 8.4 depicts the effect of graphene concentration on the EMI SE of graphene polymer composites while Table 8.5 shows the effect of higher cell density with increased cell density on the EMI SE of the materials. Liang et al. in their study as depicted in Fig. 8.8A revealed that graphene/epoxy composites prepared exhibited SE between 0.8 and 21 dB in the X-band for a graphene loading from 0.1 to 15 wt.% loading [16].

Table 8.4 Electromagnetic interference shielding performance of respective graphene polymer composite foams

Polymer matrix	Graphene concentration (wt.%)	Density (g/cm)	EMI SE (dB)	References
PU	5	0.027	12.4−34.7	[58]
PU	10	0.030	19.9−57.7	[58]
PS	30	0.45	29.3	[78]
PS	30	0.27	17.3	[78]
PMMA	5.0	0.79	19.0	[50]
PVDF	7.0	−	28.0	[26]
PI	16.0	0.28	17−21	[90]
PEI	10	0.28−0.40	41.5	[58]
PEI	10	0.29	11.0	[17]
PDMS	0.8	0.06	20.0	[74]

Table 8.5 Average cell size, cell density, density, and EMI SE of PS/TRGO composites

The concentration of TRGO in PS (wt.%)	Average cell size (mm)	Cell density (× 10⁸ Cells cm⁻³)	Density of the foam (g/cm)	EMI SE (dB)	References
0	19.5	19	–	2	[52]
3	10.4	13	0.24	–	[52]
5	8.4	23	0.26	7.0	[52]
10	6.0	60	0.39	16–19	[52]
15	3.7	160	0.63	22–24	[52]
20	–	–	0.74	–	[52]

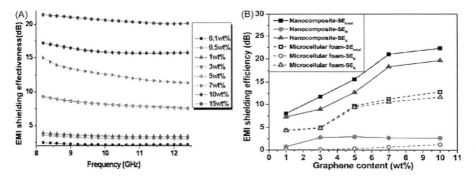

Figure 8.8 (A) EMI SE of graphene/epoxy composites with various solution-processable functionalized graphene (SPFG) loadings as a function of frequency in the X-band [16], (B) SE_{total}, SE_R, and SE_A of PEI/graphene nanocomposite and microcellular foams at 9.6 GHz [17].

8.5.3 Dispersion level of graphene

The poor dispersion of graphene in polar solvents ultimately affects its dispersion within the matrix such polymers during the processing of the composites. It is well known that poor dispersion of conductive fillers within the matrix of any polymer used for EMI SE application results in poor properties of the final composite material, while good dispersion of these fillers in different polymers matrices have shown excellent EMI SE from the literature [16,20–25,40–50].

Graphene based polymer composite foam with the amalgamation of hydroxyl and carbonyl groups functionalized graphene (f-G) and PVDF having good dispersion as a result of the functionalization of graphene for excellent dispersion within PVDF and applied for EMI shielding has been reported [16].

Properly dispersed graphene in a PVC matrix with the help of ferromagnetic Fe_3O_4 nanoparticles have been revealed to enhance fire retardancy of nanocomposites through the formation of a network-like structure as a result of the well-dispersed graphene along with improved magnetic and electrical properties resulting to high EMI SE within the X-band range of ~ 13 dB by Yao et al. [48]. Though, as a result of the dominance of the reflection mechanism in the nanocomposite, the authors reported that the EMI SE was lower in comparison with related graphene polymer-based composites: they suggested that the lower EMI SE could be improved upon by optimizing the mixing ratio between graphene and Fe3O4 nanoparticles. In another study, Gupta and coworkers modified the surface of graphene [multilayer graphene anchored with titanium dioxide (TiO_2)] to develop a novel graphene polymer polypyrrole (PPY) composite having enhanced absorption-dominated EMI SE of 53 dB along with good dispersion of graphene in PPY with the help of TiO_2 [49].

Chen et al., in their study of two polymer PS composite reinforced with TrGO and CrGO [87]. Their report revealed that much higher electrical conductivity and EMI shielding effectiveness were achieved for TGO containing composites in comparison to those of RGO composites [87]. Largely, GO has been acknowledged as a more suitable precursor for the fabrication of 3D graphene polymer framework due to its crumpled morphology with coiling edges and good dispersibility in aqueous media and most solvents [15] in comparison to graphene or rGO. Mamunya et al., have shown from work that Gr, TEG possess better distribution within the polymer matrix and lower percolation thresholds "ϕc" = 0.21 and 0.55 vol.% were observed for the composites Gr/PE and TEG/PE in comparison to thermally treated anthracite (A)/PE composite having ϕc = ~ 3 vol.% for the SPC and ϕc = ~ 25 vol.% for composite with random filler distribution These authors found from their work that the percolation threshold for Gr and A filler is 100 and 10 times lower in the segregated system was 100 and 10 times lower than that for A filler with its random distribution in the polymer matrix. Also, their report revealed noticeably enhanced EMI SE value in the SPC system due to multiple internal reflections of EM wave within the framework of the segregated architecture [91].

8.5.4   Processing conditions of composites

The low-slung price and availability of the raw graphite in large quantities along with comparatively facile industrially feasible solution process makes graphene a potential candidate as an effective conductive filler for the fabrication of conductive polymer composites for EMI SE application [26]. Graphene's high aspect ratio of supports in reinforcing the polymer matrix at lower loadings percentage. Recent research on graphene polymer composites having EMI shielding potential with the consideration of the type of polymer, the fabrication process, nanofiller content percolation threshold, and the EMI SE of the prepared composites along with the core shielding mechanism have been explored by scientist around the globe. One of the most utilized fabrication/processing techniques adopted so far is polymer foaming because it provides a greater advantage in EMI SE coupled with lightweight and the use of already convention foaming processing techniques. An increase in the thickness ($2-7$ mm) of the PANI-GNP-GO nanocomposite has been reported to provide a corresponding increase in EMI SE from 20 to 33 dB [85]. However, the foaming process presents itself as the most attractive approach because it can be used to prepare graphene polymer composites with high needed thickness but low weight.

Another process that has been used for the preparation of graphene polymer composite (GPC) is a hot compressing molding [52,53,76,78]. The EMI SE of compression molded composite depends on its thickness after processing low temperature, short compressing time, and high viscosity of the polymer melt which minimizes the migration of the conductive graphene filler particles into polymer molecules, thereby reducing the threshold concentration value.

Currently, researchers have focused on engineering graphene polymer composites with efficient EMI SE without neglecting the need for decreased energy consumption, manufacturing along with desired material properties like lightweight, excellent electromagnetic waves absorption, improved specific mechanical properties, controlled cell size/density, porosity, etc. Graphene polymer foams composites have presented themselves as potential candidates with the likelihood of controlled processing conditions for composition fit for EMI SE application [84]. The cellular structure of thermoplastic polymers which is the most common approach used for the preparation of graphene polymer composites for EMI SE by the utilization of foaming agents has been reportedly used by researchers [51,52,78]. Controllable cellular-structured graphene aerogel-epoxy composites (GA-EP) prepared using the infiltration method utilizing porous GA as a framework have been reported [88]. These groups of researchers obtained GA-EP composites having a thickness of 3 mm with an exhibited EMI SE of ~30 dB. It has been reported that graphene polymer composites containing only single-phase 3D graphene framework hardly possessed excellent EMI SE as a result of defects on the surface of the reduced graphene and the low electrical conductivity [15,89], and lack of seamless conducting networks along with the existence of interface incompatibility [15]. The aforementioned challenges have been to a large extent addressed by the adoption of template method (Fig. 8.9) by Liang et al. for the fabrication of 3D graphene nanoplatelets/reduced graphene oxide foam/epoxy (GNPs/rGO/EP) composites, in which 3D GNPs/rGO composites showed better EMI SE in comparison to GNPs/EP in their study [15].

8.5.4.1 Some other factors

Other factors affecting EMI SE are unarguably the density, porosity, crystallinity, morphology and cellular structure, etc. of the composites and the reinforcing graphene material or its hybrids. These factors consequently have a great influence on the specific EMI SE. For example, controlling the foaming process variables, the morphology along with the cellular structure of polymer graphene fabricated foams can be engineered to function effectively in EMI shielding application. This process can also adjust the graphene reinforcing filler orientation and dispersion resulting in desired architecture such as exfoliation of the graphene filler within the polymer matrix. Based on the literature, the saturation of the reinforcing filler material with

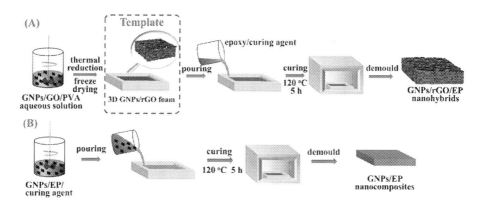

Figure 8.9 Schematic diagram of the fabrication for (A) 3D GNPs/rGO/EP nanocomposites by a template method and (B) GNPs/EP nanocomposites by the traditional blending-casting method [15].

a foaming agent can be performed either below or above the glass transition temperature (Tg) of the polymeric matrix in the batch foaming process. Again, whenever the dissolution temperature of the polymer is higher than the Tg, the release of pressure results in supersaturating along with cell nucleation and growth while the fixation of the cell structure is achieved by cooling the materials below it Tg [92]. In an event where the saturation temperature is less than the Tg, the network structured cells are not capable of nucleating and growing after release of the pressure and foaming may perhaps take place when the temperature is elevated above the Tg and the cellular structure is again secure through cooling the composite material [92]. The cellular structure which has a great influence on the morphology and consequently the absorption/reflection of EM waves is also majorly controlled by variables like the saturation graphene-polymer composite foam processing pressure, pressure drop rate, temperature along with the temperature ramp rate.

In PU based graphene composite, the cell size and cell density are among the factors influencing EMI SE has been reported pointed out by the literature on graphene polymer composites [83]. The reduction in cell size leads to an increase in cell density which in turn favors enhancement of EMI SE of GO-PU composites [83].

It has also been shown by Basavaraja et al. in their work that, increase in the thickness (2−7 mm) of the PANI-GNP-GO nanocomposite led to a corresponding increase in EMI SE (20−33 dB) [85]. Li et al. have also published a report which agrees with the report from Basavaraja et al. where they also affirmed that multilayered composites fabricated by simply stacking several layers of the materials together resulted in significant improvement of the EMI SE via enhanced multiple reflections on the internal interfaces [86]. It has been recently shown that the electrical conductivity of graphene polymer nanocomposite has a direct relationship with the EMI SE [91].

8.6   Applications focused on electromagnetic interference shielding

In the literature reports by Zhao et al., it is revealed that using two kinds of fillers namely, carbon nanotubes and graphene nanoplatelets on poly (vinylidene fluoride) with the Ni chains PVDF/CNT/Ni-chain and the PVDF/GNP/Ni-chain composites showed increased electrical conductivity and the EMI shielding properties with the increase in Ni filler content. Also, it was observed that the EMI shielding properties of PVDF-based composites were dependent on the thickness of the films [44]. Whereas Wu et al., concluded that using graphene foam (GF) which was drop coated on poly (3,4 ethylene dioxythiophene): poly(styrenesulfonate) [PEDOT: PSS] composites gave remarkable EMI shielding performance with shielding effectiveness (SE) of 91.9 dB and a specific SE (SSE) of 3124 dB cm^3/g, which were the highest among those reported in the literature for carbon-based polymer composites [45]. Reports by Modak et al., on polymer composites based on PANI and (graphene nanocomposites) GNP (PANI/GNP), inferred that EMI shielding effectiveness (SE) was found to be dependent on the amount of the graphene added to the nanocomposite [46].

Different reports published on the basis of the dispersion used for preparing graphene based nanocomposites stated that PANI-graphene films gave shielding efficiencies of around 42 and 32 dB, respectively with the C-band (4−8 GHz) and X-band frequency ranges, wherein they corresponded to more than 99.99% microwave attenuation [47]. In a report published by Yao et al., it was proven that graphene can be utilized for being dispersed into the matrix of PVC with the help of ferromagnetic Fe_3O_4 nanoparticles which enhanced the electrical and magnetic properties of the nanocomposites, showing high EMI SE in the X-band range (13 dB) [48]. There was a study reported by Gupta et al., wherein on the surface of graphene (Multiwalled Nanotubes) MWNTs where anchored which showed improved absorption-dominated EMI SE, specifically when multilayer graphene was anchored with TiO_2 which was combined with conductive PPY, it gave total SE of 53 dB in the high frequency range 12.4−18 GHz (Ku band) [49]. The EMI SE of graphene (at varying wt.%) based polymer nanocomposites and its X-band frequency range has been reported [50−54].

In a study reported by Shahzad and coworkers with the use of (reduced graphene oxide) along with polymer of sulfur nanocomposites (rGO-PS) gave higher electrical conductivity (150%higher) with improved EMI SE (24.5 dB when compared to the 21.4 dB of undoped rGO-PS nanocomposites) [54].

Lately, a research conducted by Singh et al., revealed that with the combination of both types of conductive carbon-based nanoparticles which are based on different morphologies of the polymer-like tubular CNTs and layered rGO had high influence on the alignment of the nanotubes on the polymer surface, it was then modified with magnetic Fe_3O_4 nanoparticles which were sandwiched between the rGO layers as depicted in Fig. 8.10. These showed a higher EMI shielding efficiency of 37 dB in the Ku-band (as shown in Fig. 8.10) [55].

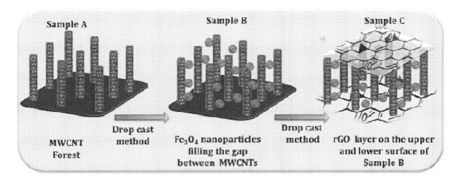

Figure 8.10 Schematic representation exhibits the infiltrated MWCNT forest (Sample B) and sandwiching the sample B in rGO sheets (Sample C) [55].

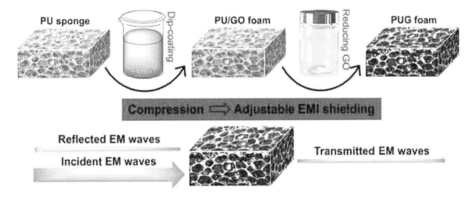

Figure 8.11 Pictorial representation of polyurethane sponge process for obtaining adjustable electromagnetic interference shielding [58].

Maiti and Khatua worked on optimizing the ratio of (Graphene Nanoplates) GnP and MWNTs with a polycarbonate matrix for preparation of the nanocomposites, they were successful to reach the EMI shielding efficiency of around 21.6 dB at a relatively low concentration of conductive nanofillers which was 4 wt.% and it was higher than that of individually used graphene or only CNTs [56].

Studies on nanocomposites done by Yousefi et al., comprised of rGO and epoxy nanocomposites showed high performance on EMI shielding with a remarkable shielding efficiency of 38 dB [57].

An article published by Shen bin et al. confirmed that PU sponge can be dip-coated with graphene oxide (GO) dispersion to produce polymer/graphene—(PU/graphene foam) nanocomposite and with the compression can have adjustable EMI shielding property. It is pictorially well depicted in Fig. 8.11 [58].

In a report published by Kar and coauthors, a dispersion of MWNTs with the blends of (Polyvinylidene fluoride) PVF and (acrylonitrile butadiene styrene) ABS

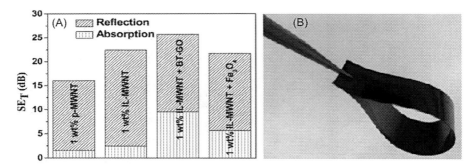

Figure 8.12 (A) Reflection and absorption components in the blends, and (B) twisted film of the blend with 1 wt.% IL modified MWNTs and BT—GO showing the flexibility of the film [59].

was designed and it showed enhanced EMI shielding efficiencies. Particularly, they modified the MWNTs with an amine-terminated ionic liquid which proved to improve the interfacial interaction with PVDF and smooth the process for the formation of a MWNT network structure.

Also, a ferroelectric phase (barium titanate (BT) nanoparticles (Fig. 8.12) chemically grafted onto GO) or a ferromagnetic phase (Fe_3O_4 nanoparticles) added to the MWNTs has been reported to possess a very high SMI SE with increase concentration of conductive filler material [59].

Li et al., in his research fabricated GF with poly(dimethylsiloxane) (PDMS) with a variable thin-layer concentration of GF contents ranging from 15.9% to 31.7% and found that it had excellent EMI shielding effectiveness (up to 36.1 dB) over a broad frequency range of 8.2—18 GHz [60]. On similar lines, Shen et al. worked on the fabrication of a PEI/graphene@Fe_3O_4 as composite foam. He varied the amount of graphene with $Fe_3O_4$to the range of 10 wt.% and came up with some excellent EMI shielding effectiveness (EMI SE) of ~ 41.5 dB/(g/cm^3) at 8—12 GHz [61].

In the study reported by Jia et al., for producing a Superhydrophobic EMI shielding textile material they used a combination of an original textile, a silver nanowire (AgNW) which acted as a conductive layer and a carbon nanotube (CNT) for its Superhydrophobic nature were utilized, along with polytetrafluoroethylene (PTFE) nanoparticles, and fluoroacrylic polymer (Capstone ST-110). This new modified textile material exhibited an extraordinary EMI shielding effectiveness (EMI SE) of 51.5 dB with a thickness of only 0.6 mm [62].

A report by Li et al. disclosed the preparation of thermally annealed anisotropic graphene aerogel (TAGAs) which were reinforced using composites of epoxy. They discovered that with low graphene loadings with 0.8 wt.% exhibited highly anisotropic mechanical and electrical properties and had radial EMI shielding efficiency 32 dB and along the axial direction, it was 25 dB [63]. Whereas, a report by Ling et al., stated fabrication of a microcellular polyetherimide (PEI)/graphene nanocomposite foams with a density of about 0.3 g/cm^3 using a phase separation process. It was concluded that with the utilization of the foaming process there was a significant increase in the specific EMI shielding effectiveness from 17 to 44 dB/ (g/cm^3) [18].

A research work published by Bagotia et al. stated the preparation of in situ reduced graphene oxide (IrGO) using melt blending of polycarbonate/ethylene-methyl acrylate [PC/EMA]in the ratio 95/5 (wt/wt) blend and graphene oxide to achieve high electromagnetic shielding effectiveness for the nanocomposites. Different matrix to filler ratio revealed enhanced activity for electromagnetic shielding effectiveness (-30 dB) over the frequency range of X-band (8.2$-$12.4 GHz). They concluded that the prepared PC/EMA$-$IrGO nanocomposites can be a great candidate for achieving good electromagnetic shielding effects [64].

8.7   Current status, challenges, and patents

8.7.1   Present status in industrial scale/lab scale

Yet, the area of graphene-based polymer nanocomposites research is still at the budding stage of growth. However, based on the progress made in the field of polymer nanocomposite based on graphene it can be said that it can be a promising candidate as a nanocomposite for achieving better electromagnetic shielding. Taking into consideration the fact that graphene-based polymer nanocomposites that exhibited high electrical conductivity also showed high EMI shielding efficiencies broadens their way to be used as lightweight EMI shielding materials to protect electronic devices and components from electromagnetic radiation. [46,50,57,66] Some studies also discussed the use of graphene based composite films can be utilized for fabrication of high-efficiency EMI shielding devices which requires rapid heat dissipation like charge storing devices [44,59]. In some reports, it was seen that EMI shielding was mainly reflection-dominated; it can be exploited more with certain modifications with MWCNT's which would help in shifting it to more absorption dependent shielding mechanism [47,49]. Furthermore, with the recent advancement in fabrication techniques graphene based polymer nanocomposites can be successful candidates for applications based on electrical conduction which requires EMI shielding in postprocessing techniques, which can be based on their high performance and versatility in three-dimensional (3D) printing and electrospinning [48,56,59,61]. Lastly, it can be said that there is massive scope for graphene based polymer nanocomposites for EMI applications, chiefly in microwave absorbers for electronic devices. There are several studies reported in the literature pertaining to the excellent electromagnetic shielding efficiencies due to distinctively designed graphene based foams which promises for inevitable development in the field of hybrid architectures [45,58,59,64].

We foresee a dire need for more exploration in research activities recently and therefore, anticipate a new direction in near future towards the development of graphene based polymer nanocomposites for EM shielding interfaces. Though the researches are at a nascent stage multidisciplinary study can enlarge the scope of research. Challenges observed in most of the nanocomposites were overcome with the employment of Fe_3O_4 nanoparticles or MWCNTs; though the dilemma still lies in designing a highly efficient, ultra-light and flexible EM interfaces for various microelectronics and devices. With this perspective, we possibly will be able to address the design aspects of modern EM shields with excellent flexibility and cost-effectiveness as well. Furthermore, graphene based nanocomposite materials possess excellent thermal conductivity that endows high heat dissipation ability. Therefore, excellent integrated heat dissipation for the EM shielding materials can further extend their application potential in optoelectronic devices.

8.7.2   Patents related to electromagnetic interference for graphene based polymer composites

Patent United States 10,340,054 B2 was filed by Ghosh et al., for a carbon-based polymer matrix containing CuO particles dispersed in the matrix for EMI mitigation properties. Another one was filed by Hong et al., United States 10,306,818 B2 was based on a multilayer graphene-metal-polymer sheet for shielding electromagnetic waves. Jung et al. filed a patent United States 9,942,952 B2 for the method of manufacturing graphene electromagnetic wave blocking plate and microwave oven. Al Ghamdi et al. produced a patent United States 10,490,314 B2 based on graphene oxide free-standing film and methods for shielding electromagnetic radiation at microwave frequencies. A patent filed by Morreale United States 9,793, 437 for graphene based solid-state devices which were capable of emitting electromagnetic radiation was published in 2017 and then in the same year(2017) filed another patent United States 10, 283, 674 for graphene based solid-state devices capable of emitting electromagnetic radiation and its improvements. Whereas Sarto et al., claimed patent United States 9717170B2 for graphene based nanoplatelets or graphite nanoplatelets based nanocomposites which were responsible for EMIs. Also, Avouris et al., filed a patent United States 9,413,075 and United States 8,610, 617 B1 for graphene based structures and methods for broadband electromagnetic radiation absorption at microwave and terahertz frequencies, United States 8,805, 148 B2 for the generation of terahertz electromagnetic waves in graphene by coherent photon mixing further United States 9,210, 835 B2, United States 9,174, 413 B2 and United States 9,215,835 B2 for graphene based structures and methods for shielding electromagnetic radiation have been reported. Sundara et al., patented for the production of graphene using electromagnetic radiation. In a patent United States 9,620, 597. Tzeng et al. reported graphene optoelectronic detector and method for detection photonic and electromagnetic energy. Kim et al., filed patent United States 9,924, 619 B2 for a passive layer for attenuation of near-field electromagnetic waves and heat dissipation including graphene, and electromagnetic device [65−73].

8.8   Future perspectives

In this chapter, we have presented the recent research advancements in the preparation of graphene-polymer based nanomaterials for EMI shielding applications. A brief introduction to the EMI shielding has been presented. Afterward, graphene-polymers nanomaterials, their polymer nanocomposites along their hybrids have been reviewed. Even with the noteworthy successes in this niche, rapid growth demands of succeeding generation EMI shielding materials require thinner, lighter, and highly effective polymeric matrix based hybrids. Along with the optimization of the material properties of nanofiller reinforced nanocomposite, facile novel processes are required for the development of new EMI shielding nanomaterials that need to be designed, engineered, and developed for the purpose of exploring them to fulfill the state of the art demand in material science and engineering. At the present time, the development of 2D and 3D materials with novel conducting properties have been reportedly used as filler materials for the preparation of polymer nanocomposites for EMI shielding applications. Additionally, foundational knowledge and understanding of EMI shielding based electronic materials and structural properties are necessary to optimized SE of graphene-polymer nanocomposite materials. To attain the objectives with respect to future research, certain areas in this respect are expected to be explored in the next decades:

- Graphene-polymer nanocomposites and their hybrids having suitably engineered porous structured morphology are essential to augment their SE performance. Uniformly engineered pore morphology with the help of theoretical prediction and simulation will enhance the internal reflections and overall absorption of the composite materials.
- The design and engineering of graphene-polymer nanocomposites/hybrids with titanium carbide or titanium oxide nanofillers can also produce materials with effective SE performance.
- The functional groups present on the surface of graphene can be utilized to initiate polymerization on the surface of the fillers instead of polymer intercalation which is a bulk mixing process resulting in minimum use of polymer and better distribution of the graphene filler in the nanocomposite.
- Through novel modification of graphene via the creation of defects and doping will have an immense influence on the final graphene-polymer nanocomposite nanostructure and also greatly control the properties of two dimensional nanomaterials like graphene, MXene for efficient SE.
- With proven effective SE in boron nitride materials, graphene-boron nitride materials with a predetermined engineered structure can be a promising area for the engineering of EMI shielding materials. Also, a hybrid of graphene-single layer antimony based polymeric nanocomposite materials is a promising niche to be utilized/explored for the design, engineering and fabrication of new EMI shielding materials very high SE efficiency.

8.9   Conclusions

In this chapter, we have focused on preparation techniques, properties characterization, and EMI shielding capacity of the polymer-graphene composites. So far there have been but a few theoretical studies on the mechanism of EMI shielding with regards to polymer-graphene nanocomposites. A few models have been adopted by researchers to predict the EMI shielding effectiveness of graphene for a specific application. Considering the negligible mass of graphene, ultrathin, flexible, and transparent materials hold an excellent potential for EMI shielding in portable electronic devices like transparent electronics, automobiles, along with EMI isolation of 3D integrated circuits, and so on which can be further explored.

Acknowledgment

The authors wish to thank the Centre for Research in Nanoscience and Nanotechnology, and the Department of Polymer Science and Technology, University of Calcutta for their financial and technical support.

References

[1] P. Saini, M. Arora, Microwave absorption and EMI shielding behavior of nanocomposites based on intrinsically conducting polymers, graphene and carbon nanotubes, New Polym. Spec. Appl. (2012) 71−112.

[2] G. Jose, P.V. Padeep, Electromagnetic shielding effectiveness and mechanical characteristics of polypropylene based CFRP, Int. J. Theo. Appl. Res. Mech. Eng. 3 (2014) 47−53.

[3] H.W. Ott, Carbon Nanotube-Reinforced Polymers: From Nanoscale to Macroscale: Electromagnetic Compatibility Engineering, John Wiley & Sons, 2009. P. Saini, V. Choudhary, B. P. Singh, R. B. Mathur, S. K. Dhawan, Mat. Chem. Phys. 113 919−926.

[4] S.A. Schelkunoff, Ultrashort electromagnetic waves IV-guided propagation, Elec. Eng. 62 (6) (1943) 235−246.

[5] R.B. Schulz, V.C. Plantz, D.R. Brush, Shielding theory and practice, IEEE Trans. 30 (1988) 187−201.

[6] A. Joshi, S. Datar Pramana, Carbon nanostructure composite for electromagnetic interference shielding, J. Phys. 84 (6) (2015) 1099−1116.

[7] S. Sankaran, K. Deshmukh, M.B. Ahamed, S.K. Khadheer Pasha, Recent advances in electromagnetic interference shielding properties of metal and carbon filler reinforced flexible polymer composites: a review, Compos. Part A Appl. Sci. Man. 114 (2018) 49−71.

[8] K. Jagatheesan, A. Ramasamy, A. Das, A. Basu, Electromagnetic shielding behaviour of conductive filler composites and conductive fabrics—a review, Indian. J. Fib. Text. Res. 39 (2014) 329−342.

[9] P.P. Pathak, V. Kumar, Harmful electromagnetic environment near transmission tower, Indian. J. Radio Space Phys. 32 (2003) 238−241.

[10] S. Kumar, P.P. Pathak, Effect of electromagnetic radiation from mobile phones towers on human body, Indian. J. Radio Space Phys. 40 (2011) 340−342.

[11] V. Udmale, D. Mishr, R. Gadhave, D. Pinjare, R. Yamgar, Development trends in conductive nano-composites for radiation shielding, Orient. J. Chem. 29 (3) (2013) 927−936.

[12] S. Geetha, K.K.S. Kumar, C.R. Rao, M. Vijayan, D.C. Trivedi, EMI shielding: methods and materials-A review, J. Appl. Polym. Sci. 112 (2009) 2073−2086.

[13] P. Saini, V. Choudhary, B.P. Singh, R.B. Mathur, S.K. Dhawan, Enhanced microwave absorption behavior of polyaniline-CNT/polystyrene blend in 12.4−18.0 GHz range, Synth. Met. 161 (2011) 1522−1526.

[14] G. Gedler, M. Antunes, J.I. Velasco, R. Ozisik, Enhanced electromagnetic interference shielding effectiveness of polycarbonate/graphene nanocomposites foamed via 1-step supercritical carbon dioxide process, Mat. Des. 90 (2016) 906−914.

[15] C. Liang, H. Qiu, Y. Han, H. Gu, P. Song, L. Wang, et al., Superior electromagnetic interference shielding 3D graphene nanoplatelets/reduced graphene oxide foam/epoxy nanocompos0ites with high thermal conductivity, J. Mater. Chem. C. 7 (2019) 2725−2733.

[16] J. Liang, Y. Wang, Y. Huang, Y. Ma, Z. Liu, J. Cai, et al., Electromagnetic interference shielding of graphene/epoxy composites, Carbon 47 (2009) 922−925.

[17] J. Ling, W. Zhai, W. Feng, B. Shen, J. Zhang, W.G. Zheng, Facile preparation of light-weight microcellular polyetherimide/graphene composite foams for electromagnetic interference shielding, ACS Appl. Mater. Inter. 5 (2013) 2677−2684.

[18] J. Wu, Z. Ye, H. Ge, J. Chen, W. Liu, Z. Liu, Modified carbon fiber/magnetic graphene/epoxy composites with synergistic effect for electromagnetic interference shielding over broad frequency band, J. Coll. Inter. Sci. 506 (2017) 217−226.

[19] Y. Li, Y. Zhao, J. Sun, Y. Hao, J. Zhang, X. Han, Mechanical and electromagnetic interference shielding properties of carbon fiber/graphene nanosheets/epoxy composite, Polym. Compos. 37 (8) (2016) 2494−2502.

[20] S. Chhetri, P. Samanta, N.C. Murmu, S.K. Srivastava, T. Kuila, Electromagnetic interference shielding and thermal properties of non-covalently functionalized reduced graphene oxide/epoxy composites, AIMS Mater. Sci. 4 (1) (2016) 61−74.

[21] A.F. Ahmad, S.A. Aziz, Z. Abbas, S.J. Obaiys, A.M. Khamis, I.R. Hussain, et al., Preparation of a chemically reduced graphene oxide reinforced epoxy resin polymer as a composite for electromagnetic interference shielding and microwave-absorbing applications, Polymers 10 (11) (2018) 1180.

[22] H.B. Zang, Q. Yan, W.G. Zheng, Z. He, Z.Z. Yu, Tough graphene-polymer microcellular foams for electromagnetic interference shielding, ACS Appl. Mater. Interfaces 3 (3) (2011) 918−924.

[23] H. Yu, T. Wang, B. Wen, M. Lu, Z. Xu, C. Zhu, et al., Graphene/polyaniline nanorod arrays: synthesis and excellent electromagnetic absorption properties, J. Mater. Chem. 22 (2012) 21679−21685.

[24] R. Rohini, S. Bose, Extraordinary improvement in mechanical properties and absorption-driven microwave shielding through epoxy-grafted graphene "Interconnects.", ACS Omega 3 (2018) 3200−3210.

[25] T.A. Zoubi, B. Albiss, M.A.A. Akhras, H. Qutaish, E. Alabed, S. Nazrul, NiO-nanofillers embedded in graphite/PVA-polymer matrix for efficient electromagnetic radiation shielding, AIP Conf. Proc 2083 (2019) 020002.

[26] V. Eswaraiah, V. Sankaranarayanan, S. Ramaprabhu, Functionalized graphene−PVDF foam composites for EMI shielding, Macro. Mater. Eng. 296 (2011) 894−898.

[27] M.M. Bernal, I. Molenberg, S. Estravis, M.A. Rodriguez-Perez, I. Huynen, M.A. Lopez-Manchado, et al., Comparing the effect of carbon-based nanofillers on the physical properties of flexible polyurethane foams, J. Mater. Sci. 47 (2012) 5673−5679.

[28] H. Hu, J.C. Wang, L. Wan, F.M. Liu, H. Zheng, R. Chen, et al., Preparation and properties of graphene nanosheets-polystyrene nanocomposites via in situ emulsion polymerization, Chem. Phy. Letts 484 (2010) 247−253.

[29] W. Zheng, X. Lu, S.C. Wong, Electrical and mechanical properties of expanded graphite-reinforced high-density polyethylene, J. Appl. Polym. Sci. 91 (2004) 2781−2788.

[30] L. Ye, X.Y. Meng, X. Ji, Z.M. Li, J.H. Tang, Synthesis and characterization of expandable graphite-poly(methyl methacrylate) composite particles and their application to flame retardation of rigid polyurethane foams, Polym. Degrad. Stab. 94 (2009) 971−979.

[31] G. Chen, D. Wu, W. Weng, C. Wu, Development of Cu-exfoliated graphite nanoplatelets (xGnP) metal matrix composite by powder metallurgy route, Carbon 41 (2003) 619−621.

[32] W.D. Lee, S.S. Im, Thermomechanical properties and crystallization behavior of layered double hydroxide/poly(ethylene terephthalate) nanocomposites prepared by in-situ polymerization, J. Polym. Sci. Pt. B Polym. Phys. 45 (2007) 28−40.

[33] J. Liang, Y. Huang, L. Zhang, Y. Wang, Y. Ma, T. Guo, et al., Molecular-level dispersion of graphene into poly(vinyl alcohol) and effective reinforcement of their nanocomposites, Adv. Funct. Mater. 19 (2009) 2297−2302.

[34] F. Hussain, M. Hojjati, M. Okamoto, R.E. Gorga, Review article: polymer-matrix nanocomposites, processing, manufacturing, and application: an overview, J. Compos. Mater. 40 (2006) 1511−1575.

[35] J.W. Shen, W.Y. Huang, S.W. Zuo, J. Hou, Polyethylene/grafted polyethylene/graphite nanocomposites: preparation, structure, and electrical properties, J. Appl. Polym. Sci. 97 (2005) 51−59.

[36] H.B. Hsueh, C.Y. Chen, Preparation and properties of LDHs/epoxy nanocomposites, Polymer 44 (18) (2003) 5275−5283.

[37] W.P. Wang, C.Y. Pana, Preparation and characterization of polystyrene/graphite composite prepared by cationic grafting polymerization, Polymer 45 (2004) 3987−3995.

[38] K. Kalaitzidou, H. Fukushima, L.T. Drzal, New compounding method for exfoliated graphite-polypropylene nanocomposites with enhanced flexural properties and lower percolation threshold, Compos. Sci. Technol. 67 (2007) 2045−2051.

[39] M. Zammarano, M. Franceschi, S. Bellayer, G.W. Gilman, S. Meriani, Preparation and flame resistance properties of revolutionary self-extinguishing epoxy nanocomposites based on layered double hydroxides, Polymer 46 (2005) 9314−9328.

[40] G. Broza, K. Piszczek, K. Schulte, T. Sterzynski, Nanocomposites of poly(vinyl chloride) with carbon nanotubes (CNT), Compos. Sci. Technol. 67 (2007) 890−894.

[41] S.K. Kim, N.H. Kim, J.H. Lee, Effects of the addition of multiwalled carbon nanotubes on the positive temperature coefficient characteristics of carbon-black-filled high density polyethylene nanocomposites, Scr. Mater. 55 (2006) 1119−1122.

[42] S. Kim, I. Do, L.T. Drzal, Thermal stability and dynamic mechanical behavior of exfoliated graphite nanoplatelets-LLDPE nanocomposites, Polym. Compos. 31 (2009) 755−761.

[43] G. Chen, C. Wu, W. Weng, D. Wu, W. Yan, Preparation of polystyrene/graphite nanosheet composite, Polymer 44 (2003) 1781−1784.

[44] B. Zhao, S. Wang, C. Zhao, R. Li, S.M. Hamidinejad, Y. Kazemi, et al., Synergism between carbon materials and Ni chains in flexible poly (vinylidene fluoride) composite films with high heat dissipation to improve electromagnetic shielding properties, Carbon 127 (2018) 469−478.

[45] Y. Wu, Z. Wang, X. Liu, X. Shen, Q. Zheng, Q. Xue, et al., Ultralight graphene foam/conductive polymer composites for exceptional electromagnetic interference shielding, ACS Appl. Mater. Interf. 9 (10) (2017) 9059−9069.

[46] P. Modak, S.B. Kondawar, D.V. Nandanwar, Synthesis and characterization of conducting polyaniline/graphene nanocomposites for electromagnetic interference shielding, Proc. Mater. Sci. 10 (2015) 588−594.

[47] R.R. Mohan, S.J. Varma, M. Faisal, S. Jayalekshmi, Polyaniline/graphene hybrid film as an effective broadband electromagnetic shield, RSC Adv. 5 (8) (2015) 5917−5923.

[48] K. Yao, J. Gong, N. Tian, Y. Lin, X. Wen, Z. Jiang, H. Na, T. Tang, Flammability properties and electromagnetic interference shielding of PVC/graphene composites containing Fe3O4 nanoparticles, RSC Adv. 5 (40) (2015) 31910−31919.

[49] A. Gupta, S. Varshney, A. Goyal, P. Sambyal, B.K. Gupta, S.K. Dhawan, Enhanced electromagnetic interference shielding behaviour of multi-layer graphene anchored luminescent TiO_2 in PPY matrix, Mater. Lett. 158 (2015) 167−169.

[50] H.B. Zhang, W.G. Zheng, Q. Yan, Z.G. Jiang, Z.Z. Yu, The effect of surface chemistry of graphene on rheological and electrical properties of polymethylmethacrylate composites, Carbon 50 (14) (2012) 5117−5125.

[51] S.T. Hsiao, C.C.M. Ma, H.W. Tien, W.H. Liao, Y.S. Wang, S.M. Li, et al., Using a non-covalent modification to prepare a high electromagnetic interference shielding performance graphene nanosheet/water-borne polyurethane composite, Carbon 60 (2013) 57−66.

[52] C. Li, G. Yang, H. Deng, K. Wang, Q. Zhang, F. Chen, et al., The preparation and properties of polystyrene/functionalized graphene nanocomposite foams using supercritical carbon dioxide, Poly. Int. 62 (7) (2013) 1077−1084.

[53] D.X. Yan, H. Pang, B. Li, R. Vajtai, L. Xu, P.G. Ren, et al., Structured reduced graphene oxide/polymer composites for ultra-effi cient electromagnetic interference shielding, Adv. Funct. Mat. 25 (4) (2015) 559−566.

[54] F. Shahzad, S. Yu, P. Kumar, J.W. Lee, Y.H. Kim, S.M. Hong, et al., Sulfur doped graphene/polystyrene nanocomposites for electromagnetic interference shielding, Compos. Struct. 133 (2015) 1267−1275.

[55] A.P. Singh, M. Mishra, D.P. Hashim, T.N. Narayanan, M.G. Hahm, P. Kumar, et al., Carbon 85 (2015) 79−88.

[56] S. Maiti, B.B. Khatua, Graphene nanoplate and multiwall carbon nanotube−embedded polycarbonate hybrid composites: high electromagnetic interference shielding with low percolation threshold, Polym. Compos. 37 (7) (2016) 2058−2069.

[57] N. Yousefi, X. Sun, X. Lin, X. Shen, J. Jia, B. Zhang, et al., Highly aligned graphene/polymer nanocomposites with excellent dielectric properties for high-performance electromagnetic interference shielding, Adv. Mater. 26 (2014) 5480−5487.

[58] B. Shen, Y. Li, W. Zhai, W. Zheng, Compressible graphene-coated polymer foams with ultralow density for adjustable electromagnetic interference (EMI) shielding, ACS Appl. Mater. Interf. 8 (12) (2016) 8050−8057.

[59] G.P. Kar, S. Biswas, R. Rohini, S. Bose, Tailoring the dispersion of multiwall carbon nanotubes in co-continuous PVDF/ABS blends to design materials with enhanced electromagnetic interference shielding, J. Mater. Chem. A 3 (2015) 7974−7985.

[60] L. Hongling, L. Jing, Z.L. Ngoh, R.Y. Tay, J. Lin, H. Wang, et al., Engineering of high-density thin-layer graphite foam-based composite architectures with superior compressibility and excellent electromagnetic interference shielding performance, ACS Appl. Mater. Interf. 10 (48) (2018) 41707−41716.

[61] B. Shen, W. Zhai, M. Tao, J. Ling, W. Zheng, Lightweight, multifunctional polyetherimide/graphene@Fe_3O_4 composite foams for shielding of electromagnetic pollution, ACS Appl. Mater. Interf 5 (21) (2013) 11383−11391.

[62] L.C. Jia, G. Zhang, L. Xu, W.J. Sun, G.J. Zhong, J. Lei, et al., Robustly superhydropho-
     bic conductive textile for efficient electromagnetic interference shielding, ACS Appl.
     Mater. Interf 11 (1) (2018) 1680−1688.
[63] X.H. Li, X.F. Li, K.N. Liao, P. Min, T. Liu, A. Dasari, et al., Thermally annealed ani-
     sotropic graphene aerogels and their electrically conductive epoxy composites with
     excellent electromagnetic interference shielding efficiencies, ACS Appl. Mater. Interf 8
     (2016) 33230−33239.
[64] N. Bagotia, V. Choudhary, D.K. Sharma, Superior electrical, mechanical and electro-
     magnetic interference shielding properties of polycarbonate/ethylene-methyl acrylate-in
     situ reduced graphene oxide nanocomposites, J. Mater. Sci 53 (23) (2018)
     16047−16061.
[65] D. Ghosh, B.K. Roy, S. Satarkar, Polymer composites with electromagnetic interfer-
     ence mitigation properties. United States 10,340,054 B2 (2020).
[66] S.H. Hong, H.J. Ryu, J.Y. Oh, H.J. Im, Multi-layer graphene-metal-polymer sheet for
     shielding electromagnetic wave. United States 10 306 818 B2 (2019).
[67] M. Jung, J. Moon, Method for manufacturing graphene electromagnetic wave blocking
     plate and microwave oven using same. United States 9942952B2 (2020).
[68] A.A. Al-Ghamdi, Y. Al-Turki, F. Yakuphanoglu, F. El-Tantawy, Graphene oxide free-
     standing film and methods for shielding electromagnetic radiation at microwave fre-
     quencies. United States 10 490 314B2 (2019).
[69] J.P. Morreale, Graphene based solid state devices capable of emitting electromagnetic
     radiation and improvements thereof. United States 9,793, 437 (2017).
[70] J.P. Morreale, Graphene-based solid state devices capable of emitting electromagnetic
     radiation and improvements thereof. United States 20180019379 A1 (2018).
[71] M. Sabrina, S. De Bellis, A. Tamburrano, A.G. D'Aloia. Graphene nanoplatelets- or
     graphite nanoplatelets-based nanocomposites for reducing electromagnetic interfer-
     ences. United States 9717170 B2 (2019).
[72] P. Avouris, A.V. Garcia, C.-Y. Sung, F. Xia, H. Yan. Graphene based structures and
     methods for broadband electromagnetic radiation absorption at the microwave and tera-
     hertz frequencies. United States 9413075B2 (2016).
[73] E. Varrla, J.A.S. Sasikaladevi, R. Sundara, Production of graphene using electromag-
     netic radiation. United States 8 828 193 B2 (2014).
[74] Z. Chen, C. Xu, C. Ma, W.C. Ren, H.M. Cheng, Lightweight and flexible graphene
     foam composites for high-performance electromagnetic interference shielding, Adv.
     Mater 25 (9) (2013) 1296−1300.
[75] Y. Li, X.L. Pei, B. Shen, W.T. Zhai, L.H. Zhang, W.G. Zheng, Polyimide/graphene
     composite foam sheets with ultrahigh thermostability for electromagnetic interference
     shielding, RSC Adv 5 (31) (2015) 24342−24351.
[76] D.X. Yan, H. Pang, L. Xu, Y. Bao, P.G. Ren, J. Lei, et al., Electromagnetic interference
     shielding of segregated polymer composite with an ultralow loading of in situ thermally
     reduced graphene oxide, Nanotechnology 25 (14) (2014) 145705.
[77] K. Tian, H. Wang, Z. Su, J. He, X. Tian, W. Huang, et al., Few-layer graphene sheets/
     poly(vinylidene fluoride) composites prepared by a water vapor induced phase separa-
     tion method, Mater. Res. Exp 4 (4) (2017) 045603.
[78] D.-X. Yan, P.-G. Ren, H. Pang, Q. Fu, M.-B. Yang, Z.-M. Li, Efficient electromagnetic
     interference shielding of lightweight graphene/polystyrene composite, J. Mater. Chem
     22 (36) (2012) 18772.
[79] J. Joseph, A.K. Koroth, D.A. John, A.M. Sidpara, J. Paul, Highly filled multilayer ther-
     moplastic/graphene conducting composite structures with high strength and thermal

stability for electromagnetic interference shielding applications, J. Appl. Polym. Sci 136 (29) (2019) 47792.

[80] A. Gupta, S. Varshney, A. Goyal, P. Sambyal, B. Kumar Gupta, S.K. Dhawan, Enhanced electromagnetic shielding behaviour of multilayer graphene anchored luminescent TiO2 in PPY matrix, Mater. Lett 158 (2015) 167−169.

[81] P. Sambyal, S.K. Dhawan, P. Gairola, S.S. Chauhan, S.P. Gairola, Synergistic effect of polypyrrole/BST/RGO/Fe$_3$O$_4$ composite for enhanced microwave absorption and EMI shielding in X-Band, Curr. Appl. Phys 18 (5) (2018) 611−618.

[82] N. Agnihotri, K. Chakrabarti, A. De, Highly efficient electromagnetic interference shielding using graphite nanoplatelet/poly(3,4-ethylenedioxythiophene)-poly(styrenesulfonate) composites with enhanced thermal conductivity, RSC Adv 5 (54) (2015) 43765−43771.

[83] X.Y. Chen, A. Romero, A. Paton-Carrero, M.P. Lavin-Lopez, L. Sanchez-Silva, J.L. Valverde, et al., Functionalized graphene−reinforced foams based on polymer matrices, Functionalized Graphene Nanocomposites and their Derivatives. (2019) 121−155.

[84] J. Liu, H.-B. Zhang, Y. Liu, Q. Wang, Z. Liu, Y.-W. Mai, et al., Magnetic, electrically conductive and lightweight graphene/iron pentacarbonyl porous films enhanced with chitosan for highly efficient broadband electromagnetic interference shielding, Compos. Sci. Technol 151 (2017) 71−78.

[85] C. Basavaraja, W.J. Kim, Y.D. Kim, D.S. Huh, Synthesis of polyaniline-gold/graphene oxide composite and microwave absorption characteristics of the composite films, Mater. Lett 65 (19−20) (2011) 3120−3123.

[86] Y. Li, B. Shen, D. Yi, L. Zhang, W. Zhai, X. Wei, et al., The influence of gradient and sandwich configurations on the electromagnetic interference shielding performance of multilayered thermoplastic polyurethane/graphene composite foams, Compos. Sci. Technol 138 (2017) 209−216.

[87] Y. Chen, Y. Wang, H.B. Zhang, X. Li, C.X. Gui, Z.Z. Yu, Enhanced electromagnetic interference shielding efficiency of polystyrene/graphene composites with magnetic Fe$_3$O$_4$ nanoparticles, Carbon 82 (2015) 67−76.

[88] Y.-J. Wan, S.-H. Yu, W.-H. Yang, P.-L. Zhu, R. Sun, C.-P. Wong, et al., Tuneable cellular-structured 3D graphene aerogel and its effect on electromagnetic interference shielding performance and mechanical properties of epoxy composites, RSC Adv 6 (61) (2016) 56589−56598.

[89] S.-H. Lee, D. Kang, I.-K. Oh, Multilayered graphene-carbon nanotube-iron oxide three-dimensional heterostructure for flexible electromagnetic interference shielding film, Carbon 111 (2017) 248−257.

[90] H. Yang, Z. Li, H. Zou, P. Liu, Preparation of porous polyimide/in-situ reduced graphene oxide composite films for electromagnetic interference shielding, Polym. Adv. Technol 28 (2017) 233−242.

[91] Y. Mamunya, L. Matzui, L. Vovchenko, O. Maruzhenko, V. Oliynyk, S. Pusz, et al., Influence of conductive nano- and microfiller distribution on electrical conductivity and EMI shielding properties of polymer/carbon composites, Compos. Sci. Technol 170 (2019) 51−59.

[92] L. Chen, D. Rende, L.S. Schadler, R. Ozisik, Polymer nanocomposite foams, Compos. Sci. Technol 65 (15−16) (2005) 2344−2363.

Thermal conductivity of graphene-polymer composites

Subhadip Mondal[1], Haradhan Kolya[2], Srinivas Pagidi[3], Chun-Won Kang[2] and Changwoon Nah[1]

[1]Department of Polymer-Nano Science and Technology, Jeonbuk National University, Jeonju, Republic of Korea, [2]Department of Housing Environmental Design and Research Institute of Human Ecology, College of Human Ecology, Jeonbuk National University, Jeonju, Republic of Korea, [3]Institute of Quantum Systems and Department of Physics, Chungnam National University, Daejeon, Republic of Korea

9.1 Introduction

Thermal management has become a crucial factor to control the reliability, service life, and performance of next-generation electronic devices [1,2]. With more integration, functionalization, and miniaturization of high-power density microelectronics and electronic devices and the development of modern applications including light-emitting diodes, microchips, complex circuits, flexible and wearable electronics, heat dissipation from these working devices has become a challenging problem [3]. Generally, thermal dissipation is simply the thermal energy transfer within the material that depends on the thermal conductivity of materials, as the thermal conductivity of the material demonstrates the speed of heat energy transfer within the material. Highly conductive materials have been reported to be used for the reliable heat sink whereas low thermal conductivity materials are needed for thermal insulation [4]. Therefore it is very crucial to develop and design robust thermally conductive materials to solve the emerging problem of heat generation. Since ancient times, people have been looking for metal or metal-based materials (copper, silver, iron, gold, and aluminum), which are used for good heat dissipation. But they have suffered from heavyweight, high-cost, easy to process, inferior thermal and chemical stability [2]. In recent years, polymer-based materials have received great attention because of their outstanding combination of flexibility, light-weight, low-cost, excellent processability, excellent chemical/thermal stability, any shape formability, high strength, high insulating properties, and most importantly high voltage breakdown strength [5]. High thermal conductivity is needed in electronic circuits for dissipating heat quickly from the conductor. Typically, most of the polymers are insulating in nature and exhibit intrinsically inferior thermal conductivity between 0.05 and 0.5 W/m/K, which is far from meeting industrial standards [4]. This inferior thermal conductivity is attributed to numerous phonon scattering, low atomic density, complicated crystal structure, high level of anharmonicity, and

Polymer Nanocomposites Containing Graphene. DOI: https://doi.org/10.1016/B978-0-12-821639-2.00003-3

weak atomic interactions in the molecular vibrations, and lots of defects in the amorphous phase. Moreover, polymers with insufficient thermal stability and low thermal conductivity have some limits in exploring the broader applications [3,4]. The only way to increase the thermal conductivity of polymers is to incorporate fillers into the polymer matrix. In recent years, numerous studies have been conducted on polymer-based composites containing different thermally conductive fillers such as carbon nanotubes, boron nitride, graphite, graphene nanoplatelets, graphene, aluminum oxide, and diamonds [2]. Among them, graphene can be regarded as one of the more efficient and economic nanofillers for constructing thermally conductive polymer nanocomposites because of its unique layered 2D structure, zero-gap band structure, extraordinary thermal conductivity (5000 W/m-K), and high electron mobility [6,7]. This chapter outlines the basic physical concepts of the heat conduction phenomenon, the theoretical model for thermal conductivity, the extrinsic parameters for thermal conduction behavior, and illustrates the advances in thermal conductivity of graphene-based polymer composites.

9.2   Definition of thermal conductivity and mechanism of thermal conduction

Thermal conductivity of any material is a unique intrinsic property that demonstrates its intrinsic ability to thermal dissipation or heat conduction. It is denoted by "λ" or often by "κ" and measured in watts per meter per Kelvin (W/mK) according to Fourier's law mechanism. According to Debye equation, thermal conductivity can be theoretically calculated using the following equation,

$$\lambda = \frac{C_p v l}{3} \tag{9.1}$$

where v and l are the average phonon velocity and the phonon mean free path.

From fundamental quantum theory, it is also well-defined as the intrinsic ability to transfer vibrational energy of particles to its adjacent particles through microscopic collisions of particles and movement of electrons. In terms of second law of thermodynamics, heat dissipation in material mainly occurs from a high temperature region to a low temperature region of the material. Thermal conduction phenomenon of solid materials is achieved by either electron or phonon transport. In the case of the metal type of material, heat conduction occurs through electron transport whereas in polymer and its composites, phonon transport is the dominant mechanism of thermal conductivity. In general, it is believed that phonon acts as quantized lattice vibrational energy that propagates from lattice vibration in polymeric material. When polymeric material contacts the heat source, heat reaches the first surface layer of molecular atoms in the form of random vibrations and rotations, and then transfers to the adjacent atoms, and then the next, and so on [2], as shown in Fig. 9.1. Heat cannot dissipate like waves, but it diffuses very slowly.

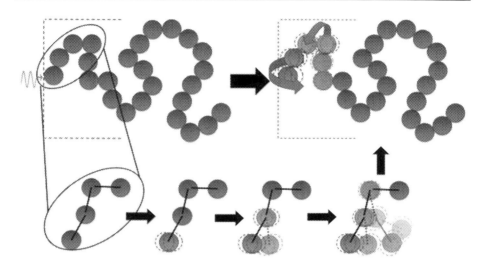

Figure 9.1 Schematic illustration of thermal conductivity mechanism of polymer.
Source: From N. Burger, A. Laachachi, M. Ferriol, M. Lutz, V. Toniazzo, D. Ruch, Review of thermal conductivity in composites: mechanisms, parameters and theory, Prog. Polym. Sci. 61 (2016) 1–28. Copyright (2016), with permission from Elsevier.

Basically, polymeric materials have disordered and irregular packing structure of atoms that significantly reduce the thermal conductivity of the polymer. To improve the heat transfer ability of polymeric materials, different types of thermally conductive fillers and additives have been used. Such fillers are incorporated in polymer matrix through various mixing methods to prepare good thermally conductive composites. Mainly, thermal conductivity of polymeric materials is dominated by thermally conductive filler. Thermal conductivity of polymer composites depends on some factors such as filler type, filler structure, filler loading, filler dispersion, interconnected filler networks, and intrinsic thermal conductivity of filler [2]. However, it is very crucial to manipulate the above-mentioned factors during processing to make excellent thermally conductive materials. In the case of graphene-based composites, the heat transfer of graphene is attributed to vibrations of phonon waves and electrons because of its metallic property, as schematically illustrated in Fig. 9.2. Concomitantly, the optimization of graphene loading, graphene layer structures, functionalization of graphene, and dispersion of graphene in the polymer matrix can enhance the thermal conductivity of graphene composites by tuning thermally conductive continuous networks inside the polymer matrix.

9.3 Measurement of thermal conductivity of composites

There are two different useful techniques that are available in the literature for determining the effective thermal conductivity of the bulk and thin-film composites,

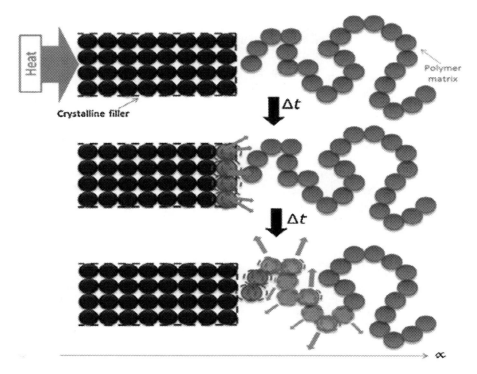

Figure 9.2 Schematic illustration of thermal conductivity mechanism of crystalline material and polymer matrix.
Source: From N. Burger, A. Laachachi, M. Ferriol, M. Lutz, V. Toniazzo, D. Ruch, Review of thermal conductivity in composites: mechanisms, parameters and theory, Prog. Polym. Sci. 61 (2016) 1−28. Copyright (2016), with permission from Elsevier.

that is, steady-state measurement method and nonsteady state measurement (Transient) method.

9.3.1 Steady-state measurement method

In the steady-state method, the thermal conductivity can be measured under the equilibrium condition by establishing a steady-state temperature gradient across the sample thickness and the controlled heat flow through. This method consists of two methods that include the guarded hot plate method [8] and the heat flow meter method [9], where the heat dissipation rate is equal to the heat transfer rate. Among the two methods, the guarded hot plate method is the most representative steady-state method that is described in ISO-8302. In this method, a rectangular or cylindrical shape sample is positioned between two parallel plates consisting of two separate sources of temperature, one of which is the heat source and the other is the heat sink, as shown in Fig. 9.3A. A steady-state power is given as an input to the

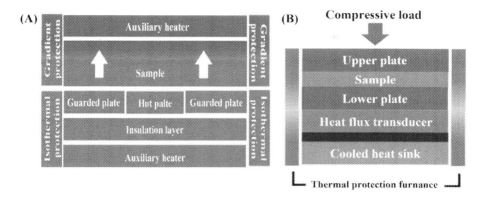

Figure 9.3 Schematic representation of thermal conductivity mechanism of (A) the guarded hot plate method and (B) guarded heat flow meter method.
Source: From X. Yang, C. Liang, T. Ma, Y. Guo, J. Kong, J. Gu, et al., A review on thermally conductive polymeric composites: classification, measurement, model and equations, mechanism and fabrication methods, Adv. Compos. Hybrid. Mater. 1 (2) (2018) 207−230. Copyright (2018), with permission from Springer.

heat source to generate unidirectional heat flow across the sample specimen under different gas environments or vacuum conditions. The resulting temperature gradient was developed across the sample which can be detected using the temperature sensor that is thermocouple. For guarded heat flow meter method, the sample is kept between two polished metal plates under a compressive load. Here, upper metal plate acts as a heat source whereas lower plate acts as heat flux transducer (Fig. 9.3B). The thermal conductivity of the sample can be estimated by Fourier's law of heat conduction expressed as,

$$\lambda_c = \frac{QL}{A\Delta T} \tag{9.2}$$

$$Q = p - Q_{\text{loss}} \tag{9.3}$$

Where Q, A, and L are the heat, cross-sectional area and length of the sample, ΔT is the temperature difference between temperature sensors, p is the applied heating power at the heat source and Q_{loss} is the heat losses due to various heat transfer methods to the ambient temperature.

Moreover, the experimental setup placed inside the vacuum chamber to avoid the effects of heat transfer and this setup also allows us to measure various environmental conditions. These two established methods are quite essential for determining the thermal conductivity of the polymer composites at a wide temperature range. But these methods have some disadvantages like time-consuming, required specific measuring conditions (adiabatic condition and controlled temperature) for systems, required known shape and size of the sample, and temperature control during the measurements. ASTM C177 [10] and ISO 8302 [9] can provide detailed information about the guarded hot plate method. The corresponding schematic diagrams for guarded hot plate method and guarded heat flow meter method is shown in Fig. 9.3.

9.3.2 Nonsteady state measurement method

The nonsteady state is well known as a transient method. The nonsteady-state method is classified into four main types: laser flash [11], transient plane source (TPS) [12], transient line source [13], hot wire [14]. These methods have been introduced to control the issues related to the steady-state method such as heat losses, time-consuming, contact resistance of temperature sensor, and environmental conditions.

9.3.2.1 Laser flash method

The laser flash method is commonly used because it works with noncontact, nondestructive temperature sensing to obtain higher precision measurements [15]. This method is a very convenient for small size samples (5−12 mm diameter) and can be measured in a matter of seconds (for 1−2 s solids) [16]. In this method, a laser flash is used as an optical source to generate heat on the sample's front side and time-dependent temperature changes are measured from the other side, as shown in Fig. 9.4A. Herein, the infrared radiation thermometer is used as a sensor to collect generated high-speed data of the temperature change. This method is not a potential method for directly measuring the thermal conductivity of the composite rather it measures only the thermal diffusivity. Using the thermal diffusivity, density, and specific heat value, the thermal conductivity of material can be estimated by the following equation,

$$\lambda_c = \alpha \rho C_p, \tag{9.4}$$

where Cp is the specific heat of the sample. Further detailed information about this method and specific apparatus, test sample and protocols for measuring the system is available in ASTM E11461 [17] and ISO 22007−4 [11].

9.3.2.2 Transient plane source method

The TPS process is a hot disk process. As shown in Fig. 9.4B, a thin metal strip or disk is used as both a temperature sensor and a heat source to determine the thermal conductivity of the material [19]. First, the metal probe disk is sealed using the electrical insulation tape on both side and inserted between two identical thin slab-shaped test samples. The available surfaces of the test sample also remain thermally insulated. A small amount of electrical current is applied to raise the temperature of the metal disk while the experiment is being conducted. The temperature variation in the test sample is recorded for the specified measurement time. The thermal

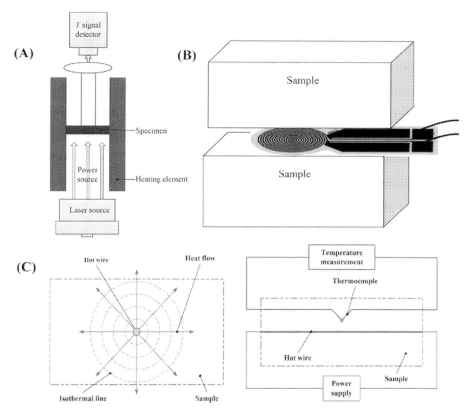

Figure 9.4 Schematic of (A) laser flash, (B) transient plane source, (C) hot wire measurement methods for determining thermal conductivity.
Source: From W. Zhao, Y. Yang, Z. Bao, D. Yan, Z. Zhu, Methods for measuring the effective thermal conductivity of metal hydride beds: a review, Int. J. Hydrog. Energy 45 (11) (2020) 6680−6700. Copyright (2020), with permission from Elsevier.

conductivity of various materials can be measured using this method which includes liquids, aerogels, and solids ranging from 0.5 to 500 W/mK with a cryogenic temperature range of 500K.

9.3.2.3 Hot wire method

In 1931, Stalhane and Pyk employed the hot wire method for measuring the thermal conductivity of the solids and isotropic samples [20]. The method usually works based on the temperature change which is measured at a known distance from the hot wire that is from the source incorporated in the sample. This method was expected to be ideal for one-dimensional radial heat flow within the homogeneous and isotropic test samples [21]. In this method, the change in temperature can be detected either by a wire acting as a temperature sensor or by placing a

thermocouple near the wire. Thermal conductivity of material can be measured by
the temperature change at a known distance from the hot wire over a known period
of time while the constant intensity of electrical current passes through the hot
wire, as shown in Fig. 9.4C. The resulting temperature change from t_1 to t_2 is written as follows [22],

$$\Delta T = T(t_2) - T(t_1) = \frac{p}{4\pi kL} \ln\left(\frac{t_2}{t_1}\right),$$
(9.5)

The thermal conductivity of the sample can be calculated from the following
equation as written below,

$$\lambda_c = \frac{p}{4\pi[T(t_2) - T(t_1)]L} \ln\left(\frac{t_2}{t_1}\right),$$
(9.6)

where p is the power supply per unit length of the heating line, W/m, and $T(t_1)$ and
$T(t_2)$ are the temperatures at t_1 and t_2, respectively. Further details and data analysis
of this method are available in ISO 8894-1 and ISO 8894-2 [23].

9.3.2.4 Transient line source method

The transient line source process is a rapid measurement technique and suitable for
measuring the thermal conductivity of molten and solid-state samples [18].
The design of this method is similar to that of the hot wire method, in which the
probe with an electrical heater and a temperature sensor replaces the hot wire. Also,
the transient line source method is referred to as the needle probe method. In this
method, the line source is placed at the center of the sample and takes the form of a
needle-sensor probe of finite length and diameter. The size of the probes is between
50−100 mm in length and 1.5−2 mm in diameter. The line source temperature
changes linearly with the logarithm of time. More details of this method are available in ASTM D5930 [18].

9.4 Theoretical models of thermal conductivity of nanocomposites

Theoretical modeling is a promising tool to predict the thermal conductivity of
composites before preparation and gives knowledge about the relation between thermal transport properties with microstructures. Thus theoretical modeling can also
deliver ideas for design and development of new composites. So far, several theoretical and mathematical models have been addressed in the literature to estimate
the thermal transport properties of the two-phase systems. The predicted models
gained considerable attention and it paves an easy pathway for better understanding
the fundamental properties like structural design and effective thermal transport

properties of the polymer composites. Based on the literature survey, a few theoretical models that have been discussed are as follows.

Maxwell was the first person to investigate the thermal conductivity of the polymer composites [24,25], the Series and Parallel model was perceived as the easiest one among all for estimation of thermal conductivity of the two-phase systems. According to established heat flow, the model has maximum and minimum limitations of thermal conductivity with the filler orientation. The thermal conductivity (λ_c) of the composite medium related to the series Eq. (9.7) and parallel Eq. (9.8) mathematical models can be determined as [3],

$$\lambda_c = \frac{\lambda_p \lambda_f}{\lambda_f(1 - \phi) + \lambda_p \phi} \tag{9.7}$$

$$\lambda_c = (1 - \phi)\lambda_p + \phi\lambda_f \tag{9.8}$$

where λ_c and λ_f are the thermal conductivity of the polymer matrix and dispersed fillers, respectively, and ϕ is the volume fraction of the dispersed fillers in a continuous polymer matrix. The series model predicts the thermal conductivity of the lower limit whereas the parallel model gives the upper limit [26,27]. Practically, the filler is randomly distributed in a continuous medium and is not aligned in the direction of heat flow. Therefore the effective thermal conductivity anticipated using the series and parallel model may not be a reasonable one for the polymer composites. In recent years, Gu et al. [28] made a small modification in the series Eq. (9.3) and parallel Eq. (9.4) models for obtaining the accurate and suitable thermal conductivity of the composites.

$$\lambda_{c,s} = \lambda_s + \alpha_p \varphi_f (\lambda_f - \lambda_m) \tag{9.9}$$

$$\lambda_{c,p} = \lambda_p + \alpha_s \varphi_f (\lambda_f - \lambda_m) \tag{9.10}$$

where $\lambda_{c,s}$ and $\lambda_{c,p}$ are the incorporated λ_c expressions of the series and parallel condition, respectively. α_s and α_p are the series and parallel connection factors vary in the range ($0 \leq \alpha_s$ or $\alpha_p \leq 1$). $\alpha_p \varphi_f$ and $\alpha_s \varphi_f$ are the parallel fillers and the series fillers in the composite are ignored in the series and parallel models.

Maxwell's model of thermal conductivity of the two-phase composite system was proposed with the use of potential theory and Laplace equations. Maxwell's mathematical model can be expressed as [25],

$$\lambda_c = \lambda_p \frac{2\lambda_p + \lambda_f + 2\phi(\lambda_p - \lambda_f)}{2\lambda_p + \lambda_f + \phi(\lambda_p - \lambda_f)} \tag{9.11}$$

Later, Eucken modified the equation of thermal conductivity by maintaining the original Maxwell model as a backbone of the system. It is named the Maxwell-Eucken model [29,30].

Bruggeman et al. [31] has developed a modified Maxwell's equation for measuring thermal conductivity of the polymer composite system, in which plausible assumptions are made related to field strength and permeability in the predicted model of a two-phase system consisting of fillers (<30vol.%) in a continuous polymer matrix. After all modifications, the Bruggeman equation was applied to estimate the thermal conductivity of the polymer composite and the model can be expressed as [32],

$$1 - \phi = \frac{\lambda_f - \lambda_c}{\lambda_f - \lambda_p} \left(\frac{\lambda_p}{\lambda_c} \right)^{\frac{1}{3}} \tag{9.12}$$

Based on Maxwell's and Frieke's models, Hamilton and Crosser [33,34] have developed a mathematical model for measuring the thermal conductivity of polymer composite when the fillers are in irregular shape in a continuous phase. Hamilton and Crosser's modified model can have determined as,

$$\lambda_c = \lambda_m \left(\frac{\lambda_f + (n-1)\lambda_p + (n-1)(\lambda_f - \lambda_p)\phi}{\lambda_f + (n-1)\lambda_p - (\lambda_f - \lambda_p)\phi} \right) \tag{9.13}$$

where n is the empirical constant, that is, $n = \psi/3$, ψ is the sphericity of the particle and it varies in the range of $0.58 < \psi < 1.0$ [33]. The sphericity ψ is defined as the surface area of the filler with the same volume as the particle divided by the surface area of the particle.

Cheng and Vachon et al. [35] was proposed a model that is an extension to Tsao's model of thermal conductivity [36]. Herein, the proposed model assumes that the discontinuous phase obeys a parabolic distribution of Tsao's model that is discrete phase in a continuous polymer matrix ($\lambda_f > \lambda_p$). According to Cheng and Vachon model, the thermal conductivity of the polymer composite described as,

$$\frac{1}{\lambda_c} = \frac{1 - B}{\lambda_p} + \frac{1}{\sqrt{K \lambda_d (\lambda_f + \lambda_d)}} . \ln \left[\frac{\sqrt{\lambda_p + B\lambda_d} + \frac{B}{2}.\sqrt{K.\lambda_d}}{\sqrt{\lambda_p + B\lambda_d} - \frac{B}{2}.\sqrt{K.\lambda_d}} \right] \tag{9.14}$$

where B and C are related factors of ϕ and can be defined as

$$B = \sqrt{\frac{3\phi}{2}}, \ K = -4\sqrt{\frac{2}{3\phi}} \ \text{and} \ \lambda_d = \lambda_f - \lambda_p$$

Russell et al. [37] derived a model for measuring the thermal conductivity of composites. The Russell model of thermal conductivity can be described as,

$$\lambda_c = \frac{\left[\phi^{\frac{2}{3}} + \frac{\lambda_p}{\lambda_f} \left(1 - \phi^{\frac{2}{3}} \right) \right]}{\left[\phi^{\frac{2}{3}} - \phi + \frac{\lambda_p}{\lambda_f} \left(1 - \phi^{\frac{2}{3}} \right) \right]} \tag{9.15}$$

Nielsen and Lewis et al. [38,39] proposed a model that is a modified version of the Halpin-Tsai's one [40,41]. The Nielsen and Lewis model mainly associates with the parameters include aggregate type, shape, and arrangement of dispersed fillers in a continuous polymer matrix. According to Nielsen and Lewis model of thermal conductivity can be written as,

$$\lambda_c = \frac{1 + AB\phi}{1 - B\eta\phi} \tag{9.16}$$

$$A = \kappa_E - 1, \ B = \frac{\frac{\lambda_f}{\lambda_p} - 1}{\frac{\lambda_f}{\lambda_p} + A}, \ \eta = 1 + \phi\frac{(1 - \phi_m)}{\phi^2_m}$$

where A is the constant that related to the generalized Einstein coefficient κ_E [42] and η is the packing fraction ϕ_m of the dispersed fillers in a matrix and thus adds the mutual interactions between the fillers. The model applicable for a wide range of fillers in a composite medium and the filler capacity is limited to <20vol.% of a composite part.

Agari and Uno et al. [43] have predicted the thermal conductivity of the polymer composite based on the series and parallel model and the model is applicable for higher loading of filler in a polymer matrix. Also, the influence of the different type of particle interactions and their sizes are considered for the formation of high thermal conductivity of the polymer composites and the model can be expressed as,

$$\lambda_c = \lambda_f^{C_2\phi}\left(C_1\lambda_p\right)^{(1-\phi)} \tag{9.17}$$

where C_1 is the characterization factor that can affect the crystallinity and crystal size and C_2 is the factor of forming the thermal conduction path inside the polymer composite. In addition to this, Agari also proposed the thermal conductivity model for polymer composite systems with different types of fillers [44]. The modified model can be determined as,

$$\lambda_c = \left(C_1\lambda_p\right)^{(1-\phi)}\lambda_{f2}^{C_2X_2\phi}\lambda_{f3}^{C_3X_3\phi} \tag{9.18}$$

where $X_{1,2,3}...$ are the statistical fraction of conductive filler in the mixed fillers dispersed in the polymer composite.

Hatta and Taya et al. [45] proposed a model for the polymer composite that is applicable to measure the conductivity of the system where the fillers having unidirectional orientation or high aspect ratio in a continuous medium. The model assumes that all the dispersed fillers in the composite system are stacked and aligned parallel or perpendicular to the established direction of heat flow. The predicted thermal conductive model can be written as,

$$\lambda_{c,i} = \lambda_p\left[1 + \frac{\phi}{S_i(1 - \phi) + \frac{\lambda_p}{\lambda_f - \lambda_p}}\right] \tag{9.19}$$

where $\lambda_{c,i}$ is the thermal conductivity of the composite system with respect to specified direction and S_i is the factor corresponding to dispersed filler orientational direction and their shape, respectively. The $S_a = S_b = d/4\ T$ and $S_c = 1 - d/4\ T$, is written for the stacked fillers aligned in parallel or perpendicular to the plane, and d and T are represented as diameter and thickness of the fillers used in the system.

9.5 Factors affecting the thermal conduction behavior of graphene and its polymer composites

Thermal management of graphene and graphene-based polymeric nanocomposites is quite interesting from both fundamental academic and practical applications point of view. Graphene has unique thermal management properties that could inspire to investigate graphene and graphene-based materials in coatings, thermal composites, and thermal interface material applications. In this section, we will discuss the effect of morphology of graphene in the polymer matrix, graphene loading, functionalizations of graphene, preparation methods, and novel architectures on the thermal conductivity of graphene-based polymer nanocomposites. Mainly, the graphene sheet size, thickness, quality, orientations, surface function nalizations, shape, defects, and graphene loading are the key characteristics for controlling the thermal conductivity of graphene-based polymeric materials. Overall, the thermal conductivity of polymer nanocomposites is correlated to the intrinsic thermal conductivity of nanofiller. The thermal conductivity of polymer nanocomposites increases with increasing the nanofiller content. The dependency of these key parameters on the thermal conductivity of graphene-based polymer nanocomposites is discussed in the following paragraph.

9.5.1 Effect of graphene sheet size and shape

The size and shape of the graphene sheets (GS) or flakes have utmost importance for enhancing the thermal transport. As we have already known that phonon transfer has a significant impact on the thermal conductivity of polymer nanocomposites. The mean free path of phonon transfer depends on the shape and size of the graphene which can control the thermal management. When phonon scattering is small enough, thermal conductivity depends on increasing the mean-free path of phonon transfer [46,47]. Large size of the graphene sheet has been stated to minimize the interfacial contact areas between the polymer matrix and the sheets, thereby reducing the resistance to phonon transport. So, that the thermal conduction networks and connected paths have been completely formed when the formation of a huge number interfaces between the polymer matrix and the graphene flakes is appropriate and results in a more favorable enhancement in thermal conductivity [4].

Kumar et al. prepared large and small size graphene flakes through low temperature chemical reduction and compared their thermal conductivity behavior [48]. The thermal conductivity of large-area graphene is around 1390 W/mK which was 54% higher than that of the thermal conductivity of small size of graphene (900 W/mK) (Fig. 9.5). This increment is mainly attributed to the presence of more acoustic phonons with longer wavelength which are available for more heat conduction on the large area graphene without thermal resistance as compared to the small area graphene. Cao et al. developed a nonequilibrium molecular dynamics method to evaluate the thermal conductivity of trilayer armchair and zig-zag graphene nanoribbons with variation of temperature and length [49]. With an increasing length from 10 to 20 nm and temperature from 150K to 310K, the thermal conductivity of both trilayer armchair and zig-zag graphene nanoribbons gradually increases. In addition, the lamellar shape of the graphene can be directly related to the contact areas which, in turn, affect the thermal conductivity of graphene-based polymer nanocomposites. Finite element analyses proved that phonons transport in lamellar fillers based on graphene easily facilitate thermal conduction pathways in polymer nanocomposites [50]. Moreover, the aspect ratio of the 2D-graphitic filler material has significant effects on the thermal transport behavior of polymer nanocomposites. For instance, Yu et al. prepared the GNP based epoxy nanocomposites and reported on the effect of GNP aspect ratio on thermal conductivity behavior [51]. Thermal conductivity of larger size of GNP having an aspect ratio 200, was compared to lower GNP having aspect ratio of 30. At the highest aspect ratio, the thermal conductivity value of GNP/epoxy composites reached 1.45 W/mK with 5.4vol.% GNP loading whereas the lowest aspect ratio GNP based nanocomposites showed 1.08 W/mK. However, the thermal conductive fillers do reveal noticeable changes with increasing aspect ratio, until its value reached a certain threshold value. Then, the high aspect ratio exhibits minimal changes with thermal conductivity of polymer nanocomposites, as also reported by Kim et al. [52] and Roy et al. [53].

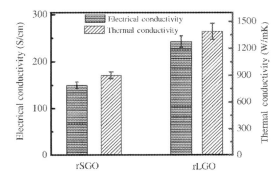

Figure 9.5 Electrical and thermal conductivity of rSGO and rLGO thin films.
Source: From P. Kumar, F. Shahzad, S. Yu, S.M. Hong, Y.-H. Kim, C.M. Koo, Large-area reduced graphene oxide thin film with excellent thermal conductivity and electromagnetic interference shielding effectiveness, Carbon 94 (2015) 494−500. Copyright (2015), with permission from Elsevier.

9.5.2 Effect of defects

For graphene and graphene-based polymer nanocomposites, defects in graphene are the crucial limiting factor to efficiently promote heat transfer via introducing phonons propagation. The transfer of phonons is the most favorable and responsible method to manipulate thermal conductivity of carbon-based materials. Therefore defects in crystal can also originate from incompleteness and irregularity such as surface defects (grain boundary and phase interfaces), line defects (dislocations), body defects (bubbles and cavities), point defects (vacancy and interstitial atoms), edge roughness, and contact disorders and wrinkles, respectively [54,55]. These defects would promote phonon transmission and short the mean free path of phonon. However, the reduction in the mean free path of phonon not only caused by atomic interactions but also by the small size defects like heteroatoms, the structure of substrate, and dislocations. At low temperature region, phonon scattering through graphene interfaces caused by bubbles and grain boundary defects where it can be neglected at high temperature region [54,55]. The smaller grain size of graphene exhibits a low thermal transport behavior due to more frequent phonon scattering and grain boundary effects on thermal conduction [56]. The thermal conductivity of suspended graphene increases with the increasing density of defects in graphene upto a certain level [57]. In graphene/epoxy composites study, the defects in graphene such as single vacancy (SV), double vacancy (DV), multivacancy (MV) and stone-wales defects (SW) can be calculated by the number of defects divided by the number of atoms in pristine grapheme (Fig. 9.6) [58]. SW and MV defects improved the thermal conductivity of epoxy composites to some extent. In comparison with the three types of defects (SV, DV, and SW), Zhang et al. also reported that SW defects exhibit a descending order in thermal conductivity followed by SV and DV defects [6]. However, the thermal conductivity on the SV defect density show the same descending trend with increasing temperature due to presence of phonon-boundary and phonon-phonon scatterings induced by the higher temperature

In another study, Zhang et al. also observed similar findings of the reduction of thermal conductivity in graphene with different vacancy defect concentrations and obtained maximum of a thousand fold reduction of thermal conductivity with a high defect density [59]. Chen et al. studied the impact of wrinkles and cracks on thermal conductivity of CVD grown suspended graphene [60]. These defects led to more phonon scattering and thus reduction in thermal conductivity. The decrease in thermal conductivity with higher temperature is also attributed to enhanced phonon-phonon scattering. Yang et al. introduced the folding pattern and changing the interlayer couplings in graphene [7]. It has been reported that the folding pattern is a responsible factor for the reduction of thermal conductivity due to anharmonic phonon-phonon scattering when the interlamellar space is reduced from 0.64 to 0.44 nm

9.5.3 Effect of surface functionalization

Surface functionalization of graphene is a unique strategy to improve its dispersion and distribution within the polymer matrix. In polymer nanocomposites, both the

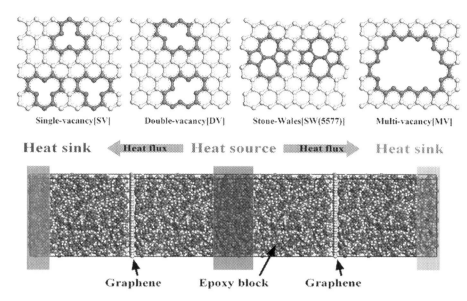

Figure 9.6 Types of graphene defects studied in graphene/epoxy composites.
Source: From M. Li, H. Zhou, Y. Zhang, Y. Liao, H. Zhou, Effect of defects on thermal conductivity of graphene/epoxy nanocomposites, Carbon 130 (2018) 295—303. Copyright (2018), with permission from Elsevier.

thermally conductive filler and polymer matrix form many contact interface areas between them. The phonon scattering at the available contact interfaces would decrease the efficiency of thermal transport and increase interfacial thermal resistance in polymer nanocomposites. This increment depends on some factors such as the poor interfacial performance, poor compatibility of thermally conductive fillers with polymer matrix, difficulty in dispersion of filler, high surface energy of nanofillers, and nanofiller agglomeration, and the presence of defects or voids at interfaces, respectively. Therefore several researchers have carried out various surface functionalizations on thermally conductive fillers via acid-base treatment, chemical modification, surface coating, and mechanochemical method to improve the polarity and wettability between nanofillers and polymer chains [2]. The surface functionalization treatments can also enhance the mechanical, thermal, and flame-retardant properties of polymer nanocomposites to some extent. Moreover, covalent or noncovalent surface functionalization on the thermally conductive fillers can improve the efficiency of thermal conduction. Especially, these treatments can not only promote the dispersion of graphene in solvents but also improve the wettability in polymer chains. For example, You et al. established a new strategy like plasma treatment and mechanochemical method to improve the compatibility between graphene nanoplatelets (GNPs) and PA66 as well as thermal conductivity of nanocomposites [1]. They found that the thermal conductivity value of PA66 nanocomposites with 20wt.% GNPs reached 3.77 W/mK which is a higher value than that of thermal conductivity value (1.32 W/mK) of PA66 nanocomposites prepared by conventional technique. In another study, Zhang et al. treated graphene-based film (GBF) with silane [61]. Silane-functionalized GBF exhibits higher thermal conductivity and heat spreading performance compared to nonsilane functionalized GBF. The thermal conductivity of functionalized graphene increased by between 15% and 56% as a function of the number density of functionalized silane molecules. However, some studies have shown that glycidyl methacrylate grafted graphene oxide (g-GO) could improve the thermal conductivity efficiency of polymer composites [5,62]. For example, Tseng et al. reported that the g-GO can lead huge improvement on thermal conductivity compared to non-g-GO [5]. The thermal conductivity of PI/g-GO nanocompoites exhibited a linear ascending trend with higher filler concentration, as shown in (Fig. 9.7).

The compatible phase between g-GO with polymer chain molecules results in a more efficient transfer of the phonons from graphene phase to PI matrix, which effectively improves the thermal conductivity performance. Teng et al. functiona-lized graphene nanosheets (GNS) with poly (glycidyl methacrylate) containing localized pyrene groups, which effectively enhanced the interfacial interaction between GNS and epoxy composites via formation of covalent bonds [63], and thus increased the thermal conductivity values of the epoxy composites. In addition, the presence of π-π stacking of pyrene filled PGMA polymer chain on GNS surface helps both the well dispersion of GNS in epoxy matrix as well as improves the polymer—filler interactions that can assist in better thermal transportation. The ther-mal conductivity value of Py-PGMA-GNS/epoxy composite containing 4 phr filler

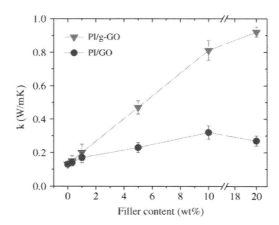

Figure 9.7 Thermal conductivity (k) of PI/GO and PI/g-GO nanocomposites as a function of filler loading. *GO, Graphene oxide.*
Source: From I.H. Tseng, J.C. Chang, S.L. Huang, M.H. Tsai, Enhanced thermal conductivity and dimensional stability of flexible polyimide nanocomposite film by addition of functionalized graphene oxide, Polym. Int. 62 (5) (2013) 827—835. Copyright (2018), with permission from Elsevier.

loading increased more than 800% and 16.4% compared with pure epoxy matrix and epoxy composites with same loading of pristine GNS.

9.5.4 Effect of novel architectures

Most researchers believe that graphene with a specific orientation can provide effective channels for phonon transport within the polymer matrix [64−68]. The thermal conductivity value in the oriented direction is much better than that of graphene with random oriented directions. For example, Li et al. compared thermal transport performances with both aligned and interconnected multilayer graphene (MLG) in epoxy nanocomposites [65]. The aligned MLG epoxy composite shows the increment in thermal conductivity from 16.75 to 33.54 W/mK with only 11.8wt.% graphene as a function of temperature ranging from 40°C to 90°C, as shown in Fig. 9.8A. This enhancement of thermal conductivity is mainly attributed to the unique alignment in the structure of MLG in epoxy composite. Zhang et al. have investigated the effect of vertically aligned graphene film based polydimethylsiloxane (PDMS) nanocomposites on thermal conductivity behavior [64]. It can be shown that the thermal conductivity of PDMS nanocomposites reaches 614.85 W/mK, which is higher than copper at room temperature (Fig. 9.8B). This enhancement is attributed to effective overall phonon transport and formation of interconnected networks by oriented graphene.

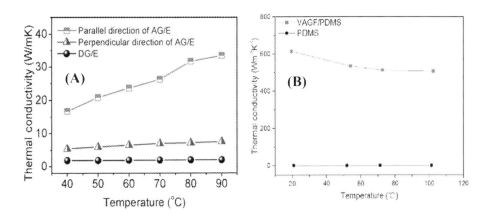

Figure 9.8 Thermal conductivity of (A) aligned multilayer graphene/epoxy composites and (B) vertically aligned graphene film/PDMS composites.
Source: From Q. Li, Y. Guo, W. Li, S. Qiu, C. Zhu, X. Wei, et al., Ultrahigh thermal conductivity of assembled aligned multilayer graphene/epoxy composite, Chem. Mater. 26(15) (2014) 4459−4465; Y.-F. Zhang, D. Han, Y.-H. Zhao, S.-L. Bai, High-performance thermal interface materials consisting of vertically aligned graphene film and polymer, Carbon 109 (2016) 552−557. [Copyright (2016), with permission from Elsevier] as a function of temperature.

Kumar et al. [67] found that the self-aligned reduced graphene oxide (rGO)/ PVDF-HFP polymer composites showed high thermal conductivity whereas Renteria et al. reported that aligned magnetically-functionalized graphene-based epoxy composites also exhibited strong increment in thermal conductivity value than that of random oriented graphene/epoxy composites [69]. Recently, 3D graphene foam (GF) has attracted immense potential due to its flexibility and excellent thermal conductivity. Liu et al. reported that GF assembled polymer composites exhibit exceptional thermal conductivity of 1.52 W/mK at 5wt.% GF [70]. This increment in thermal conductivity might be due to the easy formation of interconnected heat conduction path in 3DGF. Zhao et al. also compared the thermal conductivity of GF and GS-based polydimethylsiloxane nanocomposites [71].

It is observed that at 0.7wt.% nanofiller loading level, the thermal conductivity of GF-based PDMS nanocomposites is 0.56 W/mK, and it is nearly 300% higher than that of pure PDMS, and 20% greater than that of GS/PDMS nanocomposites. The unique interconnected architecture of GF in PDMS allows the composites with more effective heat transport. Basically, the interconnected network structure of GF reduces the interface thermal resistance and improves the overall phonon transport by minimizing the interfacial phonon scattering compared to normal GS flakes. In the case of GS/PDMS nanocomposites, double-face interfaces are formed through each GS flake and PDMS chains results complicated heat transfer process and lower thermal conductivity than GF/PDMS nanocomposites. Recently, graphene aerogels have been studied widely in thermal management applications due to its available large surface area, unique architecture, excellent mechanical properties, and outstanding thermal conductivity. For example, Yang et al. [72] prepared high-density graphene/octadecanol phase change aerogel composites. It is observed that the aerogel exhibited an exceptional thermal conductivity of ∼5.92 W/mK which is higher than that of pure 1-octadecanol. Hence, we can conclude that orientation of graphene and formation of heat transfer pathways can explicitly improve thermal conductivity of the polymer/graphene nanocomposites.

9.5.5 Effect of hybrid graphene filler

Hybrid graphene nanofillers have been widely employed as thermal conductive nanofiller, that is an interesting approach to enhance the thermal transport performance in polymer composites and as well as improved interaction between fillers and polymer chains, as compared to single thermally conductive nanofiller [4]. In general, hybrid nanofillers are composed of two or more different types nanofillers, as they are different in size, shape, and types. Basically, these binary or ternary hybrid thermally conductive nanofillers are prepared by several methods such as chemical bonding, physical bonding, direct blending, and physical adsorption, etc [4]. Several researchers have found that the incorporation of hybrid nanofiller to the polymer matrix could easily achieve higher thermal conductivity of the polymer composites than that of single types of thermally conductive fillers. The hybrid nanofillers could also improve the distribution or dispersion behavior inside polymer matrix, or reduce the surface defects and interfaces, or help reduce the voids in polymer matrix, or easily create bridge between adjacent thermally conductive nanofillers, or improve polymer-filler interaction, which actually forms interconnected thermal conduction networks for better transport of phonons, resulting in the excellent enhancement of thermal conductivity.

Liang et al. investigated the role of 3D porous GNPs/rGO hybrid foam to prepare epoxy-hybrid nanocomposites for high thermal conductivity composites [73]. A high thermal conductivity value of 1.56 W/mK was achieved for GNPs/rGO/ epoxy composite with 20.4wt.% GNPs loading, benefiting from the 3D nanohybrid framework, where the GNP/epoxy composites, prepared by ordinary blending, achieved a thermal conductivity value of only 1.01 Wm/K with same loading of filler, as shown in Fig. 9.9A. The formation of electrostatic adsorption or π-π interaction between GNPs and rGO facilitates good dispersion of hybrid filler inside epoxy matrix as well as enhances the polymer-filler interactions. Also, small sized fillers can easily fill the gaps of the large-sized fillers, results in increasing contact area and formation of compact interconnected thermal conduction networks, improving thermal conductivity of epoxy composites. Song et al. also fabricated the polymer-functionalized CNT (PCNT)/rGO/styrene butadiene rubber (SBR) composites and investigated the thermal conductivity behavior as a function of hybrid filler loading [74]. They have prepared PCNT@rGO using reversible addition-fragmentation chain transfer polymerization, esterification reaction and reduction process. PCNT filler was grafted onto the GO sheets and reduced by reducing agent to fabricate PCNT/rGO hybrid thermally conductive fillers. Fig. 9.9B depicts the

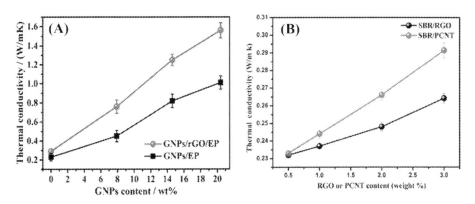

Figure 9.9 Thermal conductivity of (A) 3DGNPs/rGO/epoxy composites and (B) PCNT@RGO/SBR composites. *PCNT, polymer-functionalized carbon nanotubes. Source:* From C. Liang, H. Qiu, Y. Han, H. Gu, P. Song, L. Wang, et al., Superior electromagnetic interference shielding 3D graphene nanoplatelets/reduced graphene oxide foam/epoxy nanocomposites with high thermal conductivity, J. Mater. Chem. C. 7 (9) (2019) 2725−2733S; Song, Y. Zhang, Carbon nanotube/reduced graphene oxide hybrid for simultaneously enhancing the thermal conductivity and mechanical properties of styrene-butadiene rubber, Carbon 123 (2017) 158−167. [Copyright (2017), with permission from Elsevier] as a function of filler content.

thermal conductivity behavior of PCNT/rGO/SBR composites with variation of filler content. The thermal conductivity of hybrid nanocomposites at 3wt.% PCNT/rGO loading was 0.45 W/mK where PCNT/SBR and rGO/SBR nanocomposites were showed only 0.29 and 0.26 W/mK, respectively. The synergistic effect of PCNT and rGO in PCNT@rGO hybrid and formation of continuous filler networks inside the SBR matrix are responsible for the improved thermal conductivity.

9.5.6   Effect of processing method, filler loading and dispersion of filler

Besides the structure, orientation and functionalization of graphene, the main factors affecting the thermal conductivity of graphene-based materials are filler loading, processing methods, and dispersion of filler in polymer matrix [4]. The loading of graphene has a great influence on the thermal conductivity of the graphene-polymer composites. Generally, the thermal conductivity of composites increases with increase in graphene loading which indicates the positive correlation between them. On low loading of graphene in polymer matrix, GS are difficult to formation of good thermal conduction pathways because of GS are isolated each other inside the polymer matrix, which results in the ineffective enhancement of thermal conductivity values. Thus the high graphene loading in the composites is essential to achieve high thermal conductivity value which implies the formation of efficient thermal conductive pathways inside the polymer composites. Li et al. [75] and Michael et al. [76] established the positive correlation between thermal conductivity and graphene loading in graphene-based polymer composites. Fazel et al. reported that the thermal conductivity of graphene/stearyl alcohol gradually increases with increasing graphene content and there is no optimum threshold of graphene loading [77]. Also, Khan et al.'s group found a similar finding on the thermal conductivity of graphene-based epoxy composites [78]. As we have already known that high graphene loading is not always preferred and has an adverse effect on the mechanical properties, expensive in cost, poor processing performance, etc. Thus incorporation of graphene in polymer matrix has the limitation. So that, researchers have found that the good dispersion of graphene is beneficial to the great improvement of the thermal conductivity of polymer composites. Mainly, dispersion of fillers in polymer matrix depends on the several processing methods like powder blending, melt mixing, solution mixing, in-situ polymerization, vacuum-assisted impregnation, electrospinning, and many other methods [4]. Powder blending and melt compounding are an economical and environmentally friendly method for fabricating graphene-polymer composites. These methods are typically favorable to fabricate composites at low filler loading. Solution mixing methods is a promising method to prepare graphene-based composites up to high filler loading because of good dispersion and distribution of filler inside the polymer matrix are easily achieved. This method requires a solvent as dispersing medium for dispersing graphene into polymer matrix but most of the solvents are not ecofriendly. Many researchers try to construct efficient graphene conduction path inside the polymer matrix and increasing thermal conductivity values of polymer composites through above mentioned methods. For example,

Feng et al. developed the graphite flake wrapped PP powder based segregated composites and compared with graphite flake/PP composites prepared by conventional blending [79]. The thermal conductivity value of segregated composites was 5.4 W/mK with 21.2vol.% graphite flake, much greater than that of conventional composites (1.65 W/mK), as show in Fig. 9.10A [79]. Similar way, Wu et al. [80] prepared segregated dual-network structure of (MWCNT/PS)@GNPs composites. The thermal conductivity of segregated double network based (MWCNT/PS)@GNPs composites was achieved (1) eightfold that of randomly dispersed PS/MWCNT/GNPs composites, and (2) twofold than that of (GNPs/MWCNT)/PS segregated composites. Qin et al. [81] prepared 3D graphene on melamine sponge (MF) through infiltration and successfully prepared graphene/PDMS composite, as shown in Fig. 9.10B. Due to the interconnected double continuous network of graphene, the thermal conductivity of the RGO@MF/PDMS composite with a graphene loading of only 4.82wt.% reached 2.91 W/mK. Peng et al. used two types fused deposition 3D printing methods to fabricate vertical alignment of graphite flakes based PA6 composites and also compared

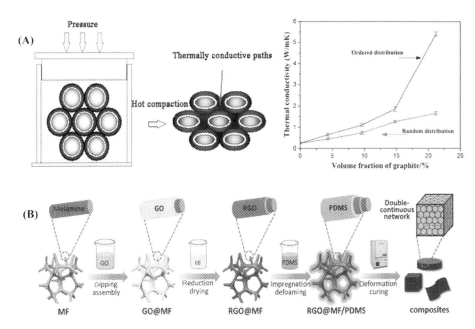

Figure 9.10 Schematic illustration of (A) the segregated structure fabrication process and (B) overall preparation procedures of 3D thermally conductive nanocomposites with double-continuous networks.

Source: From C. Feng, H. Ni, J. Chen, W. Yang, Facile method to fabricate highly thermally conductive graphite/PP composite with network structures, ACS Appl. Mater. Interfaces 8 (30) (2016) 19732–19738; M. Qin, Y. Xu, R. Cao, W. Feng, L. Chen, Efficiently controlling the 3D thermal conductivity of a polymer nanocomposite via a hyperelastic double-continuous network of graphene and sponge, Adv. Funct. Mater. 28 (45) (2018) 1805053. Copyright (2018), with permission from Wiley.

thermal conductivity with injection molding (IM) fabricated composites [82]. The thermal conductivity of stand-3D printing (SP) fabricated composites reached 5.5 W/mK with 50wt.% graphite flakes, much higher than that of IM fabricated composites (2.4 W/mK) and flat-3D printing fabricated composites (~0.5 W/mK). This enhancement is mainly due to the unique orientation of graphite flakes inside the PA-6 matrix. In SP and IM fabricated composites, the graphite flakes are aligned along the through-plane direction whereas the graphite flakes are oriented in the in-plane direction in FP fabricated composites. With an increase in the content of graphite flakes, graphite flakes are fully stacked together completely in through-plane direction for IM fabricated composites.

Yang et al. used self-assembly, freeze drying, and thermal annealing method to prepare CuNWs-thermally annealed graphene aerogel (TAGA) aerogel and then prepared 3D CuNWs-TAGA/epoxy composites [83]. The thermal conductivity coefficient of CuNWs-TAGA/epoxy composites reached 0.51 W/mK with 7.2wt.% CuNWs-TAGA loading. Magnetic field has unique advantages to get good orientation of thermal conductive fillers inside the polymer matrix in a certain direction. Chung et al. used magnetic field to induce the orientation of graphite platelets (GP) inside the poly(vinyl pyrrolidone) (PVP) matrix [84]. The thermal conductivity value of GP/PVP composite films with 80wt.% filler reached 2.4 W/mK which is 330% enhancement of through-plane thermal conductivity than that of pure PVP film.

9.6 Recent advances in thermal conductivity of graphene-polymer nanocomposites

Recently, graphene-based polymer nanocomposites have gained tremendous attention acting as effective thermal dissipation materials due to the excellent heat transport of graphene through polymer matrix. In recent years, numerous researchers have fabricated graphene-polymer nanocomposites and investigated the importance of thermal conductivity behavior of graphene-polymer nanocomposites. For example, Wu et al. made aligned graphene nanosheets and 3D interconnected GF-based natural rubber nanocomposites and found excellent thermal conductivity enhancement (TCE) of natural rubber nanocomposites [85]. The GNs/GF/natural rubber nanocomposite exhibits TCE of 8100% with 6.2vol.% graphene loading at room temperature, which is the highest ever reported value. The excellent improvement in thermal conductivity value is mainly attributed to the significant synergistic effect between aligned GNs and 3D interconnected GF. This unique hybrid structure could facilitate the formation of thermal percolation networks for remarkably improved thermal transport through natural rubber matrix [85]. Guo et al. prepared functional Al_2O_3 grafted graphene-based nanofibrillated cellulose (NFC) composite films via facile vacuum assisted filtration [86]. The in-plane thermal conductivity of composite film reached 8.3 W/mK at 30% RGO loading, which is 1975% greater than that of the NFC. The improvement of thermal conductivity behavior of composites is attributed to three factors: (1) bridging effect of functional Al_2O_3 and GS, (2) graphene nanosheets are well aligned in the plane direction during filtration process, (3) good compatibility between nanofibrillated cellulose and graphene, respectively. Cui et al. also prepared high thermally conductive NFC composite films based on nanodiamonds and GS [87]. Ren et al. investigated that the effect of double mixing process like solution cum melt blending on thermal conductivity of graphene sheet-based epoxy nanocomposites [88]. The proposed mixing technique prevents agglomeration of GSs at higher loading and helps good dispersion and distribution of GSs within the epoxy matrix, resulting in excellent thermal conductivity. The thermal conductivity of the nanocomposites reaches 3.55 W/mK with 20wt.% GS, which is 1167% higher than the pure PA6 matrix. It is mainly due to the uniform dispersion of GSs with high loading facilitates the formation of efficient interconnected thermally conductive networks through PA6 matrix for efficient heat dissipation. Liu et al. found that GF can make efficient three dimensional interconnected networks in epoxy composites than that of GNP/epoxy composites [89]. Basically, 3D architecture of GF enhances the overall heat dissipation by minimizing the interfacial phonon scattering compared to GNP nanofiller. The thermal conductivity of GF-based epoxy composites is 8.04 W/mK at a low graphene content of 6.8wt.% and it is nearly 4473% greater than neat epoxy matrix. Zhang et al. prepared highly thermally conductive but electrically insualting GNP-hBN based PA6 nanocomposites via melt compounding process. The thermal conductivity of epoxy composites becomes 3.25 W/mK at 412vol.% GNP loading. This enhancement in thermal conductivity is mainly attributed to the continuous double segregated network between GNP and h-BN nanofiller. Yuan et al. investigated thermal conductivity of polydopamine (PDA) coated graphene oxide (GO) based polystyrene nanocomposites and reported that composites exhibited both high in-plane (4.13 W/mK) and through-plane (4.56 W/mK) thermal conductivities at 0.96vol.% GO-PDA loading with excellent electrically insulating behavior of about >1014 Ωcm [90]. Chen et al. reported the first time that the thermal conductivity of epoxy composites exhibited 13.3 W/mK and 33.4 W/mK in axial and radial direction with the binary alumina-GF with 12.1wt.% graphene and 42.4wt.% alumina. This binary alumina-GF forms a pea-pod like architecture that can lead an efficient heat transport channel through the epoxy matrix [91]. Zhang et al. investigated the effect of vertically aligned graphene tubes on thermal conductivity of polydimethylsiloxane composites. The thermal conductivity of composites is 1.7 W/mK at 4.5wt.% of graphene tubes and it is 1082% greater than that of pure PDMS matrix [92]. Cui et al. obtained graphene-polymer composite with a remarkable thermal conductivity of 21.83 W/mK at 30wt.% graphene loading. This high thermal conductivity could be contributed to the bilayer structure composed by graphene [93]. Li et al. prepared 3D interconnected graphene microspheres-based epoxy composites and investigated their thermal conductive behavior. The thermal conductivity of epoxy composites shows 0.96 W/mK at 1wt.% graphene loading, which is 437% higher than that of pure epoxy [94].

9.7   Conclusions and future outlooks

In this chapter, a brief discussion on thermal conductivity of graphene-based polymeric composites is reported and attempted to elucidate the progress of research into the thermal management behavior on several factors. Thermally conductive graphene-based polymer composites are tentatively prepared by the incorporation of graphene. It is found that the types of polymer, filler size and shape, filler loading, filler surface morphology, filler surface treatment, and processing methods have great influences on the thermal conductivity coefficient of graphene-based polymer composites. Mainly, the addition of graphene in polymer matrix can help to achieve high thermal conductivity. Because of their high intrinsic thermal conductivity and unique structure, graphene has a great potential to solve the heat dissipation problems in polymeric composites. In fact, the higher graphene loading has a great effect on better heat dissipation. Excessive addition of graphene will therefore interfere with the density, mechanical properties, cost, and processing performance of the final polymeric composites and will therefore inevitably reduce the performance of the product. There is still no ultimate processing method for fabrication of thermally conductive material in a large-scale production. Significant attention should be paid to the scientific problems of thermally conductive polymer composites and their technical difficulties. In the meantime, different models have been discussed explicitly to understand the thermal conductivity behavior of graphene-based composites. Although the present thermally conductive models are still insufficient, which are needed for further modification as well as advanced and efficient techniques, they should be explored and developed to understand the precise relationship among thermal conduction networks of filler, polymer chain motions and thermal conductivity values.

Acknowledgments

This review was financially supported by the Korea Evaluation Institute of Industrial Technology, Republic of Korea, and Grant Number-20012558.

References

[1] J. You, H.-H. Choi, J. Cho, J.G. Son, M. Park, S.-S. Lee, et al., Highly thermally conductive and mechanically robust polyamide/graphite nanoplatelet composites via mechanochemical bonding techniques with plasma treatment, Compos. Sci. Technol. 160 (2018) 245−254.
[2] N. Burger, A. Laachachi, M. Ferriol, M. Lutz, V. Toniazzo, D. Ruch, Review of thermal conductivity in composites: mechanisms, parameters and theory, Prog. Polym. Sci. 61 (2016) 1−28.
[3] X. Yang, C. Liang, T. Ma, Y. Guo, J. Kong, J. Gu, et al., A review on thermally conductive polymeric composites: classification, measurement, model and equations, mechanism and fabrication methods, Adv. Compos. Hybrid. Mater. 1 (2) (2018) 207−230.
[4] Y. Guo, K. Ruan, X. Shi, X. Yang, J. Gu, Factors affecting thermal conductivities of the polymers and polymer composites: a review, Compos. Sci. Technol. 193 (2020) 108134.

[5] I.H. Tseng, J.C. Chang, S.L. Huang, M.H. Tsai, Enhanced thermal conductivity and dimensional stability of flexible polyimide nanocomposite film by addition of functionalized graphene oxide, Polym. Int. 62 (5) (2013) 827—835.

[6] Y.-Y. Zhang, Y. Cheng, Q.-X. Pei, C. Wang, Y. Xiang, Thermal conductivity of defective graphene, Phys. Lett. A 376 (47—48) (2012) 3668—3672.

[7] N. Yang, X. Ni, J.-W. Jiang, B. Li, How does folding modulate thermal conductivity of graphene? Appl. Phys. Lett. 100 (9) (2012) 093107.

[8] I.O.F. Standardization, ISO 8302: thermal insulation. Determination of Steady State Thermal Resistance and Related Properties. Guarded Hot Plate Apparatus, ISO, 1991.

[9] E.-. ASTM, Standard Test Method for Evaluating the Resistance to Thermal Transmission of Materials by the Guarded Heat Flow Meter Technique, ASTM International, West Conshohocken, 2011.

[10] A. Standard, Standard Test Method for Steady-State Heat Flux Measurements and Thermal Transmission Properties by Means of the Guarded-Hot-Plate Apparatus, Designation: C177—13, 2004.

[11] I. 22007—4, IV. Plastics—Determination of Thermal Conductivity and Thermal Diffusivity—Part 4: Laser Flash Method, 2017.

[12] I. 22007—2, Plastics—Determination of Thermal Conductivity and Thermal Diffusivity—Part 2: Transient Plane Heat Source (Hot Disc) Method, 2008.

[13] B. Shin, S. Mondal, M. Lee, S. Kim, Y.-I. Huh, C. Nah, et al., Flexible thermoplastic polyurethane-carbon nanotube composites for electromagnetic interference shielding and thermal management, Chem. Eng. J. 418 (2021) 129282. Available from: https://doi.org/10.1016/j.cej.2021.129282.

[14] A. -1, Refractory Materials—Determination of Thermal Conductivity—Part 1: Hot-Wire Methods (Cross-Array and Resistance Thermometer), AFNOR Éditions Paris, 2010.

[15] S. Min, J. Blumm, A. Lindemann, A new laser flash system for measurement of the thermophysical properties, Thermochim. Acta 455 (1—2) (2007) 46—49.

[16] X.C. Tong, Advanced Materials for Thermal Management of Electronic Packaging, Springer Science & Business Media, 2011. Available from: https://doi.org/10.1007/978-1-4419-7759-5.

[17] A. E-13, Standard Test Method for Thermal Diffusivity by the Flash Method, ASTM International, West Conshohocken, PA, 2013.

[18] W. Zhao, Y. Yang, Z. Bao, D. Yan, Z. Zhu, Methods for measuring the effective thermal conductivity of metal hydride beds: a review, Int. J. Hydrog. Energy 45 (11) (2020) 6680—6700.

[19] T. Log, S. Gustafsson, Transient plane source (TPS) technique for measuring thermal transport properties of building materials, Fire Mater. 19 (1) (1995) 43—49.

[20] B. Stalhane, S. Pyk, New method for determining the coefficients of thermal conductivity, Tek. Tidskr. 61 (28) (1931) 389—393.

[21] D. Zhao, X. Qian, X. Gu, S.A. Jajja, R. Yang, Measurement techniques for thermal conductivity and interfacial thermal conductance of bulk and thin film materials, J. Electron. Packag. 138 (4) (2016).

[22] A. Franco, An apparatus for the routine measurement of thermal conductivity of materials for building application based on a transient hot-wire method, Appl. Therm. Eng. 27 (14—15) (2007) 2495—2504.

[23] U. Hammerschmidt, J. Hameury, R. Strnad, E. Turzó-Andras, J. Wu, Critical review of industrial techniques for thermal-conductivity measurements of thermal insulation materials, Int. J. Thermophys. 36 (7) (2015) 1530—1544.

[24] J.C. Maxwell, in: W.D. Niven, J.J. Thomson (Eds.), A Treatise on Electricity and Magnetism With Prefaces, Reprint of the third (1891) edition, Dover Publications, Inc., New York, N. Y, 1954.

[25] R. Kochetov, Thermal and Electrical Properties of Nanocomposites, Including Material Properties, 2012.

[26] I. Tavman, Effective thermal conductivity of isotropic polymer composites, Int. Commun. Heat. Mass. Transf. 25 (5) (1998) 723−732.

[27] S. Agarwal, M.M.K. Khan, R.K. Gupta, Thermal conductivity of polymer nanocomposites made with carbon nanofibers, Polym. Eng. Sci. 48 (12) (2008) 2474−2481.

[28] Y. Li, G. Xu, Y. Guo, T. Ma, X. Zhong, Q. Zhang, et al., Fabrication, proposed model and simulation predictions on thermally conductive hybrid cyanate ester composites with boron nitride fillers, Compos. A Appl. Sci. Manuf. 107 (2018) 570−578.

[29] J. Felske, Effective thermal conductivity of composite spheres in a continuous medium with contact resistance, Int. J. Heat. Mass. Transf. 47 (14−16) (2004) 3453−3461.

[30] C.-W. Nan, R. Birringer, D.R. Clarke, H. Gleiter, Effective thermal conductivity of particulate composites with interfacial thermal resistance, J. Appl. Phys. 81 (10) (1997) 6692−6699.

[31] V.D. Bruggeman, Berechnung verschiedener physikalischer Konstanten von heterogenen Substanzen. I. Dielektrizitätskonstanten und Leitfähigkeiten der Mischkörper aus isotropen Substanzen, Ann. d. Phys. 416 (7) (1935) 636−664.

[32] M.-x Shen, Y.-x Cui, J. He, Y.-m Zhang, Thermal conductivity model of filled polymer composites, Int. J. Miner. Metall. Mater. 18 (5) (2011) 623.

[33] R.L. Hamilton, O. Crosser, Thermal conductivity of heterogeneous two-component systems, Ind. Eng. Chem. Fundamentals 1 (3) (1962) 187−191.

[34] R. Hamilton, Thermal Conductivity of Two-phase Materials (Dissertation), University of Oklahoma, 1960.

[35] S.C. Cheng, R. Vachon, The prediction of the thermal conductivity of two and three phase solid heterogeneous mixtures, Int. J. Heat. Mass. Transf. 12 (3) (1969) 249−264.

[36] G.T.-N. Tsao, Thermal conductivity of two-phase materials, Ind. Eng. Chem. 53 (5) (1961) 395−397.

[37] H. Russell, Principles of heat flow in porous insulators, J. Am. Ceram. Soc. 18 (1−12) (1935) 1−5.

[38] T.B. Lewis, L.E. Nielsen, Dynamic mechanical properties of particulate-filled composites, J. Appl. Polym. Sci. 14 (6) (1970) 1449−1471.

[39] L.E. Nielsen, Thermal conductivity of particulate-filled polymers, J. Appl. Polym. Sci. 17 (12) (1973) 3819−3820.

[40] J. Halpin, Stiffness and expansion estimates for oriented short fiber composites, J. Compos. Mater. 3 (4) (1969) 732−734.

[41] J.H. Affdl, J. Kardos, The Halpin-Tsai equations: a review, Polym. Eng. Sci. 16 (5) (1976) 344−352.

[42] A. Einstein, On the movement of small particles suspended in stationary liquids required by the molecularkinetic theory of heat, Ann. Phys. 17 (549−560) (1905) 1.

[43] Y. Agari, T. Uno, Estimation on thermal conductivities of filled polymers, J. Appl. Polym. Sci. 32 (7) (1986) 5705−5712.

[44] Y. Agari, A. Ueda, S. Nagai, Thermal conductivity of a polyethylene filled with disoriented short-cut carbon fibers, J. Appl. Polym. Sci. 43 (6) (1991) 1117−1124.

[45] H. Hiroshi, T. Minoru, Equivalent inclusion method for steady state heat conduction in composites, Int. J. Eng. Sci. 24 (7) (1986) 1159−1172.

[46] S.V. Kidalov, F.M. Shakhov, Thermal conductivity of diamond composites, Materials 2 (4) (2009) 2467−2495.

[47] A.M. Abyzov, S.V. Kidalov, F.M. Shakhov, High thermal conductivity composites consisting of diamond filler with tungsten coating and copper (silver) matrix, J. Mater. Sci. 46 (5) (2011) 1424−1438.

[48] P. Kumar, F. Shahzad, S. Yu, S.M. Hong, Y.-H. Kim, C.M. Koo, Large-area reduced graphene oxide thin film with excellent thermal conductivity and electromagnetic interference shielding effectiveness, Carbon 94 (2015) 494–500.

[49] H.-Y. Cao, Z.-X. Guo, H. Xiang, X.-G. Gong, Layer and size dependence of thermal conductivity in multilayer graphene nanoribbons, Phys. Lett. A 376 (4) (2012) 525–528.

[50] B. Mortazavi, M. Baniassadi, J. Bardon, S. Ahzi, Modeling of two-phase random composite materials by finite element, Mori–Tanaka and strong contrast methods, Compos. B Eng. 45 (1) (2013) 1117–1125.

[51] A. Yu, P. Ramesh, M.E. Itkis, E. Bekyarova, R.C. Haddon, Graphite nanoplatelet – epoxy composite thermal interface materials, J. Phys. Chem. C. 111 (21) (2007) 7565–7569.

[52] X. Shen, Z. Wang, Y. Wu, X. Liu, Y.-B. He, J.-K. Kim, Multilayer graphene enables higher efficiency in improving thermal conductivities of graphene/epoxy composites, Nano Lett. 16 (6) (2016) 3585–3593.

[53] V. Varshney, J. Lee, J.S. Brown, B.L. Farmer, A.A. Voevodin, A.K. Roy, Effect of length, diameter, chirality, deformation, and strain on contact thermal conductance between single-wall carbon nanotubes, Front. Mater. 5 (2018) 17.

[54] S.H. Lo, J. He, K. Biswas, M.G. Kanatzidis, V.P. Dravid, Phonon scattering and thermal conductivity in p-type nanostructured PbTe-BaTe bulk thermoelectric materials, Adv. Funct. Mater. 22 (24) (2012) 5175–5184.

[55] K. Watari, K. Ishizaki, F.J. Tsuchiya, Phonon scattering and thermal conduction mechanisms of sintered aluminium nitride ceramics, J. Mater. Sci. 28 (14) (1993) 3709–3714.

[56] I. Vlassiouk, S. Smirnov, I. Ivanov, P.F. Fulvio, S. Dai, H. Meyer, et al., Electrical and thermal conductivity of low temperature CVD graphene: the effect of disorder, Nanotechnology 22 (27) (2011) 275716.

[57] H. Malekpour, P. Ramnani, S. Srinivasan, G. Balasubramanian, D.L. Nika, A. Mulchandani, et al., Thermal conductivity of graphene with defects induced by electron beam irradiation, Nanoscale 8 (30) (2016) 14608–14616.

[58] M. Li, H. Zhou, Y. Zhang, Y. Liao, H. Zhou, Effect of defects on thermal conductivity of graphene/epoxy nanocomposites, Carbon 130 (2018) 295–303.

[59] H. Zhang, G. Lee, K. Cho, Thermal transport in graphene and effects of vacancy defects, Phys. Rev. B 84 (11) (2011) 115460.

[60] S. Chen, Q. Li, Q. Zhang, Y. Qu, H. Ji, R.S. Ruoff, et al., Thermal conductivity measurements of suspended graphene with and without wrinkles by micro-Raman mapping, Nanotechnology 23 (36) (2012) 365701.

[61] Y. Zhang, H. Han, N. Wang, P. Zhang, Y. Fu, M. Murugesan, et al., Improved heat spreading performance of functionalized graphene in microelectronic device application, Adv. Funct. Mater. 25 (28) (2015) 4430–4435.

[62] J. Gu, X. Yang, Z. Lv, N. Li, C. Liang, Q. Zhang, Functionalized graphite nanoplatelets/epoxy resin nanocomposites with high thermal conductivity, Int. J. Heat. Mass. Transf. 92 (2016) 15–22.

[63] C.-C. Teng, C.-C.M. Ma, C.-H. Lu, S.-Y. Yang, S.-H. Lee, M.-C. Hsiao, et al., Thermal conductivity and structure of non-covalent functionalized graphene/epoxy composites, Carbon 49 (15) (2011) 5107–5116.

[64] Y.-F. Zhang, D. Han, Y.-H. Zhao, S.-L. Bai, High-performance thermal interface materials consisting of vertically aligned graphene film and polymer, Carbon 109 (2016) 552–557.

[65] Q. Li, Y. Guo, W. Li, S. Qiu, C. Zhu, X. Wei, et al., Ultrahigh thermal conductivity of assembled aligned multilayer graphene/epoxy composite, Chem. Mater. 26 (15) (2014) 4459–4465.

[66] N. Song, D. Jiao, S. Cui, X. Hou, P. Ding, L. Shi, Highly anisotropic thermal conductivity of layer-by-layer assembled nanofibrillated cellulose/graphene nanosheets hybrid films for thermal management, ACS Appl. Mater. Interfaces 9 (3) (2017) 2924—2932.

[67] P. Kumar, S. Yu, F. Shahzad, S.M. Hong, Y.-H. Kim, C.M. Koo, Ultrahigh electrically and thermally conductive self-aligned graphene/polymer composites using large-area reduced graphene oxides, Carbon 101 (2016) 120—128.

[68] F.E. Alam, W. Dai, M. Yang, S. Du, X. Li, J. Yu, et al., In situ formation of a cellular graphene framework in thermoplastic composites leading to superior thermal conductivity, J. Mater. Chem. A 5 (13) (2017) 6164—6169.

[69] J. Renteria, S. Legedza, R. Salgado, M. Balandin, S. Ramirez, M. Saadah, et al., Magnetically-functionalized self-aligning graphene fillers for high-efficiency thermal management applications, Mater. Des. 88 (2015) 214—221.

[70] Z. Liu, D. Shen, J. Yu, W. Dai, C. Li, S. Du, et al., Exceptionally high thermal and electrical conductivity of three-dimensional graphene-foam-based polymer composites, RSC Adv. 6 (27) (2016) 22364—22369.

[71] Y.-H. Zhao, Z.-K. Wu, S.-L. Bai, Study on thermal properties of graphene foam/graphene sheets filled polymer composites, Compos. A Appl. Sci. Manuf. 72 (2015) 200—206.

[72] J. Yang, X. Li, S. Han, Y. Zhang, P. Min, N. Koratkar, et al., Air-dried, high-density graphene hybrid aerogels for phase change composites with exceptional thermal conductivity and shape stability, J. Mater. Chem. A 4 (46) (2016) 18067—18074.

[73] C. Liang, H. Qiu, Y. Han, H. Gu, P. Song, L. Wang, et al., Superior electromagnetic interference shielding 3D graphene nanoplatelets/reduced graphene oxide foam/epoxy nanocomposites with high thermal conductivity, J. Mater. Chem. C. 7 (9) (2019) 2725—2733.

[74] S. Song, Y. Zhang, Carbon nanotube/reduced graphene oxide hybrid for simultaneously enhancing the thermal conductivity and mechanical properties of styrene-butadiene rubber, Carbon 123 (2017) 158—167.

[75] A. Li, C. Zhang, Y.F. Zhang, Graphene nanosheets-filled epoxy composites prepared by a fast dispersion method, J. Appl. Polym. Sci. 134 (36) (2017) 45152.

[76] M. Shtein, R. Nadiv, M. Buzaglo, K. Kahil, O. Regev, Thermally conductive graphene-polymer composites: size, percolation, and synergy effects, Chem. Mater. 27 (6) (2015) 2100—2106.

[77] F. Yavari, H.R. Fard, K. Pashayi, M.A. Rafiee, A. Zamiri, Z. Yu, et al., Enhanced thermal conductivity in a nanostructured phase change composite due to low concentration graphene additives, J. Phys. Chem. C. 115 (17) (2011) 8753—8758.

[78] K.M. Shahil, A.A. Balandin, Graphene—multilayer graphene nanocomposites as highly efficient thermal interface materials, Nano Lett. 12 (2) (2012) 861—867.

[79] C. Feng, H. Ni, J. Chen, W. Yang, Facile method to fabricate highly thermally conductive graphite/PP composite with network structures, ACS Appl. Mater. Interfaces 8 (30) (2016) 19732—19738.

[80] K. Wu, C. Lei, R. Huang, W. Yang, S. Chai, C. Geng, et al., Design and preparation of a unique segregated double network with excellent thermal conductive property, ACS Appl. Mater. Interfaces 9 (8) (2017) 7637—7647.

[81] M. Qin, Y. Xu, R. Cao, W. Feng, L. Chen, Efficiently controlling the 3D thermal conductivity of a polymer nanocomposite via a hyperelastic double-continuous network of graphene and sponge, Adv. Funct. Mater. 28 (45) (2018) 1805053.

[82] Y. Jia, H. He, Y. Geng, B. Huang, X. Peng, High through-plane thermal conductivity of polymer based product with vertical alignment of graphite flakes achieved via 3D printing, Compos. Sci. Technol. 145 (2017) 55—61.

[83] X. Yang, S. Fan, Y. Li, Y. Guo, Y. Li, K. Ruan, et al., Synchronously improved electromagnetic interference shielding and thermal conductivity for epoxy nanocomposites by constructing 3D copper nanowires/thermally annealed graphene aerogel framework, Compos. A Appl. Sci. Manuf. 128 (2020) 105670.

[84] S.-H. Chung, H. Kim, S.W. Jeong, Improved thermal conductivity of carbon-based thermal interface materials by high-magnetic-field alignment, Carbon 140 (2018) 24—29.

[85] Z. Wu, C. Xu, C. Ma, Z. Liu, H.M. Cheng, W. Ren, Synergistic effect of aligned graphene nanosheets in graphene foam for high-performance thermally conductive composites, Adv. Mater. 31 (19) (2019) 1900199.

[86] S. Guo, R. Zheng, J. Jiang, J. Yu, K. Dai, C. Yan, Enhanced thermal conductivity and retained electrical insulation of heat spreader by incorporating alumina-deposited graphene filler in nano-fibrillated cellulose, Compos. B Eng. 178 (2019) 107489.

[87] S. Cui, N. Song, L. Shi, P. Ding, Enhanced thermal conductivity of bioinspired nanofibrillated cellulose hybrid films based on graphene sheets and nanodiamonds, ACS Sustain. Chem. Eng. 8 (16) (2020) 6363—6370.

[88] Y. Ren, Y. Zhang, H. Guo, R. Lv, S.-L. Bai, A double mixing process to greatly enhance thermal conductivity of graphene filled polyamide 6 composites, Compos. A Appl. Sci. Manuf. 126 (2019) 105578.

[89] Z. Liu, Y. Chen, Y. Li, W. Dai, Q. Yan, F.E. Alam, et al., Graphene foam-embedded epoxy composites with significant thermal conductivity enhancement, Nanoscale 11 (38) (2019) 17600—17606.

[90] H. Yuan, Y. Wang, T. Li, Y. Wang, P. Ma, H. Zhang, et al., Fabrication of thermally conductive and electrically insulating polymer composites with isotropic thermal conductivity by constructing a three-dimensional interconnected network, Nanoscale 11 (23) (2019) 11360—11368.

[91] Y. Chen, X. Hou, M. Liao, W. Dai, Z. Wang, C. Yan, et al., Constructing a "pea-pod-like" alumina-graphene binary architecture for enhancing thermal conductivity of epoxy composite, Chem. Eng. J. 381 (2020) 122690.

[92] Y.-F. Zhang, Y.-J. Ren, H.-C. Guo, S.-l Bai, Enhanced thermal properties of PDMS composites containing vertically aligned graphene tubes, Appl. Therm. Eng. 150 (2019) 840—848.

[93] S. Cui, F. Jiang, N. Song, L. Shi, P. Ding, Flexible films for smart thermal management: influence of structure construction of a two-dimensional graphene network on active heat dissipation response behavior, ACS Appl. Mater. Interfaces 11 (33) (2019) 30352—30359.

[94] C. Li, X.-L. Zeng, L.-Y. Tan, Y.-M. Yao, D.-L. Zhu, R. Sun, et al., Three-dimensional interconnected graphene microsphere as fillers for enhancing thermal conductivity of polymer, Chem. Eng. J. 368 (2019) 79—87.

Dispersion of graphene in polymer matrices

10

Subhendu Bhandari and Prashant Gupta
Department of Plastic and Polymer Engineering, Maharashtra Institute of Technology, Aurangabad, India

10.1 Introduction

Effective application of polymer composite depends on effective filler dispersion to a large extent. Graphene is widely used as filler in polymer matrix. Owing to its mechanical, electrical, morphological, and other characteristics, good dispersion is solicited depending on the end use requirement. Graphene layers in graphite are bonded together by weak Van der Waals force and separated by ~ 0.335 nm only [1]. Total-energy method-based theoretical investigation reveals the presence of strong covalent bond between the layers existing on the top of each other, viz. AA type stacking, having interlayer distance approximately half of that existing between alternating AB-type stacks [2]. However, stacked layers of graphene contribute to low extent of polymer-filler interaction, which is an essential requirement for improvement of mechanical strength by incorporation of reinforcing filler. Graphene may be used to prepare polymeric composite by using it individually, as well as in combination with other fillers for further improvement in desired properties [3,4]. Since, in agglomerated form, the effective specific surface area available for graphene, separation of layers (i.e., dispersion) and its homogeneous distribution are the utmost consideration for nanocomposites for improvement of electrical conductivity [5,6] gas barrier property etc. [7,8]. Relative positioning among graphene layers play important role in such cases. For example, interconnection among conductive layers are the primary requirement for electrically conductive composites to reach percolation threshold; whereas, nearly parallel orientation of graphene layers in composites exhibit improved gas barrier property, since path of gas passing through the bulk of the composite becomes more distorted (tortuous path) [8], as shown in Fig. 10.1. Few models for determination of gas barrier property have been shown in Table 10.1 where P_0 and P represent the gas permeability through the pristine polymer and composite respectively, φ is the volume fraction of filler, α is the aspect ratio of filler flakes/ layers, and S′ is the orientation factor. Typical values of S′ may be considered as 1, 0 and -0.5 for perpendicular, random, and parallel orientation with respect to the path of gas permeation [9]. Since delamination of graphene layers from stacked aggregates and its orientation in composite depend on dispersion and distribution during processing/manufacturing the composite, therefore, the processing method also play a significant role in this regard.

Polymer Nanocomposites Containing Graphene. DOI: https://doi.org/10.1016/B978-0-12-821639-2.00020-3

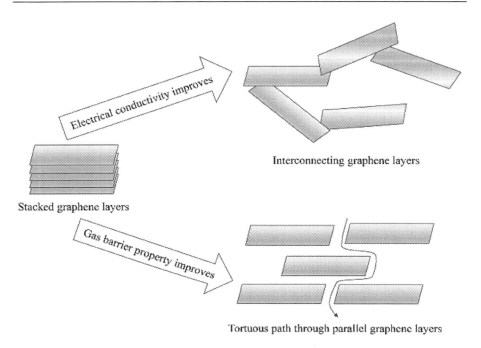

Figure 10.1 Dispersion and distribution requirement for improvement of electrical conductivity and gas barrier property.

Table 10.1 Different models of gas permeability for composites having flake/layered fillers correlating to dispersion characteristics [9].

Models	Formula	References
Nielson	$\frac{P}{P_0} = \frac{1-\varphi}{1+\left(\frac{a}{2}\right)\varphi}$	[10]
Bharadwaj	$\frac{P}{P_0} = \frac{1-\varphi}{1+\left(\frac{a}{2}\right)\left(\frac{2}{3}\right)\left(S'+\frac{1}{2}\right)\varphi}$	[11]
Cussler (Regular array)	$\frac{P_0}{P}(1-\varphi) = 1 + \frac{(\alpha\varphi)^2}{4}$	[12−14]
Cussler (Random array)	$\frac{P_0}{P}(1-\varphi) = \left(1+\frac{\alpha\varphi}{3}\right)^2$	[12−14]

Three methods are mostly used for preparation of polymer-graphene composite: (1) melt intercalation, (2) solution mixing (or, solution intercalation) and (3) in situ intercalative polymerization [15]. During intercalation interlayer distance between parallel layers of graphene increases, which can be identified from the shifting of graphitic peak to lower value of 2θ in X-ray diffraction (XRD) analysis. This may occur (1) by shearing of polymers having good polymer-filler interaction on outer surfaces of the aggregates, as observed in the case of melt mixing, or (2) application of ultrasonication for the case of both solution mixing and in situ intercalative polymerization. Intercalation of graphitic layers may be carried out by using different intercalants like AsF_5, sulfur, inorganic salts like LiCl, KCl, [16] NaCl, $CuCl_2$, [17] $FeCl_2$ [18], molten ternary salt $KCl-NaCl-ZnCl_2$ [19] etc. These may be used for either of the processes mentioned for preparation of polymer-graphene composite.

For the melt intercalation, molten polymer matrix can penetrate inside the interlayer gap, or gallery spacing which again facilitates to exfoliate upon further processing to have layers oriented in different directions isotropically (Fig. 10.2). Therefore, polymer-filler interface, that is the effective specific surface area increases with layer separation facilitating improvement of polymer-filler interface interaction which is observed in intercalation, and to the highest extent for complete exfoliation, which can be identified from complete diminishing of graphitic peak for the composite in x-ray diffractogram. For solution mixing, graphene is dispersed in a solvent medium by ultrasonication, followed by its mixing with the solution of the polymer matrix. Before drying, polymer chains enter into the interlayer spacing for intercalation, and have sufficient interaction with the layers for exfoliation. For in situ intercalative polymerization, monomers are soaked and allowed to enter inside intercalated layers of graphene followed by soaking with initiator. Therefore, polymerization occurs inside the gallery spacing too. Similar methods are also followed for other nano-layered fillers, viz., nanoclay [34], layered double hydroxide (LDH) [20] etc.

Apart from different preparation techniques of graphene and its composites, the type of functionalization also plays significant role in its dispersion characteristics. Surface modification of graphene may be carried out by either noncovalent attachment of molecules or covalent attachment of functional groups on the surface of graphene [21,22]. Chemical modification of graphene by means of bond formation results in improvement of matrix-filler interface which, as a consequence results in improvement of mechanical properties.

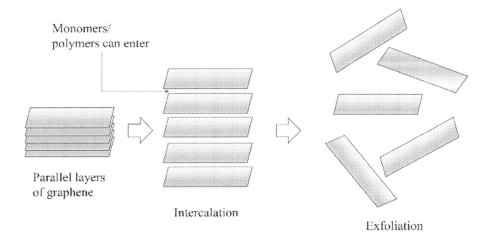

Monomers/
polymers can enter

Parallel layers
of graphene

Intercalation

Exfoliation

Figure 10.2 Exfoliation and intercalation of graphene layers.

10.1.1   Composites with covalently functionalized graphene

Song et al. [23] covalently functionalized graphene oxide (GO) with bis (3-amino-propyl)-terminated poly(ethylene glycol)(NH$_2$-PEG-NH$_2$) followed by grafting with maleic anhydride grafted polypropylene (MAPP) oligomer matrix via reactive compatibilization method. Uniformity in dispersion was attributed to augmented compatibilization. Wan et al. [24] covalently functionalized reduced graphene oxide by grafting diglycidyl ether of bisphenol-A (DGEBA) onto it which exhibited improved dispersion in epoxy matrix through ball milling. Modified graphene oxide was found to be well embedded in the epoxy matrix without remarkable de-bonding and agglomeration because of possible strong covalent bonding, π-π stacking as well as similarity of chemical structure between the matrix and the grafting molecule. Wang et al. [25] functionalized graphene nanosheets covalently with 3-aminopropyl triethoxysilane (APTS) exhibiting better dispersion in epoxy resin compared to the unmodified counterpart.

10.1.2   Composites with noncovalently functionalized graphene

Wang et al. [26] used poly(sodium 4-styrenesulfonate) (PSS) for noncovalent functionalization of graphene exhibiting improved dispersion in polyvinyl alcohol owing to H-bonding, π- π and CH- π interaction between polymer and nanofiller. Song et al. [27] synthesized poly(methyl methacrylate)-block-polydimethylsiloxane (Py-PMMA-b-PDMS) copolymers with pyrene at chain termination via activators regenerated by an electron transfer atom transfer radical polymerization (ARGET ATRP) protocol. The pyrene moiety led to improved interaction with the block copolymer via π−π stacking during covalent functionalization of graphene oxide with it. The functionalized graphene oxide facilitated in improvement of its dispersion in PMMA matrix which led to maximum tensile strength at 0.2 wt.% filler loading. Damlin et al. [28] electrochemically synthesized composite of poly(3,4-ethylene dioxythiophene) (PEDOT) with graphene oxide and reduced graphene oxide supported by ionic liquids 1-butyl-3-methyl-imidazolium tetrafluoroborate (BMIMBF4) and 1-hexyl-3-methyl-imidazolium tetrafluoroborate (HMIMBF4) (IoLiTec) without using any modifier for stabilizing dispersion. Teng et al. [29] functionalized graphene nanosheets noncovalently through p-p stacking with localized pyrene moiety existing in functional segmented poly(glycidyl methacrylate), prepared through atom transfer radical polymerization. The modification of graphene resulted in improvement of dispersion with remarkable increase of thermal conductivity in contrast to unmodified graphene nanosheet and multiwalled carbon nanotube.

10.2   Polymer-graphene composite prepared via melt intercalation

Melt processing of polymers is favored over processing in solution medium because of faster processing and requirement of no solvents. Dispersion of graphene in hydrophobic commodity polymers, viz., polyethylene, polypropylene etc. is a challenge. Even though melt intercalation of graphene for preparation of its composite with polymers using industrially feasible instruments like extruder, it still encounters the problem of inferior dispersion owing to high melt viscosity of polymers in comparison to the solution casting medium [30]. Therefore, apart from conventional direct mixing of graphene via single step melt intercalation, two-step mixing by using master batch is also an important strategy in this regard. A concentrated blend (master batch) of filler and matrix is prepared initially which is then diluted to the desired concentration by addition of pristine polymer to it [31]. Achaby et. al. [32] investigated dispersion of graphene nanosheets in polypropylene at varying filler loading (0.08−1.2 vol.% and 0.2−3 wt.%) via melt intercalation method using twin-screw co-rotating extruder at 200°C for 10 min maintaining screw speed at 100 r.p.m. Disappearance of graphitic peak (2θ = 26.23 degrees) in the composites signifies complete dispersion of graphene nanosheets in the matrix. Rheological percolation threshold, as the consequence of the appearance of graphene nanosheets in close vicinity of each other owing to proper dispersion, was achieved at 0.2−0.4 vol.% loading of graphene exhibiting liquid-like to solid-like transition at low frequency range. Melt mixing of graphene of specific surface area of 555 m^2/g (BET) in polyethylene terephthalate wars carried out by Zhang et al. [33] at 285°C using Brabender mixer. Homogeneity of dispersion and separation of layers by exfoliation at 3 vol.% was evident from bulk morphology images (Fig. 10.3). Efficient dispersion resulted in achieving electrical percolation threshold at 0.47 vol.%. Perfect dispersion of graphene in polypropylene matrix using different pre/post mixing protocols may be corroborated by achieving electrical percolation threshold in the range of 6%−15% (w/w) [35−37] which is significantly higher than the theoretically predicted value [38]. Hassouna et al. [39] varied time of melt

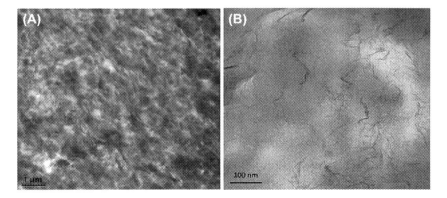

Figure 10.3 Transmission electron micrographs at (A) lower and (B) higher magnification for PET composite filled with 3 vol.% of graphene.
Source: Reproduced (adapted) with permission from H. Bin Zhang, W.G. Zheng, Q. Yan, Y. Yang, J.W. Wang, Z.H. Lu, et al., Electrically conductive polyethylene terephthalate/graphene nanocomposites prepared by melt compounding, Polymer 51 (2010) 1191−1196.

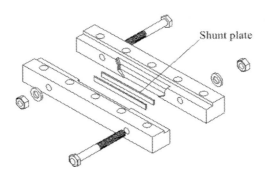

Figure 10.4 Schematic diagram of sheet die design using shunt plate.
Source: Reproduced (adapted) with permission from J. Zhang, S. He, P. Lv, Y. Chen, Processing–morphology–property relationships of polypropylene–graphene nanoplatelets nanocomposites, J. Appl. Polym. Sci. 134 (2017).

mixing and rotor speed for preparation of the expanded graphite-filled polylactic acid composite. Increase of residence time and screw speed was found to improve better dispersion of expanded graphite into the matrix. Zhang et al. [36] used shunt plates in sheet die, as schematically illustrated in Fig. 10.4, during processing of polypropylene/ graphene nanocomposite which sliced the melt flow in three different layers. Use of shunt plate was found to facilitate in the increase of pressure, shear stress, velocity, and the nucleation sites during processing which resulted in improvement of crystallinity and lowering of electrical percolation threshold signifying better dispersion.

Gkourmpis et al. [40] suggested a simplified method of preparation of composite of graphene in highly isotactic (> 90%) polypropylene matrix prepared through melt mixing method using

Brabender Plasticorder where all pre/post processing stages were eliminated. Superior dispersion of graphene was confirmed (Fig. 10.5) from very low electrical percolation threshold (0.94 wt.%). However, agglomeration of filler was observed from morphology analysis. Very less cluster formation of nanofiller was observed for 1 wt.% filler loading, whereas larger clusters were found to be present in 2.5 at % filler loading. Raef et al. [9] followed in situ reduction of graphene oxide dispersed in poly(styrene butadiene) rubber latex followed by coagulation using sulfuric acid (5 at%). Compounding of the coagulated and dried rubber nanocomposite was carried out in two-roll mill at friction ratio 1:1.2 which was further vulcanized at 160°C. Even though improvement of rubber-filler interaction was achieved by in situ reduction process, it did not facilitate significantly in improvement of dispersion of graphene oxide in the matrix. Gedler et al. [41] investigated dispersion of graphene in polycarbonate foam. Melt compounding of pelletized polycarbonate was carried out in internal mixer Brabender Plasticorder at 30, 60, and 120 r.p.m. screw speed and 180°C for 1, 2, and 3 min respectively. Exfoliation was not found in the case of polycarbonate/ graphene composite, whereas remarkable exfoliation in composite foam was evident from disappearance of graphene peak at

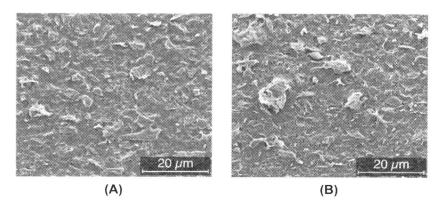

Figure 10.5 Scanning electron micrographs of solvent-etched surface of isotactic polypropylene/ graphene nanocomposite exhibiting the presence of (A) few isolated agglomerates at 1 wt.% of filler loading and (B) larger agglomerates at 2.5 wt.% of filler loading.
Source: Reproduced (adapted) from Ref. T. Gkourmpis, K. Gaska, D. Tranchida, A. Gitsas, C. Müller, A. Matic, et al., Melt-mixed 3D hierarchical graphene/polypropylene nanocomposites with low electrical percolation threshold, Nanomaterials. 9 (2019) 1−19.

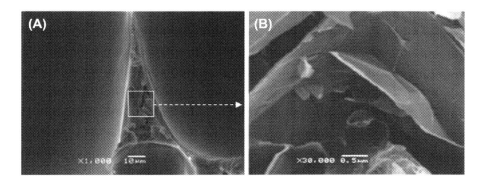

Figure 10.6 SEM image of graphene-reinforced polycarbonate foam showing the presence of (A) cell strut and (B) dispersion of graphene nanoplatelets in a cell strut.
Source: Reproduced (adapted) with permission from G. Gedler, M. Antunes, V. Realinho, J.I. Velasco, Thermal stability of polycarbonate-graphene nanocomposite foams, Polym. Degrad. Stab. 97 (2012) 1297−1304.

$2\theta = 26.5$ degrees in x-ray diffractogram. Surface morphology (Fig. 10.6) revealed the presence of cell strut of the nanocomposite foam at $\times 1000$ magnification, whereas the presence of a group of discrete graphene nanoplatelets was clearly identified at $\times 30,000$ magnification.

10.3   Polymer-graphene prepared composite via solution mixing

For mixing of graphene with polymers in solution medium, graphene layers are ultrasonicated in suitable dispersion medium prior to addition of polymer. Most of the reported works are based on the use of polar solvents like N,N-dimethyl formamide (DMF), N-methyl-2-pyrrolidone (NMP), water, etc. Moreover, many researchers adopt the strategy of functionalization of graphene to improve its dispersion in solvent media. However, many a times this strategy is not favored because it may adversely affect the electrical conductivity and interaction with polymer matrix [42]. Efficient dispersion of graphene-polymer composite depends on various factors viz. selection of solvent medium, ultrasonication system, improvement of stabilization of dispersion by using suitable surfactant or functionalization of graphene, nature of polymer matrix, etc. However, in spite of the possibility of good dispersion, this technique has some limitations like utilization of toxic solvents, removal of solvents, and limitations of thickness of solvent cast polymer composites [43,44].

Selection of suitable solvent for dispersion of graphene or its derivatives is very crucial for preparation of its polymeric composites. Ideal solvents can overcome the interlayer interaction of graphene owing to the Van der Waals force provided the interlayer thickness does not exceed 0.34 nm [45,46]. Konios et al. [47] extensively studied the dispersion behavior of graphene in 18 different solvents. For all solvents, Hansen and Hilderbrand solubility parameter values were calculated. The Hansen parameters viz. cohesive energy parameters owing to dispersion (δ_D), polarity (δ_P) and hydrogen bonding (δ_H) of graphene oxide were determined as c.17.1, 10, and 15.7 $MPa^{1/2}$ respectively, whereas the values were 17.9, 7.9, and 10.1 respectively for reduced graphene oxide. The cohesive energy parameters of graphene oxide owing to polarity and hydrogen bonding are significantly higher than in the case of reduced graphene oxide because of the presence of oxygen containing groups. Therefore, graphene oxide is less soluble in chlorinated solvents like chloroform, o-DCB, DCM etc. in contrast to graphene oxide. The Hilderbrand solubility parameter of reduced graphene oxide was found to be 22 $MPa^{1/2}$ which is somewhat less than that of graphene oxide (25.4 $MPa^{1/2}$) (Table 10.2).

Graphene oxide exhibited stable dispersion in nonpolar solvents in spite of low solubility. Its dispersion in water, NMP and ethylene glycol were found to exhibit long-term stability for 3 weeks where the solubility value reaches the maximum of c.8.7 μg/mL in NMP medium. However, interactions with solvents like CN or o-DCB were found to improve after reduction of graphene oxide which is evident from the solubility values of c.8.1 and 9 μg/mL respectively.

Apart from the solvent selection as a major concern in preparation of graphene dispersion medium, the method of ultrasonication also plays a vital role in this context. The concentration of graphene in the dispersion medium depends on both the type and the time of ultrasonication. The ultrasonication method of dispersion may be carried out by using (1) tip sonicator or (2) bath sonicator. During the

Table 10.2 Dipole moments, surface tensions, and Hildebrand parameters of solvents and the solubility values of graphene oxide (GO), reduced graphene oxide (rGO) [47].

Solvents	Dipole moment	Surface tension (mN/m)	Hilderbrand solubility parameter (δ_T) (MPa$^{\frac{1}{2}}$)	GO solubility (μg/mL)	rGO solubility (μg/mL)
Deionized water	1.85	72.8	47.8	6.6	4.74
Acetone	2.88	25.2	19.9	0.8	0.9
Methanol	1.70	22.7	29.6	0.16	0.52
Ethanol	1.69	22.1	26.5	0.25	0.91
2-propanol	1.66	21.66	23.6	1.82	1.2
Ethylene glycol	2.31	47.7	33	5.5	4.9
Tetrahydrofuran (THF)	1.75	26.4	19.5	2.15	1.44
N,N-dimethylformamide (DMF)	3.82	37.1	24.9	1.96	1.73
N-methyl-2-pyrrolidone (NMP)	3.75	40.1	23	8.7	9.4
n-hexane	0.085	18.43	14.9	0.1	0.61
Dichloromethane (DCM)	1.60	26.5	20.2	0.21	1.16
Chloroform	1.02	27.5	18.9	1.3	4.6
Toluene	0.38	28.4	18.2	1.57	4.14
Chlorobenzene	1.72	33.6	19.6	1.62	3.4
o-dichlorobenzene (o-DCB)	2.53	36.7	20.5	1.91	8.94
1-chloronaphthalene	1.55	41.8	20.6	1.8	8.1
Acetylaceton	3.03	31.2	20.6	1.5	1.02
Diethyl ether	1.15	17	15.6	0.72	0.4

ultrasonication process, the following steps occur in sequence: (1) Preparation of graphite dispersion in specific solvent medium, (2) ultrasonication-assisted exfoliation of graphitic layers and (3) purification of graphene. This graphene dispersion is further added to the solution of polymeric matrix followed by stirring for increasing homogeneity of dispersion. During ultrasonication cavitation-induced pressure pulsation leads to growth and collapse of the liquid microbubbles. The resulting high-speed microjets and shock waves generate normal and shear force on graphitic layers which facilitate in layer separation by exfoliation [48].

Surfactants are widely used for preparation of stable dispersion of graphene with lower extent of structural defects. Preparation of defect-free aqueous dispersion of graphene to the level of 0.1 mg/mL is possible by using different ionic surfactants [49], as comparatively illustrated with various nonionic surfactants in Fig. 10.7, which includes sodium dodecyl sulfate (SDS), sodium dodecylbenzene sulfonate (SDBS) [50,51], sodium cholate [52], pyrene derivatives like 1,3,6,8-pyrenetetrasulfonic acid (Py-SO$_3$) tetrasodium salt hydrate, 1-pyrene-methylamine (Py-NH2) hydrochloride [53] etc. It is noteworthy that apart from performing the role of surfactant, pyrene also facilitates in healing graphene during annealing by acting as

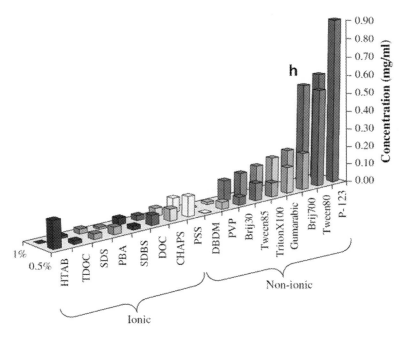

Figure 10.7 Comparative solubility of graphene at 0.5 and 1 wt.% addition of graphene in aqueous medium using different ionic and nonionic surfactants.
Source: Reproduced (adapted) with permission from L. Guardia, M.J. Fernández-Merino, J.I. Paredes, P. Solís-Fernández, S. Villar-Rodil, A. Martínez-Alonso, et al., High-throughput production of pristine graphene in an aqueous dispersion assisted by non-ionic surfactants, Carbon 49 (2011) 1653–1662.

nanographene molecule-based "electric glue" and joining the adjacent graphene layers [54].

Apart from the above-mentioned factors, chemical modification of graphene also plays important role in its dispersion behavior. Well dispersed aqueous colloid may be formed by exfoliated graphite oxide [55–58] which retains negative charge on to the surface of its nanosheets owing to ionization of phenolic hydroxyl or carboxylic acid groups [59,60]. Apart from hydrophilicity, the electrostatic repulsion too influences its dispersion significantly. Stabilization of lyophobic colloid through electrostatic repulsion may be explained by Derjaguin–Landau–Verwey–Overbeek theory [61,62]. By controlling the dispersed particle content, electrolyte concentration and pH, electrostatically stabilized dispersion may be formed. Li and coworkers [61] focused on aqueous phase dispersion of graphene nanosheets (as shown schematically in Fig. 10.8) with expected large scale producibility. In their investigation ammonia was used to increase pH to the level of 10 to achieve maximal charge density. Moreover, water-immiscible liquid (e.g., mineral oil) was used to avoid the tendency of graphene sheets to agglomerate upon evaporation of water by eliminating the air/water interface where the agglomeration takes place. Ionic species are formed by dissociation of ammonia and hydrazine in aqueous medium during the reduction process which act as electrolyte. The overuse of both chemicals adversely affects stabilization of graphene oxide with the addition of NaCl. The optimum ratio of hydrazine: graphene oxide was suggested as 7:10 (w/w) where inferior dispersion occurs with the increase of the proportion of hydrazine. Therefore, it was suggested to remove excess hydrazine immediately for stabilization of the dispersion. Moreover, concentration of graphene oxide beyond 0.5 mg/mL results in gelation over time. Guardia et al. [49] have investigated the comparative dispersion characteristics of graphene using a wide range of surfactants including both nonionic surfactants: Pluronic P-123, Tween 80, Brij 700, Gum arabic from acacia tree, Triton X-100, Tween 85, Brij 30, polyvinylpyrrolidone (PVP) and

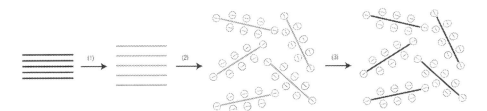

Figure 10.8 Schematic illustration of chemical modification of graphene oxide to achieve aqueous dispersion. (1) Oxidation of graphite (darker black) to form graphite oxide with increased interlayer distance (lighter black), (2) ultrasonication-assisted exfoliation of graphite oxide to form graphene oxide colloid stabilized by electrostatic repulsion, (3) reduction of graphene oxide to graphene colloid by reduction with hydrazine.
Source: Reproduced (adapted) with permission from D. Li, M.B. Müller, S. Gilje, R.B. Kaner, G.G. Wallace, Processable aqueous dispersions of graphene nanosheets, Nat. Nanotechnol. 3 (2008) 101–105.

n-dodecyl b-D-maltoside (DBDM); as well as ionic surfactants: SDS, SDBS, PSS, 3-[(3-cholamidopropyl)dimethyl ammonio]-1-propanesulfonate (CHAPS), sodium deoxycholate (DOC), 1-pyrenebutyric acid (PBA), sodium taurodeoxycholate hydrate (TDOC) and hexadecyltrimethylammonium bromide (HTAB) at comparatively high concentration (500 and 1000 mg/mL). During sample preparation ultrasonication was carried out at 40 kHz for 2 h followed by allowing settle-down period of 15 days. Nonionic surfactants were found to be more efficient in dispersing higher concentration of graphene in aqueous medium. The highest concentration of graphene dispersion as high as ~ 1 g/mL was achieved by using Pluronic P-123 surfactant.

For preparation of graphene-polymer composite, stacked graphene sheets of expanded graphite, produced by heating of intercalated sulfuric acid can be used. To overcome the limitation of insufficient interaction with matrix, breaking of expanded graphite into thinner graphite nanoplatelets followed by improvement of dispersion by using high-speed shearing methods within solution of polymeric matrix may be considered as an effective approach. In the work reported by Ramanathan et al. [63], expanded graphite was prepared by using the 1:4 (v/v) mixture of conc. HNO_3 and H_2SO_4. Graphite nanoplatelets were obtained from the suspension of the expanded graphite by ultrasonication (600 W) for 1 h. It was dispersed in polymethyl methacrylate (PMMA) matrix using tetrahydrofuran (THF) medium via bath sonication (335 W) for 30 min followed by shear mixing at 6000 r.p.m. Effective dispersion was evident from the enhancement of modulus by 133% and increase of T_g by 30°C for 5% nanofiller loading. Villar-Rodil et al. [42] prepared highly stable dispersion (Fig. 10.9) of graphene oxide in organic media. N, N dimethylhydrazine (DMH) was used to reduce the dispersions of graphene oxide in NMP and DMF which remained stable for months. The composite of PMMA prepared with 1% (w/w) addition of graphene dispersion in NMP medium resulted in slower thermal degradation and increased the process by 10°C.

High dispersion of graphene platelets in epoxy matrix by addition of expanded graphite and followed by ultrasonication was found to facilitate in enhancement of thermal conductivity by 230% and reduction of oxygen permeability about 99.6%, as reported by Corcione et al. [64]. Functionalization of graphene by using polycaprolactone diol (PCL) and 4, 4'- methylenebis(phenyl isocyanate) (MDI), as investigated by Jing et al. [65], was found to improve graphene dispersion in polyurethane (PU) matrix. Solution of MDI in anhydrous DMF was added to the dispersion of colloidal suspension of graphene oxide in the same solvent medium and kept for 24 h in inert atmosphere after heating at 80°C. Then it was solidified by coagulation in anhydrous acetone medium which was further filtered sequentially in anhydrous acetone and anhydrous DMF medium. It was again redispersed in anhydrous DMF medium ultrasonically to which the solution of PCL in same solvent medium was added. Subsequent coagulation of the product in anhydrous acetone followed by subsequent washing in DMF medium yielded the functionalized graphene oxide. Polyurethane was synthesized by using MDI, PCL and 1,4-butanediol (chain extender) at 6:1:5 molar ratio using dibutyltin dilaurate catalyst. The composite was prepared by solubilizing polyurethane in the dispersion of functionalized graphene

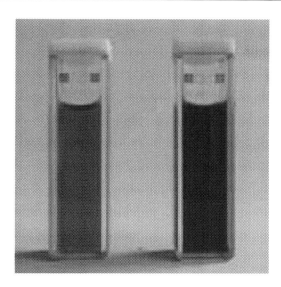

Figure 10.9 Stable dispersion of pristine (left) and reduced (right) graphene oxide in DMF medium.
Source: Reproduced (adapted) with permission from S. Villar-Rodil, J.I. Paredes, A. Martínez-Alonso, J.M.D. Tascón, Preparation of graphene dispersions and graphene-polymer composites in organic media, J. Mater. Chem. 19 (2009) 3591−3593.

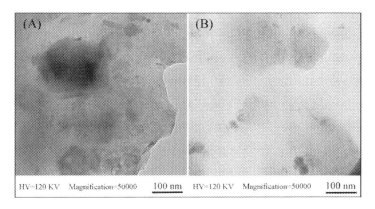

Figure 10.10 TEM images of polyurethane composite with 0.4 wt.% loading of (A) unmodified graphene oxide and (B) functionalized graphene oxide [65].

oxide in DMF medium followed by magnetic stirring and subsequent drying of the mixture. The addition of 0.4 wt.% of unmodified graphene oxide graphene oxide in polyurethane was found to exist in the bulk in stacked layers, whereas, the functionalized graphene oxide layers were found to exist in polyurethane without formation of any layer-stacking which signified the improved dispersion of graphene oxide after functionalization (Fig. 10.10). For that extent of filler loading, improved filler dispersion by using functionalized filler facilitated in improvement of elongation at break and toughness by 21.8% and 19.6% respectively in comparison to the unmodified one. However, the plasticizing effect of the surface grafter covalently grafted soft and small molecules on the filler surfaces led to lowering of modulus.

Instead of any treatment of graphene, Kim et al. [66] followed a different method of functionalization of polymer matrix. Ring-opening metathesis polymerization (ROMP) of isocyanate, cyano and amino functionalized cyclooctene were used to prepare linear low-density polyethylene with respective functionality. Dispersion of thermally reduced graphene in 1, 2-dichlorobenzene (DCB) medium to which commercial and the functionalized LLDPE as well as unmodified and methyl acrylate-grafted ethylene glycol were solubilized to carry out solution mixing. Functionalized LLDPE polymers were found to facilitate augmentation of dispersion with better exfoliation in contrast to localization of graphene clusters in the unmodified LLDPE. As an example, improved dispersion of graphene in amine-functionalized LLDPE has been represented in Fig. 10.11. Similarly, dispersion of graphene in methyl acrylate-grafted polyethylene glycol was also found to be homogeneous, whereas partial exfoliation was revealed in the unmodified matrix.

10.4 Polymer-graphene composite prepared via in situ intercalative polymerization

In this technique of preparation of graphene-polymer composite, graphene is first dispersed ultrasonically in liquid monomer or in monomer solution, followed by polymerization [67]. In another way, graphene nanosheets are swollen with monomers. After addition of initiator, polymerization is triggered by either radiation or heat [68,69]. During polymerization the interlayer spacing of graphene increases. In this method monomer or its suitable solution is soaked by graphene layers and polymerization is carried out by addition of initiator. Several researchers followed this method to prepare graphene-polymer composites in different dispersion medium using different monomers.

In situ polymerization of poly(3,4-ethylenedioxythiophene)/poly(styrenesulfonate) (PEDOT:PSS) in aqueous dispersion medium was investigated by Wan et al. [70] for dye sensitized solar cell (DSSC) application. After preparation of graphene dispersion using reducing agent hydrazine and ammonia, their presence in trace amounts was removed by adding 1M hydrochloric acid. Graphene was first stabilized with PSS so as to avoid the coagulation of graphene in the presence of HCl, followed by addition of PEDOT to a molar ratio 1:6. After stabilizing the mixture by bath sonication for 30 min, ammonium peroxydisulphate and ferric chloride was added to the mixture for a final molar ratio of 1:1 to EDOT. Color change from black to blue indicated successful polymerization. After stirring for overnight, addition of anion exchange resin 717 and cationic exchange resin 723 yielded water soluble PEDOT:PSS/ graphene composite by removing organic impurities.

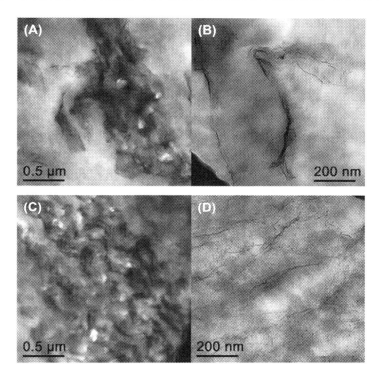

Figure 10.11 Transmission electron micrographs of solvent-cast graphene-filled nanocomposite in (A, B) unmodified polyethylene and (C, D) amine-functionalized polyethylene.
Source: Reproduced (adapted) with permission from H. Kim, S. Kobayashi, M.A. Abdurrahim, M.J. Zhang, A. Khusainova, M.A. Hillmyer, et al., Graphene/polyethylene nanocomposites: Effect of polyethylene functionalization and blending methods, Polymer 52 (2011) 1837–1846.

Unlike the previous reported work, reduction of graphene oxide to graphene may also be done after polymerization. In situ polymerization of methylmethacrylate (MMA) in dimethyl formamide (DMF) medium using benzoyl peroxide initiator was carried out by Potts et al. [71]. Reduction of reduced graphene oxide was carried out using hydrazine, followed by coagulation with addition of ethanol in 5:1 ratio to DMF. In this work, a combination of solution casting and melt processing was followed. The solvent dried PMMA/graphene composite was further processed by using twin-screw microcompounder at 100 r.p.m. screw speed and 220°C temperature, followed by injection molding to prepare the testing samples. Applying Mori-Tanaka theory [72] in dynamic mechanical analysis revealed good dispersion at filler loading up to 1 wt.%, whereas agglomeration occurs at higher filler loading owing to restacking of layers or incomplete exfoliation. As illustrated in Fig. 10.12, significant agglomeration is not observed at 0.5 wt.% loading, whereas crumpled

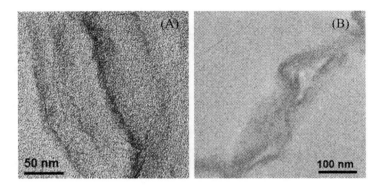

Figure 10.12 TEM images of PMMA composite with (A) 0.5 wt.% reduced graphene oxide and (B) 4 wt.% reduced graphene oxide, showing the presence of multilayer nanoplatelets with a crumpled morphology.
Source: Reproduced (adapted) with permission from J.R. Potts, S.H. Lee, T.M. Alam, J. An, M.D. Stoller, R.D. Piner, et al., Thermomechanical properties of chemically modified graphene/poly(methyl methacrylate) composites made by in situ polymerization, Carbon 49 (2011) 2615–2623.

morphology of the stacks of nanoplatelets are prominent at 4 wt.% loading of reduced graphene oxide in PMMA.

In situ intercalative polymerization inside separated graphene layers was followed by Liao and coworkers [73] for polyurethane-acrylate copolymer. Prior to the use of graphene, urethane-acrylate oligomer (UAO) was synthesized. Hexyl ethyl methacrylate (HEMA) and methylene diphenyl diisobcyanate (MDI) at molar ratio 1:1.05 were reacted together to form HEMA-isocyanate, which was further reacted with propylene oxide tri-functional polyol (PPO) at molar ratio 3.15:1 along with N,N-dimethylcyclohexylamine-based catalyst Polycat8 yielding UAO. Tripropyleneglycol diacrylate (TPGDA) comonomer in liquid state was added to it in 4:6 molar ratio to form UA, where TPGDA still remains unreacted. UA was then solubilized in tetrahydrofuran (THF) and thermally reduced graphene (TRG) was added at varying weight fraction up to 2%. After bath ultrasonication, followed by magnetic stirring, 4 wt.% of free radical initiator azobisisobutyronitrile (AIBN) was added to initiate in situ intercalative copolymerization. Curing by TPGDA exhibited electrical percolation threshold achieved at 0.5 wt.% less than the uncured mixture. Very high dispersion and uniform distribution of graphene was evident from the ultra-low percolation threshold at 0.15 wt.%, that is 0.07 vol.%. Yang et al. used reduced graphene oxide as initiator during in situ ring opening polymerization of lactide for preparation of poly(L-lactic acid)/ thermally reduced graphene oxide nanocomposite [74].

In situ intercalative polymerization may be carried out even in latex stage. Etmimi et al. [75] followed mini emulsion polymerization technique where graphene oxide was dispersed in the latex stage during synthesis of poly (styrene-co-butyl acrylate) via mini emulsion technique using SDBS stabilizing agent and 2,2′-Azobis(isobutyronitrile) (AIBN) initiator.

10.5   Polymer-graphene composite prepared via miscellaneous methods

Majority of the works on reduction of graphene oxide to graphene involves the reduction process prior to the addition of polymeric matrix with the filler. However, reports are also available on reduction of graphene oxide after mixing with polymer, for example, in the case of in situ thermal reduction method. Li et al. [76] followed a different approach for improvement of dispersion of graphene. Polyvinylidene fluoride (PVDF) powder was mixed with the suspension of GO, ultrasonicated in deionized water prior to mixing, along with mild stirring at 80°C for 3 h. Vacuum dried powder was hot pressed for 2 h at 200°C to form the composite. Since PVDF powder particles separate exfoliated graphene oxide layers, its dispersion within the bulk becomes sufficiently high so that electrical percolation threshold was achieved due to formation of segregated network structure at merely 0.105 vol.% and critical exponent $t = 1.1$. It is noteworthy that the investigation did not involve dissolution of polymer matrix in any solvent. However, relevant prior work using PVDF matrix was investigated by Tang et al. [77] by dissolving PVDF in the ultrasonicated suspension of GO in DMF medium. Percolation threshold in this work was achieved at ~ 1.6 vol.%.

Another method, named as slurry compounding, was proposed by Yang et al. [78] combining both melt mixing and suspension involving solvent exchange method. According to this method, as prepared tea polyphenol reduced graphene (TPG) was dispersed in water. It was then diluted with ethanol to prepare a slurry at TPG concentration of 1 mg/mL in a solvent mixture of ethanol and water containing 2% (v/v) water. Further dilution with ethanol followed by centrifugation yielded a slurry with 50 mg/mL solid content. The TPG slurry was directly added at 1%, 3% and 7% (w/w) loading to the matrix while compounding chlorosulfonated polyethylene (CSM) in two-roll mill. The resulting composite containing 1% TPG exhibited remarkable improvement of tensile strength to 19.2 MPa in contrast to 10 MPa as reported for the pristine polymer owing to improved dispersion of modified graphene in the CSM matrix.

Combination of the conventional three techniques of preparation of polymer-graphene composite with modifications may also be found effective. Li et al. [79] followed both high-shear mixing as well as ultrasonication method, using bromine as intercalant because of ease of intercalation, less toxicity and higher thermal stability. Vapor-phase bromination was found to improve dispersion in epoxy matrix. Interestingly, they have found the existence of both ionic and covalent bond in vapor-phase brominated graphene. According to their work, expanded graphite was produced from sulfur-intercalated (2.8 wt.%) graphite was undergone thermal shock at 1050°C for 30 s to produce exfoliated graphene nanoplatelets, which was further ultrasonicated in acetone medium at 42 kHz and 70 W for 8 h resulting in exfoliation. Exposure of the exfoliated graphene nanoplatelets to Br_2 vapor was given at room temperature for time varying between 1 and 168 h. As-formed brominated graphene nanoplatelets were dispersed in epoxy matrix at 3000 r.p.m. using a

high-shear mixer. The epoxy-brominated graphene composite was further ultrasonicated in previous condition at 80°C for 0.5 h, followed by degassing at the same
temperature for 2 h. 1,3-phenylenediamine (mPDA) was used for curing epoxy
matrix. Exposure to bromine for 3 h facilitated the fraction of ionic bond to reach
the maximum and the epoxy composite exhibited 52.9% increase of electrical conductivity in contrast to the untreated graphene composite. However, vapor-phase
bromination was found not influencing percolation threshold significantly.

Although in situ intercalative polymerization may be used for preparation of
polymeric composite of graphene in latex stage, Li et al. [80] prepared natural rub-
ber/ graphene composite using ammonia-stabilized graphene and prevulcanized natural rubber latex exhibiting improved filler dispersion.

Wang et al. [81] followed a combination of latex-stage mixing and melt mixing
methods. Reduction of graphene oxide and its functionalization with N,N-bis(3-
aminopropyl)methylamine (APMEL) were combined to simplify the preparation of
composite. Master batch was prepared by dispersing the modified filler in natural
rubber latex. The use of master batch exhibited improved dispersion in neat natural
rubber using Banbury mill as compared to direct dispersion of the modified filler in
the same matrix.

A different approach of slurry compounding was proposed by Tang and coworkers [82] to reduce the environmental hazards of solvents used in solvent casting
method for preparation of nanocomposites filled with graphene oxide. Unlike the
conventional intercalation of graphene in solvent medium, in this case dispersion of
graphene oxide in acetone medium was not carried out directly because entry of
acetone molecules is hindered by the hydrogen bonding of the graphene oxide
layers. In this method, graphene oxide was initially dispersed in water. After centrifugation, the sediment was diluted with acetone, thus forming a slurry of graphene
oxide in acetone medium by solvent exchange method. Modification of graphene
was carried out by addition of commonly used silane coupling agent Bis(3-triethox-
ysilylpropyl)-tetrasulfide (TESPT) into the dispersion medium which was further
added during compounding of acrylonitrile butadiene rubber (NBR) using two-roll
mill to prepare the nanocomposite. Multilayered stacking of stack thickness
c.10 nm signified nanoscale dispersion.

Instead of conventional preparation techniques, Kalaitzidou et al. [83] followed
coating method for dispersing graphene nanoplatelets in polypropylene matrix.
Powder particles of polypropylene was ultrasonicated in the dispersion of graphene
in isopropyl alcohol medium followed by evaporation and subsequent compression
or injection molding. The combination of coating and compression molding was
found to be the most efficient in filler dispersion which was evident from the lowest
percolation threshold achieved at 0.1 vol.% in contrast to c.7 vol.% for the injection
molded counterpart. Inferior dispersion for the combination of

Coating and injection molding was ascribed to the preferential alignment of graphene nanoplatelets along the direction of flow during injection molding. Surface
morphology revealed about the existence of (1) big agglomerates as well as deformed
and buckled graphene platelets (Fig. 10.13A and B), and (2) very small or no platelets away from the sample edge, signifying insufficient shearing to downsize the

Figure 10.13 ESEM images of the of fracture surfaces of 10 vol.% graphite nanoflake (~15 nm diameter)-filled polypropylene composites prepared by (A, B) melt mixing and (C) coating method.
Source: Reproduced (adapted) with permission from K. Kalaitzidou, H. Fukushima, L.T. Drzal, A new compounding method for exfoliated graphite-polypropylene nanocomposites with enhanced flexural properties and lower percolation threshold, Compos. Sci. Technol. 67 (2007) 2045−2051.

agglomerates. Fang et al. [84]. Followed ATRP protocol for in situ intercalative polymerization of polystyrene on to the surface of graphene, covalently bonded with monomer molecules. The tailoring of polymer-filler interface improved dispersion significantly with 70% improvement of tensile strength at 0.9 wt.% graphene content in the nanocomposite. Improvement of filler dispersion was also observed by Huang et al. [85] in the composite prepared by synthesis of polypropylene by using graphene-oxide-supported C_4H_9MgCl-based Zieglar-Natta catalyst.

10.6   Conclusions

Preparation of polymeric composites with graphene in different techniques differ in dispersion characteristics. Various aspects of dispersion behavior of graphene and its derivatives using different materials and methods have been discussed in this chapter. The composites may be prepared in single or multiple steps via polymerization, swelling or postpolymerization mixing of polymer with graphene. Conventional techniques of preparation of such composites including melt mixing, solution mixing and in situ intercalative polymerization. Apart from the techniques used most often, different modified or unconventional routes of preparation of such composites like slurry compounding, combination of coating, and melt mixing etc. have been found effective in improvement of dispersion. Polymer matrices may be functionalized for improvement of interaction with graphene. On the contrary, graphene may also be modified by suitable functionalization strategies. Covalent or noncovalent interaction between graphene and matrix has a significant effect on its dispersion. Moreover, the strategies of stabilizing graphene dispersion in liquid medium by selecting suitable ones also play a key role in optimization of dispersion characteristics.

References

[1] F. RozpŁoch, J. Patyk, J. Stankowski, Graphenes bonding forces in graphite, Acta Phys. Pol. A 112 (2007) 557–562.

[2] P.L. De Andres, R. Ramírez, J.A. Vergés, Strong covalent bonding between two graphene layers, Phys. Rev. B Condens. Matter Mater. Phys 77 (2008) 1–5.

[3] A. Das, G.R. Kasaliwal, R. Jurk, R. Boldt, D. Fischer, K.W. Stöckelhuber, et al., Rubber composites based on graphene nanoplatelets, expanded graphite, carbon nanotubes and their combination: a comparative study, Compos. Sci. Technol. 72 (2012) 1961–1967.

[4] T.K. Das, S. Prusty, Graphene-based polymer composites and their applications, Polym. Plast. Technol. Eng. 52 (2013) 319–331.

[5] L. He, S.C. Tjong, Low percolation threshold of graphene/polymer composites prepared by solvothermal reduction of graphene oxide in the polymer solution, Nanoscale Res. Lett. 8 (2013) 2–8.

[6] A.J. Marsden, D.G. Papageorgiou, C. Vallés, A. Liscio, V. Palermo, M.A. Bissett, et al., Electrical percolation in graphene-polymer composites, 2D Mater 5 (2018). aac055.

[7] B.M. Yoo, H.J. Shin, H.W. Yoon, H.B. Park, Graphene and graphene oxide and their uses in barrier polymers, J. Appl. Polym. Sci. 131 (2014) 1–23.

[8] Y. Cui, S.I. Kundalwal, S. Kumar, Gas barrier performance of graphene/polymer nanocomposites, Carbon 98 (2016) 313–333.

[9] M. Raef, M. Razzaghi-Kashani, The role of interface in gas barrier properties of styrene butadiene rubber-reduced graphene oxide composites, Polymer 182 (2019).

[10] L.E. Nielsen, Journal of macromolecular science: Part A - chemistry models for the permeability of filled polymer systems, J. Macromol. Sci. Part A Chem 5 (1967) 37–41.

[11] P.S. Nanocomposites, Modeling the barrier properties of polymer-layered silicate nanocomposites, Macromolecules 34 (2001) 9189–9192.

[12] N.K. Lape, E.E. Nuxoll, E.L. Cussler, Polydisperse flakes in barrier films, J. Membr. Sci. 236 (2004) 29–37.

[13] W.R. Falla, M. Mulski, E.L. Cussler, Estimating diffusion through flake-filled membranes, J. Membr. Sci. 119 (1996) 129–138.

[14] E.L. Cussler, S.E. Hughes, W.J. Ward, R. Aris, Barrier membranes, Chem. Eng. 38 (1988) 161–174.

[15] T.B. Rouf, J.L. Kokini, Biodegradable biopolymer–graphene nanocomposites, J. Mater. Sci. 51 (2016) 9915–9945.

[16] S. Wang, C. Wang, X. Ji, Towards understanding the salt-intercalation exfoliation of graphite into graphene, RSC Adv. 7 (2017) 52252–52260.

[17] L. Niu, M. Li, X. Tao, Z. Xie, X. Zhou, A.P.A. Raju, et al., Salt-assisted direct exfoliation of graphite into high-quality, large-size, few-layer graphene sheets, Nanoscale 5 (2013) 7202–7208.

[18] X. Liu, M. Zheng, K. Xiao, Y. Xiao, C. He, H. Dong, et al., Simple, green and high-yield production of single- or few-layer graphene by hydrothermal exfoliation of graphite, Nanoscale 6 (2014) 4598–4603.

[19] K.H. Park, B.H. Kim, S.H. Song, J. Kwon, B.S. Kong, K. Kang, et al., Exfoliation of non-oxidized graphene flakes for scalable conductive film, Nano Lett. 12 (2012) 2871–2876.

[20] G. Zhang, T. Wu, W. Lin, Y. Tan, R. Chen, Z. Huang, et al., Preparation of polymer/clay nanocomposites via melt intercalation under continuous elongation flow, Compos. Sci. Technol. 145 (2017) 157–164.

[21] F. Leroux, J. Besse, Polymer interleaved layered double hydroxide: a new emerging class of nanocomposites, Chem. Mater. 13 (2001) 3507–3515.

[22] V. Georgakilas, M. Otyepka, A.B. Bourlinos, V. Chandra, N. Kim, K.C. Kemp, et al., Functionalization of graphene: covalent and non-covalent approach, Chem. Rev. 112 (2012) 6156–6214.

[23] I.A. Vacchi, C. Ménard-Moyon, A. Bianco, Chemical functionalization of graphene family members, Phys. Sci. Rev. 2 (2017) 1–18.

[24] N. Song, J. Yang, P. Ding, S. Tang, Y. Liu, L. Shi, Effect of covalent-functionalized graphene oxide with polymer and reactive compatibilization on thermal properties of maleic anhydride grafted polypropylene, Ind. Eng. Chem. Res. 53 (2014) 19951–19960.

[25] Y.J. Wan, W.H. Yang, S.H. Yu, R. Sun, C.P. Wong, W.H. Liao, Covalent polymer functionalization of graphene for improved dielectric properties and thermal stability of epoxy composites, Compos. Sci. Technol. 122 (2016) 27–35.

[26] X. Wang, W. Xing, P. Zhang, L. Song, H. Yang, Y. Hu, Covalent functionalization of graphene with organosilane and its use as a reinforcement in epoxy composites, Compos. Sci. Technol. 72 (2012) 737–743.

[27] X. Wang, X. Liu, H. Yuan, H. Liu, C. Liu, T. Li, et al., Non-covalently functionalized graphene strengthened poly(vinyl alcohol), Mater. Des. 139 (2018) 372–379.

[28] S. Song, C. Wan, Y. Zhang, Non-covalent functionalization of graphene oxide by pyrene-block copolymers for enhancing physical properties of poly(methyl methacrylate), RSC Adv. 5 (2015) 79947–79955.

[29] P. Damlin, M. Suominen, M. Heinonen, C. Kvarnström, Non-covalent modification of graphene sheets in PEDOT composite materials by ionic liquids, Carbon 93 (2015) 533–543.

[30] C.C. Teng, C.C.M. Ma, C.H. Lu, S.Y. Yang, S.H. Lee, M.C. Hsiao, et al., Thermal conductivity and structure of non-covalent functionalized graphene/epoxy composites, Carbon 49 (2011) 5107–5116.

[31] J. Bian, X.W. Wei, H.L. Lin, I.T. Chang, E. Sancaktar, PP/PP-g-MAH/layered expanded graphite oxide nanocomposites prepared via masterbatch process, J. Appl. Polym. Sci. 128 (2013) 600–610.

[32] B. Lepoittevin, N. Pantoustier, M. Devalckenaere, M. Alexandre, C. Calberg, R. Jérôme, et al., Polymer/layered silicate nanocomposites by combined intercalative polymerization and melt intercalation: a masterbatch process, Polymer 44 (2003) 2033—2040.

[33] M. El Achaby, F.-E. Arrakhiz, S. Vaudreuil, A. el Kacem Qaiss, M. Bousmina, O. Fassi-Fehri, Mechanical, thermal, and rheological properties of graphene-based polypropylene nanocomposites prepared by melt mixing, Polym. Compos. 33 (2012) 733—744.

[34] H. Bin Zhang, W.G. Zheng, Q. Yan, Y. Yang, J.W. Wang, Z.H. Lu, et al., Electrically conductive polyethylene terephthalate/graphene nanocomposites prepared by melt compounding, Polymer 51 (2010) 1191—1196.

[35] E.V. Kuvardina, L.A. Novokshonova, S.M. Lomakin, S.A. Timan, I.A. Tchmutin, Effect of the graphite nanoplatelet size on the mechanical, thermal, and electrical properties of polypropylene/exfoliated graphite nanocomposites, J. Appl. Polym. Sci. 128 (2013) 1417—1424.

[36] J. Zhang, S. He, P. Lv, Y. Chen, Processing—morphology—property relationships of polypropylene—graphene nanoplatelets nanocomposites, J. Appl. Polym. Sci. 134 (2017).

[37] Y.S. Jun, J.G. Um, G. Jiang, G. Lui, A. Yu, Ultra-large sized graphene nano-platelets (GnPs) incorporated polypropylene (PP)/GnPs composites engineered by melt compounding and its thermal, mechanical, and electrical properties, Compos. Part B Eng. 133 (2018) 218—225.

[38] J. Li, M.L. Sham, J.K. Kim, G. Marom, Morphology and properties of UV/ozone treated graphite nanoplatelet/epoxy nanocomposites, Compos. Sci. Technol. 67 (2007) 296—305.

[39] F. Hassouna, A. Laachachi, D. Chapron, Y. El Mouedden, V. Toniazzo, D. Ruch, Development of new approach based on Raman spectroscopy to study the dispersion of expanded graphite in poly(lactide), Polym. Degrad. Stab. 96 (2011) 2040—2047.

[40] T. Gkourmpis, K. Gaska, D. Tranchida, A. Gitsas, C. Müller, A. Matic, et al., Melt-mixed 3D hierarchical graphene/polypropylene nanocomposites with low electrical percolation threshold, Nanomaterials 9 (2019) 1—19.

[41] G. Gedler, M. Antunes, V. Realinho, J.I. Velasco, Thermal stability of polycarbonate-graphene nanocomposite foams, Polym. Degrad. Stab. 97 (2012) 1297—1304.

[42] S. Villar-Rodil, J.I. Paredes, A. Martínez-Alonso, J.M.D. Tascón, Preparation of graphene dispersions and graphene-polymer composites in organic media, J. Mater. Chem. 19 (2009) 3591—3593.

[43] O. Monticelli, S. Bocchini, A. Frache, E.S. Cozza, O. Cavalleri, L. Prati, Simple method for the preparation of composites based on PA6 and partially exfoliated graphite, J. Nanomater. 2012 (2012).

[44] Y. Liu, Z. Chen, G. Yang, Synthesis and characterization of polyamide-6/graphite oxide nanocomposites, J. Mater. Sci. 46 (2011) 882—888.

[45] R. Narayan, S.O. Kim, Surfactant mediated liquid phase exfoliation of graphene, Nano Converg. 2 (2015).

[46] Y. Xu, H. Cao, Y. Xue, B. Li, W. Cai, Liquid-phase exfoliation of graphene: an overview on exfoliation media, techniques, and challenges, Nanomaterials. 8 (2018).

[47] D. Konios, M.M. Stylianakis, E. Stratakis, E. Kymakis, Dispersion behaviour of graphene oxide and reduced graphene oxide, J. Colloid Interface Sci. 430 (2014) 108—112.

[48] J.T. Han, J.I. Jang, H. Kim, J.Y. Hwang, H.K. Yoo, J.S. Woo, et al., Extremely efficient liquid exfoliation and dispersion of layered materials by unusual acoustic cavitation, Sci. Rep. 4 (2014) 1—7.

[49] L. Guardia, M.J. Fernández-Merino, J.I. Paredes, P. Solís-Fernández, S. Villar-Rodil, A. Martínez-Alonso, et al., High-throughput production of pristine graphene in an aqueous dispersion assisted by non-ionic surfactants, Carbon 49 (2011) 1653−1662.

[50] J.H. Lee, D.W. Shin, V.G. Makotchenko, A.S. Nazarov, V.E. Fedorov, J.H. Yoo, et al., The superior dispersion of easily soluble graphite, Small 6 (2010) 58−62.

[51] M. Lotya, Y. Hernandez, P.J. King, R.J. Smith, V. Nicolosi, L.S. Karlsson, et al., Liquid phase production of graphene by exfoliation of graphite in surfactant/water solutions, J. Am. Chem. Soc. 131 (2009) 3611−3620.

[52] M. Lotya, P.J. King, U. Khan, S. De, J.N. Coleman, High-concentration, surfactant-stabilized graphene dispersions, ACS Nano 4 (2010) 3155−3162.

[53] M. Zhang, R.R. Parajuli, D. Mastrogiovanni, B. Dai, P. Lo, W. Cheung, et al., Production of graphene sheets by direct dispersion with aromatic healing agents, Small 6 (2010) 1100−1107.

[54] L. Liu, K.T. Rim, D. Eom, T.F. Heinz, G.W. Flynn, Direct observation of atomic scale graphitic layer growth, Nano Lett. 8 (2008) 1872−1878.

[55] T. Szabó, A. Szeri, I. Dékány, Composite graphitic nanolayers prepared by self-assembly between finely dispersed graphite oxide and a cationic polymer, Carbon 43 (2005) 87−94.

[56] N.I. Kovtyukhova, Layer-by-layer assembly of ultrathin composite films from micron-sized graphite oxide sheets and polycations, Chem. Mater. 11 (1999) 771−778.

[57] T. Cassagneau, F. Guérin, J.H. Fendler, Preparation and characterization of ultrathin films layer-by-layer self-assembled from graphite oxide nanoplatelets and polymers, Langmuir 16 (2000) 7318−7324.

[58] M. Hirata, T. Gotou, M. Ohba, Thin-film particles of graphite oxide. 2: preliminary studies for internal micro fabrication of single particle and carbonaceous electronic circuits, Carbon 43 (2005) 503−510.

[59] T. Szabó, O. Berkesi, P. Forgó, K. Josepovits, Y. Sanakis, D. Petridis, et al., Evolution of surface functional groups in a series of progressively oxidized graphite oxides, Chem. Mater. 18 (2006) 2740−2749.

[60] A. Lerf, H. He, M. Forster, J. Klinowski, Structure of graphite oxide revisited, J. Phys. Chem. B 102 (1998) 4477−4482.

[61] D. Li, M.B. Müller, S. Gilje, R.B. Kaner, G.G. Wallace, Processable aqueous dispersions of graphene nanosheets, Nat. Nanotechnol. 3 (2008) 101−105.

[62] J. Yao, H. Han, Y. Hou, E. Gong, W. Yin, A method of calculating the interaction energy between particles in minerals flotation, Math. Probl. Eng. (2016))(2016).

[63] T. Ramanathan, S. Stankovich, D.A. Dikin, H. Liu, H. Shen, S.T. Nguyen, et al., Graphitic nanofillers in PMMA nanocomposites—an investigation of particle size and dispersion and their influence on nanocomposite properties, J. Polym. Sci. Part B Polym. Phys. 45 (2007) 2097−2112.

[64] C. Esposito Corcione, A. Maffezzoli, Transport properties of graphite/epoxy composites: thermal, permeability and dielectric characterization, Polym. Test. 32 (2013) 880−888.

[65] Q. Jing, W. Liu, Y. Pan, V.V. Silberschmidt, L. Li, Z.L. Dong, Chemical functionalization of graphene oxide for improving mechanical and thermal properties of polyurethane composites, Mater. Des. 85 (2015) 808−814.

[66] H. Kim, S. Kobayashi, M.A. Abdurrahim, M.J. Zhang, A. Khusainova, M.A. Hillmyer, et al., Graphene/polyethylene nanocomposites: effect of polyethylene functionalization and blending methods, Polymer 52 (2011) 1837−1846.

[67] Y. Guo, F. Peng, H. Wang, F. Huang, F. Meng, D. Hui, et al., Intercalation polymerization approach for preparing graphene/polymer composites, Polymers 10 (2018) 1−28. Basel.

[68] W. Zheng, X. Lu, S.C. Wong, Electrical and mechanical properties of expanded graphite-reinforced high-density polyethylene, J. Appl. Polym. Sci. 91 (2004) 2781−2788.

[69] H. Hu, X. Wang, J. Wang, L. Wan, F. Liu, H. Zheng, et al., Preparation and properties of graphene nanosheets-polystyrene nanocomposites via in situ emulsion polymerization, Chem. Phys. Lett. 484 (2010) 247−253.

[70] L. Wan, B. Wang, S. Wang, X. Wang, Z. Guo, B. Dong, et al., Well-dispersed PEDOT: PSS/graphene nanocomposites synthesized by in situ polymerization as counter electrodes for dye-sensitized solar cells, J. Mater. Sci. 50 (2015) 2148−2157.

[71] J.R. Potts, S.H. Lee, T.M. Alam, J. An, M.D. Stoller, R.D. Piner, et al., Thermomechanical properties of chemically modified graphene/poly(methyl methacrylate) composites made by in situ polymerization, Carbon 49 (2011) 2615−2623.

[72] T. Mori, K. Tanaka, Average stress in matrix and average elastic energy of materials with misfitting inclusions, Acta Metall. 21 (1973) 571−574.

[73] K.H. Liao, Y. Qian, C.W. MacOsko, Ultralow percolation graphene/polyurethane acrylate nanocomposites, Polymer 53 (2012) 3756−3761.

[74] J.H. Yang, S.H. Lin, Y. Der Lee, Preparation and characterization of poly(l-lactide)-graphene composites using the in situ ring-opening polymerization of PLLA with graphene as the initiator, J. Mater. Chem. 22 (2012) 10805−10815.

[75] H.M. Etmimi, P.E. Mallon, In situ exfoliation of graphite oxide nanosheets in polymer nanocomposites using miniemulsion polymerization, Polymer 54 (2013) 6078−6088.

[76] M. Li, C. Gao, H. Hu, Z. Zhao, Electrical conductivity of thermally reduced graphene oxide/polymer composites with a segregated structure, Carbon 65 (2013) 371−373.

[77] H. Tang, G.J. Ehlert, Y. Lin, H.A. Sodano, Highly efficient synthesis of graphene nanocomposites, Nano Lett. 12 (2012) 84−90.

[78] Z. Yang, Z. Xu, L. Zhang, B. Guo, Dispersion of graphene in chlorosulfonated polyethylene by slurry compounding, Compos. Sci. Technol. 162 (2018) 156−162.

[79] J. Li, L. Vaisman, G. Marom, J.K. Kim, Br treated graphite nanoplatelets for improved electrical conductivity of polymer composites, Carbon 45 (2007) 744−750.

[80] C. Li, C. Feng, Z. Peng, W. Gong, L. Kong, Ammonium-assisted green fabrication of graphene/ natural rubber latex composite, Polym. Compos. 34 (2013) 88−95.

[81] J. Wang, K. Zhang, S. Hao, H. Xia, M. Lavorgna, Simultaneous reduction and surface functionalization of graphene oxide and the application for rubber composites, J. Appl. Polym. Sci. 136 (2019) 1−11.

[82] Z. Tang, X. Liu, Y. Hu, X. Zhang, B. Guo, A slurry compounding route to disperse graphene oxide in rubber, Mater. Lett. 191 (2017) 93−96.

[83] K. Kalaitzidou, H. Fukushima, L.T. Drzal, A new compounding method for exfoliated graphite-polypropylene nanocomposites with enhanced flexural properties and lower percolation threshold, Compos. Sci. Technol. 67 (2007) 2045−2051.

[84] M. Fang, K. Wang, H. Lu, Y. Yang, S. Nutt, Covalent polymer functionalization of graphene nanosheets and mechanical properties of composites, J. Mater. Chem. 19 (2009) 7098−7105.

[85] Y. Huang, Y. Qin, Y. Zhou, H. Niu, Z.Z. Yu, J.Y. Dong, Polypropylene/graphene oxide nanocomposites prepared by in situ ziegler-natta polymerization, Chem. Mater. 22 (2010).

Structure-property relationship in polymer-graphene composites

Purabi Bhagabati[1] and Mostafizur Rahaman[2]
[1]School of Chemistry, The Centre for Research on Adaptive Nanostructures and Nanodevices (CRANN), Trinity College Dublin, Dublin, Ireland, [2]Department of Chemistry, College of Science, King Saud University, Riyadh, Saudi Arabia

11.1   Introduction

Graphene is an allotrope of carbon that has sp^2 hybridization forming planar structure. Each layer of planar structure of carbon is called graphene and these graphene sheets are connected with each other by Van der Waals force of attraction known as graphite. Graphene is also known as "wonder material" as it shows several exceptional properties due to high current density, ballistic transport, chemical inertness, high thermal conductivity, optical transmittance, and super hydrophobicity at nanometer scale [1]. The monoatomic single layer of graphene which has the sp^2 hybridized C atom provides great importance in developing several material properties critically far higher than similar conventional materials. The 2p orbitals of C atom form the π state bands that delocalize all over the single layer sheet of individual graphene. It is the reason why graphene exhibits exceptionally high strength like Young's modulus ~ 1100 GPa, fracture strength ~ 125 GPa, excellent electrical conductivity ~ 106 S/cm, thermal conductivity ~ 5000 W/m K, very high charge mobility of nearly $200,000$ cm^2/V s, nearly zero effective mass, impermeable to several gases, optically transparent, and a very large specific area of ~ 2630 m^2/g as per a theoretical calculation [2–4]. Details on the structure of graphene and graphite are available in Chapter 2, Preparation/processing of polymer-graphene composites by different techniques. Graphene monolayers of nanometer range in dimension are unique in its properties but the major difficulty lies at its availability in the desired form. In general, graphene nanosheets are intimately connected with each other through a strong Van der Waals force of attraction parallel to each other. Technically it is very difficult to separate these graphene nanosheets, also known as the "wonder material", from each other without affecting the closed hexagonal structure of carbon loops in each nanosheet. In other words, in the process of separating the graphene nanosheets, some major defects get inculcated within the layer and it becomes inevitable in many instances. Formation of such defects creates a huge impact in lowering the abovementioned properties of graphene. Hence, structural arrangement of graphene is an essential criterion to look into while synthesizing or fabricating the material. There are many ongoing debates that are specific to the process of manufacturing of graphene nanosheets without

Polymer Nanocomposites Containing Graphene. DOI: https://doi.org/10.1016/B978-0-12-821639-2.00016-1

any single defect. However, this chapter does not deal with those aspects, rather a detailed discussion will be made on the effect of different graphene structures including the defects into the final material properties. Polymeric materials have found great interest from multiple sectors in the materials world and more specifically in commodity, high performance, light weight electrical equipment, super capacitors, etc. not only due to its properties but also because of the flexibility in the fabrication process, low cost, and easy availability. Nonetheless, polymers are not capable enough to enter into many such high end sophisticated applications until they are modified with additives particularly for the specified applications.

The exceptionally high electrical, thermal, mechanical properties make it the most promising filler material in polymer composite applications [5]. Enormous researches were carried out to study the role of graphene in enhancing the mechanical, electrical, and thermal properties of biocomposites. As we can readily realize that the addition of graphene in various forms into polymers is primarily to improve multiple functional properties that the polymer itself cannot attain. In this case, the polymer/graphene composites show characteristics of both the constituents. However, it may be possible to observe synergistic properties of the composites that depend upon the type of graphene chemical structure, morphology, base polymer matrix and its physiochemical characteristics, interaction with the added graphene platelets, and also largely on the composite processing conditions.

In this chapter in-depth discussion on various factors affecting the properties of polymer/graphene composites are done. Further, the effect of base polymer matrix and the type of graphene as nanofiller in polymer composites are discussed in terms of various functional properties. Nevertheless, the effect of graphene nanostructure, its functionality, and its physical and chemical behavior are more relevant to final property determination. Therefore the effect of graphene structure on polymer composite properties is discussed in detail. The relationship between structure of the polymer composite and its constituents with the final properties of the composites are discussed in more detail in every segment of individual properties.

11.2 Effect of graphene structure on polymer composite properties

In this chapter, the term "graphene" is associated with graphite, graphene oxide (GO) and reduced graphene oxide (RGO). While graphite is a reasonably low cost ubiquitous raw material for graphene, a defect-free graphene nanosheet with high surface is very difficult to fabricate and detailed information on the graphene nanosheet is already discussed in Chapter 2, Preparation/processing of polymer-graphene composites by different techniques. A free standing graphene nanosheet is one of the strongest materials on Earth and research represents a tremendous increment in electrical and thermal conductivity and mechanical strength of polymer composites upon addition of minor amount of such graphene. However, it is not always the case and there is a compromise in properties due to the molecular and atomic level structural criteria. Graphene nanosheets without any defect in its structure are mostly nonreactive and stable due to its strong C−C covalent bonds with stable sp^2 hybridization. Hence, in most cases these defect-free graphene nanosheets do not impart much improvement in certain functional properties due to unavailability of any interaction between the graphene and polymer matrix. However, physical and chemical modification of these nanosheets at its surface cause development of interaction between graphene and polymer matrix and the effect of such changes can well be reflected upon mechanical, thermal, and electrical properties of the resultant polymer composites.

There are several techniques that have gained attention time to time from the sectors of material scientists to fabricate graphene from graphite. Fabrication of defect-free graphene nanosheet with high aspect ratio is expensive and does not hold the utmost importance in the area of polymer/graphene nanocomposites. An intermediate of property and cost is much relevant here. In fact, chemical or physical defect in graphene may be helpful in the polymer composite system.

11.2.1 Modification of graphene

The commonly used techniques for the fabrication of graphene from graphite powder are presented below.

11.2.2 Micromechanical cleavage

This is a sophisticated technique to develop high quality, defect-free graphene with typically $10-100 \, \mu m$ graphene particles. Adhesive tapes are used to peel off the graphene nanosheets from the graphite on the silicon dioxide (SiO_2) surface. Further, as compared to tape exfoliation, the polydimethyl siloxane (PDMS) stamping offers better morphology of the obtained graphene. The micromechanical exfoliation does not affect the chemical nature of the graphene nanosheets [6].

11.2.3 Chemical vapor deposition

In this process, graphene layers are grown on top of transition metal surfaces through thermal decomposition of hydrocarbons. The number of graphene layer deposited onto the transition metal could range from 1 to 10. Based on the type of transition metal surface, the hydrocarbon source and the operation procedure, high quality graphene can be fabricated using this process.

11.2.4 Solution exfoliation of graphite

The chemical vapor deposition can provide high quality monolayer graphene, and the micromechanical exfoliation provides physically defected mono or multilayer graphene. Here the Van der Waals forces of attraction between the graphene sheets are weakened by the inculcation of some small molecules in solution either by chemical functionalization or by intercalation. Ideally, this technique provides high quality single layer graphene nanosheet in solution. Other techniques are available to exfoliate the graphite into monolayer to few layered graphene in solution to obtain stable colloids of graphene. In this case, the structural changes through chemical functionalization of the graphite leads to formation of chemically modified graphene where certain chemical groups will be attached onto its surface. Graphene oxides are one of the most researched and applied forms of graphene being used in the polymer matrix.

11.2.5 *Oxidation and functionalization*

This is primarily due to the exfoliation process that involves flexibility in oxidizing graphite followed by further chemical treatment of the oxidized graphene with other molecules that lead to enhanced interaction with the base polymer matrix. Graphite flakes are oxidized to graphite oxide and these oxides of graphite contains a large number of reactive functional oxygen containing groups like carboxylic acid, hydroxyl, aldehyde, epoxy groups etc., which render a high degree of reactivity for further interaction with the base polymer material. These graphite oxides that possess several sites of defect can well be separated from each other through a mechanical process like mechanical stirring, thermal shock, or ultrasonication. Hence, the resulting graphene oxide nanosheet contains a large number of oxygen containing functional groups making it hydrophilic and defect as physical sites. The graphene oxide nanosheet obtaining from the graphite oxide can be monolayer to a few layered depending upon the type of oxidation and exfoliation process adopted. The received graphene oxide nanosheets are capable of altering the van der Waals interaction between polymer matrix and graphene nanosheets forming chemical interaction with functional groups of base polymers either primary covalent or secondary interaction like hydrogen bonding. Involvement of such an interaction of graphene nanosheets with base polymer matrix generally leads to enhanced mechanical strength, thermal stability, as well as electrical conductivity. Hence, it is essential to understand the behavior of the graphene oxide functional groups. Many a times researchers are capable to further react these graphene oxide functional groups with other chemicals in order to dangle new functional groups onto the surface that may specifically be targeted to interact with polymer matrix. In that case the compatibility between the graphene nanosheet and the polymer matrix rises and subsequently served functional properties at multiple levels. A good interfacial interaction between graphene nanosheet and polymer matrix eases stress transfer in polymer composites across the graphene oxide nanofillers, thereby causing increment in mechanical strength like ultimate tensile strength at break, hardness, toughness and modulus. Nevertheless, it must be kept in mind while selecting any functionalization of graphene that inculcation of functional groups onto the surface of graphene nanosheet definitely gives rise to structural defects that obstruct electronic conduction across the nanosheet. Hence, addition of functionalized graphene nanosheet may enhance mechanical and thermal degradation stability but at the cost of reduction in electrical and thermal conductivity. Thereby, it is up to the choice of functional properties that can be preferentially chosen for the final application. Some of the important properties of polymer/graphene composites based on the preferential selection of graphene nanosheet are presented below.

11.3 Mechanical properties of polymer/graphene composites

Mechanical strength of polymer composites primarily depends upon three factors:

1. Inherent strength of polymer matrix and the filler. This section includes the inherent chemical structure of polymer matrix that leads to superior physicomechanical properties including the mechanical strength of the base polymer matrix. Characteristics like chemical structure, molecular weight, crystallinity and crystal morphology, etc. are highly important in dictating the composites properties. Detailed discussion on the effect of polymer on the composite properties will be discussed in relevant sections of this chapter. On the other hand, the chemical structure of the fillers, electronic configuration of atoms involved in the fillers, morphology and aspect ratio, overall modulus and hardness of the fillers greatly influence the mechanical strength of the composites. Defect-free graphene nanosheet does not interact with polymer matrix and is commonly found to be ineffective in improving the mechanical strength like ultimate tensile strength, toughness, compressive strength, etc. Though modulus of the polymer composites may enhance at significant degree as because modulus being an additive property of materials. Whereas, graphene oxides with hydroxyl, carboxylic, ketone, ester, and epoxy groups at surfaces develop sufficient hydrophilicity making the nanosheets much more polar. Thus graphene oxide nanosheets are relatively compatible with polar polymers like polyvinyl alcohol, poly(ethylene oxide), amides, different esters, urethanes etc. [7,8]. However, these graphene oxides are not well soluble or compatible in many of organic nonpolar polymers like polyethylene, polypropylene, etc. It is even a good choice to tailor the oxygen containing functional groups in graphene nanosheet through chemical modification based on the type of polymers matrix to be used in the application. For example, amine functionalized graphene nanosheets are very effective in improving the mechanical strength of epoxy resin based composites [9]. Similarly, amidation, esterification, isocyanate modification, and wrapping the nanosheet with polymer molecules similar to the base polymer matrix help gain better properties [10−13]. Many times, covalent modification of graphene nanosheet is carried out using long chain organometal or surfactant compounds in order to incorporate hydrophobic characteristic in the graphene nanosheet. These compounds contain hydrophilic and hydrophobic components. While the polar hydrophilic component interacts with the polar functional groups of graphene oxide nanosheet, the nonpolar hydrophobic long alkyl chains generate certain degree of compatibility with the organic polymer matrix. In this procedure, it is possible to generate compatibility or even possible to develop homogeneity between polymer matrix and the graphene nanosheet. Such compatibility in polymer matrix and filler leads to enhancement in failure mechanical properties like ultimate tensile strength, lower negative change in percentage elongation at break etc. These surface modifying agents of graphene oxide nanosheet reduces the hydrophilicity and enhances the solubility within the polymer matrix. The underlying mechanism of the improved mechanical performance of the composite includes high interfacial interaction between the polymer and functionalized graphene nanosheet. Hence, such chemical functionalization of graphene nanosheet improves the solubility in polymers and enhances the processability.

2. Uniformity in dispersion of filler within the polymer matrix makes a large shift in mechanical performance of polymer composites. Graphene nanosheet after modification may show great improvement in its solubility. The enhanced solubility of the functionalized graphene nanosheet leads to uniform filler dispersion within polymer matrix. Further, there are certain factors that play an effective role in enhancing the degree of dispersion of the graphene nanosheet in polymer matrix. In the case of solution processing parameters like solvent used, solubility parameter of solvent, mixing processes like mechanical stirring, speed, time of mixing, temperature of mixing, drying condition like speed of solvent evaporation, vacuum drying, etc. While in case of the common industrial practice of melt mixing or melt compounding process, the mixing screw type and its design, mixing temperature, mixing time, rpm of the screw, torque, etc. are a few parameters that greatly influence the final composite properties. As a rule of thumb, higher mechanical shear and torque generation in polymer solution or melt during mixing breaks down any possible aggregates of filler and leads to uniform dispersion. At the same time, it is essential to understand the adverse effect of high mechanical shearing upon morphology of fillers like graphene nanosheet. Moreover, chemically modified graphene nanosheet like graphene oxide are thermally unstable and undergo shrinkage or morphological deformation upon exposure to high temperature while melt processing of polymers. Besides, the significant low bulk density of graphene nanosheet creates difficulty while feeding in the extruder or blender. Hence, the melt processing process is more relevant to thermally reduced graphene nanosheet.

3. Morphology acts as an important criterion in graphene nanosheet while using as additive in polymer and it is always desirable to have a larger surface area with minimal defect within polymer matrix. It is the reason why functionalized graphene nanosheet provides a higher mechanical strength of polymer composites under met state like melt strength, melt viscosity, storage modulus (G') in polymer melt in comparison to functionalized graphite or graphite oxides. Several times, these nanosheets are broken or deformed upon application of high thermomechanical shear during the processing of polymer composite. In that case, mechanical, electrical, and thermal properties undergo drastic deterioration. Hence, optimization of processing parameters is required in which a balance between the two factors would work effectively.

4. Another promising technique is to develop high quality well-dispersed graphene nanosheet-based polymer composite is via in-situ intercalative polymerization of respective monomers. In this process, the solution polymerization of monomer is carried out in the presence of graphene nanosheet in multiple forms. The polymerization system contains monomers that were dissolved in the solvent and the graphene nanosheet in different form are dispersed within the solution through mechanical stirring or ultrasound technology. The resulting polymer generally contains the graphene nanosheet uniformly well dispersed, examples include successful polymerization of Poly(vinyl acetate), Poly(methyl methacrylate), poly(arylene disulfide), silicon rubber etc. [14−17]. In certain cases, graphene oxide nanosheets were initially taken and mixed with the monomer and then reducing agents were added in the reaction mix. Hence, during polymerization the graphene oxide nanosheets were reduced by reducing agents and form graphene nanosheet dispersed uniformly in polymer matrix. There is another advantage that comes into effective when the functional groups attached in the graphene nanosheet interact with the monomer, and during in-situ polymerization, these polymer chains start growing from the surface of the graphene nanosheet. Examples like, in the case of strained ring structure, the opening of ring for polymerization requires active hydroxide groups to initiate the reaction, and graphene nanosheet containing similar hydroxyl group acts as reactive site for the ring opening polymerization. In other words, the polymer chains

are grafted onto the surface of graphene nanosheet. The literature demonstrates numerous polymer grafting reactions available, and it is obvious to obtain superior polymer-filler interfacial interaction and better uniformity in dispersion. However, most of the trials in bulk polymerization were not successful because of high viscosity generated, and it mostly leads to nonuniform and agglomerated dispersion of the nanosheet. Therefore, at low scale this insitu polymerization can work efficiently for developing high quality composite.

11.4 Electrical properties of polymer-graphene composites

As already discussed, defect-free, graphene nanosheet with large surface area possesses high electrical and thermal conductivity due to the structural characteristics. Successful addition of graphene nanosheet into polymer matrix with excellent homogeneity between polymer and filler, uniform dispersion of graphene nanosheet all across the polymer matrix increases the electrical conductivity of the polymer. Hence, electrical conductivity of insulating polymers can be increased by addition of very small amount of graphene in the polymer matrix. I particular, addition of graphene in olefins have gained more attention in the area of sensors, antistatic materials in electrostatic discharge (ESD), electromagnetic interference (EMI) shielding materials and various conductive coatings applications [18]. There are certain factors that dictates the change in electrical conductivity of polymers upon addition of the graphene. Other than the form of graphene used, the interfacial interaction with polymer matrix, uniformity in dispersion and concentration of loaded graphene are important for consideration. Structural modification in graphene nanosheet increases surface roughness, which can further enhance the mechanical interlocking of polymer chains in the graphene and also enhances the adsorption of polymer onto its surface. For example, thermally exfoliated graphene show higher interfacial interaction with polymethyl methacrylate (PMMA) due to enhanced mechanical interlocking and strong adhesion with the wrinkled and rough surface of the graphene causing enhanced properties of polymer composites. Effect of oxygen content in graphene has great influence in resulting polymer composite properties, in particular to electrical properties. The percolation threshold commonly increases with increasing the oxygen content in the graphene nanosheet. In other words, it is realized that the lowering of electrical conductivity in presence of large number of oxygen at the graphene surface is structural disruption of graphene. The oxygen containing groups disrupt the graphitic sp2 hybridized network and thereby decrease its conductivity. In general, electrical conductivity exhibits a nonlinear rise in electrical conductivity as the function of graphene concentration. In this case, electrical conductivity of polymer/graphene composites catastrophically increases with increasing the concentration of graphene and attains saturation at critical filler concentration, commonly known as electrical percolation threshold concentration, where the graphene forms a three-dimensional conductive network within the polymer matrix. The percolation concentration of graphene in different polymers varies with the interaction between graphene and polymer matrix, the nature of dispersion of graphene in polymer matrix. Further, based on the solubility and interaction of graphene with polymer matrix, and also the intrinsic conductivity of the graphene used, the percolation threshold may reduce or increase. The percolation threshold of graphene many a time is lower compared to carbon nanotubes when reasonably similar chemical structure of fillers is involved due to higher surface area of graphene nanosheets compared to carbon nanotubes. The mechanism for electrical conductivity of polymer/graphene composites involve three steps. At the beginning low concentration of graphene in polymer matrix show poor conductivity, as each of graphene nanosheets are surrounded by electrically insulating polymer matrix and there is close proximity among graphene nanosheets to pass the electrons from one to another. Upon slight increase in the concentration, bigger clusters of graphene begin to form and cause filler-to-filler closeness developing tunneling effects between neighbor graphene nanosheets. As per the classical percolation theory, the interparticle tunneling effect is the conduction of electron across the thin layer of polymer that surrounds the graphene nanosheets in polymer composites. In other words, the graphene nanosheets are physically not connected to each other, nevertheless, neighboring graphene nanosheets are connected electrically by the tunneling effect. This effect is primarily dependent upon the type of graphene nanosheet, its interaction with polymer matrix and the concentration. In this second stage, the conductivity rises sharply and attains nearly a maximum value. Further increase in concentration, a complete conducting path forms across the graphene nanosheet being in contact with each other, which is called as percolation threshold. This is the minimum concentration of the graphene nanosheets that are in contact with each other and capable to transmit electrons across the composite sample. Further increase in the concentration of graphene will increase the number of conducting paths. However, the overall conductivity does not show significant increase beyond the percolation threshold. This is also the case for most of the conducting nanoparticles like carbon black, carbon nanoparticles etc.

The processing behavior greatly influences the conductivity of polymer/graphene composites. Solvent assisted mixing or in-situ polymerization has turned out to be more effective in comparison to the conventional melt mixing or bulk polymerization technique at similar graphene concentration. Morphology of graphene nanosheet is important in displaying the electrical conductivity of the polymer composites. Large 2D surface of the graphene sheet without any defect in its hexagonal structure of Carbon atoms lead to high electrical conductivity. However, oxidation of graphene lead to sharp reduction in electrical properties, at the same time the graphene oxides have better compatibility with most of the organic polymers. Therefore, reduction of these graphene oxide is necessary to create good deal of electronic conduction in the resultant polymer composites. Graphene upon direct addition may not be readily dispersed in organic polymers. Therefore, use of functionalized graphene like graphene oxide is better option, and the electrical conductivity can be restored by chemically reducing the graphene oxide in-situ within the polymer composite system. For example, sulfonated polystyrene in combination with hydrazine hydrate as reducing agent of graphene oxide and the resulting polymer composite has got high electrical conductivity with uniformly dispersed graphene [19]. The in-situ reduction of graphene involves varying reducing agents with respect to the type of polymer matrix and the resulting composite show improved electrical conductivity at the cost of certain degree of polymer degradation. Thermally reduced graphene nanosheet has got better electrical conductivity and thermal stability compared to chemically reduced graphene nanosheet, making it the best candidate for melt mixing process. However, based on the chemical as well as physical nature of polymer the thermally exfoliated graphene has different state of dispersion. For example, dimethylformamide (DMF) solution cast films of thermoplastic polyurethane with thermally reduced graphene show high electrical conductivity at minimum critical concentration of graphene. In-situ polymerization of conducting polymers like polyaniline (PANI), polypyrol (PPy) and polyacetylene in presence of graphene oxide show high electrical conductivity. As expected, these polymer composites have better electrical conductivity compared to conventional nonconducting polymer based graphene composites. However, the major drawback of poor formability, processability, mechanical strength, flexibility and toughness limits is application widely.

Further, composite morphology like orientation of the anisotropic 2D graphene nanosheet in polymer matrix result high electrical conductivity along the direction of alignment. The alignment of the graphene can be made by inducing flow during the melt processing of the composites. For example, annealing of melt processed polymer/graphene composites are electrically more conducting than compression molded crash cooled samples. The surface resistivity of a material is defined as the resistance for electrical current leakage across the surface of the insulator and graphene provide good electrical surface resistivity of polymer composites. For example, unsaturated polyester resin/graphene based polymer composites show decreasing trend in surface resistivity with increase in the concentration of graphene nanosheet concentration, which is attributed to larger surface area of graphene owing to its 2D structure. This results enhanced free path for free electron to move around within the polymer composite sample surface.

11.5   Thermal conductivity of polymer/graphene composites

Thermal transport properties of graphene nanosheet are relatively high and incorporation of graphene can increase the thermal conductivity of polymer/graphene composites and thereby make it a deserving candidate for applications that require high thermal conductivity for example, miniaturized tough and flexible electronic devices, thermal pastes, heat actuated shape memory polymer applications, printed circuit boards, connectors, thermal interface materials, heat sinks, power electronics, electric motors, generator, heat exchangers etc. [5,18]. The 2D graphene nanosheet reportedly show better thermal conductivity in polymers compared to 1D carbon nanotubes (CNT) or carbon nanofibers (CNF) [20,21]. Though organic polymers are thermally insulating while graphene is thermally conducting in nature, but there exists relatively smaller difference in thermal conductivity between polymers (0.1−1 W/m K) and graphene Carbon atoms (5000 W/m K) compared to larger difference in electrical conductivity between nonconducting polymers (10^{-18t} to 10^{-13} S/cm) and graphene

(6000 S/cm). It is the reason that the rise in thermal conductivity of polymer/graphene composites is not as sharp and significant like the electrical conductivity of at similar graphene concentration. Furthermore, there is a difference in the mechanism of thermal conduction from electrical conduction mechanism in materials and so in polymer composites. Thermal energy primarily transmits in the form of lattice vibration that is phonons, in contrast to electrons. There is a poor coupling in vibration modes at the interfaces between polymer-graphene and graphene-graphene nanosheets and this mismatch in vibrations phonon scattering at the interfaces which generate resistance to its transmission across heterogeneous phases and results thermal resistance also known as "Kapitza resistance" [22]. So, practically improvising the interfaces of heterogeneous phases in polymer composites can reduce the thermal resistance. Bringing the surface chemistry of polymer and graphene nanosheet closer at the interfaces either by chemical modification of the graphene nanosheet or adding interfacial coupling agent between polymer and filler can minimize the interfacial phonon scattering. In case of graphene oxides, the oxygen containing functional groups can cause good interaction with the polymer matrix and cause rise in thermal conductivity in polymer composites. However, excessive functionalization of graphene may disrupt the hexagonal planar structure of graphene and reduce the intrinsic thermal conductivity. An optimum of functionalization is necessary which is based on the type of polymer matrix and the composite processing technique. Several thermally nonconducting polymers like polyethylene (PE), polypropylene (PP), polyamides (PA), epoxy resin, and paraffin wax are some of the commonly used polymers attaining high degree of thermal conductivity upon addition of graphene nanosheet. Though solution technique of composite preparation is more effective compared to melt processing and there is little commercial importance in solution technique. Other than the interfacial interaction, the degree of dispersion and orientation of graphene nanosheet is important in determining the thermal conductivity of polymer composites. Higher the infinite particle aspect ratio (A_f) of graphene nanosheet, higher is the thermal conductivity. It is observed that thermally reduced graphene nanosheet can readily be tailored for its A_f value and thermal conductivity responses reciprocally with the A_f value of graphene [23]. as expected, morphology of polymer/graphene composites is important in determining the thermal conductivity. For example, conductivity increases with along the direction of oriented graphene nanosheet in melt extruded or solvent casted composites compared to the direction perpendicular to the graphene alignment, implying thermal conductivity as anisotropic macroscopic property of polymer/graphene composites. In other words, parameters that influences the interfacial interaction between polymer and graphene nanosheet, dispersion and alignment in graphene nanosheet within polymer matrix increases overall thermal conductivity of the polymer composites.

11.6   Gas barrier properties of polymer/graphene composites

Gas barrier property is important in the case of packaging and certain electronic application where humidity and oxygen gas may affect the quality of materials. In that case, these gas barrier films are used to coat or wrap those material. For example, in certain food items, rust prone electronic equipment etc. needs good safety from these gases in order to make its service or shelf life longer. Defect-free graphene nanosheet are impermeable to all gas molecules including water molecules, and this is the advantage of the graphene as 2D conducting filler over other counterparts like carbon nanotubes, carbon nanofibers, carbon blacks, etc. Graphene is reported to have higher gas barrier property compared to 2D nanosheets of layered silicates. The difficulty existing in the production of defect-free graphene nanosheet and further its uniform dispersion in polymer matrix without affecting the aspect ratio of graphene structure allows to consider treated graphene. Incorporation of graphene in polymer matrix reduces the gas permeability to significant extent and the mechanism is very simple. The incorporating of these flat nanosheet of graphene in polymer matrix creates as blockage along the path through which gases are meant to pass through the polymer samples. In other words, these graphene nanosheet develops tortuosity in the path of gas molecules passing through the sample and delays the passage of gas molecules significantly.

There are several factors that affect the overall gaps barrier characteristics of polymer/graphene composites, out of which good interfacial interaction between the polymer matrix and the graphene, uniform agglomerate free dispersion of graphene nanosheet within the polymer matrix, retaining high aspect ratio of graphene nanosheet within polymer matrix are important. Further improvement in the gas barrier property can be enhances by customarily aligning these 2D nanosheet along the direction perpendicular to the direction of gas passage. Normally, most of the gases like carbon-dioxide, oxygen, nitrogen etc. show similar transport properties with only difference in the dimension of gas molecules, but the moisture barrier property of polymer/graphene composites behave differently. Other than the dimension of water molecule, there is the factor like hydrophilicity of polymer and graphene nanosheets. Further, polar oxygen containing molecules like hydroxyl ($-OH$), acid ($-COOH$) etc. either in polymer matrix or functionalized graphene nanosheet may form hydrogen bond with water molecules and result polymer plasticization leading easy passage of moisture molecules along the polymer composite samples. Therefore, special care in terms of chemical structure has to be taken while considering the polymer and functionalized graphene nanosheet in specific to attain moisture barrier properties.

11.7   Electrical properties of polymer/graphene

11.7.1   Dieclectric properties

Nonconducting polymers have poor dielectric constant and in general practice, electrically conductive fillers like graphene is added to increase the dielectric constant of these polymer composites by formation conducting network. Graphene in polymer composites can induce interfacial charge polarization of dipoles and cause increase capacitance. The major advantage of using polymer based dielectric

material over ceramic based dielectric materials like barium titanate, strontium titanate is ability to fabricate complex shape of dielectrics. Moreover, ceramic based
dielectrics suffer from inherent brittleness, while polymer based materials are tough
enough to sustain impacts like falling to hard surface from a height. In polymer
composites factors like electrical conductivity, specific surface area and alignment
in certain direction of graphene plays important role in determining the dielectric
properties of composites [24]. Further, a good polymer-graphene interaction always
leads to improvement in properties. Though graphene increases conductivity in
polymers but at the same time causes increase in current leakage and dielectric loss
lowers breakdown strength. Hence, the primary challenge for material scientists is
to deal with dielectric loss at the same time obtain high dielectric constant and this
limits its application in energy storage devices [25]. modification of graphene is the
most favorable process for improving the dielectric loss of polymer/graphene composites. At low concentration graphene gets exfoliated easily and exfoliated morphology favours low dielectric loss due to the absence of direct contact between
neighboring graphene nanosheets within polymer matrix. Hence, at percolation
threshold low dielectric loss is favoured and dielectric constant is also optimum but
can be enhanced upon increasing the concentration due to formation of conductive
networks. In case of composite made of poly(vinyl pyrrolidone) (PVP) grafted graphene nanosheet as filler in PVDF showed high dielectric constant at the same time
the dielectric loss was minimum at a percolation threshold of the PVP-grafted-
graphene nanosheet. The PVP coated graphene acted as microcapacitors within
PVDF matrix and increasing the percolation threshold of the composite facilitated
increase in number of microcapacitors and the insulated coating of PVP on graphene nanosheet led to low dielectric loss and energy dissipation [26]. Highly
mobile free electrons and the delocalized π electrons of sp^2 hybridized orbital of
graphitic carbon atom causes current leakage. In case of functionalized or grafted
graphene/polymer composites (e.g., PVP-g-graphene/polymer), the low dielectric
constant and energy dissipation is attributed to restriction developed in the mobility
of these free electrons [25]. Due to the difference in electrical conductivity between
polymer matrix and graphene nanosheet, the breakdown strength goes down significantly. Further, tracking failure is associated with thermal localized accumulation
that reduced the breakdown strength. Functionalization or grafting of graphene
nanosheet brings down the conductivity of graphene and the difference with polymer matrix reduces. It is the choice and optimization to attain best possible combination of material type and its properties.

11.7.2 *Piezoelectrics*

Piezoelectric materials or piezoelectrics are the materials that can produce electric
energy upon application of mechanical stress. A commonly known piezoelectric
material is quartz. The mechanism involves development of electric charge due to
movement of electron upon application of stress. Graphene provide piezoelectric
properties to its polymer composites and used as actuator material. Factors like
degree of homogeneity in filler dispersion in polymer matrix, its interfacial

interaction, and type of graphene based on its chemical and physical structural characteristics play major role in determining the properties. Piezoelectricity is a property of material that is often engineered into certain nonpiezoelectric materials like flexible polymer/graphene composites [27]. polymer/graphene composite based piezoelectric materials are used in pressure sensors, acoustic transducers, high voltage generators etc. For example, polyvinylidine fluoride (PVDF)/reduced graphene oxide composites show higher signal amplitude and better response in output voltage against applied frequency in stress compared to neat PVDF [28]. Higher concentration of graphene is required for the purpose of improved piezoelectricity. Polar groups present in graphene nanosheet may change the piezoelectricity in polymer composites.

11.7.3   Actuators

Electroactive polymers and polymer composites that are capable to undergo mechanical change in response to applied electrical stimuli are known as actuators. Polymer/graphene composites with high electroactivity are better actuators compared to its ceramic counterparts. Further these are light in weight, low cost and easy to process into intricate complex structures. In polymer composite system, an actuator works as a combination of two units; the "energy transfer unit" that absorbs different form of external energy stimuli such as light, thermal, electrical or chemical energy and transfer to the "molecular switch unit" that undergo mechanical change. In case of polymer/graphene composites, the polymers act as energy transfer unit, while the graphene act as molecular switch unit. PDMS/graphene based composites upon subjected to heat by infrared illumination cause dimensional change in the composite. The graphene nanosheet absorb light and optical energy and transduced into thermal energy through phonons [29]. the percolation resulting from the high thermal conductivity through phonon conduction and uniform dispersion of graphene in polymer matrix cause polymer chain contraction. The large amplitude in photo-thermal actuation in PDMS/graphene composite is primarily due to efficient heat transfer by graphene nanosheet and heat transduction to PDMS matrix [30].

11.7.4   The electromagnetic induction shielding

EMI is defined as electromagnetic radiation emitted by electrical circuits under current operation and in current era large electromagnetic radiation generates to our surroundings from the enormous numbers of electronic equipment [31]. This radiation at high frequency interrupts smooth functioning of other electrical equipment and also causes damage to humans and other animals [32]. A cover made up of electrically conducting materials serves as the purpose of shielding the EMI radiation, and metal fabrics were used for this purpose at the beginning. But eventual progress in science and technology made conducting polymer composites has outperformed metal due to its light weight, low cost, ease of processibility as per required shape and size, flexibility in choice of materials, etc. Polymer/graphene composites have emerged as one of the most desirable composites for EMI shielding applications. As already discussed, electrical conductivity of polymer/graphene composites depends on the intrinsic conductivity of graphene as well as the polymer, aspect ratio and concentration of graphene, etc. A polymer composites of graphene with electrically conducting polymer could be the best candidate for EMI shielding application. In-situ polymerization of polyacetylene, PANi, or PPy in the presence of graphene can present the best EMI shielding efficiency, although formability, processability, and poor mechanical properties are the few disadvantages here [33]. Most of the available polymers can be fabricated into foam and conductive foam of polymer composites provide better mechanical strength, ductility, lighter in weight, and capable to absorb more energy. Therefore, conductive foams of polymer composites are used as high EMI shielding component in aircrafts, spacecrafts, and automobiles [34].

11.7.5 Fuel cell applications

Fuel cells provide feasible energy conversion due to their high power density. Nafion/graphene composites offer a great advantage of thermal and mechanical stability and high degree of selectivity to certain molecules over other conductive materials [35]. In a fuel cell graphene other than functioning as a window and counter electrode, it acts as catalyst support for oxygen reduction and the methanol oxidation. Functionalized GOs exhibit unique ion-exchange capacity, phosphate adsorption ability, electrochemical detection of H_2O_2, the onset potential, electron transfer number in oxygen reduction reaction process, electrochemiluminescence activity, good solubility and stability. Such functionalized graphene is used to synthesize low humidifying polymer electrolyte membrane fuel cells with Nafion using the microwave reduction method [36]. The hexagonal planar structure of graphene oxide with high surface area provides more proton transport channels and captures larger moisture molecules, which could be beneficial for the improvement of the proton conductivity and mechanical properties of fuel cell membranes [37].

11.8   Conclusion

In summary, polymer/graphene composites can be termed as one of the most efficient and versatile materials in polymer composites and its performance is better compared to many other of its counterparts. The excellent characteristics of graphene can be inculcated into polymers through carefully looking at the structure-property relationships between polymer matrix and the graphene nanosheet. Graphene nanosheet show high electrical conductivity, making it suitable for several applications in electrical equipment. Existence of good polymer-filler interaction, uniformity in degree of dispersion, and alignment of graphene nanosheet within polymer matrix provide the utmost electrical conductivity. It also possesses high thermal conductivity, also acts as reinforcing filler to polymer matrix and thereby improves dimensional stability, thermal stability, dielectric properties for energy harvesting applications. Further, the toughness and ease of processibility of polymer is also reflected upon the composites of graphene. Hence, these composites can be used to fabricate intricate and highly complex structures.

References

[1] J.H. Chen, C. Jang, S. Xiao, M. Ishigami, M.S. Fuhrer, Intrinsic and extrinsic performance limits of graphene devices on SiO_2, Nat. Nanotechnol. 3 (2008) 206–209.

[2] Y. Pan, N.G. Sahoo, L. Li, The application of graphene oxide in drug delivery, Expert Opin. Drug Deliv. 9 (11) (2012) 1365–1376.

[3] R.F. Service, Carbon sheets an atom thick give rise to graphene dreams, Am. Assoc. Adv. Sci. (2009).

[4] D.E. Sheehy, J. Schmalian, Optical transparency of graphene as determined by the fine-structure constant, Phys. Rev. B 80 (19) (2009) 193411.

[5] H. Kim, A.A. Abdala, C.W. Macosko, Graphene/polymer nanocomposites, Macromolecules 43 (16) (2010) 6515–6530.

[6] Z.U. Khan, A. Kausar, H. Ullah, A. Badshah, W.U. Khan, A review of graphene oxide, graphene buckypaper, and polymer/graphene composites: properties and fabrication techniques, J. Plastic Film Sheet. 32 (4) (2016) 336–379.

[7] Y. Matsuo, K. Tahara, Y. Sugie, Structure and thermal properties of poly (ethylene oxide)-intercalated graphite oxide, Carbon 35 (1) (1997) 113–120.

[8] M. Hirata, T. Gotou, S. Horiuchi, M. Fujiwara, M. Ohba, Thin-film particles of graphite oxide 1: high-yield synthesis and flexibility of the particles, Carbon 42 (14) (2004) 2929–2937.

[9] S. Chatterjee, J. Wang, W. Kuo, N. Tai, C. Salzmann, W. Li, et al., Mechanical reinforcement and thermal conductivity in expanded graphene nanoplatelets reinforced epoxy composites, Chem. Phys. Lett. 531 (2012) 6–10.

[10] H.J. Salavagione, M.A. Gomez, G. Martinez, Polymeric modification of graphene through esterification of graphite oxide and poly (vinyl alcohol), Macromolecules 42 (17) (2009) 6331–6334.

[11] K.A. Worsley, P. Ramesh, S.K. Mandal, S. Niyogi, M.E. Itkis, R.C. Haddon, Soluble graphene derived from graphite fluoride, Chem. Phys. Lett. 445 (1–3) (2007) 51–56.

[12] S. Niyogi, E. Bekyarova, M.E. Itkis, J.L. McWilliams, M.A. Hamon, R.C. Haddon, Solution properties of graphite and graphene, J. Am. Chem. Soc. 128 (24) (2006) 7720–7721.

[13] S. Stankovich, R.D. Piner, S.T. Nguyen, R.S. Ruoff, Synthesis and exfoliation of isocyanate-treated graphene oxide nanoplatelets, Carbon 44 (15) (2006) 3342–3347.

[14] P. Liu, K. Gong, P. Xiao, M. Xiao, Preparation and characterization of poly (vinyl acetate)-intercalated graphite oxide nanocomposite, J. Mater. Chem. 10 (4) (2000) 933–935.

[15] J.Y. Jang, M.S. Kim, H.M. Jeong, C.M. Shin, Graphite oxide/poly (methyl methacrylate) nanocomposites prepared by a novel method utilizing macroazoinitiator, Compos. Sci. Technol. 69 (2) (2009) 186–191.

[16] S. Wang, M. Tambraparni, J. Qiu, J. Tipton, D. Dean, Thermal expansion of graphene composites, Macromolecules 42 (14) (2009) 5251–5255.

[17] X. Du, M. Xiao, Y. Meng, A. Hay, Direct synthesis of poly (arylenedisulfide)/carbon nanosheet composites via the oxidation with graphite oxide, Carbon 1 (43) (2005) 195–197.

[18] S.N. Tripathi, G.S. Rao, A.B. Mathur, R. Jasra, Polyolefin/graphene nanocomposites: a review, RSC Adv. 7 (38) (2017) 23615–23632.

[19] S. Stankovich, D.A. Dikin, R.D. Piner, K.A. Kohlhaas, A. Kleinhammes, Y. Jia, et al., Synthesis of graphene-based nanosheets via chemical reduction of exfoliated graphite oxide, Carbon 45 (7) (2007) 1558–1565.

[20] S. Xie, Y. Liu, J. Li, Comparison of the effective conductivity between composites reinforced by graphene nanosheets and carbon nanotubes, Appl. Phys. Lett. 92 (24) (2008) 243121.

[21] A. Yu, P. Ramesh, X. Sun, E. Bekyarova, M.E. Itkis, R.C. Haddon, Enhanced thermal conductivity in a hybrid graphite nanoplatelet–carbon nanotube filler for epoxy composites, Adv. Mater. 20 (24) (2008) 4740–4744.

[22] H. Zhong, J.R. Lukes, Interfacial thermal resistance between carbon nanotubes: molecular dynamics simulations and analytical thermal modelling, Phys. Rev. B 74 (12) (2006) 125403.

[23] A. Yu, P. Ramesh, M.E. Itkis, E. Bekyarova, R.C. Haddon, Graphite nanoplatelet – epoxy composite thermal interface materials, J. Phys. Chem. C. 111 (21) (2007) 7565–7569.

[24] N. Yousefi, X. Sun, X. Lin, X. Shen, J. Jia, B. Zhang, et al., Highly aligned graphene/polymer nanocomposites with excellent dielectric properties for high-performance electromagnetic interference shielding, Adv. Mater. 26 (31) (2014) 5480–5487.

[25] D. Wang, T. Zhou, J.-W. Zha, J. Zhao, C.-Y. Shi, Z.-M. Dang, Functionalized graphene–BaTiO$_3$/ferroelectric polymer nanodielectric composites with high permittivity, low dielectric loss, and low percolation threshold, J. Mater. Chem. A 1 (20) (2013) 6162–6168.

[26] H. Li, Z. Chen, L. Liu, J. Chen, M. Jiang, C. Xiong, Poly (vinyl pyrrolidone)-coated graphene/poly (vinylidene fluoride) composite films with high dielectric permittivity and low loss, Compos. Sci. Technol. 121 (2015) 49–55.

[27] K.S. Novoselov, A.K. Geim, S.V. Morozov, D. Jiang, Y. Zhang, S.V. Dubonos, et al., Electric field effect in atomically thin carbon films, Science 306 (5696) (2004) 666–669.

[28] J. Xue, L. Wu, N. Hu, J. Qiu, C. Chang, S. Atobe, et al., Evaluation of piezoelectric property of reduced graphene oxide (rGO)–poly (vinylidene fluoride) nanocomposites, Nanoscale 4 (22) (2012) 7250–7255.

[29] A.A. Balandin, Thermal properties of graphene and nanostructured carbon materials, Nat. Mater. 10 (8) (2011) 569–581.

[30] S. Choi, Z. Zhang, W. Yu, F. Lockwood, E. Grulke, Anomalous thermal conductivity enhancement in nanotube suspensions, Appl. Phys. Lett. 79 (14) (2001) 2252–2254.

[31] K.K. Sadasivuni, D. Ponnamma, S. Thomas, Y. Grohens, Evolution from graphite to graphene elastomer composites, Prog. Polym. Sci. 39 (4) (2014) 749–780.

[32] C.E. Banks, A. Crossley, C. Salter, S.J. Wilkins, R.G. Compton, Carbon nanotubes contain metal impurities which are responsible for the "electrocatalysis" seen at some nanotube—modified electrodes, Angew. Chem. Int. (Ed.) 45 (16) (2006) 2533–2537.

[33] M.D. Stoller, S. Park, Y. Zhu, J. An, R.S. Ruoff, Graphene-based ultracapacitors, Nano Lett. 8 (10) (2008) 3498–3502.

[34] R. Verdejo, F. Barroso-Bujans, M.A. Rodriguez-Perez, J.A. De Saja, M.A. Lopez-Manchado, Functionalized graphene sheet filled silicone foam nanocomposites, J. Mater. Chem. 18 (19) (2008) 2221–2226.

[35] J.R. Kim, S. Cheng, S.-E. Oh, B.E. Logan, Power generation using different cation, anion, and ultrafiltration membranes in microbial fuel cells, Environ. Sci. Technol. 41 (3) (2007) 1004−1009.

[36] J. Wang, Y. Wang, D. He, H. Wu, H. Wang, P. Zhou, et al., Influence of polymer/fullerene-graphene structure on organic polymer solar devices, Integr. Ferroelectr. 137 (1) (2012) 1−9.

[37] A.C. Mayer, S.R. Scully, B.E. Hardin, M.W. Rowell, M.D. McGehee, Polymer-based solar cells, Mater. Today 10 (11) (2007) 28−33.

Graphene as a reinforcement in thermoset resins

12

Sanjay Remanan, Tushar Kanti Das and Narayan Chandra Das
Rubber Technology Centre, Indian Institute of Technology Kharagpur, Kharagpur, India

12.1   Introduction

An editorial article published in the Carbon journal redefines the nomenclature of the various graphene-derived materials. Bianco et al. [1] defined graphene as a single atom thick hexagonally arranged, sp^2 bonded carbon atoms that is not an integral part of carbon material but is freely suspended or adhered on a foreign substrate. The editorial board suggests that other members of the graphene family simply cannot be termed as "graphene" but must be called using unique multiword, for example, graphene layer, bilayer graphene, trilayer graphene, or multilayer graphene, etc.

This 2D-material is widely studied and investigated in the scientific community owing to the diversity in applications such as water purifications, energy storage, sensors, and in various automobile fields. This can be attributed to its inherent electrical conductivity and very high in-plane tensile strength which makes the graphene an ideal candidate for various engineering applications. This is driven by the emergence of lightweight and flexible devices which decrease the amount of filler required for attaining the mechanical properties as compared to the conventional fillers like carbon black (CB). The impressive mechanical properties of this 2D material are due to its stability of the sp^2 bonds that can form a hexagonal lattice and oppose a variety of in-plane deformations [2]. This high stability imparts excellent stiffness to the graphene and the values may vary more than 100 N/m. However, the stiffness value of this filler may change with the presence of defects (such as Stone-Wales defects) and wrinkles present over the nanosheet. These wrinkles can be the stress-concentrations sites due to the uneven distribution of stress and eventually leads to material failure under loading. Defect-free graphene is the strongest material ever tested with an intrinsic strength in the range of 130 GPa. This 2D material can sustain higher breaking strength and stiffness even at higher densities of sp^3 defects but drops significantly in presence of vacancy defects. The stability may be varied in the case of multilayer graphene due to the interlocking of graphene layers. This 2D material stiffness and strength were measured with the help of nanoindentation and atomic force microscopy analysis.

Graphene/polymer nanocomposites are being used for various applications and to get the high mechanical properties which can improve the dynamic loading ability [3,4]. The sheet-like structure has the advantage of higher surface area that increases the polymer chain wetting and thereby increases interfacial adhesion. Improvement in interfacial adhesion facilitates the effective load transfer from

Polymer Nanocomposites Containing Graphene. DOI: https://doi.org/10.1016/B978-0-12-821639-2.00012-4

matrix to the graphene and decreases the structural failure. Resins are low viscous polymers compared to the elastomers and it is necessary to introduce reinforcing fillers for getting high mechanical properties. As mentioned, conventional fillers such as CB pose the problem of the requirement of high filler loading, agglomeration, and inferior physicomechanical properties. The introduction of graphene can increase the interfacial adhesion facilitate the effective load transfer from the matrix to fillers which reinforce the polymer.

Herein, the preparation of graphene/polymer nanocomposites and its reinforcing mechanisms are discussed. In the first section, different resins and their synthesis are discussed followed by reinforcing the mechanism of graphene in the resin matrix is explained. Various resins such as phenol-formaldehyde, epoxy, and silicones are considered and different reinforcing effects of graphene in these resin matrices are also discussed. In the second section, the reinforcing mechanism in the elastomeric matrices and the effect of graphene reinforcement in different elastomeric matrices are discussed.

12.2   Basic concepts

12.2.1   Elastic modulus

Modulus is the measure of material stiffness and its unit is in Pa. It is determined from the slope of the stress—strain curve in the Hookean region. Stiffness is defined as the degree of resistance to deformation.

12.2.2   Tensile strength and elongation at break

Tensile strength is defined as the stress at which sample fracture occur in a stress-strain experiment. And the strain at which sample fractured is called elongation at break. Higher elongation at break of a material indicates the ductility and lower EB is the brittle characteristics (Fig. 12.1).

12.2.3   Toughness

The area under the stress—strain curve is called the toughness. Toughness is measured in terms of energy per unit volume [5].

For liquid-like materials, shear testing can be used instead of tensile. Viscoelastic properties of the liquid-like materials can be analyzed by shearing force. During shearing of material, force is acting tangentially over the surface as a function of distance. Hence, shear stress is measured as,

$$\text{Shear stress} = \frac{\text{Force}}{\text{Surface area}}$$

$$\text{Shear strain} = \frac{\Delta U_x}{\Delta y}$$

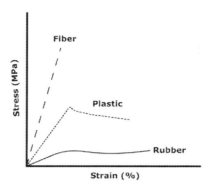

Figure 12.1 Schematic representation of the mechanical properties of various polymers in a universal testing machine.

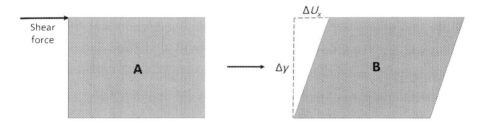

Figure 12.2 Application of shear force on a bottom fixed surface.

Were ΔU_x is the change in displacement and Δy is the vertical displacement (Fig. 12.2).

12.2.4 Shear modulus

Shear modulus is similar to the tensile modulus and it is the slope of the straight line portion of the stress-strain graph.

$$E = 2G(1 + \nu)$$ (12.1)

where ν is the Poisson's ratio and $E = 3\,G$ for noncompressible and elastic polymers.

12.2.5 Reinforcement

The term reinforcement is broadly described as the mechanical properties of the polymer. A polymer is said to be reinforced indicate its mechanical properties such as modulus, strength, and toughness reaches its optimum values.

12.2.6   Interfacial adhesion

It is the adhesive interaction between the additives and polymer matrix. The additives can be fillers, another polymer, or other particles that play a significant role in improving the mechanical properties and polymer is said to be reinforced. The interfacial adhesion between the polymer phases governs the reinforcement (ex: improvement in mechanical properties in a nanoblend) in the polymer blend while as interfacial adhesion between the polymer and filler governs the reinforcement in polymer nanocomposites.

12.3   Polymeric resins

12.3.1   Thermosetting resin

Thermosetting resin is a group of polymer that upon heating irreversibly converted into three-dimensional cross-linked rigid materials from viscous liquid and when the material is cured, further remolding and recycling are prevented. These groups of polymers distinguished themselves from thermosetting elastomers based on the cross-linked three-dimensional covalent linkages formed during the curing reaction. In the case of thermosetting elastomers, network structures are formed by a finite number of covalent bonds which restricts the mobility of the polymer chains but local molecular chain mobility is still observed. On the other hand, thermosetting resins are cross-linked by high numbers of covalent bonds which impede all the molecular chain movements resulting in infusible and insoluble polymeric materials [6,7]. The commercial thermosetting resins are phenolic resins, amino resins, epoxy resins, polyester resins, and silicone resins.

12.3.1.1   Phenolic resin

At the beginning of the twentieth century, phenolic resins are widely used and considered as one of the most important thermosetting synthetic polymers that has a major market share. Phenolic resins are prepared through either acid or base-catalyzed reaction by the mole ratio variation of phenol with formaldehyde. The first commercially synthetic plastic known as "Bakelite" is a phenolic resin prepared in 1907 by Leo Hendrik Baekeland.

The base-catalyzed phenolic resin is called resol and acid-catalyzed resin is called novolac (Fig. 12.3). In the case of resol, the mole ratio of formaldehyde to phenol is greater than unity whereas for novolac it is less than unity. Novolac is thermoplastic in nature before it is cross-linked by hexamethylenetetramine (HEXA). At elevated temperature, HEXA is fragmented into formaldehyde and ammonia and this formaldehyde reacts with phenol terminated novolac resin to convert crosslinked thermosetting resin. On the other hand, resol is self-cured by excess formaldehyde to generate three-dimensional thermosetting resins [8,9]. Compared to resol, novolac has a long shelf life. Resoles are usually liquids while novolac are solids and dimensional stability of the latter is greater than that of the former [10].

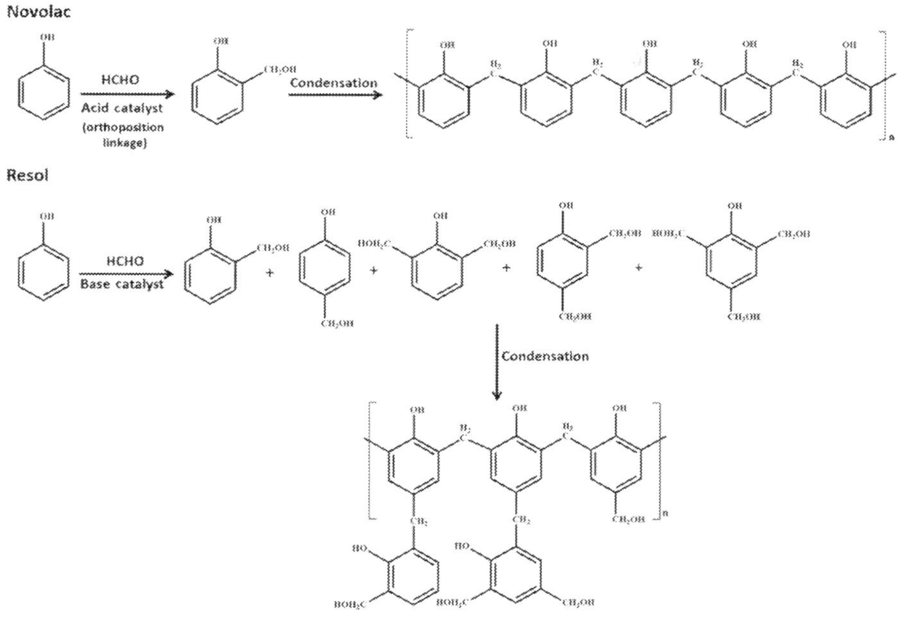

Novolac

Resol

Figure 12.3 Reaction scheme for preparing the novolac and resol resin.

12.3.1.2 Amino resins

Amino resin, next to phenolic resin, is another important class of synthetic thermosetting polymer synthesized by reacting formaldehyde with compounds containing amino ($-NH_2$) groups. Two types of amino resins namely urea-formaldehyde (UF) resin and melamine-formaldehyde (MF) resin are primarily consumed the whole market of amino resins. UF resin is prepared by a two-stage reaction (Fig. 12.4). In the first stage, under slightly alkaline conditions, urea and formaldehyde at equal molar ratio reacted with each other to produce a prepolymer of methylol derivatives of urea. In the next stage, under acidic condition (pH ≤ 5.5) the low molecular weight prepolymer undergo condensation reaction to produced three dimensional crosslinked network structure. On the other hand, MF resin (Fig. 12.5) is also produced in two stages similar to that of UF resin. In the first stage, melamine and formaldehyde reacted with each other to produce melamine derivative which undergoes curing to produce network structure in presence of heat in the next stage [11,12]. Although, both UF and MF resins have their commercial market with respect to amino resins, UF resin is consumed 80% of the total market.

12.3.1.3 Epoxy resins

Epoxy resins are prepared by condensation reaction between epoxy compounds with the multifunctional monomeric unit. The most common epoxy resin is synthesized by the condensation reaction of epichlorohydrin and bisphenol A (Fig. 12.6).

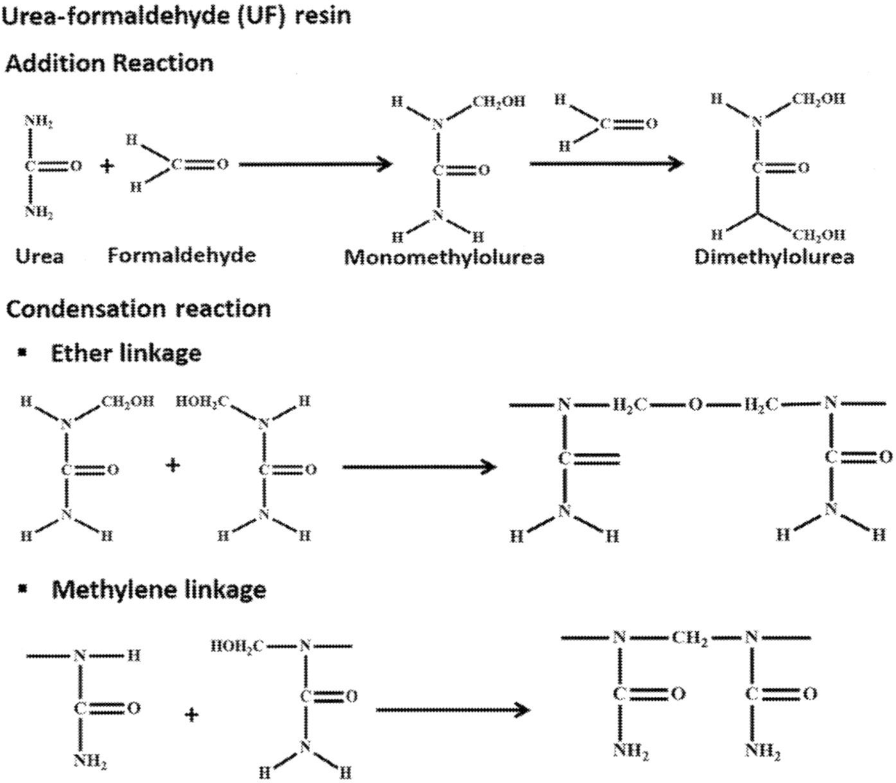

Urea-formaldehyde (UF) resin

Addition Reaction

Urea   Formaldehyde   Monomethylolurea   Dimethylolurea

Condensation reaction

- **Ether linkage**

- **Methylene linkage**

Figure 12.4 Scheme of preparing and curing reaction of urea-formaldehyde resin.

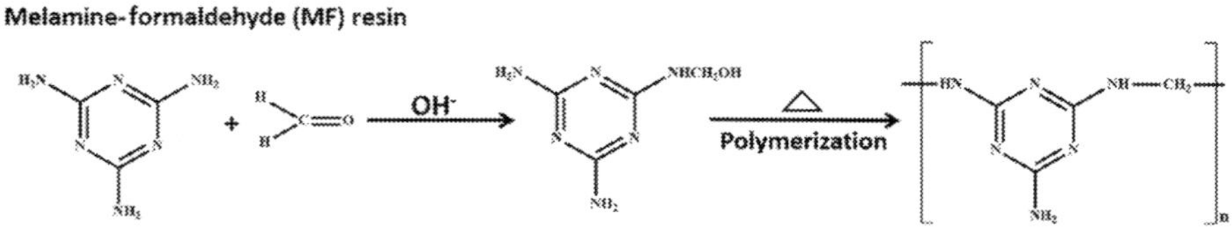

Melamine-formaldehyde (MF) resin

Figure 12.5 Reaction mechanism of preparing the melamine-formaldehyde resin.

This resin is also prepared by two steps method: in the 1st step bisphenol A and epichlorohydrin (ratio of 1:3) are reacted at around 100 °C to produce low-molecular-weight diglycidyl ether of bisphenol A. The reaction is conducted with the continuous drop-wise addition of NaOH to neutralize the reaction medium. In the 2nd step, the low-molecular-weight liquid resins are crosslinked by various crosslinking agents termed as hardening agents. The commonly used hardening

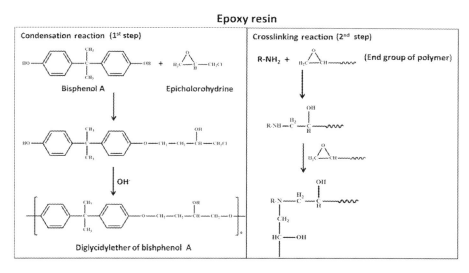

Figure 12.6 Scheme of preparing and curing reaction of epoxy resin.

agents are monoamine, diamine, polyamine, urea resin, phenolic resin, acids, and various acid anhydrides. Depending upon the property requirements, various types of hardening agents are employed. Besides this resin, other types of epoxy resin are also available such as novolac epoxy resin, aliphatic epoxy resins, halogenated epoxy resin, and glycidyl amine resin [13,14]. However, these resins have less commercial value than bisphenol A based epoxy resins which have the market consumption above 90%.

12.3.1.4 Polyester resins

Polyester resins are classified into two categories namely saturated and unsaturated polyesters. Saturated polyesters resins are synthesized by reacting multifunctional acid (or anhydride) groups containing compounds with polyols (Fig. 12.7). This reaction leads to the formation of low-molecular-weight viscous liquid which is then reacted in the presence of heat to convert into three-dimensional thermosetting resins. The most commonly used saturated polyester resin is polyethylene terephthalate (PET). PET is prepared by reacting terephthalic acid with excess ethylene glycol in the presence of base catalyst followed by heating the formed prepolymer at a higher temperature in the next stage to convert three-dimensional network structures. On the other side, unsaturated polyester resin (Fig. 12.8) is prepared by reacting either unsaturated multifunctional acid groups containing compounds with polyols or unsaturated polyols with multifunctional acid groups containing compounds or both having unsaturation. This unsaturation helps in crosslinking reactions to produce a three-dimensional network structure. The crosslinking of unsaturated polyesters includes various curing agents such as styrene, methyl acrylate, diallyl phthalate, and vinyl toluene and triallyl cyanurate. Unsaturated

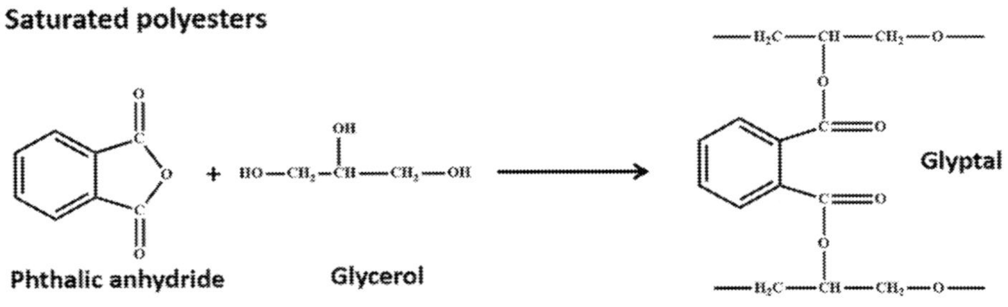

Saturated polyesters

Figure 12.7 Reaction scheme of preparation of the saturated polyester resin.

Unsaturated polyesters

- **Polymerization**

- **Crosslinking reaction**

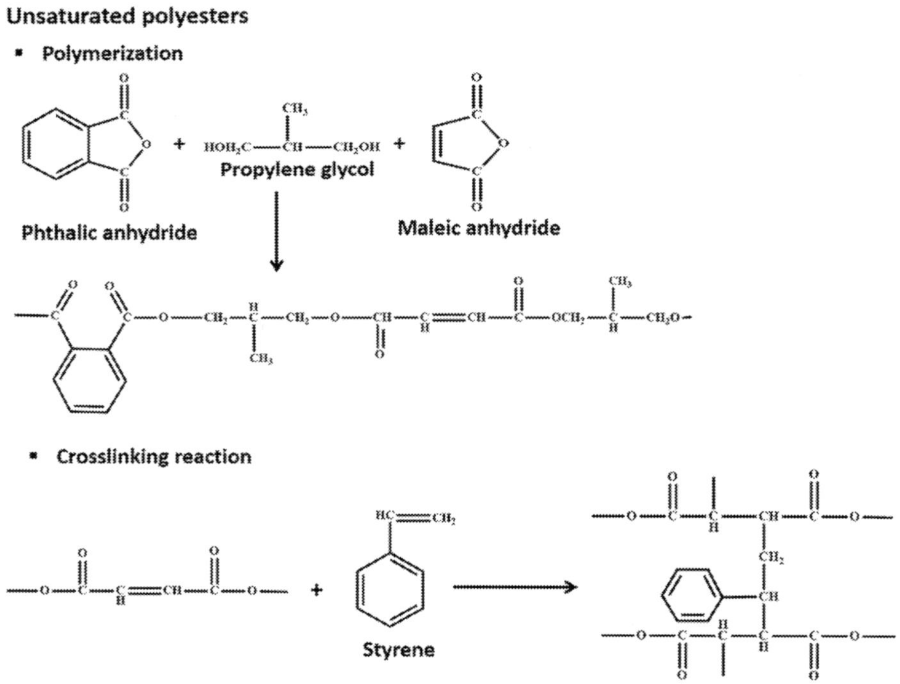

Figure 12.8 Scheme of preparing and curing reaction of the unsaturated polyester resin.

polyester resin has a wide range of applications due to its ease of handling and rapid curing behavior [15,16].

12.3.1.5 Silicone resins

Silicone resins are highly crosslinked inorganic polymers which are composed of either branched or cage-like structure of siloxanes. These resins have unique high-temperature resistance properties due to the high bond strength of the

Si-O bond. Silicone resins are manufactured by hydrolysis of organo-chlorosilanes in the presence of a catalyst. These resins are also cured by heating at high temperature in the presence of catalysts. Various types of silicone resin are also available in the market and the primary application of silicone resin is in the coating industry [17].

12.3.2 Resin reinforcing mechanism

Graphene, when it is uniformly dispersed in the resin matrix can effectively distribute the load across the nanocomposite. Uniform dispersion allows the effective load transfer from the resin to the graphene filler phase. This increases the mechanical properties such as strength, modulus, and toughness of the matrix. Hence, load transfer to the graphene depends upon the interfacial adhesion or interfacial interaction between the nanosheets and resin matrix.

The addition of filler beyond the threshold concentration leads to the filler agglomeration and these agglomerated structures act as stress concentration sites inside the nanocomposite. This increases poor load distribution across the filler system and imparts inferior property to the graphene/resin composite. As mentioned, interfacial adhesion has a crucial role in improving mechanical properties. Good mechanical properties are the result of uniform dispersion inside the resin or polymer matrix. Several strategies can be adopted for the modification of graphene and thereby increase the interfacial adhesion to get good mechanical properties [18]. The most common example is the preparation of graphene oxide (GO), thermal reduction of the graphene and covalent or noncovalent functionalization of graphene imparts higher interfacial adhesion and increase the reinforcement of resin matrix. A list of modification strategies is given in Table 12.1.

12.3.2.1 Double crosslinking effect

Functionalized graphene or fillers can also impart crosslinking sites with the functional groups of matrix polymers which imparts an additional crosslinking. This additional crosslinking in the polymer due to the modified filler is called the double crosslinking effect. Due to the double crosslinking, degree of freedom of polymer chains reduced and the matrix tends to be stiffer (Fig. 12.9). An increase in stiffness is the indication of improvement in material modulus and hence, matrix reinforcement is improved.

12.4 Resin/graphene nanocomposites

12.4.1 Phenolic resin

Phenolic resin can be reinforced with the introduction of the graphene nanosheets. Ren et al. [24] observed that the addition of graphene and modified graphene reduces both the wear rate and coefficient of friction of the nanocomposite. Modified graphene shows a lower wear rate and coefficient of friction indicates the

Table 12.1 List of functionalizing agents used for graphene in resin or elastomer matrices.

Sl. no	Functionalizing agent	System	Reference
1.	Poly(butadieneco-acrylonitrile) (ATBN)	ATBN/GnP/Epoxy	[19]
2.	3-Aminopropyltriethoxysilane (APTES, KH550)	KH550-G/boron phenolic resin	[20]
3.	Triton X-100	Triton X-100-G/boron phenolic resin	[20]
4.	Benzyl glycidyl ether (BGE)	BGE-G/epoxy	[21]
5.	4′-Allyloxy-biphenyl-4-ol (AOBPO)	AOBPO-GnP/silicone resin	[22]
6.	Polyurethane-imide (PUI)	PUI-GnP/silicone resin	[23]
7.	Polystyrene (PS)	PS-G/ Nomex fabric/ phenolic	[24]
8.	1-Pyrenebutyric acid (PBA)	PBA-G/epoxy	[25]
9.	Ortho-quinone	Ortho-quinone-G/SBR	[26]
10.	Aminopropyltriethoxysilane (APTES)	APTES-GnP/ silicone rubber	[27]
11.	Vinyltrimethoxysilane (VTMS)	VTMS-GnP/silicone rubber	[27]
12.	Triton X-100	Triton X-100-GnP/silicone rubber	[27]

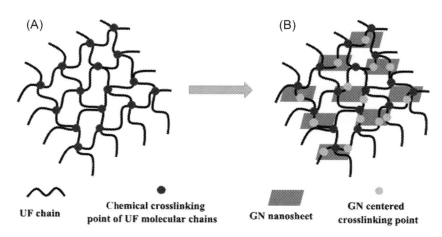

Figure 12.9 Crosslinking network structure of (A) neat urea-formaldehyde (UF) and (B) graphene (GN)-silicon coupling agent with terminal amino groups (SA)/UF nanocomposites. *Source:* Reproduced with permission from B. Wu, L. Ye, Y. Liu, X. Zhao, Intercalation structure and toughening mechanism of graphene/urea-formaldehyde nanocomposites prepared via in situ polymerization. Polym. Int. 67 (2018) 330−339, John Wiley and Sons.

better stiffness arise from the graphene. Hence, modification of graphene plays a significant role in improving its dispersion and thereby composite mechanical properties. Graphene can be functionalized to improve the dispersion in the resin matrix; 3-aminopropyltriethoxysilane (APTES, KH550) and Triton X-100 can be used for functionalization which improves the dispersion and facilitates better load transfer. This functionalization increases the flexural strength of the composite through uniform dispersion compared to the unmodified graphene [20]. Similar results are obtained when graphene is converted into GO, in which an increment in mechanical properties such as bending, tensile and impact strength were observed with the addition of even small percentage of GO [28]. The increment is possibly due to the better dispersion in the phenolic matrix and stronger bonding between the resin and GO which facilitate the efficient stress transfer through the interfacial area and distribute the effective area of strain in the matrix. GO and functionalized graphene can take part in the crosslinking reaction which further increases the entanglements between the polymer and filler, increases the nanocomposite reinforcement. The addition of functionalized graphene improves the mechanical properties by improving the interfacial adhesion. For example, the introduction of GO into the phenolic matrix improved Young's modulus and tensile strength attributed to the chemical reaction between the hydroxyl groups of resol and carboxyl groups of GO, and hence an efficient load transfer and reinforcement can be achieved in the phenolic nanocomposites [29].

12.4.2   Urea-formaldehyde resin

As discussed, functionalized graphene has a better reinforcing effect as compared to the unmodified graphene. When modified, graphene nanosheets found to have better dispersion and increase the interaction with the UF resin, transfer the stress from the resin to graphene. Hence, the interface adhesion plays a critical role in improving mechanical properties. The addition of 1% modified graphene imparts higher tensile strength in the UF matrix. Increment in storage modulus during dynamic mechanical analysis shows that graphene increases the stiffness of the resin and decreases the molecular motion [30]. The presence of functionalized graphene can also increase the crosslink density, which is also a reason for the improved mechanical properties of the graphene/UF nanocomposites. Such crosslinking due to the addition of nanofillers in the polymer matrix is called a double crosslinking network structure [31]. Wu et al. [32] recently studied the mechanical properties of GO in the mixture of UF and melamine-formaldehyde (MF) matrix which is reinforced with the r-glass fiber. With the addition of GO, an improvement in mechanical properties was observed and optimum tensile strength and modulus were achieved even at 0.6% addition of GO in the glass-reinforced polymer matrix. The increment is due to the efficient load transfer through the matrix to GO. The decrement in properties was observed at higher filler concentration is due to the filler agglomeration leads to the inefficient load distribution.

12.4.3 *Epoxy resin*

Epoxy/graphene resin nanocomposite exhibits improved mechanical properties when the graphene nanosheets are suitably modified. The presence of graphene improved the stiffness and thereby modulus of the resin. However, when it is not modified, the other mechanical properties such as modulus and elongation at break are inferior due to the poor interface formed between the graphene and epoxy resin. When the graphene is functionalized with various agents such as pyrenebutynic acid (PBA) and benzyl glycidyl ether (BGE) imparts better dispersion and thereby efficient load transfer across the interface. Hence, higher mechanical properties such as increased modulus during the dynamic loading can be achieved [25]. For instance, at 1% loading of functionalized graphene in epoxy matrix increase, the storage modulus to 1 GPa and percentage increment was 100% compared to pure epoxy and 57% compared to epoxy with 1% GO. The tan δ of the functionalized graphene/epoxy composite is lower than that of pure epoxy (Fig. 12.10). This is due to the carboxylic functional group of PBA on the modified graphene interacts with the functional groups present in the epoxy. A similar increment in mechanical properties was observed when the graphene was modified using the BGE. BZE also acts as a dispersing agent which increases the graphene dispersion inside the epoxy matrix and π-π interaction of BZE and graphene improve the mechanical properties of the epoxy nanocomposite. This interaction increases the storage modulus to a very high value (~3 GPa) and lowers the tan δ due to the higher dispersion and interfacial interaction [21]. Wang et al. [33] observed that acid and amine

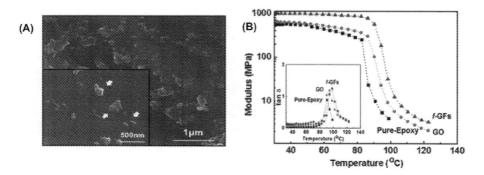

Figure 12.10 (A) Scanning electron microscope images of fracture surfaces of the functionalized-graphene flakes (f-GFs) nanocomposites. Isolated f-GFs were homogeneously dispersed in the polymer matrix. (B) dynamic mechanical analysis storage modulus curves and tanδ curves of the epoxy nanocomposites with GO and f-GFs at 1wt.%. The mechanical properties of f-GFs composites are enhanced by improving the interfacial interaction between f-GFs and polymer matrix and better dispersion of f-GFs within a polymer matrix.
Source: Reproduced with permission from S.H. Song, K.H. Park, B.H. Kim, Y.W. Choi, G.H. Jun, D.J. Lee, et al., Enhanced thermal conductivity of epoxy—graphene composites by using non-oxidized graphene flakes with non-covalent functionalization, Adv. Mater. 25 (2013) 732–737, John Wiley and Sons.

functionalization of the graphene considerably increases Young's modulus of the nanocomposite compared to the unfilled epoxy system.

Hence, the interaction of the functional groups between the polymer and modified filler plays a crucial role in increasing the reinforcement through improving the interfacial adhesion. The dramatic increment in mechanical properties is due to the efficient load transfer between in a uniformly distributed graphene and epoxy matrix. This indicates the modified graphene in the epoxy resin can improve the uniform dispersion and reduce the agglomeration as in the case of unmodified graphene nanocomposites [34,35]. From the morphology study, one can observe that graphene can act as a stress concentration site which efficiently absorbs the large amount of external energy transmitted from the epoxy matrix. Similar kinds of literature on epoxy reinforcement are widely studied and reported elsewhere [36–40].

12.4.4 Polyester resin

Graphene loading reinforces the polyester matrix when it is added to the threshold concentration. Higher filler concentration leads to the agglomeration in the resin matrix and imparts inferior mechanical properties. Also, unmodified graphene may pose a dispersion problem in the resin matrix and not improves the tensile strength and storage modulus to a great extent [41]. Functionalized graphene, for example, GO improves the mechanical properties greatly. Even at lower loading of 3%, a considerable increment in tensile strength and Yong's modulus were observed. The presence of thermally reduced graphene is another approach to produce high mechanical properties in polyester resin. Strong interaction at the interface and interlocking due to the wrinkled surface of the multilayer graphene facilitates the effective load transfer and increases the tensile modulus or reinforcing effect. In contrast elongation at break was found reduced due to the alteration in molecular rearrangement due to the interlocked graphene particles [42].

12.4.5 Silicone resin

Mechanical properties such as elastic modulus and tensile strength significantly increased even at the 1% concentration of the graphene in the silicone resin. Gui et al. [23] observed this increment in mechanical properties when the graphene was modified with polyurethane-imide. The graphene surface was functionalized through the covalent bond and π-π interaction. So, the functionalized graphene imparts higher reinforcement in the resin matrix compared to the GO or unmodified graphene. In another study, graphene nanoplatelets can be modified using the 4′-allyloxy-biphenyl-4-ol (AOBPO) and incorporated in the silicone matrix. The functionalized graphene/silicone nanocomposite shows a tremendous increment in tensile strength and elastic modulus (Fig. 12.11) compared to neat silicone resin [22]. In general, the silicone resin matrix can be reinforced by using the graphene nanosheets which can increase the interfacial adhesion and thereby improve the mechanical properties [43].

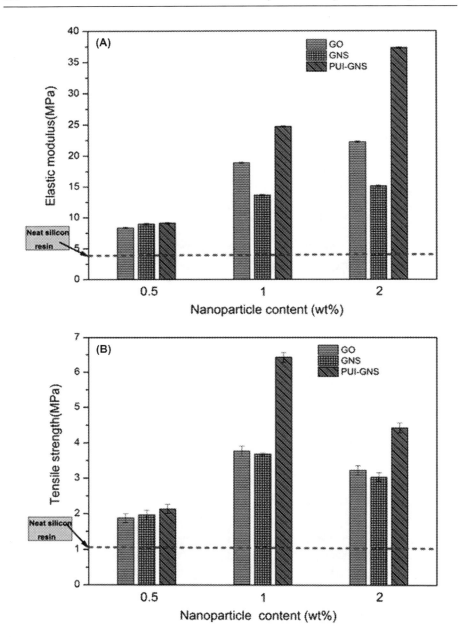

Figure 12.11 Mechanical properties of neat silicon resin and various types of nanocomposites (A) elastic modulus (B) tensile strength.
Source: Reproduced with permission from D. Gui, S. Yu, W. Xiong, X. Cai, C. Liu, J. Liu, Liquid crystal functionalization of graphene nanoplatelets for improved thermal and mechanical properties of silicone resin composites, RSC Adv. 6 (2016) 35210−35215, Royal Society of Chemistry.

12.5 Graphene elastomer nanocomposites

12.5.1 Preparation methodologies

Elastomer nanocomposites can be prepared by various processing techniques such as solution mixing, melt-mixing and in situ polymerization methods.

12.5.1.1 Solution mixing

Solution mixing involves the preparation of dissolved polymer in a suitable solvent followed by the addition of fillers in the polymer solution, mixing, and then drying. Initially, predried polymer pellets are added to the solvents at a known concentration. Polymer chains are initially wetted by the solvents and dissolved. Dissolution varies with the solubility parameter of both the solvent and the polymer. Graphene is introduced to the polymer solution by either direct addition or dispersion in a particular solvent. The dispersing solvent miscibility with the polymer solution must be closely matched for the better filler distribution in the polymer solution.

Polymer/graphene solution then stirred for sufficient time for the better mixing of the graphene in a polymer solution. Bath-sonication can be adopted for the better graphene distribution in the polymer matrix and is necessary for the preparation of uniform graphene dispersion. The mechanically stirred and sonicated mixture is then cast over a suitable substrate and dried to get the film. In the solution mixing, polymer chains are diffuse between the graphene sheets (Fig. 12.12A) and preventing the possible agglomeration [44]. For a better dispersion in the polymeric matrix, especially in higher molecular weight materials like elastomers, the dried films were further processed in a two-roll mill for homogeneous distribution of graphene or nanoparticles.

Solution mixing is a simple and straight forward method for the polymer nanocomposite preparation. The process does not require complex types of machinery rather simple stirring that forms the well-dispersed matrix. Hence, the method is facile and fast. However, the use of hazardous solvent and its release to the surrounding environment reduce the chances of upscaling of this method.

12.5.1.2 Melt-mixing

Melt-mixing is an industrially favorable processing method in which polymer nanocomposite can be prepared by using various processing instruments like internal-mixers and extruders. Melt-mixing is devoid of any multistep processing and use of any hazardous solvents like in the case of solution mixing. This method can be used for the batch/continuous line with a large production volume. During the mixing in the internal-mixers, polymer and fillers undergo distributive and dispersive mixing. Polymer pellets melt under the shearing force experienced by the rotors and barrel walls. When the viscosity of the thermoset decreases under the shearing stress, polymer chains start wet the filler particles and slowly incorporated into the bulk. The nanocomposite is prepared under the combined action of the rotor, processing temperature, and other parameters like the viscosity of the polymer matrix

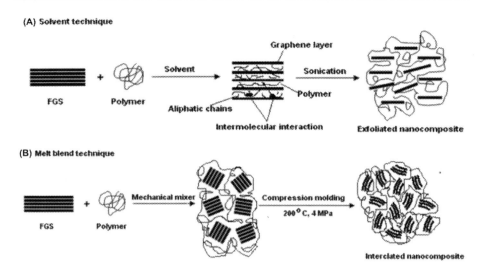

Figure 12.12 The mechanism of foliated graphene sheets incorporation into polymer by (A) solvent technique and (B) melt blend technique.
Source: Reproduced with permission from W.E. Mahmoud et al., Morphology and physical properties of poly(ethylene oxide) loaded graphene nanocomposites prepared by two different techniques, Eur. Polym. J. 47 (2011) 1534−1540, Elsevier.

and the concentration of filler loading. A uniform graphene/polymer nanocomposite in the internal mixture is prepared by the introduction of a master batch. This could reduce the possible agglomeration (Fig. 12.12B) of graphene fillers during the conventional mixing and form a uniform graphene/polymer nanocomposite.

12.5.1.3   In situ polymerization

Graphene/polymer nanocomposite is prepared in another route called in situ polymerization. Graphene flakes are introduced into the monomer or prepolymer and polymerization proceeds to yields nanocomposites. The nanocomposite has the advantage of the strong interaction between the graphene and polymer matrix through good dispersion. Additionally, functionalized graphene imparts strong interaction with polymer matrix which facilitates the load transfer and imparts high modulus and tensile strength even at small filler concentrations. Modification of graphene with suitable functional groups increases the interfacial adhesion in a polymer matrix which helps to gain high reinforcement. However, a reduction in the polymerization rate, necessary solvent conditions due to the use of graphene, and high viscosity generation during bulk polymerization are the major limitations. Different in situ polymerization techniques such as in situ polycondensation, in situ emulsion polymerization, in situ melt-polycondensation, and in situ emulsion polymerizations are reported in the literature [45−47].

12.5.2   Reinforcing mechanism in elastomers

Reinforcing characteristics of the rubber largely depends upon the filler characteristics such as intrinsic property and geometry. As mentioned, graphene has
high in-plane tensile strength, modulus, and specific surface area compared to
the conventional fillers. This increases the end-use property requirement for
optimum physicomechanical property achievement. As mentioned, interfacial
interaction or adhesion is a crucial factor for better load transfer between the
graphene and elastomer matrix. This interaction can be improved by tuning
the surface characteristics of the graphene or by suitable modification with the
functional groups, which reinforces the elastomer matrix.

Graphene derived elastomer composite shows an increment in mechanical properties. As in the case of natural rubber (NR), reinforcing mechanism is arising from
the strain-induced crystallization (SIC) and vulcanization of the elastomer matrix.
SIC increases the mechanical properties of the NR. When stretched, NR chains
undergo SIC as the polymeric chains align in the direction of stress and increase
the mechanical property. When nanosheet like graphene is introduced, it is aligned
in the direction of stress and observed that SIC is shifted to low-stress values. Thus
an increase in modulus is observed even with the small percentage of filler incorporation compared to the corresponding modulus with the high loading of CB. This is
due to the sheet-like structure and high specific surface area of the graphene which
reinforces the rubber in the direction of applied stress. Graphene sheets align in the
direction of applied strain while CB does not and aligned graphene sheets act as
load-bearing medium and shifts the SIC to lower strain. This can be analyzed by
studying the anisotropy of the graphene composite using grazing-incidence small-
angle X-ray scattering technique.

As in the case of styrene butadiene rubber (SBR) that do not have a SIC, graphene dispersion is improved when the vinyl content in the solution SBR (SSBR) is
increased. An increase in vinyl content improves the T_g, limits the chain movement,
and increases the activation energy (E_a). Activation energy is the indication of
internal friction and it is measured from the slope of the modulus versus temperature curve. Hence, graphene/SSBR imparts better reinforcement when the vinyl
concentration is increased [48]. As mentioned, interfacial adhesion or interaction
can significantly influence the mechanical properties. Modified graphene in the
SBR matrix is found to be a better candidate compared to the conventional CB fillers. Conventional reinforcing fillers currently pose the problem of nonuniform dispersion in the elastomeric matrix that leads to the energy loss (of tyres) during
service. The problem of filler agglomeration can reduce the hysteresis loss and
increase energy loss under dynamic loading. Hence, the uniform dispersion of nanoparticles inside the rubber matrix reduces the dynamic energy loss.

Graphene nanosheets when suitably modified, can increase the uniform dispersion in the elastomer matrix and increase the interfacial interaction. Lower filler
percolation threshold required for the optimum tensile strength and lower energy
loss (Fig. 12.13) during the dynamic loading can be achieved with graphene nanofiller. Modified graphene increases the mechanical properties of the composites.

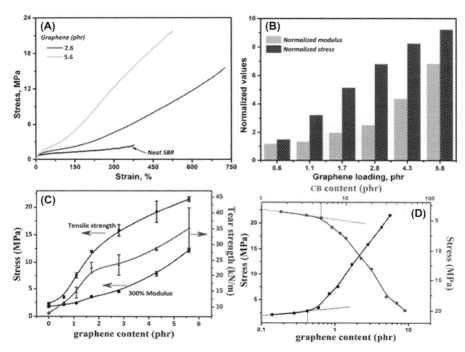

Figure 12.13 (A) Stress−strain curves of SBR/graphene nanocomposites, (B) modulus and stress values for graphene/SBR nanocomposites, normalized to the values for neat SBR, (C) tensile modulus, strength and tear strength of graphene/SBR nanocomposites and (D) percolation phenomena of SBR nanocomposites, the intersections of dash lines represent the percolation points. *SBR, styrene-butadiene rubber.*
Source: Reproduced with permission from Z. Yang, J. Liu, R. Liao, G. Yang, X. Wu, Z. Tang, et al., Rational design of covalent interfaces for graphene/elastomer nanocomposites, Compos. Sci. Technol. 132 (2016) 68−75, Elsevier.

The increment in reinforcement is due to the reduction in the nanofriction of the polymeric matrix [26].

In situ reduction of the GO in isoprene rubber matrix imparts higher interfacial interaction and thereby improved mechanical properties. Raman mapping of the prepared composite showed a thick layer of rubber molecules present over the surface of the modified graphene imparts high bound rubber value. Higher bound rubber is an indication of improved load transfer due to the increase in interfacial interaction between modified graphene and isoprene rubber. Also, modified graphene imparts higher crosslink density than the unmodified graphene in the elastomeric matrix. The modification further improves the interfacial interaction and better load transfer. A wide-angle X-ray diffraction spectroscopy can be used for the visualization of SIC with various strain in both unfilled and filled rubber samples (Fig. 12.14A) [49]. Modified graphene facilitate the early orientation of polymer chains and thereby nucleation leads to the SIC at lower strain (Fig. 12.14B and C).

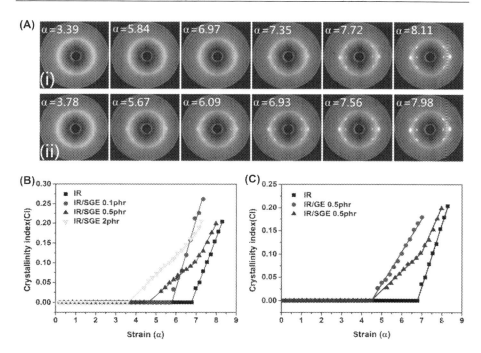

Figure 12.14 (A) WAXD patterns of unfilled isoprene rubber (IR) (i) and IR/ surface modified graphene (SGE) (ii) in various strains; (B) crystallinity index of IR/SGE with various filler contents as a function of strains; (C) crystallinity index of unfilled IR, IR/SGE and IR/GE with 0.5 phr filler as a function of strains. *WAXD, wide-angle X-ray diffraction.* *Source:* Reproduced with permission from Z.-T. Xie, X. Fu, L.-Y. Wei, M.-C. Luo, Y.-H. Liu, F.-W. Ling, et al., New evidence disclosed for the engineered strong interfacial interaction of graphene/rubber nanocomposites. Polymer 118 (2017) 30−39, Elsevier.

Li et al. [50] reported that introduction of graphene in the NR latex through ammonium assisted dispersion of graphene increase the interfacial interaction at lower phr, crosslink density and tensile strength without compromising the elongation at break. The increase in mechanical strength compared to the conventional CB filled system is due to the higher specific surface area of the graphene fillers [51].

12.5.3 Characterization of the composites

The reinforcement effect due to the graphene can be characterized using a number of instruments. Mechanical properties can be measured using the universal testing machine and dynamic properties were analyzed using a dynamic mechanical analyzer (DMA). However, the influence of monolayer graphene on reinforcement can be quantified using the Raman spectroscopy [52]. D and G bands are the important characteristic peaks of the graphene and are in the range of $1580-1360 cm^{-1}$. G band represents the in-plane bond stretching motion of the sp^2 carbon atoms while D band is the indication of breathing modes of the ring. Gong et al. [53] observed a shift in the G band when the monolayer graphene is strained and its value increased with the strain. During unloading, the G band shifts to another higher wavenumber (relaxing). This is due to the slippage in the composite during the initial tensile deformation and then subjected to unloading. The shift in the G band is an indication of the transfer of load or strain to the graphene sheet. As an additional proof, Xie et al. [49] observed that modified graphene exhibited a higher shift in G band wavenumber compared to the unmodified graphene in the elastomer matrix indicates that former one facilitates the better interfacial adhesion and thereby better load transfer compared to the pristine graphene.

12.6 Miscellaneous: different thermosetting polymer/graphene nanocomposites

Frasca et al. found that the addition of multilayered graphene increases the reinforcement in the chloroprene rubber matrix [54]. With the increase in concentration there observed an increase in storage modulus, an increase in Young's modulus, a decrease in elongation at break, and a decrease in tan δ peak. A reduction in the tan δ is an indication of a decrease in energy loss during the dynamic loading. As discussed, the increase in reinforcement is due to the strong interaction between the filler and rubber. This interaction can arrest the polymer chain mobility and result in a decrease of tan δ of the nanocomposite.

Polyimide/graphene nanocomposite shows increased modulus, lower tan δ value, and a lower coefficient of friction with the addition of graphene filler. However, more than 1% filler loading found to have a negative impact on the nanocomposite. This is due to the agglomeration of graphene during the mixing and inefficient and nonuniform load transfer while dynamic loading [55]. The excessive filler loading can also cause inferior coefficient of friction due to agglomeration of the modified filler which deteriorates the crosslinked network.

Silicone rubber has high thermal stability attributed to the presence of Si-O bond that is having higher bond dissociation energy (110 kcal/mol) compared to the C-C bonds (83 kcal/mol). Silicone rubber is also used in wide temperature range and have application in biomedical and space. However, the material poses poor mechanical properties which can be improved by the incorporation of various nanofillers. Especially, silicone/graphene composite shows improved mechanical properties [43,56]. Graphene is modified with silane coupling agents and nanocomposites exhibit higher mechanical properties. The prepared elastomeric composite exhibits a higher tensile strength of ∼0.5 MPa with the addition of 2 phr of silane-modified graphene. Few surface functionalizing agents such as aminopropyltriethoxysilane (APTES), vinyltrimethoxysilane (VTMS), and Triton X-100 (Fig. 12.15) can be used for graphene modification [27]. The increment in mechanical properties in a silicone rubber matrix is observed when the interfacial interaction between the silane coupling agents/graphene/silicone rubber is established. A similar improvement in mechanical

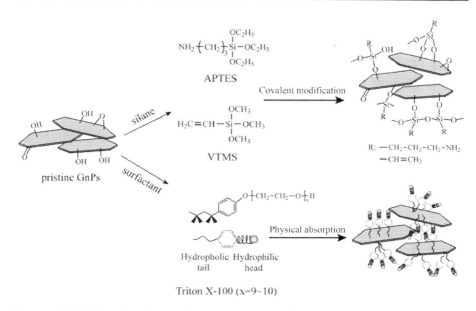

Figure 12.15 Schematic of the surface treatment of graphene nanoplatelets (GnP) particles with different modifiers.
Source: Reproduced with permission from G. Zhang, F. Wang, J. Dai, Z.J.M. Huang, Effect of functionalization of graphene nanoplatelets on the mechanical and thermal properties of silicone rubber composites, Materials 9 (2016) 92, MDPI.

properties was observed when the silicone rubber is mixed with a heterogeneous filler system by introducing SiO_2 into the graphene/silicone interface. Developed compounds exhibit higher mechanical properties such as improved hardness, tensile strength, elongation at break, and tear strength. The increment is attributed to the improved interfacial adhesion of the heterogeneous system [57].

12.7 Conclusion and outlook

Graphene/polymer composites show excellent improvement in mechanical properties when the nanosheet is added at optimum concentrations and is suitably modified. Modified graphene increases the filler polymer interaction, improves the interfacial adhesion, and thereby increases the uniform dispersion in the resin or polymer matrix. Increased uniform dispersion helps the effective transfer of the load from the resin or polymer matrix to filler and equally distributing the applied strain in the nanocomposite system. Various resin matrices such as phenol-formaldehyde, MF, epoxy, and polyester resin synthesis and graphene reinforcement in these resin matrices are discussed. This reinforcement is also achieved through the double-crosslinking effect which can further arrest the movement of polymer chains and improve mechanical strength. Reinforcement in the high molecular weight polymer/graphene nanocomposites, especially polymers that have the SIC followed a similar reinforcing mechanism. Finally, the increment in mechanical properties is much better than that of conventional fillers like CB which poses the problem of higher filler agglomeration. Hence, this two-dimensional filler serves as an excellent nanomaterial for increasing the mechanical properties and thereby reinforcing the polymer matrix.

References

[1] A. Bianco, H.-M. Cheng, T. Enoki, Y. Gogotsi, R.H. Hurt, N. Koratkar, et al., All in the graphene family—a recommended nomenclature for two-dimensional carbon materials, Carbon 65 (2013) 1−6.
[2] D.G. Papageorgiou, I.A. Kinloch, R.J. Young, Mechanical properties of graphene and graphene-based nanocomposites, Prog. Mater. Sci. 90 (2017) 75−127.
[3] S. Ganguly, S. Ghosh, P. Das, T.K. Das, S.K. Ghosh, N.C. Das, Poly(N-vinylpyrrolidone)-stabilized colloidal graphene-reinforced poly(ethylene-co-methyl acrylate) to mitigate electromagnetic radiation pollution, Polym. Bull. 77 (2020) 2923−2943.
[4] S. Ganguly, D. Ray, P. Das, P.P. Maity, S. Mondal, V.K. Aswal, et al., Mechanically robust dual responsive water dispersible-graphene based conductive elastomeric hydrogel for tunable pulsatile drug release, Ultrason. Sonochem. 42 (2018) 212−227.
[5] A. Kumar, R.K. Gupta, Fundamentals of Polymer Engineering, third ed., CRC Press, Boca Raton, FL, 2018.
[6] W. Peng, B. Riedl, Thermosetting resins, J. Chem. Educ. 72 (1995) 587.
[7] A.R. Smith, A review of the chemistry of thermosetting resins and related compounds and their application to textiles, J. Soc. Dyers Colour. 77 (1961) 416−434.
[8] A. Noparvar-Qarebagh, H. Roghani-Mamaqani, M. Salami-Kalajahi, Novolac phenolic resin and graphene aerogel organic-inorganic nanohybrids: High carbon yields by resin modification and its incorporation into aerogel network, Polym. Degrad. Stab. 124 (2016) 1−14.
[9] Y. Huang, J. Yang, H. Cai, Y. Zhai, D. Feng, Y. Deng, et al., A curing agent method to synthesize ordered mesoporous carbons from linear novolac phenolic resin polymers, J. Mater. Chem. 19 (2009) 6536−6541.
[10] A. Pizzi, C.C. Ibeh, 2—Phenol−formaldehydes, in: H. Dodiuk, S.H. Goodman (Eds.), Handbook of Thermoset Plastics, third ed., William Andrew Publishing, Boston, MA, 2014, pp. 13−44.
[11] R. Ghafari, K. DoostHosseini, A. Abdulkhani, S.A. Mirshokraie, Replacing formaldehyde by furfural in urea formaldehyde resin: effect on formaldehyde emission and physical−mechanical properties of particleboards, Eur, J. Wood Wood Prod. 74 (2016) 609−616.
[12] S. Sharma, V. Choudhary, Poly(melamine-formaldehyde) microcapsules filled with epoxy resin: effect of M/F ratio on the shell wall stability, Mater. Res. Express 4 (2017) 075307.
[13] F.-L. Jin, X. Li, S.-J. Park, Synthesis and application of epoxy resins: a review, J. Ind. Eng. Chem. 29 (2015) 1−11.

[14] S. Sprenger, Epoxy resin composites with surface-modified silicon dioxide nanoparticles: A review, J. Appl. Polym. Sci. 130 (2013) 1421—1428.

[15] W. Smith, Resin Systems, Delaware Composites Design Encyclopedia, first ed., Routledge, 2017, pp. 13—86.

[16] Y. Gao, H. Zhang, M. Huang, F.J.C. Lai, Unsaturated polyester resin concrete: a review, Constr. Build. Mater. 228 (2019) 116709.

[17] C. Robeyns, L. Picard, F. Ganachaud, Synthesis, characterization and modification of silicone resins: an "Augmented Review," Prog. Org. Coat. 125 (2018) 287—315.

[18] D. Cai, M. Song, Recent advance in functionalized graphene/polymer nanocomposites, J. Mater. Chem. 20 (2010) 7906—7915.

[19] F. Wang, L.T. Drzal, Y. Qin, Z. Huang, Effects of functionalized graphene nanoplatelets on the morphology and properties of epoxy resins, High. Perform. Polym. 28 (2016) 525—536.

[20] J. Dai, C. Peng, F. Wang, G. Zhang, Z.J.J.O.N. Huang, Effects of functionalized graphene nanoplatelets on the morphology and properties of phenolic resins, J. Nanomater. 2016 (2016) 3485167.

[21] W. Duan, Y. Chen, J. Ma, W. Wang, J. Cheng, J. Zhang, High-performance graphene reinforced epoxy nanocomposites using benzyl glycidyl ether as a dispersant and surface modifier, Compos. B 189 (2020) 107878.

[22] D. Gui, W. Xiong, G. Tan, S. Li, X. Cai, J. Liu, Improved thermal and mechanical properties of silicone resin composites by liquid crystal functionalized graphene nanoplatelets, J. Mater. Sci.—Mater. Electron. 27 (2016) 2120—2127.

[23] D. Gui, S. Yu, W. Xiong, X. Cai, C. Liu, J. Liu, Liquid crystal functionalization of graphene nanoplatelets for improved thermal and mechanical properties of silicone resin composites, RSC Adv. 6 (2016) 35210—35215.

[24] G. Ren, Z. Zhang, X. Zhu, B. Ge, F. Guo, X. Men, et al., Influence of functional graphene as filler on the tribological behaviors of Nomex fabric/phenolic composite, Compos. A 49 (2013) 157—164.

[25] S.H. Song, K.H. Park, B.H. Kim, Y.W. Choi, G.H. Jun, D.J. Lee, et al., Enhanced thermal conductivity of epoxy—graphene composites by using non-oxidized graphene flakes with non-covalent functionalization, Adv. Mater. 25 (2013) 732—737.

[26] Z. Yang, J. Liu, R. Liao, G. Yang, X. Wu, Z. Tang, et al., Rational design of covalent interfaces for graphene/elastomer nanocomposites, Compos. Sci. Technol. 132 (2016) 68—75.

[27] G. Zhang, F. Wang, J. Dai, Z.J.M. Huang, Effect of functionalization of graphene nanoplatelets on the mechanical and thermal properties of silicone rubber composites, Materials 9 (2016) 92.

[28] X. Yi, A. Feng, W. Shao, Z. Xiao, Synthesis and properties of graphene oxide—boron-modified phenolic resin composites, High. Perform. Polym. 28 (2015) 505—517.

[29] J. Zhou, Z. Yao, Y. Chen, D. Wei, Y. Wu, T. Xu, Mechanical and thermal properties of graphene oxide/phenolic resin composite, Polym. Compos. 34 (2013) 1245—1249.

[30] B. Wu, L. Ye, Y. Liu, X. Zhao, Intercalation structure and toughening mechanism of graphene/urea-formaldehyde nanocomposites prepared via in situ polymerization, Polym. Int. 67 (2018) 330—339.

[31] B. Wu, Y. Liu, Y. Shu, L. Ye, X. Zhao, Intrinsic flame-retardant urea formaldehyde/graphene nanocomposite foam: Structure and reinforcing mechanism, Polym. Compos. 40 (2019) E811—E820.

[32] C. Wu, Z. Chen, F. Wang, Y. Hu, E. Wang, Z. Rao, et al., Mechanical and drilling properties of graphene oxide modified urea-melamine-phenol formaldehyde composites reinforced by glass fiber, Compos. B 162 (2019) 378—387.

[33] T.-Y. Wang, P.-Y. Tseng, J.-L. Tsai, Characterization of Young's modulus and thermal
      conductivity of graphene/epoxy nanocomposites, J. Compos. Mater. 53 (2018)
      835−847.
[34] Y. Sun, L. Chen, L. Cui, Y. Zhang, X. Du, Molecular dynamics simulation of cross-
      linked epoxy resin and its interaction energy with graphene under two typical force
      fields, Comput. Mater. Sci. 143 (2018) 240−247.
[35] V.B. Mohan, K.-T. Lau, D. Hui, D. Bhattacharyya, Graphene-based materials and their
      composites: a review on production, applications and product limitations, Compos. B
      142 (2018) 200−220.
[36] J. Sun, J. Ji, Z. Chen, S. Liu, J. Zhao, Epoxy resin composites with commercially avail-
      able graphene: toward high toughness and rigidity, RSC Adv. 9 (2019) 33147−33154.
[37] H.B. Kulkarni, P. Tambe, G.M. Joshi, Influence of covalent and non-covalent modifica-
      tion of graphene on the mechanical, thermal and electrical properties of epoxy/graphene
      nanocomposites: a review, Compos. Interfaces 25 (2018) 381−414.
[38] M.-Y. Shen, T.-Y. Chang, T.-H. Hsieh, Y.-L. Li, C.-L. Chiang, H. Yang, et al.,
      Mechanical properties and tensile fatigue of graphene nanoplatelets reinforced polymer
      nanocomposites, J. Nanomater. 2013 (2013) 565401.
[39] H. Saleem, A. Edathil, T. Ncube, J. Pokhrel, S. Khoori, A. Abraham, et al., Mechanical
      and thermal properties of thermoset−graphene nanocomposites, Macromol. Mater.
      Eng. 301 (2016) 231−259.
[40] X. Wang, L. Song, W. Pornwannchai, Y. Hu, B. Kandola, The effect of graphene pres-
      ence in flame retarded epoxy resin matrix on the mechanical and flammability proper-
      ties of glass fiber-reinforced composites, Compos. A 53 (2013) 88−96.
[41] M.-T. Le, S.-C.J.M. Huang, Thermal and mechanical behavior of hybrid polymer nano-
      composite reinforced with graphene nanoplatelets, Materials 8 (2015) 5526−5536.
[42] D.R. Son, A.V. Raghu, K.R. Reddy, H.M. Jeong, Compatibility of thermally reduced
      graphene with polyesters, J. Macromol. Sci. B Phys. 55 (2016) 1099−1110.
[43] S. Pan, I.A. Aksay, R.K. Prud'homme, Multifunctional Graphene-Silicone Elastomer
      Nanocomposite, Method of Making The Same, and Uses Thereof, United States
      Patents, United States 9441076B2, 2016.
[44] W.E. Mahmoud, Morphology and physical properties of poly(ethylene oxide) loaded
      graphene nanocomposites prepared by two different techniques, Eur. Polym. J. 47
      (2011) 1534−1540.
[45] M. Zhang, Y. Li, Z. Su, G. Wei, Recent advances in the synthesis and applications of
      graphene−polymer nanocomposites, Polym. Chem. 6 (2015) 6107−6124.
[46] S. Das, F. Irin, L. Ma, S.K. Bhattacharia, R.C. Hedden, M.J. Green, Rheology and mor-
      phology of pristine graphene/polyacrylamide gels, ACS Appl. Mater. Interfaces 5
      (2013) 8633−8640.
[47] M.A. Milani, D. González, R. Quijada, N.R.S. Basso, M.L. Cerrada, D.S. Azambuja,
      et al., Polypropylene/graphene nanosheet nanocomposites by in situ polymerization:
      synthesis, characterization and fundamental properties, Compos. Sci. Technol. 84
      (2013) 1−7.
[48] Y. Luo, R. Wang, S. Zhao, Y. Chen, H. Su, L. Zhang, et al., Experimental study and
      molecular dynamics simulation of dynamic properties and interfacial bonding charac-
      teristics of graphene/solution-polymerized styrene-butadiene rubber composites, RSC
      Adv. 6 (2016) 58077−58087.
[49] Z.-T. Xie, X. Fu, L.-Y. Wei, M.-C. Luo, Y.-H. Liu, F.-W. Ling, et al., New evidence
      disclosed for the engineered strong interfacial interaction of graphene/rubber nanocom-
      posites, Polymer 118 (2017) 30−39.

[50] C. Li, C. Feng, Z. Peng, W. Gong, L. Kong, Ammonium-assisted green fabrication of graphene/natural rubber latex composite, Polym. Compos. 34 (2013) 88−95.
[51] M.M. Möwes, F. Fleck, M. Klüppel, Effect of filler surface activity and morphology on mechanical and dielectric properties of nbr/graphene nanocomposites, Rubber Chem. Technol. 87 (2013) 70−85.
[52] L. Bokobza, M. Couzi, J.-L. Bruneel, Raman spectroscopy of polymer−carbon nano-material composites, Rubber Chem. Technol. 90 (2016) 37−59.
[53] L. Gong, I.A. Kinloch, R.J. Young, I. Riaz, R. Jalil, K.S. Novoselov, Interfacial stress transfer in a graphene monolayer nanocomposite, Adv. Mater. 22 (2010) 2694−2697.
[54] D. Frasca, D. Schulze, M. Böhning, B. Krafft, B. Schartel, Multilayer graphene chlorine isobutyl isoprene rubber nanocomposites: influence of the multilayer graphene concen-tration on physical and flame-retardant properties, Rubber Chem. Technol. 89 (2016) 316−334.
[55] X. Wu, Y. Zhang, P. Du, Z. Jin, H. Zhao, L. Wang, Synthesis, characterization and properties of graphene-reinforced polyimide coatings, N. J. Chem. 43 (2019) 5697−5705.
[56] L. Tian, E. Jin, H. Mei, Q. Ke, Z. Li, H. Kui, Bio-inspired graphene-enhanced ther-mally conductive elastic silicone rubber as drag reduction material, J. Bionic Eng. 14 (2017) 130−140.
[57] H. Wang, C. Yang, R. Liu, K. Gong, Q. Hao, X. Wang, et al., Build a rigid−flexible graphene/silicone interface by embedding SiO_2 for adhesive application, ACS Omega 2 (2017) 1063−1073.

Electrical and electronic applications of polymer-graphene composites

13

Krishnendu Nath, Suman Kumar Ghosh and Narayan Chandra Das
Rubber Technology Centre, Indian Institute of Technology Kharagpur, Kharagpur, India

13.1 Introduction

Nowadays, material science has drawn impactful attention in to the world of nanomaterials and nanoscience. Various electronic equipment such as computer, mobile phone, laptops, etc. play a significant part in our day to day life and polymers along with conducting nanofillers (such as carbon nanotubes, graphenes, carbon nanofillers, etc.) are raw materials of which the electronic devices are made of. The modern era of nanoscience has already taught us the useful exploitation of polymer by blending it with conducting fillers in the electrical and electronics fields through the application in field effect transistors (FET) [1], gas sensors [2], electromagnetic interference shielding materials [3], photovoltaic cells [4], etc. When polymers are blended with nanofillers (or nanoparticles) they produce polymer nanocomposites. Polymer nanocomposites are basically a composite material produced from incorporating different additives (one of which should have a size in the range of nanometer) in a polymer matrix. The nanosized particles could be one dimensional (e.g., carbon nanotubes, carbon nanofiber), two dimensional (e.g., various graphene based nanosheets), or three dimensional (e.g., various conductive carbon nanoparticles) in geometrical shape. When these nanoscale additives are blended with the polymer matrix the resulting polymer nanocomposite exerts an impressive optical, mechanical [5], thermal [6], and most importantly electrical [7] properties when added even in a very small quantity. The fact of achievement of optimized mechanical, electrical, or thermal properties on the addition of a very little amount of nanofillers is due to larger surface area compared to the volume of the filler particles. Presently, the most common nanofiller that has been extensively used to produce nanocomposites are carbon nanotubes (CNT) but there is a burden of high cost and difficulty in dispersion [8] which lead to imposing restrictions on the use of CNTs. The other most common and popular conductive nanomaterial is graphene which was discovered in 2004 [9] by Novoselov et al. [10]. Compared to CNT two-dimensional graphene sheets exert a cluster of properties such as extreme mechanical durability along with the impressive electrical and thermal conductivity [11] due to its higher surface area compared to the particle volume when incorporated in a polymer or polymer blend. Graphene is a lightweight, structurally two-dimensional

Polymer Nanocomposites Containing Graphene. DOI: https://doi.org/10.1016/B978-0-12-821639-2.00002-1

honeycomb-like arrangement of carbon atoms. There are different carbon allotropes, for example, fullerene, graphite (Fig. 13.1). etc. where graphene structures are main building blocks of those allotropes. The surface area of graphenes are higher compared to CNT as the inner surface of the CNT cylinders remains inaccessible to the polymer chains.

This leads to higher accessibility of surface area of graphene due to its sheet-like structure when compared to cylindrical structure of CNT. In general CNT and fullerenes are synthesized by wrapping up of the basic graphene nanosheet [12] while the graphite has been prepared by stacking up of the graphene nanosheets [13] and interlayer distance of each sheet is approximately 3.37 Å. After graphene was discovered scientists had started to improve the compatibility and dispersion properties of the 2D graphene nanosheets with respect to their application in the polymer [14]. In this regard graphene oxide (GO) [15] along with the reduced graphene oxide (RGO) [16,17] were discovered and gradually they became the most desired material regarding their applications in the nanotechnology [18,19], biology [20,21], nanomedicine [22,23] fields, etc. Electrical and electronic applications of both graphene, RGO, and GO are enormous such as electromagnetic interference shielding material [24], actuators, supercapacitors [25], etc. Besides all of the advantages of incorporation of graphene in polymer nanocomposite, there is a certain disadvantage of agglomeration (which is irreversible) of the nanomaterial while incorporating it in the polymer matrix. Surfactant-assisted modification of the filler [26] could

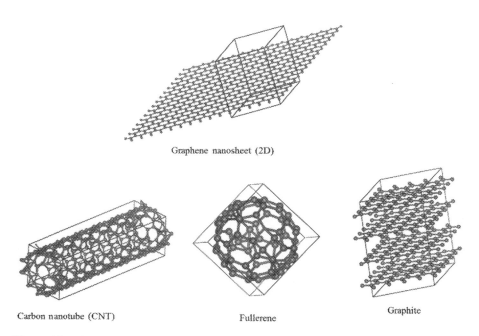

Graphene nanosheet (2D)

Carbon nanotube (CNT)   Fullerene   Graphite

Figure 13.1 Graphene nanosheet and its different derivatives (carbon nanotubes, fullerene, and graphite).

help in this regard. In a typical process of surfactant-assisted graphene modification, the graphene powder is oxidized first and then the GO sample was dispersed in a surfactant solution above its CMC (critical micelle concentration). In this way the surfactant will get physically absorbed on the surface of the nanofilller which helps in the dispersibility of it in the polymer matrix leading to optimized mechanical properties on the resulting polymer nanocomposite. Numerous methods of dispersion of graphene into the polymer matrix will be discussed later in this chapter along with some exciting features of the graphene-based polymer nanocomposite.

13.2   Synthesis of graphene

Some earlier attempts were made in the 1970s [27] by synthesizing thin graphitic film upon the transition metal surfaces [28] and in 1975 to prepare a mixture of single and multilayer graphene by thermal decomposition of ethylene upon the crystal faces of platinum [29]. The process was known as surface graphitization on which the main motive was to study the formation of a graphitic layer upon the crystal structure of the metal to examine the type of arrangement of carbon contaminants (much like graphitic layer). Upon the metal crystal surface, although those processes did not produce any fruitful identification of the product, even the electronic applications were not studied. However, the most effective attempt was made by Novoselov et. al. in 2004 [9]. They had synthesized high quality monocrystalline, two-dimensional semimetallic graphitic films of thickness of a few atoms yet stable under ambient conditions. The films showed an ambipolar electric field due to a small scale overlaping of valence bands with conductance bands. The concentrations of electrons/holes were $\sim 10^{13}/cm^2$ with mobility $\sim 10,000\ cm^2/V$ S at room temperature on applying the gate voltage. There are some other feasible techniques that were developed recently to prepare graphenes, for example − Thermal-assisted chemical vapor deposition technique [30], chemical exfoliation technique [31], mechanical exfoliation technique [32] synthesis based on microwave technique [33], etc. Electrolytic exfoliation technique was also used to produce graphene from graphite using an electrolyte such as poly(sodium-4-styrenesulfonate) [34] and high purity graphite rods (diameter $\sim$ 6 mm) were used as electrodes. In a typical process, 0.001 M solution of Poly(sodium-4-styrenesulfonate) (Mw $\sim 70,000$) was prepared using deionized (DI) water as solvent. The electrolyte solution was then placed inside of an electrolysis cell containing two graphite rods as electrodes. The potential difference was kept constant at 5 V between the electrodes. The electrolysis was continued for 20 min after which appearance of black precipitate at the anode was observed and the exfoliation was continued for around 4 h. After the exfoliation process was completed the dispersion of the product was centrifuged at 1000 rpm for the purpose of the removal of the large agglomerates and hence decanted from the top of the dispersion. Later the dispersion was vacuum dried at oven around 80°C to get the graphene powder. The procedure follwed is portrayed in Fig. 13.2.

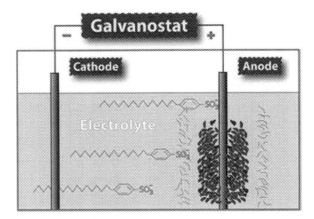

Figure 13.2 Diagram of graphite synthesis using electrolytic exfoliation technique.
Source: Reproduced with the permission from G. Wang, B. Wang, J. Park, Y. Wang, B. Sun, J. Yao, Highly efficient and large-scale synthesis of graphene by electrolytic exfoliation, Carbon 47 (2009) 3242−3246.

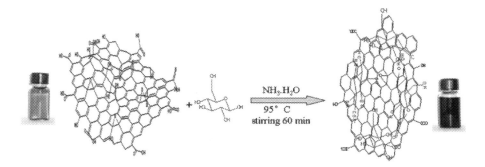

Figure 13.3 Green synthesis of graphene.
Source: Reproduced with the permission from C. Zhu, S. Guo, Y. Fang, S. Dong, Reducing sugar: new functional molecules for the green synthesis of graphene nanosheets, ACS Nano, 4 (2010) 2429−2437.

    Chengzhou Zhu et al. [35] approached a green method to synthsize graphene from the reducing sugar (e.g., glucose, sucrose etc). They have used a precursor that is exfoliated graphite oxide for that purpose. In a typical method, they had prepared a dispersion of GO using graphite as the raw material following modified Hummers method [36]. After the dispersion was prepared glucose along with ammonia solution was added to it. Ammonia was used probably to accelarate the process of reduction by the glucose. Fig. 13.3 represents the proceedure diagramatically.
    In a typical two stage chemical exfoliation process, the initial step involves the formation of graphene intercalated compound by reducing the interlayer van der

Waals forces to increase the spacing between the graphite layers [37]. Employment of rapid heating or ultra-sonication further separates the graphenes to a single or few layers. For single layer graphene synthesis ultrasonication is done [38] and to get few layer thick graphene Density Gradient Ultracentrifugation [39] method is employed. Graphene could be modified to further exfoliate, or to optimize dispersion with the polymer matrix for that reasons graphenes could be oxidized to GO which is synthesized exploiting the Hummers method [38] to oxidize the graphite powder. To increase the interlayer spacing, graphite powders are ultrasonicated in a DMF (dimethyl formamide)/water mixture (9:1) is followed. Present day researchers follow modified hummers method (15 to achieve more efficient oxidation of graphite by excluding $NaNO_3$ along with excess $KMnO_4$ in a 9:1 mixture of H_2SO_4/H_3PO_4 solution). The method improves the hydrophilicity of the product GO when compared to GO synthesized following Hummers' method. On the other-hand, electrical conductivity could be further increased by reducing GO to RGO. Shin et al. [40] Performed comparative study of reducing GO with the aid of $NaBH_4$ (sodium borohydride) and N_2H_4 (hydrazine) and it was found out that the former treatment increased the electrical conductivity more effectively than in the later case. Let us discuss some applications along with the neccesary modification of graphenes.

13.3 Applications of polymer-graphene composites in electrical and electronic fields

The applications of the nanocomposites in the field of electronics are vast. Graphene shows high elecrical mobility around 25 $m^2/V/s$ [41] along with a good electrical conductivity property of around 6500 S/m [42]. Graphene showed good electrical conductivity even with a minimum filler content in the polymer/polymer blend [43]. Due to those above mentioned electrical properties, graphene is becoming a very important nanomaterial. Graphene and its modified versions (GO or RGO) have various applications which were discussed earlier in the introduction section and will be elaborated on in the present section.

13.3.1   Field effect transistors

FET exploits an electric field to manipulate the flow of current. Mainly, semiconductors (such as silicon) are used to produce FETs. The first graphene based FETs in the integrated circuits were produced by IBM in 2011 [44]. There are some biological applications such as biosensor [45], protein detectors [46], protein sensor [47], real time DNA detecting device [48], electronic olfactory sensor [49], etc. RGO could be exploited as a transducer in the field emission transducers as H_2O_2 biosensor when modified with polypyrrole nanotube [50] where the scheme given below (Fig. 13.4) had been for the modification of the GO nanosheets.

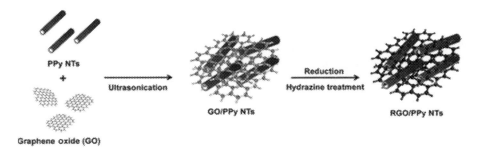

Figure 13.4 Modification of reduced graphene oxide.
Source: Reproduced with the permission from J.W. Park, S.J. Park, O.S. Kwon, C. Lee, J. Jang, Polypyrrole nanotube embedded reduced graphene oxide transducer for field-effect transistor-type H_2O_2 biosensor, Anal. Chem. 86 (2014) 3, 1822—1828.

After the particle was modified, a glass substrate was designed by a microarray containing almost 80 pairs of gold impregnated microelectrodes. The microelectrode substrate was then subjected to a treatment of 0.1% aqueous solution of the modified graphene nanoparticle composite (RGO/PPy NT) and dried in oven. Electrical features of the RGO/PPY-NT composite particles were characterized using a current-voltage ($I–V$) plots applying a range of -0.4 to $+0.4$ V in the Fig. 13.5A where the RGO/PPy NT gave an improved result when compared to RGO or PPy NTs in terms of electrical conductivity. The experiment suggests that there was a favorable electron conduction between the PPy NT and RGO. The transistors containing rGO/PPy NT channels were subjected to the source-drain current versus source drain voltage ($I_{SD} – V_{SD}$) measurement under various gate biases to evaluate their electrical properties. The ranges of the V_g (gate voltage) employed was from -2.0 to $+0.4$ V, with a gate voltage sweep rate of 0.2 V/s. The whole device was put in an electrolyte solution (phosphate-buffered saline), from the Fig. 13.6B. we can observe that I_{SD} increased on the application of more negative bias. So it was noticeably behaving like a p-channel transistor due to PPy modification on the RGO because conjugated polymers behave like semiconductors.

The FET was immersed in a PBS electrolyte as shown in the Fig. 13.6A. we can see that the sensitivity refers to the change $\Delta I_{SD}/I_0 = (I_{SD} - I_0)/I_0$, here I_0 equals to the initially applied current and I_{SD} equals to the current measured after stabilization on different H_2O_2 concentration. the value of I_{SD} is directly proportional to the concentration of H_2O_2.

In another example Cai et al. had produced FET biosensor device based on RGO [45] which was ultrasensitive to detect DNA via PNA — DNA hybridization. The RGO suspension was drop casted on the sensor surface to modify the same. The schematic method is given in Fig. 13.7 below.

The capture probe is PNA instead of DNA, and DNA was detected by observing the PNA-DNA hybridization in the FET biosensor. The lowest detection limit was 100 fM. Here the RGO-based FET biosensor easily detected the target DNA

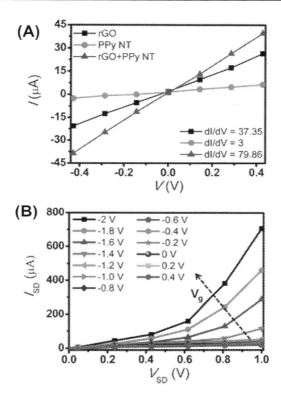

Figure 13.5 (A) I versus V curve for different experimental channels (B) I_{SD} versus V_{SD} curves.
Source: Reproduced with the permission from J.W. Park, S.J. Park, O.S. Kwon, C. Lee, J. Jang, Polypyrrole nanotube embedded reduced graphene oxide transducer for field-effect transistor-type H_2O_2 biosensor, Anal. Chem. 86 (2014) 3, 1822−1828.

keeping aside the noncomplementary DNA as shown in the above picture. The fabricated DNA biosensor had the capability to regenerate the results, had a potential application to diagnose diseases based on its ultrasensitivity and exquisite specificity. RGO-FET could also be used to detect strain sensing property [51] which could observe the slow or rapid physical movement in the human body, sensing variation in temperature and infrared radiation [52], presence of toxic Pb^{2+} ion in the aqueous solution where RGO had been used as a semiconducting channel material [53] where in a typical process reduced GO/l-glutathione reduced gold nanoparticle (NP) hybrid structure was developed for applications to sense water following the process given below (Fig. 13.8).

The hybrid structure was successfully applied as a water sensing FET device as depicted in the following figure (Fig. 13.9). There are other various applications of graphene as the biosensor material [54] but in the present section we had just covered only a fraction of it.

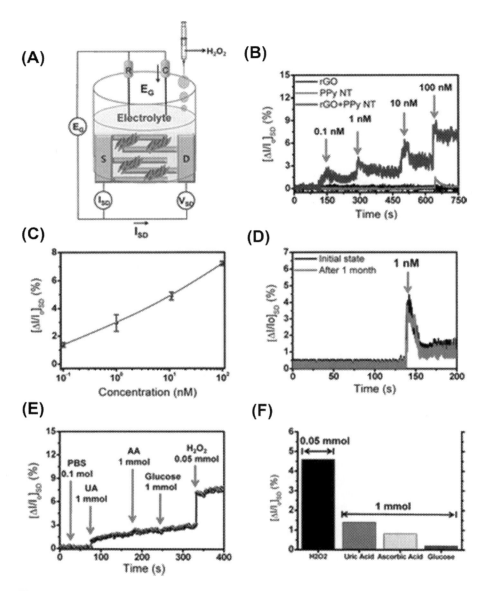

Figure 13.6 (A) The field effect transistors setup inside the PBS electrolyte (B) real time H_2O_2 sensing response (C) a calibration curve ($V_g = 0.1$ V) at different H_2O_2 concentrations varied from 0.1 to 100 nM (D) long time storage stability of the of the produced sensor based on its sensing performance (E) Responses of the RGO/PPy-NT based sensor toward different substances (F) a histogram depicting real time response of sensor toward the mentioned substances.

Source: Reproduced with the permission from J.W. Park, S.J. Park, O.S. Kwon, C. Lee, J. Jang, Polypyrrole nanotube embedded reduced graphene oxide transducer for field-effect transistor-type H_2O_2 biosensor, Anal. Chem. 86 (2014) 3, 1822−1828.

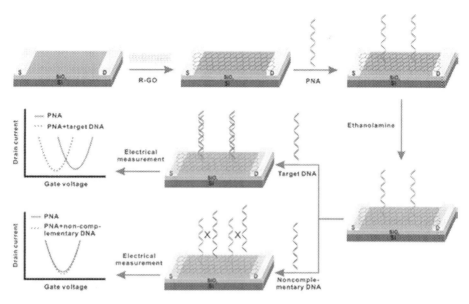

Figure 13.7 Method of reduced graphene oxide-based field effect transistors preparation. *Source*: Reproduced with the permission from B. Cai, S. Wang, L. Huang, Y. Ning, Z. Zhang, G.J. Zhang, Ultrasensitive label-free detection of PNA-DNA hybridization by reduced graphene oxide field-effect transistor biosensor, ACS Nano, 8 (2014), 2632–2638.

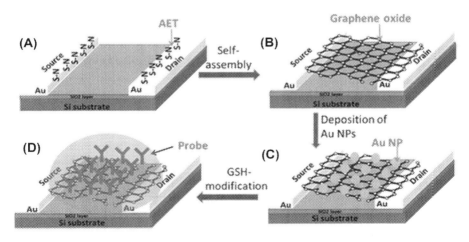

Figure 13.8 Fabrication of reduced graphene oxide/l-glutathione reduced gold nanoparticle hybrid structure. *Source*: Reproduced with the permission from B. Cai, S. Wang, L. Huang, Y. Ning, Z. Zhang, G.J. Zhang, Ultrasensitive label-free detection of PNA-DNA hybridization by reduced graphene oxide field-effect transistor biosensor, ACS Nano, 8 (2014), 2632–2638.

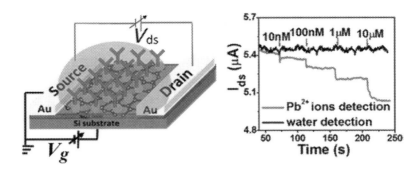

Figure 13.9 Detection of the Pb^{2+} in water.
Source: Reproduced with the permission from B. Cai, S. Wang, L. Huang, Y. Ning, Z. Zhang, G.J. Zhang, Ultrasensitive label-free detection of PNA-DNA hybridization by reduced graphene oxide field-effect transistor biosensor, ACS Nano, 8 (2014), 2632–2638.

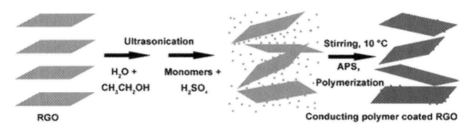

Figure 13.10 Process of reduced graphene oxide /conductive polymer composite synthesis.
Source: Reproduced with the permission J. Zhang, X.S. Zhao, Conducting polymers directly coated on reduced graphene oxide sheets as high-performance supercapacitor electrodes, J. Phys. Chem. C. 116 (2012) 5420–5426.

13.3.2 Supercapacitors

Supercapacitors are the special grade of capacitors which has high capacitance value compared to the normal capacitors with very low voltage limit. Super capacitors can store about 20–90 times more electricity per unit mass or volume compared to that of the normal capacitors. The charge delivery rate of supercapacitors is much faster than that of batteries. Graphene has tons applications in the field of supercapacitance; for example RGO/conductive polymer composite could be used as high performance super capacitor material [55]. In a typical process, RGO sheets were ultrasonicated in acidic solution, then conducting polymers (poly aniline, poly (3,4-ethylenedioxythiophene) (PEDOT)) are synthesized in presence of the RGO particles. The diagram is given below (Fig. 13.10).

Cyclic voltametric diagrams of the prepared composites (Fig. 13.11) showing a typical rectangular shape suggesting about good capacitance of the supercapacitor.

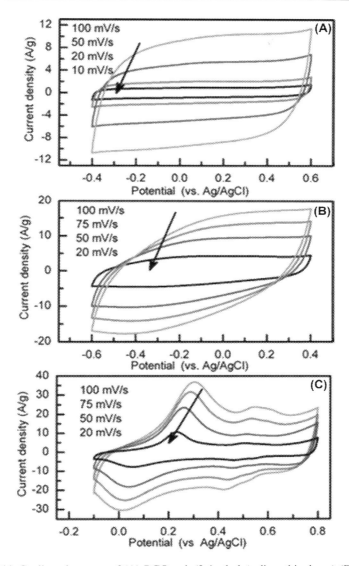

Figure 13.11 Cyclic voltagrams of (A) RGO-poly(3,4-ethylenedioxythiophene) (B) RGO-polypyrrole (PPy) and (C) RGO-Polyaniline (PANi) composites. *RGO*, reduced graphene oxide.
Source: Reproduced with the permission from J. Zhang, X.S. Zhao, Conducting polymers directly coated on reduced graphene oxide sheets as high-performance supercapacitor electrodes, J. Phys. Chem. C. 116 (2012) 5420−5426.

In another example Zhang et al. produced GO/polypyrrole composites which exhibited high electrocapacitive performance [56]. This composite material possessed specific capacitance over 500 F/g. Here layered graphene oxide sandwiched

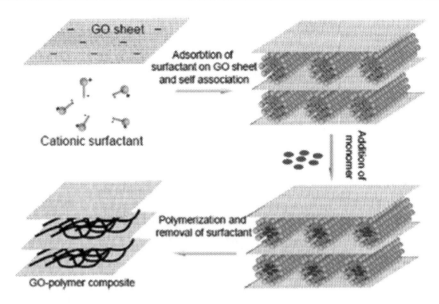

Figure 13.12 Schematic of GOPPy composite formation. *GOPPy*, Graphene oxide-polypyrrole.
Source: Reproduced with the permission from L.L. Zhang, S. Zhao, X.N. Tian, X.S. Zhao, Layered graphene oxide nanostructures with sandwiched conducting polymers as supercapacitor electrodes, Langmuir, 26 (2010) 17624–17628.

conducting polymer composite preparation was approached based on the concept of electrostatic interaction between positively charged surfactant molecules with graphene oxide sheet having negative charge. Fig. 13.12 demonstrates the formation procedure of GO-polypyrrole (GOPPy) composite whereas Fig. 13.13 shows the electrochemical properties of prepared composites and also a comparison of this properties for different sets of sample such as GOPPy-F, GOPPy-S and PPy-F. Fig. 13.13D clearly shows excellent capacitive performance of the prepared composites. The effective capacitance of the GOPPy-F is due to the contribution of the conductive PPy fibers in the corresponding composite used as electrode. Sample GOPPy-F exhibits an effective specific capacitance of 510 F/g. Also, this sample showed a good capacitance retention ratio on increment of 17 times of current density.

Single layered two dimensional structure of graphene has various technological advantages compared to other nanomaterials. But sometimes it loses its identity and other advantages due to restacking of layers in most of the solution-based scalable synthesis processes. In a typical synthesis process melamine resin (Mr) monomer was used to inhibit the restaking of graphene molecules. Here, GO dispersion was added to the melamine resin monomer solution [57]. Then Mr-GO solution was dried and finally reduced thermally. The final cMR-rGO$_{th}$ carbonaceous composite

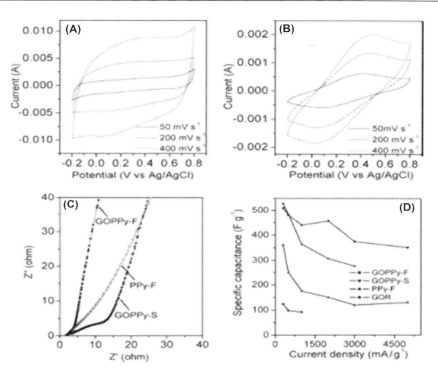

Figure 13.13 CV profiles of (A) GOPPy-F and (B) PPy-F measured at different sweep rates. (C) Nyquist plots of GOPPy-F and PPy-F. (D) Specific capacitance as a function of current density. *GOPPy*, Graphene oxide-polypyrrole.
Source: Reproduced with the permission from L.L. Zhang, S. Zhao, X.N. Tian, X.S. Zhao, Layered graphene oxide nanostructures with sandwiched conducting polymers as supercapacitor electrodes, Langmuir, 26 (2010) 17624–17628.

exhibited exquisite performance due to its high surface area (1040 m^2/g) as an electrode material (Fig. 13.14) [57].

The specific capacitance of the composite was found to be 210 F/g with good retention of capacitance after 20,000 cycles. Here, galvanostatic measurement was done keeping the current density same around 5 A/g which was clearly shown from Fig. 13.15A. cMR-rGO$_{th}$ showed more effective capacitance compared to rGO$_H$ as indicated by higher charging and discharging period. Also, cMR-rGO$_{th}$ possesses higher capacitance than that of CNT-rGO$_{th}$. The specific capacitance of this composite was two times higher than that of rGO$_H$. This increased capacitance might be due to morphological effect as well as Mr-assisted restaking inhibition on the capacitance.

Fig. 13.16 represents the cyclic performance of the composites at two different current density such as 0.5 and 2 A/g. From the figure it was clear that this composite material exhibited excellent cyclic performance for both current densities. For

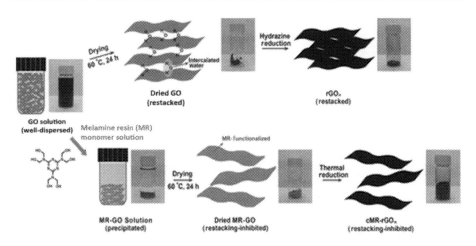

Figure 13.14 Schematic representation showing synthesis pathway of rGO$_H$ and cMR-rGO$_{th}$. *Source:* Reproduced with the permission from J.H. Lee, N. Park, B.G. Kim, D.S. Jung, K. Im, J. Hur, et al., Restacking-inhibited 3D reduced graphene oxide for high performance supercapacitor electrodes, ACS Nano 7 (2013) 9366–9374.

the first case, in the first 5000 cycles the specific capacitance increased to 120% of its initial value whereas for the latter case cMR-rGO$_{th}$ showed good durance of capacitance (96%) with almost no loss after 20,000 cycles.

13.3.3 Electromagnetic interference shield materials

Excessive utilization of commercially available electronic gadgets and communication instruments causes electromagnetic pollution and has a hazardous effect on humans [3]. These electrical and electronic devices emit harmful electromagnetic radiation in wavy form which not only deteriorates normal functioning of nearby other electronic gadgets but also causes human health issues [58]. This interference with other electronic devices normally affecting hardware malfunctioning and inaccurate signaling are generally named as electromagnetic interference or simply EMI. Reducing electromagnetic pollution is a big challenging task in recent years. For this shielding of electromagnetic radiation by some substances is an effective way to mitigate this pollution. Electrically conductive metals are utilized as EMI shielding material. But they suffer some disadvantages such as high specific gravity, corrosive, low flexibility and poor processability. Conducting polymer can be used as efficient EMI shields but they possess some disadvantageous properties which limit their application in such field [59]. In this respect, polymers with conducting carbonaceous fillers play an important role. Among several carbonaceous fillers RGO, GO and graphene nanoplatelets (GNP) are extensively used to prepare flexible, light weight polymer composite which could be utilized as efficient EMI shielding material. These graphene material imparts conductivity in composite materials due to their exceptional

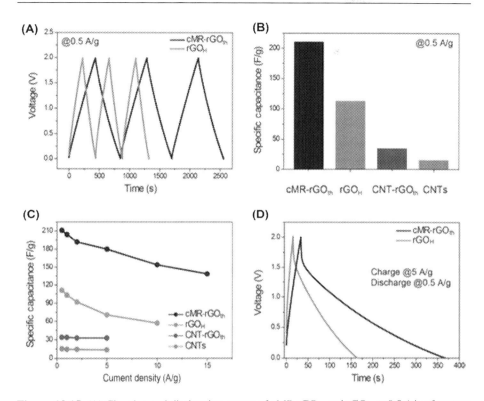

Figure 13.15 (A) Charging and discharging curves of cMR-rGO$_{th}$ and rGO$_H$ at 0.5 A/g of current density. (B) The specific capacitances of the samples measured at 0.5 A/g. (C) Gravimetric capacitances at different current density. (D) Fast charging and slow discharging tests of cMR-rGO$_{th}$ and rGO$_H$ at current densities of 5.0 and 0.5 A/g respectively.
Source: Reproduced with the permission from J.H. Lee, N. Park, B.G. Kim, D.S. Jung, K. Im, J. Hur, et al., Restacking-inhibited 3D reduced graphene oxide for high performance supercapacitor electrodes, ACS Nano 7 (2013) 9366–9374.

electrical properties [60]. These polymer-graphene nanocomposites can be prepared by three different processes like solution blending, melt mixing and in situ polymerization techniques. In a typical process Bhawal et al. prepared polymer-graphene composites via melt mixing of polyethylene-co-methyl acrylate (EMA) and graphene oxide [61]. Here GO was in situ reduced during melt mixing at high temperature (200°C). The pausible mechanism of in situ reduction of GO to in situ reduced graphene oxide (IRGO) via "oxidation debris" theory is represented pictorially in Fig. 13.17.

The resultant nanocomposites (EIRGO) exhibited improved EMI shielding effectiveness along with controlled electromechanical properties. IRGO content was varied from 1 to 7wt.%. The composite material possessed a very percolation threshold of 1.98% IRGO. With increase in IRGO content total EMI shielding efficiency

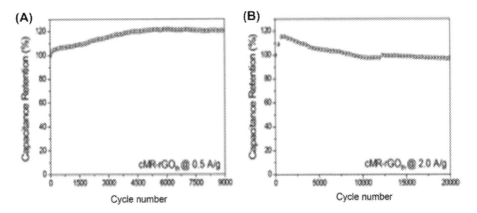

Figure 13.16 The cycling performance of cMR-rGO$_{th}$ measured at (A) 0.5 and (B) 2.0 A/g. *Source*: Reproduced with the permission from J.H. Lee, N. Park, B.G. Kim, D.S. Jung, K. Im, J. Hur, et al., Restacking-inhibited 3D reduced graphene oxide for high performance supercapacitor electrodes, ACS Nano 7 (2013) 9366−9374.

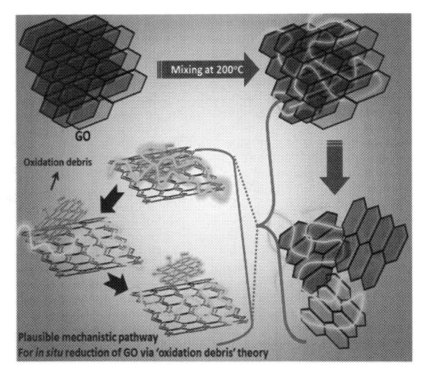

Figure 13.17 Schematic in situ reduction of grapheme oxide in polyethylene-co-methyl acrylate during melt mixing. *Source*: Reproduced with the permission from P. Bhawal, S. Ganguly, T.K. Das, S. Mondal, S. Choudhury, N.C. Das, Superior electromagnetic interference shielding effectiveness and electro-mechanical properties of EMA-IRGO nanocomposites through the in-situ reduction of GO from melt blended EMA-GO composites, Compos. B Eng. 134 (2018) 46−60.

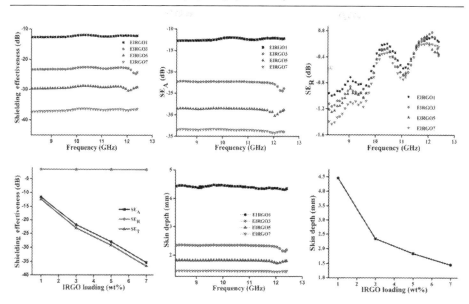

Figure 13.18 (A) Total SE (SE_T) with X-band frequency, (B) microwave absorption (SE_A) with frequency, (C) microwave reflection (SE_R) with frequency, (D) comparison of SE_T, SE_A, SE_R as a function of IRGO loading measured at fixed frequency of 10 GHz, (E) variation of skin depth with frequency, (F) skin depth as a function of IRGO loading.
Source: Reproduced with the permission from P. Bhawal, S. Ganguly, T.K. Das, S. Mondal, S. Choudhury, N.C. Das, Superior electromagnetic interference shielding effectiveness and electro-mechanical properties of EMA-IRGO nanocomposites through the in-situ reduction of GO from melt blended EMA-GO composites, Compos. B Eng. 134 (2018) 46−60.

increases. Composite containing 7wt.% of IRGO showed SE of -34.9 dB measured in the X-band of radiation frequency arena (8.2−12.4 GHz). EMI shielding route was governed by absorption mechanism rather than reflection as graphene molecules from effective conductive network due compactly situated graphene sheets. The results of the EIRGO composites are portrayed below in Fig. 13.18.

Sometimes graphene surface was modified using some surfactants. Hsiao et al. synthesized water bome-polyurethane (WPU)/GO composites by following the layer-by-layer (L-b-L) stacking of intrinsic negatively charged graphene oxide and positively charged cationic surfactants (didodecyldimethylammonium bromide, DDAB) imbibed graphene oxide [62].

These two oppositely charged GO were deposited upon the polyurethane fibers (negatively charged). Then WPU fiber matrix was wrapped completely by GO bilayers and finally fine connections were revealed by electrospum WPU fibers. Finally hydroiodic acid (HI) was used to reduce GO to rGO. The whole preparation procedure was represented in Fig. 13.19. Periods of L-b-L stacking were varied up to 20 times. The resultant rGO/WPU-15 composite portrayed effective electrical conductivity around 17 S/m.

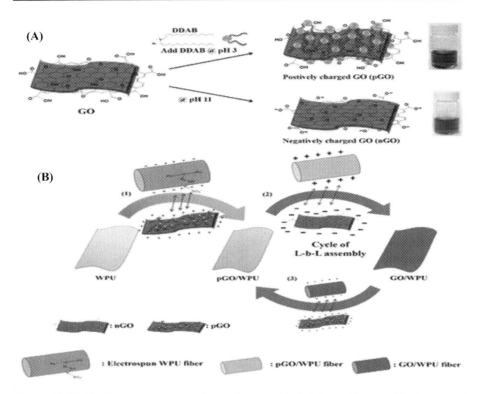

Figure 13.19 (A) Strategy to prepare the portions used in L-b-L stacking: positively charged portion, pGO, and negatively charged portion, nGO. (B) Strategic procedure of L-b-L stacking of pGO and nGO on the WPU fiber. The development order included three steps: (1) accumulation of WPU and pGO and synthesis of pGO/WPU; (2) accumulation of pGO/WPU and nGO. The first cycle of L-b-L stacking is over when the GO/WPU composite is produced. (3) The GO/WPU composite is continually used in the next period of L-b-L stacking, which is imitated as important to obtain the aimed times. *GO*, Graphene oxide; *WPU*, water bome-polyurethane.
Source: Reproduced with the permission from S.T. Hsiao, C.C.M. Ma, W.H. Liao, Y.S. Wang, S.M. Li, Y.C. Huang, et al., Lightweight and flexible reduced graphene oxide/waterborne polyurethane composites with high electrical conductivity and excellent electromagnetic interference shielding performance, ACS Appl. Mater. Interfaces 6 (2014) 10667−10678.

With 20 periods of L-b-L stacking the nanocomposite showed a total EMI SE of 34 dB for only 1 mm sample thickness. With increase in number of periods of L-b-L stacking shielding mechanism was governed by absorption of EM waves. EMI shielding results are shown in Fig. 13.20.

Ganguly et al. prepared polyvinyl pyrrolideone (PVP)-stable graphene dispersion and used this stabilized graphene dispersion as conducting particle in EMA matrix to produce highly conductive, efficient EMI shielding material via wet mixing

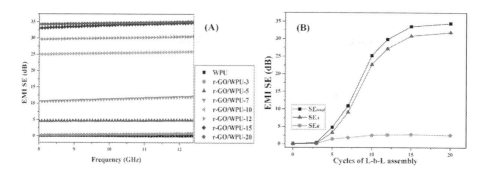

Figure 13.20 (A) Electromagnetic interference shielding effectivity of rGO/WPU composites measured over the frequency range of 8.2−12.4 GHz. (B) Change in total shielding effectivity, shielding by absorption and shielding by reflection mechanism with periods of L-b-L stacking.
Source: Reproduced with the permission from S.T. Hsiao, C.C.M. Ma, W.H. Liao, Y.S. Wang, S.M. Li, Y.C. Huang, et al., Lightweight and flexible reduced graphene oxide/waterborne polyurethane composites with high electrical conductivity and excellent electromagnetic interference shielding performance, ACS Appl. Mater. Interfaces 6 (2014) 10667−10678.

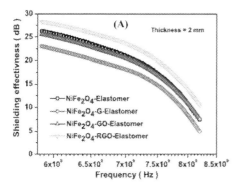
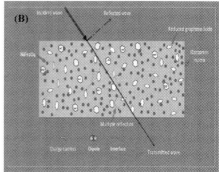

Figure 13.21 (A) Total electromagnetic interference shielding effectivity of graphene-polyethylene-co-methyl acrylate nanocomposite sample of 2 mm thickness. (B) Strategic attenuation of electromagnetic interference shielding in $NiFe_2O_4$-RGO-elastomer nanocomposite sheet.
Source: Reproduced with the permission from R.S. Yadav, I. Kuřitka, J. Vilcakova, D. Skoda, P. Urbánek, M. Machovsky, et al., Lightweight $NiFe_2O_4$-reduced graphene oxide-elastomer nanocomposite flexible sheet for electromagnetic interference shielding application, Compos. B Eng. 166 (2019) 95−111.

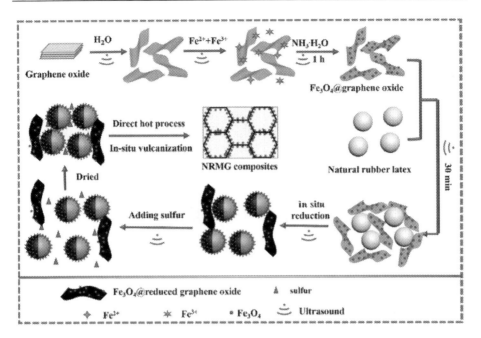

Figure 13.22 Strategic representation of production of NR/Fe$_3$O$_4$-Rgo composite.
Source: Reproduced with the permission from Y. Zhan, J. Wang, K. Zhang, Y. Li, Y. Meng, N. Yan, et al., Fabrication of a flexible electromagnetic interference shielding Fe$_3$O$_4$@reduced graphene oxide/natural rubber composite with segregated network, Chem. Eng. J. 344 (2018) 184—193.

method [63]. This light weight, highly flexible composite material possessed a DC conductivity of 0.259 S/cm and frequency independent electroconducting behavior at higher concentration. As graphene content in nanocomposite increased EMI shielding effectiveness increased. Highest content of graphene filled EMA composite material showed EMI SE of −30 dB deduced in X-band frequency arena (8.2—12.4 GHz) for only 1 mm of sample thickness. Here also EMI shielding attenuation was dominated by absorption rather than transmission with filler loading. This nanocomposite material could be an alternate for heavy weighted EMI shields.

To improve the electromagnetic properties along with EMI shielding efficiency of the polymer nanocomposites some magnetic NP such as Fe$_3$O$_4$, NiFe$_2$O$_4$ were added along with graphene molecules. These materials enhanced the magnetic contribution of the EMI shielding. These magnetic materials were added directly with graphene or prepared via in situ method on graphene surface. Yadav et al. prepared polypropylene-based elasomeric nanocomposites by using prepared NiFe$_2$O$_4$ spinel ferrite NP along with GO and rGO via melt mixing in extruder [64]. The resultant flexible, light weight nanocomposite sheets exhibited ferromagnetic behavior with high EMI shielding efficiency. NiFe$_2$O$_4$-RGO-Elastomer nanocomposite of 2 mm thick sheet portrayed EMI SE of −28.5 dB measured at frequency arena of

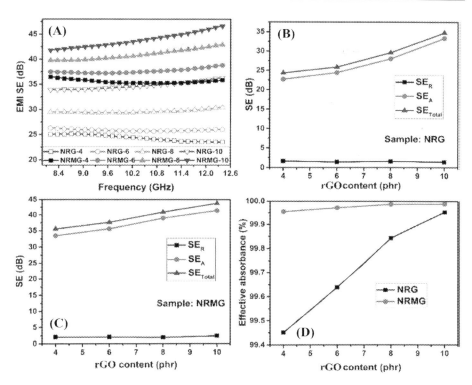

Figure 13.23 (A) Change in EMI shielding effectivity of the NRMG and NRG nanocomposites RGO amount and frequency, (B) Different parts of EMI shielding and total shielding of NRG nanocomposites. (C) Different parts of EMI shielding and total shielding of NRMG nanocomposites (D) Effective absorbance of 1.8 mm thick the NRMG and NRG nanocomposites. *EMI*, electromagnetic interference; *RGO*, reduced graphene oxide.
Source: Reproduced with the permission from Y. Zhan, J. Wang, K. Zhang, Y. Li, Y. Meng, N. Yan, et al., Fabrication of a flexible electromagnetic interference shielding Fe_3O_4@reduced graphene oxide/natural rubber composite with segregated network, Chem. Eng. J. 344 (2018) 184−193.

5.8−8.2 GHz. The results showed that absorption attenuation was a dominant process of EMI shielding. The overall EMI shielding of the prepared nanocomposites and strategic attenuation of EMI shielding in the nanocomposites were shown in Fig. 13.21.

In a typical process Fe_3O_4@GO was prepared, blended with natural rubber latex and then the GO was reduced by in situ and finally composite was vulcanized in a hot press by adding sulphur to the latex [65]. The whole composite preparation was shown in Fig. 13.22. The resultant flexible nanocomposite sheet was named as NRMG. NR/rGO (NRG) Composite without magnetic NP was also prepared and EMI shielding effectiveness for both types of samples were examined and compared.

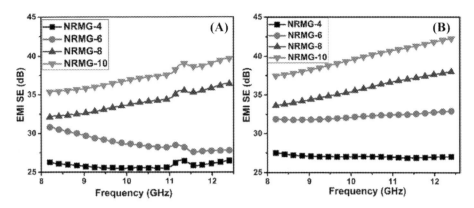

Figure 13.24 EMI shielding effectivity of NRMG nancomposites before (A) and after (B) heat treatment. *EMI*, electromagnetic interference.
Source: Reproduced with the permission from Y. Zhan, J. Wang, K. Zhang, Y. Li, Y. Meng, N. Yan, et al., Fabrication of a flexible electromagnetic interference shielding Fe_3O_4@reduced graphene oxide/natural rubber composite with segregated network, Chem. Eng. J. 344 (2018) 184−193.

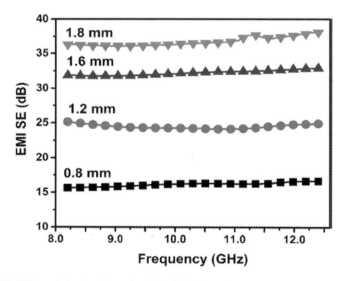

Figure 13.25 Effect of the depth on the EMI shielding effectivity of NRMG-6 composites after heat treatment.
Source: Reproduced with the permission from Y. Zhan, J. Wang, K. Zhang, Y. Li, Y. Meng, N. Yan, et al., Fabrication of a flexible electromagnetic interference shielding Fe_3O_4@reduced graphene oxide/natural rubber composite with segregated network, Chem. Eng. J. 344 (2018) 184−193.

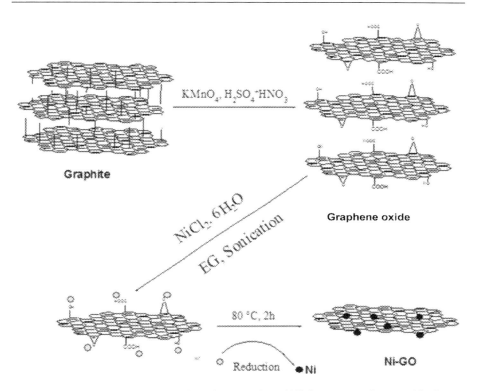

Figure 13.26 Schematic representation of preparation of Ni-decorate grapheme oxide sheets.
Source: Reproduced with the permission from P.K.S. Mural, S.P. Pawar, S. Jayanthi, G. Madras, A.K. Sood, S. Bose, Engineering nanostructures by decorating magnetic nanoparticles onto graphene oxide sheets to shield electromagnetic radiations, ACS Appl. Mater. Interfaces 7, (2015) 16266−16278.

In the former case decoration of Fe_3O_4 magnetic nanaoparticles facilitated higher magnetic property of the nanocomposite material. Electrical activities and EMI shielding effectivity of NRMG nanocomposites were much improved compared to that of NRG nanocomposites. 10 phr (part per hundred parts of rubber) of RGO portrayed EMI shielding effectivity of −48 dB, almost 1.4 times larger than that of NRG nanocomposite with similar RGO amount measured over X-band (8.2−12.4 GHz). Presence of Fe_3O_4 governs EMI shielding absorption mechanism. Effective magnetic permeability exerted by Fe_3O_4 present in the nanocomposite samples leads to high absorption and low reflection loss of penetrated EM waves by magnetic and ohmic losses. Also, thermally treated NRMG composites showed higher conductivity and better EMI SE than untreated composites. The EMI shielding results of both NRG and NRMG composites were demonstrated in Fig. 13.23 whereas EMI Shielding effectivity of nanocomposites next to heat treatment was shown in Fig. 13.24.

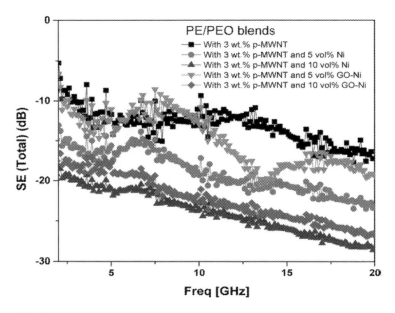

Figure 13.27 Total electromagnetic interference Shielding Effectiveness of Polyethylene/
poly(ethylene oxide) blend systems with various filler loading.
Source: Reproduced with the permission from P.K.S. Mural, S.P. Pawar, S. Jayanthi, G. Madras,
A.K. Sood, S. Bose, Engineering nanostructures by decorating magnetic nanoparticles onto
graphene oxide sheets to shield electromagnetic radiations, ACS Appl. Mater. Interfaces 7,
(2015) 16266−16278.

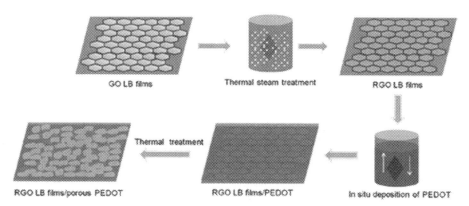

Figure 13.28 Representation of preparation method for reduced graphene oxide /porous
PEDOT nanocomposite.
Source: Reproduced with the permission from Y. Yang, S. Li, W. Yang, W. Yuan, J. Xu, Y.
Jiang, In situ polymerization deposition of porous conducting polymer on reduced graphene
oxide for gas sensor, ACS Appl. Mater. Interfaces 6 (2014) 13807−13814.

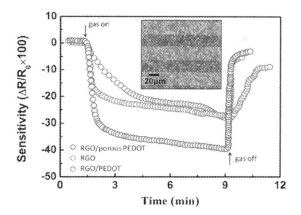

Figure 13.29 Gas sensitiveness of different equipment with RGO, RGO/PEDOT and RGO/alveolar PEDOT gas sensing layers to 20 parts per million of NO$_2$. The infused picture is the RGO/alveolar PEDOT covered interdigitated electrode 9. *PEDOT*, poly (3,4-ethylenedioxythiophene); *RGO*, reduced graphene oxide.
Source: Reproduced with the permission from Y. Yang, S. Li, W. Yang, W. Yuan, J. Xu, Y. Jiang, In situ polymerization deposition of porous conducting polymer on reduced graphene oxide for gas sensor, ACS Appl. Mater. Interfaces 6 (2014) 13807−13814.

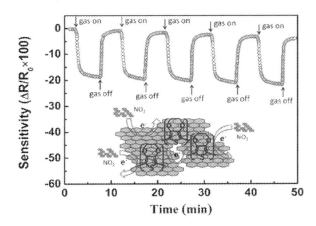

Figure 13.30 Regeneratable characteristic of RGO/alveolar PEDOT sensor to 2 parts per million of NO$_2$ gas and strategic explanation of cooperation between NO$_2$ and nanocomposites.
Source: Reproduced with the permission from Y. Yang, S. Li, W. Yang, W. Yuan, J. Xu, Y. Jiang, In situ polymerization deposition of porous conducting polymer on reduced graphene oxide for gas sensor, ACS Appl. Mater. Interfaces 6 (2014) 13807−13814.

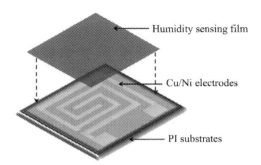

Figure 13.31 Strategic representation of the humidity sensor.
Source: Reproduced with the permission from D. Zhang, J. Tong, B. Xia, Q. Xue, Ultrahigh
performance humidity sensor based on layer-by-layer self-assembly of graphene oxide/
polyelectrolyte nanocomposite film, Sens. Actuators B Chem. 203 (2014) 263−270.

EMI SE of the composite sample strongly relies on sample thickness. If sample
thickness increases than EMI SE also increases. Here variation of EMI SE of
NRMG-6 with sample thickness was represented in Fig. 13.25. As NRMG-6 film
depth promoted from 0.8 to 1.8 mm, EMI shielding effectivity increased from 18 to
37 dB. This was due to fact that higher thickness implies higher amount of
Fe_3O_4@rGO interacts with incident EM waves resulting better shielding efficiency.

Now a days instead of single polymer, blends of two polymers are extensively
used to fabricate polymer-graphene composites. This composites exhibit higher
conductivity and better EMI SE at a lower density of NP loading facilitating double
percolation phenomenon by strategic distribution of NP in one particular phase of
polymer blend systems. Sometimes hybrid filler (mixture of two filler) are utilized
to produce polymer nanocomposites to achieve better overall properties of the
same. In a typical process Mural et al. synthesized Ni-doped GO through following
process represented in Fig. 13.26 [66]. This Ni-GO was utilized as conducting filler
in combination with multiwalled carbon nanotubes (MWNTs) to improve electrical
as well as EMI shielding efficiency of the polyethylene/poly(ethylene oxide) blend
material.

When 10vol.% of graphene oxide-Ni was used together with 3wt.% of p-MWNT
the resultant nanocomposite showed EMI SE of more than 25 dB. The graphene
oxide-Ni sheets not only enhanced permeability in the blend system but also simplified in a susceptible charge transfer as pronounced from efficient electrical conductivity in the polymer blends. This polyethylene/poly(ethylene oxide) blend with
MWNTs/GO-Ni decayed the entering electromagnetic radiation mostly by absorption (Fig. 13.27) [66].

13.3.4 Sensor

Initially, Novoselov et al. [67] discovered the application of graphene in the field of
gas sensors. High chemical reactivity of RGO due to excellent electrical

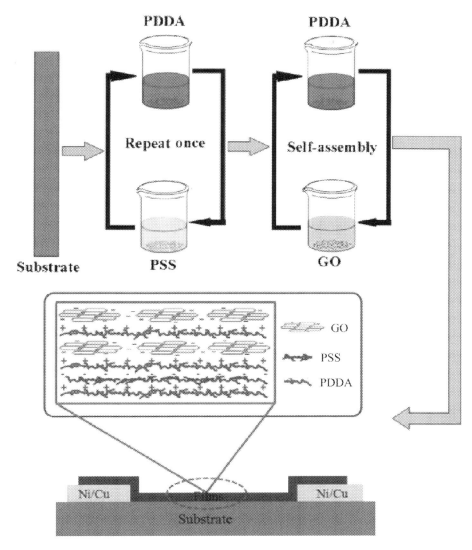

Figure 13.32 Schematic representation of layer-by-layer stacking of graphene oxide/poly (diallyldimethylammoniumchloride) film.
Source: Reproduced with the permission from D. Zhang, J. Tong, B. Xia, Q. Xue, Ultrahigh performance humidity sensor based on layer-by-layer self-assembly of graphene oxide/ polyelectrolyte nanocomposite film, Sens. Actuators, B Chem. 203 (2014) 263−270.

performance and defective nature compared to graphene oxide and graphene makes them suitable for gas sensing with high sensitivity [68]. Conducting polymers with high reliability and low cost properties are extensively used as gas sensor due to their high sensitivities. Compared to pure GO and RGO sensors hybridization of

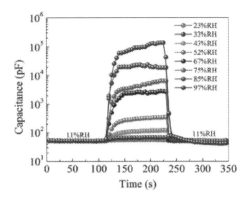

Figure 13.33 Typical reaction and recovery plots of the grapheme oxide/poly (diallyldimethylammonium chloride) film sensor to a RHpulse from 11% to various RH levels.
Source: Reproduced with the permission from D. Zhang, J. Tong, B. Xia, Q. Xue, Ultrahigh performance humidity sensor based on layer-by-layer self-assembly of graphene oxide/ polyelectrolyte nanocomposite film, Sens. Actuators, B Chem. 203 (2014) 263–270.

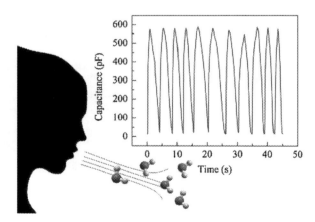

Figure 13.34 Reaction characteristic of graphene oxide/poly (diallyldimethylammoniumchloride) film sensor to human breath.
Source: Reproduced with the permission from D. Zhang, J. Tong, B. Xia, Q. Xue, Ultrahigh performance humidity sensor based on layer-by-layer self-assembly of graphene oxide/ polyelectrolyte nanocomposite film, Sens. Actuators, B Chem. 203 (2014) 263–270.

polymers especially conducting polymers with RGO and GO lead to better sensor material with enhanced sensitivity [69]. Sometimes conducting polymer with porous structure was utilized for fabrication of graphene based gas sensor. Yang et al. produced alveolar conjugated polymer poly(3,4-ethylenedioxythiophene)/

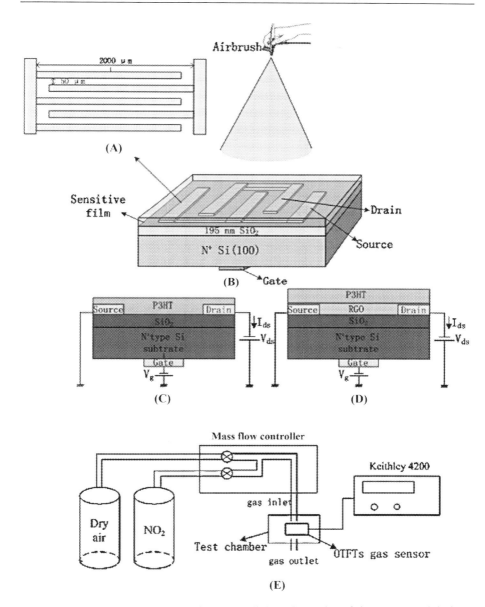

Figure 13.35 Schematic diagram of (A) interdigital electrodes of the source and drain, (B) OTFT gas sensor and airbrush process, (C) the single layer film as sensitive film, (D) the bilayerfilm as sensitive film and (E) the strategic diagram of the test system for the gas sensor.
Source: Reproduced with the permission from D. Zhang, J. Tong, B. Xia, Q. Xue, Ultrahigh performance humidity sensor based on layer-by-layer self-assembly of graphene oxide/polyelectrolyte nanocomposite film, Sens. Actuators, B Chem. 203 (2014) 263–270.

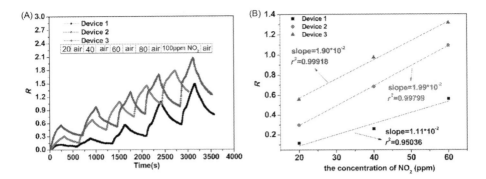

Figure 13.36 The gas sensing of equipment (A) Real-time reaction of OTFT gas sensors to NO_2. (B) The sensitive characteristics of OTFT gas sensors to NO_2.
Source: Reproduced with the permission from T. Xie, G. Xie, Y. Zhou, J. Huang, M. Wu, Y. Jiang, et al., Thin film transistors gas sensors based on reduced graphene oxide poly(3-hexylthiophene) bilayer film for nitrogen dioxide detection, Chem. Phys. Lett. 614 (2014) 275−281.

RGO-based gas sensing material. The synthesis process was schematically represented in Fig. 13.28 [70]. Here single layered graphene oxide sheet was prepared GO Langmuir − Blodgett deposition followed by thermal treatment. The resultant nanocomposite showed excellent gas sensitivity for NO_2 detection.

The gas sensitivity of various film base sensor devices was tested depending on the variation in resistance of the corresponding devices after exposure to analyte gas (NO_2). Following equation was used to calculate the sensing response of the devices for NO_2 detection.

$$R(\%) = \Delta R/R \times 100 = (R_I - R_e) \times 100/R_e$$

Where R_e and R_I are the measured resistance of alveolar PEDOT/RGO nanocomposite before and later exposition to NO_2 gas, gradually. Fig. 13.29 gives a comparison of gas sensitivity of different sensor devices toward 20 parts per million of NO_2 gas. Compared to pure RGO based sensor. Alveolar PEDOT/rGO sensor exhibited enhanced sensitivity (approximately 2 order of magnitude) along with response and recovery performance. This is due to cooperative effect of huge surface area, electrical performance of RGO and porous PEDOT and opening porous structure of the nanocomposites.

Fig. 13.30 shows the response of alveolar PEDOT/RGO based gas sensor under repeatable cycles. From the figure it is clear that this sensor exhibited an excellent repeatable characteristics.

In recent times, humidity sensors have gained attention followed by their usage in several areas such as medicinal science, agricultural science, and industry. Graphene-filled polymer nanocomposite based sensors are extensively utilized for humidity sensing. Fig. 13.31 shows a film-based humidity sensor. In a typical procedure multilayer film depending on GO/poly(diallyldimethylammoniumchloride) nanocomposite was fabricated on a substrate of polyimide with the help of layer-by-layer stacking technique. Fig. 13.32 represents the fabrication process of GO/poly(diallyldimethylammoniumchloride) multilayer film along with its hierarchical structure [71]. This multilayer films were utilized as humidity sensor.

The sensitivity of the sensor material with RH was measured by exposing it to different RH levels (11%−97%) in a closed vessel. The normalized response (R) and sensitivity (S) were calculated using the following equations:

$$R = \Delta C / C_0 = (C_x - C_{11})/C_{11} \times 100\% \text{ and } S = (C_x - C_{11})/(RH_x - RH_{11})$$

Here C_x and C_{11} represents the capacitance at the x% and 11% RH levels, gradually for the sensing material. During exposure to various humidity level the sensor showed novel reaction of up to 265,640% much better than that of other sensors. This sensor was also capable of monitoring human breath exhibiting ultrafast recovery and response time. The reaction and recovery profile of the GO/PDDSA sensor at different RH levels were represented in Fig. 13.33 whereas Fig. 13.34 shows the reaction characteristic of GO/poly(diallyldimethylammonium chloride) based film sensor to human breath.

Xie et al. produced poly(3-hexylthiophene)/Reduced graphene oxide nanocomposite based multilayer film and utilized this film as effective layers in organic thin film transistor (OTFT) gas sensors to detect nitrogen dioxide (NO_2) [72]. Fig. 13.35 represents the corresponding OTFT gas sensor and the test method for the sensor.

The resultant sensor device exhibited transistor behavior and exquisite gas sensing properties at 25°C. The real-time reaction of OTFT sensor and sensing characteristics of the sensor for NO_2 detection were shown in Fig. 13.36.

13.4 Conclusion

Although discovered more than a decade ago, the research regarding graphene and graphene-based polymer nanocomposites is still a burning topic among the society of material scientists and still a lot of things to come to illuminate the unnoticed applications on several other fields which were not discussed in the above sections. The important applications along with the corresponding research analysis are well discussed and documented in this chapter.

References

[1] R. Stine, J.T. Robinson, P.E. Sheehan, C.R. Tamanaha, Real-time DNA detection using reduced graphene oxide field effect transistors, Adv. Mater. 22 (2010) 5297−5300.
[2] S.K. Mishra, S.N. Tripathi, V. Choudhary, B.D. Gupta, SPR based fibre optic ammonia gas sensor utilizing nanocomposite film of PMMA/reduced graphene oxide prepared by in situ polymerization, Sens. Actuat. B Chem. 199 (2014) 190−200.

[3] R. Ravindren, S. Mondal, K. Nath, N.C. Das, Prediction of electrical conductivity, double percolation limit and electromagnetic interference shielding effectiveness of copper nanowire filled flexible polymer blend nanocomposites, Compos. B Eng. 164 (2019) 559−569.

[4] E. Kymakis, K. Savva, M.M. Stylianakis, C. Fotakis, E. Stratakis, Flexible organic photovoltaic cells with in situ nonthermal photoreduction of spin-coated graphene oxide electrodes, Adv. Funct. Mater. 23 (2013) 2742−2749.

[5] S. Srivastava, M. Haridas, J.K. Basu, Optical properties of polymer nanocomposites, Bull. Mater. Sci. 31 (2008) 213−217.

[6] R. Gulotty, M. Castellino, P. Jagdale, A. Tagliaferro, A.A. Balandin, Effects of functionalization on thermal properties of single-wall and multi-wall carbon nanotube-polymer nanocomposites, ACS Nano 7 (2013) 5114−5121.

[7] R.M. Mutiso, K.I. Winey, Electrical properties of polymer nanocomposites containing rod-like nanofillers, Prog. Polym. Sci. 40 (2015) 63−84.

[8] H.H. So, J.W. Cho, N.G. Sahoo, Effect of carbon nanotubes on mechanical and electrical properties of polyimide/carbon nanotubes nanocomposites, Eur. Polym. J. 43 (2007) 3750−3756.

[9] K.S. Novoselov, A.K. Geim, S.V. Morozov, D. Jiang, Y. Zhang, S.V. Dubonos, et al., Electric field in atomically thin carbon films, Science (80-) 306 (2004) 666−669.

[10] X. Zhao, Q. Zhang, D. Chen, P. Lu, Enhanced mechanical properties of graphene-based polyvinyl alcohol composites, Macromolecules 43 (2010) 2357−2363.

[11] M. Kole, T.K. Dey, Investigation of thermal conductivity, viscosity, and electrical conductivity of graphene based nanofluids, J. Appl. Phys. 113 (2013).

[12] M. Antonietti, K. Müllen, Carbon: the sixth element, Adv. Mater. 22 (2010) 787.

[13] A. Fasolino, J.H. Los, M.I. Katsnelson, Intrinsic ripples in graphene, Nat. Mater. 6 (2007) 858−861.

[14] Z. Sekhavat Pour, M. Ghaemy, Polymer grafted graphene oxide: for improved dispersion in epoxy resin and enhancement of mechanical properties of nanocomposite, Compos. Sci. Technol. 136 (2016) 145−157.

[15] D.C. Marcano, D.V. Kosynkin, J.M. Berlin, A. Sinitskii, Z. Sun, A. Slesarev, et al., Improved synthesis of graphene oxide, ACS Nano 4 (2010) 4806−4814.

[16] S. Abdolhosseinzadeh, H. Asgharzadeh, H.S. Kim, Fast and fully-scalable synthesis of reduced graphene oxide, Sci. Rep. 5 (2015) 10160.

[17] G. Williams, B. Seger, P.V. Kamt, TiO$_2$-graphene nanocomposites. UV-assisted photocatalytic reduction of graphene oxide, ACS Nano 2 (2008) 1487−1491.

[18] J. Pyun, Graphene oxide as catalyst: application of carbon materials beyond nanotechnology, Angew. Chemie - Int. (Ed.), 50, 2011, pp. 46−48.

[19] G. Eda, G. Fanchini, M. Chhowalla, Large-area ultrathin films of reduced graphene oxide as a transparent and flexible electronic material, Nat. Nanotechnol. 3 (2008) 270−274.

[20] M. Li, Q. Liu, Z. Jia, X. Xu, Y. Cheng, Y. Zheng, et al., Graphene oxide/hydroxyapatite composite coatings fabricated by electrophoretic nanotechnology for biological applications, Carbon N. Y. 67 (2014) 185−197.

[21] J. Bai, X. Jiang, A facile one-pot synthesis of copper sulfide-decorated reduced graphene oxide composites for enhanced detecting of H$_2$O$_2$ in biological environments, Anal. Chem. 85 (2013) 8095−8101.

[22] S.Y. Wu, S.S.A. An, J. Hulme, Current applications of graphene oxide in nanomedicine, Int. J. Nanomed. 10 (2015) 9−24.

[23] J. Liu, K. Liu, L. Feng, Z. Liu, L. Xu, Comparison of nanomedicine-based chemotherapy, photodynamic therapy and photothermal therapy using reduced graphene oxide for the model system, Biomater. Sci. 5 (2017) 331−340.

[24] D.X. Yan, H. Pang, B. Li, R. Vajtai, L. Xu, P.G. Ren, et al., Structured reduced graphene oxide/polymer composites for ultra-efficient electromagnetic interference shielding, Adv. Funct. Mater. 25 (2015) 559−566.

[25] M. Ji, N. Jiang, J. Chang, J. Sun, Near-infrared light-driven, highly efficient bilayer actuators based on polydopamine-modified reduced graphene oxide, Adv. Funct. Mater. 24 (2014) 5412−5419.

[26] Y.J. Wan, L.C. Tang, D. Yan, L. Zhao, Y.B. Li, L. Bin, et al., Improved dispersion and interface in the graphene/epoxy composites via a facile surfactant-assisted process, Compos. Sci. Technol. 82 (2013) 60−68.

[27] M. Eizenberg, J.M. Blakely, Carbon monolayer phase condensation on Ni(111), Surf. Sci. 82 (1979) 228−236.

[28] M. Eizenberg, J.M. Blakely, Carbon interaction with nickel surfaces: monolayer formation and structural stability, J. Chem. Phys. 71 (1979) 3467.

[29] B. Lang, A LEED study of the deposition of carbon on platinum crystal surfaces, Surf. Sci. 53 (1975) 317−329.

[30] L. Meng, Q. Sun, J. Wang, F. Ding, Molecular dynamics simulation of chemical vapor deposition graphene growth on Ni (111) surface, J. Phys. Chem. C. 116 (2012) 6097−6102.

[31] L.H. Viculis, J.J. Mack, R.B. Kaner, A chemical route to carbon nanoscrolls, Science (80-.) 299 (2003) 1361.

[32] B. Jayasena, S. Subbiah, A novel mechanical cleavage method for synthesizing few-layer graphenes, Nanoscale Res. Lett. 6 (2011) 95.

[33] G. Xin, W. Hwang, N. Kim, S.M. Cho, H. Chae, A graphene sheet exfoliated with microwave irradiation and interlinked by carbon nanotubes for high-performance transparent flexible electrodes, Nanotechnology 21 (2010) 405201.

[34] G. Wang, B. Wang, J. Park, Y. Wang, B. Sun, J. Yao, Highly efficient and large-scale synthesis of graphene by electrolytic exfoliation, Carbon N. Y. 47 (2009) 3242−3246.

[35] C. Zhu, S. Guo, Y. Fang, S. Dong, Reducing sugar: new functional molecules for the green synthesis of graphene nanosheets, ACS Nano 4 (2010) 2429−2437.

[36] R. Muzyka, M. Kwoka, Ł. Smędowski, N. Díez, G. Gryglewicz, Oxidation of graphite by different modified Hummers methods, Xinxing Tan. Cailiao/New Carbon Mater. 32 (2017) 15−20.

[37] Y.H. Wu, T. Yu, Z.X. Shen, Two-dimensional carbon nanostructures: fundamental properties, synthesis, characterization, and potential applications, J. Appl. Phys. 108 (2010) 071301.

[38] W.S. Hummers, R.E. Offeman, Preparation of graphitic oxide, J. Am. Chem. Soc. 80 (1958) 1339.

[39] A.A. Green, M.C. Hersam, Solution phase production of graphene with controlled thickness via density differentiation, Nano Lett. 9 (2009) 4031−4036.

[40] H.J. Shin, K.K. Kim, A. Benayad, S.M. Yoon, H.K. Park, I.S. Jung, et al., Efficient reduction of graphite oxide by sodium borohydride and its effect on electrical conductance, Adv. Funct. Mater. 19 (2009) 1987−1992.

[41] K.S. Novoselov, V.I. Fal'Ko, L. Colombo, P.R. Gellert, M.G. Schwab, K. Kim, A roadmap for graphene, Nature 490 (2012) 192−200.

[42] S. Park, R.S. Ruoff, Chemical methods for the production of graphenes, Nat. Nanotechnol. 4 (2009) 217−224.

[43] S. Stankovich, D.A. Dikin, G.H.B. Dommett, K.M. Kohlhaas, E.J. Zimney, E.A. Stach, et al., Graphene-based composite materials, Nature 442 (2006) 282−286.

[44] Y.M. Lin, A. Valdes-Garcia, S.J. Han, D.B. Farmer, I. Meric, Y. Sun, et al., Wafer-scale graphene integrated circuit, Science (80-.) 332 (2011) 1294−1297.

[45] B. Cai, S. Wang, L. Huang, Y. Ning, Z. Zhang, G.J. Zhang, Ultrasensitive label-free detection of PNA-DNA hybridization by reduced graphene oxide field-effect transistor biosensor, ACS Nano 8 (2014) 2632−2638.

[46] D.J. Kim, I.Y. Sohn, J.H. Jung, O.J. Yoon, N.E. Lee, J.S. Park, Reduced graphene oxide field-effect transistor for label-free femtomolar protein detection, Biosens. Bioelectron. 41 (2013) 621−626.

[47] S. Mao, K. Yu, G. Lu, J. Chen, Highly sensitive protein sensor based on thermally-reduced graphene oxide field-effect transistor, Nano Res. 4 (2011) 921.

[48] Z. Yin, Q. He, X. Huang, J. Zhang, S. Wu, P. Chen, et al., Real-time DNA detection using Pt nanoparticle-decorated reduced graphene oxide field-effect transistors, Nanoscale 4 (2012) 293−297.

[49] M. Larisika, C. Kotlowski, C. Steininger, R. Mastrogiacomo, P. Pelosi, S. Schütz, et al., Electronic olfactory sensor based on A. *mellifera* odorant-binding protein 14 on a reduced graphene oxide field-effect transistor, Angew. Chem. - Int. (Ed.) 54 (2015) 13245−13248.

[50] J.W. Park, S.J. Park, O.S. Kwon, C. Lee, J. Jang, Polypyrrole nanotube embedded reduced graphene oxide transducer for field-effect transistor-type H_2O_2 biosensor, Anal. Chem. 86 (2014) 1822−1828.

[51] T.Q. Trung, N.T. Tien, D. Kim, M. Jang, O.J. Yoon, N.E. Lee, A flexible reduced graphene oxide field-effect transistor for ultrasensitive strain sensing, Adv. Funct. Mater. 24 (2014) 117−124.

[52] T.Q. Trung, N.T. Tien, D. Kim, J.H. Jung, O.J. Yoon, N.E. Lee, High thermal responsiveness of a reduced graphene oxide field-effect transistor, Adv. Mater. 24 (2012) 5254−5260.

[53] G. Zhou, J. Chang, S. Cui, H. Pu, Z. Wen, Real-time, selective detection of Pb^{2+} in water using a reduced graphene oxide/gold nanoparticle field-effect transistor device, J. Chem. ACS Appl. Mater. Interfaces 6 (2014) 19235−19241.

[54] M. Hasegawa, Y. Hirayama, Y. Ohno, K. Maehashi, K. Matsumoto, Characterization of reduced graphene oxide field-effect transistor and its application to biosensor, Jpn. J. Appl. Phys. 53 (2014) 05FD05.

[55] J. Zhang, X.S. Zhao, Conducting polymers directly coated on reduced graphene oxide sheets as high-performance supercapacitor electrodes, J. Phys. Chem. C 116 (2012) 5420−5426.

[56] L.L. Zhang, S. Zhao, X.N. Tian, X.S. Zhao, Layered graphene oxide nanostructures with sandwiched conducting polymers as supercapacitor electrodes, Langmuir 26 (2010) 17624−17628.

[57] J.H. Lee, N. Park, B.G. Kim, D.S. Jung, K. Im, J. Hur, et al., Restacking-inhibited 3D reduced graphene oxide for high performance supercapacitor electrodes, ACS Nano 7 (2013) 9366−9374.

[58] M.H. Al-Saleh, G.A. Gelves, U. Sundararaj, Copper nanowire/polystyrene nanocomposites: lower percolation threshold and higher EMI shielding, Compos. A Appl. Sci. Manuf. 42 (2011) 92−97.

[59] S. Sankaran, K. Deshmukh, M.B. Ahamed, S.K. Khadheer Paha, Recent advances in electromagnetic interference shielding properties of metal and carbon filler reinforced flexible polymer composites: a review, Compos. A Appl. Sci. Manuf. 114 (2018) 49−71.

[60] B. Shen, W. Zhai, M. Tao, J. Ling, W. Zheng, Lightweight, multifunctional polyetherimide/graphene@Fe$_3$O$_4$ composite foams for shielding of electromagnetic pollution, ACS Appl. Mater. Interfaces 5 (2013) 11383−11391.

[61] P. Bhawal, S. Ganguly, T.K. Das, S. Mondal, S. Choudhury, N.C. Das, Superior electromagnetic interference shielding effectiveness and electro-mechanical properties of EMA-IRGO nanocomposites through the in-situ reduction of GO from melt blended EMA-GO composites, Compos. B Eng. 134 (2018) 46−60.

[62] S.T. Hsiao, C.C.M. Ma, W.H. Liao, Y.S. Wang, S.M. Li, Y.C. Huang, et al., Lightweight and flexible reduced graphene oxide/water-borne polyurethane composites with high electrical conductivity and excellent electromagnetic interference shielding performance, ACS Appl. Mater. Interfaces 6 (2014) 10667−10678.

[63] S. Ganguly, S. Ghosh, P. Das, T.K. Das, S.K. Ghosh, N.C. Das, Poly(N-vinylpyrrolidone)-stabilized colloidal graphene-reinforced poly(ethylene-co-methyl acrylate) to mitigate electromagnetic radiation pollution, Polym. Bull. 77 (2020) 2923−2943.

[64] R.S. Yadav, I. Kuřitka, J. Vilcakova, D. Skoda, P. Urbánek, M. Machovsky, et al., Lightweight NiFe$_2$O$_4$-reduced graphene oxide-elastomer nanocomposite flexible sheet for electromagnetic interference shielding application, Compos. B Eng. 166 (2019) 95−111.

[65] Y. Zhan, J. Wang, K. Zhang, Y. Li, Y. Meng, N. Yan, et al., Fabrication of a flexible electromagnetic interference shielding Fe$_3$O$_4$@reduced graphene oxide/natural rubber composite with segregated network, Chem. Eng. J. 344 (2018) 184−193.

[66] P.K.S. Mural, S.P. Pawar, S. Jayanthi, G. Madras, A.K. Sood, S. Bose, Engineering nanostructures by decorating magnetic nanoparticles onto graphene oxide sheets to shield electromagnetic radiations, ACS Appl. Mater. Interfaces 7 (2015) 16266−16278.

[67] F. Schedin, A.K. Geim, S.V. Morozov, E.W. Hill, P. Blake, M.I. Katsnelson, et al., Detection of individual gas molecules adsorbed on graphene, Nat. Mater. 6 (2007) 652−657.

[68] G. Lu, L.E. Ocola, J. Chen, Gas detection using low-temperature reduced graphene oxide sheets, Appl. Phys. Lett. 94 (2009) 083111.

[69] X. Huang, N. Hu, R. Gao, Y. Yu, Y. Wang, Z. Yang, et al., Reduced graphene oxide-polyaniline hybrid: preparation, characterization and its applications for ammonia gas sensing, J. Mater. Chem. 22 (2012) 22488−22495.

[70] Y. Yang, S. Li, W. Yang, W. Yuan, J. Xu, Y. Jiang, In situ polymerization deposition of porous conducting polymer on reduced graphene oxide for gas sensor, ACS Appl. Mater. Interfaces 6 (2014) 13807−13814.

[71] D. Zhang, J. Tong, B. Xia, Q. Xue, Ultrahigh performance humidity sensor based on layer-by-layer self-assembly of graphene oxide/polyelectrolyte nanocomposite film, Sens. Actuat., B Chem. 203 (2014) 263−270.

[72] T. Xie, G. Xie, Y. Zhou, J. Huang, M. Wu, Y. Jiang, et al., Thin film transistors gas sensors based on reduced graphene oxide poly(3-hexylthiophene) bilayer film for nitrogen dioxide detection, Chem. Phys. Lett. 614 (2014) 275−281.

Structural/load bearing characteristics of polymer-graphene composites

14

Pankaj Tambe, Monica Tanniru and B.L.N. Krishna Sai
School of Mechanical Engineering, VIT-AP University, Amravati, India

14.1   Introduction

Composite materials have been under intense investigation for many decades [1−4]. In the case of composite materials, polymer matrix composites are of great interest. Polymeric material have limitations in terms of mechanical properties enhancement, as high molecular weight limits the processing of polymer in melt state due to increased viscosity of polymer melt. In this regard, the development of glass fiber and carbon fiber technology gives a giant leap toward the development of composite materials for aerospace, automobile, and other applications [1,4]. Based on the requirements, proper fiber can be chosen for reinforcement in polymer matrix for the load bearing applications [1]. In addition, the orientation of fiber in polymer matrix gives the flexibility to the designer to design composite material for intended applications [3].

The emergence of nanomaterials in the early 1990s has boosted research in the area of composite materials [5,6]. Nanomaterials like carbon nanotubes (CNTs) and clay were widely reported in the literature. These nanomaterials were dispersed in the long fiber reinforced polymer matrix composites to explore the possibility of its use in aerospace and automobile industries [1]. CNTs were coated over glass/carbon fiber by various methods and used to make epoxy composites for load bearing applications. The CNTs coated over glass/carbon fiber epoxy composites show the enhanced mechanical properties and other properties of interest [1,2]. In 2004, Geim and Novoselov [7] discovered the graphene by scotch tape method and measured its remarkable electrical properties. Subsequently, the measurements of remarkable thermal and mechanical properties of graphene have leads toward the emergence of graphene as superior reinforcing filler [8,9]. Synthesis of graphene is one of the promising areas of research. Top down and bottom up approaches for synthesis of graphene have been studied extensively. There are research effort toward the functionalization of graphene with covalent and noncovalent strategies [10]. Graphene-filled polymer composites have been under investigation for various applications in the last decade. Thermoplastics and thermosets-based composites filled with graphene were well studied [10]. In this chapter, synthesis of graphene and its dispersion in polymer matrix is discussed. In addition, graphene as one of

Polymer Nanocomposites Containing Graphene. DOI: https://doi.org/10.1016/B978-0-12-821639-2.00005-7

the filler along with short or long fiber reinforcement in polymer composites for load bearing application is discussed.

14.2 Synthesis of graphene

The two types of synthesis route of graphene that are popular are liquid phase exfolition of graphene and oxidation of graphene follwed by its reduction. In liquid phase exfoliation route, graphite powder is added in a suitable solvent followed by sonication or shear exfoliation to exfoliate graphite into graphene nanosheets and by decanting the top 80% solution. The choice of solvents plays a key role in achieving high concentration dispersion of graphene. Hansen and Hildebrand solubility parameters calculations are used to find the suitable solvent. It has to be noted that the solubility parameters of solvent and graphene must be nearer in order to achieve good dispersability of graphene. As per the literature, N-methyl pyrrolidone (NMP) and dimethyl formamide (DMF) are the good solvent for the graphene dispersion. A few authors also studied the influence of surfactant and solvent for graphene dispersion. Hernandez et al. [11] used NMP as a solvent to disperse graphene as a solvent using sonication for 30 min and centrifugation at 500 rpm for 90 min. The graphene flakes have up to five layers and lateral dimension is in microns. Alaferdov et al. [12] used DMF as a solvent to disperse graphene using sonication for 240 min and centrifugation at 90 min for 10,000 rpm. The graphene flakes' thickness is $11-14$ nm. Loyata et al. [13] used surfactant and solvent to disperse graphene using sonication for 30 min and centrifugation at 90 min for 500 rpm. The 48% of graphene flakes has less than five layers. Sonication energy too played an important role in exfoliation of graphene in solvent [14].

The graphene oxidation method is one of the widely used methods to oxidized graphene. There are five methods for the oxidation of graphite powder to prepare graphene oxide. They are Boride method [15], Staudenmaier method [16], Hummers method [17], modified Hummers method [18] and improved method [19]. In Bordie method [15], graphite powder was mixed with fumic nitric acid followed by addition of potassium chlorate. In the case of Staudenmaier method [16], graphite powder was mixed with fumic nitric acid and sulfuric acid in the ratio of 1:3. Hummers in 1958 [17] came up with a method of oxidizing graphene. He mixed graphite powder in concentrated sulfuric acid followed by potassium permanganate and sodium nitrate. At last at the end of reaction, hydrogen peroxide was added. In the case of modified Hummers method [18], in order to increase the extent of oxidation of graphene, potassium permanganate is added in twice the concentration as compared to Hummers method. In the case of Hummers and modified Hummers method, toxic gas like nitrogen oxide was formed. An improved method [19] was developed by Tour group which does not produce toxic gas like nitrogen oxide during synthesis of graphene oxide. In this method, sodium nitrate is not used; instead concentrated sulfuric acid to phosphoric ratio is 9:1 with an addition of same concentration of potassium permanganate used in modified Hummers method.

Graphene oxide nanosheets are reduced with suitable reducing agent in order to prepare graphene nanosheets. The graphene oxide nanosheets are exfoliated in queaous solution using sonication. Park et al. [20] used hydrazine as a reducing agent for graphene oxide reduction. Upon reduction, reduced graphene oxide is stable in aqueous solution with appropriate p^H value. Shin et al. [21] used sodium borohydrate ($NaBH_4$) as a reducing agent for graphene oxide reduction. The obtained reduced graphene oxide is purer as the C:O ratio is more as compared to the hydrazine used as a reducing agent and exhibiting more sheet resistance. Pham et al. [22] used in-situ gaseous hydrogen to reduce the graphene oxide and is effective to achieve high C:O ratio. Wang et al. [23] used hydroquinone solution a reducing agent for graphene oxide reduction, but it is not as effective as hydrazine and $NaBH_4$. Wan [24] used alkaline solution as a reducing agent for graphene oxide reduction, but it is not as effective as hydrazine and $NaBH_4$.

14.3   Dispersion of graphene in polymer matrix

The dispersion of graphene in polymer matrix is challenging due to strong Van der Waal's forces of attraction between graphene nanosheets. To overcome the strong Van der Waal's forces of attraction between graphene nanosheets, graphene is finctionalized covalently and noncovalently. Since, it is well known that the reduction of graphene oxide results in graphene with functional group at edges. The presence of functional group on graphene is utilized to covalently graft the molecule with chemical processing. Graphene synthesized with chemical vapor deposition (CVD) does not have the functional group at the edges of graphene. To add functional group at the edges, graphene is chemically treated to add hydroxyl functional group followed by reaction with functional molecule. But, covalent functionalization comes at the cost of creation of defects. The creation of defects results in deteoriating the superior properties of graphene. Thus, noncovalent functionalization of graphene attracted researchers' attention. The noncovalent functionalization of graphene is achieved using cation-π, anion- π type of interaction, hydrogen bonding, π-π interaction, and surfactant-assisted modification. A briew review of graphene dispersion in polymer matrix is presented in this section.

14.3.1   Epoxy/graphene nanocomposites

Epoxy is one of the thermoset polymer that has applications in aerospace and automobile industries. Epoxy resin processing is easier, as it is avaliable in liquid state and combined with hardener forms a crosslinked chains and gives it a stable solid state. Graphene is added in epoxy resin using solvent-assisted route. The required quantity of graphene is mixed in a solvent such as ethyl alcohol or acetone using sonication. The advantage of use of low boiling point solvent is that it can be easily evoporated from epoxy resin. Solvent assistant mixing of graphene in epoxy do not result in good dispersion, so further processing is required to achieve the good

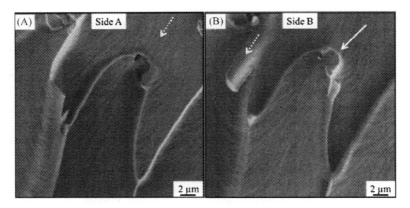

Figure 14.1 Crack pinning failure in epoxy/graphene nanocomposites.
Source: Reproduced with permission from S. Chandrasekaran, S. Narumichi, T. Folke, R. Mülhaupt, B. Fiedler, K. Schulte, Fracture toughness and failure mechanism of graphene based epoxy composites, Compos. Sci. Technol. 97 (2014) 90−99.

dispersion of graphene in epoxy. Epoxy/graphene mixure is further probe sonicated or ball milled to achieve good dispersion of graphene in epoxy matrix. Few researcher have use three-roll mill for the dispersion of graphene in epoxy matrix [10].

Failure mechanism of graphene filled epoxy mechanism was studied by Chandrasekaran et al. [25]. Fig. 14.1 shows the fracture surface SEM images of graphene filled epoxy nanocomposites. It shows two sides of fracture surface named as Side A and Side B respectively. The two surface formed concave and convex surface respectively. The arrow indicates the flow direction of the crack propogation. It can be seen clearly that the flow pattern wraps and curved around the graphene presence in an epoxy matrix. The fracture mechanism deduced from the SEM images are categorized in three modes. First by crack pinnig occurs at the visinity of the graphene in epoxy matrix. The sencond mechanism is seperation between the graphene nanosheets. The third mechanism is shear failure due to the difference in height on the fracture surface [25]. The fracture mechanism were concluded from the low concentration dispersion of graphene in epoxy matrix, which shows the improved fracture toughness. At high concentration (> 0.2 wt.%) of graphene, the fracture toughness decreases for epoxy nanocomposites.

The mechanical properties of epoxy nanocomposites depends on the graphene dispersion in epoxy matrix. Tang et al. [26] studied the dispersion of graphene in epoxy matrix using sonication and ball milling process strategies. Fig. 14.2A shows the transmission optical microscope (TOM) image of sonicated graphene dispersion in ethyl alcohol and subsequent dispersion in epoxy matrix. While, Fig. 14.2B shows the TOM image of ball milled graphene dispersion in epoxy matrix. TOM image in Fig. 14.2A shows the more aggregates of graphene in epoxy matrix as compared to Fig. 14.2B. Similar observation is confirmed with transmission electron microscope (TEM) of sonicated and ball milled graphene dispersion in epoxy matrix in Fig. 14.2C ans D respectively. The improved disperrsion resulted in

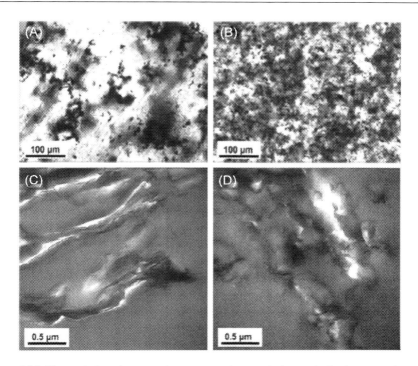

Figure 14.2 Transmission electron microscope and transmission optical microscope images of epoxy/graphene nanocomposites (A, C) without (B, D) with ball milling.
Source: Reproduced with permission from L.C. Tang, Y.J. Wan, D. Yan, B.P. Yong, L. Zhao, Y.B. Li, et al., The effect of graphene dispersion on the mechanical properties of graphene/epoxy composites, Carbon 60 (2013) 16−27.

enhanced flexural, tensile modulus and fracture toughness for ball milled graphene filled epoxy nanocomposites.

Shen et. [27] studied the influence of cryogenic temperature on mechanical properties of epoxy/graphene nanocomposites. Due to clamping effect of epoxy chains associated with thermal shrinkage results in an enhancement in mechanical properties of epoxy/graphene nanocomposites for cryogenic treatement as compared to graphene seperation observed in epoxy/graphene nanocomposites at room temperature. Liang et al. [28] showed that graphene dispersion at a concentrtion above the percolation threshold results in higher EMI shielding effectiveness for epoxy/graphene nanocomposites. Mani et al. [29] suggest the partially reduced graphene oxide results in improved dispersion of graphene and enahncement in mechanical properties of epoxy/graphene nanocomposites. Bindu et al. [30] reduced the graphene oxide with microwave followed by epoxy nonocomposites processing. The microwaved reduced graphene oxide shows the enhancement in flexural and tensile properties at 0.2 wt.% concentration of microwaved reduced graphene oxide in epoxy matrix.

Rafiee et al. [31] mixed 1 wt.% of graphene concentration in epoxy matrix. They tested the nanocomposites specimen to find out the critical buckling load. The critical buckling load is enhanced by 52% for 1 wt.% of graphene concentration in epoxy matrix. Zandiatashabar et al. [32] studied the creep behavior of graphene reinforced epoxy nanocomposites. The creep testing was performed at constant load of 20 MPa and 40 MPa respectively. The creep resistance at higher load of 40 MPa for epoxy nanocomposites are lower as compared to creep load of 20 MPa. Ma et al. [33] synthesized graphene using graphite powder. They intercalated graphite with compound followed by heat treatement at 700°C and reaction with a diamine based surfactant to exfoliate the graphene. The synthesized graphene was used to prepare epoxy nanocomposites. At 0.98 vol.% of diamine modified graphene in epoxy matrix improve the Young's modulus by 14% as compared to pure epoxy.

Covalent functionalization of graphene is done to improve the compatibility between graphene and epoxy matrix. Naebe et al. [34] covalently functionalized the graaphene using the Bingel reaction. Fig. 14.3 shows the FTIR specta of graphene oxide, graphene, and functionalized graphene using Bingel reaction. In the case of graphene oxide, the chatacteristic band of -C = O is present at 1711 cm^{-1}, while reduction of graphene oxide shows the disappeareance of peak at 1711 cm^{-1}. Further, the functionlized graphe shows the -C = O peak at 1711 cm^{-1} of −COOH group. The incorporation of 0.1 wt.% of functionalized graphene in epoxy matrix results in an increase in flexural propeties. It shows 22% increase in flexural stength for 1 wt.% of functionalized graphene in epoxy matrix as compared to pure epoxy.

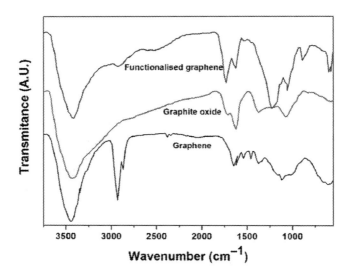

Figure 14.3 FTIR spectro of synthesized graphene oxide, thermally reduced graphene oxide and functionalized graphene.
Source: Reproduced with permission from M. Naebe, J. Wang, A. Amini, H. Khayyam, N. Hameed, L.H. Li, et al., Mechanical property and structure of covalent functionalized graphene/epoxy nanocomposites, Sci. Rep. 4 (4375) (2014) 1−7.

Ahmadi-Moghadam et al. [35] functionalized graphene with silane coupling agent. The addition of 0.5 wt.% of silanized graphene in epoxy matrix enhanced the tensile modulus and tensile strength. Tang et al. [36] modified graphene with diamine and disperesed in rubbery epoxy matrix. The 2.7 vol.% addition of modified graphene with diamine in rubbery epoxy matrix results in 536% and 269% enhancement in tensile modulus and tensile strength respectively as compared to pure rubbery epoxy. This composition shows good electrical conductivity.

Noncovalent functionalization of graphene using surfactant is widely used in literature to disperse graphene in epoxy matrix. Wan et al. [37] used Triton-X100 as a surfactant to noncovalently modify the graphene. Fig. 14.4 shows the temerature profile of the curing process and optical microscope images during the curing process. With prstine graphene, it shows the graphene reaggregates in the epoxy resin as time progresses during curing. The agglomerate formation is due to high Van der Waals forces of attraction between graphene nanosheets and the elevated temperature plays an additional drriving force for agglomeration of graphene nanosheets. The optical micosope experiments were performed over a hot stage optical microscope. With Triton, modified graphene show more stable dispersion during curing of epoxy matrix, due to the interaction between Triton and epoxy matrix. The influence of Triton modified graphene on mechanical properties epoxy nanocomposites is shown in Fig. 14.5. With an increase in graphene concentration in epoxy matrix,

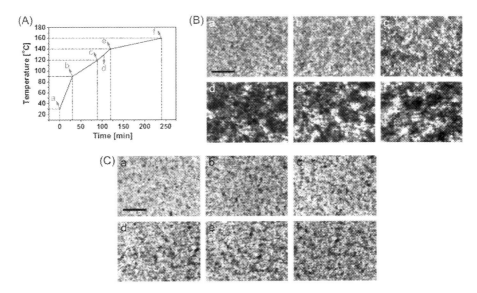

Figure 14.4 (A) Curing process thermal cycle. (B) Reagglomeration of graphene in epoxy matrix. (C) Reagglomeration of surfactant modified graphene in epoxy matrix.
Source: Reproduced with permission from Y.J. Wan, C.T. Long, Y. Dong et al., Improved dispersion and interface in the graphene/epoxy composites via a facile surfactant-assisted process, Compos. Sci. Technol. 82 (2013) 60−68.

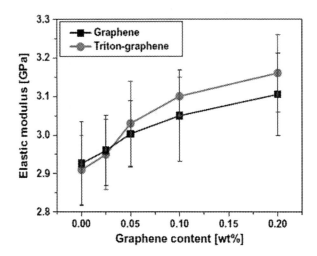

Figure 14.5 Elastic modulus of epoxy/graphene nanocomposites with different concentration of graphene and trition modified graphene.
Source: Reproduced with permission from Y.J. Wan, C.T. Long, Y. Dong et al., Improved dispersion and interface in the graphene/epoxy composites via a facile surfactant-assisted process, Compos. Sci. Technol. 82 (2013) 60−68.

the tensile modulus increases up tp 0.2 wt.% concenteration in epoxy matrix. Further, modification of graphene with Triton resulted in a further enhancement in tenisle modulus in the concentration range of 0.05−0.2 wt.% of graphene in a epoxy matrix.

Teng et al. [38] used poly(glycidyl methacrylate) (Py-PGMA) for modification of graphene. The Py-PGMA contains localized pyrene groups. Fig. 14.6 shows the protocol of noncovalent modification of graphene with Py-PGMA. It shows the synthesis of graphene oxide from graphite powder using modified Hummers method followed by thermal reduction at a higher temperature to obtain graphene. The modifier Py-PGMA interacts with graphene due to π-π stacking. The unmodified and Py-PGMA modified graphene are dispersed in epoxy matrix. The measured thermal conductivity shows 588% enhancement with an addition of Py-PGMA modified graphene in epoxy matrix as compare to epoxy matrix. The high thermal conductivity is achieved due to enhanced dispersion of graphene in an epoxy matrix using Py-PGMA as a modifier. Saha et al. [39] utilized commercially available polyvinylpyrrolidone (PVP) surfactant for non-covalent modification of graphene. The estimated Hansen and Hildebrand solubility parameters of PVP help to achieve good dispersion of graphene in epoxy matrix. The enhanced dispersion of graphene using PVP shows enhancement in flexural modulus and flexural strength of epoxy/PVP-graphene nanocomposites as compared to epoxy and epoxy/graphene nanocomposites.

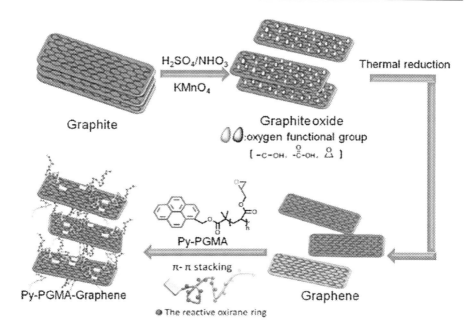

Figure 14.6 Schematic of protocal of functionalization of graphene.
Source: Reproduced with permission from C.C. Teng, C.M. Chen, H.L. Chu et al., Thermal conductivity and structure of non-covalent functionalized graphene/epoxy composites, Carbon 9 (2011) 5107−5116.

14.3.2   Thermoplastic/graphene nanocomposites

In this subsection, graphene reinforced thermoplastic nanocomposites research is discussed.

Thayumanavan et al. [40] used sodium dodecylsulfate (SDS) to disperse the graphene in aqueous solution. In a typical experiment, SDS was dissolved in different concenteration in aqueous solution followed by addition of graphene and sonication subsequently. Upon sonication, the graphene aqueous solution was centrifuged to sediment the large particles and top 80% solution was decanted. The UV-Vis-NIR spectra abosorbance value at a constant wavelength and calculated concentration of graphene in solution using Beer Lambert's law for different concentration of SDS in aqueous solution. Above the critical micelle concenteration, the absorbance value is highest depicting the highest dispersion of graphene achieved in the aqueous solution. This SDS concentration was used to dipserse graphene in different concenterations in polyvinyl alcohol (PVA) matrix. The measured tensile properties of the PVA/graphene nanocomposites shows the significant enhancement as compared to pure PVA film. It show the 151% enhancement in tensile modulus at 0.5 wt.% of graphene in PVA matrix. Further modification of 0.5 wt.% of graphene with sodium alginate shows enahncement in tensile modulus to 385% as compred to pure PVA films. The signifant increase in mechanical properties is due to the hydrogen bonding in between sodium alginate and PVA. Crystallinity of PVA increased with an addition of SDS and sodium alginate modified graphene in PVA matrix.

Ishtrate et al. [41] processed graphene-filled polyethylene terpthlate (PET) nanocomposites using melt mixing method. An internal mixture followed by compression molding was used to process PET/graphene nanocomposites. TEM image of PET/graphene nanocomposites shows the random distribution of graphene in PET matrix. With 0.07 wt.% of graphene addition in PET matrix shows the 10% enahancement in tensile properties as compared to the pure PET (Fig. 14.7). Fig. 14.7 suggest that at very low level of loading of graphene reinforcement influences the fracture process. Fang et al. [42] covalently grafted graphene with diazonium addition. The synthesized graphene was used to prepare polystyrene nanocomposites through atom transfer radical polymerization. Young's modulus increased by 57% for 0.9 wt.% of graphene in polystyrene matrix as compared to pure polystyrene. Glass transition temperature increased by 15°C for 0.9 wt.% of graphene in polystyrene matrix as compared to pure polystyrene. Wang et al. [43] used graphene with low functional defect to coat over the polyamide 6 (PA6) powder and process PA6 nanocomposites with 3D network of graphene in PA6 matrix. The 3D interconnected graphene in PA6 matrix shows high electrical conductivity at low volume fraction. In addition at low volume fraction, mechanical properties increases for 3D interconnected graphene in PA6 matrix. This method is industrially important for commercial utilization. Pang et al. [44] used thermally treated graphene to prepared

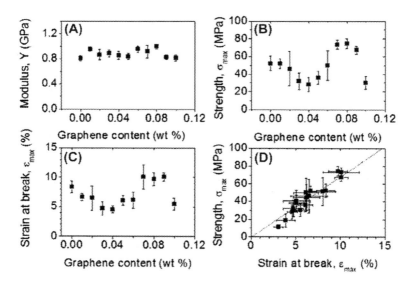

Figure 14.7 Mechanical properties of PET/graphene nanocomposites.
Source: Reproduced with permission from O.M. Istrate, K.R. Paton, U. Khan, A. O'Neill, A.P. Bell, J.N. Coleman, Reinforcement in melt-processed polymer−graphene composites at extremely low graphene loading level, Carbon 78 (2014) 243−249.

the polyehtylene (PE) film by sol-gel extrusion followed by freezing of polyemr chains and mechanical drawing. With 1 wt.% of graphene in PE increases the mechanical properties of PE nanocomposites tapes with increasing the drawing ratio. Molecular dynamics simulation shows the thermally treated graphene has good interaction with PE matrixes compared to PE-PE polymer chain interaction. This has resulted in enhanced stress transfer from graphene to PE matrix of tapes. Wu et al. [45] functionalized the graphene with hyperbranched aromatic polyamide. They dispersed hyperbranched aromatic polyamide functionlized graphene in thermoplastic polyurethane marix. The hydrogen bond is formed in between hyperbranched aromatic polyamide functionalized graphene and thermoplastic polyurethane matrix, which has resulted in enhanced stress transfer from graphene to polymer matrix. The dielectric performace is also enhanced due to formation of small microcapacitors for hyperbranched aromatic polyamide functionlized graphene in thermoplastic polyurethane marix.

14.4   Glass fiber reinforced polymer graphene composites

Glass fibers reinforced with epoxy matrix containing graphene are discussed in this section. Graphene is generally mixed with polymer matrix or coated over the glass fiber to be utilized in various applications.

14.4.1   Fiber-bundle pull out test

The mechanical properties of the polymer/long fiber composites with graphene as reinforcement depend on the fiber-matrix interaction and graphene. Yang et al. [46] coated the graphene over the glass fiber bundle using high pressure spraying of dispersed graphene in isopropyl alcohol followed by drying at 70°C. They calculated the interfacial shear stress (IFSS) of composites using the measured value of pull out force at fiber-matrix interface and embedded area of fiber. The IFSS of graphene-coated glass fiber in epoxy matrix is 12.34% higher as compared to uncoated glass fiber in epoxy matrix. The coating of graphene over glass fiber and subsequently embedded in the epoxy matrix exhibits toughening. In addition, there is a decrease in IFSS of glass fiber and graphene reinforced epoxy composites with increase in temperature, ascribed to epoxy chain relaxation at elevated temperature.

14.4.2   Flexural properties of graphene filled glass fiber/epoxy composites

Two different sizes of graphene were used to prepare epoxy/glass fiber composites by Wang et al. [47]. The impregnation of epoxy-graphene resin in the bundles of glass fiber was observed during composite processing. The flexural testing result shows the increase in both flexural strength and flexural modulus at 3 wt.% of graphene addition in epoxy matrix with lower diameter glass fiber. Further, addition of 5 wt.% of graphene in epoxy/glass fiber composites shows a decrease in flexural strength exhibiting more aggregation of graphene in epoxy matrix. The 3 wt.% of graphene in epoxy/glass fiber composites shows better dispersion of graphene than 5 wt.% of graphene in epoxy/glass fiber composites. It is to be noted that with an increase in concentration of graphene in epoxy resin, the viscosity of epoxy increases. The increased viscosity resulted in difficulty in removing the entrapped air bubbles which results in more void formation. Void formation resulted in weaker interfacial strength of graphene in epoxy/glass fiber composites. During the fracture of composites, crack initiates and grows with void coalescence which can result in catastrophic failure. It is important to understand the flexural behavior of composites material in order to efficiently design the product. In flexural loading, the fiber will carry most of the load during deformation of composite material. The presence of graphene in epoxy improves the interfacial interaction between epoxy/graphene and fiber. The combined effect of both fiber properties and interfacial interaction is con-tributing the flexural properties of graphene-filled epoxy/glass fiber composites.

14.4.3   Graphene functionalization and its influence

The functionalization of graphene has influence over the dispersion quality of gra-phene in epoxy matrix containing glass fiber. Zaheer et al. [48] functionalized the graphene with ozone. The ozone functionalized graphene addition in epoxy with glass fiber shows lower void content. The reason for lower void content is that the well dispersed graphene imporved the thermal conductivity and enhances the heat transfer which results in removig entrapped bubble from non-percolated graphene in epoxy matrix. Thus air bubble mover from epoxy resin to surfce from the path where there are gaps between graphene nanosheets. The incorporation of ozone functionalized graphene in epoxy with glass fiber show that graphene sticking to the glass fiber. With the improvved adhesion of graphene with other materials of composites, they found that the tenisle, compressive and flexural properties incre-seas up to 0.5 wt.% addition of ozone functionalized graphene in epoxy with glass fiber. This observation highlights the quality of functionalization of graphene which helps in superior dispersion of graphene in epoxy matrix.

Kamar et al. [49] treated graphene with amine-based hardener and coated over glass fiber. The coated glass fiber epoxy composites show an increase in flexural modulus up to concentration of 0.25 wt.%. The oxidation of graphitic lattice is referred to as a graphene oxide and was used as filler to process epoxy/glass fiber composites [50]. Up to 0.5 wt.% of graphene oxide in epoxy/glass fiber composites show an enhancement in the flexural properties.

14.4.4   Coating of graphene over glass fiber

Chen et al. [51] modified glass fiber with graphene oxide. Fig. 14.8 shows the protocol for the modification of glass fiber with graphene oxide. It involves the amine functional group addition over the glass fiber using silane coupling agent. Further, the amine

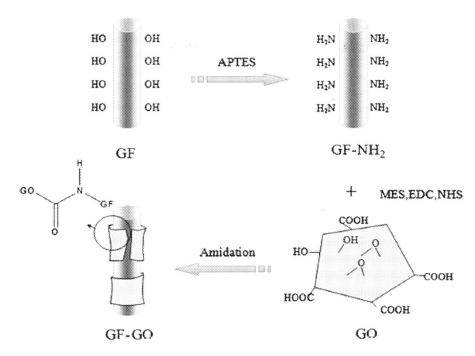

Figure 14.8 Schematic of coating of glass fiber with graphene oxide.
Source: Reproduced with permission from J. Chen, D. Zhao, X. Jin, C. Wang, D. Wang, H. Ge, Modifying glass fiber with graphene oxide: towards high-performance polymer composites, Compos. Sci. Technol. 97 (2014) 41–45.

modified glass fiber is treated with graphene oxide and other chemical so that the graphene oxide is attached to glass fiber through amidation reaction. The influence of the modification of graphene oxide over glass fiber results in an improvement in ILSS of graphene oxide grafted over glass fiber epoxy composites as compared to unmodified glass fiber epoxy composites. The ILSS enhancement is 41.3% for graphene oxide grafted over glass fiber epoxy composites as compared to epoxy/glass fiber composites. It was noted form the SEM image of the graphene oxide grafted glass fiber show rough surface. The rough surface results in mechanical interlocking between glass fiber and epoxy. In addition, the covalent bonding of graphene oxide with glass fiber also contributes to the ILSS enhancement. Thus, mechanical interlocking and chemical bonding are contributed to the ILSS enhancement of for graphene oxide grafted over glass fiber epoxy composites as compared to epoxy/glass fiber composites.

14.4.5 Graphene as a strain sensor in glass fiber/epoxy composites

Structural health monitoring is an important area where grahene high electrical conductivity was utilized to develop strain sensor. Mahmood et al. [52] coated glass fibers with graphene. They synthesized graphene oxide using modified Hummers method and coated glass fiber with graphene oxide using electrophoretic deposition technique. The glass fiber coated with graphene oxide was further treated with hydrazine vapor to reduce the graphene oxide. The electrical conductivity mesurements of reduced graphene oxide coated over glass fiber with epoxy matrix show volume resistivity of 4.5×10^2 ohm · cm. The achieved an electrically conductive nature of composites which were probed further for strain monitoring. Fig. 14.9A shows the piezoelectric response of the tenisle loaded condition monitored on the composite specimen. It shows the steady resistance response until 2.5% of strain and thereafter a dramitic influence over resistance response due to breaking of fibers. Fig. 14.9B shows the resistance of network of reduced graphene oxide coated over glass fiber under fatigue loading. It shows the reversible piezoelectricity. These results indicate strain monitoring possibility due to the control of electrical resistance. In addition, flexural, fracture toughness, and ILSS properties were enhanced for reduced graphene oxide coated glass fiber with epoxy matrix. Creep test results show the enahced creep resistance for reduced graphene oxide coated glass fiber with epoxy matrix.

14.4.6 Tribological study of graphene filled glass fiber/epoxy composites

Tribological properties of glass fiber epoxy composites needs to be studied in view of sliding contact because of abrasion. Wang et al. [53] studied the inflence of graphene dispersed in epoxy matrix and graphene coating over glass fabric of epoxy/graphene/glass fiber composites. Fig. 14.10A shows the variation of

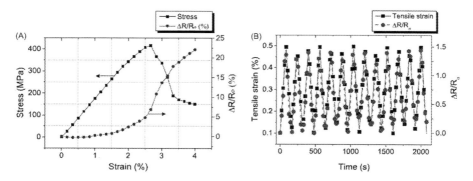

Figure 14.9 (A) Tensile loaded and (B) Piezoelectric response of tensile loaded condition of composites.
Source: Reproduced with permission from H. Mahmood, L. Vanzetti, M. Bersani, Alessandro Pegoretti, Mechanical properties and strain monitoring of glass-epoxy composites with graphene-coated fibers, Compos. Part A 107 (2018) 112–123.

coefficient of friction (COF) as a function of time for various compositions of composites. All samples show COF is divided into three stages viz initial run-in period, slow growth period, and steady state period. In initial run-in period the COF increases rapidly as there is a production of debris, which hinders the sliding movement. Followed by run-in peroid, there is a transformation of slow growth period due to formation of transfer film. Thereafter, steady state COF is reached. In the presence of graphene in composites, the length of run-in period and slow growth period reduces due to increase intercation between graphene and its sorrounding thereby decreaseing the formation of debris (Fig. 14.10B). The self lubricating properties of bucky paper do not withstand the contact pessure as it shows the similar COF profile as epoxy/glass fiber composites (Fig. 14.10C). Graphene coating oveer the glass fiber in epoxy/glass fiber composites show the gradual increase in COF till it stabilizes to steady state (Fig. 14.10D). This indicates the formation of high quality transfer fillm.

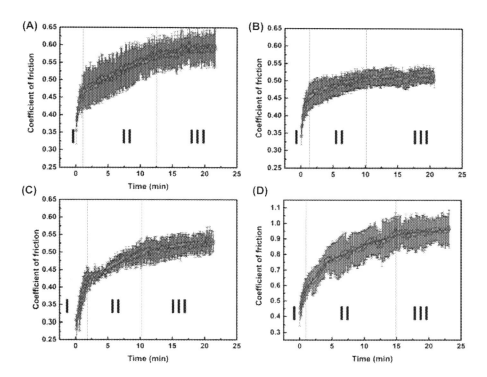

Figure 14.10 Variation of coefficient of friction of composites (A) Raw (B) With graphene (C) With bucy paper (D) With graphene coating.
Source: Reproduced with permission from B. Wang, W. Han, X. Zhang, Y. Zhu, Y. Duan, H. Wang, et al., Preparation and tribological study of graphene coating on glass fiber-reinforced composite using modified percolating assisted resin film infusion method, Materials 13 (2020) 851−866.

14.4.7 EMI shielding study of graphene filled glass fiber/epoxy composites

Glass fiber reinforced composites are light weight but the addition of graphene can improve EMI shielding performance for advanced aerspace applications. These nanomaterials filled composites requires both mechanical properites improvement and high EMI shielding effectiveness. In this regard, Marra et al. [54] coated glass fiber with graphene-epoxy mixuture with air-sprayed printing technology. After preparing graphene expoxy coated film over glass fiber, the layered coated glass fiber filled with liquid resin infusion method for preparation of composites. They found that the wide band reflection coefficent has value less than -10 dB in the range of 8–16 GHz. In addition, flexural properties of composites were improved. Wan et al. [55] coated the graphene oxide over glass fiber followed by redcution of graphene oxide (RGO). The 40 wt.% concentration of RGO coated glass fiber in epoxy matrix improved the EMI shielding effectiveness with SE value of $\sim$ 21 dB at 8 GHz. Yuchang et al. [56] processed graphene from exfoliation of expanded graphite uisng chemical oxidation followed by rapid heat treatement and sonication. The obtaine gaphene mixed with epoxy resin was used for processing of glass fiber composites. They found enhancement in flexural modulus and elastic modulus by incorporation of 2 wt.% of graphene. For same composition, reflection loss is $-$ 10 dB in Ku band range absorbing 95% EM radiation.

14.5 Glass fiber reinforced thermoplastics/graphene composites

Thermoplastics like polypropylene (PP) and its composites were widely studied in literature due to their application in the automobile industry. These composites are manufactured using the twin screw extruder operated in different temperature zone and with a high rotational speed of the screw. The shear forces applied during melt mixing process by the screw leads to the dispersion of filler in melt matrix. Followed by melt mixing of composite material, injection molding machine has to be used to give the desired shape to the composite material. Zeng et al. [57] modified the short glass fiber (SGF) with graphene oxide followed by processing PP composites using melt processing. Graphene was synthesized using modified Hummers method and used to prepare the solution of graphene oxide (GO) using sonication followed by addition of SGF. The solvent is vaporized to obtain the graphene oxide coated SGF. The graphene oxide coated SGF was melt-mixed using twin screw extrusion at a screw speed of 25 rpm and palletized further to prepare injection molding specimens. Additionally, Zeng et al. [57] added aluminum (Al) in the PP composites. The tensile test of GO coated SGF filled PP composites as a function of varying concentration of fillers was performed. It shows that with an increase in concentration of GO coated SGF in PP matrix, tensile modulus increases. But addition of fixed concentration of Al (0.5 wt.%) in PP along with

GO coated SGF in PP matrix shows significant enhancement at higher concentration of GO coated SGF in PP matrix. The GO hydroxyl functional group acts as a bridge between the SGF and PP, thus enhancing the stress transfer across fiber matrix interface.

Pan et al. [58] incorported graphene in PA6 matrix in presence of flame retardant. The flame retardant used was Al hypophosphite. The graphene incorporation helps in improving both mehanical properties, while Al hypophosphite helps in improving fire retardancy of PA6 composites. Papageorgiou et al. [59] simultaneoly added SGF and graphene in PP matrix. The simultaneous addition improves the mechancal and thermal properties. It is due to the interfacial stress transfer from graphe to glass fiber and PP matrix. SEM observation reveal the aggregation of graphene presence at high concentaton of graphene in PP/glass fiber composites, while at low concentration do not show aggregation of graphene in PP/glass fiber composites. In addition, SEM images of fractured PP composites shows the pull out of fibers, which shows the orientation of glass fiber during PP composites. The Raman spectroscopy investigation suggest the enhanced heat transfer form graphene to PP matrix. Addition of glass fiber in PP matrix decrease the glass transiton temperature and crystallinity. While addition of graphene along with glass fiber in PP matrix increases the glass transiton temperature and crystallinity. It is due to stiffening effect of graphene presence in PP/glass fiber composites. Papageorgiou et al. [60] simultaneoly added SGF and graphene in PP matrix to enhance the fire retardancy. Addition of SGF in PP matrix decrease the thermal stability of PP composites, while the presence of graphene in PP/SGF composites increse the thermal stability. It is due to nanoconfinement effect of PP in graaphene nanosheets where the gaseous PP molecule during combustion is difficult to escape. The presnce of graphen acts a barrier layer for PP/SGF composites and thus improving the concentration of char formation and thereby improving the flamme retardancy. Zhang and Han [61] incorporated graphene and SGF in PTFE marix to improve the mechanical properties.

14.6   Carbon fiber reinforced epoxy/graphene composites

Carbon fiber is a strategic load bearing material in view of its superior mechanical properties. Coating of glass fiber with graphene can create the opportunity to design the advanced composite materials for aerospace applications and in specific to load bearing applications. Gangineni et al. [62] used graphene and its derivative to coat over the carbon fiber using electrophoretic deposition technique. Graphene-coated carbon fiber epoxy composites show the rough surface. ILSS test was performed over the various composites specimens. It shows that the functionalized graphene coated over carbon fiber and its composites show the high ILSS as compared to neat composites. The reasons are due to the roughness of coated graphene results in mechanical interlocking, and chemical bonding in between functionalized graphene and epoxy matrix. Fractography images of composites show the fiber pull out in case of neat carbon fiber and graphene-coated fiber and its epoxy composites. While functionalized graphene has shown good wettability of functionalized over carbon fiber, no pull out was observed. From DMA measurements, it was noticed that from the loss modulus curves that the peak height was reduced in the case of graphene filled carbon fiber epoxy composites. It is due to viscoelastic deformation of epoxy chain at the interface. Ashori et al. [63] did functionalization of graphene oxide using different diamines. There is a chemical bonding between amine functional groups of diamine with carboxylic and epoxy functional group of graphene oxide. The *p*-phenylenediamine functionalized graphene oxide incorporated in epoxy and its carbon fiber composites shows the highest enhancement in mechanical properties. The remaining amine groups on graphene modified with diamine reacts with the epoxy matrix and thus increases the mechanical properties. Pathak et al. [64] mixed graphene oxide with epoxy matrix and prepared carbon fiber composites. There exists a hydrogen bonding between graphene oxide and epoxy. Due to interaction between graphene oxide and epoxy, the mechanical properties of composites increase. The ILSS is increased by 25% with an incorporation of graphene oxide in carbon fiber epoxy composites.

14.7 Conclusions

In this chapter, a brief review of graphene filled with polymer matrix was done. The influence of modification of graphene for dispersion of graphene in polymer matrix was discussed in detail. In addition, graphene and graphene oxide filled along with short or long fiber polymer composites were done. The research work in this area is in the initial stage. There is still huge amount of research need to be carried out in this area for successful realization of products for load bearing application in aerospace and automobile industries. Glass fabric reinforced graphene filled epoxy matrix is widely researched for load bearing to strain monitoring and other applications. The electrically conducting nature of graphene opens the opportunity of multifunctional composites materials. In addition, SGF along with graphene as filler in thermoplastics is a promising material for automobile industries. There are few research papers available on graphene reinforced carbon fiber epoxy composites which has a potential application in aerospace industries.

References

[1] G. Lubineau, A. Rahaman, A review of strategies for improving the degradation properties of laminated continuous-fiber/epoxy composite with carbon-based nanoreinforcements, Carbon 50 (7) (2012) 2377–2395.

[2] M. Saha, P. Tambe, S. Pal, G. Manivasagan, A. Xavier, V. Umashankar, Effect of nonionic surfactant assisted modification of exfoliated hexagonal boron nitride nanosheets

on the mechanical and thermal properties of epoxy nanocomposites, Compos. Interface 22 (7) (2015) 611−627.

[3] T. Siddiqui, M. Shukla, P. Tambe, Optimisation of surface finish in abrasive water jet cutting of Kevlar composites using hybrid Taguchi and response surface method, Int. J. Mach. Mach. Mater. 3 (4) (2008) 384−402.

[4] M. Shukla, P. Tambe, Predictive modeling of surface roughness and kerf width in abrasive waterjet cutting of Kevlar composites using artificial neural network, Int. J. Mach. Mach. Mater. 8 (1) (2010) 226−246.

[5] P. Tambe, A.R. Bhattacharyya, S. Kamath, A.R. Kulkarni, T.V. Sreekumar, A. Srivastav, et al., Structure property relationship studies in amine functionalized multiwall carbon nanotubes filled polypropylene composite fiber, Polym. Eng. Sci. 52 (6) (2012) 1183−1194. 6.

[6] P. Tambe, A.R. Bhattacharyya, A.R. Kulkarni, The influence of melt-mixing conditions on electrical conductivity of polypropylene/multiwalled carbon nanotubes composites, J. Appl. Polym. Sci. 127 (2) (2013) 1017−1026.

[7] K.S. Novoselov, A.K. Geim, S.V. Morozov, D. Jiang, Y. Zhang, S.V. Dubonos, et al., Electric field effect in atomically thin carbon films, Science 5696 (306) (2004) 666−669.

[8] C.G. Lee, X.D. Wei, J.W. Kysar, J. Hone, Measurement of the elastic properties and intrinsic strength of monolayer graphene, Science 5887 (321) (2008) 385−388.

[9] A.A. Balandin, G. Suchismita, B. Wenzhong, C. Irene, T. Desalegne, M. Feng, et al., Superior thermal conductivity of single-layer graphene, Nano Lett. 8 (3) (2008) 902−907.

[10] H. Kulkarni, P. Tambe, G. Joshi, Influence of processing and modification of graphene on the mechanical, thermal and electrical properties of epoxy/graphene nanocomposites: a review, Compos. Interfaces 25 (5−7) (2018) 381−414.

[11] Y. Hernandez, M. Lotya, D. Rickard, S.D. Bergin, J.N. Coleman, Measurement of multicomponent solubility parameters for graphene facilitates solvent discovery, Langmuir 26 (5) (2010) 3208−3213.

[12] A.V. Alaferdov, A. Shirazi, M.A. Chanesqui, Y. Danilov, S.A. Moshkalev, Size-controlled synthesis of graphite nanoflakes and multi-layer graphene by liquid phase exfoliation of natural graphite, Carbon 69 (2014) 525−535.

[13] M. Lotya, P.J. King, U. Khan, S. De, J.N. Coleman, High-concentration, surfactant-stabilized graphene dispersions, ACS Nano 4 (6) (2010) 3155−3162.

[14] R. Durge, R.V. Kshirsagar, P.B. Tambe, Effect of sonication energy on the yield of graphene nanosheets by liquid-phase exfoliation of graphite, Procedia Eng. 97 (2014) 1457−1465.

[15] B.C. Brodie, On the atomic weight of graphite, Philos. Trans. R. Soc. Lond. 149 (1859) 249−259.

[16] L. Staudenmaier, Verfahren zur Darstellung der Graphitsäure, Berichte der deutschen chemischen Ges. 31 (1898) 1481−1487.

[17] S. Hummers, R.E. Offeman, Preparation of graphitic oxide, J. Am. Chem. Soc. 80 (6) (1958) 1339. 1339.

[18] J. Chen, B. Yao, C. Li, G. Shi, An improved Hummers method for eco-friendly synthesis of graphene oxide, Carbon 64 (2013) 225−229.

[19] D.C. Marcano, D.V. Kosynkin, J.M. Berlin, A. Sinitskii, Z. Sun, A. Slesarev, et al., Improved synthesis of graphene oxide, ACS Nano 4 (8) (2010) 4806−4814. 6.

[20] S. Park, J. An, J.R. Potts, A. Velamakanni, S. Murali, R.S. Ruoff, Hydrazine-reduction of graphite- and graphene oxide, Carbon (49)(2011) 3019−3023.

[21] H. Shin, K. Kim, A. Benayad, Y. Seon-Mi, H. Park, I. Jung, et al., Efficient reduction of graphite oxide by sodium borohydride and its effect on electrical conductance, Adv. Funct. Mater. 19 (12) (2009) 1987−1992. 12.

[22] V.H. Pham, H.D. Pham, T.T. Dang, Chemical reduction of an aqueous suspension of graphene oxide by nascent hydrogen, J. Mater. Chem. 22 (2012) 10530−10536.

[23] G. Wang, B. Wang, J. Park, J. Yang, X. Shen, J. Yao, Synthesis of enhanced hydrophilic and hydrophobic graphene oxide nanosheets by a solvothermal method, Carbon 47 (8) (2009) 68−72.

[24] J.B. Wan, M. Sookhakian, S. Baradaran, M.R. Mahmoudian, M. Ebadi, Solid-phase electrochemical reduction of graphene oxide films in alkaline solution, Nanoscale Res. Lett. 8 (2013) 1−9.

[25] S. Chandrasekaran, S. Narumichi, T. Folke, R. Mülhaupt, B. Fiedler, K. Schulte, Fracture toughness and failure mechanism of graphene based epoxy composites, Compos. Sci. Technol. 97 (2014) 90−99.

[26] L.C. Tang, Y.J. Wan, D. Yan, B.P. Yong, L. Zhao, Y.B. Li, et al., The effect of graphene dispersion on the mechanical properties of graphene/epoxy composites, Carbon 60 (2013) 16−27.

[27] X.J. Shen, L. Yu, H.M. Xiao, Q.P. Feng, The reinforcing effect of graphene nanosheets on the cryogenic mechanical properties of epoxy resins, Compos. Sci. Technol. 72 (13) (2012) 1581−1587.

[28] J. Liang, Y. Wang, Y. Huang, Y. Ma, Z. Liu, J. Cai, et al., Electromagnetic interference shielding of graphene/epoxy composites, Carbon 47 (3) (2009) 922−925.

[29] A. Mani, P. Tambe, A. Rahaman, Flexural properties of multiscale nanocomposites containing multiwalled carbon nanotubes coated glass fabric in epoxy/graphene matrix, Compos. Interfaces 26 (11) (2019) 935−962.

[30] S.T.K. Bindu, A.B. Nair, B.T. Abraham, S. Beegum, E. Thomas, Microwave exfoliated reduced graphene oxide epoxy nanocomposites for high performance applications, Polymer 55 (16) (2014) 3614−3627.

[31] M.A. Rafiee, J. Rafiee, Z.Z. Yu, N. Koratkar, et al., Buckling resistant graphene nanocomposites, Appl. Phys. Lett. 95 (22) (2009) 223103.

[32] A. Zandiatashbar, C.R. Picu, N. Koratkar, Control of epoxy creep using graphene, Small 8 (11) (2012) 1675−1681.

[33] J. Ma, Q. Meng, I. Zaman, S. Zhu, A. Michelmore, N. Kawashima, et al., Development of polymer composites using modified, high-structural integrity graphene platelets, Compos. Sci. Technol. 91 (2014) 82−90.

[34] M. Naebe, J. Wang, A. Amini, H. Khayyam, N. Hameed, L.H. Li, et al., Mechanical property and structure of covalent functionalized graphene/epoxy nanocomposites, Sci. Rep. 4 (4375) (2014) 1−7.

[35] B. Ahmadi-Moghadam, M. Sharafimasooleh, S. Shadlou, F. Taheri, Effect of functionalization of graphene nanoplatelets on the mechanical response of graphene/epoxy composites, Mater. Des. 66 (2015) 142−149.

[36] G. Tang, Z.G. Jiang, L. Xiaofeng, H. Zhang, S. Hong, Z. Yu, Electrically conductive rubbery epoxy/diamine-functionalized graphene nanocomposites with improved mechanical properties, Compos. Part B Eng. 67 (2014) 564−570.

[37] Y.J. Wan, C.T. Long, Y. Dong, et al., Improved dispersion and interface in the graphene/epoxy composites via a facile surfactant-assisted process, Compos. Sci. Technol. 82 (2013) 60−68.

[38] C.C. Teng, C.M. Chen, H.L. Chu, et al., Thermal conductivity and structure of non-covalent functionalized graphene/epoxy composites, Carbon 9 (2011) 5107−5116.

[39] M. Saha, P. Tambe, S. Pal, Thermodynamic approach to enhance the dispersion of graphene in epoxy matrix and its effect on mechanical and thermal properties of epoxy nanocomposites, Compos. Interface 23 (3) (2016) 255−272.

[40] N. Thayumanavan, P. Tambe, G. Joshi, Effect of surfactant and sodium alginate modification of graphene on mechanical and thermal properties of polyvinyl alcohol (PVA) nanocomposites, Cellulose Chem. Technol. 49 (1) (2015) 69−80.

[41] O.M. Istrate, K.R. Paton, U. Khan, A. O'Neill, A.P. Bell, J.N. Coleman, Reinforcement in melt-processed polymer−graphene composites at extremely low graphene loading level, Carbon 78 (2014) 243−249.

[42] M. Fang, K. Wang, H. Lu, Y. Yang, S. Nutt, Covalent polymer functionalization of graphene nanosheets and mechanical properties of composites, J. Mater. Chem. 38 (2009) 7098−7105.

[43] P. Wang, H. Chong, J. Zhang, H. Lu, Constructing 3D graphene networks in polymer composites for significantly improved electrical and mechanical properties, ACS Appl. Mater. Interfaces 9 (26) (2017) 22006−22017.

[44] Y. Pang, J. Yang, T.E. Curtis, S. Luo, D. Huan, Z. Feng, et al., Exfoliated graphene leads to exceptional mechanical properties of polymer composite films, ACS Nano 13 (2) (2019) 1097−1106.

[45] C. Wu, X. Huang, G. Wang, X. Wu, K. Yang, S. Li, et al., Hyperbranched-polymer functionalization of graphene sheets for enhanced mechanical and dielectric properties of polyurethane composites, J. Mater. Chem. 22 (2012) 7010−7019.

[46] B. Yang, X. Tang, K. Yang, F. Xuan, Y. Xiang, L. He, et al., Temperature effect on graphene-filled interface between glass-carbon hybrid fibers and epoxy resin characterized by fiber-bundle pull-out test, J. Appl. Polym. Sci. 135 (19) (2018) 1−10.

[47] F. Wang, L.T. Drzal, Y. Qin, Z. Huang, Size effect of graphene nanoplatelets on the morphology and mechanical behaviour of glass fiber/epoxy composites, J. Mater. Sci. 51 (2016) 3337−3348.

[48] U. Zaheer, A. Khurram, T. Subhani, A treatise on multiscale glass fiber epoxy matrix composites containing graphene nanoplatelets, Adv. Compos. Hybrid. Mater. 1 (2018) 705−721.

[49] N. Kamar, M.M. Hossain, A. Khomenko, M. Haq, L.T. Drzal, A. Loos, Interlaminar reinforcement of glass fiber/epoxy composites with graphene nanoplatelets, Compos. Part A Appl. Sci. Manuf. 70 (2015) 82−92.

[50] R.K. Prusty, S.K. Ghosh, D.K. Rathore, B.C. Ray, Reinforcement effect of graphene oxide in glass fibre/epoxy composites at in-situ elevated temperature environments: an emphasis on graphene oxide content, Compos. Part A Appl. Sci. Manuf. 95 (2017) 40−53.

[51] J. Chen, D. Zhao, X. Jin, C. Wang, D. Wang, H. Ge, Modifying glass fiber with graphene oxide: towards high-performance polymer composites, Compos. Sci. Technol. 97 (2014) 41−45.

[52] H. Mahmood, L. Vanzetti, M. Bersani, A. Pegoretti, Mechanical properties and strain monitoring of glass-epoxy composites with graphene-coated fibers, Compos. Part A 107 (2018) 112−123.

[53] B. Wang, W. Han, X. Zhang, Y. Zhu, Y. Duan, H. Wang, et al., Preparation and tribological study of graphene coating on glass fiber-reinforced composite using modified percolating assisted resin film infusion method, Materials 13 (2020) 851−866.

[54] F. Marra, J. Lecini, A. Tamburrano, L. Pisu, M.S. Sarto, Electromagnetic wave absorption and structural properties of wide-band absorber made of graphene-printed glass-fibre composite, Sci. Rep. 8 (2018) 12029−12038.

[55] X. Wan, H. Lu, J. Kang, S. Li, Y. Yue, Preparation of graphene-glass fiber-resin composites and its electromagnetic shielding performance, Compos. Interfaces 25 (10)
(2018) 883−900.

[56] Q. Yuchang, J. Wang, H. Wang, F. Luo, W. Zhou, Graphene nanosheets/E-glass/epoxy
composites with enhanced mechanical and electromagnetic performance, RSC Adv. 6
(2016) 80424−80430.

[57] L.Z. Xiao, C. Wu, B. Tang, X.J. Shen, K. Liu, B. Ye, et al., Spray free polypropylene
composite reinforced by graphene oxide@short glass fiber, Polym. Compos. 41 (4)
(2020) 1215−1223.

[58] Y. Pan, N. Hong, J. Zhan, B. Wang, L. Song, Y. Hu, Effect of graphene on the fire and
mechanical performances of glass fiber-reinforced polyamide 6 composites containing
aluminum hypophosphite, Polym. Technol. Eng. 53 (14) (2014) 1467−1475.

[59] D.G. Papageorgiou, I.A. Kinloch, R.J. Young, Hybrid multifunctional graphene/glass-
fibre polypropylene composites, Compos. Sci. Technol. 137 (2016) 44−51.

[60] D.G. Papageorgiou, Z. Terzopoulou, A. Fina, F. Cuttica, G.Z. Papageorgiou, D.N.
Bikiaris, et al., Enhanced thermal and fire retardancy properties of polypropylene reinforced with a hybrid graphene/glass-fibre filler, Compos. Sci. Technol. 156 (2018)
95−102.

[61] B. Zhang, J. Han, Strong nanocomposites reinforcement effects in PTFE/glass fabric
composites modified with graphene, Compos. Interfaces 26 (6) (2019) 525−535.

[62] P.K. Gangineni, S. Yandrapu, S.K. Ghosh, A. Anand, R.K. Prusty, B.C. Ray,
Mechanical behavior of Graphene decorated carbon fiber reinforced polymer compo-
sites: an assessment of the influence of functional groups, Compos. Part A Appl. Sci.
Manuf. 122 (2019) 36−44.

[63] A. Ashori, H. Rahmani, R. Bahrami, Preparation and characterization of functionalized
graphene oxide/carbon fiber/epoxy nanocomposites, Polym. Test. 48 (2015) 82−88.

[64] A.K. Pathak, M. Borah, A. Gupta, T. Yokozeki, S.R. Dhakate, Improved mechanical
properties of carbon fiber/graphene oxide-epoxy hybrid composites, Compos. Sci.
Technol. 135 (2016) 28−38.

Polymer-graphene composites as sensing materials

Prashant Gupta
Department of Plastic and Polymer Engineering, Maharashtra Institute of Technology, Aurangabad, India

15.1 Introduction

Sensors are integral to our daily life. In fact, we live in a world full of sensors that can be found in our homes, offices, cars, etc., which make our lives easier by employing smarter means for manual work such as turning on/off the light, detecting a vehicle in front while driving forward and reverse, detecting smoke, cooking the food in a microwave, setting up a burglar alarm, automatically opening/closing glass doors at shopping malls, etc., and turning it to automatic via their deployment. A sensor can be defined as an input device programmed to detect events or changes in its environment to send the detected information to a processing system or simply a device to convert signals from one energy domain to the electrical domain. To understand the functioning of a sensor, let us take the example of autopilot mode in Tesla's automobile. An automatic drive control system comprises of various sensors for tasks such as speed control, positioning, obstacles, lane number, visibility through heavy rain, fog, dust, energy consumption, etc. A set of predesigned data is used by a computer for comparing the sensor data in order to process it for asserting control over various parameters such as speed, acceleration, braking effectiveness, fuel efficiency, etc. that aid in a smooth ride which is only possible through a combined effort of sensors, computers, and mechanics involved. As shown in Fig. 15.1, these parameters such as, the sensors (input), computer (processing system) and mechanics (output) are just as important as other factors in order to successfully build and run an automatic system. Furthermore, the sensors present in the Tesla car records its perceptive opinion of the surroundings and feeds it to a massive database which processes it and helps to make the autopilot system smarter for all associated automobiles made by them [1].

Figure 15.1 Auto drive system components for a four-wheeler automobile.

Polymer Nanocomposites Containing Graphene. DOI: https://doi.org/10.1016/B978-0-12-821639-2.00017-3

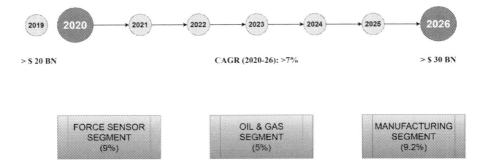

Figure 15.2 Sensors market growth data [3].
Source: Adapted from P. Wadhwani, S. Yadav, Industrial Sensors Market Size By Product (Level Sensor, Temperature Sensor, Flow Sensor, Position Sensor, Pressure Sensor, Force Sensor, Image Sensor, Gas Sensor), By Application (Energy & Power, Oil & Gas, Mining, Chemical, Manufacturing, Pharmaceutical), Industry Analysis Report, Regional Outlook, Application Potential, Competitive Market Share & Forecast, 2020—2026, 2020. https://www.gminsights.com/industry-analysis/industrial-sensors-market (accessed December 21, 2020).

The global sensor-based market is expected to grow at a compounded annual growth rate of 11.4% from $193.9 billion in 2020 to $332.8 billion by 2025 [2]. The industrial sensor market accounts for around 10% of the total market share with around $20 billion in 2019 and an expectancy of a 7% growth over the next 6-year cycle as shown in Fig. 15.2. Amongst the various industrial verticals, the growth is expected to be in the oil and gas, manufacturing, healthcare, energy, and power sectors. The market size is further driven by the implementation of advanced technologies such as cloud computing, robotics, Internet of Things (IoT) along with industrial sensors, etc. Furthermore, the emergence of Industry 4.0 methodology will further enhance the use of IoT for enhancement of productivity, performance, and efficacy in manufacturing by the deployment of smart and wireless sensors in order to collect important information related to the process of manufacturing and supply chain. A 51% market share is owned by the Asia Pacific due to the initiatives taken by local governance. India, for instance, has enjoyed good growth in the manufacturing sector via. the make in India initiative laid out by the government which has inspired local manufacturing for several products within the country for the economy and market growth [3].

Graphene is a functional material with intrinsically high thermal and electrical conductivities. These properties combine with its high aspect ratio to attain the percolation threshold at a low volume fraction of filler loading in the polymeric material which makes it suitable for use as sensing materials. Graphene sheet atoms are surface-based and responsible for a highly sensitive nature towards adsorbed molecules in terms of molecular interaction that is, electron transport through it [4,5]. It is a two-dimensional nanomaterial that provides good reproducibility for devices using it on a large scale. Also, its monolayer's whole volume adds to the sensing effect as it is totally exposed to the environment. Graphene functions as electron donors or acceptors as the device's conductance changes due to adsorption (chemical/biological) on the surface [6]. For instance, due to graphene's remarkable electronic transportation property along with high electrocatalytic activity, analyte's electrochemical reactions are greatly promoted on graphene film leading to an enhanced voltammetric response. Furthermore, additional functional materials such as nano catalysts can improve the electrochemical properties of graphene resulting in the versatility of electrochemical sensory performance.

Graphene offers superior properties to that of polymers alone, thus prompting its use with them. This has given resulted in the fabrication/synthesis of graphene-polymer nanocomposites as they potentially have better electrical, thermal. mechanical, gas bar-rier, flame retardance, etc. compared to the pristine polymer [7,8]. The improvement in mechanical and electrical properties is in fact much better than fillers such as clay or other carbon-based fillers. Carbon nanotubes (CNTs) offer comparable properties in terms of mechanical strength but certainly lacks in electrical and thermal aspects against graphene [9,10]. Also, they exhibit high conductivity, production ease, biocom-patibility, and abundance of the precursor which is used as the starting material for its synthesis [5,11]. It thus boasts of the right features that make it a suitable candidate for advanced electrode materials and can be integrated with other functional materials to enable fabrication of sensing interface for electroanalysis [12,13].

The physicochemical nature of graphene-polymer nanocomposite is reliant on the layer distribution of graphene within the matrix and bonding of the interface between them. Graphene family materials consist of ultrathin graphite, few-layer graphene, graphene oxide, reduced graphene oxide, and graphene nanosheets [14]. They have varying layer number, size, and surface properties [15]. The compatibil-ity of neat graphene is not good with organic polymers that result in heterogeneous composites. Out of all the materials listed above, graphene oxide is a very important graphene derivative that is produced via the Hummers method using energetic oxi-dation reaction of graphite. The reactive surface functional sites facilitate the con-nection with a host of materials such as biomolecules, proteins, DNA, and polymers [16−20]. This has resulted in a significant attraction toward graphene oxide as a nanofiller for polymer composites in sensory applications [21].

15.2 Classification of sensors

There are various classifications that sensors may be classified based on the source of power requirement into active and passive types. Active sensors, also known as the self-generating type, do not require an external power signal for excitation of the sen-sor as they generate power within themselves. For instance, a piezoelectric crystal is capable of self-generating the charge as an electrical output when put under

Table 15.1 Properties measured by sensors with various forms of energy [22].

Energy type	Properties measured
Mechanical	Pressure, length. Force, area, volume, acoustics
Chemical	Concentration, rate of reaction, pH, redox potential
Radiant	Wavelength, reflectance, transmittance, refractive index
Thermal	Temperature, specific heat, heat flux
Magnetic	Field intensity, magnetic moment
Electrical	Voltage, current, charge, resistance, capacitance

Source: Adapted from F. Regan, Sensors: Overview, in: P. Worsfold, C. Poole, A. Townshend, M. Miró (Eds.), Encyclopedia of Analytical Science, third edition, Academic Press, Oxford, 2019, pp. 172–178. https://doi.org/10.1016/B978-0-12-409547-2.14540-8.

acceleration thereby ensuring that energy generation is done from the quantity being measured. Alternatively, passive sensors require an external power source as they are not capable of generating the power response on their own for, for example, a host of inductive and resistive sensors are of the passive type. The second classification is based on whether the output interface is analog or digital. Analog sensors are capable of providing a continuous output signal with respect to the parameter under observation in the form of analog output for, for example, thermocouple, strain gauge, etc. whereas digital sensors function with digital data are employed for transmission and conversion and produce an output in the form of a pulse for, for example, encoders. Thirdly, as an alternative to true sensors, there are inverse sensors that are based on the phenomena of conversion of input and output. They are capable of sensing a physical parameter and convert it into another form along with the ability to sense the output signal form to get back the parameter in the original form. Some of the common examples include electrochemical, electromagnetic, piezoelectric, thermoelectric, etc.

Sensors respond to an employed stimulus by generating an electrical signal which must correspond in an expected manner to the stimulus that might be mechanical, radiation, thermal, electrical, optical, biological, acoustic, chemical, and so on. The conversion of a signal from a physical form (Table 15.1) to another is facilitated by transducers as they have a pivotal role to play in sensing technology for the 21st century. Hence, the most relevant classification for the understanding of this chapter will be based on mechanical, thermal, and chemical changes that cause a change in energy/electrical resistivity within an electrochemical cell. We will discuss a variety of specific application-based sensors under the broad categories wherein graphene polymer composites find their use.

15.3 Mechanical strain-based sensors

The material of choice for fabrication of strain sensors have been metals and semiconductors post the reports of resistance change due to elongation in iron and copper by Thompson in 1856 [23–26]. They work by noticing the variation in voltage/resistance of a material in response to changes in its length [27]. These strain gauges exhibiting high sensitivity are cheaper in comparison with polymeric strain sensors. On the other hand, they cannot be integrated into the structural modules effectively and offer a low nanoscale resolution [28].

The strain sensing behavior of conductive thermoplastic polyurethane (TPU) matrix composite filled with graphene is reported by Liu et al. A 2D flake shaped graphene structure enabled a plane-plane contact leading to a conductive network formation that resulted in good sensing stability and high sensitivity along with good recoverability and reproducibility for varying strain patterns after stabilization through the cyclic load. They concluded that a host of applications require varying demands to be fulfilled with regard to strain sensing patterns and strain response patterns that can be engineered on the basis of application requirements are desired [29]. Upon application of external strain, the strain response is simply addressed by quantifying the resistance variation, which arises due to tunneling or simply change of local contacts between the conducting fillers that are uniformly spread in the matrix. In addition to filler which is conductive in nature, its morphology also plays a pivotal effect in strain sensing. The addition of graphene as a bifiller along with CNT into polyester-based thermoplastic polymer matrix was studied by Liu et al. Graphene worked as a spacer to individualize the entwined CNTs wherein CNT played the role of a link in between single graphene leading to better CNT dispersion. A conductive path formation was observed due to which electrical conductivity was better at lower filler loading. In comparison with CNT/TPU based strain sensors, the bifiller based sensors exhibited a single peak response pattern under small strain against that of dual peak response pattern of the former indicating engineering of the sensor properties on the basis of strain requirements. Under higher strain, the conductive network was prestrained and tunable single peak response patterns were obtained. Furthermore, the composites also exhibited good reversible and reproducible nature under cyclic extension [30].

Eswaraiah et al. synthesized and demonstrated a novel real-time strain sensor-based response of functionalized graphene oxide-polyvinylidene fluoride (f-GO-PVDF). The functionalization of graphene was done enabling the introduction of hydroxyl and carboxyl groups to aid dispersion in the polymer matrix. They used a shear-based solvent casting method that enabled a 2D crosslinked graphene network within the PVDF matrix. As the distance between two layers of graphene determines the current conduction methodology, Fig. 15.3 demonstrates the tunneling effect resulting in current flow between two electrodes from one secluded graphene layer to another, and an exponential change of resistance is hypothesized with the change in distance. A 2% filler loading in f-GO-PVDF films exhibited better strain sensing behavior than polymer composites based on CNT. The change in voltage followed the cyclic behavior which indicates the bending and relaxation of graphene on a continual basis. The flexibility of graphene along with the intrinsic changes in its contact and tunneling resistance within the PVDF polymer matrix is largely responsible for this level of strain sensing performance [31,32].

Qin et al. reported light, elastic and flexible graphene/polyimide films for strain sensing by altering reduced graphene oxide (rGO) aerogel from fragile into super

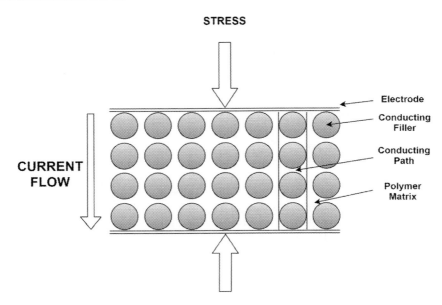

Figure 15.3 A representation of tunneling effect observed in graphene polymer composites. *Source*: Adapted from X.-W. Zhang, Y. Pan, Q. Zheng, X.-S. Yi, Time dependence of piezoresistance for the conductor-filled polymer composites, J. Polym. Sci. Part B Polym. Phys. 38 (2000) 2739−2749. https://doi.org/10.1002/1099-0488(20001101)38:21 < 2739:: AID-POLB40 > 3.0.CO;2-O.

extensible 3D architecture with the introduction of the water-soluble polyimide polymer matrix. The nanocomposites were fabricated using freeze casting and thermal annealing techniques. The nanocomposites were durable and possessed the required electrical conductivity and compression stability. The scanning electron microscopy images of rGO aerogel, PI monolith, and rGO/PI nanocomposite is shown in Fig. 15.4A−C. Fig. 15.4A reveals that rGO aerogel had a porous structure that was randomly oriented throughout the material made via. interlinked rGO sheets by the fractional overlapping of pi-pi interactions in a 3D configuration. A higher magnification image below shows thin pore boundaries, where the wrinkles/ folds of its sheets are seen. Fig. 15.4B shows lamellar structures of PI monoliths with struts linking the adjoining layers which were due to the structural changes happening after the synthesis was done by freeze casting technique. The spatial gap within the layers was observed to be in between nano to microscale. Fig. 15.4C shows the micrographs for rGO/PI nanocomposite which exhibited an unusual layer-strut bracing architecture resulting from an increase in single layer thickness and its internal distance. A magnified image of the strut shown below revealed the wrinkles, folds, and ripples which are alike the rGO sheets indicating the strong interconnection between them and PI matrix blocks. This, in turn, promotes efficient load transfer under mechanical deformation that steers improvement of mechanical properties. The fabricated sensors were light, super elastic, flexible, and thus had a high recovery rate

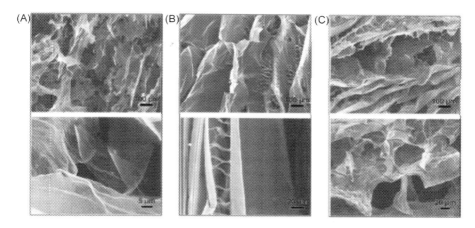

Figure 15.4 Scanning electron micrographs of (A) rGO aerogel, (B) PI monolith, and (C) rGO/PI nanocomposite at low (above) and high (below) magnifications [33].
Source: Adapted from Y. Qin, Q. Peng, Y. Ding, Z. Lin, C. Wang, Y. Li, et al., Lightweight, superelastic, and mechanically flexible graphene/polyimide nanocomposite foam for strain sensor application, ACS Nano 9 (2015) 8933–8941.

along with amazing reversible compressibility which makes them suitable for use in compression, flexure, tension, and torsion as hypothesized in Fig. 15.5. The unloading curves indicated that rGO/PI composites completely recover without any plastic deformation which is distinct and is comparable to the recovery of an open cellular foam material. The hysteresis curves for 2000 cycles of fatigue that is, cyclic load application support the structural robustness as no significant plastic deformation is observed and the energy loss coefficient initially decreases before eventual stabilization. However, there is a small decrease in the stiffness over cyclic load but it retained over 88% of maximum stress and just a 3% change in strain which is considered to be stable over 2000 cycles [33].

Wearable strain sensors are the core of flexible electronics. However, the requirements of wearable strain sensors are not satisfied with usual strain measuring sensors as they need to be light, compliant, and should be sensitive enough to detect processes like breathing, heartbeat, etc. Also, they should be able to function accurately at high strain motions along with functionality to track fast intended and unintentional motions at high speeds and subsequent strain rates. Boland et al. demonstrated the use of graphene-based high strain, high rate, sensitive body wearable motion sensors in an elastomeric matrix. They reported them to be light, cheap, mechanically compatible along with reasonable sensitivity at high strain values and subsequent rates. A simplified method to infuse graphene exfoliated in a liquid into a natural rubber matrix for making conductive composites. They experimented with the graphene-based band (G-Band) for its use for body-oriented motion sensors to observe joint and muscle motion along with pulse and breathing rate. The G-bands were tested to ascertain their broad dynamic range as strain sensors for fingers,

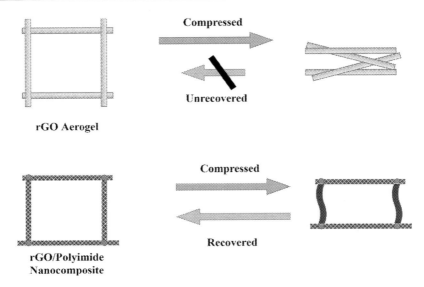

Figure 15.5 Schematic diagram for compressive recovery rGO aerosol and rGO/PI nanocomposite [33].
Source: Adapted from Y. Qin, Q. Peng, Y. Ding, Z. Lin, C. Wang, Y. Li, et al., Lightweight, superelastic, and mechanically flexible graphene/polyimide nanocomposite foam for strain sensor application, ACS Nano 9 (2015) 8933−8941.

hand, and throat motion which exhibited 40%, 4%, and very minor strain motions, respectively. The sensitivity of the throat-based sensor, which was attached to basically measure speech-related changes was so high that it also sensed the neck strain movement which was involuntary. The Gauge Factors (GF) of around 35 were measured thereby indicating reasonably high sensitivity. Furthermore, the study showed a 10,000-fold increase in resistance along with exhibiting its performance above 800%. They could track dynamic strain effectively at vibration frequencies of 160 Hz on the lower side. The monitoring of at least 6% strains could be done at a strain rate of 6000% per second or higher at a frequency of 60 Hz [27].

Liu et al. in another study approached the fabrication of strain-based sensor in a nontraditional manner wherein the styrene acrylate emulsion polymer layer was sandwiched in between the graphene layer and the rubber substrate. The GF was measured to be around 35, which was significantly higher than commercially available conductive metal-based strain gauges. To ascertain the long-term stability of the sensory device, reliability tests were conducted and the strain was identified to be around 5% during cyclic tests. The GF was almost similar during the 100 cycles over which the standard deviation was measured to be 2.3% which could be attributed to lower abrasion between sliding layers. Upon studying the GF changes under varying strain levels of 5%, 10%, 15%, and 25% as a function of the initial level of resistance, an increase in GF with an increase in sheet resistance was observed. Also, the varying behavior of the curve indicated three sections wherein the first is continually changing with the increase in sheet resistance, the second section is observed as power-law fulfilling range in log-log plot for 5% strain, and the third section of higher resistivity samples showing a larger sensitivity to change in strain values. This is a major change over the traditional metallic sensors where GF is a constant. The tunability of GF can fulfill the requirements of conventional gauges along with making new ones modified to factor the individualized use such as high GF for applications having low strain or low GF for applications having higher strain factors [34].

Costa et al. developed styrene ethylene butylene styrene-graphene composites for highly sensitive stretchable piezoresistive strain sensors based on GO, rGO, and graphene nanoplatelets (GNP). The percolation threshold for GO and rGO were three times lower than that of GNP. Also, GO and rGO based composites exhibited a higher variation in resistivity upon application of strain with GF of 15−120 and 10−90, respectively up to 10% strain values as shown in Fig. 15.6 with 4% filler loading. Also, a suitable response was observed over 1000 cycles with a slight decrease in the first 50 cycles and further stabilization in values for the further cycles with no effect of initial decrease on GF. The sensors were successfully tested on hand gloves with good sensory activity, that is, readability and repeatability [35].

D'elia et al. exhibited a self-healing graphene-polyborosiloxane (rGO-PBS) composite which has strain sensing capabilities. The self-healing capabilities were owing to the solid-liquid behavior of PBS polymer which is reported to be retained even after high loadings of fillers like clay, graphene, etc. They were reportedly tested for self-repair over 6−8 cycles without any loss of structural healing capabilities. The

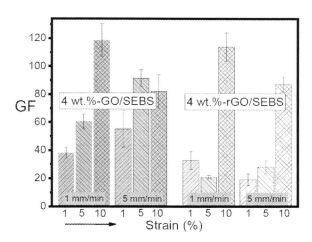

Figure 15.6 GF of 4% GO and rGO filled SEBS composites for 1%, 5% and 10% of strain at 1 and 5 mm/min strain rate [35].
Source: Adapted from P. Costa, S. Gonçalves, H. Mora, S.A.C. Carabineiro, J.C. Viana, S. Lanceros-Mendez, Highly sensitive piezoresistive graphene-based stretchable composites for sensing applications, ACS Appl. Mater. Interfaces 11 (2019) 46286−46295. https://doi.org/10.1021/acsami.9b19294.

filler and polymer complement each other as the graphene scaffold deforms well after the addition of PBS due to rearrangement and sliding of graphene surface walls in the polymer composite during tension. There was a linear relationship between pressure applied and voltage measured up to 500 kPa pressures with reproducible results. The sensors could be made into flexible films of 500 microns or lower thickness for flexion sensing applications with a fully repeatable cycle [36].

15.4   Chemical sensors

Chemical sensors are sensory devices that transform a chemical/physical property of a specific analyte, component, or chemical species into a quantifiable signal, whose magnitude is proportional to the concentration of the material under detection [37]. They are usually based on one molecule in which there are two key components: first functioning as a receptor/input which selectively interacts with the analyte and the second is responsible for varying a chemical/physical property as output/read-out reaction to the binding event. These molecules may have a spacer, thereby working as a connector in between the receptor and read-out [38]. In short, it consists of a physical transducer and a chemically sensitive layer. The chemically sensitive layer reacts with the chemical environment to generate a measurable signal through transduction. The composition is critical as it controls various factors such as selectivity, sensitivity, time of response, and durability. A good chemical sensor is identified by signal-to-noise ratio and quick response which is both sensitive and selective [22]. These sensors emit response signals upon detecting the presence of certain chemicals for which they are designed. There are various mechanisms on which sensing may depend upon, such as, charge carrier concentrations, electroactive properties of materials, chemical reactions (redox), etc. A variety of chemical sensors are used in biological, environmental, and industrial monitoring of the presence of proteins, gases, vapors, enzymes, etc., and hence have drawn great attention of the research fraternity to improve their effectiveness and efficiencies along with improving the product range [39].

The polymer matrix in graphene-polymer composite swells upon detection of organic vapors/liquids which, in turn, changes the connectivity between the graphene, thereby resulting in an increase in sample resistance which serves as the basis of the manufacture of vapor and liquid sensors for detection of changes in nearby environmental conditions. Olean-Olivera et al. developed rGO-poly(azo-Bismarck Brown Y) (rGO-BBY), a π-conjugated azo polymer-based chemiresistor sensor for detection of dissolved oxygen due to the redox properties of the azo polymer. The sensor showed excellent ability to detect molecular oxygen in acidic and slightly alkaline solution due to the affinity of the hydrazine group to molecular oxygen. The resistance of the sensor electrode based on fluoride-doped tin oxide coated with rGO-BBY linearly increases with decreasing oxygen content in the electrolyte. The sensitivities were determined using two calibration curves and were 0.204 and 0.133 kΩ cm^{2}L/g for oxygen detection at varying pH of 1 and 7.4, respectively. Furthermore, based on the electrochemical impendence spectroscopy data, the variation was as little as 4.1% in the overall response of the chemiresistor per cycle of regeneration [40]. There are reports for polymer-graphene chemiresistive sensors for the detection of chemical warfare agents [41], graphene-polyethylene amine/polyethylene glycol for the detection of carbon dioxide via charge transfer mechanism [42], trimethyl silane propyl methacrylate modified GO/2-hydroxy methacrylate/methyl methacrylate [TMSPMA-GO-co-HEMA/ MMA] for sensing creatinine in blood and urine samples [43], graphene/polyaniline based ultra-sensitive sensor for the detection of urea in food, water, dairy items, and environmental monitoring [44].

Wang et al. synthesized pH-responsive hydrogels, based on GO doped acrylic acid-hydroxyethyl acrylate copolymer to fabricate a flexible sensor for pH-responsive environment that was tested between a pH of 2−8, resulting in increasing diameter and gradual change in the ionization degree of the carboxyl functional group along with the ion concentration in the hydrogel [45]. Paek et al. reported the development of a colorimetric pH sensor based on polyacrylic acid/poly 2 vinyl pyridine-quantum dot doped GO. The sensing system was based on a responsive polymer and quantum dot hybrids put together on one GO sheet with the composite providing high water dispersibility, excellent reversibility, and high signal-to-noise ratio making it a suitable material for use in bioenvironmental applications [46]. Liu et al. has reported pH-sensitive graphene modified with pyrene-terminated poly (2-N,N'-(dimethyl amino ethyl acrylate)) (PDMAEA) and poly(acrylic acid) (PAA) via $\pi - \pi$ stacking that demonstrated phase transfer behavior between aqueous and organic media at pH values varying from 4 to 11 [47].

According to Wang et al., the chemical sensory systems can be divided into three different categories on the basis of the method of operation and working principle that is, potentiometric, amperometric, and impedimetric transducers converting the chemical information into a measurable amperometric signal. A host of measurement and conversion techniques for, for example, cyclic voltammetry, square-wave voltammetry, differential pulse voltammetry, electrochemical impedance spectroscopy, etc. to name a few for detection organic molecules, liquid chemicals, gases or vapors, biomaterials, etc. as given in Table 15.2 [58−61].

15.5 Thermal sensors

Thermal sensors are construction elements to measure temperature and engage functional dependence of a particular physical property of the sensory material on the temperature, which is routinely identified and well defined [62]. The need for these sensors is imminent in manufacturing and process monitoring industries such as chemical plants, agriculture, automotive, power plants, aerospace, logistics, medical, health, safety, etc. [63].

Kong et al. reported the development of rGO-based temperature-sensitive electrode via. inkjet printing of rGO onto Polyethylene Terephthalate (PET) which served as a temperature-sensitive substrate with an electrical resistance of 0.3 MΩ/square

Table 15.2 Recent reported literature on developments in chemical sensors for sensing of liquid, gas, vapor, biomaterials etc.

S. no.	Filler-polymer sensory system	Chemical detected	Observations	Reference
1	PMMA supported graphene film on polymer optical fiber	$(CH_3)_2$ CO Vapor	Capable of sensing acetone vapor concentrations as low as 44 ppm	[48]
2	Polydiacetylene-graphene film	VOC such as $CHCl_3$, CH_3OH, DMF and THF	Extremely sensitive response to low concentrations ($\sim 0.01\%$) of VOC	[49]
3	rGO/MnO_2 based PANI	NH_3	Composite sensory devices made up of RGO–PANI hybrid device showed a much better response to NH_3 gas than those of the PANI nanofiber sensor and bare graphene device (3.4 and 10.4 times/50 ppm of NH_3)	[50]
4	White graphene reinforced water soluble polypyrrole/PVOH	Liquified petroleum gas C_6H_6, $CHCl_3$, C_2H_5OH and $(CH_3)_2$ CO vapors	A maximum sensitivity $S = 0.25\%$ LPG/600 ppm concentration at room temperature with response times of around 30–32 min in a PVA/WPPy matrix. The nanocomposite with 6 wt.% filler loading shows good selectivity for LPG over vapors of related chemicals	[51]
5	Graphene-Sodium polystyrene sulfonate	NO_2, C_2H_5OH vapor and NH_3	Reversible color change of film from blue to yellow as it is sensitive to the detection of nitrogen dioxide with good selectivity as tested for nine other volatile chemical compounds	[52]
6	Graphene-polyurethane	C_6H_{12} and CCl_4	Novel negative vapor coefficient for tested vapors along with fast response, good reversibility and reproducibility for cyclohexane and chloroform	[53]

7	rGO-Cellulose	CH_3OH	The increase in resistance in the range of 7%–40% methanol with increasing signal sensing with increasing methanol vapor concentration with fast reaction, good reproducibility, high sensitivity, and good differentiating ability between various organic vapors	[54]
8	GO-PANI	CH_3OH	GF for methanol sensing is about 21–37 with good reversibility and response time in comparison with 3.77 and 3.1 for ethanol and propanol, respectively	[55]
9	Poly (3, 4-ethylenedioxythiophene) (PEDOT) doped with graphene oxide (GO)	Dopamine	The fabricated electrode exhibited lower electrochemical impedance and excellent electrochemical ability towards oxidation of dopamine and capability of sensitive and selective detection of dopamine free from uric and ascorbic acid interferences	[56]
10	GO-PANI	Deoxyribonucleic Acid (DNA)	Vertical array-based PANI-GO nanocomposites exhibited the highest sensitivity which is two order of magnitudes lower than others offering good balance for DNA immobilization and hybridization	[57]

and optical transparency of 86% which is tunable to the respective needs. The electrode behaved as a negative temperature coefficient (nonmetallic behavior) material, showing a sharp deterioration in electrical resistance with an increase in temperature. There was a faster response time observed by an order of magnitude along with sensitivity, similar to those of conventional materials where the dR/dT is negative. The potential of such a composite electrode can be used as a writable, flexible, extremely thin, and transparent temperature sensor. Fig. 15.7 exhibits the results for the evaluation of temperature sensory role upon tapping the electrode with a finger-tip at room temperature. Due to human touch, a change in temperature over the sensor is observed with a decrease in resistance. However, when tapped with any other nonliving objects which were at equilibrium with room temperature, no change was observed for the electrode resistance (not shown in the figure) which indicated that these changes were a result of heat transfer between the tip of the finger and the electrode due to variation in temperature between both [64].

A positive temperature coefficient effect was reported by Tian et al. They developed a temperature sensor based on GNP/silicon rubber/silver paste/interdigital electrode over polyimide substrate construction fabricated via. preheat process (150°C) to improve the stability and sensitivity of sensors. The experimental data collected of tests carried out from −40°C to 150°C exhibits that an initial decrease in resistance is observed with increasing temperature followed by a positive temperature coefficient effect above 30°C. The temperature sensing tests exhibited good linearity ($\sim$0.996%) and high temperature coefficient of resistance or TCR ($\sim$0.0371/°C) in the region of 10°C−60°C which is much higher than MWCNT and other fillers as per published literature. They reported great repeatability and small hysteresis with a time constant of 5.1 s, respectively which is much shorter

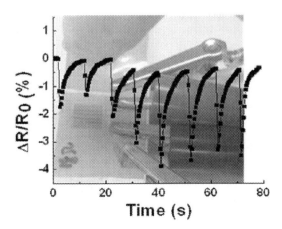

Figure 15.7 Responses in relative electrical resistance upon repeated fingertip tapping with experimental setup for finger tapping in the background [64].
Source: Adapted from D. Kong, L.T. Le, Y. Li, J.L. Zunino, W. Lee, Temperature-dependent electrical properties of graphene inkjet-printed on flexible materials, Langmuir 28 (2012) 13467−13472.

than 12 s reported for MWCNTs and comparable to that of 3.6 s for mercury thermometer and 5 s for Pt100 [65].

Liu et al. reported the fabrication of lightweight, low-cost, flexible temperature sensors based on rGO, single and multiwalled carbon nanotubes with PET for use in the relevant environments of IoT. The graphene sensors reported the most balanced performance in terms of linearity, sensitivity, and repeatability with minimal resistance change upon bending into different angles. Fig. 15.8 represents the experimental characterizations of rGO-PET sensors with a, b, and c figures showing a linear change in resistance with a change in temperature, relative resistance changes in rGO vs CNT based sensors showing variation in resistance decreasing linearly in rGO based ones along with repeatability and stability in resistance values tested for three cycles of heating and cooling which makes them a suitable material for temperature sensors. Also, the sensitivity was reported to be $0.6345\%/^\circ C$ which was almost 10 times that of CNT-based sensor values of $0.068\%/^\circ C$. Under varying conditions of humidity, and the presence of some gases (water vapor, oxygen, carbon dioxide, ammonia, vapors of ethanol, formaldehyde, methanol. and acetone), the performance of these sensors remained stable and reproducible as tested for three cycles of heating and cooling [66].

Yang et al. reported the development of P7AC-b-PNIPAM-b-PSN3 triblock copolymers that comprise of poly(7-(4-(acryloyloxy)butoxy)coumarin) (P7AC) as the fluorescent part, poly(N-isopropylacrylamide) (PNIPAM) as the thermally responsive polymer, and a short poly(azidostyrene) (PSN3) block for thermally responsive sensor/switch. PNIPAM has extensively found its use in such systems due to its biocompatibility, solubility in water, and a Low Critical Solution Temperature (LCST) of $32^\circ C$. In the reported graphene-polymer composite, covalent alteration of PNIPAM was achieved with fluorescent chemical moieties. A strong photoluminescent effect was observed below the LCST as swelling took place in the polymeric chains of PNIPAM. On the other hand, contraction of PNIPAM chains occurred above LCST, thereby reducing the gap between fluorescent moiety and the GO sheet surface which brought about a sort of quenching

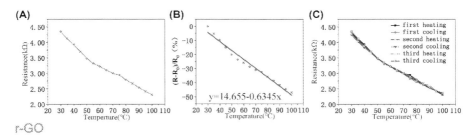

Figure 15.8 Experimental characterizations of rGO-PET temperature sensor for (A) the resistance change; (B) the relative resistance change in the rGO versus CNT based sensors for $30^\circ C-100^\circ C$; (C) the resistance responses of sensor for three heating-cooling cycles [66]. *Source*: Adapted from G. Liu, Q. Tan, H. Kou, L. Zhang, J. Wang, W. Lv, et al., A flexible temperature sensor based on reduced graphene oxide for robot skin used in Internet of Things, Sensors 18 (2018) 1400. https://doi.org/10.3390/s18051400.

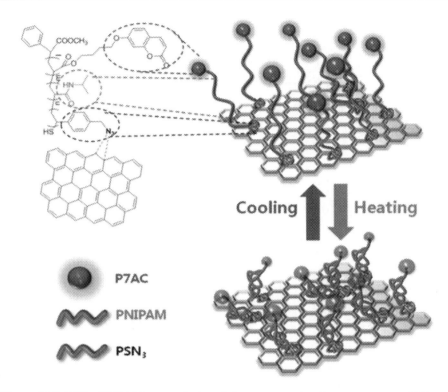

Figure 15.9 FGO−PNIPAM composite based thermoresponsive fluorescent sensor [67]. *Source*: Adapted from H. Yang, K. Paek, B.J. Kim, Efficient temperature sensing platform based on fluorescent block copolymer-functionalized graphene oxide, Nanoscale 5 (2013) 5720. https://doi.org/10.1039/c3nr01486j.

effect for photoluminescence. Upon lowering the temperature by cooling the solution, the photoluminescence effect could be recovered completely which may be translated as the on-off switch effect upon temperature change. The mechanism explained above is shown in Fig. 15.9 [67]. There is more work reported on the use of graphene with PNIPAM, P(NIPAM 95%-r-butyl methacrylate 5%) and P (NIPAM 90%-r-dimethylaminopropyl acrylamide 10%) for adjustment in LCST in temperature ranges of 25°C−45°C, the design of which is shown in Fig. 15.10 [68].

15.6   Miscellaneous sensors

Graphene has a good near infra-red (NIR) response that can be utilized to make light-responsive materials that are the basis for the sensing and triggering mechanism in mechanical actuators, self-healing materials, photothermal therapy, and controlled drug delivery [69]. They offer advantages in wireless actuation and remote

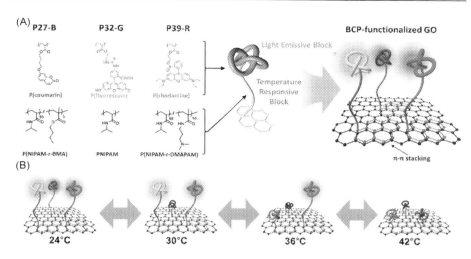

Figure 15.10 (A) Schematic illustration of the Block copolymer-GO based thermal sensor and (B) color changes of polymer-GO composite sensor as a function of temperature [68]. *Source:* Adapted from J. Lee, H. Yang, C.H. Park, H.-H. Cho, H. Yun, B.J. Kim, Colorimetric thermometer from graphene oxide platform integrated with red, green, and blue emitting, responsive block copolymers, Chem. Mater. 28 (2016) 3446−3453. https://doi.org/10.1021/acs.chemmater.6b00913.

controlling for any object or device. There are reports of the blending of graphene into an elastic polymeric matrix to form a graphene-polymer composite. The filler here is capable of absorbing and passing NIR energy at lower percolation threshold values. Loomis et al. used GNP in polydimethyl siloxane matrix for the absorption of light and corresponding transduction of energy to the polymer chains for actuator operation via. elastic entropy of the prestrained concrete [70,71]. Ling et al. reported the use of thermally reduced graphene oxide in TPU for the development of photo-mechanical actuators which can be made tunable by varying the applied prestrain values along with the volume fraction of the filler. A lower loading of 2% filler exhibited 50% photomechanical strain with stress values of 1680 kPa at a prestrain of 220% at a lower intensity of 16 mW/cm^2 which serve as a potential material for bio-medical devices and robotics [72]. The same polymer matrix has been reported at 1% sulfonated-graphene filler loading which is capable of lifting 21.6 g of weight by 3.1 cm with 0.21 N of force when exposed to infra-red light along with an increase in the mechanical properties such as young's modulus [73]. Furthermore, self-healing polymers benefit from the addition of graphene which has been experimented in TPU [74] and poly(N,N-dimethylacrylamide) [75] via the use of graphene component for converting light energy to heat, resulting in entanglement and diffusion of polymer chains at the rupture/crack for self-healing which can be employed in artificial tissues, biomedical devices, surgical dressing and so on.

The introduction of magnetic/superparamagnetic materials into graphene-based polymer gives magnetic responsiveness to the composite material. Lee et al proposed a faster response of shape memory polymers with enhanced stiffness and thermal conductivity with the use of graphene-ferromagnetic (nano-Fe_2O_3) hybrid 3-D structure in a PU matrix. The distribution was key in the magnetic control performance of such composites as they were uniformly placed over the walls of a 3D graphene framework. This hybrid network was inducted in shape memory PU matrix which resulted in rapid actuation under the effect of an alternating magnetic field. The ferromagnetic nanoparticles functioned as remote heat transferring medium under the field but the acceleration of the actuation rate was due to the 3D network put together by graphene-ferromagnetic structural network also resulting in increasing the thermal conductivity and stiffness of the composite [76].

There are instances where multiple sensing parts are integrated into one sensor to achieve different functions in a single unit. Zhao et al. attempted the fabrication of flexible and bifunctional sensors by assembling the temperature and humidity sensing layers based on graphene woven fabrics (GWF) in the PDMS polymer matrix. The coefficient of thermal expansion on PDMS has an influence over its layers expanding with an increase in the temperature which causes deformation thus resulting in an increase of the resistance of GWF giving rise to the PTC effect. Also, cellulose acetate butyrate, a humidity sensitive material was incorporated as the humidity sensing part which increases the capacitance upon sensing moisture and eventually absorption of the same [77].

15.7 Conclusion

Graphene is said to be as flexible as an elastomer and possesses good surface area while having good efficiency at conducting heat and electricity. Also, the 2D nature of the atom which is really thin gives it better light and water-oriented properties. The properties of graphene when considered with the versatile nature of polymers make the polymer-graphene nanocomposites use as interdisciplinary and multidirectional materials. Their use in various domains such as chemistry, physics, biology, chemical, civil, mechanical, electrical, etc. have certainly lived up to the mark and have led to the commercial scale-up of GO and GNP in the onset of the primary use of graphene in polymer nanocomposites. A host of graphene family materials such as graphene, GO, rGO, GNP, etc. have been discussed for applications in various sensors such as strain, thermal, chemical, optical, multifunctional types for detection of temperature, liquids, gases, humidity, organic vapors, pollutants, etc. The role of matrix-filler combination is pivotal in the outcome of sensory activities. It was observed that chemical modification of graphene using various functionalizing agents has a role to play in exhibiting improved sensing behavior. Their performance is better in terms of accuracy, sensitivity, response time, selectivity, detection limits, rapidness, reversibility, etc. and are capable to perform at room temperature with lower consumption of electricity. There can be upcoming fabrication technologies such as 3D graphene polymer composites [78] and electrospinning which can be used to develop graphene impregnated nanofiber sensory materials that can bridge the gap between the laboratory to the industry [79]. A large number of new generation graphene-based nanocomposites are expected to be developed based on various thermoplastic (especially commodity) and thermosetting polymer matrices along with a variety of graphene nanoelements capable to have different sizes, shapes, and functionalities such as an imaginative design of polymeric surfactants to solve the issues regarding solubility, film formation, and dispersion of graphene in polymers. Also, future experiments on graphene-polymer sensing platforms, in combination with versatile sensing strategies, are expected to reduce the detection threshold on a continual basis. However, in spite of the scale-up regarding the fabrication of graphene-based nanomaterials, its cost acceptability in the market is still commercially challenging. Also, the potential health risks especially the cytotoxicity of graphene-based materials need to be thoroughly researched and evaluated before large-scale production and utilization as sensory materials for fabrication of polymer-graphene composites for sensory and responsive effect.

References

[1] Tesla Inc, Autopilot, Tesla. https://www.tesla.com/autopilot, 2020 (accessed 21.12.20).
[2] BCC Publishing, Sensors market size, share, trend & industry analysis report, BCC Research. https://www.bccresearch.com/market-research/instrumentation-and-sensors/sensors-technologies-markets-report.html, 2020 (accessed 21.12.20).
[3] P. Wadhwani, S. Yadav, Industrial Sensors Market Size By Product (Level Sensor, Temperature Sensor, Flow Sensor, Position Sensor, Pressure Sensor, Force Sensor, Image Sensor, Gas Sensor), By Application (Energy & Power, Oil & Gas, Mining, Chemical, Manufacturing, Pharmaceutical), Industry Analysis Report, Regional Outlook, Application Potential, Competitive Market Share & Forecast, 2020−2026. https://www.gminsights.com/industry-analysis/industrial-sensors-market, 2020 (accessed 21.12.20).
[4] L. Chen, Y. Tang, K. Wang, C. Liu, S. Luo, Direct electrodeposition of reduced graphene oxide on glassy carbon electrode and its electrochemical application, Electrochem. Commun. 13 (2011) 133−137.
[5] M.S. Artiles, C.S. Rout, T.S. Fisher, Graphene-based hybrid materials and devices for biosensing, Adv. Drug. Deliv. Rev. 63 (2011) 1352−1360.
[6] W. Wu, Z. Liu, L.A. Jauregui, Q. Yu, R. Pillai, H. Cao, et al., Wafer-scale synthesis of graphene by chemical vapor deposition and its application in hydrogen sensing, Sens. Actuat. B: Chem. 150 (2010) 296−300.
[7] S. Stankovich, D.A. Dikin, G.H.B. Dommett, K.M. Kohlhaas, E.J. Zimney, E.A. Stach, et al., Graphene-based composite materials, Nature 442 (2006) 282−286. Available from: https://doi.org/10.1038/nature04969.
[8] T. Ramanathan, A.A. Abdala, S. Stankovich, D.A. Dikin, M. Herrera-Alonso, R.D. Piner, et al., Functionalized graphene sheets for polymer nanocomposites, Nat. Nanotechnol. 3 (2008) 327−331.
[9] A. Das, G.R. Kasaliwal, R. Jurk, R. Boldt, D. Fischer, K.W. Stöckelhuber, et al., Rubber composites based on graphene nanoplatelets, expanded graphite, carbon nanotubes and their combination: a comparative study, Compos. Sci. Technol. 72 (2012) 1961−1967.

[10] N. Bagotia, V. Choudhary, D.K. Sharma, A review on the mechanical, electrical and EMI shielding properties of carbon nanotubes and graphene reinforced polycarbonate nanocomposites, Polym. Adv. Technol. 29 (2018) 1547−1567.

[11] Y. Fan, J.-H. Liu, C.-P. Yang, M. Yu, P. Liu, Graphene−polyaniline composite film modified electrode for voltammetric determination of 4-aminophenol, Sens. Actuat. B Chem. 157 (2011) 669−674.

[12] F. Li, J. Li, Y. Feng, L. Yang, Z. Du, Electrochemical behavior of graphene doped carbon paste electrode and its application for sensitive determination of ascorbic acid, Sens. Actuat. B Chem. 157 (2011) 110−114.

[13] Y. Wang, Y. Li, L. Tang, J. Lu, J. Li, Application of graphene-modified electrode for selective detection of dopamine, Electrochem. Commun. 11 (2009) 889−892.

[14] N. Chatterjee, H.-J. Eom, J. Choi, A systems toxicology approach to the surface functionality control of graphene−cell interactions, Biomaterials. 35 (2014) 1109−1127.

[15] S.-Y. Wu, S.S.A. An, J. Hulme, Current applications of graphene oxide in nanomedicine, Int. J. Nanomed. 10 (2015) 9.

[16] H. Wang, H. Gu, N. Xiao, L. Ye, Q. Xu, Chlorotoxin-conjugated graphene oxide for targeted delivery of an anticancer drug, Int. J. Nanomed. 9 (2014) 1433.

[17] H.A. Becerril, J. Mao, Z. Liu, R.M. Stoltenberg, Z. Bao, Y. Chen, Evaluation of solution-processed reduced graphene oxide films as transparent conductors, ACS Nano 2 (2008) 463−470.

[18] D.A. Dikin, S. Stankovich, E.J. Zimney, R.D. Piner, G.H. Dommett, G. Evmenenko, et al., Preparation and characterization of graphene oxide paper, Nature 448 (2007) 457−460.

[19] J.L. Vickery, A.J. Patil, S. Mann, Fabrication of graphene−polymer nanocomposites with higher-order three-dimensional architectures, Adv. Mater. 21 (2009) 2180−2184.

[20] P. Bhagabati, M. Rahaman, S. Bhandari, I. Roy, A. Dey, P. Gupta, et al., Synthesis/preparation of carbon materials, in: M. Rahaman, D. Khastgir, A.K. Aldalbahi (Eds.), Carbon-Containing Polymer Composites, Springer, Singapore, 2019, pp. 1−64. Available from: https://doi.org/10.1007/978-981-13-2688-2_1.

[21] T.K. Das, S. Prusty, Graphene-based polymer composites and their applications, Polym. Technol. Eng. 52 (2013) 319−331. Available from: https://doi.org/10.1080/03602559.2012.751410.

[22] F. Regan, Sensors: overview, in: P. Worsfold, C. Poole, A. Townshend, M. Miró (Eds.), Encyclopedia of Analytical Science (Third Edition), Academic Press, Oxford, 2019, pp. 172−178. Available from: https://doi.org/10.1016/B978-0-12-409547-2.14540-8.

[23] W. Thomson, On the electro-dynamic qualities of metals:Effects of magnetization on the electric conductivity of nickel and of iron, Proc. R. Soc. Lond. (1857) 546−550.

[24] K.E. Petersen, Silicon as a mechanical material, Proc. IEEE 70 (1982) 420−457.

[25] J.C.F. Millett, N.K. Bourne, Z. Rosenberg, On the analysis of transverse stress gauge data from shock loading experiments, J. Phys. D Appl. Phys. 29 (1996) 2466.

[26] A.A. Barlian, W.-T. Park, J.R. Mallon, A.J. Rastegar, B.L. Pruitt, Semiconductor piezoresistance for microsystems, Proc. IEEE 97 (2009) 513−552.

[27] C.S. Boland, U. Khan, C. Backes, A. O'Neill, J. McCauley, S. Duane, et al., Sensitive, high-strain, high-rate bodily motion sensors based on graphene−rubber composites, ACS Nano 8 (2014) 8819−8830. Available from: https://doi.org/10.1021/nn503454h.

[28] L. Lin, H. Deng, X. Gao, S. Zhang, E. Bilotti, T. Peijs, et al., Modified resistivity−strain behavior through the incorporation of metallic particles in conductive polymer composite fibers containing carbon nanotubes, Polym. Int. 62 (2013) 134−140.

[29] H. Liu, Y. Li, K. Dai, G. Zheng, C. Liu, C. Shen, et al., Electrically conductive thermoplastic elastomer nanocomposites at ultralow graphene loading levels for strain sensor applications, J. Mater. Chem. C 4 (2016) 157−166.

[30] H. Liu, J. Gao, W. Huang, K. Dai, G. Zheng, C. Liu, et al., Electrically conductive strain sensing polyurethane nanocomposites with synergistic carbon nanotubes and graphene bifillers, Nanoscale. 8 (2016) 12977−12989.

[31] X.-W. Zhang, Y. Pan, Q. Zheng, X.-S. Yi, Time dependence of piezoresistance for the conductor-filled polymer composites, J. Polym. Sci. Part B Polym. Phys. 38 (2000) 2739−2749. https://doi.org/10.1002/1099-0488(20001101)38:21 < 2739::AID-POLB40 > 3.0.CO;2-O.

[32] V. Eswaraiah, K. Balasubramaniam, S. Ramaprabhu, Functionalized graphene reinforced thermoplastic nanocomposites as strain sensors in structural health monitoring, J. Mater. Chem. 21 (2011) 12626−12628.

[33] Y. Qin, Q. Peng, Y. Ding, Z. Lin, C. Wang, Y. Li, et al., Lightweight, superelastic, and mechanically flexible graphene/polyimide nanocomposite foam for strain sensor application, ACS Nano 9 (2015) 8933−8941.

[34] Y. Liu, D. Zhang, K. Wang, Y. Liu, Y. Shang, A novel strain sensor based on graphene composite films with layered structure, Compos. Part A Appl. Sci. Manuf. 80 (2016) 95−103.

[35] P. Costa, S. Gonçalves, H. Mora, S.A.C. Carabineiro, J.C. Viana, S. Lanceros-Mendez, Highly sensitive piezoresistive graphene-based stretchable composites for sensing applications, ACS Appl. Mater. Interfaces. 11 (2019) 46286−46295. Available from: https://doi.org/10.1021/acsami.9b19294.

[36] E. D'Elia, S. Barg, N. Ni, V.G. Rocha, E. Saiz, Self-healing graphene-based composites with sensing capabilities, Adv. Mater. 27 (2015) 4788−4794. Available from: https://doi.org/10.1002/adma.201501653.

[37] T. Ohashi, L. Dai, C60 and carbon nanotube sensors, in: L. Dai (Ed.), Carbon Nanotechnology, Elsevier, Amsterdam, 2006, pp. 525−575. Available from: https://doi.org/10.1016/B978-044451855-2/50018-8.

[38] A. Chen, W. Wu, W.E. Jones, Polymers and molecular wires as chemical sensors, in: J. L. Atwood (Ed.), Comprehensive Supramolecular Chemistry II, Elsevier, Oxford, 2017, pp. 179−195. Available from: https://doi.org/10.1016/B978-0-12-409547-2.12615-0.

[39] S. Bhandari, Polymer/carbon composites for sensor application, in: M. Rahaman, D. Khastgir, A. K. Aldalbahi (Eds.), Carbon-Containing Polymer Composites, Springer, Singapore, 2019, pp. 503−531. Available from: https://doi.org/10.1007/978-981-13-2688-2_14.

[40] A. Olean-Oliveira, M.F.S. Teixeira, Development of a nanocomposite chemiresistor sensor based on π-conjugated azo polymer and graphene blend for detection of dissolved oxygen, Sens. Actuat. B Chem. 271 (2018) 353−357. Available from: https://doi.org/10.1016/j.snb.2018.05.128.

[41] M.S. Wiederoder, E.C. Nallon, M. Weiss, S.K. McGraw, V.P. Schnee, C.J. Bright, et al., Graphene nanoplatelet-polymer chemiresistive sensor arrays for the detection and discrimination of chemical warfare agent simulants, ACS Sens. 2 (2017) 1669−1678. Available from: https://doi.org/10.1021/acssensors.7b00550.

[42] M. Son, Y. Pak, S.-S. Chee, F.M. Auxilia, K. Kim, B.-K. Lee, et al., Charge transfer in graphene/polymer interfaces for CO_2 detection, Nano Res. 11 (2018) 3529−3536.

[43] T.S. Anirudhan, J.R. Deepa, N. Stanly, Fabrication of a molecularly imprinted silylated graphene oxide polymer for sensing and quantification of creatinine in blood and urine samples, Appl. Surf. Sci. 466 (2019) 28−39.

[44] R. Sha, K. Komori, S. Badhulika, Graphene−Polyaniline composite based ultrasensitive electrochemical sensor for non-enzymatic detection of urea, Electrochim. Acta 233 (2017) 44−51.

[45] T. Wang, X. Zhang, Z. Wang, X. Zhu, J. Liu, X. Min, et al., Smart composite hydrogels with pH-responsiveness and electrical conductivity for flexible sensors and logic gates, Polymers. 11 (2019) 1564.

[46] K. Paek, H. Yang, J. Lee, J. Park, B.J. Kim, Efficient colorimetric pH sensor based on responsive polymer–quantum dot integrated graphene oxide, ACS Nano 8 (2014) 2848–2856.

[47] J. Liu, L. Tao, W. Yang, D. Li, C. Boyer, R. Wuhrer, et al., Synthesis, characterization, and multilayer assembly of pH sensitive graphene-polymer nanocomposites, Langmuir. 26 (2010) 10068–10075.

[48] H. Zhang, A. Kulkarni, H. Kim, D. Woo, Y.-J. Kim, B.H. Hong, et al., Detection of acetone vapor using graphene on polymer optical fiber, J. Nanosci. Nanotechnol. 11 (2011) 5939–5943.

[49] X. Wang, X. Sun, P.A. Hu, J. Zhang, L. Wang, W. Feng, et al., Colorimetric sensor based on self-assembled polydiacetylene/graphene-stacked composite film for vapor-phase volatile organic compounds, Adv. Funct. Mater. 23 (2013) 6044–6050. Available from: https://doi.org/10.1002/adfm.201301044.

[50] X. Huang, N. Hu, R. Gao, Y. Yu, Y. Wang, Z. Yang, et al., Reduced graphene oxide–polyaniline hybrid: preparation, characterization and its applications for ammonia gas sensing, J. Mater. Chem. 22 (2012) 22488–22495.

[51] J. Gounder Thangamani, K. Deshmukh, K.K. Sadasivuni, D. Ponnamma, S. Goutham, K. Venkateswara Rao, et al., White graphene reinforced polypyrrole and poly(vinyl alcohol) blend nanocomposites as chemiresistive sensors for room temperature detection of liquid petroleum gases, Microchim. Acta 184 (2017) 3977–3987. Available from: https://doi.org/10.1007/s00604-017-2402-1.

[52] Y. Li, Preparation of graphene/polymer composite film and its response to organic pollutants, Qilu University of Technology. http://www.theseus.fi/handle/10024/151981, 2018 (accessed 26.12.20).

[53] H. Liu, W. Huang, X. Yang, K. Dai, G. Zheng, C. Liu, et al., Organic vapor sensing behaviors of conductive thermoplastic polyurethane–graphene nanocomposites, J. Mater. Chem. C 4 (2016) 4459–4469.

[54] Y. Chen, P. Pötschke, J. Pionteck, B. Voit, H. Qi, Aerogels based on reduced graphene oxide/cellulose composites: preparation and vapour sensing abilities, Nanomaterials 10 (2020) 1729. Available from: https://doi.org/10.3390/nano10091729.

[55] S. Konwer, A.K. Guha, S.K. Dolui, Graphene oxide-filled conducting polyaniline composites as methanol-sensing materials, J. Mater. Sci. 48 (2013) 1729–1739. Available from: https://doi.org/10.1007/s10853-012-6931-z.

[56] W. Wang, G. Xu, X.T. Cui, G. Sheng, X. Luo, Enhanced catalytic and dopamine sensing properties of electrochemically reduced conducting polymer nanocomposite doped with pure graphene oxide, Biosens. Bioelectron. 58 (2014) 153–156.

[57] T. Yang, L. Meng, J. Zhao, X. Wang, K. Jiao, Graphene-based polyaniline arrays for deoxyribonucleic acid electrochemical sensor: effect of nanostructure on sensitivity, ACS Appl. Mater. Interfaces. 6 (2014) 19050–19056. Available from: https://doi.org/10.1021/am504998e.

[58] M.H. Naveen, N.G. Gurudatt, Y.-B. Shim, Applications of conducting polymer composites to electrochemical sensors: a review, Appl. Mater. Today. 9 (2017) 419–433. Available from: https://doi.org/10.1016/j.apmt.2017.09.001.

[59] W. Lei, W. Si, Y. Xu, Z. Gu, Q. Hao, Conducting polymer composites with graphene for use in chemical sensors and biosensors, Microchim. Acta 181 (2014) 707–722. Available from: https://doi.org/10.1007/s00604-014-1160-6.

[60] T. Wang, D. Huang, Z. Yang, S. Xu, G. He, X. Li, et al., A review on graphene-based gas/vapor sensors with unique properties and potential applications, Nano-Micro Lett. 8 (2016) 95−119. Available from: https://doi.org/10.1007/s40820-015-0073-1.

[61] D.K. Mahla, S. Bhandari, M. Rahaman, D. Khastgir, Morphology and cyclic voltamme-try analysis of in situ polymerized polyaniline/graphene composites, J. Electrochem. Sci. Eng. 3 (2013) 157−166. Available from: https://doi.org/10.5599/jese.2013.0038.

[62] J. Šesták, Thermophysical examination and temperature control, in: J. Šesták (Ed.), Science of Heat and Thermophysical Studies, Elsevier Science, Amsterdam, 2005, pp. 378−411. Available from: https://doi.org/10.1016/B978-044451954-2/50013-8.

[63] M. Kobayashi, C.-K. Jen, Transducers for non-destructive evaluation at high tempera-tures, in: K. Nakamura (Ed.), Ultrasonic Transducers, Woodhead Publishing, 2012, pp. 408−443. Available from: https://doi.org/10.1533/9780857096302.3.408.

[64] D. Kong, L.T. Le, Y. Li, J.L. Zunino, W. Lee, Temperature-dependent electrical properties of graphene inkjet-printed on flexible materials, Langmuir. 28 (2012) 13467−13472.

[65] M. Tian, Y. Huang, W. Wang, R. Li, P. Liu, C. Liu, et al., Temperature-dependent electrical properties of graphene nanoplatelets film dropped on flexible substrates, J. Mater. Res. 29 (2014) 1288−1294. Available from: https://doi.org/10.1557/jmr.2014.109.

[66] G. Liu, Q. Tan, H. Kou, L. Zhang, J. Wang, W. Lv, et al., A flexible temperature sensor based on reduced graphene oxide for robot skin used in Internet of Things, Sensors 18 (2018) 1400. Available from: https://doi.org/10.3390/s18051400.

[67] H. Yang, K. Paek, B.J. Kim, Efficient temperature sensing platform based on fluores-cent block copolymer-functionalized graphene oxide, Nanoscale 5 (2013) 5720. Available from: https://doi.org/10.1039/c3nr01486j.

[68] J. Lee, H. Yang, C.H. Park, H.-H. Cho, H. Yun, B.J. Kim, Colorimetric thermometer from graphene oxide platform integrated with red, green, and blue emitting, responsive block copolymers, Chem. Mater. 28 (2016) 3446−3453. Available from: https://doi.org/10.1021/acs.chemmater.6b00913.

[69] X. Yu, H. Cheng, M. Zhang, Y. Zhao, L. Qu, G. Shi, Graphene-based smart materials, Nat. Rev. Mater. 2 (2017) 17046. Available from: https://doi.org/10.1038/natrevmats.2017.46.

[70] J. Loomis, B. King, B. Panchapakesan, Layer dependent mechanical responses of gra-phene composites to near-infrared light, Appl. Phys. Lett. 100 (2012) 073108.

[71] J. Loomis, X. Fan, F. Khosravi, P. Xu, M. Fletcher, R.W. Cohn, et al., Graphene/elasto-mer composite-based photo-thermal nanopositioners, Sci. Rep. 3 (2013) 1900.

[72] Q. Meng, S. Han, S. Araby, Y. Zhao, Z. Liu, S. Lu, Mechanically robust, electrically and thermally conductive graphene-based epoxy adhesives, J. Adhes. Sci. Technol. 33 (2019) 1337−1356. Available from: https://doi.org/10.1080/01694243.2019.1595890.

[73] M.N. Muralidharan, S. Ansari, Thermally reduced graphene oxide/thermoplastic polyurethane nanocomposites as photomechanical actuators, Adv. Mater. Lett. 4 (2013). 1−0.

[74] L. Huang, N. Yi, Y. Wu, Y. Zhang, Q. Zhang, Y. Huang, et al., Multichannel and repeatable self-healing of mechanical enhanced graphene-thermoplastic polyurethane composites, Adv. Mater. 25 (2013) 2224−2228.

[75] C. Hou, Y. Duan, Q. Zhang, H. Wang, Y. Li, Bio-applicable and electroactive near-infrared laser-triggered self-healing hydrogels based on graphene networks, J. Mater. Chem. 22 (2012) 14991−14996.

[76] S.-H. Lee, J.-H. Jung, I.-K. Oh, 3D networked graphene-ferromagnetic hybrids for fast shape memory polymers with enhanced mechanical stiffness and thermal conductivity, Small. 10 (2014) 3880−3886.

[77] X. Zhao, Y. Long, T. Yang, J. Li, H. Zhu, Simultaneous high sensitivity sensing of temperature and humidity with graphene woven fabrics, ACS Appl. Mater. Interfaces 9 (2017) 30171–30176.

[78] N. Baig, T.A. Saleh, Electrodes modified with 3D graphene composites: a review on methods for preparation, properties and sensing applications, Microchim. Acta. 185 (2018) 283. Available from: https://doi.org/10.1007/s00604-018-2809-3.

[79] A.M. Al-Dhahebi, S.C.B. Gopinath, M.S.M. Saheed, Graphene impregnated electrospun nanofiber sensing materials: a comprehensive overview on bridging laboratory set-up to industry, Nano Converg. 7 (2020) 27. Available from: https://doi.org/10.1186/s40580-020-00237-4.

The use of polymer-graphene composites in fuel cell and solar energy

Yasir Qayyum Gill, Umer Abid, Umer Mehmood, Abdulrehman Ishfaq and Muhammad Baqir Naqvi
Department of Polymer and Process Engineering (PPE), University of Engineering & Technology (UET), Lahore, Pakistan

16.1 Introduction

The graphene is a wonderful material as it is a thinnest and the strongest material universally. Its high mobility is due to the charge carriers and zero effective mass that makes it capable to travel in micron-distance without the effect of scattering at ambient temperature. The current density sustainability of graphene is dominantly higher as compare to the copper, which presents the best-conducting properties and strength [1].

Many studies have shown that the graphene is very active towards conduction properties and catalytic action, which makes it able to be the part of energy conversion and storage applications. The replacement of graphene with platinum for the application of DSSC counter electrode have shown the capabilities of graphene towards solar cell energy conversion. In a research, it is concluded that the honeycomb structured graphene counter electrode converted 7.8% of the light energy in the electrical energy which is very comparable with conventional platinum based solar cell with the conversion efficiency of 8%. The researchers have found the synthesis of three-dimensional honeycomb graphene, easier and economical as compare to the platinum. They also identified that there are no challenges to use honey comb graphene to fabricate the counter electrode [1].

Generally, the graphene has exhibited the best chemical inertness and mechanical properties with high mobility and best optical transparency properties. Single-layered graphene allows the light of broad range and wavelength due to its high optical transmissivity up to 98%. For that good reason, the graphene is recorded as an optical conducting window. Moreover, the featured properties of graphene-like optical, electrical and electrolytic are highly tunable. These properties of graphene are optimized or tuned as per the requirements through functionalization of graphene. Chemical functionalization of graphene with desired properties makes it the best-suited candidate for the applications in the areas of optoelectronics and energy harvesting devices [2].

Polymer Nanocomposites Containing Graphene. DOI: https://doi.org/10.1016/B978-0-12-821639-2.00004-5

For many centuries, the researchers are in continuous movement to dig out new technologies, methods and devices to efficiently store, covert and utilize the solar energy, or to use it in a pure and clean energy production processes. The sun is considered a natural energy source that provides us with sustainable and clean energy. This energy from light can be converted and transmitted into the required electrical form of energy without any challenging environmental or other problems. By using the concept of the photoelectric effect, the light is easily converted or transformed in electricity with the facilities of photovoltaic substances and required facilitations. According to a report by the European Photovoltaic Industry Association, in 2017, the photovoltaic systems having a capacity of 420 GW were installed, that can produce electricity of 160 TW annually [3].

The dye-sensitized solar cells (DSSCs) have replaced the conventional solar cells rapidly because DSSCs provide high conversion efficiency, economic ease, and easy fabrication [4,5]. The light conversion followed by the charge transport is entirely different in DSSC as compare to the conventional solar cells. In DSSC, the transportation of charge and light conversion are considered as distinct phases. In a DSSC, the photoelectrode is deposited by a crystalline porous nanostructure of a semiconductor oxide, providing the large contact or surface area for molecules of dye. The dye molecules are absorbed in conduction substrate, electrolytes and dye-sensitizers. Usually, the organic electrolytic solution is sandwiched between semiconductor oxide deposited photoelectrodes and platinized counter electrode [6].

Graphene is particularly considered as a perfect material to be used for the applications of energy conversion and storage for solar cells, supercapacitors, fuel cells and batteries [7−9], due to its highly attractive properties as its high young's modulus or stiffness (1.0 TPa thermal conductivity (5000 W/m/K),), specific surface area (2630 m^2/g), transparency up to 97% and excellent charge mobility (20,000 cm^2/V/S) [10−12].

The fabrication of zero-dimensional carbon structured graphene quantum dots (GQDs) have created a boom for researchers because of its amazing properties including good water solubility [13,14], fluorescence emission due to quantum yield [15−17], lower cytotoxicity [14,18], excellent biocompatibility [19,20] and photostability [13,19]. These properties make graphene feasible for the applications of photovoltaic devices [21−24] bio-imaging [25,26], photocatalysis [27,28], LEDs [13,29−31] and phototherapy [32,33].

The technique of Chemical Vapor Deposition (CVD) have made researchers able to study more about the graphene-based hetero-junction devices and large surface graphene films [34−36]. The conductivity of thin films of graphene is becoming a key area of research nowadays. The graphene film conductivity is optimized typically by doping [37−39]. Generally, there are two methods to dope graphene. First method includes the chemical modification of the graphene films by affixing the reagents or metals on the active graphene films surface. This method is more efficient for the provision of charge carriers for high conductivity. The second method of graphene doping was introduced by Xinming Li et al. in which the graphene is doped chemically using acid for the optimization of graphene and silicon-based solar device efficiency [40].

Recently, the nitrogen-doped graphene is one of the key advancements in research sciences and is becoming one of the most popular materials for electronic devices in the coming years. In recent researches, few methods have been established for the fabrication of N-doped graphene. One approach is via chemical reactions [41] or plasma treatment [42−44] processing after the graphene synthesis. The other method is the direct production of N-doped graphene using precursors of nitrogen and carbon such as pyridine [45], ammonia and methane [46,47]. However, another effective double step technique for larger-scale production of N-graphene was adopted. The production was done using CVD method for graphene synthesis followed by Radio Frequency discharge treatment of synthesized graphene films [48].

Graphene is a highly crystalline nanostructure that makes it a favorable material for electronic applications as graphene possess potential electronic characteristics. Graphene structure is highly flexible towards change adaptation as its functional sites, and active surface area are boosted through nitrogen doping ultimately to optimize its electronic, chemical and morphological properties. The nitrogen doping of in graphene has resulted in various configurations including Graphitic N, Pyrrolic N and Pyridinic N. However; via these modifications graphene is well suited for nano-electronics, energy harvesting and electrochemical bio-sensing [49].

Yanyan Zhu et al. researched and reported the fabrication of nitrogen-doped graphene frameworks, through a single step rapid pyrolysis using fumaric acid and sodium carbonate with urea incorporating as nitrogen precursor. The synthesis was done, resulting in very effective specific surface area of value 1149 m^2/g [50]. The synthesis graphene frameworks are very efficient for counter electrode applications in perovskite solar cells (PSCs).

A new approach is noted under recent researches to dope nitrogen in graphene through plasma treatment. In this technique, the graphene or oxide of graphene is placed in a medium of nitrogen plasma. The concentration of nitrogen is regulated via the strength of plasma and time of exposure [49]. In research work, the synthesis of nitrogen-doped graphene by plasma treatment is verified [51].

16.2 Graphene

16.2.1 History

Graphene word is a combination of two words "graph" which comes from graphite and "ene" comes from the carbon double bond [52]. Graphene is a 2D layer of the carbon atoms which is separated from graphite [53]. Each carbon atom is attached with other 3 neighboring atoms through σ bonds to form a honeycomb structure and the remaining P electrons form a delocalized π bond [54]. Graphene is the building block of the other form of the carbon like carbon nanotubes, fullerene, and graphite [55]. The structure of graphene and its relation with other forms of carbon is shown in Fig. 16.1 [56].

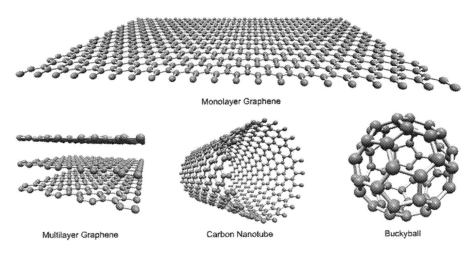

Figure 16.1 Graphene structure and relation with other carbon forms [56].

Research history of the graphene dates back to 1895 when British scientist Benjamin Brodie made graphene oxide (GO) using graphite and strong acid, which is graphene containing hydroxyl and epoxide bond attached to the carbon atoms. After this discovery, several other researchers tried to observe the thickness and structure of GO. In 1962 Boehm and Hofmann, regarded as first observers of the graphene sheet, identified reduced Graphite oxide (rGO) as a monolayer. Followed by this observation monolayer graphite layer was grown on metallic and nonmetallic substrates. Graphene term was coined by the Boehm et al. in 1986 to describe the single layer of carbon, which was formalized by the IUPAC in 1997 [57].

Fast forward to 2004 when Prof. Andre Geim and Prof. Constantine Novoselov were able to successfully separate monolayer of Graphene from using mechanical exfoliation or the popularly known scotch tape method. They were later awarded Nobel Prize in 2010 for this ground-breaking work [58]. Previously it was believed that graphene cannot exist in a free State and termed as academic material. Historical events in the history of Graphene research are shown in Fig. 16.2 [52].

After the successful isolation of the graphene from graphite in 2004, an exponential rise was observed in research publications and patent filings related to the graphene. This sudden growth was attributed to the astonishing electrical, thermal, optical, and mechanical properties of the graphene which are of great importance in composites, energy storage, electronics, biology, and other applications [59].

According to Prof. Andre Geim, this burst was not due to the isolation and observation of the graphene but it was the result of the outstanding electronic properties of which graphene offers [57]. Another major contributor to this growth rise was the fact that many researchers previously working on other carbon forms shifted their attention toward graphene research. The first sharp rise in graphene-related publications and patents was observed in 2005, which is still on its path to

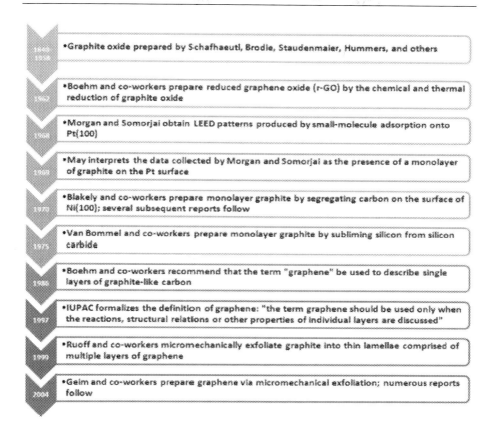

Figure 16.2 Historical events in history of graphene research.

grow further. After 2010, the strong trend was seen [59]. The graphene publication trimline from 2004 to 2013 is shown in Fig. 16.3 [60].

Fig. 16.4A reveals the number of publications on graphene in a year from 2004–17 [52]. On average 1500 research articles increase per year are being observed from different fields, which include chemistry, materials, science and technology, engineering, energy, and polymers [55]. Mainly topics like synthesis, properties, and application of graphene have been studied in research publications [52]. Fig. 16.4B reveals the different areas contributing to the graphene research [61].

Keeping in view the important properties, that graphene offers every country is trying to position itself among the leaders in graphene research. Most of the research done on graphene comes from 30 countries and the top five producers are China, the United States, South Korea, Japan, and India. Similarly, the top five leaders in the patent filing are China, South Korea, the United States, Japan, and Germany [59].

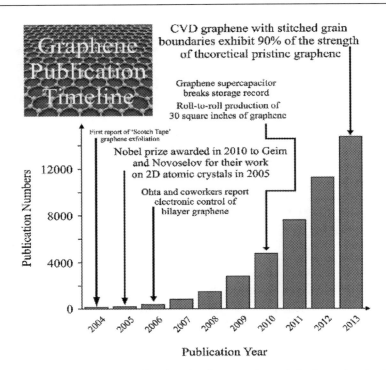

Figure 16.3 Graphene research publication timeline over the period of past decades [60].

16.2.2   Synthesis method of graphene

The methods available for the synthesis of graphene can be classified into two major groups (1) Top-down methods and (2) bottom-up methods. Top-down which is also known as the destructive technique involves the isolation or exfoliation of graphene sheets from graphite and its derivatives. Top-down methods have been further classified into mechanical exfoliation, chemical exfoliation, and chemical synthesis.

On the other hand, bottom-up methods are constructive in which graphene is grown on a different substrate using hydrocarbons present in gaseous and liquid states. The bottom-up methods include epitaxial growth, pyrolysis, CVD, and other methods. The main advantage of these methods is that high quality and large size graphene sheets can be produced [62]. Fig. 16.5 [52] summarizes the methods which are mostly used for the synthesis of graphene.

Some of the commonly used methods for graphene synthesis are described below briefly.

16.2.2.1   Mechanical exfoliation

Mechanical exfoliation is one of the simplest and first method which made free existing graphene a reality when Prof. Andre Geim and Prof. Constantine Novoselov used simple scotch tape to exfoliate the graphene from graphite. This method produced

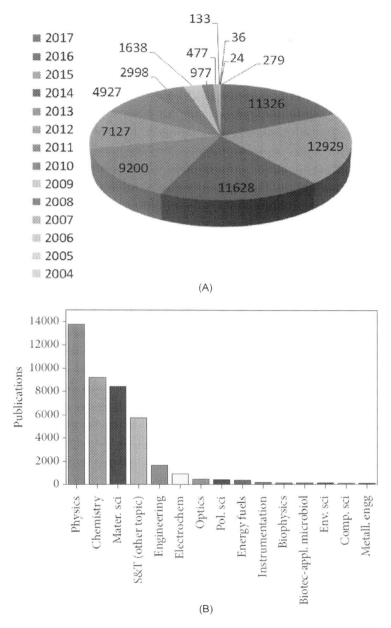

(A)

(B)

Figure 16.4 (A) Number of publications in a year from 2004 to 2017 [52] (B) Different fields contributing to the graphene research [61].

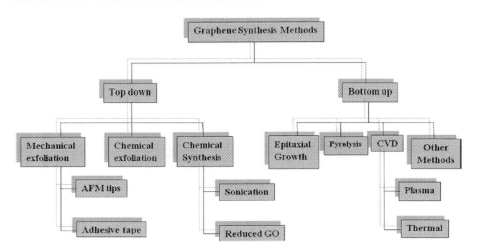

Figure 16.5 Synthesis methods for graphene [52].

high-quality graphene but this is only suitable for lab-scale experimentation as it cannot be scaled up [63]. Other mechanical means to exfoliate the graphene from graphite are ultra-sonication, electric field, and transfer printing technique [62].

16.2.2.2 Chemical vapor deposition

CVD is one of the most important methods to produce large-area graphene in which different substrates are exposed to the gaseous and liquid hydrocarbon in the presence of a very high temperature of 700°C−850°C. Gases and liquid mostly used for this purpose are methane, ethylene, acetylene, hexane, and pentane. Apart from these, biological materials like an insect, food, and waste can also be used as a precursor of graphene. Several metal substrates have been used in this method which include copper, nickel, iridium, palladium, and ruthenium [63]. Copper and nickel are mostly used as a substrate for the development of graphene in the CVD method. The main advantages of copper are its low cost and readily etching capability without disturbing the deposited graphene layer on its surface. The CVD method can be used to make an inch size large graphene sample but the main disadvantage of this technique is the difficulty in the transfer of graphene from a growth substrate to another substrate due to the inert nature of grapheme [62].

Plasma enhanced CVD technique (PECVD) is modified from CVD in which dissociation of hydrocarbon precursor is achieved using plasma at low temperatures. The main advantage of this method is the absence of any metal catalyst and different metallic and nonmetallic growth substrate can be used.

16.2.2.3 Liquid phase exfoliation

In this technique mono and few-layer graphene can be produced from graphite directly in the presence of liquid media without the involvement of any chemical oxidation step. This technique gaining quite a lot of attention due to its low cost, efficiency, wide range of available methods, and being a viable option for large scale production of graphene. Graphene in its pure form shows poor colloidal stability and weak solubility which is not favorable in graphene reinforced polymers. This technique provides a solution for this problem by giving the option of functionalization of graphene during the exfoliation process for better solubility in polymer composites.

Liquid Phase exfoliation can be achieved using different available routes which include, ultrasonic exfoliation using a combination of different solvents, surfactants, and salts, electrochemical exfoliation, shear exfoliation, and functionalization-assisted exfoliation [64]. Energy input by the ultra-sonication and choice of the solvent are two critical factors that directly control the yield of graphene in ultrasonic exfoliation [65]. In the case of electrochemical exfoliation type of electrodes, electrolyte and applied voltages are the key factors. Shear exfoliation has been achieved by using different techniques including wet-ball milling, micro fluidization, and homogenization [64].

16.2.2.4 Thermal degradation of Silicon Carbide and other materials

Thermal degradation another popular method of producing graphene by decomposing the Silicon Carbide (SiC) at a high temperature of 1250°C−1450°C for a short period of time that is 1−20 min. As shown in Fig. 16.6 [52].

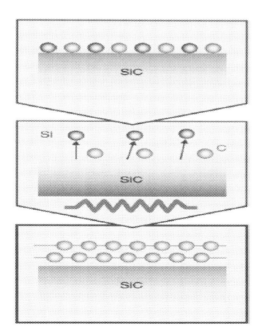

Figure 16.6 Thermal degradation of SiC to form grapheme.

When SiC is subjected to high temperature, Si atoms leave the surface layer of SiC leaving behind graphene film. Graphene with a $1-3$ layer has been grown using this technique by changing the decomposition temperature. This is a decent method to produce a continuous layer of graphene. Thickness control and reproducibility of graphene are some key issues in this practice [52].

Similarly, other materials like single crystal Ru (0001) have been used to grow monolayer graphene by the decomposition of pre absorbed ethylene on the surface of the crystal at ultra-high vacuum conditions. Graphene with high purity and large surface area has been grown using this technique. Other transition a metal which could be used for the method are Ir, Ni, Co, and Pt [66].

16.2.3   Characterization of graphene

Graphene is a multifunctional thinnest, lightest, toughest, and most flexible material discovered [58]. It has shown excellent electronic (250000 cm^2/vs), mechanical (modulus ~ 1 TPa), thermal (5000 W/m/K), and various other properties [67]. This multifunctionality enables the characterization of graphene possible using a wide range of already available characterization techniques. Being an optically active material, it can be observed with an optical microscope and different layers of graphene can be differentiated. To analyze, the surface and any wrinkles on graphene scanning electron microscopy (SEM) is used which can also be used as in situ SEM during the growth of graphene on a metallic substrate in the CVD method to study the growth dynamics. Transmission electron microscopy (TEM) has been used to observe the atomic structure of the graphene which gives information about the number of layers, bond rotations, grain boundaries, and other properties.

In order to observe the chemical reaction of dispersion of graphene in different solvents UV–Vis spectroscopy is used. X-ray diffraction (XRD) technique is also used to evaluate the exfoliation and number of layers in graphene. By analyzing the Bragg peak of graphite at 2 h $\approx$ 26 estimates can be made about the number of layers of graphene in the sample [68].

Raman spectroscopy is one of the most important and convenient techniques to characterize graphene. The number of layers and Bernal stacking order of graphite in multilayer graphene can be obtained by analyzing the shifting behavior of typical intensities of 2D (at 2700 cm^{-1}) and G(1560 cm^{-1}) band [63]. To study the intermolecular forces in graphene reinforced polymer nanocomposites FTIR has also been used [69].

16.2.4   Surface graphene for polymer nanocomposites modification

As mentioned earlier that Graphene in its pure form shows poor colloidal stability and weak solubility which is not favorable in graphene reinforced polymers. That's why the functionalization of graphene is an important step for better solubility in polymer composites [64].

GO is a derivative of graphene which is easy to process, viable to large scale production, and relatively less expensive. It has the same structure as the GO but

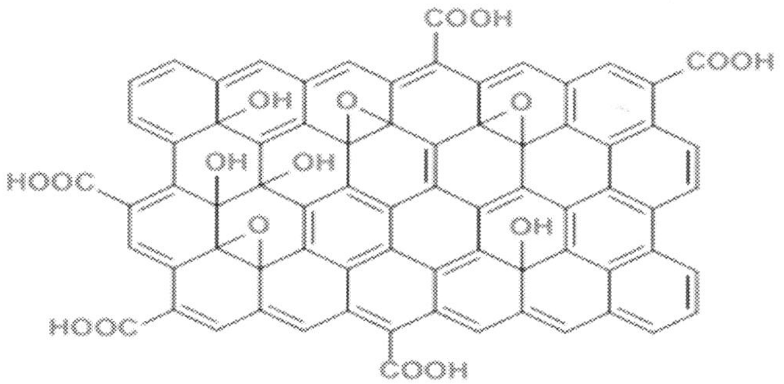

Figure 16.7 Graphene oxide molecule chemical structure [70].

the main difference is the presence of oxygen atom connected with some carbon atoms. According to the K-L model hydroxyl and epoxy groups are bonded to randomly while carboxyl and carbonyl groups are present at the boundary of the GO layer. it is also a viable option to do big scale surface modification of graphene by functionalization [54].

There are three main routes through which graphene and GO can be functionalized that is covalent bonding, noncovalent bonding, and element doping [54].

Covalent bonding is used when strong interactions between polymer and graphene are desired. But the main difficulty in graphene modification is the lack of presence of functional which can be used for modification. So, the only way to modify the graphene is to disturb the π bond network of carbon atoms which ultimately affect the natural conductivity of graphene.

On the other side covalent modification of GO Is the comparatively easy due presence of oxygen-containing functional groups [71]. Covalent modification of graphene is achieved using carbon skeleton modification while Hydroxyl functionalization and carboxyl functionalization is used to modify the GO surface [54].

Noncovalent bonding of graphene and GO is achieved with the help of electrostatic forces and hydrogen bonding between graphene and functional groups. This technique not only preserves the chemical structure of the graphene but also improves the solubility and stability of graphene and GO [54]. There are different interactions through which noncovalent modification can be achieved which include van der Waals force, electrostatic interaction, Hydrogen bonding, and π−π stacking interaction. (Fig. 16.7)

16.3   Polymer/graphene nanocomposites

16.3.1   Fabrication of graphene/polymer composites (synthesis methods)

A nanocomposite is a material comprising of more than single solid phase both structurally and compositionally where at the minimum one dimension is in nanometer

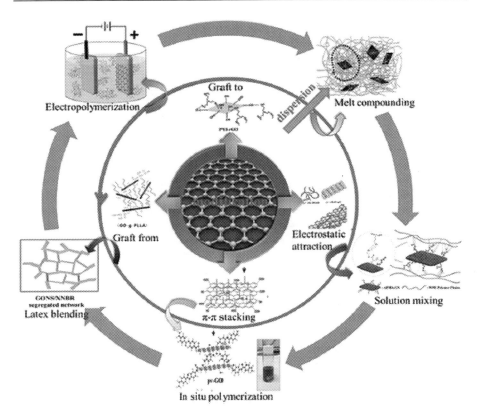

Figure 16.8 Typical graphene/polymer composite fabrication and graphene modification techniques [74].

range. The combination of nanomaterial with polymeric material is quite attractive not only for the development of nanocomposites but also for the reinforcement of polymeric material for potential applications. To achieve a homogeneous dispersion of graphene in the polymer matrix and to develop high potential polymer/graphene nanocomposites, numerous fabrication strategies have been developed. Typically, solution mixing, latex mixing, melt compounding, in situ polymerization, and electropolymerization are various extensively utilized synthesis techniques for the fabrication of polymer/graphene nanocomposites as shown in Fig. 16.8 [72–74] High industrial acceptability of melt blending as compared to solution blending is attributed to its cost-effectiveness and high scalability whereas, solution blending is attractive owing to better mixing which is achievable by solution blending technique [75,76].

16.3.1.1 Solution blending

In solution blending technique, polymeric material and nanomaterials (graphene or its derivatives) are firstly homogeneously dissolved in solvent. The selection of solvents for both polymer and nanomaterial is dependent of the miscibility of the both materials respectively. However, most of the time, both nanocomposite constituents are dissolved in same solvent depending on the miscibility. After dissolving both materials in solvents, the prepared solutions are homogeneously mixed together to acquire homogeneously dispersed graphene in the polymer matrix assisted by ultra-sonication, mechanical agitation, or high-speed shear mixing [77]. Finally, the polymer/graphene nanocomposites are acquired by adopting various procedures to remove solvent such as filtration [78], evaporation [79], lyophilization [80], and precipitation [81]. Transparent films of polymer/graphene nanocomposites can be evaporated or spin coated to manufacture composite films that have potential application as counter electrode material in the fabrication of solar cells owing to the high conductivity features of polymer/graphene nanocomposites [82]. The solution blending technique to fabricate nanocomposites can lead to comparatively better dispersion and distribution of graphene in the polymer matrix than that of melt processing. However, this polymer/graphene nanocomposite fabricating technique is less attractive for industrial application owing to extensive utilization of organic solvents to dissolve polymers [76].

In solution blending strategy, thermally or chemically reduced graphene can be uniformly dispersed in organic solvents owing to the availability of oxygen-containing functional groups which are not entirely removed. Solution blending technique has been reported in the literature for the fabrication of numerous types of polymer/graphene nanocomposites including polystyrene nanocomposites [83,84], poly(methyl methacrylate) nanocomposites [85,86], Nafion nanocomposites [87,88], polyurethane nanocomposites [89,90], polyethylene nanocomposites [91,92], and the like.

16.3.1.2 Melt blending

Owing of versatility, cost effectiveness, and environmental friendliness of melt mixing process, it is most commonly adopted polymer/graphene nanocomposites fabrication technique [93]. The feature of uniform dispersion of graphene in the polymer matrix by using melt blending cannot be compared with uniformity in dispersion of graphene achieved by solution blending or in situ polymerization but still, the commercial standard nanocomposites can be achieved by melt blending [76]. Currently, melt blending technique is more compatible with available industrial production processes. For melt blending processes to fabricate nanocomposites, there is requirement of strong shear forces and high melt temperature under extreme conditions. In this method, at high temperature, a thermoplastic polymeric material in molten state can be mixed with graphene or modified graphene by utilizing injection molding or extrusion processes [94,95]. The process basically involves the melting of pellets of polymeric material for the formation of viscous liquid and then the utilization of high shear mixing to disperse the graphene in the polymer matrix. Various other equipment namely internal mixer, two-roll mill, and tri-roll mill can be used for the melt blending to acquire polymer/graphene nanocomposites [96,97]. It does not require any solvent for the dispersion of nanomaterial hence, it is considered as environmentally friendly process to fabricate polymer/graphene nanocomposites. All polymeric materials can be easily processed by utilizing this technique expect ultra-high molecular weight polymers owing to their high viscosity and high molecular weight [98].

However, severe shearing action during melt blending for the structural distortion and size reduction can affect adversely but reducing the graphene reinforcement efficiency for polymeric materials [76]. Furthermore, the utilization of this technique can only be successful for the thermally rGO (TRG) due to the high temperature instability of most of the chemically modified graphene. Another downside is that the graphene density becomes too small after thermally reducing GO. This can lead to the feeding difficulties and nonuniform dispersion of graphene nanomaterial in the polymer matrix which adversely affect the mechanical properties of nanocomposites [99].

Kim et al. [76] fabricated polyurethane/graphene nanocomposites by utilizing three extensively utilized nanocomposite fabrication techniques namely solvent blending, in situ polymerization, and melt compounding. Fig. 16.9 illustrates the processing routes of polyurethane/graphene nanocomposites where (A) shows the oxidation of graphite in the presence of concentrated acids to acquire functionalized layers of graphene via (B) thermal expansion in argon environment or via (C) organic modification of graphite oxide with isocyanate by utilizing anhydrous N,N-dimethylformamide (DMF). Fig. 16.9D shows the incorporation of graphite and TRG in thermoplastic polyurethane (TPU) by adopting melt blending technique at 180°C whereas, (E) shows the solution blending of nanocomposite constituents in the presence of DMF, followed by removal of excess solvent and (F) displays the fabrication of TPU/graphene nanocomposite by incorporating monomer rather than polymeric material by utilizing in situ polymerization technique. In Fig. 16.9 black lines are representing graphite reinforcements. Hard segments of TPU are represented by short blue blocks and soft segments of TPU are denoted by thin red curves. Study revealed the more effective distribution of thin graphene-based exfoliated sheets in the polymeric material by solution blending as compared to melt processing which was indicated by the morphological characterization of nanocomposites.

16.3.1.3 In situ chemical polymerization

In situ polymerization is one of the convenient processing techniques for the fabrication of polymer/graphene nanocomposites and it has also been widely utilized for the development of nanocomposites based on graphene derivatives namely GO, rGO or TRG by incorporating graphene or graphene derivatives during the polymerization stage of monomer to acquire uniform dispersion between both polymer and nanomaterial phases (graphene or graphene derivatives) [98]. The major advantage lies in the utilization of in situ polymerization technique to fabricate polymer/graphene nanocomposites is that it enables the polymerization of monomer resided in and out of the interlayers of graphene to fabricate polymer/graphene nanocomposites in which the platelets of graphene are delaminated at nano scale. Furthermore, the dispersion of ability of nanomaterial is much better in in situ polymerization technique as compared to other mixing strategies [100]. Besides its superior

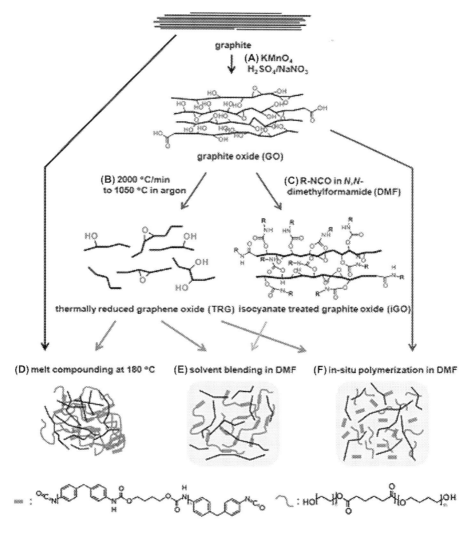

Figure 16.9 (A) Oxidation of graphite, (B) thermal expansion of graphite in argon environment, (C) organic modification of graphite oxide, (D) graphite and thermally reduced graphite addition in TPU, (E) solution blending of nanocomposite constituents, and (F) TPU/graphene nanocomposite fabrication by in situ polymerization technique [76]. *TPU*, thermoplastic polyurethane.

features, there are some defects of this nanocomposite fabrication strategy. The major drawback is the loss of polymerization rate during in situ polymerization at the latter stages of the polymer/graphene nanocomposite processing. Another foremost drawback, which is observable in most of the instances, is that the monomers have been polymerized only in the presence of solvent or at least equilibration

solution of GO in the distilled water [101]. In this case, the considerably high viscosity of even dilute graphene nanosheets dispersion becomes a prominent con-
straint related to the bulk-phase in situ polymerization.

For the development of polymer/graphene nanocomposites my utilizing in situ
polymerization, the surface modified graphene is comparatively more advantageous
over GO owing to its superior interfacial bonding features and better dispersion
properties. A. Tchernook et al. [102] generated graphene based polyethylene nanocomposites by in situ polymerization of ethylene monomer. Fig. 16.10 displays the
fabrication of polyethylene/graphene nanocomposites by the in situ polymerization
and postpolymerization mixing. Graphene was ultrasonicated for 60 min in DMF
prior to incorporation in stainless steel mechanically stirrer polymerization reactor
with volume of 500 mL for the in situ polymerization. Ethylene monomer and surfactant along with catalyst was then incorporated in the polymerization reactor for
the uniform dispersion of graphene. The mixture was vigorously stirred in a pressure reactor to yield graphene/polyethylene dispersions. The acquired nanocomposite yields suggested some partial deactivation of catalyst to take place at higher
concentration of graphene loading. The solid generated after polymerization was filtered and washed with methanol solution before drying in vacuo for time of 24 h at
40°C. However, in postpolymerization mixing technique as shown in Fig. 16.10,
sodium dodecyl sulfate was incorporated in the preformed dispersion of polyethylene (10−20 mL) for the prevention of coagulation during the incorporation of graphene dispersion in it. The graphene dispersion in DMF was then incorporated
dropwise during stirring and the mixture was stirred for 10 min. The solids were

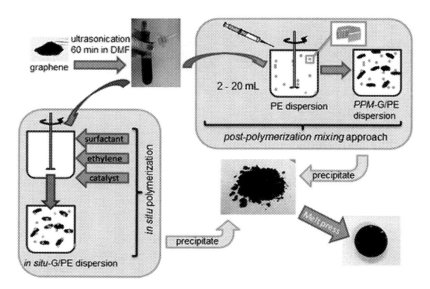

Figure 16.10 Fabrication of nanocomposite of polyethylene and graphene by in situ
polymerization method and by using postpolymerization mixing approach [102].

filtered and washed with methanol solution prior to drying in vacuo for time of 24 h at 40°C.

16.3.1.4   Latex mixing

Latex mixing is another strategy for the fabrication of polymer/graphene nanocomposites. Nanocomposites produced by blending polymer latex with graphene are environmentally more preferable to nanocomposites prepared by solution-blending owing to no release of harmful volatile substances during the nanocomposite film formation. Moreover, enhanced properties can be acquired at lower graphene loading owing to creation of excluded volume by the particles of polymer that forces the graphene into the interstitial spaces available between the particles of polymer during the drying process. Unlike various other processing techniques, by utilizing latex mixing method, graphene can be distributed along the specific paths available due to interstitial spacing for the formation of segregated structure [103]. This segregated network is a special arrangement of graphene in the polymeric material for the fabrication of polymer/graphene nanocomposites. The procedure followed in this technique is quite simple. Graphene or modified graphene is firstly coated on the polymer particle surface and then hot-pressing or any other corresponding techniques can be adopted for the fabrication of final nanocomposite. The morphology of the nanocomposites developed by this method is inherently dependent on the relative polymer latex dimensions and the dimensions of graphene nanofiller. The foremost drawback of the latex mixing technique is its narrow application range and less industrial attractiveness which is only limited to the conductive materials [104].

16.3.1.5   In situ electrochemical polymerization

Electrochemical polymerization (electropolymerization) has been extensively used to fabricate graphene-based polymer composites for various energy applications. In this technique, the electropolymerization of adsorbed monomer of the graphene sheet surface is accomplished to form polymer between the interfacial layers of the surrounding graphene sheets [105]. Graphene/polymer films are fabricated on the surface of the glassy electrode via single step in situ electrochemical polymerization in a reaction solution [106]. This technique is utilized widely for the fabrication of conductive composite films. The electropolymerization can be accurately controlled by the time of polymerization, applied electric potential, and current density [107]. It encompasses the electrochemical oxidation of monomer in the reaction solution in the accompany of its counterion. The requirement of the monomer for this polymerization technique to develop graphene-based polymer composites is in very small amounts and the rate of deposition is quite fast. The deposition on the conducting substrate is localized on the designated area of deposition. After the electropolymerization, the layered structure of graphene sheet remains maintained. The electropolymerization can be perceived by the changings in the coloring of the working electrode (WE) on its surface. After the deposition, washing of the WE is carried out in a distilled water and then electrode is dried in oven for 1 day at

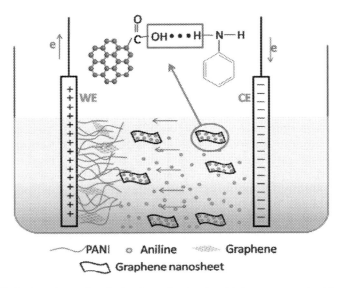

PANI   ∘ Aniline   Graphene
Graphene nanosheet

Figure 16.11 In situ electrochemical polymerization in which monomer of aniline is adsorbed on nanosheets of graphene through H-bonding [109].

around 60°C temperature. A resulting film of the polymer/graphene composite acquired by the in situ electrochemical polymerization is highly dependent on the type and magnitude of the oxidation potential [108]. The characteristics of the acquired composite films are reliant on the type of solvent selected for the reaction.

Liwen Hu et al. [109] fabricated graphene based nanorod-polyaniline composites by utilizing in situ electrochemical polymerization technique. Fig. 16.11 describes the scheme of the fabrication technique in which part of the aniline monomer has been adsorbed on the graphene nanosheet surface through H-bonding after sonication and magnetic stirring. In the Fig. 16.11, it can be observed that the availability of pulse current moved the water pool in the direction of WE via electromigration process. After colliding of water pool with WE, the deposition of aniline monomer after oxidation was perceived. Concurrently, graphene was observed to be embedded in the nanorods of polyaniline signifying the interfacial interconnection of graphene with the polymer network.

16.3.2 Graphene-based different polymer composites

S. Goswami et al. [110] synthesized graphene/polyaniline composites by utilizing two different approaches. (1) In first approach, the in situ polymerization was carried out in which precooled ammonium peroxodisulfate solution containing GO was incorporated in the β-CSA solution which was also dissolved in GO aqueous dispersion. The molar ratio of both aniline/ammonium peroxodisulfate and aniline/β-CSA was kept same and the reaction was carried out for 5 h. Collection of

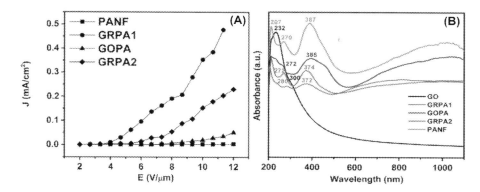

Figure 16.12 Field emission behavior (A) and UV-vis characterization (B) of composite specimens GOPA, PANF, GRPA1, GRPA2, and graphene oxide (GO) [110].

powder of GO and polyaniline/β-CSA nanofibers (PANF) composite, named as GOPA, at the end of the reaction was accomplished by filtration and repetitive washing with distilled water and methanol. The synthesis of graphene/polyaniline composite (GRPA2) was accomplished after the mil stirring of GOPA with hydrazine hydrate and ammonia solution, resulting composite was acquired through vacuum filtration process. (2) In second approach, mixing of the PANF, preprepared by oxidative polymerization of monomer by ammonium peroxodisulfate in the presence of β -CSA, and the presynthesized GO in 15:1 weight ratio was carried out. Prepared graphene/polyaniline composite (GRPA1) specimens were also characterized for their field emission behavior and the J-E (current density-field strength) features were examined as shown in Fig. 16.12A. The acquired value of highest current density for GRPA2 (specimen fabricated via in situ polymerization) was 0.8 mA/cm^2 with corresponding highest electric field value of 15.96 V/μm. In UV-vis characterization of the graphene/polyaniline composites, the peaks observed at 272−280 and 207 nm are most likely the π-π interaction peaks in-between different constituents of composite material namely PANF, GO, and graphene (GR) as shown in Fig. 16.12(b).

V. Loryuenyong et al. [111] fabricated graphene based polypyrrole (PPy) composites for potential application in the solar cells. GO was synthesized prior to the composite fabrication. Graphene/PPy composite films were acquired by utilizing polymerization method and GO reduction process using ascorbic acid. Composites were developed with 0, 5, 10, 15, 20, and 25 wt.% of rGO. Powders of both GO and PPy were homogeneously mixed in 10 mm of distilled water and afterwards, 0.088 g of ascorbic acid was incorporated as binding and reducing agent. Doctor-blade method was utilized to coat prepared suspension on FTO glass substrate followed by the drying of film for 6 h at 70°C. SEM micrographs revealed the uniformity in dispersion of spherical-like monodispersed particles having 120 nm particle size in average. For rGO, no apparent XRD patterns were observed which indicated the intercalation of PPy between reduce GO sheets (GOSs) which disallowed its chemical structure to collapse back. TGA thermograms manifested the main PPy weight loss at around 250°C but an appreciable improvement is observed with the incorporation of rGO with the increase in residual content, as shown in Fig. 16.13A, owing to the high thermal stability of rGO. Cyclic voltammograms curves were also acquired as shown in Fig. 16.13B and the charge transfer processes occurring in the composite films were studied in them along with their electrocatalytic activity which showed the occurrence of redox peaks for pure PPy specimens with highest current density. The interfacial interaction signifies the homogeneity of the composite. Solid state shear milling technique in preparation of HDPE/graphene composite was utilized by P. Wei and S. Bai [112] to overcome the problems associated with the re-aggregation and re-stacking of graphene sheets. HDPE/graphene composites were prepared by using melt processing technique utilizing the uniformly dispersible and exfoliated superfine HDPE/graphene compounded superfine powdered prepared by shear milling technique in solid state. TEM and SEM micrographs along with wide-angle XRD manifested the exfoliation of graphene sheets in the individual polymer matrix sheets. Considerable improvement the mechanical properties of composites were observed by utilizing aforementioned solid-state shearing milling technique. Impact strength of composites acquired with conventional blending technique with 0.1% of graphene was 22.6 kJ/m^2 whereas with the same graphene content, the impact strength increased to 27.6 22.6 kJ/m^2, as shown in Fig. 16.14, by using solid state shear milling technique. Yield strength acquired for composite with solid state processing was about 25.25 MPa at 2 wt. % of graphene whereas yield strength acquired for the same specimen with melt processing was 19.25. Y. Qing et al. [113] prepared modified GO based polystyrene composite with high flame-retardant properties. Suspension polymerization of styrene was carried out to prepare GO (ODOPM treated)/polystyrene composites. ODOPM is the hydroxylation product of 9, 10-dihydro-9-oxygen-heterooxy-10-phosphoro-10-oxygen

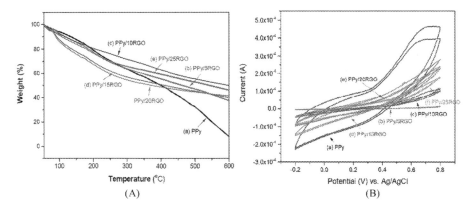

Figure 16.13 TGA thermograms (A) and cyclic voltammograms curves (B) of pure polypyrrole (PPy) and composite with 5 wt.% (PPy/5RGO), 10 wt.% (PPy/10RGO), 15 wt.% PPy/15RGO), 20 wt.% (PPy/20RGO), and 25 wt.% (PPy/25RGO) of reduced graphene oxide (RGO) in polypyrrole [111].

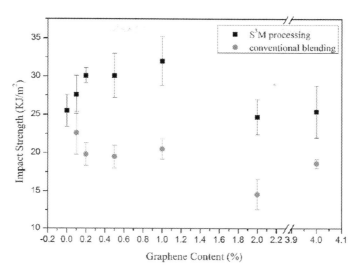

Figure 16.14 Comparison of impact strength of nanocomposite fabricated via S3M technique and by conventional blending method [112].

(DOPO) which is a DOPO derivative. ODOPM, an excessively utilized flame retardant, utilization for GO modification resulted in a composite with high flame retardancy properties. Thermal analysis revealed more than 33.6% and 36.2% decrease in total and peak heat release rate, THR and HRR respectively, for developed composites microspheres with 3.0 wt.% of additives.

Zeng et al. [114] fabricated spray-free polypropylene composites reinforced with GO modified short glass fibers (GO@SGF) by adopting melt compounding technique via extrusion and injection molding. Analysis revealed the improvement in mechanical properties by incorporating GO@SGF in polypropylene. Al powder along with GO@SGF was incorporated in 200 g polypropylene as filler. Tensile strength increase was detected from 32.32 MPa (pure polypropylene) to 39 MPa (3phr content of GO@SGF). Moreover, the tensile strength of composites with GO coated on short glass fiber (GO@SGF/Al/PP composite) was considerably higher as compared to the composites without GO as shown in Fig. 16.15A. This improvement was observed owing to coated GO, PP, and short glass fiber interfacial interaction. Fig. 16.15B manifested the increase in elongation at break up to 3 phr of GO@SGF based composite, as compared to decrease in elongation at break for composite without GO, and this increase was attributed to the fact that the yielding of matrix around GO will take place during the stretching of composite and the tensile deformation energy will be absorbed which will resultantly increase the stress concentration and therefore tensile ductility will increase. N. Song et al. [115] developed a PP/graphene composite with significantly high thermal conductivity for an application in electronic devices as a thermal control material. Composite with 3D graphene framework were developed by using in situ building technique.

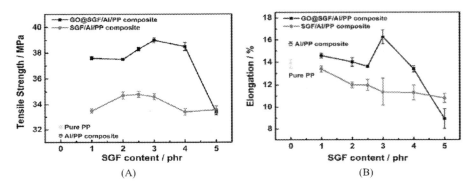

Figure 16.15 (A) Tensile strength and (B) elongation at break comparison of composites prepared with and without graphene oxide [114].

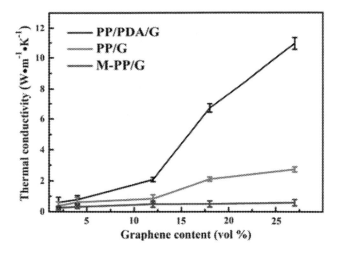

Figure 16.16 Through-plane thermal conductivity of graphene based PP composites [115].

The developed PP/graphene composites exhibited 10.93 W/m/K of through-plane thermal conductivity value, as shown in Fig. 16.16, which is a significantly high value and is about 55 times greater than that of virgin polypropylene. Thermal conductivity observed for composite specimen with graphene (G) and polydopamine (PDA) content was observed to be higher as compared to other specimens with only graphene (PP/G) and specimen prepared by melt mixing technique (M-PP/G). The utilization of PDA was advantageous owing to its richness in aromatic rings, oxygen functional groups, and nitrogen functional groups which promote the inter-facial interaction with graphene.

C.C.N. de Melo et al. [116] prepared GO reinforced nylon 6 nanocomposites and characterized the prepared nanocomposites for their nonisothermal nano-crystallization behavior through kinetics by varying the amount of GO. Nanocomposites were developed by adopting two processing techniques. Firstly, production of master batch was accomplished by solution mixing technique and then master batch was diluted in nylon 6 matrix through melt blending in extruder. Mo's theory that describes the polymer crystallization mechanism was observed to be fitted well with the experimental results. Owing to the presence of GO in nanocomposites, the secondary crystallization was highly favored. However, the presence of GO as a reinforcing nanomaterial resulted in the decrease of degree of crystallinity in the nylon 6 matrix. N. Aranburu et al. [117] studied the improvement in dispersion of graphene in polyurethane matrix by the modification of graphene with ionic liquid (IL). Polyurethane/unmodified graphene and polyurethane/IL modified graphene nanocomposites were fabricated by melt mixing technique in twin screw extruder. Electrical conductivity measurement of both nanocomposite materials approved the improvement in the dispersion of graphene material after modifying with IL as the nanocomposite material with IL modified graphene manifested considerably low electrical percolation threshold value at 1.99 wt.% graphene whereas the percolation threshold value was quite higher for unmodified graphene-based nanocomposite which appeared at 3.21 wt.% of graphene as shown in Fig. 16.17. 65% improvement in Young's modulus was observed with nanocomposite specimen with 4 wt.% of graphene. Mechanical properties maintained a slight increase with Il modified graphene

Blend of acrylonitrile butadiene and polycarbonate with graphene as a reinforcing filler was prepared by V. Tambrallimath et al. [118] by fabricating composites with fused deposition modeling (FDM) technique. Both polymers and graphene platelets were blended and extruded out in the filament shape of 1.75 mm diameter.

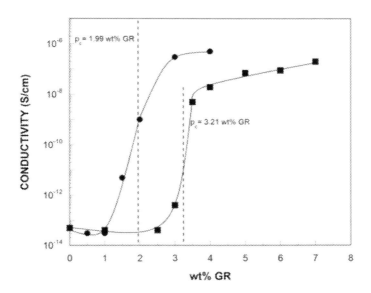

Figure 16.17 Electrical percolation threshold value determination of nanocomposite based on modified and unmodified nanocomposites through conductivity experiment [117].

Test specimens were developed from FDM technique by utilizing the prepared nanocomposite filament as a feed material. Nanocomposites were characterized for mechanical and thermal properties improvement with the addition of graphene. 0.8% of graphene resulted in the substantial increase in the Young's modulus value to 4.038 GPa. TGA thermograms revealed the increase in both degradation initiation temperature and complete decomposition temperature with the increase in graphene loading. Decomposition temperature was increased to 575.1°C (0.8 wt.% of graphene) from 558.5°C (0.2 wt.% of graphene). The increase in graphene content resulted in increase in Tg of nanocomposites and reduction of mass at higher temperature.

16.3.3 Properties of graphene/polymer nanocomposites

Thermal, rheological, barrier, electrical, and mechanical properties of graphene polymer nanocomposites are highly interlinked with the intrinsic features of and level of dispersion of graphene material and its derivatives in the polymer matrix. Rheological studies of nanocomposites are significantly important to analyze the microstructure of graphene/polymer nanocomposites and to optimize the processing of the nanocomposites [119,120]. During the measurement of viscoelastic rheology of nanocomposite materials, the formation of storage modulus plateau (low frequency G') is the sign of the rheological percolation owing to the generation of graphene elastic network in the polymer matrix [121]. The strongest material with an in-plane elastic modulus (G'') value of about 1.1 TPa has been reported to be pristine defect-free graphene [122]. These high modulus values of graphene and graphene derivatives along with their large platelets surface areas allow them to be the chief load bearing constituents of graphene-based polymer nanocomposites. The enhancement of mechanical properties of the graphene/polymer nanocomposites has been reported to be correlated with the improvement in the dispersion of graphene [123] and the alignment of graphene in the polymeric matrix may improve the reinforcement [124]. However, it is also evident that the graphene or graphene derivate-based nanofiller which, on short length scale, appeared to be uniformly dispersed in the polymer matrix may actually be aggregated while observing at micron level scale [125]. On the contrary, some studies also suggested the improvement in reinforcement owing to the presence of large scale nanofiller aggregates in polymer matrix [126,127].

Graphene and the derivatives of graphene are considered to be promising fillers for the fabrication of nanocomposites with high gas barrier properties. It is observed that defect free graphene shows high gas impermeable properties. Graphene repels gas molecules and is also able to repel atoms therefore graphene show low gas solubility [128]. The incorporation of 1% w/w content of GO in poly(methyl methacrylate) matrix reduced the O_2 permeability of nanocomposites by 50% and the incorporation of 10% w/w content of GO resulted in almost complete impermeability of nanocomposites to O_2 as shown in Fig. 16.18 [129]. The thermal conductivity of pristine graphene measured at room temperature when suspended [130] and when supported on a substrate of SiO_2 [131] has been reported to be exceeding

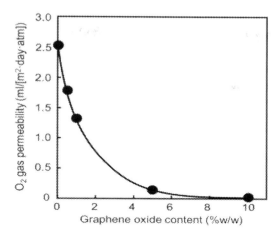

Figure 16.18 O_2 permeability of graphene oxide based poly(methyl methacrylate) nanocomposites [129].

3000 and 600 W/mK, respectively. The coefficient of thermal expansion (CTE) of polymer in nanocomposite material has been reported to be significantly decreased attributed to the negative CTE of graphene nanofiller [132,133]. The increase in thermal stability, noticed by the maximum value of weight loss rate acquired from thermogravimetric analysis, has been reported in literature by the incorporation of graphene and its derivates [134]. Even despite of thermally unstable of GO, it can also improve the thermal stability of the nanocomposites [73,135].

16.3.4 Factors affecting the performance of graphene/polymer nanocomposites

16.3.4.1 Dispersion of graphene into polymers

The thermal, mechanical, electrical, conductive, and various other properties required for the utilization of graphene-based nanocomposites in fuel cells, supercapacitors, hydrogen storage, and solar energy applications are highly dependent on the stable dispersion of graphene in polymeric material and the optimization of nanocomposites microstructure. Generation of polymer based functionalization on the surface of the graphene is often essential, especially when host polymeric material is of nonpolar nature [136]. Polarity of the polymeric material is crucial for the efficient and successful graphene stabilization in the polymer matrix. Prevention from agglomeration of graphene nanolayers and their restacking is attributed to the availability of colloidal particles on the graphene nanolayers surface which results in the fine and uniform dispersion of nanolayers throughout the matrix of polymeric material [137].

GO as a precursor material to the graphene is extensively utilized and is easy to disperse in polar polymeric materials [138]. The chemically rGO dispersion in

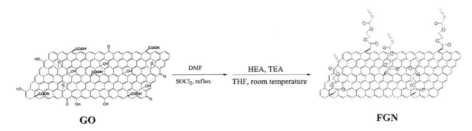

GO   **FGN**

Figure 16.19 Modification of GO to prepare functionalized graphene (FGN) [140]. *GO*, graphene oxide.

polymer aqueous medium fails to provide a stable dispersion which is attributed to the fact that during the reduction reaction, severe agglomeration of rGO occurs while neutralizing and elimination of charged group of oxygen [139]. Therefore, the focus in various researches is usually on the preservation of charged groups during the production of stable rGO dispersion. This preservation can be done by controlling the reduction reaction or by creating charged functional sites in the structure of rGO. Fig. 16.19 displays the graphene functionalization method adopted by Jiang et al. [140] by utilizing PMMA for improvement of graphene dispersion via melt blending technique and by latex technique. The PMMA presence on the surface of the graphene was found to efficiently prevent the agglomeration of graphene and the graphene dispersion improvement was obvious. Fig. 16.19 represents the modification of GO with hydroxyethyl acrylate (HEA) and concurrent reduction of GO. Functionalized GO (FGO) resulted, with double bonds, owing to the covalent bonding in-between GO and HEA. Emulsion copolymerization technique was adopted for the pretreatment of FGO with methyl methacrylate by bonding between monomer and double bonds of FGO. Finally, melt blending was accomplished to acquire graphene/PMMA composites

Solution processing for the fabrication of graphene-based nanocomposites involves long time drying of the graphene/polymer nanocomposites which mostly result in graphene layers restacking thus affects the dispersion of graphene in the polymer matrix [141]. A solvent free fabrication process is therefore usually induced for the preparation of nanocomposites with good graphene particle dispersion. Presently, the dispersion of graphene in the polymer matrix can be improved by physical methods for dispersion, covalent, and noncovalent bonding methods for dispersion [142]. In physical methods, mechanical techniques for the dispersion of agglomerated nanosheets and nanoparticles of graphene are adopted. In covalent bonding methods, GO material is combined with organic molecules (polymers) with excellent dispersibility via chemical reactions for the improvement in dispersion of graphene [74]. Noncovalent bonding dispersion methods involve the graphene combination with organic functional materials (polymers) through hydrogen bonding, π-π bonding, an ionic bonding, and various other noncovalent bonds [143]. Graphene dispersion strategy has been proposed by Iqbal et al. [123] in which they improved the graphene dispersibility in PE matrix for the improvement in conductivity of nanocomposites. Solution blending method was adopted to mix PE, oxidized PE, and graphene. The dispersion of graphene in the PE and oxidized PE matrix was observed to be improved greatly.

16.3.4.2   Bonding structure of graphene/polymer interface

The improved mechanical properties of the graphene/polymer nanocomposites are attributed to the strong GO and polymer interfacial interactions owing to the GO nanosheets dispersion and strong hydrogen bonding interaction in between GO and polymer matrix. The interfacial interactions or bonding of graphene and polymer matrix is dependent of interfacial area [144]. Fig. 16.20 revealed the breakage of covalent bond between polymer matrix and graphene, pulling out of graphene, and functionalization of graphene which improved the interfacial bonding structure between polymer matrix and graphene to developed a nanocomposite with improved mechanical properties. The greater the interfacial area, the greater would be the bonding. The interface has a critical role to determine the mechanical as well as functional features of nanocomposites. Previously, the studies on the interface of the graphene/polymer focused on the use of rGO which usually had low structural integrity limitations. J. Ma et al. [145] utilized 4,4'-diaminophenylsulfone to modify the chemical bonding structure of covalently bonded structure of graphene platelets (GnPs) with polymer matrix. The stacking of the GnPs was observed to be prevented by this modification. The modified GnPs further reacted with the molecules of diglycidyl ether of bisphenol A (DGEBA) and this modification in two steps resulted in the creation of covalently bonded epoxy and GnPs nanocomposite interface. The atomic structure of created bonding interface between polymer matrix and GnPs is displayed in Fig. 16.21.

The GO bonding with polymer matrix is usually by electrostatic forces and van der Waal forces. Polymers and GO then aggregate to create a compact composite sandwiched structure. Hydrogen bonding also play a vital role in the interfacial

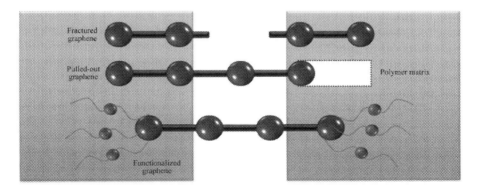

Figure 16.20 Breakage of covalent bond, pulling out of graphene, and improved interfacial interactions between graphene and polymer matrix due to functionalization of graphene [144].

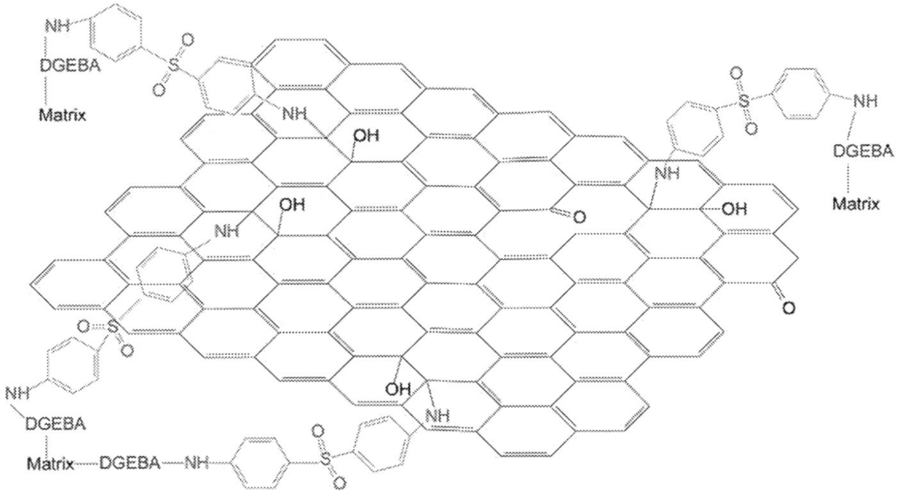

Figure 16.21 Covalently bonded atomic structure of interface in-between matrix and GnPs [145]. *GnPs*, graphene platelets.

bonding and adhesion of graphene and its derivative to polymer matrix in the highly ordered nanocomposite layered by layered structure [146]. The amino groups available on the amine terminated polystyrene and the carboxylic acid based residual groups available on the graphene are useful to undergo electrostatic interactions and to acquire graphene based polymer composite materials [147]. Cano et al. [148] prepared polyvinyl alcohol grafted GO surface by utilizing covalent modification mechanism to acquire composite material with improved mechanical features. The functionalization of surface of GO was accomplished by utilizing carbodiimide esterification reaction as shown in Fig. 16.22.

16.4 Applications of polymer/graphene nanocomposites in fuel cells

16.4.1 Polymer/graphene nanocomposite membrane

Polymer and graphene-based nanocomposite have been presented themselves as promising materials for membranes applications in fuel cells, providing best electronic properties and anticrossover properties of methanol. There is still a huge challenge to combine pristine graphene with the matrix of polymer to attain a hybrid structure with high disparity at the atomic level. The leading cause of this problem is the presence of aromatic groups in the structure. GO nanosheets in a combination of multiple groups such as hydroxyl and carboxyl play an essential role to disperse the polymer electrolyte. Generally, the GO is compatible with the polymeric surfaces as compared to the hydrophobic graphene. The chemistry of GO

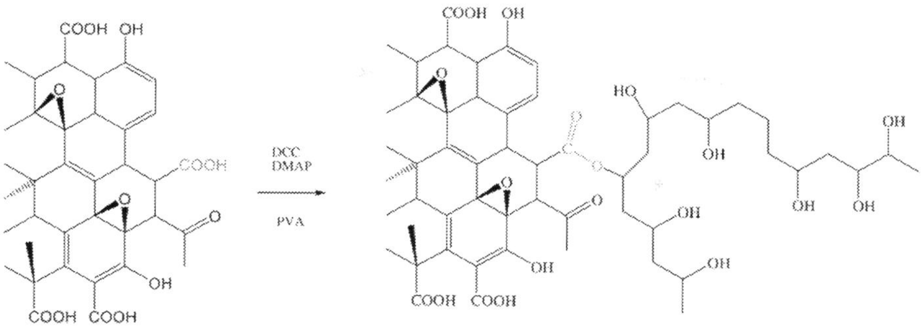

Figure 16.22 Functionalization of surface of graphene oxide by adopting covalent modification technique [148].

surface is quite flexible, tunable and easy to modify for the sake of interaction with ionic functionalized groups [149].

The applications of the polymers are in sensors, storage devices, biomedical, batteries, composites and electronics [150,151]. Several naturally occurring and synthetic polymers have introduced for these applications. The polymers are utilized in applications in multiple forms such as fibers, sheets, powder and membranes. Among all of these forms with particular and suitable applications, the membranes have achieved a great position in thousands of application because of their attractive properties and selectivity as well [152,153]. The mostly used polymers for membrane fabrications are polyamide, polyphenylene sulfate, polyimide, polysulfone, polyethersulfone, polyethylene terephthalate, polycarbonate, polyethylene, polyphenylene oxide, polyurethane, polyether ketone, vinyl polymers, cellulose, nitrocellulose, cellulose acetate and many other polymers [154–156] The fabrication of the polymer membrane is done by in situ polymerization, solution technique, melt technique, spin coating, layer by layer formation and coagulation method of solvent/nonsolvent. Membrane nanocomposites and hybrid structural composite are hot topics if membrane technology. The polymer membranes are highly flexible, and its properties rely on multiple factors such as type of fabrication method, pore size, pore size distribution, additive type and concentration, and concentration of polymer matrix. Several types of research on nanostructures have innovated for hundreds of applications related to the electronics, aeronautical and medical field [157]. Firstly, the carbon nanotubes (CNTs) were the main focus for the fabrication of nanocomposites. But after the innovation of graphene, the researchers moved towards better material for membranes as compared to the other nanotubes or nanofillers [158,159]. The graphene plays a significant role in the matrix to maintain the homogenous nature of the surface and to improve the dispersion without aggregation of particles. The graphene explored a new world of applications for membrane technology. The efficiency and performance of the polymer composite based on graphene are highly dependent on the structure if polymer matrix and graphene, a system of interaction, chemical and physical properties. Apart from them, the stacking, functionalization of graphene and amount of nanomaterial also affect the performance. The overall properties of the nanocomposite are determined by the compatibility of polymer and graphene which may cause, hydrogen and covalent bonding or week interactive forces, the weak π-π interactive forces between the delocalized electrons (graphene) and polymer aromatics. The presence of covalent bond increases the compatibility of graphene and polymer in nano composite. While physical forces are week and doesn't provide enough energy to form a strong link. In a nutshell type composite, the desired properties of membranes such as moisture content, permeation, surface porosity, hydrophilicity, flexibility and selectivity can be obtained by optimization of design and structure of composite, and type of polymer. Proceeding with these approaches, the nano composite membranes can be achieved with required properties as per the end use application demand [160].

Graphene is very considerable reinforcement to improve the qualities of a nano-composite based on polymer. The main characteristics of the polymer/graphene composite are its mechanical and thermal stability, and a number of applications are linked with these properties of polymer composite [161−165].

As the graphene has excellent material properties, so, it is highly recommended as reinforcement for polymer composite (Fig. 16.23). The composites are always found with the better strength and modulus resulting in the increase of glass transition temperature, which is also directly linked with the nanofiller loading. Many types of research are done to explain the interfacial properties of the polymer and nanofiller [128,166]. The modification and optimization of interface chemistry directly improve the mechanical properties of the nanocomposite, for example, polymer/GO nano-composite [167−169]. The nanocomposite membranes based on graphene shows

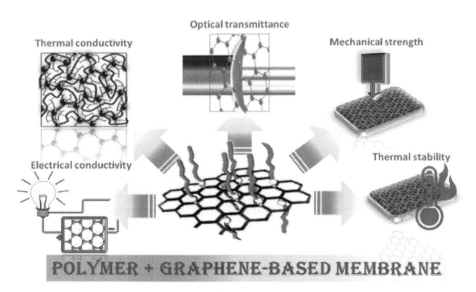

Figure 16.23 Effect of graphene nanomaterials on the properties of the polymeric membrane [160].

exceptional properties comprising the electrical conductivity, temperature stability, mechanical strength, thermal conductivity and optical transmittance.

The simplest way to get graphene is cutting by length along the longitudinal direction of CNT. After cutting the graphene is collected as web [170] with robust properties [171]. Structural Model of graphene and its exclusive characteristics are presented in Fig. 16.24A and B.

The functionalization of graphene is done to obtain the effect of hydrophilicity. The zigzag corrugated structure of graphene is itself hydrophobic in nature. The conductivity is achieved through holes heterogeneously. The platinum doped graphene was introduced in Nafion using its dispersion in water and isopropyl alcohol mixture. Eventually, at around 70°C, the casted membrane is obtained after the solvent of evaporation [172]. The reason behind the addition of graphene is to lower down the humidity level for the improvement in membrane working. In this study, the membrane properties are investigated at a relative humidity of 40%, 60% and 100%. An increase in efficiency was recorded from 0.435 A/cm^2 to 1.27 A/cm^2 with the addition of platinum-based GO in Nafion by 3% of the weight of the polymer at RH of 100% as compared to the raw Nafion [173]. The other modified characteristics were improved flexible, ductile and barrier properties with better chemical, thermal and mechanical stability. The proposed research also concluded the cost-effectiveness of this experiment.

16.4.2   Graphene-polymer nanocomposites in polymer membrane fuel cell

High power density and low greenhouse gas release have made the Polymer electrolyte membrane fuel cells (PEMFC) as the best-suited materials for energy conversion and storage devices [174]. Due to the best properties such as rapid startup and excellent efficiency, low-temperature PEMFC are in huge consideration [175]. Firstly, to attain the properties of chemical resistance and conductivity, Nafion, which is a perfluorinated polymer, is used in membranes. But the high cost of Nafion membrane based polymer electrolytes overcome the utilization of these materials for energy storage and conversion [176]. Thus, PEMFC membranes based on polyfluorosulfonic acid or acid doping have been developed. Graphene and its relative derivatives are considered a hot research topic relevant to this scope of study [177−179]. In graphene derivatives, the GOs are mostly used and experimented materials for energy applications [180]. The best characteristics of GO are its proton transport channels and water holding capability. GO is obtained after chemical modification and these characteristics are due to the excess presence of oxygen-based functional groups present in its structure with a larger surface area. Therefore, GO is a considerable nanomaterial in order to attain the best mechanical strength to membranes and also the proton conductivity [181]. However, the purest form of GO shows electronically nonconductive properties presenting a gradient conductivity of range from 1 to 5×10^{-3} S/cm at biasing of 10 V [182].

As Nafion based polymer electrolyte membrane suffers with low efficiency at high temperature. This decline in the performance has made researchers active to improve this declining efficiency [183]. A nanocomposite of GO and Nafion is

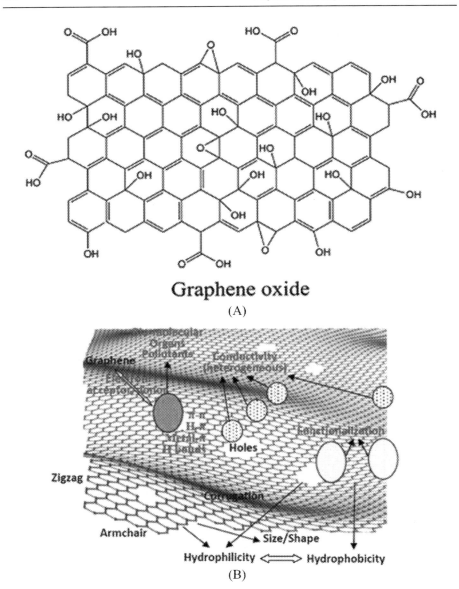

Figure 16.24 (A) Graphene structure and microstructural variation by the addition of carboxyl groups of GO [170]. (B) Exclusive characteristics of graphene [171]. *GO*, graphene oxide.

used for the fabrication of membranes for the applications of PEMFC [184]. The interaction of Nafion with GO at the interface is improved by a chemical reaction known as atom transfer radical addition, in which C-C group of graphene and CRF group from the chains of Nafion undergo a chemical change to enhance the compatibility at the interface. The conductivity of the GO-Nafion nanocomposite is very high, ultimately increasing the performance of the cell as compared to the Nafion based membrane. NM/GO-0.05 based MEA appears as a material with the highest fuel cell performance with the power and current density of 886 mWcm^2 and 1376 mAcm^2 at biasing of 0.6 V as shown in Fig. 16.25.

In a research study, Feng et al. described a method to develop nonhybrid polymer electrolyte membrane by the rolling the GOSs on the surface of aluminum foil by evaporation through their disperse solution (Fig. 16.25) [185]. These structurally modified sheets support the development of GO/Nafion based membranes. This membrane structure provides excellent proton conductivity and even at low humidity of 40% RH. A dominant variation is observed between the activation energy of reformed Nafion and GO/Nafion composite. The activation energy of 22.82 kJ/mol at 40% RH and 14.80 kJ/mol at 40% RH respectively for recast Nafion and GO/Nafion composite is explained with massive variation in Fig. 16.26.

As mentioned above that there are number applications of Nafion for PEMFCs as it is highly mechanically and chemically stable with excellent conductivity of ions. However, there is a big challenge of its commercialization because of its cost, but many kinds of research are ongoing to develop the solution to this problem. Cao et al. prepared a membrane of nanocomposite PEO/GO using solution mixing technique [186]. The developed poly(ethylene oxide)/GO (PEO/GO) nanocomposite are used for the low-temperature fuel cell membrane. The 0.5 wt.% nanofiller loaded membrane has a thickness of 80 μm (Fig. 16.27). The membrane possesses high tensile strength, flexibility, and ion-transport behavior. A considerable increase in the power density from 21 to 53 mW/cm^2 was recorded even at a high-temperature range from 30°C to 60°C with the relative humidity of 100% as shown in (Fig. 16.28). At 60 °C, the open-circuit voltage value is recorded as 760 mV. The value of limiting current is also improved from 90 to 180 mA. When the operating temperature is increased from 30 °C to 60 °C, the ionic conductivity was also increased from 0.086 to 0.134 S/cm. The tensile strength was observed with a

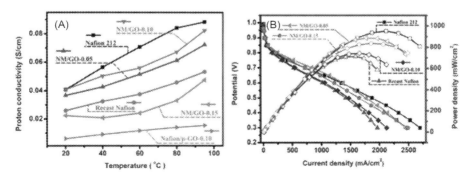

Figure 16.25 (A) Effect of temperature conductivity (proton) for the membranes based on Nafion (B) MEAs are polarizing curves for membranes based on Nafion using air as the oxidant [184].

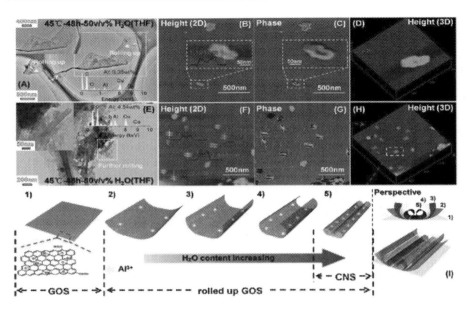

Figure 16.26 (A−H) TEM, EDS, and AFM characterizations of GOSs (rolled up) fabricated at 50/80 vol./vol.% H₂O under 45°C for 48 h (H₂O/THF); (1) schematic quantity (of Al³⁺ ions)-dependent rolling process of the "evaporated" GOSs on the aluminum foil surface [185].

Figure 16.27 PEO/GO membrane image with 0.5% addition by weight of nano filter [186]. *GO*, graphene oxide. *PEO*, poly(ethylene oxide).

massive increment at 52.22 MPa and modulus at 3.21 GPa. The overall thickness of 0.5% nanofiller embedded membrane was 80 μm (Fig. 16.27) [186].

The PEMFCs were run at a high temperature of above 100°C, showing the best and optimized performance because of the improved reaction kinetics of electrode,

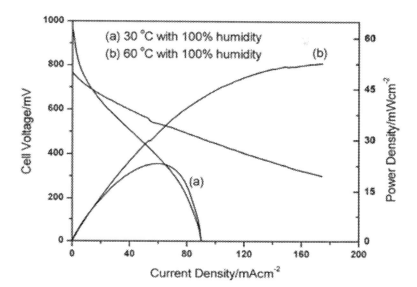

Figure 16.28 Power density and polarization curves of fuel cell membranes in a test at altered operation temperature. Loading of Pt was around 0.7 mg/cm^2, and O$_2$, as well as H2, was used in the test without back pressure [186].

minimizing the usage of expensive metals as a catalyst, that is, platinum. This modification also increases CO tolerance. The Nafion based membranes have a limitation of 80°C temperature for their usage because, at a temperature greater than 80°C, the amount of water content declines dramatically. Consequently, the protic ILs (PILs) and complexes based on polymer acids are now widely considered for high-temperature applications in order to replace the Nafion [187]. Strong research has been done on the composite membrane materials of PILs and IL polymer-modified graphene (PIL(NTFSI)-G) [188]. At the temperature of 160°C, the conductivity was recorded as 7.5 × 10^{-3} S/cm with the 0.5% weightage of the loading of graphene. This value of conductivity is four times greater than a membrane of sulfonated polyimide/PIL (SPI/PIL). Considering the comparison with the SPI/PIL membranes, the composite membrane showed an increase of 345% and 127% for the modulus and tensile strength respectively with the 0.9% weightage of graphene loading.

A research was made possible to study the GO functionalization with 3-mercaptopropyltrimethoxysilane by Zarrin et al. In this study, the effect of oxidation of the thiol group with sulfonic acid is also considered as filler for Nafion membranes at elevated temperature in the application of polymer fuel cells [189]. The results of conductivity and one cell testing were obtained for both recast Nafion and GO/Nafion functionalized composite, with four times improvement in GO/Nafion functionalized composite at the temperature of 120°C with a relative humidity of 25%.

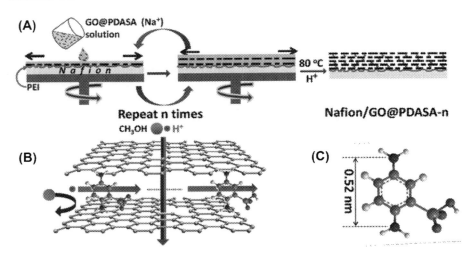

Figure 16.29 (A) Crosslinked GO film based on phenylenediamine-2-sulfonic acid (PDASA) fabrication through a technique of spin-coating; (B) crosslinked GO film based on PDASA transport mechanism; and (C) PDASA morphological structure [190]. *GO*, graphene oxide.

In recent research, Nafion /GO composite membranes are fabricated through the method of spin coating, with a curing agent of 1,4-phenylenediamine-2-sulfonic acid by HE et al. as in Fig. 16.29. It was claimed that synergistic combination transport of methanol and protons in the films of GO had decreased the permeability of methanol up to 93%, conserving the higher value of conductivity by the proton [190].

In 2014, Ravi at el. [191] developed a composite using Sulfonated GO/Sulfonated polyether ether ketone nanocomposite membrane for PEMFCs. The properties, such as the performance, efficiency, and ionic conductivity, are studied in this research with the effect of SGO on the characteristics of SPEEK. The interconnection of the proton transfer pathway was done by introducing the nanosheets of GO functionalized with sulfonic acid in the structural layers of sulfonated polyether ether ketone. Sulfonated GO/Sulfonated polyether ether ketone nanocomposite membrane showed excellent properties for energy conversion and storage. These results were far better than the recast SPPEK. Acid-base composites have presented themselves as promising candidates for the innovations in PEM under high temperatures and limited humid conditions to attain the charge transport properties. In these composites, the acid and base group represent themselves as donor and acceptor, respectively, and are linked very effectively. Protons are transported through the Grotthuss mechanism from the donor to accepter without the presence of water. The developed composite presented a very great conductivity up to 0.055 S/cm and a fuel cell performance of 378 mW/cm^2 at a relative humidity of 30% at 80°C. These results were far better than the recast SPPEK [191]. Sahu et al. developed a Sulfonated Graphene−Nafion Composite Membranes for PEFCs Operating under Reduced Relative Humidity and found that the composite membrane of Nafion−S-graphene might be used to discuss

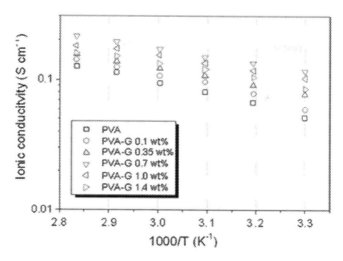

Figure 16.30 Effect of temperature on conductivities of VA and PVA/graphene composite membranes [193].

various critical complications related with commercial Nafion membranes in the fuel cell applications [192].

In a research, a very typical blending method was utilized to synthesize the nanocomposite based on PVA and graphene by Ye et al. A well-established conductivity value was achieved through the best-connected transport channels of graphene inside the membrane structure. A small weightage up to 0.7% increased the conductivity up to 126% and reduced the permeability of methanol up to 55%, as shown in Fig. 16.30. The brilliant structural arrangement of PVA and graphene composite presented the most potent strength and robust adhesion characteristics. This arrangement also increased the power density to 148% to improve the performance and efficiency of the polymer fuel cells [193].

16.4.3 Graphene-polymer nanocomposites in direct methanol fuel cell

The PEMFCs have mostly used polymer fuel cells. After them, the Direct Methanol Fuel Cells are making theirs in the applications of energy stooge and conversions. DMFCs provide a special tow-dimensional flexible structural arrangement with functionalization of considerable and high-temperature stability as per the requirements of the polymer electrolyte membrane. In comparison with inorganic materials, they possess a high range of properties due to the flexible nature of functionalized groups and their presence [194]. GO applications for energy purposes are increasing day by day. GO-based on Nafion composites are studied and researched extensively nowadays in recent researches. Thus, they are extensively used among the applications of PEM. Moreover, Nafion is well known for its good methanol permeability, which makes it a considerable material for methanol transport with water through conductive channels full of charge carriers, so it is also referred to as methanol restrictor. The commercialization of direct methanol fuel cells is very challenging practically because of the lowering in performance due to the poisoning of catalysts and potential mixing [195,196]. For the applications of DMFCs, the crossover of methanol is controlled by the membrane thickness. The recommended thickness of Nafion is 175 mm (Nafion 117). The disadvantage of adopting this strategy increases the current resistance and ultimately decreases the power density of the fuel cell [197]. Multiple researches are being done and are ongoing to overcome the issues and problems related to the methanol crossover. Many nanofillers and polymers are introduced to modify the Nafion, such as zeolite, zirconium, PPy, and SiO_2 [198−204]. In the previous studies, two modifications are adopted to improve the properties of Nafion. The first one is the addition of nanofillers dispersion [205]. The second approach is the nanocomposite membrane formation by the in situ synthesis of the inorganic nanofillers [206]. The Nafion based composite membranes show low methanol crossover and high-water acceptance as compare to the recast Nafion. A huge value of loading of inorganic stuff upon membranes makes them highly brittle and poor conductor to current. These effects are high, which are developed through the above-mentioned methods. In this study, a novel approach was adopted to low crossover methanol electrolyte using SGO [207].

Great results were obtained by the introduction of GO in Nafion for DMFCs applications by the Choi group [208]. The robust advantages of GO are not realized yet to improve fuel cell efficiency. In a study, the GO/Nafion composite is revealed to have a nonuniform distribution of GO. Graphene absorbed with surfactants also improved the proton conductively and decreased the permeability of methanol [209]. The matrix of polymer is the main cause of GO aggregation, which is a considerable challenge to overcome. This problem was addressed by Hung at el. [210] via determination of viscosity variation of SGO/Nafion blend. The study concluded and provided the optimized percentage of loading of sulfonated GO to Nafion up to 1% by weightage approximately. The analysis clearly concluded that the introduction of SGO to Nafion improves the strength, proton conductivity, mechanical properties and decreases the methanol crossover and methanol permeability.

In the latest research, the Nafion/GO nanocomposite is developed for the application of DMFC as PEM. The composite is fabricated using LBL methodology, utilizing the phenyl diamine hydrochloride (PDHC) as a curing agent [211]. The developed composite presented improved selectivity with lower methanol permeability than Nafion 117. In research by Yuan et al. it was explained that the multilayer Go incorporation doesn't only increase the strength but also improves the crossover reduction by up to 67%. Yuan et al. used poly (diallyl dimethylammonium chloride) (PDDA) as a crosslinking agent [212].

In a research study, Choi et al. modified the structure of Nafion with the addition of SGO. The electronic properties of the modified composite structure were better than the recast Nafion and GO/Nafion nanocomposite. The properties such as proton conductivity and methanol crossover were improved due to the confined water in order to increase fuel efficiency [208]. The limited elevated temperature and high cost are the limitations of Flemion, Nafion, and perfluoro polymers membranes. Moreover, the temperature limitation of the Nafion membrane is 80°C because of the decrement in water content [107].

Under a low relative humid environment, the Nafion is of minor function [213]. The temperature limitation is one of the reasons for hurdles in large scale production and applications. To overcome this limitation, many other polymeric materials have gained dominant attention, such as PBI, PSF, and PEEK [191]. Mishra et al. demonstrated the introduction of SPEEK into the Nafion with increased proton conductivity. The two-dimensional structure of GO has the capability to improve the mechanical properties up to a very high extent [214].

In a research study, the SGO and SPES were used to develop a composite membrane structure using the process of sonication and membrane casting, providing improved strength, conductivity, and transport capacity due to increased interfacial bonding because of the larger lamination surface area of SPES and SGO. The best suitable results were obtained with 5% weightage of SGO, providing the optimized electronic properties suitable for fuel cell applications [215]. The intercalation method was also utilized to develop a Nafian based composite with the addition of organo-modified sulfonate-based GO [216]. The plate-like structure with a dispersive pattern-oriented structure of GO was observed using characterization techniques, especially, NMR and electrochemical analysis. The structure provides a high barrier, mechanical and electronic properties. This attractive structure improves the blocking effect and also widens the temperature range for fuel cell applications (Fig. 16.31).

The layer by layer arrangement is a very simple and typical method to be used [217,218]. This approach is beneficial for the applications of polymer electrolyte membranes and direct methanol fuel cells [219−221]. The deposition technique of LBL technology is prevalent for the formation of GO/PDDA composite onto the Nafian membranes to reduce the methanol permeability and improve the conductivity and power density. The blocking characteristics of methanol are obtained when bilayer develops a fragile sheet on the surface to suppress the diffusion efficiency of methanol up to 67% [222].

The developed structure of the bilayers membranes shows the blocking property of methanol up to 63%, decrement in methanol permeability up to 60%, and improved power density than pristine membrane. In research by Jia et al., the rGO fused Nafion is formed using the vacuum filtration method [223]. The 3D structure with improved electronic properties empowers the long-routed transport channels with improved hydrogen bonding between the layers with rod structured nanomaterials (Fig. 16.32).

SPEEK/PVA based blend of polymer and nanosheets of SGO/Fe_3O_4 were used to develop a nanocomposite for the application of direct methanol fuel cells by Beydaghi et al. The 5% addition of SGO/Fe_3O_4 reduced the permeability of the methanol from 1.78 × 10^{-6} cm^2/s to 8.83 × 10^{-7} cm^2/s because of the effective barrier properties of graphene. However, the conductivity of the proton transportation is increased due to the enhanced interaction between the SGO and Fe_3O_4 with the water through Grotthus and Vehicle mechanism, respectively [224]. The nanoscaled materials, such as GO, are combined with PVA or SSA are very popular for the application of polymer electrolytes and membranes [194]. In order to overcome the limitations of cost and functional performance, the functionalized groups of nanoparticles are incorporated with the synergistic

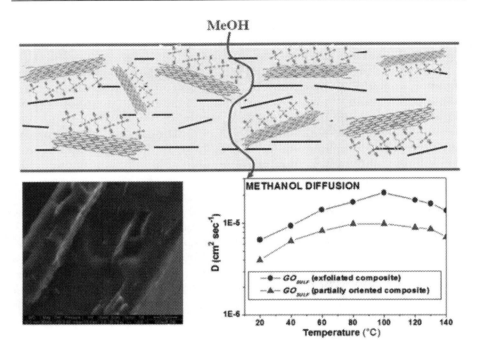

Figure 16.31 Illustration of the methanol molecules diffusion pathway through an exfoliated/aligned nanocomposite membrane, respectively SEM images of the Naf/GOSULF-kn self-diffusion coefficients of methanol for nanocomposite membranes prepared [216].

effect of inorganic material. The synergistic effect provides high thermal, physical stability, and polymer related characteristics, such as flexibility and conductivity [225]. While in the case of elevated temperature, the inorganic materials are added to the polymeric matrix to improve the proton conductivity and as a bridge to link consecutive ionic groups in the hydrated form of polymeric matrix [226,227]. This is the only cause of cluster formation of inorganic materials in the way of hydrophilic channels resulting in the chocking or blockage. However, the addition of nanoparticles reduces the crossover of fuel. The diffusion through the charge transport path (holes and pores) causes the permeation of fuel. The incorporation of inorganic material nanoparticles, causing the blockage resulting in a decrease in fuel permeability and crossover as well. In this way, a well-established balanced can be achieved among the high electronic properties and low permeation properties [225]. (Table 16.1)

16.4.4 Graphene-polymer nanocomposites as oxidation reduction reaction catalysts

Graphene functionalized with polymer, specifically the polymers having nitrogen in them, are found best materials as metal-free electrocatalysts for the oxidation-reduction reaction at the cathode for the applications of lithium-ion batteries and

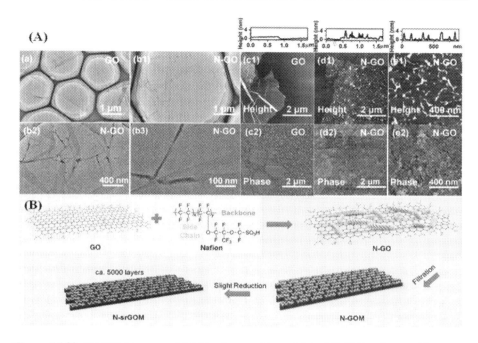

Figure 16.32 (A) TEM images of (a) Graphene oxide and (b1–b3) N-Graphene oxide nanocomposite. AFM height images and height profiles of (c1) GO and (d1, e1) N-GO nanocomposites. AFM phase images of (c2) GO and (d2, e2) N-GO nanocomposites. (B) Graphic illustration of the preparation of N-srGOM [223].

fuel cells. As platinum-based catalysts are expensive and less stable so, these ORR catalysts are suitable for overcoming these problems.

Platinum has wide applications in fuel cells as a catalyst [253]. Conversely, the expense of platinum, platinum cluster formation, and poisoning have restricted its applications practically. Moreover, the applications of platinum as electrocatalyst are also restricted due to the effect of crossover in direct methanol fuel cells. Till now, much researches have been made possible to astounded the effect of these challenging problems [254–259], nonnoble metals [260–262], and enzymes [263] to supplant electrocatalysts based on platinum. In many kinds of research, the high resilience, stability, and durability are reported by graphene doped with nitrogen because of its highly conductive two-dimensional morphological structure [46]. Along with the graphene doped with nitrogen, the nanocomposites based on graphene and polymeric materials based on nitrogen are also tested and considered as a great catalyst for oxidization reduction reaction (ORR) [264].

According to quantum physics, the atoms of nitrogen having the ability to accept electrons carries a highly positive charge on the consecutive atoms of carbons through the intramolecular transfer of charge. The delocalization of the charge rapidly creates an attractive force between the O_2 and the anode, and to complete the

Table 16.1 Literature review for application of graphene polymer nano composites in fuel cells.

Polymer electrolyte membrane fuel cells				
Method of solution casting				
Sr.	**Type of graphene**	**Amount of graphene (wt.%)**	**Type of polymer**	**Suitable solvent**	**Modified/improved properties**
1	Graphene oxide	4	Nafion	Dimethylacetamide (DMAc)	• Power density • Proton Conductivity [228]
2	Graphene oxide (functionalized)	5 and 10	Nafion	Ethanol	• Water retention • Mechanical & chemical stability • Ion transport • Proton Transport [189]
3	Graphene oxide (coupled with polyoxometalate)	1	Nafion	Deionized water	• Water uptake • Proton Conductivity • Resistance Fall [229]
4	Graphene oxide sheets (rolled)	—	Nafion	Dimethylformamide (DMF)	• Activation energy reduction • Retention of water • Proton conductivity [185]

#	Material	Value	Polymer	Solvent	Properties
5	Graphene oxide (Pt-graphene)	0.5–4.5	Nafion	Isopropyl alcohol (aqueous solution)	Tensile strength [173]
6	Graphene/SiO2 (platinum Based)	0.5–3	Nafion	Isopropyl alcohol (aqueous solution)	• Water uptake • Proton conductivity [230]
7	Graphene oxide	2.3, 3 and 5	Nafion	Dimethylacetamide (DMAc)	• Water retention • Tensile properties • Proton & electronic conductivity [231]
8	Graphene oxide	0.05, 0.10 and 0.15	Nafion	Isopropyl alcohol (aqueous solution)	• Water retention • Mechanical properties • Ionic conductivity [184]
9	Graphene oxide	1	Nafion (Pt-TiO2)	Isopropyl Alcohol (Aqueous Solution)	• Proton Conductivity • Ion Exchange Capacity [232]
10	Sulfonated graphene oxide with graphite oxide	2	Polybenzimidazole (PBI)	Dimethylacetamide (DMAc)	Proton & ionic conductivity [233]
11	Graphene oxide (modified)	GO: 1–5; iGO: 5–15	Polybenzimidazole (PBI)	Dimethylacetamide (DMAc)	• Chemical stability • Proton conductivity [234]
12	Graphene oxide	0.5	Polyethylene oxide (PEO)	Distilled water	• Mechanical properties • Reduced resistance [186]

(Continued)

Table 16.1 (Continued)

Polymer electrolyte membrane fuel cells				
Method of solution casting				
Sr.	Type of graphene	Amount of graphene (wt.%)	Type of polymer	Suitable solvent	Modified/improved properties
13	Graphene sheets (modified)	10	Sulfonated polyimide (SPI)	Dimethyl sulfoxide (DMSO)	• Ionic conductivity • Mechanical properties [188]
14	Graphene oxide	0.75	Nafion–sulfonated polyetherether ketone (SPEEK)	Ethanol (aqueous solution)	• Ionic conductivity • Current & power density [235]
15	Sulfonated graphene oxide	2	Chitosan (CS)	Acetic acid (aqueous solution)	• Hydrous & anhydrous conductivity • Proton conductivity • Mechanical & thermal stability [236]
16	Sulfonated graphene oxide	—	Sulfonated polyether-ether ketone (SPEEK)	Dimethylacetamide (DMAc)	Proton conductivity [236]

Sr.	Type of graphene	Amount of graphene (wt.%)	Type of polymer	Suitable solvent	Modified/improved properties
17	Sulfonated graphene oxide (silica grafted)	1, 2 and 5	Chitosan/polyvinyl alcohol	Deionized water	• Proton conductivity • Thermal & mechanical stability • Elastic properties [237]
18	Poly(2,5-benzimidazole) grafted graphene oxide & graphene oxide	0.5, 1, 2 and 4	Polyarylene ether sulfone (sulfonated)	Dimethylacetamide (DMAc)	• Proton conductivity • Dimensional & mechanical stability • Young's modulus • Elongation [238]

Other methods

Sr.	Type of graphene	Amount of graphene (wt.%)	Type of polymer	Suitable solvent	Modified/improved properties
1	Graphene oxide (modified)	2.5, 5, 7.5, and 10	Sulfonated polyether-ether ketone (SPEEK)	Dimethylformamide (DMF)	• Proton conductivity • Current & power density [239]
2	Graphene oxide (functionalized)	5	Polybenzimidazole (PBI)	Dimethylacetamide (DMAc)	• Proton & ionic conductivity [240]
3	Graphene oxide (functionalized)	—	Polybenzimidazole (PBI)	Dimethylacetamide (DMAc)	• Tensile strength • Proton conductivity [241]

(Continued)

Table 16.1 (Continued)

Polymer electrolyte membrane fuel cells					

Method of solution casting

Sr.	Type of graphene	Amount of graphene (wt.%)	Type of polymer	Suitable solvent	Modified/improved properties
4	Sulfonated graphene oxide	–	Nafion 115	–	• Ion exchange capacity • Proton conductivity [181]

Method of solution casting

Sr.	Type of graphene	Amount of graphene (wt.%)	Type of polymer	Suitable solvent	Modified/improved properties
1	Sulfonated graphene oxide	0.5	Nafion	Dimethylformamide	• Power density • Methanol crossover reduction • Activation energy reduction [210]
2	Graphene oxide	0.1–2	Nafion	Dimethylformamide	• Ionic conductivity • Methanol crossover reduction • Thermal properties • Mechanical properties [208]

#	Material	Concentration	Polymer	Solvent	Properties improved
3	Sulfonated graphene oxide	0.05–0.5	Nafion	Dimethyl acetamide (DMAc)	• Water uptake • Swell ratio reduction • Permeability of methanol • Methanol crossover reduction • Proton conductivity [185]
4	Sulfonated graphene oxide	0–10	Sulfonated polyether ether ketone	Dimethyl acetamide (DMAc)	• Mechanical properties • Methanol crossover reduction • Selectivity • Retention of water • Proton conductivity [242]
5	Sulfonated graphene oxide /Fe$_2$O$_3$	3, 5 and 7	SPEEK/PVA	Dimethyl acetamide (DMAc)	• Methanol crossover reduction • Mechanical properties • Proton conductivity [224]

(Continued)

Table 16.1 (Continued)

Polymer electrolyte membrane fuel cells					
Method of solution casting					
Sr.	**Type of graphene**	**Amount of graphene (wt.%)**	**Type of polymer**	**Suitable solvent**	**Modified/improved properties**
6	Graphene oxide	0–2	115 Nafion	Deionized water	• Permeability of methanol • Selectivity • Tensile properties • Proton conductivity [243]
7	Sulfonated propyl-9 silane graphene oxide	0–8	Sulfonated polyimide	Dimethyl acetamide (DMAc)	• Water retention • Proton conductivity • Internal self-humidification • Bound water content • Thermal, mechanical & chemical stability [244]
8	Sulfonated imidized graphene oxide	0–15	Sulfonated polyimide	M-cresol	• Proton conductivity • Water retention • Ion exchange capacity

#	Material	Loading	Solvent	Properties
				• Thermal, mechanical & chemical stability [245]
9	Graphene oxide (modified)	0–8	N-o-sulfonic acid benzyl chitosan	Deionized water	• Mechanical stability • Water retention • Proton conductivity [246]
10	Graphene (functionalized)	0.10 and 0.25	Sulfonated polyether ether ketone	Dimethylformamide	• Internal self-humidification • Water retention • Methanol crossover reduction • Proton conductivity [247]
11	Graphene oxide (functionalized)	5	Sulfonated polyether ether ketone	Dimethylformamide	• Water retention • Proton conductivity • Methanol permeability • Internal self-humidification [209]

(Continued)

Table 16.1 (Continued)

Polymer electrolyte membrane fuel cells				
Method of solution casting				
Sr.	**Type of graphene**	**Amount of graphene (wt.%)**	**Type of polymer**	**Suitable solvent**	**Modified/improved properties**
12	Graphene oxide (functionalized)	GO: 5; SSi-GO: 3–8	Sulfonated polyether ether ketone	Dimethylformamide	• Ion exchange capacity • Water uptake • Proton conductivity [248]
13	Graphene oxide (zwitterion-coated)	5–25	Polybenzimidazole (PBI)	Dimethyl acetamide (DMAc)	• Water uptake • Swell ratio reduction • Proton conductivity • Methanol permeability [249]
14	Graphene oxide	1, 2, 4 and 6	Sulfonated polyether ether ketone	Dimethylformamide	• Selectivity • Methanol crossover reduction • Proton conductivity [250]

| 15 | Sulfonated graphene oxide | 0.5, 5 and 10 | Chitosan/sulfonated chitosan | Deionized water | • Selectivity
• Proton conductivity
• Methanol permeability
• Thermal, mechanical & chemical stability [251] |

Other methods

Sr.	Type of graphene	Amount of graphene	Type of polymer	Suitable solvent	Modified/improved properties
1	Graphene (monolayer)	–	Nafion	Deionized water	• Proton conductivity • Methanol permeability reduction [88]
2	Graphene oxide (laminates)	3 mg/L	Membrane (self-supported)	Deionized water	• Power density • Methanol permeability reduction • Proton conductivity [252]
3	Graphene oxide	–	Nafion 115	Deionized water	• Selectivity • Proton conductivity • Methanol permeability [243]

(Continued)

Table 16.1 (Continued)

Polymer electrolyte membrane fuel cells				
Method of solution casting				
Sr.	**Type of graphene**	**Amount of graphene (wt.%)**	**Type of polymer**	**Suitable solvent**	**Modified/improved properties**
4	Graphene oxide	2.8 wt.%	Nafion	1,4-Phenyldiamine hydrochloride	• Water uptake • Swell ratio reduction • Oxidation stability • Ion exchange capacity • Methanol permeability [211]

process of ORR; it readily transports the oxygen from one side to another [265,266]. Nitrogen-carbon interactive modification tools are being studied for some of the previous years. By optimization of this interaction, many ways related to the applications of electrocatalyst based on Carbon, will open.

In a research study by Wang et al., the poly (diallyldimethylammonium) (PDDA) was reduced by using sodium boro-hydrate (NaBH₄). As a result of that reduction, a nano-hybrid structure was formed called PDDA-functionalized graphene. In this action, the PDDA shows its ability to accept an electron to support the oxidation-reduction reaction by increasing the performance of electrocatalyst. The developed hybrid material showed better stability, selectivity, and antipoisoning ability compared to the commercialized platinum-based catalysts. These hybrid structures also presented them far better than primeval graphene by giving a better onset potential of 0.15 V in the comparison of 0.25 V [267].

Sun immobilized graphitic carbon nitride (GCN), and graphene (converted chemically) based nanocomposite was developed the absorption of graphene on GCN under the high value of temperature involving the polymerization of molecules of melamine as shown in Fig. 16.33. The reinforcement of the graphene in GCN increased the conductivity of the material and also enhanced the surface area of the nanocomposite, which ultimately improved the catalytic activity for ORR as compared to the platinum-based composite. The developed composite has low fabrication cost and flexible properties for fuel cell applications 64 [264]. The nanocomposites of 2D-graphene functionalized with O-EDOT were fabricated using a very simple one-step redox reaction, involving the reduction of GO to graphene followed by the oxidation of EDOT into the oxidized EDOT (O-EDOT). The induction of functionalized groups on the planner structure of graphene definitely improves the physical and chemical characteristics of the nanocomposite. A very strong attraction between the polymer and graphene due to compactness leads for the best electrocatalytic activity to reduce oxygen and

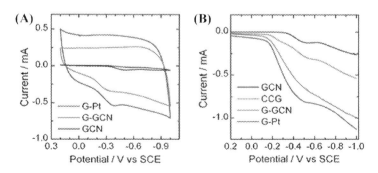

Figure 16.33 (A) CVs at 100 mV sand scan rate for G-Pt, G-GCN, and GCN electrodes (B) RDE voltammograms 100 mV sand scan rate of 10 mV/s, for G-Pt, G-GCN, CCG, and GCN electrodes with the rotation rate of 1,500 rpm in O₂-saturated 0.1 mol/L KOH solution 64. Copyright 2010. Reproduced with permission from Royal Society of Chemistry [264].

to convert I^{-3} into I^-, as compared to the structure based on the only graphene. This innovation of material structure is highly recommended for DSSCs (DSSCs) and fuel cells (Alkaline Membrane Fuel Cells) as a cathode electro catalyst [268].

Carbon nitrides are considered as promising materials as the catalyst for the oxidation-reduction reaction. Especially, the g-C_3N_4 has a very good and well-established synthetic polymer structure, which is formed through the polymerization of melamine, cyanamide, or dicyandiamide. It is observed that the higher content of N, structural, and thermal stability make g-C_3N_4 as a suitable material for the applications of metal-free ORR electro catalyst [269,270].

In research by Lyth et al. electro-active catalytic properties of pristine g-C_3N_4 were studied by the provision of the acidic medium. On analysis, it was found that the onset potential in these results was higher than the previous literature values of carbon black. But the power density of g-C_3N_4 was very low due to very low surface area, which restricted its applications in fuel cells potentially. Moreover, it was also concluded that the combination of carbon black and g-C_3N_4 would dominantly increase onset potential up to 0.76 V and current density too [271].

In multiple types of research, carbon nanomaterials with conductive nature were combined with g-C_3N_4 to enhance the electrochemical characteristics and conductivity of the material by the improvisation of the electron transport chain [272]. The use of Carbon-based material, especially the two-dimensional graphene nanosheets, is found as an excellent choice for the development of nanocomposite based on g-C_3N_4 g [273,274]. Layer by layer assembly mode is adopted in the fabrication of these composites. To provide a longer transport flow channel for electron, the nanosheets of g-C_3N_4 are forming an analogous planner structure.

A g-C_3N_4/graphene-based nanocomposites were fabricated by Sun et al. in which the melamine was polymerized at greater temperature with graphene to deposit g-C_3N_4 onto the surface of graphene. The incorporation of graphene increased the CO tolerance characteristics and electrocatalytic activity of the g-C_3N_4 for redox reactions. This study concluded that the mechanism of catalytic activity reacted to the g-C_3N_4/graphene is in the resemblance of other metal-free catalysts for redox reactions, that is, nitrogen-doped carbon nanotubes.

A novel method was introduced by Yang et al. to fabricate a graphene-based carbon nitride nanosheets as in Fig. 16.34. The hard templates based on GM-silica nanosheets were used with carbon tetrachloride as CN precursors and ethylenediamine. The incorporation of graphene added value for the improvement in catalytic activity, selectivity, and long-term performance as compare to the carbon nitride without graphene. The best catalytic activity performance revealed the presence of higher surface area, the content of nitrogen, electrical conductivity, and also the thin layer thicknesses. Moreover, the content of pyridinic N also affected the electrocatalytic activity potentially [275].

According to the theoretical results of the multiple research simulations, the form and content of nitrogen directly affect the activity of the redox-active sites, especially in the nitrogen-carbon catalytic system. The nitrogen content formed a positive charge cloud on atoms of Carbon, especially on the foursome sites on the edges of graphene [273,274,276,277].

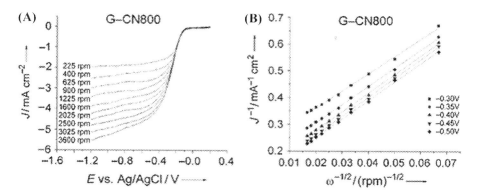

Figure 16.34 (A) At 5 mV/s of scan rate various rotation rates at a scan rate, RDE linear sweep voltammograms of G-CN800 in O_2-saturated 0.1 M KOH. (B) Koutecky–Levich plots of G–CN800 extracted from (A) [275].

16.5 Applications of polymer/graphene nanocomposites in solar energy

16.5.1 Application of polymer/graphene nano composite in perovskite solar cells

16.5.1.1 Transparent conductive electrodes

In polymer solar cells, the use of graphene and its derivatives is increasing day by day as transport conductive electrode replacing the ITO based conventional electrodes. Currently, the research work on PET related to the polymer solar cells is on the peak. The PET is being used as a substrate. For this purpose, thermal annealing of rGO is done to deposit them on the PET substrate. The fabrication of photovoltaic cells is done using the spin coating of this plasma-treated PET/rGO for the provision of surface (hydrophilic) [278]. Possessing the thickness and transmittance of 16 nm and 65%, respectively, the films of GOs show J_{SC} of 4.39 mA/cm^2, PCE of 0.78%, and V_{oc} of 0.56 V. Consequently, the fabricated device can withstand up to 1200 bending cycles with consistence performance.

The Practical applications of the ITO is strongly prohibited by the elevated temperature, deprived mechanical stability, and the expensive costing of raw material. Due to these reasons, many types of research has been done to alternate the ITO based applications in transparent conducting electrodes with the use of nanomaterials like carbon-nanotubes and graphene [279–296]. Among these, electrodes based on graphene have been studied via liquid suspension of the graphene and macro-scale graphene produced through the CVD [282,297–301]. The graphene coatings based on suspensions of liquids are very profitable in terms of cost. They provide acceptable costs related to printing, spin coating, and roll/roll processing. Moreover, the graphene deposited using the CVD method shows the best film resistance and transmittance of up to 30 Ω/Sq and 30%, respectively, with the best physical characteristics [282].

The challenging factor is the transparency of the graphene base transport conductive electrons as compared to the transparency level of the transport electrons based on metals because of the less sheet resistance. However, the TCEs based on graphene is highly stable, easy to process, and cheap [302]. In recent studies, to overcome the disadvantages of all this, a hybrid structure based on metallic and graphene morphology has been developed for energy harvesting applications [289,303]. It's been described that the TCEs performance is enhanced by hybridizing the inorganic and organic materials, such as ITO and poly(3,4-ethylene dioxythiophene) polystyrene sulfonate (PEDOT: PSS) with grapheme [304,305].

In a research study by the Xu et al. [306], the synthesis of transport conductive electrons based on sulfonated graphene (SG)/PEDOT composites through situ polymerization was recorded (Fig. 16.35). The fabricated composite showed the best transparency, conductivity, thermal stability, and processability. For very little film thickness, the composite showed conductivity and transmittance of 0.2 S/cm and 80%, respectively, at a wavelength of range 400−1800 nm. This time the observed conductivity was far better than in PEDOT: PSS (10^{-6}−10^{-5} S/cm) used commercially. Additionally, the obtained composite, when treated with the PMMA under inward bending conditions, it shows the conductive stability excellently.

In the latest research, graphene doped with CVD with four layers is combined with the commercial PEDOT: PSS and added in polymer solar cells as a cathode. The active layer was composed of poly[(5,6-difluoro-2,1,3-benzothiadiazole-4,7-diyl)-alt-(3,3000-di

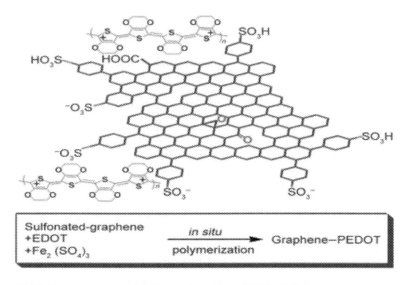

Figure 16.35 Nanocomposite of Sulfonated graphene (SG)/poly(3,4-ethylenedioxythiophene): poly(styrenesulfonate) (PEDOT) and the conditions of it synthesis-reaction conditions [306].

(2-octyl dodecyl)2,20;50,200;500,2000-quaterthiophen-5,5000-diyl)] (PffBT4T-2OD) as donor and phenyl-C71-butyric acid methyl ester (PC70BM) as acceptor. The obtained assembly showed the J_{sc} of 10.5 mA/cm^2, V_{oc} of 0.72 V, a PCE of 2.8%, and FF of 0.37, respectively, which is not comparable to the cells based on ITO. In some of the researches, the spray coating technique is used with the assistance of vibrations to fabricate the composite of graphene doped PEDOT:PSS [307,308]. The obtained results showed brilliant transparency level as compared to the ITO based solar cells. They also present the best conductivity up to 298 S/cm, around the 10 order of the un-doped PEDOT-PSS. This tremendous modification showed the high charge mobility of graphene sheets between PEDOT layers: PSS in best charge mobility channels because of the strong $\pi-\pi$ interactions. Moreover, these $\pi-\pi$ forces also remove the deficiencies are defects in PEDOT:PSS. Additionally, the doping of graphene provides a wide range of tunability of characteristics, with better strength, stability and adding the hardness and wear resistance.

16.5.1.2 Interfacial layers

The use of graphene-based nanocomposites as interfacial layers (electron and hole transporting layers) is highly recommended in polymer solar cells. They are placed between the active layer and electrodes. The main reason behind using these composites as interfacial layers is improving electrical contact and charge transport. They also improve the performance consistency with improved light absorption and distribution of radiations. Graphene and its derivatives show high corrosion resistance and best charge transport due to its advantage of quality energy-band morphology [309]. In research, graphene doped PEDOT:PSS was used in BHJ polymer solar cell based on PCDTBT:PC71BM present the PCE of up to 4.28%. The results also showed improved stability and reproducibility. The GO, due to its high value of the band gap, restricts the flow of electrons, ultimately increasing the shunt resistance value [310].

A rGO composite was developed through the grafting of polyacrylonitrile to apply the hole transport layer in PTB7-Th based polymer solar cells by Jung et al. [311]. A combination of graft polymerization and in situ synthesis was applied in this research for the fabrication of the nanocomposite. The polymerization was done of acrylonitrile with GO functionalized with styryl. The developed composite provided good weather stability, film structure, electrical conductivity, and high work function up to 4.87 eV. These all characteristics make this material well suited for the polymer solar cells with enhanced stability and cell efficiency as interracial layers. The obtained composite showed a V_{oc} of 0.76 V, PCE of 7.24%, FF of 0.64, and J_{sc} of 14.78 mA/cm^2. This developed composite was also observed as durable.

In a recent study, the spin coating method was used to deposit GO thin films into the active layers of P3HT: PCBM and ITO to overcome the current leakages and recombination of holes and electrons [312]. A layer of GO of 2 nm thickness, as a hole-transport layer, provided the best synergetic effects. The results of the studies showed FF of 0.54, PCE of 3.5%, V_{oc} of 0.57 V, and J_{sc} of 11.4 mA/cm^2. But the increase in thickness showed the decrement in the conductivity of the hole-transport layer. However, the spin coating method becomes the most difficult for such a thin film of GO. To look after this problem, carbon base nanotubes (single-walled) (SWCNTs) are developed. The combination of GO and (SWCNTs) significantly increases the conductivity of the nanocomposite and other characteristics compared to the composite based on pristine GO. The incorporation of GO sheets in few amounts into the PEDOT: PSS also increased the PCE from 3.1% to 3.7% [313].

Furthermore, a study dealing with the butylamine-modified G/PEDOT: PSS nanocomposite increased the PCE up to 0.74% from 0.42%. Additionally, an experiment was done, including GO/PEDOT: PSS based composite as a hole-transport layer. The results showed the obtained PCE up to 1.47% with an increment of 12% compared to the GO deficient cell, with enhanced conductivity and, ultimately, the charge transport [171]. As a whole, the GO addition improved the overall characteristic of the polymer solar cells.

16.5.1.3   Active layers

Typically, the structure of polymer solar cells and conventional solar cells is quite similar. A pair of electrodes bind the donor and acceptor between them like a sandwich. The donor consists of a polymer thin layer active material forming a heterojunction while the accepter is a film sheet. The responsibility of the donor and acceptor interface is to dissociate the exciton in electron-hole pairs. Liu et al. [314] firstly used the graphene functionalized with phenyl isocyanate as acceptor and poly(3-octylthiophene) (P3OT) as donor polymer solar cells. The device assembly consisting of the structure of ITO/PEDOT: PSS/P3OT: G/LIF/Al with 5% weightage of graphene showed 1.4% of PCE upon optimization through annealing at 160°C for 20 min. A similar study was done with 10% weightage of graphene using P3HT as a donor for Polymer Solar Cells (PSCs) [315] with annealing for 10 min. The obtained results showed FF of 0.38, PCE of 1.1%, V_{oc} of 0.72 V and J_{sc} of 4.0 mA/cm^2. These experiments however, improved the exciton dissociation but at temperature above 210°C, a reduction in PCE is also observed using a graphene based composite developed by Wang et al. [316]. In another research, the preparation of process able solution based on P3HT with graphene functionalized carbon based (multiwall) composite was done. In this fabrication, the P3HT acted as donor while the graphene acted as acceptor and path of percolation for holes [317]. However, the resulting assembly showed the J_{sc} of 4.7 mA/cm^2, PCE of 1.05%, V_{oc} of 0.67 V, and FF of 0.32. Furthermore, this approach of process able solution significantly lowers the value of PCE while these deficiencies can be covered with the tuning of optimization of graphene.

The bandgap of the graphene is highly tunable. The conversion of two-dimensional graphene to the GQDs (zero-dimensional) is a very operative method to improve and optimize the graphene's bandgap. In a study, the GQDs of size 3−5 nm were used in a composite structure of ITO/PEDOT: PSS/P3HT: GQDs/Al as acceptors [318], presenting the PCE and V_{oc} of 1.28% and 0.67 V, respectively, using the electrochemical approach. GQDs modified with aniline were also incorporated as acceptors in PSCs using the hydrothermal method resulting PCE of 1.14% see

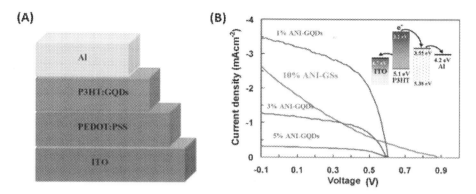

Figure 16.36 (A) Schematic representation; and (B) $J-V$ curves of ITO/PEDOT:PSS/P3HT: GQDs/Al device based on GQDs (aniline-modified) with altering content of GQDs [23,318].

Fig. 16.36 [23]. A PCE of 5.24% was recorded when GQDs obtained after the derivation from CNTs (double-walled), are introduced in P3HT: PCBM active layers [319].

Furthermore, innovative research is on the way to improve the graphene's structural characteristics, with optimal fabrication methods to develop a highly efficient and effective solar cell.

16.5.2 Application of polymer/graphene nano composite in dye-sensitized solar cells

16.5.2.1 Counter electrodes

The applications of counter electrodes in DSSC are highly explored now with PPy, polyaniline, and poly(3-alkylthiophene) as conjugated polymers in graphene nanocomposite. Counter electrodes were developed by Chen et al. [320] using GQDs with the doping in the film of polypyrol on the FTO glass, for DSSC. This formed composite showed a highly porous surface as compared to the virgin polypyrrol.

The composites electrodes based on graphene/NP are highly effective to increase the efficiency of the DSSC. Platinum NP-anchored graphene was explained as DSSC with high conductivity and electro activity with reduction of tri-iodide [321−323]. The PCE of 5.78% was obtained with the electrode composites of graphene/TiN NP for the application of electrode of DSSC by Wen et al. [324] While the use of CNTs in the fabrication of catalytic electrodes is highly effective through triiodide reduction.

The nanocomposite based on rGO and polypyrrol (PPy/rGO) were developed by Liu et al. [325] for the measurement of CV. In this study, the formed composite presented improved catalytic efficiency with a reduction of triiodide, showing a 6.45% PCE. A counter electrode based on plastic material was also formed, presenting the PCE of 4.25% high compared with the platinum-based electrode having PCE of 4.83%.

A composite is obtained by the in situ polymerization of the rGO and PPy followed by PEDOT's deposition onto the composite by Ramasamy et al. [326]. The fabricated counter electrodes showed the PCE up to 7.1% highly competitive with platinum. A liquid/liquid polymerization at interface approach was adopted by the Bore et al. [327] to fabricate a counter electrode for DSSC applications, presenting PCE up to 4.8%.

Lim et al. [328] used ITO surface for the deposition of the electrochemically polymerized PPy composite with the introduction of the rGO. The developed DSSC using this particular composite electrode provided a PCE of 2.21% compared to the platinum-based counter electrode, showing PCE 2.19%. The electrode's RCT based on rGO/PPy and electrode of sputtered platinum were obtained and compared, showing the values 14.8 and 14.3 Ω/cm^2 respectively. The graphene-based polymer composite presents a great perspective to substitute conventional platinum counter electrodes in DSSCs.

16.5.2.2 Electrolyte

Graphene/polymer nanocomposites have extensively used an electrolyte to enhance the PCE. Yuan et al. [329] fabricated the DSSC with the addition of graphene, GO, and nanographene into the poly(acrylic acid)-cetyltrimethylammonium bromide, which is a microporous conducting gel, resulting in the 7.06%, 6.35%, and 6.17% of PCE as compare to the PCE of 6.07 offered by graphene deficient fabrications. Yuan et al. [330] in a research study, added the graphene, GO, and graphite in a microporous matrix of polyacrylate—poly(ethylene glycol) see Fig. 16.37. The developed DSSCs presented the PCE of 7.74%, 6.49%, and 5.63%, respectively, which are highly comparable with platinum-based electrodes DSSC showing the

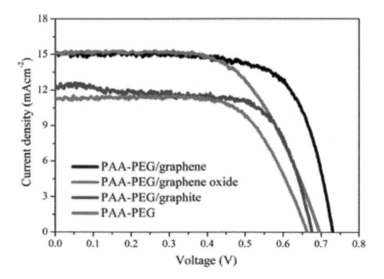

Figure 16.37 $J-V$ characteristics of DSSCs based on gel electrolytes of PAA—PEG/ graphene, PAA—PEG/graphite PAA—PEG/graphene oxide, and PAA—PEG [329].

PCE value of 5.02% for a pure PAA−PEG electrode. The addition of graphene increased the PCE up to 1.5 times showing the characteristics and improvements due to the micro porous nature of added conducting gels.

Duan et al. [331] used conduction electrolyte gels based on the graphene and polyacrylamide as microporous surface to fabricate the solar cells based on quasi solid state. The fabricated structure showed PCE of around 2.34% compared to 1.64% PCE of graphene deficient polyacrylamide. The measured value of PCE for polyacrylamide and 3%, 5%, 7%, 9% and 0% weightage of graphene was recorded as 1.67%, 1.84%, 2.24%, 1.88%, and 1.36%, respectively. Similarly, the values of RCT were recorded as 8.718, 3.381, 2.075, 3.094, and 12.57 cm^2 for polyacrylamide and 3%, 5%, 7%, 9% and 0% weightage of graphene respectively. The trends were similar for the electrocatalytic activity and charge transfer.

Chan et al. [332] used polymer hybrid electrolyte gels based on the graphene and polyacrylonitrile to fabricate the solar cells based on quasi solid-state. The clean two-dimensional graphene was added with the gel polymeric electrolyte to provide the facilitation to the diffusion of iodide/triiodide (I-/I − 3). The overall characteristics of the DSSC, especially the ionic conductivity and photovoltaic properties, are adjusted using the content of graphene in the polymeric gel of PAN. With 2% graphene weightage, PCE's value was recorded at 5.41%, which is 1.5 times more than the PCE of a cell having liquid state electrolytes (3.72%).

In a study by Akhter et al. [333], a composite was developed based on polyethylene oxide and graphene. The fabricated composite presented the best redox couple and rapid ionization. With 5% weightage of graphene, the nanocomposite showed the 3.32 mS/cm of ionic conductivity. The DSSC formed with 5% weightage of graphene provided the short circuit current up to 18.32 mA/cm^2, V_{oc} of 0.592 V, PCE of 5.23%, and FF of 0.48. Consequently, the higher photovoltaic characteristics are directly related to the effective and efficient ionization.

16.6   Key challenges

Graphene-based polymer nanocomposite materials have been previously investigated extensively owing to their increasing importance in solar and fuel energy applications. Despite of all the fascinating features and high performance of polymer/graphene nanocomposites, there are some complexities and crucial issues that are challenging for the commercial acceptability and broad scale applicability of these nanocomposites in the field of solar energy and fuel cell. Owing to the highly hydrophobic nature of raw graphene, it is rarely shows compatibility with all class of organic polymers and therefore cannot be uniformly and homogeneously dispersed in the polymer matrix. Therefore, the biggest challenge in the fabrication of graphene/polymer nanocomposite is the achievability of uniformly dispersed state of graphene sheets in the polymer [334]. The only way to tackle this compatibility issue is the functionalization of raw graphene or its chemical modification and in this context, a smart and alternative choice can be the utilization of GO and rGO [334].

Another challenging factor in the bulk scale use of graphene/polymer nanocomposites in the field of solar and fuel energy application is the lack of proper adhesion in-between polymers and graphene which is highly desired to prevent the scattering of photons through the interface of nanocomposite [335]. Furthermore, an improved adhesion can yield considerably high thermal conductivity of nanocomposites and can also reduce the free volume within the nanocomposite which is required for high gas permeability properties. The development of faster and continuous path for conductivity by development of better interaction of graphene with polymer matrix is highly required for long service life of fuel cells and solar devices.

Besides all other processing and application related challenges of graphene based polymer nanocomposites, there are some cost related challenges that are required to be confronted. Mono and bilayer sheets of graphene supported on SiO2 and fabricated by micromechanical processing are sold at a price range of $0.5-3$ £/μm^2 [336]. The economical fabrication of graphene sheets from GOs is possible but lacks the commercial applicability and will not be acceptable for commercial scale production until the graphene sheets can be yield at prices below carbon nanotubes. Along with cost related issues, difficulties faced in the handling of graphene sheets for their incorporation in polymer matrices during processing are also barrier to commercial applicability of graphene/polymer nanocomposites [337]. Use of graphene layers like thermally reduced graphene with particularly low bulk density and high surface area is challenging attributed to their feeding difficulties during melt compounding. Processing and fabrication of graphene/polymer nanocomposites may face environmental barriers due to the light weightiness of graphene sheets which can possibly lead to intake by human beings during its handling [338].

In energy applications, a critical barrier lies in the efficient fabrication of graphene/polymer nanocomposite with proper spacing of individual layer of graphene and its derivative which will permit the intercalation of ions or molecules by providing empty spaces within the nanocomposites. Moreover, the problem related to the restacking of graphene and limited exposure of actual graphene surface area in polymer matrix can be resolved by fabricating graphene/polymer nanocomposites with unique morphology and by incorporating heteroatoms that also have potential to enhance electrochemical and conductivity performance [339]. In the fuel cell applications another challenging knot is the assurance of the ideal structure of electrons and molecules of the graphene/polymer nanocomposites which is required to be known for the determination of chemical stability and electrocatalytic activities [340]. The commercialization of fuel cells based on graphene/polymer nanocomposite based membranes and its practical use still demand standard systems for the characterization and evaluation of membranes, cathode, and anode materials.

References

[1] T.N. Kumar, Solar cell made using graphene, IOSR J. Mech. Civ. Eng. 11 (2014) 71−81.
[2] S. Das, P. Sudhagar, Y.S. Kang, W. Choi, Graphene synthesis and application for solar cells, J. Mater. Res. 29 (2014) 299−319.

[3] T. Mahmoudi, Y. Wang, Y.B. Hahn, Graphene and its derivatives for solar cells application, Nano Energy 47 (2018) 51−65.

[4] B. O'Regan, M. Grätzel, A low-cost, high-efficiency solar cell based on dye-sensitized colloidal TiO_2 films, Nature. 353 (1991) 737−740.

[5] M. Grätzel, Photoelectrochemical cells, Mater. Sustain. Energy (2010) 26−32.

[6] D. Dodoo-Arhin, M. Fabiane, A. Bello, N. Manyala, Graphene: synthesis, transfer, and characterization for dye-sensitized solar cells applications, Ind. Eng. Chem. Res. 52 (2013) 14160−14168.

[7] L. Dai, D.W. Chang, J.B. Baek, W. Lu, Carbon nanomaterials for advanced energy conversion and storage, Small. 8 (2012) 1130−1166.

[8] D. Chen, H. Zhang, Y. Liu, J. Li, Graphene and its derivatives for the development of solar cells, photoelectrochemical, and photocatalytic applications, Energy Environ. Sci. 6 (2013) 1362−1387.

[9] H.J. Choi, S.M. Jung, J.M. Seo, D.W. Chang, L. Dai, J.B. Baek, Graphene for energy conversion and storage in fuel cells and supercapacitors, Nano Energy 1 (2012) 534−551.

[10] M.J. Allen, V.C. Tung, R.B. Kaner, Honeycomb carbon: a review of graphene, Chem. Rev. 110 (2010) 132−145.

[11] A.K. Geim, K.S. Novoselov, The rise of graphene, Nanoscience and Technology: A Collection of Reviews from Nature Journals, World Scientific Publishing Co., 2009, pp. 11−19.

[12] K.S. Novoselov, A.K. Geim, S.V. Morozov, D. Jiang, Y. Zhang, S.V. Dubonos, et al., Electric field in atomically thin carbon films, Science (80) 306 (2004) 666−669.

[13] M.T. Hasan, R. Gonzalez-Rodriguez, C. Ryan, N. Faerber, J.L. Coffer, A.V. Naumov, Photo-and electroluminescence from nitrogen-doped and nitrogen-sulfur codoped graphene quantum dots, Adv. Funct. Mater. 28 (2018) 1804337.

[14] L. Zhou, J. Geng, B. Liu, Graphene quantum dots from polycyclic aromatic hydrocarbon for bioimaging and sensing of Fe3 + and hydrogen peroxide, Part. Part. Syst. Charact. 30 (2013) 1086−1092.

[15] J. Sun, S. Yang, Z. Wang, H. Shen, T. Xu, L. Sun, et al., Ultra-high quantum yield of graphene quantum dots: aromatic-nitrogen doping and photoluminescence mechanism, Part. Part. Syst. Charact. 32 (2015) 434−440.

[16] F. Niu, Y. Xu, J. Liu, Z. Song, M. Liu, J. Liu, Controllable electrochemical/electroanalytical approach to generate nitrogen-doped carbon quantum dots from varied amino acids: pinpointing the utmost quantum yield and the versatile photoluminescent and electrochemiluminescent applications, Electrochim. Acta. 236 (2017) 239−251.

[17] D. Qu, M. Zheng, L. Zhang, H. Zhao, Z. Xie, X. Jing, et al., Formation mechanism and optimization of highly luminescent N-doped graphene quantum dots, Sci. Rep. 4 (2014) 1−11.

[18] M. Nurunnabi, Z. Khatun, K.M. Huh, S.Y. Park, D.Y. Lee, K.J. Cho, et al., In vivo biodistribution and toxicology of carboxylated graphene quantum dots, ACS Nano 7 (2013) 6858−6867.

[19] Q. Liu, B. Guo, Z. Rao, B. Zhang, J.R. Gong, Strong two-photon-induced fluorescence from photostable, biocompatible nitrogen-doped graphene quantum dots for cellular and deep-tissue imaging, Nano Lett 13 (2013) 2436−2441.

[20] W. Shang, X. Zhang, M. Zhang, Z. Fan, Y. Sun, M. Han, et al., The uptake mechanism and biocompatibility of graphene quantum dots with human neural stem cells, Nanoscale. 6 (2014) 5799−5806.

[21] B.J. Moon, D. Jang, Y. Yi, H. Lee, S.J. Kim, Y. Oh, et al., Multi-functional nitrogen self-doped graphene quantum dots for boosting the photovoltaic performance of BHJ solar cells, Nano Energy 34 (2017) 36−46.

[22] D. Carolan, C. Rocks, D.B. Padmanaban, P. Maguire, V. Svrcek, D. Mariotti, Environmentally friendly nitrogen-doped carbon quantum dots for next generation solar cells, Sustain. Energy Fuels 1 (2017) 1611–1619.

[23] V. Gupta, N. Chaudhary, R. Srivastava, G.D. Sharma, R. Bhardwaj, S. Chand, Luminscent graphene quantum dots for organic photovoltaic devices, J. Am. Chem. Soc. 133 (2011) 9960–9963.

[24] H. Wang, P. Sun, S. Cong, J. Wu, L. Gao, Y. Wang, et al., Nitrogen-doped carbon dots for "green" quantum dot solar cells, Nanoscale Res. Lett. 11 (2016) 27.

[25] A.P. Alivisatos, W. Gu, C. Larabell, Quantum dots as cellular probes, Annu. Rev. Biomed. Eng. 7 (2005) 55–76.

[26] S. Zhu, J. Zhang, C. Qiao, S. Tang, Y. Li, W. Yuan, et al., Strongly green-photoluminescent graphene quantum dots for bioimaging applications, Chem. Commun. 47 (2011) 6858.

[27] Z. Zeng, S. Chen, T.T.Y. Tan, F.X. Xiao, Graphene quantum dots (GQDs) and its derivatives for multifarious photocatalysis and photoelectrocatalysis, Catal. Today. 315 (2018) 171–183.

[28] Q. Lu, Y. Zhang, S. Liu, Graphene quantum dots enhanced photocatalytic activity of zinc porphyrin toward the degradation of methylene blue under visible-light irradiation, J. Mater. Chem. A. 3 (2015) 8552–8558.

[29] S. Coe, W.K. Woo, M. Bawendi, V. Bulović, Electroluminescence from single monolayers of nanocrystals in molecular organic devices, Nature. 420 (2002) 800–803.

[30] N. Tessler, V. Medvedev, M. Kazes, S.H. Kan, U. Banin, Efficient near-infrared polymer nanocrystal light-emitting diodes, Science (80−) 295 (2002) 1506–1508.

[31] D.I. Son, B.W. Kwon, D.H. Park, W.S. Seo, Y. Yi, B. Angadi, et al., Emissive ZnO-graphene quantum dots for white-light-emitting diodes, Nat. Nanotechnol. 7 (2012) 465–471.

[32] J. Ge, M. Lan, B. Zhou, W. Liu, L. Guo, H. Wang, et al., A graphene quantum dot photodynamic therapy agent with high singlet oxygen generation, Nat. Commun. 5 (2014) 1–8.

[33] T.A. Tabish, C.J. Scotton, D.C. J. Ferguson, L. Lin, A. van der Veen, S. Lowry, et al., Biocompatibility and toxicity of graphene quantum dots for potential application in photodynamic therapy, Nanomedicine. 13 (2018) 1923–1937.

[34] K.S. Novoselov, V.I. Fal'Ko, L. Colombo, P.R. Gellert, M.G. Schwab, K. Kim, A roadmap for graphene, Nature. 490 (2012) 192–200.

[35] M.P. Ramuz, M. Vosgueritchian, P. Wei, C. Wang, Y. Gao, Y. Wu, et al., Evaluation of solution-processable carbon-based electrodes for all-carbon solar cells, ACS Nano 6 (2012) 10384–10395.

[36] S. Ren, P. Rong, Q. Yu, Preparations, properties and applications of graphene in functional devices: a concise review, Ceram. Int. 44 (2018) 11940–11955.

[37] Y. Xue, B. Wu, L. Jiang, Y. Guo, L. Huang, J. Chen, et al., Low temperature growth of highly nitrogen-doped single crystal graphene arrays by chemical vapor deposition, J. Am. Chem. Soc. 134 (2012) 11060–11063.

[38] A. Guermoune, T. Chari, F. Popescu, S.S. Sabri, J. Guillemette, H.S. Skulason, et al., Chemical vapor deposition synthesis of graphene on copper with methanol, ethanol, and propanol precursors, Carbon N. Y 49 (2011) 4204–4210.

[39] Y.-F. Lu, S.-T. Lo, J.-C. Lin, W. Zhang, J.-Y. Lu, F.-H. Liu, et al., Nitrogen-doped graphene sheets grown by chemical vapor deposition: synthesis and influence of nitrogen impurities on carrier transport, ACS Nano 7 (2013) 6522–6532.

[40] X. Li, D. Xie, H. Park, T.H. Zeng, K. Wang, J. Wei, et al., Anomalous behaviors of graphene transparent conductors in graphene-silicon heterojunction solar cells, Adv. Energy Mater. 3 (2013) 1029–1034.

[41] D. Deng, X. Pan, L. Yu, Y. Cui, Y. Jiang, J. Qi, et al., Toward N-doped graphene via solvothermal synthesis, Chem. Mater. 23 (2011) 1188−1193.

[42] Y.C. Lin, C.Y. Lin, P.W. Chiu, Controllable graphene N-doping with ammonia plasma, Appl. Phys. Lett. 96 (2010) 133110.

[43] Y. Wang, Y. Shao, D.W. Matson, J. Li, Y. Lin, Nitrogen-doped graphene and its application in electrochemical biosensing, ACS Nano 4 (2010) 1790−1798.

[44] J. Moon, J. An, U. Sim, S.-P. Cho, J.H. Kang, C. Chung, et al., One-step synthesis of n-doped graphene quantum sheets from monolayer graphene by nitrogen plasma, Adv. Mater. 26 (2014) 3501−3505.

[45] Z. Jin, J. Yao, C. Kittrell, J.M. Tour, Large-scale growth and characterizations of nitrogen-doped monolayer graphene sheets, ACS Nano 5 (2011) 4112−4117.

[46] L. Qu, Y. Liu, J.-B. Baek, L. Dai, Nitrogen-doped graphene as efficient metal-free electrocatalyst for oxygen reduction in fuel cells, ACS Nano 4 (2010) 1321−1326.

[47] Y. Zhao, S. Araki, J. Wu, T. Teramoto, Y.F. Chang, M. Nakano, et al., An expanded palette of genetically encoded Ca2 + indicators, Science (80−) 333 (2011) 1888−1891.

[48] M. Rybin, A. Pereyaslavtsev, T. Vasilieva, V. Myasnikov, I. Sokolov, A. Pavlova, et al., Efficient nitrogen doping of graphene by plasma treatment, Carbon N. Y 96 (2016) 196−202.

[49] R. Yadav, C.K. Dixit, Synthesis, characterization and prospective applications of nitrogen-doped graphene: a short review, J. Sci. Adv. Mater. Devices. 2 (2017) 141−149.

[50] Y. Zhu, S. Jia, J. Zheng, Y. Lin, Y. Wu, J. Wang, Facile synthesis of nitrogen-doped graphene frameworks for enhanced performance of hole transport material-free perovskite solar cells, J. Mater. Chem. C. 6 (2018) 3097−3103.

[51] Y. Shao, S. Zhang, M.H. Engelhard, G. Li, G. Shao, Y. Wang, et al., Nitrogen-doped graphene and its electrochemical applications, J. Mater. Chem. 20 (2010) 7491−7496.

[52] N.A.A. Ghany, S.A. Elsherif, H.T. Handal, Revolution of graphene for different applications: state-of-the-art, Surfaces and Interfaces 9 (2017) 93−106.

[53] D.R. Dreyer, R.S. Ruoff, C.W. Bielawski, From conception to realization: an historial account of graphene and some perspectives for its future, Angew. Chemie—Int. (Ed.) 49 (2010) 9336−9344.

[54] W. Yu, L. Sisi, Y. Haiyan, L. Jie, Progress in the functional modification of graphene/graphene oxide: a review, RSC Adv 10 (2020) 15328−15345.

[55] D. Mudusu, K.R. Nandanapalli, S. Lee, Y.B. Hahn, Recent advances in graphene monolayers growth and their biological applications: a review, Adv. Colloid Interface Sci 283 (2020) 102225.

[56] J. Lloyd-Hughes, T.I. Jeon, A review of the terahertz conductivity of bulk and nanomaterials, J. Infrared, Millim. Terahertz Waves 33 (2012) 871−925.

[57] A.K. Geim, Graphene prehistory, Phys. Scr. (2012).

[58] V. Dhinakaran, M. Lavanya, K. Vigneswari, M. Ravichandran, M.D. Vijayakumar, Review on exploration of graphene in diverse applications and its future horizon, Mater. Today Proc. 27 (2019) 824−828.

[59] L. Zou, L. Wang, Y. Wu, C. Ma, S. Yu, X. Liu, Trends analysis of graphene research and development, J. Data Inf. Sci. 3 (2020) 82−100.

[60] E.P. Randviir, D.A.C. Brownson, C.E. Banks, A decade of graphene research: production, applications and outlook, Mater. Today. 17 (2014) 426−432.

[61] V. Dhand, K.Y. Rhee, H. Ju Kim, D. Ho Jung, A comprehensive review of graphene nanocomposites: research status and trends, J. Nanomater. 2013 (2013).

[62] A.A. Iqbal, N. Sakib, A.K.M.P. Iqbal, D.M. Nuruzzaman, Graphene-based nanocomposites and their fabrication, mechanical properties and applications, Materialia 12 (2020) 100815.

[63] D.G. Papageorgiou, I.A. Kinloch, R.J. Young, Mechanical properties of graphene and graphene-based nanocomposites, Prog. Mater. Sci. 90 (2017) 75−127.

[64] A. Amiri, M. Naraghi, G. Ahmadi, M. Soleymaniha, M. Shanbedi, A review on liquid-phase exfoliation for scalable production of pure graphene, wrinkled, crumpled and functionalized graphene and challenges, FlatChem 8 (2018) 40−71.

[65] X. Gu, Y. Zhao, K. Sun, C.L.Z. Vieira, Z. Jia, C. Cui, et al., Method of ultrasound-assisted liquid-phase exfoliation to prepare graphene, Ultrason. Sonochem. 58 (2019) 104630.

[66] W. Choi, I. Lahiri, R. Seelaboyina, Y.S. Kang, Synthesis of graphene and its applications: a review, Crit. Rev. Solid State Mater. Sci. 35 (2010) 52−71.

[67] B. Sreenivasulu, B.R. Ramji, M. Nagaral, A review on graphene reinforced polymer matrix composites, Mater. Today Proc. 5 (2018) 2419−2428.

[68] M. Han, J. Yun, H. Il Kim, Y.S. Lee, Effect of surface modification of graphene oxide on photochemical stability of poly(vinyl alcohol)/graphene oxide composites, J. Ind. Eng. Chem. 18 (2012) 752−756.

[69] A.T. Lawal, Graphene-based nano composites and their applications. A review, Biosens. Bioelectron. 141 (2019) 111384.

[70] Z. İlbay, A. Haşimoğlu, O.K. Özdemir, F. Ateş, S. Şahin, Highly efficient recovery of biophenols onto graphene oxide nanosheets: valorisation of a biomass, J. Mol. Liq. 246 (2017) 208−214.

[71] X. Ji, Y. Xu, W. Zhang, L. Cui, J. Liu, Review of functionalization, structure and properties of graphene/polymer composite fibers, Compos. Part A Appl. Sci. Manuf. 87 (2016) 29−45.

[72] S. Vadukumpully, J. Paul, N. Mahanta, S. Valiyaveettil, Flexible conductive graphene/poly(vinyl chloride) composite thin films with high mechanical strength and thermal stability, Carbon N. Y 49 (2011) 198−205.

[73] Z. Xu, C. Gao, In situ polymerization approach to graphene-reinforced nylon-6 composites, Macromolecules. 43 (2010) 6716−6723.

[74] M. Zhang, Y. Li, Z. Su, G. Wei, Recent advances in the synthesis and applications of graphene-polymer nanocomposites, Polym. Chem. 6 (2015) 6107−6124.

[75] H. Kim, S. Kobayashi, M.A. Abdurrahim, M.J. Zhang, A. Khusainova, M.A. Hillmyer, et al., Graphene/polyethylene nanocomposites: effect of polyethylene functionalization and blending methods, Polymer (Guildf) 52 (2011) 1837−1846.

[76] H. Kim, Y. Miura, C.W. MacOsko, Graphene/polyurethane nanocomposites for improved gas barrier and electrical conductivity, Chem. Mater. 22 (2010) 3441−3450.

[77] Y. Zhang, J.E. Mark, Y. Zhu, R.S. Ruoff, D.W. Schaefer, Mechanical properties of polybutadiene reinforced with octadecylamine modified graphene oxide, Polymer (Guildf) 55 (2014) 5389−5395.

[78] G. Eda, G. Fanchini, M. Chhowalla, Large-area ultrathin films of reduced graphene oxide as a transparent and flexible electronic material, Nat. Nanotechnol. 3 (2008) 270−274.

[79] S.T. Hsiao, C.C.M. Ma, W.H. Liao, Y.S. Wang, S.M. Li, Y.C. Huang, et al., Lightweight and flexible reduced graphene oxide/water-borne polyurethane composites with high electrical conductivity and excellent electromagnetic interference shielding performance, ACS Appl. Mater. Interfaces. 6 (2014) 10667−10678.

[80] W. Han, L. Ren, L. Gong, X. Qi, Y. Liu, L. Yang, et al., Self-assembled three-dimensional graphene-based aerogel with embedded multifarious functional nanoparticles and its excellent photoelectrochemical activities, ACS Sustain. Chem. Eng. 2 (2014) 741−748.

[81] K. Nawaz, M. Ayub, N. Ul-Haq, M.B. Khan, M.B.K. Niazi, A. Hussain, The effect of large area graphene oxide (LAGO) nanosheets on the mechanical properties of polyvinyl alcohol, J. Polym. Eng. 36 (2016) 399–405.

[82] A. Chunder, J. Liu, L. Zhai, Reduced graphene oxide/poly(3-hexylthiophene) supramolecular composites, Macromol. Rapid Commun. 31 (2010) 380–384.

[83] Q. qi Bai, X. Wei, J. hui Yang, N. Zhang, T. Huang, Y. Wang, et al., Dispersion and network formation of graphene platelets in polystyrene composites and the resultant conductive properties, Compos. Part A Appl. Sci. Manuf. 96 (2017) 89–98.

[84] M. Fang, K. Wang, H. Lu, Y. Yang, S. Nutt, Single-layer graphene nanosheets with controlled grafting of polymer chains, J. Mater. Chem. 20 (2010) 1982–1992.

[85] Y. Wang, X. Liao, S. Li, Y. Luo, Q. Yang, G. Li, Poly(methyl methacrylate) nanocomposites based on graphene oxide: a comparative investigation of the effects of surface chemistry on properties and foaming behavior, Polym. Int. 65 (2016) 1195–1203.

[86] R. Balasubramaniyan, V.H. Pham, J. Jang, S.H. Hur, J.S. Chung, A one pot solution blending method for highly conductive poly (methyl methacrylate)-highly reduced graphene nanocomposites, Electron. Mater. Lett. 9 (2013) 837–839.

[87] S. Scalese, I. Nicotera, D. D'Angelo, S. Filice, S. Libertino, C. Simari, et al., Cationic and anionic azo-dye removal from water by sulfonated graphene oxide nanosheets in Nafion membranes, New J. Chem. 40 (2016) 3654–3663.

[88] X.H. Yan, R. Wu, J.B. Xu, Z. Luo, T.S. Zhao, A monolayer graphene—Nafion sandwich membrane for direct methanol fuel cells, J. Power Sources. 311 (2016) 188–194.

[89] U. Khan, P. May, A. O'Neill, J.N. Coleman, Development of stiff, strong, yet tough composites by the addition of solvent exfoliated graphene to polyurethane, Carbon N. Y 48 (2010) 4035–4041.

[90] Q. Jing, W. Liu, Y. Pan, V.V. Silberschmidt, L. Li, Z.L. Dong, Chemical functionalization of graphene oxide for improving mechanical and thermal properties of polyurethane composites, Mater. Des. 85 (2015) 808–814.

[91] T. Kuila, S. Bose, C.E. Hong, M.E. Uddin, P. Khanra, N.H. Kim, et al., Preparation of functionalized graphene/linear low density polyethylene composites by a solution mixing method, Carbon N. Y 49 (2011) 1033–1037.

[92] Y. An, Z. Tai, Y. Qi, X. Yan, B. Liu, Q. Xue, et al., Friction and wear properties of graphene oxide/ultrahigh-molecular-weight polyethylene composites under the lubrication of deionized water and normal saline solution, J. Appl. Polym. Sci. 131 (2014).

[93] B. Shen, W. Zhai, M. Tao, D. Lu, W. Zheng, Enhanced interfacial interaction between polycarbonate and thermally reduced graphene induced by melt blending, Compos. Sci. Technol. 86 (2013) 109–116.

[94] Y.Q. Gill, U. Abid, M. Song, High performance Nylon12/clay nanocomposites for potential packaging applications, J. Appl. Polym. Sci. 137 (2020).

[95] J. Jin, R. Rafiq, Y.Q. Gill, M. Song, Preparation and characterization of high performance of graphene/nylon nanocomposites, Eur. Polym. J. 49 (2013) 2617–2626.

[96] B. Yuan, L. Song, K.M. Liew, Y. Hu, Solid acid-reduced graphene oxide nanohybrid for enhancing thermal stability, mechanical property and flame retardancy of polypropylene, RSC Adv 5 (2015) 41307–41316.

[97] S. Chandrasekaran, C. Seidel, K. Schulte, Preparation and characterization of graphite nano-platelet (GNP)/epoxy nano-composite: mechanical, electrical and thermal properties, Eur. Polym. J. 49 (2013) 3878–3888.

[98] S.N. Tripathi, G.S.S. Rao, A.B. Mathur, R. Jasra, Polyolefin/graphene nanocomposites: a review, RSC Adv 7 (2017) 23615–23632.

[99] W. Chen, H. Weimin, D. Li, S. Chen, Z. Dai, A critical review on the development and performance of polymer/graphene nanocomposites, Sci. Eng. Compos. Mater. 25 (2018) 1059—1073.

[100] A. Szymczyk, S. Paszkiewicz, J. Typek, Z. Spitalsky, I. Janowska, G. Żołnierkiewicz, et al., Magnetic properties of poly(trimethylene terephthalate-block-poly(tetramethylene oxide) copolymer nanocomposites reinforced by graphene oxide-Fe3O4 hybrid nanoparticles, Phys. Status Solidi. 216 (2019).

[101] D. Yuan, B. Wang, L. Wang, Y. Wang, Z. Zhou, Unusual toughening effect of graphene oxide on the graphene oxide/nylon 11 composites prepared by in situ melt polycondensation, Compos. Part B Eng 55 (2013) 215—220.

[102] A. Tchernook, M. Krumova, F.J. Tölle, R. Mülhaupt, S. Mecking, Composites from aqueous polyethylene nanocrystal/graphene dispersions, Macromolecules. 47 (2014) 3017—3021.

[103] E. Bourgeat-Lami, J. Faucheu, A. Noël, Latex routes to graphene-based nanocomposites, Polym. Chem. 6 (2015) 5323—5357.

[104] B. Yuan, B. Wang, Y. Hu, X. Mu, N. Hong, K.M. Liew, et al., Electrical conductive and graphitizable polymer nanofibers grafted on graphene nanosheets: improving electrical conductivity and flame retardancy of polypropylene, Compos. Part A Appl. Sci. Manuf. 84 (2016) 76—86.

[105] D.W. Wang, F. Li, J. Zhao, W. Ren, Z.G. Chen, J. Tan, et al., Fabrication of graphene/polyaniline composite paper via in situ anodic electropolymerization for high-performance flexible electrode, ACS Nano 3 (2009) 1745—1752.

[106] E. Asadian, S. Shahrokhian, A.I. Zad, E. Jokar, In-situ electro-polymerization of graphene nanoribbon/polyaniline composite film: application to sensitive electrochemical detection of dobutamine, Sensors Actuators, B Chem 196 (2014) 582—588.

[107] Y. Sun, G. Shi, Graphene/polymer composites for energy applications, J. Polym. Sci. Part B Polym. Phys. 51 (2013) 231—253.

[108] S. Wustoni, T.C. Hidalgo, A. Hama, D. Ohayon, A. Savva, N. Wei, et al., In situ electrochemical synthesis of a conducting polymer composite for multimetabolite sensing, Adv. Mater. Technol. 5 (2020) 1900943.

[109] L. Hu, J. Tu, S. Jiao, J. Hou, H. Zhu, D.J. Fray, In situ electrochemical polymerization of a nanorod-PANI—graphene composite in a reverse micelle electrolyte and its application in a supercapacitor, Phys. Chem. Chem. Phys. 14 (2012) 15652.

[110] S. Goswami, U.N. Maiti, S. Maiti, S. Nandy, M.K. Mitra, K.K. Chattopadhyay, Preparation of graphene-polyaniline composites by simple chemical procedure and its improved field emission properties, Carbon N. Y 49 (2011) 2245—2252.

[111] V. Loryuenyong, A. Khadthiphong, J. Phinkratok, J. Watwittayakul, W. Supawattanakul, A. Buasri, The fabrication of graphene-polypyrrole composite for application with dye-sensitized solar cells, Mater. Today Proc, Elsevier Ltd, 2019, pp. 1675—1681.

[112] P. Wei, S. Bai, Fabrication of a high-density polyethylene/graphene composite with high exfoliation and high mechanical performance via solid-state shear milling, RSC Adv 5 (2015) 93697—93705.

[113] Y. Wang, Y. Qing, Y. Sun, M. Zhu, S. Dong, A study on preparation of modified graphene oxide and flame retardancy of polystyrene composite microspheres, Des. Monomers Polym. 23 (2020) 1—15.

[114] X. Zeng, C. Wu, B. Tang, X. Shen, K. Liu, B. Ye, et al., Spray-free polypropylene composite reinforced by graphene oxide@short glass fiber, Polym. Compos. 41 (2020) 1215—1223.

[115] N. Song, D. Cao, X. Luo, Q. Wang, P. Ding, L. Shi, Highly thermally conductive polypropylene/graphene composites for thermal management, Compos. Part A Appl. Sci. Manuf. 135 (2020) 105912.

[116] C.C.N. de Melo, C.A.G. Beatrice, L.A. Pessan, A.D. de Oliveira, F.M. Machado, Analysis of nonisothermal crystallization kinetics of graphene oxide—reinforced polyamide 6 nanocomposites, Thermochim. Acta. 667 (2018) 111–121.

[117] N. Aranburu, I. Otaegi, G. Guerrica-Echevarria, Using an ionic liquid to reduce the electrical percolation threshold in biobased thermoplastic polyurethane/graphene nanocomposites, Polymers (Basel), 11, 2019, p. 435.

[118] V. Tambrallimath, R. Keshavamurthy, D. Saravanabavan, P.G. Koppad, G.S.P. Kumar, Thermal behavior of PC-ABS based graphene filled polymer nanocomposite synthesized by FDM process, Compos. Commun 15 (2019) 129–134.

[119] T. Javanbakht, E. David, Rheological and physical properties of a nanocomposite of graphene oxide nanoribbons with polyvinyl alcohol, J. Thermoplast. Compos. Mater. (2020). 089270572091276.

[120] M.J. Solomon, A.S. Almusallam, K.F. Seefeldt, A. Somwangthanaroj, P. Varadan, Rheology of polypropylene/clay hybrid materials, Macromolecules. 34 (2001) 1864–1872.

[121] J. Vermant, S. Ceccia, M.K. Dolgovskij, P.L. Maffettone, C.W. Macosko, Quantifying dispersion of layered nanocomposites via melt rheology, J. Rheol. (N. Y. N. Y) 51 (2007) 429–450.

[122] C. Lee, X. Wei, J.W. Kysar, J. Hone, Measurement of the elastic properties and intrinsic strength of monolayer graphene, Science (80–) 321 (2008) 385–388.

[123] M.Z. Iqbal, A.A. Abdala, V. Mittal, S. Seifert, A.M. Herring, M.W. Liberatore, Processable conductive graphene/polyethylene nanocomposites: effects of graphene dispersion and polyethylene blending with oxidized polyethylene on rheology and microstructure, Polymer (Guildf) 98 (2016) 143–155.

[124] D.G. Papageorgiou, Z. Li, M. Liu, I.A. Kinloch, R.J. Young, Mechanisms of mechanical reinforcement by graphene and carbon nanotubes in polymer nanocomposites, Nanoscale. 12 (2020) 2228–2267.

[125] R.A. Vaia, J.F. Maguire, ReViews Polymer Nanocomposites with Prescribed Morphology: Going beyond Nanoparticle-Filled Polymers (2007).

[126] P. Akcora, S.K. Kumar, J. Moll, S. Lewis, L.S. Schadler, Y. Li, et al., "Gel-like" mechanical reinforcement in polymer nanocomposite melts, Macromolecules. 43 (2010) 1003–1010.

[127] P. Akcora, H. Liu, S.K. Kumar, J. Moll, Y. Li, B.C. Benicewicz, et al., Anisotropic self-assembly of spherical polymer-grafted nanoparticles, Nat. Mater. 8 (2009) 354–359.

[128] M. Terrones, O. Martín, M. González, J. Pozuelo, B. Serrano, J.C. Cabanelas, Interphases in graphene polymer-based nanocomposites: achievements and challenges, Adv. Mater. 23 (2011) 5302–5310.

[129] S. Morimune, T. Nishino, T. Goto, Ecological approach to graphene oxide reinforced poly (methyl methacrylate) nanocomposites, ACS Appl. Mater. Interfaces. 4 (2012) 3596–3601.

[130] S. Ghosh, I. Calizo, D. Teweldebrhan, E.P. Pokatilov, D.L. Nika, A.A. Balandin, et al., Extremely high thermal conductivity of graphene: prospects for thermal management applications in nanoelectronic circuits, Appl. Phys. Lett. 92 (2008) 151911.

[131] J.H. Seol, I. Jo, A.L. Moore, L. Lindsay, Z.H. Aitken, M.T. Pettes, et al., Two-dimensional phonon transport in supported graphene, Science (80–) 328 (2010) 213–216. Available from: https://doi.org/10.1126/science.1184014.

[132] A. Fasolino, J.H. Los, M.I. Katsnelson, Intrinsic ripples in graphene, Nat. Mater. 6 (2007) 858–861.

[133] T.C. Mokhena, M.J. Mochane, J.S. Sefadi, S.V. Motloung, D.M. Andala, Thermal conductivity of graphite-based polymer composites, Impact of Thermal Conductivity on Energy Technologies, InTech, 2018.

[134] Q. Yang, Z. Zhang, X. Gong, E. Yao, T. Liu, Y. Zhang, et al., Thermal conductivity of graphene-polymer composites: implications for thermal management, Heat Mass Transf. Stoffuebertragung 56 (2020) 1931−1945.

[135] W.L. Zhang, B.J. Park, H.J. Choi, Colloidal graphene oxide/polyaniline nanocomposite and its electrorheology, Chem. Commun. 46 (2010) 5596−5598.

[136] A.C. Balazs, T. Emrick, T.P. Russell, Nanoparticle polymer composites: where two small worlds meet, Science (80−) 314 (2006) 1107−1110.

[137] M.M. Gudarzi, F. Sharif, Molecular level dispersion of graphene in polymer matrices using colloidal polymer and graphene, J. Colloid Interface Sci 366 (2012) 44−50.

[138] S. Stankovich, D.A. Dikin, G.H.B. Dommett, K.M. Kohlhaas, E.J. Zimney, E.A. Stach, et al., Graphene-based composite materials, Nature 442 (2006) 282−286.

[139] D. Li, M.B. Müller, S. Gilje, R.B. Kaner, G.G. Wallace, Processable aqueous dispersions of graphene nanosheets, Nat. Nanotechnol. 3 (2008) 101−105.

[140] S. Jiang, Z. Gui, C. Bao, K. Dai, X. Wang, K. Zhou, et al., Preparation of functionalized graphene by simultaneous reduction and surface modification and its polymethyl methacrylate composites through latex technology and melt blending, Chem. Eng. J. 226 (2013) 326−335.

[141] Y.J. Noh, H.I. Joh, J. Yu, S.H. Hwang, S. Lee, C.H. Lee, et al., Ultra-high dispersion of graphene in polymer composite via solvent free fabrication and functionalization, Sci. Rep. 5 (2015) 1−7.

[142] A. Liang, X. Jiang, X. Hong, Y. Jiang, Z. Shao, D. Zhu, Recent developments concerning the dispersion methods and mechanisms of graphene, Coatings 8 (2018) 33.

[143] S. Sayyar, E. Murray, B.C. Thompson, J. Chung, D.L. Officer, S. Gambhir, et al., Processable conducting graphene/chitosan hydrogels for tissue engineering, J. Mater. Chem. B 3 (2015) 481−490.

[144] R. Atif, F. Inam, Modeling and simulation of graphene based polymer nanocomposites: advances in the last decade, Graphene 05 (2016) 96−142.

[145] J. MA, Q. Meng, A. Michelmore, N. Kawashima, I. Zaman, C. Bengtsson, et al., Covalently bonded interfaces for polymer/graphene composites, J. Mater. Chem. A Mater. Energy Sustain 1 (2013) 4255−4264.

[146] Y. Zhang, Q. Zhang, D. Hou, J. Zhang, Tuning interfacial structure and mechanical properties of graphene oxide sheets/polymer nanocomposites by controlling functional groups of polymer, Appl. Surf. Sci. 504 (2020).

[147] E.Y. Choi, T.H. Han, J. Hong, J.E. Kim, S.H. Lee, H.W. Kim, et al., Noncovalent functionalization of graphene with end-functional polymers, J. Mater. Chem. 20 (2010) 1907−1912.

[148] M. Cano, U. Khan, T. Sainsbury, A. O'Neill, Z. Wang, I.T. McGovern, et al., Improving the mechanical properties of graphene oxide based materials by covalent attachment of polymer chains, Carbon N.Y 52 (2013) 363−371.

[149] J. Zhu, F. Liu, N. Mahmood, Y. Hou, Graphene polymer nanocomposites for fuel cells, Graphene-Based Polymer Nanocomposites in Electronics, Springer International Publishing, 2015, pp. 91−130.

[150] V.K. Thakur, M.R. Kessler, Self-healing polymer nanocomposite materials: a review, Polymer (Guildf) 69 (2015) 369−383.

[151] C. Chen, X. Hong, T. Xu, J. Lu, Y. Gao, Preparation and electrochemical and electrochromic properties of wrinkled poly(N-methylthionine) film, Synth. Met. 205 (2015) 175−184.

[152] A.G. Kumar, D. Bera, E. Mistri, S. Banerjee, Triphenyl amine containing sulfonated aromatic polyimide proton exchange membranes, Eur. Polym. J. 60 (2014) 235−246.

[153] J.J. Wu, H.W. Lee, J.H. You, Y.C. Kau, S.J. Liu, Adsorption of silver ions on polypyrrole embedded electrospun nanofibrous polyethersulfone membranes, J. Colloid Interface Sci 420 (2014) 145−151.

[154] A. Pappu, V. Patil, S. Jain, A. Mahindrakar, R. Haque, V.K. Thakur, Advances in industrial prospective of cellulosic macromolecules enriched banana biofibre resources: a review, Int. J. Biol. Macromol. 79 (2015) 449−458. Available from: https://doi.org/10.1016/j.ijbiomac.2015.05.013.

[155] V.K. Thakur, M.K. Thakur, Recent advances in green hydrogels from lignin: a review, Int. J. Biol. Macromol. 72 (2015) 834−847.

[156] S.L. Soto Espinoza, C.R. Arbeitman, M.C. Clochard, M. Grasselli, Functionalization of nanochannels by radio-induced grafting polymerization on PET track-etched membranes, Radiat. Phys. Chem. 94 (2014) 72−75.

[157] X. Ruan, G. He, B. Li, J. Xiao, Y. Dai, Cleaner recovery of tetrafluoroethylene by coupling residue-recycled polyimide membrane unit to distillation, Sep. Purif. Technol. 124 (2014) 89−98.

[158] M.F. Lin, V.K. Thakur, E.J. Tan, P.S. Lee, Dopant induced hollow $BaTiO_3$ nanostructures for application in high performance capacitors, J. Mater. Chem. 21 (2011) 16500−16504.

[159] K.A. Mahmoud, B. Mansoor, A. Mansour, M. Khraisheh, Functional graphene nanosheets: the next generation membranes for water desalination, Desalination. 356 (2015) 208−225.

[160] A. Kausar, Applications of polymer/graphene nanocomposite membranes: a review, Mater. Res. Innov. 23 (2019) 276−287.

[161] A. Kausar, Potential of polymer/graphene nanocomposite in electronics, Am. J. Nanosci. Nanotechnol. Res. 6 (2018) 55−63.

[162] A. Kausar, Enhanced electrical and thermal conductivity of modified poly(acrylonitrile-co-butadiene)-based nanofluid containing functional carbon black-graphene oxide, Fuller. Nanotub. Carbon Nanostructures 24 (2016) 278−285.

[163] A. Kausar, Composite coatings of polyamide/graphene: microstructure, mechanical, thermal, and barrier properties, Compos. Interfaces. 25 (2018) 109−125.

[164] A. Kausar, I. Rafique, B. Muhammad, Aerospace application of polymer nanocomposite with carbon nanotube, graphite, graphene oxide, and nanoclay, Polym. Plast. Technol. Eng. 56 (2017) 1438−1456.

[165] A. Kausar, Review on structure, properties and appliance of essential conjugated polymers, Am. J. Polym. Sci. Eng. 4 (2016) 91−102.

[166] Y.S. Lipatov, Interfaces in polymer-polymer composites, Controlled Interphases in Composite Materials, Springer, Netherlands, 1990, pp. 599−611.

[167] T.K. Das, S. Prusty, Graphene-based polymer composites and their applications, Polym. Plast. Technol. Eng. 52 (2013) 319−331.

[168] M.D. Stoller, S. Park, Z. Yanwu, J. An, R.S. Ruoff, Graphene-based ultracapacitors, Nano Lett 8 (2008) 3498−3502.

[169] Y. Xu, W. Hong, H. Bai, C. Li, G. Shi, Strong and ductile poly(vinyl alcohol)/graphene oxide composite films with a layered structure, Carbon N.Y 47 (2009) 3538−3543.

[170] G. Kucinskis, G. Bajars, J. Kleperis, Graphene in lithium ion battery cathode materials: a review, J. Power Sources 240 (2013) 66—79.

[171] X. Hu, Q. Zhou, Health and ecosystem risks of graphene, Chem. Rev. 113 (2013) 3815—3835.

[172] M. Miculescu, V.K. Thakur, F. Miculescu, S.I. Voicu, Graphene-based polymer nanocomposite membranes: a review, Polym. Adv. Technol. 27 (2016) 844—859.

[173] D.C. Lee, H.N. Yang, S.H. Park, W.J. Kim, Nafion/graphene oxide composite membranes for low humidifying polymer electrolyte membrane fuel cell, J. Memb. Sci. 452 (2014) 20—28.

[174] U. Sen, A. Bozkurt, A. Ata, Nafion/poly(1-vinyl-1,2,4-triazole) blends as proton conducting membranes for polymer electrolyte membrane fuel cells, J. Power Sources 195 (2010) 7720—7726.

[175] M. Li, K. Scott, A polymer electrolyte membrane for high temperature fuel cells to fit vehicle applications, Electrochim. Acta. 55 (2010) 2123—2128.

[176] R.K. Nagarale, W. Shin, P.K. Singh, Progress in ionic organic-inorganic composite membranes for fuel cell applications, Polym. Chem. 1 (2010) 388—408.

[177] G. Eda, M. Chhowalla, Chemically derived graphene oxide: towards large-area thin-film electronics and optoelectronics, Adv. Mater. 22 (2010) 2392—2415.

[178] D. Chen, L. Tang, J. Li, Graphene-based materials in electrochemistry, Chem. Soc. Rev. 39 (2010) 3157—3180.

[179] D.R. Dreyer, S. Park, C.W. Bielawski, R.S. Ruoff, The chemistry of graphene oxide, Chem. Soc. Rev. 39 (2010) 228—240.

[180] A. Chandan, M. Hattenberger, A. El-Kharouf, S. Du, A. Dhir, V. Self, et al., High temperature (HT) polymer electrolyte membrane fuel cells (PEMFC)-a review, J. Power Sources 231 (2013) 264—278.

[181] R. Kumar, K. Scott, Freestanding sulfonated graphene oxide paper: a new polymer electrolyte for polymer electrolyte fuel cells, Chem. Commun. 48 (2012) 5584—5586.

[182] C. Tseng, Y. Ye, M. Cheng, K. Kao, W. Shen, J. Rick, et al., Sulfonated polyimide proton exchange membranes with graphene oxide show improved proton conductivity, methanol crossover impedance, and mechanical properties, Adv. Energy Mater. 1 (2011) 1220—1224.

[183] H. Zhang, P.K. Shen, Recent development of polymer electrolyte membranes for fuel cells, Chem. Rev. 112 (2012) 2780—2832.

[184] K.J. Peng, J.Y. Lai, Y.L. Liu, Nanohybrids of graphene oxide chemically-bonded with Nafion: preparation and application for proton exchange membrane fuel cells, J. Memb. Sci. 514 (2016) 86—94.

[185] K. Feng, B. Tang, P. Wu, "Evaporating" graphene oxide sheets (GOSs) for rolled up GOSs and its applications in proton exchange membrane fuel cell, ACS Appl. Mater. Interfaces 5 (2013) 1481—1488.

[186] Y.C. Cao, C. Xu, X. Wu, X. Wang, L. Xing, K. Scott, A poly (ethylene oxide)/graphene oxide electrolyte membrane for low temperature polymer fuel cells, J. Power Sources 196 (2011) 8377—8382.

[187] M. Li, K. Scott, X. Wu, A poly(R1R2R3)-N + /H3PO4 composite membrane for phosphoric acid polymer electrolyte membrane fuel cells, J. Power Sources. 194 (2009) 811—814.

[188] Y.S. Ye, C.Y. Tseng, W.C. Shen, J.S. Wang, K.J. Chen, M.Y. Cheng, et al., A new graphene-modified protic ionic liquid-based composite membrane for solid polymer electrolytes, J. Mater. Chem. 21 (2011) 10448—10453.

[189] H. Zarrin, D. Higgins, Y. Jun, Z. Chen, M. Fowler, Functionalized graphene oxide nanocomposite membrane for low humidity and high temperature proton exchange membrane fuel cells, J. Phys. Chem. C 115 (2011) 20774—20781.

[190] G. He, X. He, X. Wang, C. Chang, J. Zhao, Z. Li, et al., A highly proton-conducting, methanol-blocking Nafion composite membrane enabled by surface-coating cross-linked sulfonated graphene oxide, Chem. Commun. 52 (2016) 2173–2176.

[191] R. Kumar, M. Mamlouk, K. Scott, Sulfonated polyether ether ketone-sulfonated graphene oxide composite membranes for polymer electrolyte fuel cells, RSC Adv 4 (2014) 617–623.

[192] A.K. Sahu, K. Ketpang, S. Shanmugam, O. Kwon, S. Lee, H. Kim, Sulfonated graphene–nafion composite membranes for polymer electrolyte fuel cells operating under reduced relative humidity, J. Phys. Chem. C. 120 (2016) 15855–15866.

[193] Y.S. Ye, M.Y. Cheng, X.L. Xie, J. Rick, Y.J. Huang, F.C. Chang, et al., Alkali doped polyvinyl alcohol/graphene electrolyte for direct methanol alkaline fuel cells, J. Power Sources. 239 (2013) 424–432.

[194] U.R. Farooqui, A.L. Ahmad, N.A. Hamid, Graphene oxide: a promising membrane material for fuel cells, Renew. Sustain. Energy Rev. 82 (2018) 714–733.

[195] Sarah C. Ball, Electrochemistry of proton conducting membrane fuel cells I Johnson Matthey technology review, Platin. Met. Rev. 49 (2005) 27–32.

[196] J. Ling, O. Savadogo, Comparison of methanol crossover among four types of Nafion membranes, J. Electrochem. Soc. 151 (2004) A1604.

[197] C. Lim, C.Y. Wang, Development of high-power electrodes for a liquid-feed direct methanol fuel cell, J. Power Sources. 113 (2003) 145–150.

[198] R. Jiang, H.R. Kunz, J.M. Fenton, Composite silica/Nafion® membranes prepared by tetraethylorthosilicate sol-gel reaction and solution casting for direct methanol fuel cells, J. Memb. Sci. 272 (2006) 116–124.

[199] P. Dimitrova, K.A. Friedrich, U. Stimming, et al., Modified Nafion®-based membranes for use in direct methanol fuel cells, Solid State Ionics, Elsevier, 2002, pp. 115–122.

[200] C.L. L.M.G. S.T. and D. da C.J.C. Pienaar, **The effect of gravity on the viscosity of a silica sol-gel—UQ eSpace, in: A Dicks (Ed.), Proc. ARCCFN Annu. Conf., Coffs Harbour, Australia, 2004, pp. 27–30.

[201] G. Vaivars, N.W. Maxakato, T. Mokrani, L. Petrik, J. Klavins, G. Gericke, et al., Zirconium phosphate based inorganic direct methanol fuel cell, Mater. Sci. 10 (2004) 1320–1392.

[202] B.P. Tripathi, V.K. Shahi, Organic-inorganic nanocomposite polymer electrolyte membranes for fuel cell applications, Prog. Polym. Sci. 36 (2011) 945–979.

[203] V. Tricoli, F. Nannetti, Zeolite-Nafion composites as ion conducting membrane materials, Electrochim. Acta. 48 (2003) 2625–2633.

[204] H. Lin, C. Zhao, W. Ma, H. Li, H. Na, Layer-by-layer self-assembly of in situ polymerized polypyrrole on sulfonated poly(arylene ether ketone) membrane with extremely low methanol crossover, Int. J. Hydrogen Energy. 34 (2009) 9795–9801.

[205] C.W. Lin, K.C. Fan, R. Thangamuthu, Preparation and characterization of high selectivity organic-inorganic hybrid-laminated Nafion 115 membranes for DMFC, J. Memb. Sci. 278 (2006) 437–446.

[206] Y.M. Kim, K.W. Park, J.H. Choi, I.S. Park, Y.E. Sung, A Pd-impregnated nanocomposite Nafion membrane for use in high-concentration methanol fuel in DMFC, Electrochem. Commun. 5 (2003) 571–574.

[207] L.-D. Tsai, H.-C. Chien, W.-H. Huang, C.-P. Huang, C. Kang, J.-N. Lin, et al., Novel bilayer composite membrane for passive direct methanol fuel cells with pure methanol, Int. J. Electrochem. Sci. 8 (2013) 9704–9713.

[208] B.G. Choi, Y.S. Huh, Y.C. Park, D.H. Jung, W.H. Hong, H. Park, Enhanced transport properties in polymer electrolyte composite membranes with graphene oxide sheets, Carbon N. Y 50 (2012) 5395–5402.

[209] Z. Jiang, X. Zhao, Y. Fu, A. Manthiram, Composite membranes based on sulfonated poly(ether ether ketone) and SDBS-adsorbed graphene oxide for direct methanol fuel cells, J. Mater. Chem. 22 (2012) 24862−24869.

[210] H.C. Chien, L.D. Tsai, C.P. Huang, C.Y. Kang, J.N. Lin, F.C. Chang, Sulfonated graphene oxide/Nafion composite membranes for high-performance direct methanol fuel cells, Int. J. Hydrogen Energy. 38 (2013) 13792−13801.

[211] L. Sha Wang, A. Nan Lai, C. Xiao Lin, Q. Gen Zhang, A. Mei Zhu, Q. Lin Liu, Orderly sandwich-shaped graphene oxide/Nafion composite membranes for direct methanol fuel cells, J. Memb. Sci. 492 (2015) 58−66.

[212] T. Yuan, L. Pu, Q. Huang, H. Zhang, X. Li, H. Yang, An effective methanol-blocking membrane modified with graphene oxide nanosheets for passive direct methanol fuel cells, Electrochim. Acta. 117 (2014) 393−397.

[213] Y.D. Lim, D.W. Seo, S.H. Lee, S.Y. Choi, S.Y. Lee, L. Jin, et al., The sulfonated poly (ether sulfone ketone) ionomers containing partial graphene of mesonaphthobifluorene for PEMFC, Electron. Mater. Lett. 10 (2014) 205−207.

[214] M. de Cazes, M.P. Belleville, M. Mougel, H. Kellner, J. Sanchez-Marcano, Characterization of laccase-grafted ceramic membranes for pharmaceuticals degradation, J. Memb. Sci. 476 (2015) 384−393.

[215] S. Gahlot, P.P. Sharma, V. Kulshrestha, P.K. Jha, SGO/SPES-based highly conducting polymer electrolyte membranes for fuel cell application, ACS Appl. Mater. Interfaces 6 (2014) 5595−5601.

[216] I. Nicotera, C. Simari, L. Coppola, P. Zygouri, D. Gournis, S. Brutti, et al., Sulfonated graphene oxide platelets in Nafion nanocomposite membrane: advantages for application in direct methanol fuel cells, J. Phys. Chem. C 118 (2014) 24357−24368.

[217] G. Decher, M. Eckle, J. Schmitt, B. Struth, Layer-by-layer assembled multicomposite films, Curr. Opin. Colloid Interface Sci. 3 (1998) 32−39.

[218] G. Decher, Fuzzy nanoassemblies: toward layered polymeric multicomposites, Science (80−) 277 (1997) 1232−1237.

[219] D.W. Kim, H.S. Choi, C. Lee, A. Blumstein, Y. Kang, Investigation on methanol permeability of Nafion modified by self-assembled clay-nanocomposite multilayers, Electrochim. Acta 50 (2004) 659−662.

[220] T.R. Farhat, P.T. Hammond, Engineering ionic and electronic conductivity in polymer catalytic electrodes using the layer-by-layer technique, Chem. Mater. 18 (2006) 41−49.

[221] S.P. Jiang, Z. Liu, Z.Q. Tian, Layer-by-layer self-assembly of composite polyelectrolyte-nafion membranes for direct methanol fuel cells, Adv. Mater. 18 (2006) 1068−1072.

[222] S. Novikova, S. Yaroslavtsev, V. Rusakov, T. Kulova, A. Skundin, A. Yaroslavtsev, $LiFe_{1-x}MII_xPO_4/C$ (M II = Co, Ni, Mg) as cathode materials for lithium-ion batteries, Electrochim. Acta 122 (2014) 180−186.

[223] W. Jia, B. Tang, P. Wu, Novel slightly reduced graphene oxide based proton exchange membrane with constructed long-range ionic nanochannels via self-assembling of Nafion, ACS Appl. Mater. Interfaces 9 (2017) 22620−22627.

[224] H. Beydaghi, M. Javanbakht, A. Bagheri, P. Salarizadeh, H.G. Zahmatkesh, S. Kashefi, et al., Novel nanocomposite membranes based on blended sulfonated poly (ether ether ketone)/poly(vinyl alcohol) containing sulfonated graphene oxide/Fe3O4 nanosheets for DMFC applications, RSC Adv 5 (2015) 74054−74064.

[225] S.P. Nunes, B. Ruffmann, E. Rikowski, S. Vetter, K. Richau, Inorganic modification of proton conductive polymer membranes for direct methanol fuel cells, J. Memb. Sci. 203 (2002) 215−225.

[226] T.S. Chung, L.Y. Jiang, Y. Li, S. Kulprathipanja, Mixed matrix membranes (MMMs) comprising organic polymers with dispersed inorganic fillers for gas separation, Prog. Polym. Sci. 32 (2007) 483−507.

[227] N. Miyake, J.S. Wainright, R.F. Savinell, Evaluation of a sol-gel derived nafion/silica hybrid membrane for proton electrolyte membrane fuel cell applications: I. Proton conductivity and water content, J. Electrochem. Soc. 148 (2001) A898.

[228] R. Kumar, C. Xu, K. Scott, Graphite oxide/Nafion composite membranes for polymer electrolyte fuel cells, RSC Adv 2 (2012) 8777−8782.

[229] Y. Kim, K. Ketpang, S. Jaritphun, J.S. Park, S. Shanmugam, A polyoxometalate coupled graphene oxide-Nafion composite membrane for fuel cells operating at low relative humidity, J. Mater. Chem. A 3 (2015) 8148−8155.

[230] D.C. Lee, H.N. Yang, S.H. Park, K.W. Park, W.J. Kim, Self-humidifying Pt-graphene/ SiO2 composite membrane for polymer electrolyte membrane fuel cell, J. Memb. Sci. 474 (2015) 254−262.

[231] Q. Li, L. Deng, J.-K. Kim, Y.Q. Zhu, S.M. Holmes, M. Perez-Page, et al., Growth of carbon nanotubes on electrospun cellulose fibers for high performance supercapacitors, J. Electrochem. Soc. 164 (2017) A3220−A3228.

[232] H.N. Yang, W.H. Lee, B.S. Choi, W.J. Kim, Preparation of Nafion/Pt-containing TiO2/graphene oxide composite membranes for self-humidifying proton exchange membrane fuel cell, J. Memb. Sci. 504 (2016) 20−28.

[233] C. Xu, Y. Cao, R. Kumar, X. Wu, X. Wang, K. Scott, A polybenzimidazole/sulfonated graphite oxide composite membrane for high temperature polymer electrolyte membrane fuel cells, J. Mater. Chem. 21 (2011) 11359−11364.

[234] C. Xue, J. Zou, Z. Sun, F. Wang, K. Han, H. Zhu, Graphite oxide/functionalized graphene oxide and polybenzimidazole composite membranes for high temperature proton exchange membrane fuel cells, Int. J. Hydrogen Energy 39 (2014) 7931−7939.

[235] A.K. Mishra, N.H. Kim, D. Jung, J.H. Lee, Enhanced mechanical properties and proton conductivity of Nafion-SPEEK-GO composite membranes for fuel cell applications, J. Memb. Sci. 458 (2014) 128−135.

[236] Y. Liu, J. Wang, H. Zhang, C. Ma, J. Liu, S. Cao, et al., Enhancement of proton conductivity of chitosan membrane enabled by sulfonated graphene oxide under both hydrated and anhydrous conditions, J. Power Sources 269 (2014) 898−911.

[237] P.P. Sharma, V. Kulshrestha, Synthesis of highly stable and high water retentive functionalized biopolymer-graphene oxide modified cation exchange membranes, RSC Adv 5 (2015) 56498−56506.

[238] T. Ko, K. Kim, M.Y. Lim, S.Y. Nam, T.H. Kim, S.K. Kim, et al., Sulfonated poly (arylene ether sulfone) composite membranes having poly(2,5-benzimidazole)-grafted graphene oxide for fuel cell applications, J. Mater. Chem. A. 3 (2015) 20595−20606.

[239] Y. He, J. Wang, H. Zhang, T. Zhang, B. Zhang, S. Cao, et al., Polydopamine-modified graphene oxide nanocomposite membrane for proton exchange membrane fuel cell under anhydrous conditions, J. Mater. Chem. A. 2 (2014) 9548−9558.

[240] E. Bakangura, L. Wu, L. Ge, Z. Yang, T. Xu, Mixed matrix proton exchange membranes for fuel cells: state of the art and perspectives, Prog. Polym. Sci. 57 (2016) 103−152.

[241] J. Yang, C. Liu, L. Gao, J. Wang, Y. Xu, R. He, Novel composite membranes of triazole modified graphene oxide and polybenzimidazole for high temperature polymer electrolyte membrane fuel cell applications, RSC Adv 5 (2015) 101049−101054.

[242] Y. Heo, H. Im, J. Kim, The effect of sulfonated graphene oxide on sulfonated poly (ether ether ketone) membrane for direct methanol fuel cells, J. Memb. Sci. 425−426 (2013) 11−22.

[243] C.W. Lin, Y.S. Lu, Highly ordered graphene oxide paper laminated with a Nafion membrane for direct methanol fuel cells, J. Power Sources 237 (2013) 187−194.

[244] R.P. Pandey, A.K. Thakur, V.K. Shahi, Sulfonated polyimide/acid-functionalized graphene oxide composite polymer electrolyte membranes with improved proton conductivity and water-retention properties, ACS Appl. Mater. Interfaces. 6 (2014) 16993−17002.

[245] R.P. Pandey, V.K. Shahi, Sulphonated imidized graphene oxide (SIGO) based polymer electrolyte membrane for improved water retention, stability and proton conductivity, J. Power Sources. 299 (2015) 104−113.

[246] R.P. Pandey, V.K. Shahi, A N-o-sulphonic acid benzyl chitosan (NSBC) and N,N-dimethylene phosphonic acid propylsilane graphene oxide (NMPSGO) based multifunctional polymer electrolyte membrane with enhanced water retention and conductivity, RSC Adv 4 (2014) 57200−57209.

[247] S.H. Lee, S.H. Choi, S.A. Gopalan, K.P. Lee, G. Anantha-Iyengar, Preparation of new self-humidifying composite membrane by incorporating graphene and phosphotungstic acid into sulfonated poly(ether ether ketone) film, Int. J. Hydrogen Energy. 39 (2014) 17162−17177.

[248] Z. Jiang, X. Zhao, A. Manthiram, Sulfonated poly(ether ether ketone) membranes with sulfonated graphene oxide fillers for direct methanol fuel cells, Int. J. Hydrogen Energy. 38 (2013) 5875−5884.

[249] F. Chu, B. Lin, T. Feng, C. Wang, S. Zhang, N. Yuan, et al., Zwitterion-coated graphene-oxide-doped composite membranes for proton exchange membrane applications, J. Memb. Sci. 496 (2015) 31−38.

[250] Y. Yin, H. Wang, L. Cao, Z. Li, Z. Li, M. Gang, et al., Sulfonated poly(ether ether ketone)-based hybrid membranes containing graphene oxide with acid-base pairs for direct methanol fuel cells, Electrochim. Acta. 203 (2016) 178−188.

[251] A. Shirdast, A. Sharif, M. Abdollahi, Effect of the incorporation of sulfonated chitosan/sulfonated graphene oxide on the proton conductivity of chitosan membranes, J. Power Sources 306 (2016) 541−551.

[252] R. Kumar, M. Mamlouk, K. Scott, A graphite oxide paper polymer electrolyte for direct methanol fuel cells, Int. J. Electrochem. 2011 (2011) 1−7.

[253] A.S. Aricò, P. Bruce, B. Scrosati, J.M. Tarascon, W. Van Schalkwijk, Nanostructured materials for advanced energy conversion and storage devices, Nat. Mater. 4 (2005) 366−377.

[254] V. Stamenković, T.J. Schmidt, P.N. Ross, N.M. Marković, Surface composition effects in electrocatalysis: kinetics of oxygen reduction on well-defined Pt3Ni and Pt3Co alloy surfaces, J. Phys. Chem. B. 106 (2002) 11970−11979.

[255] H.R. Colón-Mercado, B.N. Popov, Stability of platinum based alloy cathode catalysts in PEM fuel cells, J. Power Sources. 155 (2006) 253−263.

[256] T. Toda, H. Igarashi, H. Uchida, M. Watanabe, Enhancement of the electroreduction of oxygen on Pt Alloys with Fe, Ni, and Co, J. Electrochem. Soc. 146 (1999) 3750−3756.

[257] H. Yang, N. Alonso-Vante, J.-M. Léger, C. Lamy, Tailoring, structure, and activity of carbon-supported nanosized Pt − Cr alloy electrocatalysts for oxygen reduction in pure and methanol-containing electrolytes, J. Phys. Chem. B. 108 (2004) 1938−1947.

[258] H. Yang, W. Vogel, C. Lamy, N. Alonso-Vante, Structure and electrocatalytic activity of carbon-supported Pt − Ni alloy nanoparticles toward the oxygen reduction reaction, J. Phys. Chem. B. 108 (2004) 11024−11034.

[259] J. Zhang, H. Yang, J. Fang, S. Zou, Synthesis and oxygen reduction activity of shape-controlled Pt3Ni nanopolyhedra, Nano Lett 10 (2010) 638−644.

[260] H. Zhong, H. Zhang, G. Liu, Y. Liang, J. Hu, B. Yi, A novel non-noble electrocatalyst for PEM fuel cell based on molybdenum nitride, Electrochem. Commun. 8 (2006) 707—712.

[261] F. Charreteur, F. Jaouen, S. Ruggeri, J.P. Dodelet, Fe/N/C non-precious catalysts for PEM fuel cells: influence of the structural parameters of pristine commercial carbon blacks on their activity for oxygen reduction, Electrochim. Acta. 53 (2008) 2925—2938.

[262] Y. Liang, Y. Li, H. Wang, J. Zhou, J. Wang, T. Regier, et al., Co3O4 nanocrystals on graphene as a synergistic catalyst for oxygen reduction reaction, Nat. Mater. 10 (2011) 780—786.

[263] C. Liu, Z. Chen, C.Z. Li, Surface engineering of graphene-enzyme nanocomposites for miniaturized biofuel cell, IEEE Trans. Nanotechnol. 10 (2011) 59—62.

[264] Y. Sun, C. Li, Y. Xu, H. Bai, Z. Yao, G. Shi, Chemically converted graphene as substrate for immobilizing and enhancing the activity of a polymeric catalyst, Chem. Commun. 46 (2010) 4740—4742.

[265] K. Gong, F. Du, Z. Xia, M. Durstock, L. Dai, Nitrogen-doped carbon nanotube arrays with high electrocatalytic activity for oxygen reduction, Science (80—) 323 (2009) 760—764.

[266] C. Zhang, R. Hao, H. Liao, Y. Hou, Synthesis of amino-functionalized graphene as metal-free catalyst and exploration of the roles of various nitrogen states in oxygen reduction reaction, Nano Energy 2 (2013) 88—97.

[267] S. Wang, D. Yu, L. Dai, D.W. Chang, J.-B. Baek, Polyelectrolyte-functionalized graphene as metal-free electrocatalysts for oxygen reduction, ACS Nano 5 (2011) 6202—6209.

[268] S.M. Unni, S.N. Bhange, B. Anothumakkool, S. Kurungot, Redox-mediated synthesis of functionalised graphene: a strategy towards 2D multifunctional electrocatalysts for energy conversion applications, Chempluschem. 78 (2013) 1296—1303.

[269] A. Thomas, A. Fischer, F. Goettmann, M. Antonietti, J.O. Müller, R. Schlögl, et al., Graphitic carbon nitride materials: variation of structure and morphology and their use as metal-free catalysts, J. Mater. Chem. 18 (2008) 4893—4908.

[270] Y. Wang, X. Wang, M. Antonietti, Polymeric graphitic carbon nitride as a heterogeneous organocatalyst: from photochemistry to multipurpose catalysis to sustainable chemistry, Angew. Chemie-Int. (Ed.) 51 (2012) 68—89.

[271] S.M. Lyth, Y. Nabae, S. Moriya, S. Kuroki, M. Kakimoto, J. Ozaki, et al., Carbon nitride as a nonprecious catalyst for electrochemical oxygen reduction, J. Phys. Chem. C 113 (2009) 20148—20151.

[272] Y. Zheng, J. Liu, J. Liang, M. Jaroniec, S.Z. Qiao, Graphitic carbon nitride materials: controllable synthesis and applications in fuel cells and photocatalysis, Energy Environ. Sci. 5 (2012) 6717—6731.

[273] K.A. Ritter, J.W. Lyding, The influence of edge structure on the electronic properties of graphene quantum dots and nanoribbons, Nat. Mater. 8 (2009) 235—242.

[274] D. Deng, L. Yu, X. Pan, S. Wang, X. Chen, P. Hu, et al., Size effect of graphene on electrocatalytic activation of oxygen, Chem. Commun. 47 (2011) 10016—10018.

[275] S. Yang, X. Feng, X. Wang, K. Müllen, Graphene-based carbon nitride nanosheets as efficient metal-free electrocatalysts for oxygen reduction reactions, Angew. Chemie—Int. (Ed.) 50 (2011) 5339—5343.

[276] X. Wang, H. Dai, Etching and narrowing of graphene from the edges, Nat. Chem. 2 (2010) 661—665.

[277] X. Li, X. Wang, L. Zhang, S. Lee, H. Dai, Chemically derived, ultrasmooth graphene nanoribbon semiconductors, Science (80—) 319 (2008) 1229—1232.

[278] Q. He, S. Wu, S. Gao, X. Cao, Z. Yin, H. Li, et al., Transparent, flexible, all-reduced graphene oxide thin film transistors, ACS Nano 5 (2011) 5038−5044.

[279] E.J. López-Naranjo, L.J. González-Ortiz, L.M. Apátiga, E.M. Rivera-Muñoz, A. Manzano-Ramírez, Transparent electrodes: a review of the use of carbon-based nanomaterials, J. Nanomater. 2016 (2016).

[280] A.I. Hofmann, E. Cloutet, G. Hadziioannou, Materials for transparent electrodes: from metal oxides to organic alternatives, Adv. Electron. Mater. 4 (2018) 1700412.

[281] H. Wu, D. Kong, Z. Ruan, P.C. Hsu, S. Wang, Z. Yu, et al., A transparent electrode based on a metal nanotrough network, Nat. Nanotechnol. 8 (2013) 421−425.

[282] S. Bae, H. Kim, Y. Lee, X. Xu, J.S. Park, Y. Zheng, et al., Roll-to-roll production of 30-inch graphene films for transparent electrodes, Nat. Nanotechnol. 5 (2010) 574−578.

[283] K. Ellmer, Past achievements and future challenges in the development of optically transparent electrodes, Nat. Photonics. 6 (2012) 809−817. Available from: https://doi.org/10.1038/nphoton.2012.282.

[284] S. Ye, A.R. Rathmell, Z. Chen, I.E. Stewart, B.J. Wiley, Metal nanowire networks: the next generation of transparent conductors, Adv. Mater. 26 (2014) 6670−6687.

[285] Y. Xia, K. Sun, J. Ouyang, Solution-processed metallic conducting polymer films as transparent electrode of optoelectronic devices, Adv. Mater. 24 (2012) 2436−2440.

[286] D. Lee, H. Lee, Y. Ahn, Y. Jeong, D.Y. Lee, Y. Lee, Highly stable and flexible silver nanowire-graphene hybrid transparent conducting electrodes for emerging optoelectronic devices, Nanoscale. 5 (2013) 7750−7755.

[287] S. De, T.M. Higgins, P.E. Lyons, E.M. Doherty, P.N. Nirmalraj, W.J. Blau, et al., Silver nanowire networks as flexible, transparent, conducting films: extremely high DC to optical conductivity ratios, ACS Nano 3 (2009) 1767−1774.

[288] L. Yu, C. Shearer, J. Shapter, Recent development of carbon nanotube transparent conductive films, Chem. Rev. 116 (2016) 13413−13453.

[289] D.S. Hecht, L. Hu, G. Irvin, Emerging transparent electrodes based on thin films of carbon nanotubes, graphene, and metallic nanostructures, Adv. Mater. 23 (2011) 1482−1513.

[290] M.W. Rowell, M.A. Topinka, M.D. McGehee, H.J. Prall, G. Dennler, N.S. Sariciftci, et al., Organic solar cells with carbon nanotube network electrodes, Appl. Phys. Lett. 88 (2006) 233506.

[291] M. Luo, Y. Liu, W. Huang, W. Qiao, Y. Zhou, Y. Ye, et al., Towards flexible transparent electrodes based on carbon and metallic materials, Micromachines 8 (2017) 12.

[292] Z. Wu, Z. Chen, X. Du, J.M. Logan, J. Sippel, M. Nikolou, et al., Transparent, conductive carbon nanotube films, Science (80−) 305 (2004) 1273−1276.

[293] J.Y. Lee, S.T. Connor, Y. Cui, P. Peumans, Solution-processed metal nanowire mesh transparent electrodes, Nano Lett 8 (2008) 689−692.

[294] H. Wu, L. Hu, M.W. Rowell, D. Kong, J.J. Cha, J.R. McDonough, et al., Electrospun metal nanofiber webs as high-performance transparent electrode, Nano Lett 10 (2010) 4242−4248.

[295] L. Hu, H.S. Kim, J.Y. Lee, P. Peumans, Y. Cui, Scalable coating and properties of transparent, flexible, silver nanowire electrodes, ACS Nano 4 (2010) 2955−2963.

[296] L. Hu, H. Wu, Y. Cui, Metal nanogrids, nanowires, and nanofibers for transparent electrodes, MRS Bull 36 (2011) 760−765.

[297] J.K. Wassei, R.B. Kaner, Graphene, a promising transparent conductor, Mater. Today. 13 (2010) 52−59.

[298] J. Wu, H.A. Becerril, Z. Bao, Z. Liu, Y. Chen, P. Peumans, Organic solar cells with solution-processed graphene transparent electrodes, Appl. Phys. Lett. 92 (2008) 263302.

[299] J. Wu, M. Agrawal, H.A. Becerril, Z. Bao, Z. Liu, Y. Chen, et al., Organic light-emitting diodes on solution-processed graphene transparent electrodes, ACS Nano 4 (2010) 43−48.

[300] L. La Notte, E. Villari, A.L. Palma, A. Sacchetti, M. Michela Giangregorio, G. Bruno, et al., Laser-patterned functionalized CVD-graphene as highly transparent conductive electrodes for polymer solar cells, Nanoscale. 9 (2017) 62−69.

[301] L. Gomez De Arco, Y. Zhang, C.W. Schlenker, K. Ryu, M.E. Thompson, C. Zhou, Continuous, highly flexible, and transparent graphene films by chemical vapor deposition for organic photovoltaics, ACS Nano 4 (2010) 2865−2873.

[302] T. Sannicolo, M. Lagrange, A. Cabos, C. Celle, J.P. Simonato, D. Bellet, Metallic nanowire-based transparent electrodes for next generation flexible devices: a review, Small 12 (2016) 6052−6075.

[303] C.L. Kim, C.W. Jung, Y.J. Oh, D.E. Kim, A highly flexible transparent conductive electrode based on nanomaterials, NPG Asia Mater 9 (2017) 438.

[304] D. Yoo, J. Kim, J.H. Kim, Direct synthesis of highly conductive poly(3,4-ethylene-dioxythiophene):Poly(4-styrenesulfonate) (PEDOT:PSS)/graphene composites and their applications in energy harvesting systems, Nano Res 7 (2014) 717−730.

[305] T.H. Seo, S. Lee, K.H. Min, S. Chandramohan, A.H. Park, G.H. Lee, et al., The role of graphene formed on silver nanowire transparent conductive electrode in ultra-violet light emitting diodes, Sci. Rep. 6 (2016).

[306] Y. Xu, Y. Wang, J. Liang, Y. Huang, Y. Ma, X. Wan, et al., A hybrid material of graphene and poly (3, 4-ethyldioxythiophene) with high conductivity, Flexibility, and Transparency, 2009.

[307] F. Soltani-kordshuli, F. Zabihi, M. Eslamian, Graphene-doped PEDOT:PSS nanocomposite thin films fabricated by conventional and substrate vibration-assisted spray coating (SVASC), Eng. Sci. Technol. Int. J. 19 (2016) 1216−1223.

[308] Q. Chen, F. Zabihi, M. Eslamian, Improved functionality of PEDOT:PSS thin films via graphene doping, fabricated by ultrasonic substrate vibration-assisted spray coating, Synth. Met. 222 (2016) 309−317.

[309] X.F. Lin, Z.Y. Zhang, Z.K. Yuan, J. Li, X.F. Xiao, W. Hong, et al., Graphene-based materials for polymer solar cells, Chin. Chem. Lett. 27 (2016) 1259−1270.

[310] S. Rafique, S.M. Abdullah, M.M. Shahid, M.O. Ansari, K. Sulaiman, Significantly improved photovoltaic performance in polymer bulk heterojunction solar cells with graphene oxide /PEDOT:PSS double decked hole transport layer, Sci. Rep. 7 (2017).

[311] C.H. Jung, Y.J. Noh, J.H. Bae, J.H. Yu, I.T. Hwang, J. Shin, et al., Polyacrylonitrile-grafted reduced graphene oxide hybrid: an all-round and efficient hole-extraction material for organic and inorganic-organic hybrid photovoltaics, Nano Energy 31 (2017) 19−27.

[312] S.S. Li, K.H. Tu, C.C. Lin, C.W. Chen, M. Chhowalla, Solution-processable graphene oxide as an efficient hole transport layer in polymer solar cells, ACS Nano 4 (2010) 3169−3174.

[313] D.D. Nguyen, N.H. Tai, Y.L. Chueh, S.Y. Chen, Y.J. Chen, W.S. Kuo, et al., Synthesis of ethanol-soluble few-layer graphene nanosheets for flexible and transparent conducting composite films, Nanotechnology. 22 (2011).

[314] Z. Liu, Q. Liu, Y. Huang, Y. Ma, S. Yin, X. Zhang, et al., Organic photovoltaic devices based on a novel acceptor material: graphene, Adv. Mater. 20 (2008) 3924−3930.

[315] Q. Liu, Z. Liu, X. Zhang, L. Yang, N. Zhang, G. Pan, et al., Polymer photovoltaic cells based on solution-processable graphene and P3HT, Adv. Funct. Mater. 19 (2009) 894−904.

[316] H. Wang, D. He, Y. Wang, Z. Liu, H. Wu, J. Wang, Organic photovoltaic devices based on graphene as an electron-acceptor material and P3OT as a donor material, Phys. Status Solidi. 208 (2011) 2339—2343.

[317] Z. Liu, D. He, Y. Wang, H. Wu, J. Wang, H. Wang, Improving photovoltaic properties by incorporating both SPFGraphene and functionalized multiwalled carbon nanotubes, Sol. Energy Mater. Sol. Cells. 94 (2010) 2148—2153.

[318] Y. Li, Y. Hu, Y. Zhao, G. Shi, L. Deng, Y. Hou, et al., An electrochemical avenue to green-luminescent graphene quantum dots as potential electron-acceptors for photovoltaics, Adv. Mater. 23 (2011) 776—780.

[319] F. Li, L. Kou, W. Chen, C. Wu, T. Guo, Enhancing the short-circuit current and power conversion efficiency of polymer solar cells with graphene quantum dots derived from double-walled carbon nanotubes, NPG Asia Mater 5 (2013) e60.

[320] L. Chen, C.X. Guo, Q. Zhang, Y. Lei, J. Xie, S. Ee, et al., Graphene quantum-dot-doped polypyrrole counter electrode for high-performance dye-sensitized solar cells, ACS Appl. Mater. Interfaces. 5 (2013) 2047—2052.

[321] Handbook of Organic-Inorganic Hybrid Materials and Nanocomposites, n.d.

[322] M. Gratzel, Conversion of sunlight to electric power by nanocrystalline dye-sensitized solar cells, J. Photochem. Photobiol. A Chem. 164 (2004) 3—14.

[323] M. Grätzel, Solar Energy Conversion by Dye-Sensitized Photovoltaic Cells, 2005.

[324] S. Yun, A. Hagfeldt, T. Ma, Pt-free counter electrode for dye-sensitized solar cells with high efficiency, Adv. Mater. 26 (2014) 6210—6237.

[325] W. Liu, Y. Fang, P. Xu, Y. Lin, X. Yin, G. Tang, et al., Two-step electrochemical synthesis of polypyrrole/reduced graphene oxide composites as efficient pt-free counter electrode for plastic dye-sensitized solar cells, ACS Appl. Mater. Interfaces. 6 (2014) 16249—16256.

[326] M. Sekkarapatti Ramasamy, M.S. Ramasamy, A. Nikolakapoulou, D. Raptis, V. Dracopoulos, G. Paterakis, et al., Reduced graphene oxide/polypyrrole/PEDOT composite films as efficient Pt-free counter electrode for dye-sensitized solar cells, Electrochim. Acta 173 (2015) 276—281.

[327] C. Bora, C. Sarkar, K.J. Mohan, S. Dolui, Polythiophene/graphene composite as a highly efficient platinum-free counter electrode in dye-sensitized solar cells, Electrochim. Acta. 157 (2015) 225—231.

[328] S.P. Lim, A. Pandikumar, Y.S. Lim, N.M. Huang, H.N. Lim, In-situ electrochemically deposited polypyrrole nanoparticles incorporated reduced graphene oxide as an efficient counter electrode for platinum-free dye-sensitized solar cells, Sci. Rep. 4 (2014) 1—7.

[329] S. Yuan, Q. Tang, B. He, Y. Zhao, Multifunctional graphene incorporated conducting gel electrolytes in enhancing photovoltaic performances of quasi-solid-state dye-sensitized solar cells, J. Power Sources. 260 (2014) 225—232.

[330] S. Yuan, Q. Tang, B. Hu, C. Ma, J. Duan, B. He, Efficient quasi-solid-state dye-sensitized solar cells from graphene incorporated conducting gel electrolytes, J. Mater. Chem. A. 2 (2014) 2814—2821.

[331] J. Duan, Q. Tang, R. Li, B. He, L. Yu, P. Yang, Multifunctional graphene incorporated polyacrylamide conducting gel electrolytes for efficient quasi-solid-state quantum dot-sensitized solar cells, J. Power Sources. 284 (2015) 369—376.

[332] Y.F. Chan, C.C. Wang, C.Y. Chen, Quasi-solid DSSC based on a gel-state electrolyte of PAN with 2-D graphenes incorporated, J. Mater. Chem. A. 1 (2013) 5479—5486.

[333] M.S. Akhtar, S. Kwon, F.J. Stadler, O.B. Yang, High efficiency solid state dye sensitized solar cells with graphene-polyethylene oxide composite electrolytes, Nanoscale. 5 (2013) 5403—5411.

[334] M. Bera, P. Maji, Graphene-based polymer nanocomposites: materials for future revolution, MOJ Polym. Sci 1 (2017) 1−4.
[335] G. Anagnostopoulos, L. Sygellou, G. Paterakis, I. Polyzos, C.A. Aggelopoulos, C. Galiotis, Enhancing the adhesion of graphene to polymer substrates by controlled defect formation, Nanotechnology. 30 (2018) 15704.
[336] Graphene-industries raw graphene, devices and membranes for sale, grapheneindustries.com/?SampleþCatalog (n.d.).
[337] H. Kim, A.A. Abdala, C.W. Macosko, Graphene/polymer nanocomposites, Macromolecules. 43 (2010) 6515−6530.
[338] B.J. Panessa-Warren, J.B. Warren, S.S. Wong, J.A. Misewich, Biological cellular response to carbon nanoparticle toxicity, J. Phys. Condens. Matter. 18 (2006) S2185−S2201.
[339] C. Uthaisar, V. Barone, Edge effects on the characteristics of Li diffusion in graphene, Nano Lett 10 (2010) 2838−2842.
[340] G. Malucelli, Graphene-based polymer nanocomposites: recent advances and still open challenges, Curr. Graphene Sci 1 (2017) 16−25.

Biomedical application of polymer-graphene composites

Monalisa Adhikari[1], Jonathan Tersur Orasugh[1,2] and Dipankar Chattopadhyay[1,3]

[1]Department of Polymer Science and Technology, University of Calcutta, Kolkata, India, [2]Department of Jute and Fiber Technology, Institute of Jute Technology, University of Calcutta, Kolkata, India, [3]Center for Research in Nanoscience and Nanotechnology, Acharya Prafulla Chandra Roy Sikhsha Prangan, University of Calcutta, Kolkata, India

17.1 Introduction

Nanotechnology development was a great challenge for scientists in the nineteenth century and the improvement of nanomaterial helps humans move toward a brilliant future.

There is a large interest in multifaceted biodegradable biomaterials in biomedical applications that can improve the properties of different tissues in the human body [1]. A wide variety of natural and synthetic polymers are used for such applications.

Nanocomposite is one of the most essential types of nanomaterials while one of the basic and the key elements of nanotechnology are the "nanomaterials." Nanomaterials are materials having at least one of its dimensions <100 nm. These materials are usually 10^{-9} m in size; which means at least one of their dimensions is one-billionth of a meter. Nanomaterials also show different physicochemical properties in comparison to their bulk material which inherently depends on their size and shape [1,2]. Nanomaterials can be classified based on their structures including nanorods, nanoparticles, and nanosheets. Nanomaterials with zero-dimensional are normally nanoparticles, one dimensional are mostly nanofibers, nanorods, or nanotubes and two dimensional are generally films and layers types as schematically represented in Fig. 17.1 [3].

Among the many carbon allotropes, graphene is one of the most important allotropes depending on their properties and application. Improvement of mechanical, electrical and biological behavior of graphene nanomaterials reinforced polymeric matrix even at very low wt.% have revealed enhancement in the overall properties of the fabricated nanocomposite for any category of graphene nanomaterials polymeric nanocomposites along with their enormous application expanse.

17.2 Polymer composites/nanocomposite

The extensive use of polymers owing to their unique properties, ease of production, lightweight, and often ductile nature but they possess lower mechanical properties

Polymer Nanocomposites Containing Graphene. DOI: https://doi.org/10.1016/B978-0-12-821639-2.00011-2

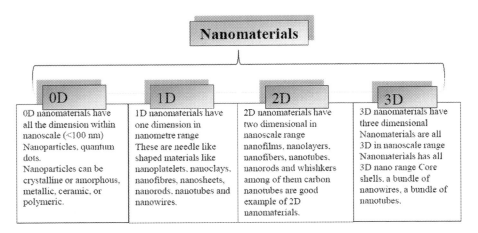

Figure 17.1 Nanomaterials dimensionally classification.

in comparison to metals or ceramics because of that the inclusion (fibers, whiskers, platelets, or particles) into a polymer matrix results into a composite [4]. Polymer nanocomposites (PNC) have been investigated for about three decades. In a recent development, polymer properties (thermal, mechanical, electrical, optical) are enhanced via the addition of novel nanoscale filler [4,5].

PNCs are in general classified into three main categories with reference to the dimensions of their dispersed reinforcing nanoscale fillers. PNCs have shown some exceptional perfunctory properties such as high elastic stiffness and strength due to the small concentration of nano-additives. The other excellent properties of PNCs possess properties such as flame retardancy, barrier resistance, wear resistance, magnetic, electrical and optical properties making them an ideal material for bio-tissue engineering, high temperature materials, advanced packaging, etc.

Depending upon the dimensions and the dispersed level of novel nanoscale reinforcing fillers PNC can be grouped into three main types. In the first instance, 2D (two dimensional) nanoscale reinforcing fillers such as layered silicate [6,7], graphene [5,8] or MXene [5] which are sheets of one to a few nanometers broad and of 100−1000 s of nm long included in polymer matrix. The equivalent PNCs can be grouped into layered PNC, secondly, two dimensions in nanometer scale and the 3rd is outsized, resulting into an elongated 1Darchitecturesuch as nanoscale fillers including nanofibers or nanotubes, for example, carbon nano-fibers (CNFs) and nano-tubes (CNTs) [3] or halloysite nano-tubes (HNTs) [3,9] as reinforcing nanofillers to obtain materials with exceptional properties. The 3rdare nanocomposites comprising of nanoscale reinforcing fillers of 3D in nm scale. These 3D nanoscale reinforcing fillers are isodimensional low aspect ratio nanoparticles like spherical nanosilica [6,7], semiconductor nanoclusters [7] and quantum dots (Qdots) [9]. Fig. 17.1 shows the schematic representation of an assortment of nanoscale reinforcing fillers.

Graphite in its atomic form consists of layered structure but its layered structure must be broken down if we can apply homogeneous dispersion in a polymeric matrix. The homogenous dispersion is much easier by using a single sheet of graphite oxide. The molecular dynamic simulations were performed for nanocomposites of graphite and polypropylene. It was reported that the nanocomposites having 3.0 nm separations between graphite layers were more effective [3].

17.2.1   Preparation method of polymer nanocomposite

Many methods have been developed for the preparation of polymer nanocomposites. The most important techniques utilized thus far are intercalation of polymer, in situ polymerization, melt intercalation method, sol−gel method, template synthesis, etc.

17.2.1.1   Intercalation of the polymer

The intercalation process is used for the preparation of polymer-based nanocomposites which generally involves the dispersion of nanoplatelets types of nanomaterials into the polymer matrix as depicted in Fig. 17.2. As a top-down approach, it is developed for the inclusion of graphene or clays (nanomaterials) into polymeric matrix of any kind aimed at improving the bulk properties like stiffness, shrinkage, and flammability of polymer matrix [1,3]. Nanocomposites may be prepared from this method starting from intercalated to exfoliate. Depending on the ability of penetration of the polymer matrix into the nanofiller material such as graphene or modified graphene is dispersed in a suitable solvent like acetone, toluene, water, tetrahydrofuran (THF), chloroform, dimethylformamide (DMF), etc. the polymer adsorbs on to the nanofiller surface while the solvent is evaporated subsequently.

Nanoplatelets can be homogeneously dispersed by the utilization of two systems.

1. **Chemical deposition**
    This procedure includes the in situ polymerization technique in which nanoparticles are dispersed in monomer and afterward the solution is subjected to polymerization.

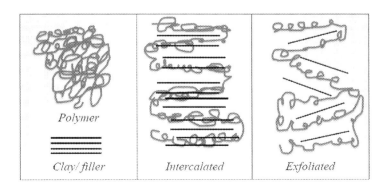

Figure 17.2 Polymer intercalation.

In this technique, nanoplatelets are dispersed into polymer continuously by additional polymerization process. The nanoplatelets are swollen in monomer solution and the polymer formation happens between the intercalated sheets and the monomer in situ.

2. **Mechanical deposition**

   In this technique, direct intercalation of polymer with nanoplatelets happens through solution mixing. The polymer is dissolved in a co solvent and nanoplatelets sheets are swollen in the solvent and then the two solvent are mixed together, the polymer chains in the final solution intercalate into the nanoplatelets layers and dislodge the dissolvable

17.2.1.2   In situ polymerization

In this method the reinforcing nanofiller must be suitably dispersed within the monomer before commencement of the polymerization process, make sure that the polymer is formed between the reinforcing nanoparticles. Polymerization can be initiated using a suitable initiator (heat or radiation use of an appropriate initiator, etc.). Using this method (as schematically shown in Fig. 17.3), a polymer-graft nanoparticle and lofty nanofiller loading devoid of aggregation is achievable. An organic modifier may be used to enhance better dispersion of the nanoparticles and take part in polymerization. It may be adopted as a different approach for the fabrication of nanocomposites with thermally unstable and insoluble polymers. This process is appropriate for the synthesis of higher performance polymeric products. Mini-emulsion polymerization is based on the synthesis of monomer droplets dispersed in a defined solution within range nanoscale. Several polymer nanocomposites are reportedly fabricated using this approach [10–20]: polystyrene

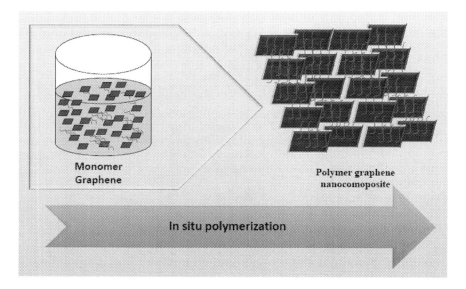

Figure 17.3 In situ polymerization graphene polymer nanocomposite.

(PS)/graphene, polyimide (PI) [10], (12), PU, polyacrylonitrile [12], PP [13], PS [14], Nylon-6 [15], poly(ethylene terephthalate)PET [16], PMMA [17,18], epoxy [19], phenolformaldehyde [20]. The application of higher nanofillers contents without agglomeration leads to enhanced performance of the final nanocomposite products via the expansion of the solvent-free form resulting in a covalent bond among the nanoparticle's functional groups and polymer chains and suitable for use for both thermoset and thermoplastic polymers. One main limitation is the ease of agglomeration.

17.2.1.3 Melt intercalation method

Melt intercalation method is mostly utilized by the industry as in extrusion and injection molding and it permits the processing of some specific polymers which are not appropriate for in situ polymerization or intercalation (See Fig. 17.4). In this approach, a thermoplastic polymer is mixed mechanically with graphene or modified graphene matrix at high temperatures with the aid of traditional techniques such as injection molding and extrusion, with exfoliation or intercalation of the polymer molecules to form nanocomposites. In the melt intercalation technique, graphene (in any of its various forms) is variegated with the thermoplastic polymer matrix in a molten state (no solvent is required in this method). Essentially, a homogeneous blending of graphene and the thermoplastic polymers is achieved by high shear mixing at raised temperatures. This approach has found widespread application for the synthesis/fabrication of thermoplastic nanocomposites/composites. The final shape of components can be fabricated by compression molding, injection molding, fiber extrusion technique, etc. Researchers have exploited this approach for the synthesis of graphene/polymer nanocomposites such as polystyrene/graphite nanocomposites [21], graphite/polypropylene nanocomposite [22], polyethylene terephthalate (PET)/graphene nanocomposites [23], polypropylene (PP)/expanded graphite (EG) [22], High dense polyethylene (HDPE)/EG [24], polyphenylenesulfide (PPS)/EG [21], polyamide (PA6)/EG [25], etc. The leading shortcoming of this process is poor dispersity and distribution of graphene fillers in the matrix in

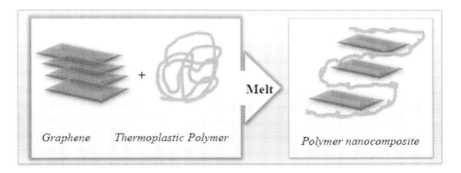

Figure 17.4 Polymer nanocomposite preparation by melt intercalation method.

comparison to other techniques. Also, the application of high shear forces creates flaws and splintering of graphene nanosheets.

17.2.1.4 Sol−gel method

The sol−gel is combined with two relative steps (Fig. 17.5), sol and gel. In a monomer solution where sol is a colloidal dispersion of solid reinforcing nanoparticles having a3D interconnecting architecture gel formed between phases. In this technique, a colloidal dispersion of solid reinforcing nanoparticle solution (sol) formed with the suspension of the solid nanoparticles in the solution of the monomer resulting in a form of an interconnecting network between the phases (gel). It is widely used in the glass and ceramic industries. At first, the metal alkoxide solution undergoes hydrolysis with water or alcohol in the presence of acid or base followed by polycondensation. Due to polycondensation, the liquid phase gets altered into the gel phase by the removal of the water molecules in the solution resulting to increase in the viscosity of the solution. Mainly the water molecules are condensed then the gel phase changes into the powder phase. Some additional heat is required to get

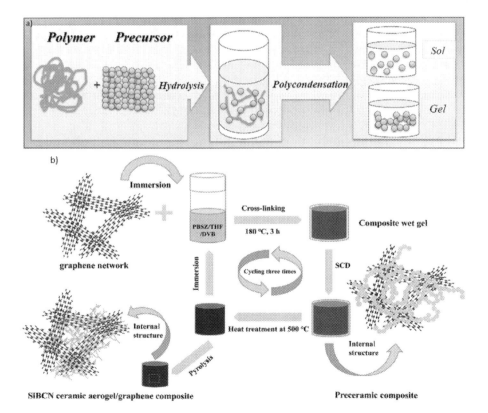

Figure 17.5 Sol−gel method preparation SiBCN ceramic aerogel/graphene composites [26].

the fine crystalline nature to the powder. This technique has been helpful in the synthesis of oxides, nanocomposites/composites or organic/inorganic hybrids. The basis of the sol−gel method presents inorganic polymerization process. Sol−gel is a simple bottom-up approach so that the purity of the final composite material is less in this method. Therefore, post-treatment is required for the purification of the sample. The adoption of this technique for the fabrication of carbon or graphene/polymer nanocomposite and its hybrids has been adopted successfully by several researchers [26−32]. Jiang and co have reported the preparation of poly (vinyl alcohol)-based hybrid nanocomposites by the sol−gel method [27].

17.2.1.5   Template synthesis

In this method where the silicates are formed in situ in an aqueous solution comprising of the polymeric matrix and the silicate building entities which has been extensively adopted for the fabrication of double-layered hydroxide based composites/nanocomposites but undeveloped for layered-silicates. In this approach, with respect to self-assembly forces, the polymer enhances the nucleation and growth of the inorganic host crystals while it is locked in within the layers as it grows [33]. The adoption of this technique for the synthesis of imprinted polymers coated magnetic graphene oxide of dual-dummy-template molecularly has been reported [33]. Only a few reports employing this method for the preparation of graphene-polymer composites/nanocomposites have been exploited [33−36]. This process has been widely used for the synthesis of silica-based polymer nanocomposites [7,37].

17.3   Graphene-based polymer nanocomposites

Graphene is a one-atom-thick, 2D layer of sp^2 bonded carbon atoms that are organized into a planner honeycomb lattice with a carbon-carbon bond length of 0.142 nm. In the beginning a German Scientist Hanns-Peter Boehm and his co-workers during their experimental work discovered graphene in 1962 and introduced the term in *Graphene* 1986. The name *"Graphene"* comes from *"Graphite + Ene = Graphene."* It was first created by Andre Geim and Konstantin Novoselov in 2004 which were awarded the Nobel Prize in Physics in 2010 for "ground-breaking experiments regarding the two dimensional material graphene" [38]. Due to outstanding strength, graphene holds promising applications in polymer composites but considering its aggregation challenges, chemically reduced graphene is commonly used for the fabrication of polymer/graphene composites. Generally, graphene family nanomaterials (GFNs) include ultrathin graphite, few-layer graphene (FLG), graphene oxide (GO; from monolayer to few layers), reduced graphene oxide (rGO), graphene nanosheets (GNS) and graphene quantum dots, etc. to be utilized for biomedical materials application should be reliable, biocompatible and must be stable in physiological environment. To adapt the exclusive physiological environment (blood, tissue, body fluid, extracellular matrix, etc.) of the body for avoidance of any bad consequences (thrombus, inflammation, bacterial infection, etc.) caused by flocculation, blocking, and fouling of materials in complex life system, unique surface modification of graphene is of significant importance. With rational design and careful synthesis, materials exhibit high stability in physiological environment and biocompatibility. Therefore, among various graphene family materials, the oxidative derivatives (e.g., GO, rGO) are superior to graphene, graphene (graphene should not be confused with graphene, a two-dimensional form of carbon alone. Graphene is a form of hydrogenated graphene. The carbon bonds in graphene are in sp^3 alignment as contrasting to graphene's sp^2 bond conformation, making graphene is a 2-dimensional analog of cubic diamond), along with graphdiyne for biomedical use due to its outstanding stability and hydrophilicity in physiological environments; also it is facially modified surface characteristics. Graphene based polymeric composites/nanocomposites are made of a wide range of reinforcing nanofillers such as EG, CNT, CNF [38−40].

17.3.1 Properties of graphene-polymer nanocomposites

Graphene-polymer nanocomposites possess enhanced mechanical strength, optical transparency, thermal stability, barrier properties, improved flexibility, and electrical properties over its conventional polymer composite material. This attribute also applies to most nanomaterials polymer matrix filled/reinforced nanocomposites materials.

Graphene has excellent mechanical, thermal, electrical and optical properties, like the high carrier mobility, quantum hall effect at ambient atmospheric condition ($\sim 250,000$ cm^2/V/s) [38,39], exceptional thermal conductivity (3000−5000 W/m/K) [40,41], good optical transparency ($\sim 97.7\%$) [42], high specific surface area (~ 2600 m^2/g) [43], and superior mechanical properties with Young's modulus of ~ 1 TPa [44]. For these excellent properties nowadays graphene is one of the most utilized nanofiller for polymer matrices to modify their virgin material properties. It is at large prepared via the reduction of its precursor known as GO. Both graphene and GO sheets show very high mechanical properties with accepted biocompatibility and also have potential applications as biomaterials. Few out of several polymeric matrices utilized with graphene for the preparation of composites/nanocomposites are depicted in Fig. 17.6.

17.3.2 Polyaniline/graphene nanocomposites

Polyaniline (PANI) is out of many conducting polymers the most utilized conducting polymer owning to its exiting characteristics such as good thermal, environmental, and chemical stability, adjustable conductivity via switching between semiconducting and insulating material, simplistic synthesis, low operational voltage, the potentiality for real-world applications and low cost [45].

Since PANI is non-biodegradable, non-flexible and non-processable, only a very low amount of PANI has been proposed and used for the preparation of conductive scaffolds. Two methods including chemical oxidation and electrochemical synthesis are employed for preparing PANI. Even though prepared PANI using

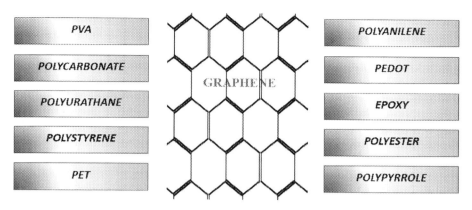

Figure 17.6 Some graphene/polymer nanocomposites.

electrochemical synthesis in general has higher conductivity, this approach is limited by the size, shape, and nature of the electrode adopted. Aimed at improving the thermal and electrical properties of PANI, conductive novel nanomaterials like CNT and graphene nanosheets are generally used.

In order to produce highly conductive polymer nanoparticles for conductive enabled scaffolds materials having good potential for neural cell proliferation, aniline has been polymerized using chemical oxidation technique through emulsion polymerization with the inclusion of chemically reduced GO nanosheets. The results revealed that spherical nanoparticles of polyaniline having sizes of 10−15 nm were successfully produced and incorporation of a small amount of graphene into a polymeric matrix dramatically improved the electrical properties of PANI. Hence, this material was proposed to be a proper candidate for different biomedical applications especially the fabrication of conductive scaffolds for nerve regeneration [46].

Yet in another report, PANI/graphene composite paper (GPCP) was prepared by two processes; first by vacuum infiltration and then in situ anodic electro-polymerization (AEP) of aniline on graphene paper [47]. Polymerization was carried out using a three-electrode anodic electro-polymerization cell. In this method, a Pt-plate, standard calomel electrode (SCE) and graphene-paper were used as the counter (reference) and the operational (working) electrode respectively. 0.05M aniline and 0.5M sulfuric acid were adopted as the electrolyte, then PANI was electro-polymerized in situ onto the graphene-paper at a constant potential of 0.75 V for altered periods [47].

For improving the energy storage performance of supercapacitors, hybrid electrode materials have been fully utilized for their advantages. Graphene-PANI nanocomposites have been proposed to be a future trend in the next decades due to their superior and unique properties [10].

Ma et al prepared a composite having a hierarchical structure, where PANi nanorods vertically oriented on sulfonated-graphene sheets thru an interfacial polymerization technique. The hierarchical nanocomposite exhibited specific capacitance of 497 F/g at 0.2 A/g and far better rate capability and cycling stability than those of pristine neat PANi nanorods [48].

For improving the energy storage performance of supercapacitors the hybrid electrode materials have been fully utilized for their advantages. Ma et al prepared a composite having a hierarchical structure, where PANi nanorods vertically aligned on sulfonated graphene sheets via an interfacial polymerization method. The hierarchical nanocomposite exhibited specific capacitance of 497 F/g at 0.2 A/g and far better rate capability and cycling stability than those of pristine neat PANi-nanorods [48].

PANI/GNS/CNT composite prepared by in situ polymerization has been reported [49]. All electrochemical measurements were performed with a three-electrode setup: Ni-foam covered with the nanocomposites as the working electrode, platinum-foil, and Hg/HgO electrode as the counter and reference electrodes. Also, PANI/GO/CNT films with an optimized PANI loading were constructed into symmetric, solid-state, flexible supercapacitors, and they were spun into helical fibers to create stretchable fiber-shaped supercapacitors [49].

PANI nanofibers, PANI/GO and PANI/GO/ZnO nanocomposites were successfully synthesized via nanoemulsion method for gas sensing applications. ZnO which is known to have high electrochemical stability, as well as good resistivity towards gas pollutants, was chosen because of its novel properties [50]. Apart from these, PANI is another molecule having excellent sensing properties. It has been broadly studied established upon its excellent reliability high sensitivity, and low cost. PANI nanofibers have been reported to be prepared with oil in water nanoemulsion. Other techniques may also be adopted for the synthesis of nanofibers such as centrifugal spinning, melt blowing, electrospinning, and force spinning, which are complex methods requiring careful solvent selection and been non-viable with regards to energy consumption [50].

17.3.3 Polypyrrole/graphene nanocomposite

The aspect of biomedical applications of conducting polypyrrole (PPy) has been considered for the preparation of controlled biosensor, drug delivery systems, artificial muscles, super batteries, corrosion protection, supercapacitor and recently applied as an antioxidant material. Among the conducting polymers, PPy is one of the most frequently studied conducting polymers due to its low cost, high stability, high conductivity, exciting redox property as well as minimum toxicity and biocompatibility which favors its use in biomedical applications [51]. PPy/graphene nanocomposite has become a subject of increasing importance due to their potential applications in optical, electrical, and sensing devices [52]. PPy can be useful in the treatment of optic nerve mainly atrophy, apoptosis or death of retinal ganglion cells (RGCs). In vitro growth of RGCs on the PPy-Graphene/PLGA nanofibrous scaffold under periodical electric stimulation has been studied. A RGC is a type of neuron located near the inner surface (the ganglion cell layer) of the retina of the eye [52], which receives visual information from photoreceptors though two intermediate neuron kinds: bipolar cells and retina amacrine cells. The optic nerve is composed of RGC axons that flock together at the optic disk and across the sclera. CPs (e.g. PPy) functionalized graphene (PPy-G) using a novel polymer polymerization enhanced ball milling (PPEBM) technique have been reported [53]. In a typical experiment, PPy was incorporated in a ball milling process to obtain well-defined PPy conjugated graphene hybrids [53]. Finally, a nanofibrous scaffold from PPy-G/PLGA was obtained and applied as the electrode for the stimulated growth of RGCs. PPy can enhance cell viability cell length and anti-aging ability after electrical stimulation (ES).

17.3.4   Chitosan/hydroxyapatite/graphene nanocomposite

Hydroxyapatite based polymer nanocomposites are fabricated and used to enhance their mechanical performance. A nanocomposite made of GO/hydroxyapatite/chitosan as potential appropriate platform for the growth of bones has been designed and fabricated [54]. Chitosan is a biopolymer made up of glucosamine and N-acetyl glucosamine linked with β 1−4 glucosidic linkage. It is synthesized by the partial deacetylation of chitin, another bioactive material that is derived from crustaceans [54]. Chitosan is considered to be one of the most used natural polymers for various tissue engineering applications. Chitosan is easily converted to the hydrogel format and molded into the desired shapes resulting in a porous scaffold-like structure upon lyophilization. For the preparation of the scaffolds graphene oxide/chitosan, chitosan/hydroxyapatite and graphene oxide/chitosan/hydroxyapatite; GO, CS, and HAp formulations have been blended at different concentrations to fabricate scaffolds in several compositions along with glutaraldehyde as a crosslinking agent for the synthesis of novel scaffolds [54].

17.3.5   Poly 3, 4-ethyldioxythiophene/graphene nanocomposite

PEDOT is a conducting polymer-based on 3,4-ethylenedioxythiophene or EDOT. According to Yoo et al., highly conductive poly(3,4-ethylenedioxythiophene): poly (4-styrenesulfonate) (PEDOT:PSS)/graphene nanocomposites fabricated using in situ polymerization for potential applications in thermoelectric devices as well as a platinum (Pt)-free dye-sensitized solar cell (DSSC) for energy harvesting systems [55].

Kim and co-workers prepared a thin film of PEDOT: PSS containing 1, 2, 3 wt.% of RGO sheets by a spin coating method and showed augmentation in the conductivity and thermoelectric properties. The highly conducting composite showed a transparency value of approx. 92% at 550 nm light wavelength [10]. Also, Xu et al. have made hybrid materials of graphene and PEDOT and analyzed the conductivity value of the order of 0.2 S/cm and transmittance value >80% [56]. Again, Lin et al. have identified the effects of graphene doping in PEDOT: PSS because of PEDOT: PSS/graphene nanocomposite which showed enhanced the performance of polymer light-emitting diodes (PLEDs). GO was first prepared and then intercalated with PEDOT: PSS solution and then thermally reduced to reduce the GO to graphene [57].

17.3.6   Polylactic acid-graphene composite

Polylactic acid or polylactide (PLA) obtained from the family of aliphatic polyesters, commonly made from alpha-hydroxy acids which is a polymer (lactic acid, cyclic di ester and lactide considered as a monomer) in which, the stereochemical structures can be modified by polymerization of the controlled mixture of L and D isomers to yield high molecular weight and amorphous or semi-crystalline polymerization has been greatly used for the fabrication of numerous nanomaterials with excellently enhanced conductivity. Market survey in 2010 showed that the PLA was in the second place at demand after starch-based plastic in the bio-based plastic category for global demand [58]. PLA is considered to be biodegradable, which is suitable for short-term packaging, and noncytotoxic/biocompatible when in contact with living tissues/cells, making it suitable for medical applications, such as tissue scaffolds, internal sutures, and implant devices, surgical implant materials drug delivery systems, guided tissue and bone regeneration platforms, such as porous scaffolds for the growth of neo-tissue [59] but also have some kind of disadvantages such as relatively poor mechanical properties, slow crystallization rate, and low thermal resistance, limiting its application. As an alternative method, adding nanofillers such as graphene based materials to improve the properties and to expand the use of PLA is highly attractive. PLA-graphene nanocomposites have been increasingly studied recently due to enhancement in their properties owing to their attractive features. Fabrication of PLA/graphene nanocomposites using different techniques like in situ intercalation polymerization, melt intercalation, solution intercalation has been widely utilized by scientist worldwide. Among all graphene allotropes, CNT have demonstrated a tendency to formulate bundles that can prove difficult to break down. This raises the need to develop new methods of obtaining good distributions of CNT including chemical modifications. Conversely, with reports from several kinds of research that graphene and GO are a potential replacement for CNT considering their excellent electronic, thermal, and mechanical properties for the synthesis of essentially novel materials in nanotechnology and various other engineering disciplines.

Kim and Jeong have reported the morphology, structure, thermal stability, mechanical, and electrical properties of PLA/exfoliated graphite nanocomposites compared to PLA/micron-sized natural graphite composites [60]. They confirmed from SEM images and XRD patterns that the exfoliated graphite with 15 nm thickness was homogeneously dispersed in the PLA matrix resulting in enhanced thermal stability and Young's moduli of the PLA/exfoliated nanocomposites [60]. Chieng et al. comparatively studied the effect of graphene nanoplatelets and rGO incorporation in virgin PLA and plasticized PLA. Their experimental results showed that the addition of rGO and graphene nanoplatelets in PLA and plasticized PLA improves its thermal property and mechanical/tensile strength without deteriorating its elasticity with the nanocomposites showing uniform and homogeneous morphology as observed by SEM and TEM [61] (Fig. 17.7).

17.3.7   Polystyrene/graphene composite

There are different methods utilized by many researchers aimed at the preparation of graphene oxide−polystyrene composites. One such method is foaming through

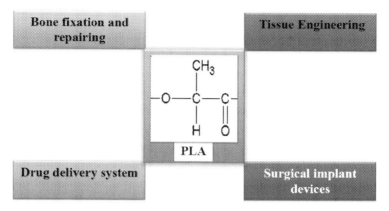

Figure 17.7 Polylactic acid biomedical application.

the blending of polystyrene followed by CO_2 supercritical drying [62]. Polystyrene and graphene oxide nanocomposite modified with a vinyl functional group, prepared by melt intercalation method and melt blending polymerization method for improved dispersion of modified GO throughout the PS matrix has been reported [63]. Some researcher identified the nanosheets of polystyrene nanocomposites were made by in situ emulsion polymerization and RGO have been used as a hydrazine hydrate [14]. Yu et al. show the first successful application of p-phenylenediamine4vinylbenzen-polystyrene and modified graphene oxide for application in corrosion protection.

17.3.8 Polyurethane/graphene composites

Among all polymers used in the biomedical field, polyurethanes (PUs) have been one of the most attractive materials. PUs has superior biocompatibility and mechanical properties and is inherently more thromboresistant than most other polymers. PU is most widely used in tissue engineering. But some drawbacks leading to limited applications of PU in tissue engineering are low conductivity thet allows cells or tissue cultured upon the biomaterials to be stimulated. This limitation can be improved by applications of electroconductive fillers like graphene, carbon-based materials, and graphene quantum dots. The large surface area of graphene provides additional advantages for integration of graphene with tissues, therefore, graphene (G), GO and RGO has been utilized as an inclusively to prepare PU based nanocomposite for diverse applications like augmentation in thermal, mechanical, and electrical properties, and also inducing shape memory effect into PU matrix [64].

PU/G composites are prepared with different methods that is melt processing, solution mixing, and in situ polymerization, etc. Liang and co-workers prepared three types of nanocomposites (such as TPU/GO, TPU/graphene, TPU/RGO) with the solution mixing process. They focused mainly on isocyanate modified graphene, sulfonated graphene sheet and reduced graphene as nanofiller and thermoplastic polyurethane (TPU) as the matrix polymer. The rate of thermal degradation of TPU/isocyanate modified G-nanocomposite was said to be considerable higher than that of TPU/sulfonated graphene and TPU/GO nanocomposites [64].

17.3.9 Polyvinyl alcohol-graphene composites

Owing to the non-toxicity, biocompatibility, hydrophilicity, and biodegradability of polyvinyl alcohol (PVA), PVA based composites have attracted great interest in theoretical and applied studies. From recent studies, researchers are focused on the reinforcement materials for PVA composites using various fillers such as CNT, GO, G, and corn fiber, etc. [65]. Nowadays, PVA/GO nanocomposite is widely applicable in the biomedical, food packaging industry. The preparation method of PVA/graphene nanocomposite in situ in general involves the reduction of GO/PVA solution in the presence of hydrazine at 100°C at about 24 h of continuous stirring. Usually, PVA/graphene nanocomposite is prepared by a direct solution intercalation method where water is used as a surfactant. Mechanical properties of the nanocomposite such as tensile strength and Young's modulus are significantly enhanced (up to 76% and 62% improvement of by addition of only 0.7 wt.% GO) by the large aspect ratio of the G-sheets, graphene sheets dispersion at molecular level in PVA matrix, and strong interfacial-adhesion owning to oxygen-containing groups attached to the surface of graphene resulting in strong interactions with the PVA matrix through H-bonding [65].The thermal stability of materials would be increased with the low quantity nanofiller such as GO and the result of the thermal property improvement for the sample is performed by thermogravimetric analysis TGA [66].

Some researcher has successfully prepared (designed by gel casting method) a Cs/PVA/GO/CuO patch that demonstrated high antibacterial activity against wound pathogens resulting to the promotion of wound healing rate [67].

17.3.10 Epoxy/graphene composites

Epoxy polymers have a superior property including good thermal stability outstanding mechanical and chemical properties good wear resistance and low cost due leading its wide application for adhesives, coatings, structural material or as a matrix in composites or nanocomposites. It has a high degree of cross-linked structure resulting in its high rigidity and strength but the cross-linked structure makes epoxy brittle and vulnerable to cracks which in many applications is considered as a limitation on its application. Hence the use of graphene as a filler which offers superb mechanical strength with excellent electrical conductivity and ability to highly improve the tensile strength and fracture toughness of polymers also resistive to corrosion so it is used commonly lubricants, additives to battery electrodes, functional textiles, as well as many other applications is comely and has been exploited for the preparation of epoxy/graphene composites [19,28]. The graphene-epoxy nanocomposite is based on the classical percolation theory to avoid the limitation of epoxy. Recently, Greece-based industrial paint producer hydroton has launched a new graphene-enhanced zinc epoxy paint—the 23303 ZINCTON GNC. This material presents enhanced cathodic protection as a result of graphene inclusion in comparison to epoxy-zinc-rich primers. Graphene also improved adhesion to the metal as zinc content is low. Due to low cost and easily available, it is recommended for applications in highly corrosive environments.

17.3.11 Polytetrafluoroethylene/graphene composite

Polytetrafluoroethylene (PTFE) is a synthetic fluoropolymer of tetrafluoroethylene that has numerous applications also most widely used for composing supercapacitor electrodes. It includes more fluorine atoms than PVDF and is dispersed in water, ethanol, and isopropanol. It is nonreactive as of the strength of carbon-fluorine bonds, and therefore it is widely used as a non-stick coating for pans and other cookware in containers and pipework for reactive and corrosive chemicals and supercapacitor industry [68,69]. PTFE matrix utilization is very well known whereas fabricating graphene nanostructured-reinforced polymeric electrodes due to low cost and water-soluble also, using an organic solvent such as the water does not lead to safety problems. This polymer matrix is habitually favored as a matrix material for preparing graphene based composite electrodes considering its two major benefits: high adhesion to active components like G-nanosheets and GO-nanosheets and low reactivity. The synthesis and proposed application of PTFE/graphene composites have in biomaterials have been reported by a few researchers [68,69].

17.3.12 Polyvinylidene fluoride/graphene composite

In recent years, there has been a growing interest in Polyvinylidene Fluoride (PVDF) polymer because it exhibits the strongest piezoelectric properties as compared to any other commercial polymer. It is a highly nonreactive thermoplastic fluoropolymer. It is mostly used in chemical, semiconductor, medical and defense industry and energy storage purposes. PVDF is often used as a binder for cathodes and anodes in lithium-ion batteries (Fig. 17.8).

17.3.13 The advantages of polymer nanocomposites

1. The enhancement in properties of the polymer matrix material can be reached by the addition of a small amount of nanofiller materials compared to conventional composites that require a high concentration of microparticles to improve properties.

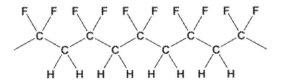

Figure 17.8 Molecular structure of polyvinylidene fluoride.

2. Due to the inclusion of a small percent of reinforcing nanofiller materials, polymeric matrix nanocomposites presents lighter weight when compared with orthodox composites.
3. PNC must have size-dependent properties that enhance thermal, chemical, mechanical, optical, magnetic and electrical properties to a much greater extent than conventional composites [70].

All the above-described nanostructured graphene based polymer composite have been used for a widely used in different biomedical applications such as controlled drug delivery, bioimaging, tissue engineering, and regenerative medicine.

17.4 Application of graphene based polymer composites in biomedical field

Within the last few decades, there has been a tremendous development of nanotechnologies and nanomaterials which has brought brilliant prospects in the biomedical field. Nanomaterials with good biocompatibility, high physiological stability, appropriate physical and chemical properties, and excellent performance can help to secure the medical field in the future [69]. Fig. 17.6 depicts a few applications of graphene based polymer composites in the biomedical field within the past few decades (Fig. 17.9).

17.4.1 Drug delivery

Drug delivery is a potential and tremendous approach for the formulation, technologies, and systems for successfully transporting a pharmaceutical compound within the body sometimes based on nanoparticles as needed to safely achieve its desired therapeutic effect. Chiefly a large surface area, chemical purity and the possibility for its easy functionalization allow graphene to provide opportunities for drug delivery. It can be of various types such as oral, injection-based transdermal or carrier-based, etc. With the aid of an atomistic simulation, Wang and Wang, 2019 proposed thet all-graphene-based novel nanostructures for advanced DDS can be developed. However, every drug delivery system should possess some novel properties like

- Biocompatibility, reliability, easy to administer and it must be inexpensive
- Inert, mechanically strong and easy to design and delivered as per the patient requirement
- Must be non-hazardous and intact within the specific disease tissue (should not affect the healthy tissue)
- Should maintain all guideline of drug delivery system (DDS)

More than a few techniques have been used for synthesizing nanomaterials targeted at drug-delivery systems such as spray-drying, solvent evaporation, physicochemical solvent/non-solvent method, electrospinning, in situ polymerization, etc. The application of graphene in drug delivery have been reported by several researchers in applications such as the delivery of 5-fluorouracil for colorectal liver metastasis [71], hydrophilic (doxorubicin, DOX) & hydrophobic (Methotrexate MTX) [72],

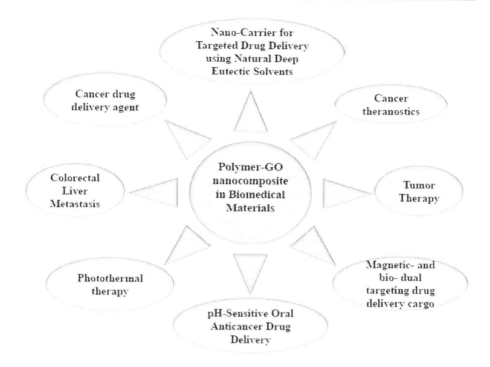

Figure 17.9 Polymer graphene nanocomposite in the medical field.

doxorubicin hydrochloride (DOX) [73,74], Acyclovir [75], ibuprofen (IBU) [76], etc. Luo et al., have shown that a hybrid system of graphene/nanocellulose presented excellent result for controlled drug delivery application. The novel properties of the proposed hybrid system may be due to the excellent sustained release of drug entities such as proven from nanocellulose based systems [77–79].

Recently, PEGylated graphene nanocomposite has shown itself to as a promising material in drug delivery. It has been shown to provide high aqueous solubility and stability in physiological solutions including cell medium and serum. Many other polymers, such as polyacrylic acid, poly(vinyl alcohol), polyethyleniminepoly(N-isopropylacrylamide) have been used for graphene surface modifications in drug delivery applications [80]. The surface modification of polymers could change the surface characteristics of materials and avoid aggregation. Among many such polymers, like polyethylene glycol (PEG) is a widely used polymer for surface modification, which can avoid the nonspecific protein adsorption and aggregation of GO in a physiological environment. Therefore, PEGylated GO exhibits improved physiological stability and enhanced biosafety [80]. The water solubility and biocompatibility of GO-PPEGMA (polyethylene glycol methyl ether methacrylate)is reported to enhance its properties, making it acceptable for biomedical application where the authors demonstrated that GO possess a high specific surface area, which is promising for carrying several anticancer drugs such as paclitaxel, cisplatin, and DOX

(Doxorubicin), etc. [64]. They observed that DOX can easily adsorb onto GO sheets via hydrophobic interaction, pi-pi stacking, hydrogen bonding and their combined action [64].

17.4.2   Tissue engineering

Tissue Engineering (TE) is an interdisciplinary field that makes a bridge between engineering and biology towards the development of tissue and organ substitutes by controlling biological, biophysical and/or biomechanical parameters in the laboratory. A commonly applied definition of tissue engineering, as stated by Langer and J. Vacanti is "An interdisciplinary field that applies the principles of engineering and life sciences toward the development of biological substitutes that restore, maintain and improve [biological tissue] function or whole organ."

Tissue engineering can be adopted as an efficient approach to repair or replace organs/glands/tissues such as skin, muscles, bone as well as the improvement or replacement of organs like heart, kidney, liver, and knee via a novel principle as depicted in Fig. 17.10.

Biodegradable polymers have been used in biomedical applications generally, and in tissue engineering due to their good physical and biological properties. We developed POSS-PCL as a nanocomposite polymer for various tissue engineering applications POSS-PCL to function as an effective construct for neural tissue

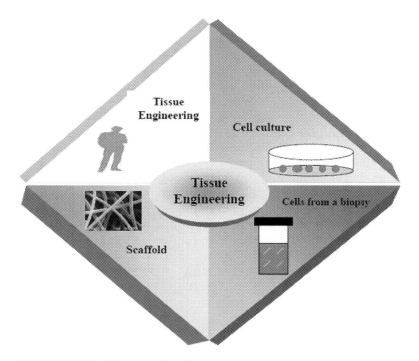

Figure 17.10 Principle of tissue engineering.

engineering: it has to possess electrical conductivity. Graphene as a filler in POSS-PCL would confer both added mechanical stability and also converts its polymer composite form into an electrically conductive material. Numerous neural tissue engineered materials are present in the market. Nevertheless, they are largely not biodegradable and lack electrically conductivity which is vital for impedance of full recovery onto proper nerve function restoration. Therefore, preparing an electrically conductive polymer matrix having robust mechanical property and yet biodegradable can form the basis of a platform technique which may be adopted in both PNS and CNS applications.

GO/hydroxyapatite/chitosan has demonstrated support for the growth of bones by Thompson et al. who used nano-GO for initiation of implantable biomaterials presenting noteworthy revolution in medical treatment, permitting the improvement in the fields of TE and bionic medical devices (e.g., cochlea implants to restore hearing, vagus nerve stimulators to control Parkinson's disease, and cardiac pacemakers) [81].

PLA is widely used as scaffolds due to its good biocompatibility however scaffold plays a key role in cell adhesion and growth and cell proliferation which can help bone tissue repairing such as bone regeneration blood vessels, and neural system [58−60,63,81]. Nowadays, PLA and PLGA are the two promising polymeric materials widely studied and used in the preparation of porous scaffolds to repair damaged bones [58−60,63,81] as an ideal and suitable bone substitute materials which possess outstanding biocompatibility and osteoinductive and osteoconductive properties. Pins, plates, and screws are those frequently used forms of polymer devices in orthopedic, neural, oral, and craniofacial surgeries owing to the excellent degradation of PLA thereby eliminating the need for the second operation and also prevents implant removal so that the pain of patients can be reduced. PLA/Graphene or PLLA/HAp (Hydroxyapatite) nanocomposite is better for bone fixation than pure PLA polymer [58−60,63,81].

17.4.3 Antibacterial study

Antibacterial nanomaterials possess the capability to inhibit or destroy the growth of the bacteria; hence, they are effectively used in biomedical devices, biomechanics, and tissue engineering applications. Conventionally, silver nanoparticles are being used as antibacterial materials in biomedical devices; however, they have certain disadvantages such as high cost, scalability, toxicity, and problems in disposal of the wastes to the environment [82]. Therefore recently graphene and its derivatives have most commonly used to make the antibacterial materials because of their attractive properties and excellent antibacterial activities against bacteria [83]. Recently, GOs have been authenticated as potential candidate for killing Gram-negative (*Escherichia coli*) and Gram-positive (*Staphylococcus aureus*) bacteria. Owning to their excellent antibacterial nature, the reinforcement of the GOs in a polymer matrix aimed at enhanced antibacterial nature also advances the thermal, mechanical, and chemical stabilities [84]. Besides, the hydrophilic functional groups such as OH, COOH, and O on GO allow homogeneous dispersion in polymer and proper alignment of GO in the polymer matrix. Such outstanding properties of GO makes it the ideal candidate for developing nanocomposites with different types of polymers having enhanced antimicrobial activity [84]. Fig. 17.11 shows the utilization of graphene in organic/inorganic composite for antibacterial activity and other applications. Liu et al. have reported that the tensile strength and antibacterial properties of the polylactic acid were improved by Ag@GO via the electrospinning technique [81].

Usman et al. reported that the antimicrobial property of polymer (polyvinyl alcohol (PVA)), polymeric carbohydrate (starch), and nanomaterial (GO, silver), that is, PVA/starch/GO/Ag nanocomposites improved greatly with the addition of GO [85].

Some researchers have analyzed Dual-polymer PEG and polyhexamethylene guanidine hydrochloride (PHGC)-functionalized GO (GO_PEG_PHGC) developed by a facile synthetic method [86]. They showed in their studies that the antibacterial activity of the composite tested against Gram-positive (*S. aureus*) and Gram-negative (*E. coli*) bacterial strains showed enhanced antimicrobial activity where GO_PEG_PHGC showed enhanced antimicrobial activity compared to GO, GO_PEG, and GO_PHGC alone. Also, Zeng et al. reported GO quantum Qdots covalently-bonded with amino-modified polyvinylidene fluoride (PVDF) for improved antifouling and bactericidal activity [87].

17.4.4 Biosensor

A biosensor is defined as a "Self-contained integrated device that is capable of providing specific qualitative and semi-quantitative analytical information using a

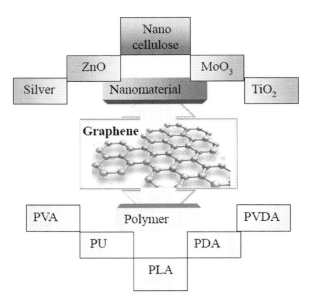

Figure 17.11 Schematic of graphene based polymer nanocomposite.

biological recognition element which is in direct spatial contact with a transduction element." **(IUPAC, 1998)**. The term "biosensor" was introduced by Clark and Lyos in 1962. In the modern age, the definition of a biosensor is a device that integrates the bioactive element with a physiochemical transducer to produce an electronic signal proportional to a single analyte which is then conveyed to a detector. A biosensor can be noninvasively or minimally invasive and we can produce reliable results within a fraction of second, so it is widely used in the healthcare industry, food and beverage industry, agricultural field, forensic study, medical science, and security purpose. Biosensors with low detection limits and high sensitivity are very vital in providing essential information for the timely diagnosis of diseases for timely suppression of the disease progression via anticipatory healthcare practice. 3D graphene composite foam consisting of CuO nanoflowers for the potential sensing of ascorbic acid has been reported [88]. Yet in another study, graphene based composite foam and cobalt oxide nanowires were proposed as effective glucose sensor substrate [89]. Also, the composite foam comprising of graphene and ZnO nanowire arrays for the sensing of uric acid, dopamine, and ascorbic acid has been proposed [90]. The biochemical considerations of graphene suggestively consider several diseases like schizophrenia, gout, and hyperuricemia, etc. [91−93].

17.4.5 Flexible supercapacitors

Nowadays electronic products are expected to be light, ultrathin, and more flexible and must be portable and may be wearable electronics finding itself in a large application area such as artificial smart skin, implantable medical devices, bendable displays, and smart card, mobile phones, military garment devices, heart-rate sensors and monitors, pedometer devices, and various military, medical, and fitness applications. Because all these applications and devices require esthetic appeal [35,44,48−50]. Graphene based polymer nanocomposite has proven effective in this niche.

17.4.6 Graphene based nanocomposite body implants

The implantation of orthopedic devices is associated with a high risk of postoperative complications that increases high chances with each revision surgery [94]. A group of researchers have shown that the cytotoxicity of Mg-GNPs nanocomposites studied using osteoblast-like $SaOS_2$ cells presents graphene as an excellent candidates for reinforcements in Mg matrices for the manufacture of biodegradable Mg-based composite implants having improved mechanical properties via synergetic strengthening modes while retaining the structural integrity of GNPs during the processing leading to improved ductility, compressive strength, and corrosion resistance of the nanocomposites [94]. The authors also reported that the cytotoxicity analyses did not show any noteworthy toxicity with the inclusion of GNPs to Mg matrices. Graphene based polymeric composite materials reportedly utilized for tissue scaffold application or medical implants must possess some basic properties as shown in Fig. 17.12.

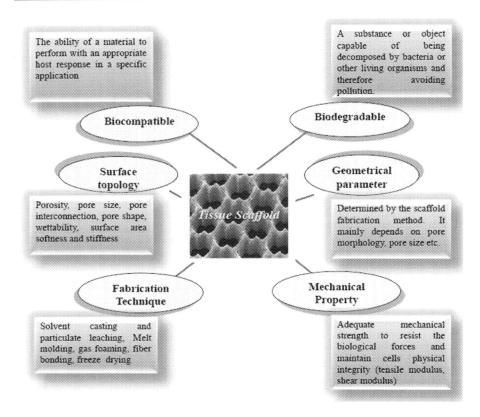

Figure 17.12 Scheme showing biomaterial property dependence of graphene scaffold materials on some factors.

17.5 Present hindrance to effective implementation of graphene based polymer composites

The incorporation of metal oxides in polymer-graphene based nanocomposites/ hybrids can be a challenging task at a large scale processing suitable for industries. Conversely, the progress in graphene based nanocomposite design and fabrication is on the progressive but prodigious defies such as poor dispersibility, obstinate contagions, extreme clump/aggregation, high synthesis/processing cost, poor regulation of surface groups chemistries, and truncated interfacial interactions between the polymeric matrix and graphene remain to be solved to a large extent. The lack of reliably facile prediction approaches for effective design and development of graphene polymer composites/hybrids also limits the engineering of these systems for effectual possible application for biomedical devices in real-life. The building of an understanding of the behavioral characteristics graphene-polymer based systems with regard to their interfacial interaction which is largely dependent on the graphene surface chemistry, the relative arrangement of constituents and its relationship in its nanocomposite properties should be a current research frontier in nanocomposites that yokes the curiosity of research scientists in materials science, physical chemistry, and engineering.

17.6 Conclusion and future perspective

We have presented the current progress on the applications of graphene-polymer composites/nanocomposites/hybrids in the biomedical field. Biomedical application of polymer-graphene composites/nanocomposites/hybrids epitomize one of the most technologically auspicious developments in polymer materials science and engineering. Despite the significant advances graphene-based polymeric nanocomposites, generous crucial research is important to give an essential comprehension of these materials to empower the full application of their nanoengineering potential. Considering the graphene chemistry, size, shape and properties, most graphene-polymer nanocomposites share conjoint features with regard to preparation methodologies, processing, morphology characterization, and fundamental physiochemical properties. This chapter on the biomedical application of polymer-graphene composites emphasized their possible applications in the next decade for biomedical applications, such as ultra-small low cost biosensors for the examination of blood and other body fluids. Graphene polymer composites can also be used as electrode materials in a range of biomedical devices. In the coming decade, it is predictable that a large number of new graphene/modified graphene polymer nanocomposites along with their hybrids using different polymer matrices (thermoplastic, and thermosetting) will be explored and implemented. There is still much room at the bottom and at the top for the utilization of graphene and its materials for application in molecular design of polymeric surfactants, solubility, dispersion in polymers, film-forming ability, electrochemical behavior, semiconductor properties, fluorescence properties of graphene quantum materials, etc. along with the understanding of the enhancement effect/mechanisms of graphene materials in polymeric nanocomposites.

Declaration

All authors have contributed equally to the writing of this chapter.

Acknowledgment

The authors wish to appreciate the Center for Research in Nanoscience and Nanotechnology, and the Department of Polymer Science and Technology University of Calcutta for their financial and technical support.

References

[1] S. Fu, Z. Sun, P. Huang, Y. Li, N. Hu, Some basic aspects of polymer nanocomposites: a critical review, Nano Mater. Sci. 1 (1) (2019) 2−30.

[2] J.T. Orasugh, S. Dutta, D. Das, J. Nath, C. Pal, D. Chattopadhyay, Utilization of cellulose nanocrystals (CNC) biopolymer nanocomposites in ophthalmic drug delivery system (ODDS), J. Nanotechnol. Res. 01 (02) (2019).

[3] A. Naz, A. Kausar, M. Siddiq, M.A. Choudhary, Comparative review on structure, properties, fabrication techniques, and relevance of polymer nanocomposites reinforced with carbon nanotube and graphite fillers, Polym. Technol. Eng. 55 (2) (2015) 171−198.

[4] L.A. Kolahalam, I.V. KasiViswanath, B.S. Diwakar, B. Govindh, V. Reddy, Y.L.N. Murthy, Review on nanomaterials: synthesis and applications, Mater. Today: Proc. 18 (2019) 2182−2190.

[5] T. Saleh, Editorial: advanced nanomaterial synthesis and their applications for engineering research, Curr. Nanomater. 2 (2) (2018) 75.

[6] F. Yang, Y. Ou, Z. Yu, Polyamide 6/silica nanocomposites prepared by in situ polymerization, J. Appl. Polym. Sci. 69 (2) (1998) 355−361.

[7] M. Alexandre, P. Dubois, Polymer-layered silicate nanocomposites: preparation, properties and uses of a new class of materials, Mater. Sci. Eng.: R: Rep. 28 (1−2) (2000) 1−63.

[8] T.K. Das, S. Prusty, Graphene-based polymer composites and their applications, Polym. Technol. Eng. 52 (4) (2013) 319−331.

[9] J. Jordan, K.I. Jacob, R. Tannenbaum, M.A. Sharaf, I. Jasiuk, Experimental trends in polymer nanocomposites—a review, Mater. Sci. Eng.: A. 393 (1−2) (2005) 1−11.

[10] D. Wang, F. Li, J. Zhao, W. Ren, Z. Chen, J. Tan, et al., Fabrication of graphene/polyaniline composite paper via in situ anodic electropolymerization for high-performance flexible electrode, ACS Nano 3 (7) (2009) 1745−1752.

[11] G.H. Kim, D.H. Hwang, S.I. Woo, Thermoelectric properties of nanocomposite thin films prepared with poly(3,4-ethylenedioxythiophene) poly(styrenesulfonate) and graphene, Phys. Chem. Chem. Phys. 14 (10) (2012) 3530.

[12] L. Yin, J. Wang, F. Lin, J. Yang, Y. Nuli, Polyacrylonitrile/graphene composite as a precursor to a sulfur-based cathode material for high-rate rechargeable Li−S batteries, Energy Environ. Sci. 5 (5) (2012) 6966.

[13] M.A. Milani, D. González, R. Quijada, N.R. Basso, M.L. Cerrada, D.S. Azambuja, et al., Polypropylene/graphene nanosheet nanocomposites by in situ polymerization: synthesis, characterization and fundamental properties, Compos. Sci. Technol. 84 (2013) 1−7.

[14] H. Hu, X. Wang, J. Wang, L. Wan, F. Liu, H. Zheng, R. Chen, C. Xu, Preparation and properties of graphene nanosheets−polystyrene nanocomposites via in situ emulsion polymerization, Chem. Phys. Lett. 484 (4−6) (2010) 247−253.

[15] Z. Xu, C. Gao, In situ polymerization approach to graphene-reinforced nylon-6 composites, Macromolecules. 43 (16) (2010) 6716−6723.

[16] W.D. Lee, S.S. Im, Thermomechanical properties and crystallization behavior of layered double hydroxide/poly(ethylene terephthalate) nanocomposites prepared by in-situ polymerization, J. Polym. Sci. Part. B: Polym. Phys. 45 (1) (2006) 28−40.

[17] J.R. Potts, S.H. Lee, T.M. Alam, J. An, M.D. Stoller, R.D. Piner, et al., Thermomechanical properties of chemically modified graphene/poly(methyl methacrylate) composites made by in situ polymerization, Carbon. 49 (8) (2011) 2615−2623.

[18] S.N. Tripathi, P. Saini, D. Gupta, V. Choudhary, Electrical and mechanical properties of PMMA/reduced graphene oxide nanocomposites prepared via in situ polymerization, J. Mater. Sci. 48 (18) (2013) 6223–6232.

[19] Y. Guo, C. Bao, L. Song, B. Yuan, Y. Hu, In situ polymerization of graphene, graphite oxide, and functionalized graphite oxide into epoxy resin and comparison study of on-the-flame behavior, Ind. Eng. Chem. Res. 50 (13) (2011) 7772–7783.

[20] M. Yang, Z. Zhang, X. Zhu, X. Men, G. Ren, In situ reduction and functionalization of graphene oxide to improve the tribological behavior of a phenol formaldehyde composite coating, Friction 3 (1) (2015) 72–81.

[21] G. Chen, C. Wu, W. Weng, D. Wu, W. Yan, Preparation of polystyrene/graphite nanosheet composite, Polymer 44 (6) (2003) 1781–1784.

[22] K. Kalaitzidou, H. Fukushima, L.T. Drzal, A new compounding method for exfoliated graphite—polypropylene nanocomposites with enhanced flexural properties and lower percolation threshold, Compos. Sci. Technol. 67 (10) (2007) 2045–2051.

[23] H. Zhang, W. Zheng, Q. Yan, Y. Yang, J. Wang, Z. Lu, et al., Electrically conductive polyethylene terephthalate/graphene nanocomposites prepared by melt compounding, Polymer 51 (5) (2010) 1191–1196.

[24] S. Kim, I. Do, L.T. Drzal, Thermal stability and dynamic mechanical behavior of exfoliated graphite nanoplatelets-LLDPE nanocomposites, Polym. Compos. 31 (5) (2009) 755–761.

[25] W. Weng, G. Chen, D. Wu, Transport properties of electrically conducting nylon 6/foliated graphite nanocomposites, Polymer 46 (16) (2005) 6250–6257.

[26] G. An, H. Liu, H. Li, Z. Chen, J. Li, Y. Li, SiBCN ceramic aerogel/graphene composites prepared via sol-gel infiltration process and polymer-derived ceramics (PDCs) route, Ceram. Int. (2019).

[27] S. Jiang, Z. Bai, G. Tang, Y. Hu, L. Song, Fabrication and characterization of graphene oxide-reinforced poly(vinyl alcohol)-based hybrid composites by the sol-gel method, Compos. Sci. Technol. 102 (2014) 51–58.

[28] C. Kuan, W. Chen, Y. Li, C. Chen, H. Kuan, C. Chiang, Flame retardance and thermal stability of carbon nanotube epoxy composite prepared from sol-gel method, J. Phys. Chem. Solids 71 (4) (2010) 539–543.

[29] S. Karamikamkar, E. Behzadfar, H.E. Naguib, C.B. Park, Insights into in-situ sol-gel conversion in graphene modified polymer-based silica gels for multifunctional aerogels, Chem. Eng. J. 392 (2019) 123813.

[30] X. Wang, W. Xing, L. Song, H. Yang, Y. Hu, G.H. Yeoh, Fabrication and characterization of graphene-reinforced waterborne polyurethane nanocomposite coatings by the sol—gel method, Surf. Coat. Technol. 206 (23) (2012) 4778–4784.

[31] L. Fang, Q. He, M. Zhou, J. Zhao, J. Hu, Electrochemically assisted deposition of sol—gel films on graphene nanosheets, Electrochem. Commun. 109 (2019) 106609.

[32] J.D. Maeztu, P.J. Rivero, C. Berlanga, D.M. Bastidas, J.F. Palacio, R. Rodriguez, Effect of graphene oxide and fluorinated polymeric chains incorporated in a multilayered Sol-gel nanocoating for the design of corrosion resistant and hydrophobic surfaces, Appl. Surf. Sci. 419 (2017) 138–149.

[33] G. An, H. Liu, H. Li, Z. Chen, J. Li, Y. Li, SiBCN ceramic aerogel/graphene composites prepared via Sol-gel infiltration process and polymer-derived ceramics (PDCs) route, Ceram. Int. 46 (6) (2020) 7001–7008.

[34] S. Wang, T. Gao, Y. Li, S. Li, G. Zhou, Fabrication of vesicular polyaniline using hard templates and composites with graphene for supercapacitor, J. Solid. State Electrochem. 21 (3) (2016) 705–714.

[35] A. El-Basaty, E. Moustafa, A. Fouda, A. El-Moneim, 3D hierarchical graphene/CNT with interfacial polymerized polyaniline nano-fibers, Spectrochim. Acta Part. A: Mol. Biomol. Spectrosc. 226 (2020) 117629.

[36] M. Masteri-Farahani, S. Mashhadi-Ramezani, N. Mosleh, Molecularly imprinted polymer containing fluorescent graphene quantum dots as a new fluorescent nanosensor for detection of methamphetamine, Spectrochim. Acta Part. A: Mol. Biomol. Spectrosc. 229 (2020) 118021.

[37] Q.T. Nguyen, D.G. Baird, Preparation of polymer—clay nanocomposites and their properties, Adv. Polym. Technol. 25 (4) (2006) 270—285.

[38] K.S. Novoselov, A.K. Geim, S.V. Morozov, D. Jiang, M.I. Katsnelson, I.V. Grigorieva, et al., Two-dimensional gas of massless Dirac fermions in graphene, Nature 438 (7065) (2005) 197—200.

[39] K.S. Novoselov, Electric field effect in atomically thin carbon films, Science 306 (5696) (2004) 666—669.

[40] H. Boehm, R. Setton, E. Stumpp, Nomenclature and terminology of graphite intercalation compounds, Carbon 24 (2) (1986) 241—245.

[41] A.A. Balandin, S. Ghosh, W. Bao, I. Calizo, D. Teweldebrhan, F. Miao, et al., Superior thermal conductivity of single-layer graphene, Nano Lett. 8 (3) (2008) 902—907.

[42] R.R. Nair, P. Blake, A.N. Grigorenko, K.S. Novoselov, T.J. Booth, T. Stauber, et al., Fine structure constant defines visual transparency of graphene, Science 320 (5881) (2008) 1308.

[43] W. Hooch, Y. Antink, K. Choi, et al., Hall of fame article: recent progress in porous graphene and reduced graphene oxide-based nanomaterials for electrochemical energy storage devices, Adv. Mater. Interfaces 5 (5) (2018) 1870023.

[44] M.D. Stoller, S. Park, Y. Zhu, J. An, R.S. Ruoff, Graphene-based ultracapacitors, Nano Lett. 8 (10) (2008) 3498—3502.

[45] J. Jiang, J. Wang, B. Li, Young's modulus of graphene: a molecular dynamics study, Phys. Rev. B 80 (11) (2009).

[46] H. Baniasadi, S.A.A. Ramazani, S. Mashayekhan, F. Ghaderinezhad, Preparation of conductive polyaniline/graphene nanocomposites via in situ emulsion polymerization and product characterization, Synth. Met. 196 (2014) 199—205.

[47] L. Al-Mashat, K. Shin, K. Kalantar-zadeh, J.D. Plessis, S.H. Han, R.W. Kojima, et al., Graphene/polyaniline nanocomposite for hydrogen sensing, J. Phys. Chem. C. 114 (39) (2010) 16168—16173.

[48] M. Moussa, M.F. El-Kady, Z. Zhao, P. Majewski, J. Ma, Recent progress and performance evaluation for polyaniline/graphene nanocomposites as supercapacitor electrodes, Nanotechnology 27 (44) (2016) 442001.

[49] B. Ma, X. Zhou, H. Bao, X. Li, G. Wang, Hierarchical composites of sulfonated graphene-supported vertically aligned polyaniline nanorods for high-performance supercapacitors, J. Power Sources 215 (2012) 36—42.

[50] Q. Jiang, Y. Shang, Y. Sun, Y. Yang, S. Hou, Y. Zhang, et al., Flexible and multi-form solid-state supercapacitors based on polyaniline/graphene oxide/CNT composite films and fibers, Diam. Relat. Mater. 92 (2019) 198—207.

[51] G. Gaikwad, P. Patil, D. Patil, J. Naik, Synthesis and evaluation of gas sensing properties of PANI based graphene oxide nanocomposites, Mater. Sci. Eng.: B 218 (2017) 14—22.

[52] C.F. Hsu, L. Zhang, H. Peng, J. Travas-Sejdic, P.A. Kilmartin, Free radical scavenging properties of polypyrrole and poly(3,4-ethylenedioxythiophene), Curr. Appl. Phys. 8 (3—4) (2008) 316—319.

[53] R. Jain, N. Jadon, A. Pawaiya, Polypyrrole based next generation electrochemical sensors and biosensors: a review, TrAC Trends Anal. Chem. 97 (2017) 363–373.

[54] L. Yan, B. Zhao, X. Liu, X. Li, C. Zeng, H. Shi, et al., Aligned nanofibers from polypyrrole/graphene as electrodes for regeneration of optic nerve via electrical stimulation, ACS Appl. Mater. Interfaces 8 (11) (2016) 6834–6840.

[55] P. Yılmaz, E. ÖztürkEr, S. Bakırdere, K. Ülgen, B. Özbek, Application of supercritical gel drying method on fabrication of mechanically improved and biologically safe three-component scaffold composed of graphene oxide/chitosan/hydroxyapatite and characterization studies, J. Mater. Res. Technol. 8 (6) (2019) 5201–5216.

[56] D. Yoo, J. Kim, J.H. Kim, Direct synthesis of highly conductive poly(3,4-ethylenedioxythiophene):poly(4-styrenesulfonate) (PEDOT:PSS)/graphene composites and their applications in energy harvesting systems, Nano Res. 7 (5) (2014) 717–730.

[57] Y. Xu, Y. Wang, J. Liang, Y. Huang, Y. Ma, X. Wan, et al., A hybrid material of graphene and poly (3,4-ethyldioxythiophene) with high conductivity, flexibility, and transparency, Nano Res. 2 (4) (2009) 343–348.

[58] C. Lin, K. Chen, J. Ho, J.J. Cheng, R.C. Tsiang, PEDOT: PSS/graphene nanocomposite hole-injection layer in polymer light-emitting diodes, J. Nanotechnol. (2012) 1–7.

[59] Protecting plastics and rubber. Ceresana analyzes the stabilizer market. Pigment Resin Technol. 40 (5) (2011).

[60] I. Bayer, Thermomechanical properties of polylactic acid-graphene composites: a state-of-the-art review for biomedical applications, Materials 10 (7) (2017) 748.

[61] I. Kim, Y.G. Jeong, Polylactide/exfoliated graphite nanocomposites with enhanced thermal stability, mechanical modulus, and electrical conductivity, J. Polym. Sci. Part. B: Polym. Phys. 48 (8) (2010) 850–858.

[62] J. Yang, M. Wu, F. Chen, Z. Fei, M. Zhong, Preparation, characterization, and super-critical carbon dioxide foaming of polystyrene/graphene oxide composites, J. Supercrit. Fluids 56 (2) (2011) 201–207.

[63] B. Chieng, N. Ibrahim, W. Yunus, M. Hussein, Y. Then, Y. Loo, Effects of graphene nanoplatelets and reduced graphene oxide on poly(lactic acid) and plasticized poly(lactic acid): a comparative study, Polymers 6 (8) (2014) 2232–2246.

[64] P. Gao, M. Liu, J. Tian, F. Deng, K. Wang, D. Xu, et al., Improving the drug delivery characteristics of graphene oxide based polymer nanocomposites through the "one-pot" synthetic approach of single-electron-transfer living radical polymerization, Appl. Surf. Sci. 378 (2016) 22–29.

[65] J. Liang, Y. Xu, Y. Huang, L. Zhang, Y. Wang, Y. Ma, et al., Infrared-triggered actuators from graphene-based nanocomposites, J. Phys. Chem. C. 113 (22) (2009) 9921–9927.

[66] C. Li, Y. Li, X. She, J. Vongsvivut, J. Li, F. She, et al., Reinforcement and deformation behaviors of polyvinyl alcohol/graphene/montmorillonite clay composites, Compos. Sci. Technol. 118 (2015) 1–8.

[67] J. Liang, Y. Huang, L. Zhang, Y. Wang, Y. Ma, T. Guo, et al., Molecular-level dispersion of graphene into poly(vinyl alcohol) and effective reinforcement of their nanocomposites, Adv. Funct. Mater. 19 (14) (2009) 2297–2302.

[68] X. Liu, Q. Zhao, S. Veldhuis, I. Zhitomirsky, Cholic acid is a versatile coating-forming dispersant for electrophoretic deposition of diamond, graphene, carbon dots and polytetrafluoroethylene, Surf. Coat. Technol. 384 (2020) 125304.

[69] M.T. Masood, E.L. Papadopoulou, J.A. Heredia-Guerrero, I.S. Bayer, A. Athanassiou, L. Ceseracciu, Graphene and polytetrafluoroethylene synergistically improve the tribological properties and adhesion of nylon 66 coatings, Carbon 123 (2017) 26–33.

[70] W. Khan, R. Sharma, P. Saini, Carbon nanotube-based polymer composites: synthesis, properties and applications, Carbon Nanotubes—Curr. Prog. Polym. Compos. (2016).

[71] B. Zhang, Y. Yan, Q. Shen, D. Ma, L. Huang, X. Cai, et al., A colon targeted drug delivery system based on alginate modificated graphene oxide for colorectal liver metastasis, Mater. Sci. Eng.: C. 79 (2017) 185−190.

[72] M. Pooresmaeil, H. Namazi, Surface modification of graphene oxide with stimuli-responsive polymer brush containing β-cyclodextrin as a pendant group: preparation, characterization, and evaluation as controlled drug delivery agent, Colloids Surf. B: Biointerfaces 172 (2018) 17−25.

[73] M. Xie, F. Zhang, H. Peng, Y. Zhang, Y. Li, Y. Xu, et al., Layer-by-layer modification of magnetic graphene oxide by chitosan and sodium alginate with enhanced dispersibility for targeted drug delivery and photothermal therapy, Colloids Surf. B: Biointerfaces 176 (2019) 462−470.

[74] M. Wei, T. Lu, Z. Nong, G. Li, X. Pan, Y. Wei, et al., Reductive response and RGD targeting nano-graphene oxide drug delivery system, J. Drug. Deliv. Sci. Technol. 53 (2019) 101202.

[75] A. HortêncioMunhoz Jr, M. Romero Filho, H. De Arruda Kleist, G. Dias Moreno, M. Oliva de Oliveira, R. Meneghetti Peres, et al., Carmelino Cardoso Sarmento, Synthesis of pseudoboehmite-graphene oxide for drug delivery system, Mater. Today: Proc. 14 (2019) 700−707.

[76] H. Luo, H. Ao, G. Li, W. Li, G. Xiong, Y. Zhu, et al., Bacterial cellulose/graphene oxide nanocomposite as a novel drug delivery system, Curr. Appl. Phys. 17 (2) (2017) 249−254.

[77] J.T. Orasugh, N.R. Saha, D. Rana, G. Sarkar, M.M. Mollick, A. Chattopadhyay, et al., Jute cellulose nano-fibrils/hydroxypropylmethylcellulose nanocomposite: a novel material with potential for application in packaging and transdermal drug delivery system, Ind. Crop. Prod. 112 (2018) 633−643.

[78] J.T. Orasugh, G. Sarkar, N.R. Saha, B. Das, A. Bhattacharyya, S. Das, et al., Effect of cellulose nanocrystals on the performance of drug loaded in situ gelling thermo-responsive ophthalmic formulations, Int. J. Biol. Macromol. 124 (2019) 235−245.

[79] J.T. Orasugh, S. Dutta, D. Das, C. Pal, A. Zaman, S. Das, et al., Sustained release of ketorolac tromethamine from poloxamer 407/cellulose nanofibrils graft nanocollagen based ophthalmic formulations, Int. J. Biol. Macromol. 140 (2019) 441−453.

[80] Z. Liu, J.T. Robinson, X. Sun, H. Dai, PEGylated nanographene oxide for delivery of water-insoluble cancer drugs, J. Am. Chem. Soc. 130 (33) (2008) 10876−10877.

[81] B.C. Thompson, E. Murray, G.G. Wallace, Graphite oxide to graphene. Biomaterials to bionics, Adv. Mater. 27 (46) (2015) 7563−7582.

[82] S. Chernousova, M. Epple, Silver as antibacterial agent: ion, nanoparticle, and metal, Angew. Chem. Int. Ed. 52 (6) (2012) 1636−1653.

[83] O. Akhavan, E. Ghaderi, Toxicity of graphene and graphene oxide nanowalls against bacteria, ACS Nano 4 (10) (2010) 5731−5736.

[84] S. Some, J.S. Sohn, J. Kim, S. Lee, S.C. Lee, J. Lee, et al., Graphene-iodine nanocomposites: highly potent bacterial inhibitors that are bio-compatible with human cells, Sci. Rep. 6 (1) (2016).

[85] A. Usman, Z. Hussain, A. Riaz, A.N. Khan, Enhanced mechanical, thermal and antimicrobial properties of poly (vinyl alcohol)/graphene oxide/starch/silver nanocomposites films, Carbohydr. Polym. 153 (2016) 592−599.

[86] P. Li, S. Sun, A. Dong, Y. Hao, S. Shi, Z. Sun, et al., Developing of a novel antibacterial agent by functionalization of graphene oxide with guanidine polymer with enhanced antibacterial activity, Appl. Surf. Sci. 355 (2015) 446−452.

[87] Z. Zeng, D. Yu, Z. He, J. Liu, F. Xiao, Y. Zhang, et al., Graphene oxide quantum dots covalently functionalized PVDF membrane with significantly-enhanced bactericidal and antibiofouling performances, Sci. Rep. 6 (1) (2016).

[88] Y. Ma, M. Zhao, B. Cai, W. Wang, Z. Ye, J. Huang, 3D graphene foams decorated by CuO nanoflowers for ultrasensitive ascorbic acid detection, Biosens. Bioelectron. 59 (2014) 384−388.

[89] X.C. Dong, H. Xu, X.W. Wang, Y.X. Huang, M.B. Chan-Park, H. Zhang, et al., 3D graphene-cobalt oxide electrode for high-performance supercapacitor and enzyme less glucose detection, ACS Nano 6 (2012) 3206−3213.

[90] H.Y. Yue, S. Huang, J. Chang, C. Heo, F. Yao, S. Adhikari, et al., ZnO nanowire arrays on 3D hierachical graphene foam: biomarker detection of Parkinson's disease, ACS Nano 8 (2014) 1639−1646.

[91] R. Aggarwal, S. Ringold, D. Khanna, T. Neogi, S.R. Johnson, A. Miller, et al., Distinctions between diagnostic and classification criteria? Arthritis Care Res. 67 (2015) 891−897.

[92] N.B. Mota, M. Copelli, S. Ribeiro, Thought disorder measured as random speech structure classifies negative symptoms and schizophrenia diagnosis 6 months in advance, NPJ Schizophr. 3 (1) (2017).

[93] T. Kawamura, I. Sato, K. Kawakami, Factors influencing the placebo effect in patients with primary open-angle glaucoma or ocular hypertension: an analysis of two randomized clinical trials, PLOS One 11 (6) (2016) e0156706.

[94] K. Munir, C. Wen, Y. Li, Graphene nanoplatelets-reinforced magnesium metal matrix nanocomposites with superior mechanical and corrosion performance for biomedical applications, J. Mag. Allo. (2020).

The use of polymer-graphene composites in catalysis

18

Haradhan Kolya[1], Subhadip Mondal[2], Chun-Won Kang[1] and Changwoon Nah[2]
[1]Department of Housing Environmental Design and Research Institute of Human Ecology, College of Human Ecology, Jeonbuk National University, Jeonju, Republic of Korea, [2]Department of Polymer-Nano Science and Technology, Jeonbuk National University, Jeonju, Republic of Korea

18.1 Introduction

As well as population growth and economic development, global energy demand will continue to rise in the foreseeable future [1]. To address the issue of renewable energy intermittency, catalytic development is an attractive approach [2]. Recently, graphene has attracted researchers' attention more significantly due to its exclusive and outstanding properties. It has been described as a planer sheet with one atom thickness of sp^2 hybridized 'C' atoms arranged in a honeycomb lattice. The carbon atoms are bound to the three adjacent atoms through two single and one double covalent bonds. The structure of graphene consists of multi-layered graphene sheets and it can be found in different configurations such as monolayers, a few layers, nanosheets, and nanoplatelets [3,4]. Various structures of graphene nanofillers are shown in Fig. 18.1. The graphene was found to be displaying remarkable mechanical strength (Young's modulus 1Ta, Fracture strength 130 GPa), electrical conductivity (10^4 S/cm), electrical mobility (250,000 $cm^2/V \cdot s$), thermal conductivity (5300 W/mK), high specific surface area (2630 m^2/g), and optical transmittance (97.7%) [5,6].

Composite materials are a class of hybrid materials consisting of two or more materials, to take advantage of the best characteristics and properties out of them [7,8]. Many graphene composites have been reported, and there is ongoing research on the development of more composites for various potential applications [9]. Graphene is a good choice for use as a safe and excellent catalyst support due to high metal/metal oxide nanoparticles (NPs) immobilization on its high surface area or within layers to prepare of more effective heterogeneous catalysts [10−12]. The transfer of electrical density from graphene to reinforced metal or metal oxides NPs deposited on graphene was mostly invoked to realize the catalytic efficiency of the as-prepared metal catalysts [13]. However, graphene sheets appear to stack together through the π-π interactions due to the presence of 2-D graphene layers [14]. In addition, the enfolding of graphene sheets and a small interlamellar spacing restrict the

Polymer Nanocomposites Containing Graphene. DOI: https://doi.org/10.1016/B978-0-12-821639-2.00013-6

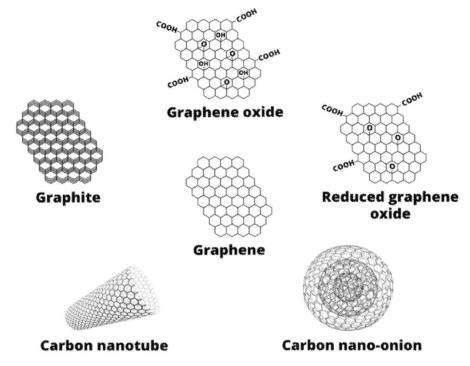

Figure 18.1 Schematic presentation of graphene nanofillers [19].
Source: D. Plachá, and J. Jampilek, Graphenic materials for biomedical applications,
Nanomaterials 9 (2019). https://doi.org/10.3390/nano9121758. Copyright 2019; reproduced
with permission from MDPI, Switzerland.

potential use of such an applied catalyst [15,16]. Therefore, stabilization and surface
modification of graphene through material change are required to avoid the undesired
aggregation, extending its catalytic applications [17]. The addition of polymers to the
graphene layers will minimize the accumulation of graphene sheets [18].

   In recent years, polymer composites displayed significantly improved properties
compared to virgin polymers or their general composites, such as improved catalytic properties, modulus and strength, gas barrier properties, electrical conductivity,
solvent and heat resistance properties, reduced flammability, and better lumines-
cence and nonlinear optical properties [20−22]. The presence of C−C bonds in the
graphene matrix along with the defect site results in a unique structure for better
catalytic efficiency [23]. Concomitantly, graphene polymer composites used in
imminent application areas include photocatalyst, electrocatalyst, drug delivery,
high-performance materials, sensors, biomedical materials and energy storage, as
shown in Fig. 18.2 [19].

   This chapter provides a comprehensive review focusing on the recent development and current polymer nanocomposites for catalytic applications. The use of

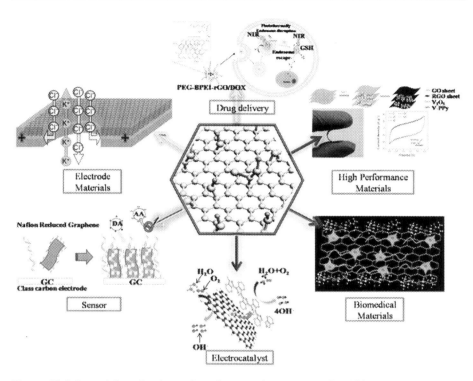

Figure 18.2 Potential applications of graphene−polymer composites [24].
Source: M. Zhang, Y. Li, Z. Su, G. Wei, Recent advances in the synthesis and applications of graphene−polymer nanocomposites, Polym. Chem. 6 (2015) 6107−6124. Copyright 2010; reproduced with permission from RSC, Great Britain.

polymer graphene composites for catalytic applications such as photocatalytic, electrocatalytic, catalytic in the hydrogenation reaction, and the lithium-air battery has been discussed. This chapter highlights the challenges and new directions for the potential growth of polymer graphene composites.

18.2   Photocatalytic activity

Photocatalysis is an essential phenomenon that deals with environmental decontamination and energy generation. Numerous photocatalysts, especially based on metal oxides photocatalysts such as titanium oxide (TiO_2), tin oxide (SnO_2), and zinc oxide (ZnO) are promising materials for the degradation of organic pollutants such as methyl orange (MO), methylene blue (MB), rhodamine B, p-chlorophenol, 2,4-dichlorophenoxyacetic acid, and reactive dyes by sunlight [25−27]. It has been well known that these metal oxide photo-catalysts alone can't be used as an effective alternative for the photodecomposition of organic pollutants, as they have certain disadvantages. Disadvantages are as follows; (1) mostly used metal oxides such as TiO_2 and ZnO, are only active in ultraviolet region (UV). It is well known that sunlight contains about 40% visible light and 10% UV light energy. (2) Visible light active metal oxide Fe_2O_3 and $BiVO_4$ have a fast recombination rate of photon-charges. (3) photo-catalysts like CdS, Fe_2O_3, and ZnO undergo photo corrosion quickly [28], respectively. In recent years, it has been reported that graphene has desirable properties for improving the overall photocatalytic activity of its corresponding composites [29]. In the presence of photocatalysis, graphene can act as an excellent electron-acceptor or transport substantially to efficiently facilitate the movement of photoinduced electrons and impede the recombination of charges in electron-transfer processes, in which photocatalytic performance is enhanced [27]. The graphene nanomaterials aid in the charge separation and transport of the composites because of their excellent electronic conductivity. In the relationship between conductivity and photocatalytic activity, conductive material exhibits higher photocatalytic activity with higher conductivity. However, this phenomenon does not always apply because there is no direct relationship between conductivity and photocatalytic activity [30−32]. Graphene can affect the optical properties and improve the specific surface area of the composites [33,34]. Graphene can also act as a structure-directing template, influencing the particle shape and size of the metal oxide NPs anchored on their surface [35]. The excellent metal adsorption properties of GO and rGO compared to graphene have been found due to the oxygen-containing functionalities [36]. Besides, rGO-based binary composite photocatalysts with their suitable light energy have been discussed in detail by Suresh et al. [37]. It has also been reported that different types of reduced graphene derivatives contain very few polar functional groups (−OH, −COOH) on the surface regularly, resulting in extremely low hydrophilicity and thus making it difficult to achieve further modification by aqueous chemical approaches. A widely applicable technique for enhancing the dispersibility of hydrophobic materials in an aqueous solution is to apply surfactants to the solution. Poly (2-acrylamide-2-methyl-1-propane sulfonic acid) is a water-soluble polymer containing lipophilic polyethylene and hydrophilic amide and sulfonic acid group that can accumulate to form aggregates in graphene layers by hydrophobic interaction [38]. It has been observed that encapsulation of various metal oxide and rGO with conductive polymer improved photocatalytic properties in a wide range of light wavelengths due to the combination effects and extended π-conjugated electron system [28,39,40]. A pictorial representation is shown in Fig. 18.3.

In recent years, it was found that graphene-based polymethylmethacrylate composites are a promising candidate in photocatalytic activity because of the reduced bandgaps (E_g^d and E_g^{ind}) of polymethylmethacrylate with the addition of graphene [41,42]. The rapid increase of absorbance values in the UV region was correlated with the optical transfers of graphene nanomaterials electrons from the valence band to the conduction band [41]. Hence, graphene based-polymethylmethacrylate exhibited excellent photocatalytic activity in the rapid degradation of amoxicillin under visible radiation. A schematic representation of the probable mechanism of amoxicillin photodegradation is shown in Fig. 18.4.

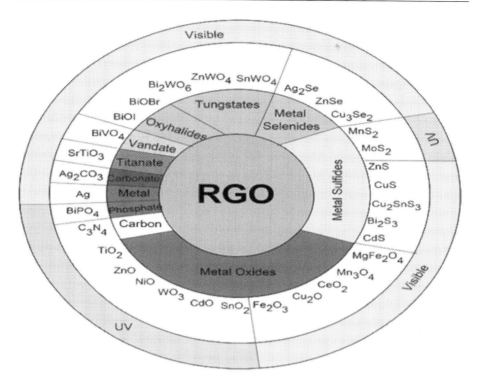

Figure 18.3 Pictorial representation of various RGO-based binary composite photo-catalysts with their suitable light energy [37].
Source: R. Suresh, R.V. Mangalaraja, H.D. Mansilla, P. Santander, J. Yáñez, Reduced graphene oxide-based photocatalysis BT: green photocatalysts, in: M. Naushad, S. Rajendran, E. Lichtfouse (Eds.), Springer International Publishing, Cham, 2020, pp. 145−166.
Copyright 2020; reproduced with permission from Springer Nature Switzerland AG.

The polymer composite of reduced graphene oxide (GO)-boron carbonitride-poly (diallyl dimethyl ammonium chloride) [PDDA] and sheets with layers of negatively charged MoS_2 and $MoSe_2$ have recently been reported to exhibit improved photocatalytic hydrogen evolution reaction (HER) activity compared to individual components, with increasing $MoS_2/MoSe_2$ content. The rGO-PDDA-MoS_2 (1:5) composites showed higher photocatalytic HER activity. The composites of rGO-PDDA-$MoSe_2$ (1:5) and rGO- boron carbonitride-MoS_2 (1:5) exhibited much lower catalytic activity 9540 and 8593 μmol/h/g, respectively. The high HER performance of the composites has attributed to some factors such as more numbers of catalytically active sites, improved charge-transfer rates between the hetero layers, and superlattice-like assemblies, and some of the 2D nanosheets (shown in Fig. 18.5) [43].

On the other hands, effective and ecofriendly photocatalytic materials such as rGO loaded to microspheres of amine-functionalized poly (styrene/glycidyl

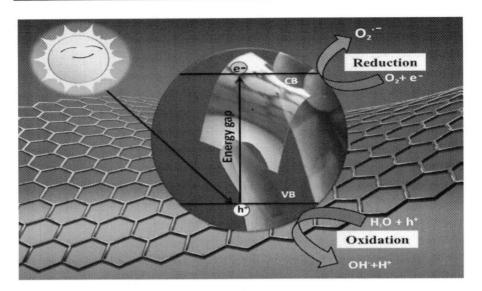

Figure 18.4 Proposed mechanism for the photocatalytic degradation of amoxicillin using
PMMA/GNP nanocomposite [41].
Source: M.S.A. Hussien, M.I. Mohammed, I.S. Yahia, Multifunctional applications of
graphene-doped PMMA nanocomposite membranes for environmental photocatalytic, J.
Inorg. Organomet. Polym. Mater. (2020). Copyright 2020; reproduced with permission from
Springer Nature.

methacrylate) (rGO/PSGM) show improved photocatalytic degradation MO dyes of
doped TiO_2 in the visible light [44]. This is because of the rapid separation of
photogenerated electrons and holes by the rGO nanosheet as electron cocatalyst
[44]. Besides, it was found that polyaniline (PANI)/rGO composites show remarkably enhanced photocatalytic activity for cationic and anionic color degradation
under visible light irradiation compared to pure PANI or rGO. The 5 wt.% of rGO
in PANI/rGO composite has emerged as the best combination degradation percentages for malachite green (99.68%), rhodamine B (99.35%), and Congo red
(98.73%) in 15, 30, and 40 min, respectively. This is because of the intercalation of
rGO which can minimize the agglomeration of PANI NPs and improve the light
absorption of the materials due to a high surface area [45,46]. In addition, PANI—
graphene nanocomposites possessed the partial hydrogen bond between the imine
(NH) of PANI and the carboxylic groups present in the graphene sheet surface [47].
The photocatalytic activity and chemical stability of graphene-polyaniline nanocomposite have great potential to purify industrial wastewater under exposure to
regular sunlight. After photolysis reaction, polymer graphene composites materials
can be recycled for repetitive use [48]. One more efficient ternary photocatalyst
with prominent OER performance under visible-light photoelectrocatalytic H_2O
oxidation has been reported utilizing PANI—graphene—TiO_2 composites materials
[49]. In addition, Cu_2O NPs loaded on the PANI and rGO nanosheets and the

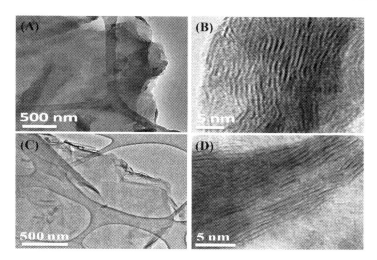

Figure 18.5 (A) TEM and (B) HRTEM images of an rGO-PDDA-MoS$_2$ composite. (C) TEM and (D) HRTEM images of a PDDA-boron carbonitride-MoS$_2$ composite [43]. *Source*: K. Pramoda, S. Servottam, M. Kaur, C.N.R. Rao, Layered nanocomposites of polymer-functionalized reduced graphene oxide and borocarbonitride with MoS2 and MoSe2 and their hydrogen evolution reaction activity, ACS Appl. Nano Mater. 3 (2020) 1792−1799. Copyright 2020; reproduced with permission from American Chemical Society.

photocatalytic activity of ternary have been improved due to the enhancement of electron-hole separation and electron transfer to the surface of the composite materials [50].

18.3 Electrocatalytic activity

Electrocatalysts are one kind of catalyst that participate in electrochemical reactions. Electrocatalysts can increase the rate of electrochemical reactions very well. At the time of the electrochemical reaction, the rate of electron transfers between the electrode and the electrocatalyst reactants increases at lower potential values. For electrochemical responses, a stronger electrocatalyst can always have lower over potential. Electrochemical reactions include overall water splitting consisting of HER at cathode and oxygen evolution reaction (OER) at anode for energy conversion, and other significant reactions such as oxygen reduction reaction (ORR) in fuel cell operation. The combination of HER and OER can transform electrical energy into hydrogen energy. In contrast, OER and ORR can convert chemical energy into electrical energy, which is used in metal-ion batteries [51]. There are most essential for two-electron transform system for HER and four or more electron transfer system for OER and ORR to carry out the reaction, which is kinetically slow. Therefore, suitable catalysts are urgently required to improve the reaction rate of OER and ORR. The platinum (Pt) and Ruthenium oxide (RuO_2) are two well-known catalysts for ORR/HER and OER, respectively [51]. However, it is not clear whether their catalytic properties can function better in overall acidic or alkaline conditions [2,51,52]. Hence, stable, effective and benchmark electrocatalysts are needed for improving the electrochemical reaction of OER, ORR and HER.

Recently, it has been reported that graphene-based electrocatalysis has attracted increasing interest worldwide due to its multifunctional properties [24,51]. N-containing graphene has been considered as a promising candidate for ORR, although significant challenges for fuel cell technology still delay its commercialization. A deep understanding of the roles played by different nitrogen states during electrocatalysis and enhancement of chemical doping procedures are some of the critical issues that need to be addressed before commercializing its feasibility [53]. Three kinds of N species containing graphitic-N, pyridinic-N and pyrrolic-N could be introduced into the graphene sheet. Each of these could affect its atomic charge distribution differently along with their catalytic properties [54]. Ruoff's group tested that N-doped graphene's electrocatalytic activity is highly dependent on graphitic-N content, whereas pyridinic-N species improved ORR's onset potential [55]. A low number of N doping (2.8%) is sufficient to achieve high ORR activity via four-electron processes due to N-doped graphene's synergistic effect [56]. Besides, porous morphological structures of materials can perhaps tune the bandgap energy of N-doped graphene [57]. There were many publications regarding the electrocatalytic application of N-doped graphene, S-doped graphene and transitional metal-doped graphene etc. [51].

It has been found that the number of chemical doping approaches reported to date are complex and conducted under harsh conditions, thus severely restricting their use on a large scale [51]. As described above, the enhanced catalytic ORR activity of N-doped graphene is due to the N atoms' ability to accept electrons. Thus, researchers extended this principle to the functionalization of graphene as metal-free electrocatalysts with polyelectrolytes [59,60]. For example, a positively charged polyelectrolyte like PDDA treated graphene exhibited improved electrocatalytic activity towards ORR. It could be attributed to the intermolecular charge transfer between PDDA and graphene surface. Schematic illustration of the electron-withdrawing or charge transfer process is presented in Fig. 18.6. These graphene-PDDA composites showed better fuel selectivity and stability than the commercially available Pt/C catalyst [58]. The results of the linear sweep voltammetry study of graphene and graphene-PDDA composites are shown in Fig. 18.7. The results show that the number of electron transfer for ORR process is more significantly influenced in graphene-PDDA composites than the pure graphene electrode. Hence, results suggesting that the graphene-PDDA composites exhibited more efficiency as electrocatalyst in ORR process. It has also been reported that the graphene-PDDA has higher fuel selectivity and durability toward ORR than the commercial Pt/C electrocatalyst. This is because the active site for ORR is still carbon atoms in graphene with adsorbed PDDA to establish somewhat delocalized positive charges on graphene's conjugated carbon surface to change the graphene's electronic properties and its oxygen adsorption actions to facilitate the ORR

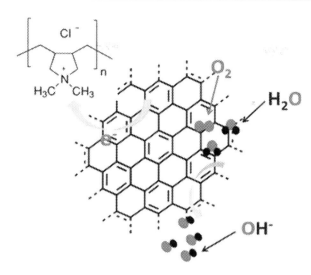

Figure 18.6 Schematic illustration of the electron-withdrawing or charge transfer from graphene in ORR process [58].
Source: S. Wang, D. Yu, L. Dai, D.W. Chang, J.-B. Baek, Polyelectrolyte-functionalized graphene as metal-free electrocatalysts for oxygen reduction, ACS Nano 5 (2011) 6202−6209. Copyright 2011; reproduced with permission from American Chemical Society.

reaction cycle. Although, graphene-PDDA electrocatalytic activity was not as good as that of CNT-N content and Pt/C [58]. These findings suggest that it can be considered an effective alternative to metal-free graphene-based polymer composite for electrochemical reactions due to its intermolecular charge transfer attitude.

Also, graphene conducting-polymer based composites exhibited improved electrocatalytic activity towards the oxidation of dopamine. The graphene-based polymer composite of graphene/poly (3,4-ethylene dioxythiophene) was prepared by a simple electrochemical reduction process [61]. Electrochemical reaction of dopamine was investigated using cyclic voltammetry study, and results showed that graphene/poly (3,4-ethylene dioxythiophene) composite has high electrochemical activity towards dopamine due to the presence of large effective surface area or many active sites which accelerates the electron transfer between the electrode and dopamine [61].

The horseradish peroxidase (HRP) immobilized in GO−Nafion composite exhibited improved electrocatalytic activity towards electrochemical reduction of oxygen and hydrogen peroxide. It could detect hydrogen peroxide and oxygen at a limit of 4.0×10^{-7} mol/L and 1.0×10^{-7} mol/L, respectively. This investigation shows that an effective enzyme enhancement associated with direct electron transfer between HRP and the electrode of GO, which can be attributed to the strong biocompatibility of GO−Nafion polymer and the large active surface area of GO [62]. As the conducting polymer, graphene and copper NPs have good electrochemical characteristics; therefore, Ehsani et al. reported hybrid composite materials such as

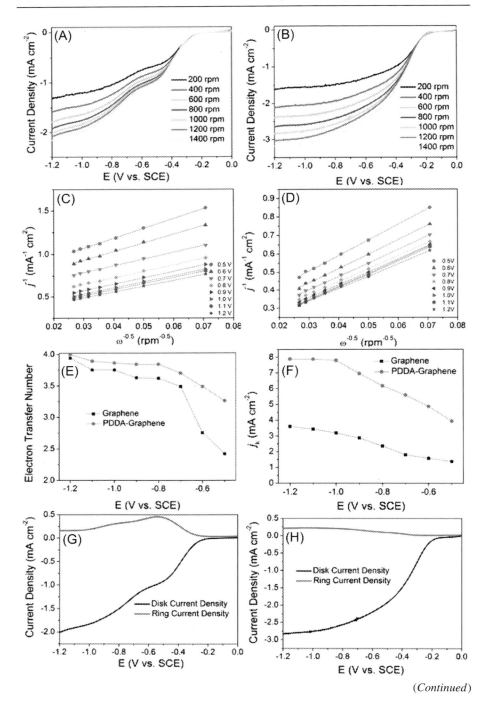

(*Continued*)

rGO-CuNPs-polytyramine for methanol oxidation [63]. It is mainly due to the extensive surface area and better electronic/ionic conductivity of polymer composite. Graphene-platinum-polyallamine composites were developed to improve the methanol electro-oxidation performance and better stability of commercial Pt/C electrocatalysts [64]. Graphene-Pt-polyallamine composite exhibited high catalytic activity for methyl oxidation reaction to the electrochemical surface of Pt/C (40.53 vs 17.61 m^2/mgPt), which may be due to the structural changes caused by the high active surface area of the polymer composite materials. The results of chronoamperometry showed that this graphene polymer composite also retains a higher current density than Pt/C in the presence of methanol [64]. Zhang et al. reported that electrochemically exfoliated graphene (EEG)-polyaniline-Pt composite exhibited higher performance in methyl alcohol oxidation reaction than rGO-polyaniline-Pt [65]. The synergistic interaction between Pt and PANI could accelerate the dispersion of Pt NPs, stabilize Pt NPs during cycling and enhanced electrical conductivity of Pt-based hybrids materials, which was favorable for improving electrocatalytic ability [65]. Besides, the combination of EEG and PANI would give more N content for the growth of Pt nanocrystals, which could further promote the electrocatalytic efficiency of Pt electrode. The CV curves measured in 0.5 M H$_2$SO$_4$ solution were studied to calculate active surface area of Pt electrode, and results are shown in Fig. 18.8A and B.

Fig. 18.8C and D showed that the two peaks appeared due to the oxidation of methanol oxidation (forward peak) and the oxidation of the residual carbonaceous content (mainly CO) formed during the forward-scan (the backward peak). The forward peak current density was found higher value in EEG-Pt compared to rGO-Pt. The polyaniline was used to modify both EEG-Pt and rGO-Pt for improving electrocatalytic activity. EEG-polyaniline-Pt represented high electrocatalytic activity in the methanol oxidation reaction [65].

18.4 Catalytic activity in the hydrogenation reaction

Hydrogenation is a chemical reaction of molecular hydrogen and substrate in the presence of a catalyst such as metal NPs, metal nanocomposites, and precious metal like silver, palladium, and platinum [19,66]. It is crucial to degrade toxic

◀

Figure 18.7 LSV curves at different rotation rates for oxygen reduction at (A) the graphene and (B) PDDA−graphene electrode in an O$_2$-saturated 0.1 M KOH solution; K−L plots of ORR on (C) the graphene and (D) PDDA−graphene electrode. (E) The dependence of the electron transfer number and (F) the kinetic current density on the potential for both the graphene and PDDA−graphene electrodes. ORR on the RRDE of (G) the graphene and (H) PDDA−graphene electrode in an O2-saturated 0.1 M KOH solution [58].
Source: S. Wang, D. Yu, L. Dai, D.W. Chang, J.-B. Baek, Polyelectrolyte-functionalized graphene as metal-free electrocatalysts for oxygen reduction, ACS Nano 5 (2011) 6202−6209. Copyright 2011; reproduced with permission from American Chemical Society.

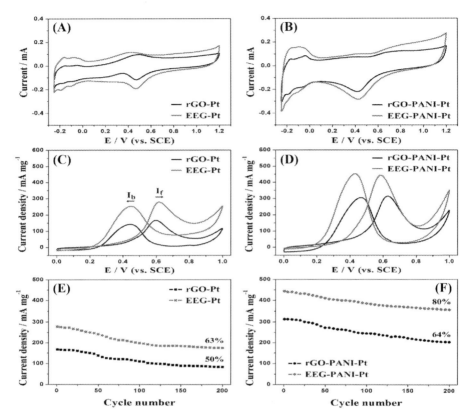

Figure 18.8 CV curves of (A) rGO-Pt, EEG-Pt and (B) rGO-PANI-Pt, EEG-PANI-Pt in
0.5 M H$_2$SO$_4$. CV curves of (C) rGO-Pt, EEG-Pt and (D) rGO-PANI-Pt, EEG-PANI-Pt in
0.5 M H$_2$SO$_4$ and 1 M CH$_3$OH. The forward peak current densities of (E) rGO-Pt, EEG-Pt
and (F) rGO-PANI-Pt, EEG-PANI-Pt as a function of the cycle number for methanol
oxidation [65].
Source: J. Zhang, L. Nan, W. Yue, X. Chen, Enhanced methanol electro-oxidation activity of
electrochemically exfoliated graphene-Pt through polyaniline modification, J. Electroanal.
Chem. 858 (2020) 113821. Copyright 2020; reproduced with permission from Elsevier B.V.

nitroaromatic and toxic azo dyes molecules from industrial and environmental perspectives. Catalytic hydrogenation reaction from various treatment technologies has
been intensively investigated [67–69]. Taking full advantage of the functional surface groups and their large specific surface area, GO nanosheets are emerging as
suitable supports for the production of hybrid nanomaterials and, in particular, for
catalytic applications. To date, a variety of metals, metal oxides, semiconducting
and magnetic NPs hosting on the surface of GO have been developed for catalytic
hydrogenation reaction [18,67,70]. The rGO-based hybrid hydrogels prepared via
vitamin C, including different metal ion NPs have been developed for catalytic

hydrogenation reaction. The rGO-Au NPs hybrid hydrogel was studied in the hydrogenation reaction of 4-nitrophenol (4-NP) and 4-nitro ammine [70]. After the reaction, the catalyst can be recovered and reused for various cycles without the significant loss of catalytic activity. In recent year, it has been reported rGO based Ag NPs nanocomposites were also an efficient catalyst for the hydrogenation reaction of 4-NP and MO dye molecules. It was found that 12 min was enough for the complete reduction of 4-NP and MO dye molecules. The degradation of azo bonds in dye molecules and the reduction of nitro groups occurred onto the surface of the in-situ formed metal hydride. The faster rate of reduction reaction held on the surface of NPs because of the greater availability of the active surface area of AgNPs [67]. A probable mechanism is shown in the following Fig. 18.9.

The graphene-polymer composites supported-NPs-based architecture to the multifunctional membrane has been developed for water purification. The casting mixture of polysulfone, N-Methyl-2-pyrrolidone and polyethylene glycol, exfoliated graphite nanoplatelets (xGnPs) decorated with Au nanoparticles (Au NPs) was used as a model hierarchical nanofiller membrane for filtration of 4-NP and reported catalytic activity in the presence of sodium borohydride [71]. The rGO-AgNPs based polyethyleneimine (PEI) hydrogel composites were found as an efficient catalyst for removing dye molecules from an aqueous medium [72].

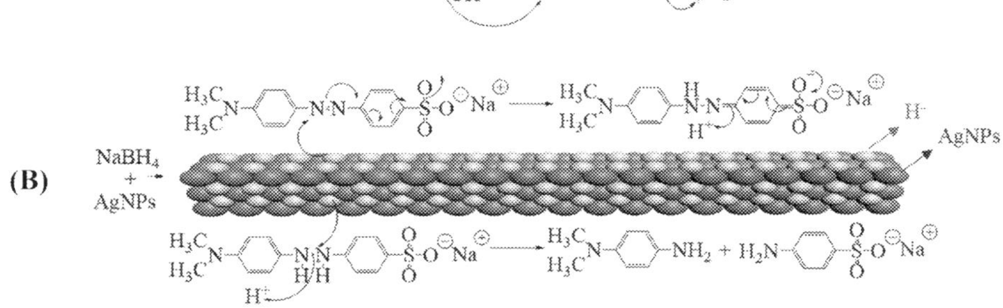

Figure 18.9 The probable mechanism of hydrogenation reaction onto the Ag NPs surface (A) 4-nitrophenol (B) azo bond degradation of MO dye molecules [67].
Source: H. Kolya, T. Kuila, N.H. Kim, J.H. Lee, Bioinspired silver nanoparticles/reduced graphene oxide nanocomposites for catalytic reduction of 4-nitrophenol, organic dyes and act as energy storage electrode material, Compos. Part B Eng. 173 (2019) 106924. Copyright 2019; reproduced with permission from Elsevier B.V.

18.5   Catalytic activity in Li-air battery

Batteries can be used as a dominant energy storage device compared to supercapacitors due to their high energy density. The high storage capacity per volume of lithium-ion (Li-ion) rechargeable batteries has made them the commercial alternative for consumer electronic goods like camcorders, cameras, and digital watches. The principle of Li-ions battery is the shuttling of Li-ions between a lithium-metal negative electrode and an intercalating compound that acts as the positive electrode or between two intercalating electrodes (the so-called rocking chair battery) [73,74]. Lithium transfers its 2s electrons to the carbon host during the intercalation process and is placed between the carbon sheets. Among the metal-air batteries (lithium-air, zinc-air batteries and aluminum-air batteries), the lithium-air battery has gained significant global attention due to its high energy capacity [75]. The critical difference between the Li-ions and Li-air batteries is the configuration of their electrodes. Similar to all the other chemical batteries, it has much higher energy densities. Li-air batteries nowadays also play a significant role in electronics, cars, driving computers, remote sensors, and robotic uses. However, Li-air battery technology is not yet mature, and the crucial issues should be adequately resolved to make it commercially viable for applications. Improving some of the Li-air battery factors such as life cycle, usable power, and round-trip performance is most important. In these batteries, oxygen (O_2) is picked up from the air, and the pure Li is used as an electrode. The energy capacity of such batteries is tentatively based only on the Li electrode, from which it can be significantly improved. Relevant chemistry of lithium-air battery involves lithium dissolution and deposition to the anodic electrode (lithium), ORR, and OER to the cathodic electrode (air). A polymer with highly electroactive functional groups and a durable skeleton may be a robust cathode contestant because of its redox kinetics which are significantly faster than inorganic cathodes. Graphene-polymer composites were specifically used as a cathode in the lithium-ion battery without binder, providing a very high discharge capacity of 156 mAh/g at 0.1°C with an ultra-high use of poly (anthraquinonyl sulfide) (94.9%) and an impressive rate of 102 mAh/g with a high current density at 20°C [76]. Multilayer assembly of GO-PDDA-poly (ethylene oxide) (PEO) was prepared by a dip-assisted layer-by-layer method based on electrostatic and epitaxial adsorption of polymers for lithium-ion electrode applications [73]. Three-dimensional GO was intercalated with polyvinyl alcohol (PVA) polymer composite has been reported as a cathode electrode material for Li-ion battery and provided discharge capacities of 1395 mAh/g at 0.1 A/g and 606 mAh/g at 5 A/g [77]. Recently, Guo et al. reported that LiI in low concentration polymer electrolyte (0.05 M) exhibited improved activity in the Li-air battery. The polymer gel electrolyte (GPE or 0.05 m LiI-GPE) includes the polymer matrix properties and the liquid electrolyte properties. Polymer membrane surface morphology shows regularly distributed elliptic micro-pores with an average diameter of 100 nm, indicating good efficiency in the load/discharge cycle, that is 3500 mAh/g (liquid electrolyte-based cell), 3100 mAh/g (GPE-based cell) and 3400 mAh/g (0,05 m LiI-GPE-based cell) [78]. Compared to those of the usually used carbon blacks and Pt/C catalysts, the graphene-rich composite catalysts (Cobalt-N-MWNTs) demonstrate significantly improved ORR efficiency in the non-aqueous lithium-ion electrolyte, as shown by both rotating disk electrode and $Li-O_2$ battery experiments. A nitrogen-doped graphene composite with CNTs provides promising electrocatalysis application for non-aqueous ORR for $Li-O_2$ battery cathodes [79].

18.6   Conclusions and future scope

In this chapter, we summarized the catalytic use of polymer graphene composites and their different applications are highlighted. The recent development of polymer graphene composites exhibits innovative catalytic activity due to good graphene properties such as high specific surface area, low mass density, good compatibility, excellent conductivity, and elegant flexibility. It has been found that polymer graphene composites are widely used as an electrocatalyst rather than a photocatalyst, catalyst in hydrogenation reaction, and lithium-air battery. It was evident that there is a considerable scope of future research to address the following points:

1. It is crucial to search for the most active polymer-graphene composites for the real world in photocatalytic, electrocatalytic, fuel cell, and lithium-air batteries.
2. More research is needed to set up polymer composite potential for various catalytic applications on an industrial scale.
3. More research is needed to be done on the fabrication of polymer graphene composites in an ecofriendly way, such as by electrochemical methods, melt compounding, and self-assembly techniques.
4. More research is required to improve the respective applications of polymer composites with all the interrelated factors being taken into consideration.

Acknowledgments

Professor Kang is thankful to the Basic Science Research Program through the National Research Foundation of Korea(NRF) funded by the Ministry of Education (NRF-2019R111A3A02059471) and Professor Nah is thankful to the Korea Evaluation Institute of Industrial Technology, Grant Number- 20012558.

References

[1] B.M. Hunter, H.B. Gray, A.M. Müller, Earth-abundant heterogeneous water oxidation catalysts, Chem. Rev. 116 (2016) 14120−14136.
[2] B. You, Y. Sun, Innovative strategies for electrocatalytic water splitting, Acc. Chem. Res. 51 (2018) 1571−1580.
[3] M. Ioniţă, G.M. Vlăsceanu, A.A. Watzlawek, S.I. Voicu, J.S. Burns, H. Iovu, Graphene and functionalized graphene: extraordinary prospects for nanobiocomposite materials, Compos. Part B Eng. 121 (2017) 34−57.

[4]   G. Yang, L. Li, W.B. Lee, M.C. Ng, Structure of graphene and its disorders: a review, Sci. Technol. Adv. Mater. 19 (2018) 613−648.

[5]   J. Sanes, C. Sánchez, R. Pamies, M.-D. Avilés, M.-D. Bermúdez, Extrusion of polymer nanocomposites with graphene and graphene derivative nanofillers: an overview of recent developments, Materials 13 (2020).

[6]   C. Ramirez, F.M. Figueiredo, P. Miranzo, P. Poza, M.I. Osendi, Graphene nanoplatelet/silicon nitride composites with high electrical conductivity, Carbon 50 (2012) 3607−3615.

[7]   R. Miranda, A.L. Vázquez de Parga, Surfing ripples towards new devices, Nat. Nanotechnol. 4 (2009) 549−550.

[8]   Y. Liu, S. Kumar, Polymer/carbon nanotube nano composite fibers: a review, ACS Appl. Mater. Interfaces 6 (2014) 6069−6087.

[9]   T. Sattar, Current review on synthesis, composites and multifunctional properties of graphene, Top. Curr. Chem. 377 (2019) 10.

[10] M.M. Ayad, W.A. Amer, M.G. Kotp, Magnetic polyaniline-chitosan nanocomposite decorated with palladium nanoparticles for enhanced catalytic reduction of 4-nitrophenol, Mol. Catal. 439 (2017) 72−80.

[11] S. Naghdi, M. Sajjadi, M. Nasrollahzadeh, K.Y. Rhee, S.M. Sajadi, B. Jaleh, Cuscuta reflexa leaf extract mediated green synthesis of the Cu nanoparticles on graphene oxide/manganese dioxide nanocomposite and its catalytic activity toward reduction of nitroarenes and organic dyes, J. Taiwan Inst. Chem. Eng. 86 (2018) 158−173.

[12] S.I. El-Hout, S.M. El-Sheikh, H.M.A. Hassan, F.A. Harraz, I.A. Ibrahim, E.A. El-Sharkawy, A green chemical route for synthesis of graphene supported palladium nanoparticles: a highly active and recyclable catalyst for reduction of nitrobenzene, Appl. Catal. A Gen. 503 (2015) 176−185.

[13] N.M. Julkapli, S. Bagheri, Graphene supported heterogeneous catalysts: an overview, Int. J. Hydrogen Energy 40 (2015) 948−979.

[14] L. Qu, Y. Liu, J.-B. Baek, L. Dai, Nitrogen-doped graphene as efficient metal-free electrocatalyst for oxygen reduction in fuel cells, ACS Nano 4 (2010) 1321−1326.

[15] Y. Li, H. Wang, L. Xie, Y. Liang, G. Hong, H. Dai, MoS_2 nanoparticles grown on graphene: an advanced catalyst for the hydrogen evolution reaction, J. Am. Chem. Soc. 133 (2011) 7296−7299.

[16] C. Nethravathi, M. Rajamathi, Chemically modified graphene sheets produced by the solvothermal reduction of colloidal dispersions of graphite oxide, Carbon 46 (2008) 1994−1998.

[17] T. Niu, G.L. Liu, Y. Liu, Preparation of Ru/graphene-meso-macroporous SiO_2 composite and their application to the preferential oxidation of CO in H_2-rich gases, Appl. Catal. B Environ. 154−155 (2014) 82−92.

[18] G.M. Scheuermann, L. Rumi, P. Steurer, W. Bannwarth, R. Mülhaupt, Palladium nanoparticles on graphite oxide and its functionalized graphene derivatives as highly active catalysts for the Suzuki − Miyaura coupling reaction, J. Am. Chem. Soc. 131 (2009) 8262−8270.

[19] D. Plachá, J. Jampilek, Graphenic materials for biomedical applications, Nanomaterials 9 (2019). Available from: https://doi.org/10.3390/nano9121758.

[20] T. Sridhar, A.U. Kini, A. Hiremath, A review on nanoparticle filled polymer nanocomposites, Int. J. Eng. Technol. 7 (2018) 169−173.

[21] H. Kim, A.A. Abdala, C.W. Macosko, Graphene/polymer nanocomposites, Macromolecules 43 (2010) 6515−6530.

[22] T. Kuilla, S. Bhadra, D. Yao, N.H. Kim, S. Bose, J.H. Lee, Recent advances in graphene based polymer composites, Prog. Polym. Sci. 35 (2010) 1350−1375.

[23] Y. Li, X. Fan, J. Qi, J. Ji, S. Wang, G. Zhang, et al., Palladium nanoparticle-graphene hybrids as active catalysts for the Suzuki reaction, Nano Res. 3 (2010) 429−437.

[24] M. Zhang, Y. Li, Z. Su, G. Wei, Recent advances in the synthesis and applications of graphene−polymer nanocomposites, Polym. Chem. 6 (2015) 6107−6124.

[25] D. Chu, Y. Masuda, T. Ohji, K. Kato, Formation and photocatalytic application of ZnO nanotubes using aqueous solution, Langmuir 26 (2010) 2811−2815.

[26] Z. Wang, B. Huang, Y. Dai, X. Qin, X. Zhang, P. Wang, et al., Highly photocatalytic ZnO/In$_2$O$_3$ heteronanostructures synthesized by a coprecipitation method, J. Phys. Chem. C 113 (2009) 4612−4617.

[27] Q.-H. Xia, H.-Q. Ge, C.-P. Ye, Z.-M. Liu, K.-X. Su, Advances in homogeneous and heterogeneous catalytic asymmetric epoxidation, Chem. Rev. 105 (2005) 1603−1662.

[28] K.R. Reddy, K. Nakata, T. Ochiai, T. Murakami, D.A. Tryk, A. Fujishima, Nanofibrous TiO$_2$-core/conjugated polymer-sheath composites: synthesis, structural properties and photocatalytic activity, J. Nanosci. Nanotechnol. 10 (2010) 7951−7957.

[29] G. Mamba, G. Gangashe, L. Moss, S. Hariganesh, S. Thakur, S. Vadivel, et al., State of the art on the photocatalytic applications of graphene based nanostructures: from elimination of hazardous pollutants to disinfection and fuel generation, J. Environ. Chem. Eng. (2019) 103505.

[30] A. Kolouch, M. Horáková, P. Hájková, E. Heyduková, P. Exnar, P. Spatenka, Relationship between photocatalytic activity, hydrophilicity and photoelectric properties of TiO$_2$ thin films, Вопросы Атомной Науки и Техники (2006).

[31] N. Chopra, J. Wu, L. Summerville, Controlled assembly of graphene shells encapsulated gold nanoparticles and their integration with carbon nanotubes, Carbon 62 (2013) 76−87.

[32] B.P. Vinayan, S. Ramaprabhu, Platinum−TM (TM = Fe, Co) alloy nanoparticles dispersed nitrogen doped (reduced graphene oxide-multiwalled carbon nanotube) hybrid structure cathode electrocatalysts for high performance PEMFC applications, Nanoscale 5 (2013) 5109−5118.

[33] X. Men, H. Chen, K. Chang, X. Fang, C. Wu, W. Qin, et al., Three-dimensional freestanding ZnO/graphene composite foam for photocurrent generation and photocatalytic activity, Appl. Catal. B Environ. 187 (2016) 367−374.

[34] H. Wang, D. Peng, T. Chen, Y. Chang, S. Dong, A novel photocatalyst AgBr/ZnO/RGO with high visible light photocatalytic activity, Ceram. Int. 42 (2016) 4406−4412.

[35] X. Li, Z. Le, X. Chen, Z. Li, W. Wang, X. Liu, et al., Graphene oxide enhanced amine-functionalized titanium metal organic framework for visible-light-driven photocatalytic oxidation of gaseous pollutants, Appl. Catal. B Environ. 236 (2018) 501−508.

[36] M. Minella, F. Sordello, C. Minero, Photocatalytic process in TiO$_2$/graphene hybrid materials. Evidence of charge separation by electron transfer from reduced graphene oxide to TiO$_2$, Catal. Today. 281 (2017) 29−37.

[37] R. Suresh, R.V. Mangalaraja, H.D. Mansilla, P. Santander, J. Yáñez, Reduced graphene oxide-based photocatalysis BT: green photocatalystsin: M. Naushad, S. Rajendran, E. Lichtfouse (Eds.), Springer International Publishing, Cham, 2020, pp. 145−166.

[38] D. Hu, C. Song, X. Jin, Q. Huang, Polymer solution-assisted assembly of hierarchically nanostructured ZnO onto 2D neat graphene sheets with excellent photocatalytic performance, J. Alloys Compd. 843 (2020) 156030.

[39] J. Wang, X. Ni, Photoresponsive polypyrrole- TiO$_2$ nanoparticles film fabricated by a novel surface initiated polymerization, Solid. State Commun. 146 (2008) 239−244.

[40] Y. Du, N. Cao, L. Yang, W. Luo, G. Cheng, One-step synthesis of magnetically recyclable rGO supported Cu@Co core−shell nanoparticles: highly efficient catalysts for

hydrolytic dehydrogenation of ammonia borane and methylamine borane, N. J. Chem. 37 (2013) 3035−3042.

[41] M.S.A. Hussien, M.I. Mohammed, I.S. Yahia, Multifunctional applications of graphene-doped PMMA nanocomposite membranes for environmental photocatalytic, J. Inorg. Organomet. Polym. Mater. (2020).

[42] J. Feng, A. Athanassiou, F. Bonaccorso, D. Fragouli, Enhanced electrical conductivity of poly(methyl methacrylate) filled with graphene and in situ synthesized gold nanoparticles, Nano Future 2 (2018) 25003.

[43] K. Pramoda, S. Servottam, M. Kaur, C.N.R. Rao, Layered nanocomposites of polymer-functionalized reduced graphene oxide and borocarbonitride with MoS_2 and $MoSe_2$ and their hydrogen evolution reaction activity, ACS Appl. Nano Mater. 3 (2020) 1792−1799.

[44] Y. Wu, H. Mu, X. Cao, X. He, Polymer-supported graphene−TiO_2 doped with nonmetallic elements with enhanced photocatalytic reaction under visible light, J. Mater. Sci. 55 (2020) 1577−1591.

[45] M. Mitra, S.T. Ahamed, A. Ghosh, A. Mondal, K. Kargupta, S. Ganguly, et al., Polyaniline/reduced graphene oxide composite-enhanced visible-light-driven photocatalytic activity for the degradation of organic dyes, ACS Omega 4 (2019) 1623−1635.

[46] K.-C. Huang, J.-H. Huang, C.-H. Wu, C.-Y. Liu, H.-W. Chen, C.-W. Chu, et al., Nanographite/polyaniline composite films as the counter electrodes for dye-sensitized solar cells, J. Mater. Chem. 21 (2011) 10384−10389.

[47] S. Ameen, H.-K. Seo, M. Shaheer Akhtar, H.S. Shin, Novel graphene/polyaniline nanocomposites and its photocatalytic activity toward the degradation of rose Bengal dye, Chem. Eng. J. 210 (2012) 220−228.

[48] G.M. Neelgund, V.N. Bliznyuk, A. Oki, Photocatalytic activity and NIR laser response of polyaniline conjugated graphene nanocomposite prepared by a novel acid-less method, Appl. Catal. B Environ. 187 (2016) 357−366.

[49] L. Jing, Z.-Y. Yang, Y.-F. Zhao, Y.-X. Zhang, X. Guo, Y.-M. Yan, et al., Ternary polyaniline−graphene−TiO_2 hybrid with enhanced activity for visible-light photo-electrocatalytic water oxidation, J. Mater. Chem. A. 2 (2014) 1068−1075.

[50] J. Miao, A. Xie, S. Li, F. Huang, J. Cao, Y. Shen, A novel reducing graphene/polyaniline/cuprous oxide composite hydrogel with unexpected photocatalytic activity for the degradation of Congo red, Appl. Surf. Sci. 360 (2016) 594−600.

[51] B. Xia, Y. Yan, X. Wang, X.W. (David) Lou, Recent progress on graphene-based hybrid electrocatalysts, Mater. Horizons 1 (2014) 379−399.

[52] K. Parvez, S. Yang, Y. Hernandez, A. Winter, A. Turchanin, X. Feng, et al., Nitrogen-doped graphene and its iron-based composite as efficient electrocatalysts for oxygen reduction reaction, ACS Nano 6 (2012) 9541−9550.

[53] H. Liu, Y. Liu, D. Zhu, Chemical doping of graphene, J. Mater. Chem. 21 (2011) 3335−3345.

[54] C.H. Choi, M.W. Chung, S.H. Park, S.I. Woo, Enhanced electrochemical oxygen reduction reaction by restacking of N-doped single graphene layers, RSC Adv. 3 (2013) 4246−4253.

[55] L. Lai, J.R. Potts, D. Zhan, L. Wang, C.K. Poh, C. Tang, et al., Exploration of the active center structure of nitrogen-doped graphene-based catalysts for oxygen reduction reaction, Energy Environ. Sci. 5 (2012) 7936−7942.

[56] D. Geng, Y. Chen, Y. Chen, Y. Li, R. Li, X. Sun, et al., High oxygen-reduction activity and durability of nitrogen-doped graphene, Energy Environ. Sci. 4 (2011) 760−764.

[57] T. Palaniselvam, H.B. Aiyappa, S. Kurungot, An efficient oxygen reduction electrocatalyst from graphene by simultaneously generating pores and nitrogen doped active sites, J. Mater. Chem. 22 (2012) 23799−23805.

[58] S. Wang, D. Yu, L. Dai, D.W. Chang, J.-B. Baek, Polyelectrolyte-functionalized graphene as metal-free electrocatalysts for oxygen reduction, ACS Nano 5 (2011) 6202−6209.

[59] V. Georgakilas, M. Otyepka, A.B. Bourlinos, V. Chandra, N. Kim, K.C. Kemp, et al., Functionalization of graphene: covalent and non-covalent approaches, derivatives and applications, Chem. Rev. 112 (2012) 6156−6214.

[60] Q. Tang, Z. Zhou, Z. Chen, Graphene-related nanomaterials: tuning properties by functionalization, Nanoscale 5 (2013) 4541−4583.

[61] W. Wang, G. Xu, X.T. Cui, G. Sheng, X. Luo, Enhanced catalytic and dopamine sensing properties of electrochemically reduced conducting polymer nanocomposite doped with pure graphene oxide, Biosens. Bioelectron. 58 (2014) 153−156.

[62] L. Zhang, H. Cheng, H. Zhang, L. Qu, Direct electrochemistry and electrocatalysis of horseradish peroxidase immobilized in graphene oxide−Nafion nanocomposite film, Electrochim. Acta. 65 (2012) 122−126.

[63] A. Ehsani, B. Jaleh, M. Nasrollahzadeh, Electrochemical properties and electrocatalytic activity of conducting polymer/copper nanoparticles supported on reduced graphene oxide composite, J. Power Sources. 257 (2014) 300−307.

[64] M. Yaldagard, Synthesis and electrochemical characterization of graphene-polyallylamine nanocomposites as a new supports of Pt catalyst for direct methanol fuel cell application, Iran. J. Hydrogen Fuel Cell. 6 (2020) 141−161.

[65] J. Zhang, L. Nan, W. Yue, X. Chen, Enhanced methanol electro-oxidation activity of electrochemically exfoliated graphene-Pt through polyaniline modification, J. Electroanal. Chem. 858 (2020) 113821.

[66] T. Bürgi, A. Baiker, Heterogeneous enantioselective hydrogenation over cinchona alkaloid modified platinum: mechanistic insights into a complex reaction, Acc. Chem. Res. 37 (2004) 909−917.

[67] H. Kolya, T. Kuila, N.H. Kim, J.H. Lee, Bioinspired silver nanoparticles/reduced graphene oxide nanocomposites for catalytic reduction of 4-nitrophenol, organic dyes and act as energy storage electrode material, Compos. Part B Eng. 173 (2019) 106924.

[68] H. Kolya, P. Maiti, A. Pandey, T. Tripathy, Green synthesis of silver nanoparticles with antimicrobial and azo dye (Congo red) degradation properties using *Amaranthus gangeticus* Linn leaf extract, J. Anal. Sci. Technol. 6 (2015) 33.

[69] Y. Choi, H.S. Bae, E. Seo, S. Jang, K.H. Park, B.-S. Kim, Hybrid gold nanoparticle-reduced graphene oxide nanosheets as active catalysts for highly efficient reduction of nitroarenes, J. Mater. Chem. 21 (2011) 15431−15436.

[70] M. Nasrollahzadeh, Z. Nezafat, M.G. Gorab, M. Sajjadi, Recent progresses in graphene-based (photo)catalysts for reduction of nitro compounds, Mol. Catal. 484 (2020) 110758.

[71] C.A. Crock, A.R. Rogensues, W. Shan, V.V. Tarabara, Polymer nanocomposites with graphene-based hierarchical fillers as materials for multifunctional water treatment membranes, Water Res. 47 (2013) 3984−3996.

[72] T. Jiao, H. Guo, Q. Zhang, Q. Peng, Y. Tang, X. Yan, et al., Reduced graphene oxide-based silver nanoparticle-containing composite hydrogel as highly efficient dye catalysts for wastewater treatment, Sci. Rep. 5 (2015) 11873.

[73] J.R. Owen, Rechargeable lithium batteries, Chem. Soc. Rev. 26 (1997) 259−267.

[74] T. Cassagneau, J.H. Fendler, High density rechargeable lithium-ion batteries self-assembled from graphite oxide nanoplatelets and polyelectrolytes, Adv. Mater. 10 (1998) 877−881.

[75] B. Chen, D.Y.C. Leung, J. Xuan, H. Wang, A mixed-pH dual-electrolyte microfluidic aluminum−air cell with high performance, Appl. Energy. 185 (2017) 1303−1308.

[76] Y. Zhang, Y. Huang, G. Yang, F. Bu, K. Li, I. Shakir, et al., Dispersion−assembly approach to synthesize three-dimensional graphene/polymer composite aerogel as a powerful organic cathode for rechargeable Li and Na batteries, ACS Appl. Mater. Interfaces. 9 (2017) 15549−15556.

[77] Y. Zhang, Q. Ma, S. Wang, X. Liu, L. Li, Poly(vinyl alcohol)-assisted fabrication of hollow carbon spheres/reduced graphene oxide nanocomposites for high-performance lithium-ion battery anodes, ACS Nano 12 (2018) 4824−4834.

[78] Z. Guo, C. Li, J. Liu, Y. Wang, Y. Xia, A. Long-Life, Lithium−air battery in ambient air with a polymer electrolyte containing a redox mediator, Angew. Chem. Int. Ed. 56 (2017) 7505−7509.

[79] G. Wu, N.H. Mack, W. Gao, S. Ma, R. Zhong, J. Han, et al., Nitrogen-doped graphene-rich catalysts derived from heteroatom polymers for oxygen reduction in nonaqueous lithium−O_2 battery cathodes, ACS Nano 6 (2012) 9764−9776.

The use of polymer-graphene composites as membrane

19

Biswajit Bera[1] and Ayan Dey[2]
[1]Department of Chemical Engineering, Indian Institute of Technology Kharagpur, Kharagpur, India, [2]Indian Institute of Packaging, Mumbai, India

19.1 Introduction

Graphene is a single layer of carbon that has a honeycomb structure. High specific surface area, remarkable elasticity, and strength make it a superior material for use as the membrane. The last decade has evidenced several developments of polymer graphene composite for the use as membrane. Many polymers are used as matrix for dispersing graphene. Some examples are polysulfone, polyethyleneimine, isophthaloyl chloride, 5-isocyanate-isophthaloyl chloride, 5-chloroformloxy-isophthaloyl chloride, m-phenylenediamine, N,O-carboxymethyl chitosan, poly ether sulfone, etc. [1]. A membrane can be described as a barrier which allows transport of specific molecules while blocking the others type of molecules; these result in a separation of different type of molecules. Membranes are majorly used in filtration, gas separation, and dialysis. The application of membrane is further extended to the fuel cell, supercapacitor, etc. In the fuel cell, polymer graphene composite is used as a proton exchange membrane. Pure graphene without pores was reported to be highly impermeable. The presence of π-electron cloud over the graphene molecules may be the reason for such impermeability. Ultrafast water flow was reported for GO membrane through the nano-channel with a size distribution as low as $3-5$ nm [2]. In this chapter, various polymer-graphene composite based membranes are discussed in mainly four applications: (1) filtration, (2) gas separation, (3) dialysis/electrodialysis, and (4) energy storage devices. Moreover, the effect of GO concentration and incorporation of various functionalized graphenes on the properties of the membrane are also discussed in this chapter.

19.2 Filtration

Membrane separation technology has been accepted to resolve the worldwide water pollution issues, for example, drinking water purification, water desalination, and wastewater treatment, by means of its high efficiency and low cost. The polymeric membranes have gained much attention owing to their mechanical stability, ease of processing, low manufacturing cost, etc. [3]. The membranes are fabricated by

Polymer Nanocomposites Containing Graphene. DOI: https://doi.org/10.1016/B978-0-12-821639-2.00024-0

different techniques, such as casting, interfacial polymerization (IP), layer-by-layer assembly, surface modification, and spinning [4]. Among different configurations, electrospun hollow fiber membranes possess a high surface to volume ratio which shows the best performance in water filtration applications. Following these methods different categories of membranes, for example, microfiltration (MF), ultrafiltration (UF), nanofiltration (NF), reverse osmosis (RO), and forward osmosis (FO) have been developed according to the membrane pore size and corresponding rejection mechanism [5]. MF membranes generally possess a symmetric microporous structure. However, UF and NF membranes show asymmetric porous structures that consist of a thin nanoporous active layer with an underlying microporous layer. Conventional phase inversion technique is applied to fabricate these porous membrane structures. In the case of RO, FO, and sometimes NF membranes, an individual nonporous, dense polyamide (PA) active layer is deposited onto porous support via the IP method to get higher selectivity. These membranes are identified as thin film composite (TFC) membranes which shows better salt rejection performance.

A variety of polymers including polysulfone (PSf) [6], poly(ether sulfone) (PES) [7], poly(acrylo nitrile) (PAN) [8], poly(vinylidene fluoride) (PVDF) [9], poly(vinyl chloride) (PVC) [10], cellulose triacetate [11], etc. are commonly used to fabricate the membranes. Even though polymeric membranes have been studied widely both in the scientific research domain as well as in industrial applications, there are some drawbacks due to their inherent hydrophobicity, severe fouling, permeability/selectivity trade-off, and shorter life span issues [12,13]. The incorporation of nanomaterials into the polymeric matrix has been proved to be very effective to enhance membrane performance by increasing the surface hydrophilicity, mechanical strength, and durability [14]. A very low loading of different nanofillers, for example, zeolites [15], silica [16], TiO_2 [17], Ag NPs [18], and carbon based nanomaterials [19], etc., have shown enhanced separation performance in comparison to the bare polymeric membrane. Among these nanomaterials, graphene and its functionalized derivatives have exhibited remarkable progress in graphene/polymer nanocomposite membrane for filtration application. Nanoporous graphene with its molecular sieving property was initially studied as a potential membrane separation material for desalination application. The simulation study showed that single-layer nanoporous graphene has superior water permeability and salt rejection capacity compared to RO membranes [20]. Incorporation of other functionalized derivatives of graphene, for example, GO, amine-functionalized GO, sulfonated GO, GO-COOH, etc. into the polymer matrix improves the permeability-selectivity trade-off, antifouling properties, and mechanical strength of the nanocomposite membranes [21−23].

1. Permeability and selectivity:

   The most important properties of membranes are permeability and selectivity in water treatment applications. However, all types of porous and nonporous membranes always show a permeability-selectivity trade off [5,24]. In the last two decades, a large number of attempts have been made by researchers to develop membranes with high permeability and solute selectivity. Hydrophilic functionalized graphene derivatives as nanofillers have attracted much attention to increasing membrane permeability. Hydrophilic functional groups promote faster water transport by forming hydrogen bonds with the water molecules. Also, the incorporation of graphene sheets into the polymer matrix increases the free volume by disrupting polymer chain packing which enhances the permeate flux [25]. Furthermore, inter-layer spacing of functionalized graphene could act as a nanochannel for the smooth transport of water molecules [26]. This interlayer spacing can be tuned by changing the functional groups and the solute rejection capacity can be increased [27].

Water permeability and solute rejection are mainly governed by the molecular sieving mechanism of a porous membrane. A uniform cylindrical pore with high density is desirable for an efficient membrane. In practice, sponge-like tortuous pores are formed during the phase inversion of polymeric membranes with a non-uniform size distribution. The addition of functionalized graphene into the polymer matrix promotes the formation of a finger-like pore structure that increases the porosity and pore size. Thus the developed nanocomposite membranes are favorable for better water permeation compared to a sponge-like structure. However, incorporation of excess graphene into the polymer matrix increases the viscosity of the casting solutions which decreases the phase inversion rate and prevents larger size finger-like pore structure.

In MF and UF membrane application the effect of graphene incorporation into the polymer matrix is directly reflected in their water permeation and solute rejection results. In the case of NF, RO, and FO membranes addition of graphene and its derivatives in the support layer of TFC membrane also influence the formation of the PA active layer during IP and the overall separation performance is affected. Furthermore, graphene derivatives are incorporated into the PA active layer to form thin film nanocomposite (TFN) membranes. This also significantly influences water permeability and solute rejection performance. In the last six years, different nanocomposite membranes, for example, MF, UF, NF, RO, and FO developed by incorporating graphene and its derivatives are presented in (Table 19.1−19.5) respectively. The tables describe different types of membrane composition and their separation performances.

According to Lai et al. with increasing GO concentration into the PSf support layer the PA active layer thickness was increased during the use of polyisoprene (PIP) in the aqueous phase [28]. They argued with increasing GO, the hydrophilicity of PSf/GO surface increases resulting in much more accumulation of PIP monomers. This large amount of PIP monomers interacted with TMC solution to form a thick membrane layer. The increased PA layer thickness exhibits more resistance to permeation. On the contrary, the use of m-phynelene diamine (MPD) monomer instead of PIP didn't show a significant increase in PA active layer on GO incorporated PSf support layer [29]. In another study, the effect of surface roughness of GO/PSf support on the PA layer was observed by Park et al. [30]. They identified that a small GO content is beneficial for IP, whereas, higher GO loading reduces the IP efficiency and results in poor solute rejection performance of the membrane. The surface roughness of the PA layer is also related to the GO content of the support layer [31].

Other than pure GO, functionalized GO also played an important role in improving water permeability when incorporated into the PA active layer of TFC membranes. Mass transport through this dense skin layer is mainly governed by the solution-diffusion mechanism. The addition of a small quantity of functionalized graphene into the dense PA layer improved the water permeability because of the increase in hydrophilicity of the composite layer. Further, the addition of functionalized graphene hinders the cross-linking process between the -C(=O)Cl group and amine group during IP. This reduces the cross-link density of the PA layer lowers the permeation resistance and consequently the flux increases [14]. However, at a higher concentration of functionalized graphene,

Table 19.1 Graphene-polymer nanocomposites as microfiltration (MF) membranes.

Filler material	Polymer matrix	Membrane configuration	Pressure kPa	Pure water flux (LMH)	Rejection	Antifouling	References
GO	PSf	Flat sheet	NA	NA	MB: 84.2%	NA	[6]
GO	Aminated-PAN	Flat sheet	NA	NA	Oil: 98.1%	FRR (Oil): 67.9 %	[49]
g-C$_3$N$_4$-GO	CA	Flat sheet	NA	957	Polystyrene (PS) (360 nm): 50 %, PS (430 nm): 94 %	NA	[50]

Table 19.2 Graphene-polymer nanocomposites as ultrafiltration (UF) membranes.

Filler material	Polymer matrix	Membrane configuration	Pressure kPa	Pure water flux (LMH)	Rejection	Antifouling	References
Cysteine-GO	PES	Flat sheet	200	82.6	BSA: 99.8%	FRR (BSA): 92.1%	[7]
PEG-GO	PVDF	Flat sheet	100	93	BSA: >94%	FRR (BSA): 78%	[51]
PFSA-g-GO	PVDF	Flat sheet	100	587.4	BSA: 93.9% HA: 79.6%	FRR (BSA): 90.8%	[52]
GO-Al$_2$O$_3$	PSf	Flat sheet	87	16.3	Oil: 74%	NA	[53]
GO & rGO	PBI	Flat sheet	NA	NA	Oil: 99.9%	FRR: 91.8%	[54]
GO	PSf	Flat sheet	100	NA	BSA: 97%	FRR BSA: 99%	[55]
Flat GO & crumpled GO	PSf	Flat sheet	100	123	(BSA): 100%(MO): 41%	FRR (BSA): 52%	[56]
Guanidyl-graphene	PSf	Flat sheet	100	217	BSA: 95%	FRR (BSA): 82.4%	[23]
GO-ND	PVC	Flat sheet	200	440	BSA: 95%	FRR (BSA): 83%	[10]
GO	PES	Flat sheet	100	340	HA: 94.5%	FRR (HA): ~95%	[57]
GO/MOF	CA	Flat sheet	150	183.5	BSA: 95.37%	FRR (BSA): 88.1 %	[58]
Fe$_3$O$_4$/ GO-COOH	PES	Hollow fiber	50	110	Lysozyme: 92.9%, trypsin: 94.5%, pepsin: 96.9%, albumin: 99.5%, γ-globulin: 100%, & fibrinogen: 100%	FRR: 97.8%	[59]
GO	PES/SPSf	Flat sheet	100	816.9	BSA: 99.5%	FRR (BSA): 94.2 %	[60]
PEI-g-gO	Nylon	Flat sheet	NA	450 (per bar)	Dye: >99%	NA	[61]

(Continued)

Table 19.2 (Continued)

GO-CNTs	PVDF	Flat sheet	100	220	BSA: ~55%	FRR: (BSA): 90 %	[62]
Zwitterion modified GO, GO-g-PMSA	PVDF	Flat sheet	700	42.9	NaCl: ~80%	FRR: 95.3%,	[63]
GO-ND	PVC	Flat sheet	200	440	BSA: 95.1%	FRR: (BSA) 83.1%	[10]
Nano GO	PSf	Flat sheet	100	620	Nitrate: 22%	FRR: $Escherichia$ $coli$: 95%	[64]
Ag NPs/GO	PSf	Flat sheet	200	66.7	NA	FRR: BSA: 92.5% $E.$ $coli$: 70.7% $Staphylococcus$ $aureus$: 61.8%	[65]
SiO$_2$/GO	PAN	Hollow fiber	10	3151	Oil: >99%	FRR: Oil: 99%	[8]
GO	PEES	Flat sheet	345	186	NA	FRR (BSA): 83.2 %	[66]
SGO	PSf	Flat sheet	200	175.2	BSA: >98%	NA	[21]
GO	PES	Flat sheet	100	245	BSA: 93.3%	FRR (BSA): 80%	[67]
PEI-g-GO	PES	Flat sheet	NA	322.8	NA	NA	[22]
GO	PAI	Flat sheet	350	77	BSA: 88% HA: 89%	BSA: 90% HA: 86%	[68]
GO	PES	Hollow fiber	100	30	NA	FRR (BSA): 93%	[69]
GO	PES	Filtration	350	150	HA: 85%	FRR (HA): 89.5%	[70]
SGO	PFSA/PVP/PVDF	Hollow fiber	100	156.2	HA: 98.4% BSA: 96.6%	FRR (HA): 94%	[71]
GO	PVP/PSF	Flat sheet	140	309.2	NA	FRR (BSA): 90.4%	[72]
GO	PVP/PEI	Flat sheet	200	101.5	HA: 77.4%, Cd^{2+}: 92.4%, Ni^{2+}: 95.5%	FRR (HA): 94%	[73]

Material	Additive/polymer	Type			Rejection	FRR	Ref.
GO	PVP/PVC	Flat sheet	100	430	BSA: 91.2%	FRR (BSA): 70.4%	[74]
Amine-graphene-PANCMI	PVP/PES	Hollow fiber	100	767	Oil: >99%	NA	[48]
GO-TiO$_2$	PVP/PVDF	Flat sheet	100	487.8	BSA: 92.5%	FRR (BSA): 71.1%	[45]
GO-Fe$_3$O$_4$	PVP/PVDF	Flat sheet	100	595.4	BSA: 92%	FRR (BSA): 86.4%	[17]
SGO	PVP/PVDF	Flat sheet	100	740	BSA: 98%	FRR (BSA): 88.7%	[36]
Crumpled GO	PAA/PES	Filtration	100	220	BSA: 91%	NA	[75]
rGO/TiO$_2$	PVP/PVDF	Flat sheet	300	221	BSA: 99.3%	FRR (protein): 94.9%	[39]
GO/Co$_3$O$_4$	PVP/PES	Flat sheet	100	365	BSA: 94%	FRR (sludge): 81.1%	[37]
Ag/Halloysite nanotube/rGO	PVP/PES	Flat sheet	100	158.4	PEG: 84%	FRR (BSA): 95%	[76]

Note: *CA*, Cellulose acetate; *GO-ND*, (graphene oxide-nanodiamond); *PAI*, poly amide imide; *PANCMI*, poly(acrylonitrile-co-maleimide); *PBI*, Polybenzimidazole; *PEES*, Polyphenylene-ether-ether-sulfone; *PEI*, Polyethylenimine; *PFSA*, perfluorosulfonic acid; *PSMA*, polystyrene maleic acid; *PVP*, Polyvinylpyrrolidone; *SGO*, Sulfonated-GO; *SPSf*, sulfonated polysulfone.

Table 19.3 Graphene-polymer nanocomposite as nanofiltration (NF) membranes.

Filler material	Active layer/ support	Membrane configuration	Pressure (kPa)	Pure water flux (LMH)	Separation performance		Antifouling/ biofouling	References
					Flux (LMH)	Rejection		
GO	PES	Flat sheet	500	45	NA	NaCl: 72% Na$_2$SO$_4$: 68%, MgSO$_4$:65%) Heavy metal (Zn: 88%, Cu: 72%, Cd: 68%); Dye (MO: 88%, MB: 70%)	FRR (BSA): 85%	[77]
GO	MPD-TMC /PES	Flat sheet	1600	NA	200	Tennary wastewater (Cr: 99.5%, COD: >99%, TDS: >96%)	NA	[78]
GO	MPD-TMC /PES	Hollow fiber	800	10.2	NA	NaCl: 97.67%	NA	[79]
PSBMA-g-GO	MPD-TMC /PES	Flat sheet	1360	NA	5.4	NaCl: 94.3%, MgSO$_4$: 97.6%	FRR (BSA): >95%	[80]
GO	MPD-TMC /PSf	Flat sheet	2070	NA	47.6	NaCl: 97%	NA	[81]
GO-EDA & GO-PEI	PIP-TMC /PSf	Flat sheet	NA	59.6 & 62.1	11.9 & 12.4	Na$_2$SO$_4$: 98.2% both MgSO$_4$: 97.4% & 97.8% MgCl$_2$: 92.1% & 93.4 NaCl: 35.9% & 38.2%	FRR (BSA): >90%	[82]
Sulfonated-GO	PIP-TMC /PSf	Flat sheet	500	11.9	NA	Na$_2$SO$_4$: 96.5% MgSO$_4$: 98.1% NaCl: 77.6%	FRR (BSA): 98%	[83]
rGO/TiO$_2$/Ag	MPD-TMC on Si$_3$N$_4$/PES	Hollow fiber	700	NA	52	Salt (Na$_2$SO$_4$): 96% Dye: 98%	E. coli: 100%	[84]
GO	PIP-TMC /PSf	Flat sheet	600	46.9	NA	Na$_2$SO$_4$: 98%	NA	[85]
GO/Ag	PVDF	Flat sheet	7	348.8	NA	NA	FRR: 88%	[86]
APTS-GO	PVDF	Flat sheet	NA	NA	6.2	NaCl: 99.9%	NA	[87]
GQD	PIP-TMC /PES	Flat sheet	200	102	NA	Dye (CR): 99.9% (Orange G II): 95%	FRR: (BSA) 91.9%	[88]

GO	PIP-TMC /PSf	Flat sheet	276	242	66.6	$MgSO_4$: 90%	NA	[89]
GO/Chitosan	PES	Flat sheet	300	NA	55	Fe: 99%,Mn: 85%	FRR (BSA): 52% (E. coli): 93%	[90]
GO	PIP-TMC /PES	Flat sheet	400	62.6	NA	Na_2SO_4: 96.6% $MgSO_4$: 90.5%	NA	[91]
GO	PES	Flat sheet	414	70	NA	TOC: 59%	FRR: >90%	[92]
GO	PIP-TMC /PSf	Flat sheet	800	39.4	33.6	Na_2SO_4: 95.8% $MgSO_4$: 97.7%	NA	[93]
GO/ZnO	MPD-TMC /PSf	Flat sheet	2070	42.5	42.4	NaCl: 97%, Na_2SO_4: 98%	FRR (HA): 97%	[94]
GOQD	TA- IPDI/ PAN	Flat sheet	200	23.3	18.2	Na_2SO_4: 66.7% NaCl: 17.2%, CR:99.8%, MB: 97.6%	FRR (BSA): 91.7%	[95]
GO	PIP-TMC/PSf	Flat sheet	800	33.04	NA	NaCl: 37.4%, Na_2SO_4: 99.2%, $MgSO_4$: 99.1%	FRR (BSA): 98.9%	[96]
SG	PVP/PES	Flat sheet	2000	190.5	NA	Dye: 91%	FRR (BSA): 87.5%	[97]
GO	MPD-TMC /PSf	Flat sheet	1500	NA	29.6	NaCl: 98%	FRR (BSA): 85%	[98]
GO	MPD-TMC /PES	Flat sheet	2070	59.4	NA	NaCl: 93.8%	NA	[99]
GO	PIP-TMC/ PSf	Flat sheet	800	5	18	Na_2SO_4: 95% $MgSO_4$: 91%	NA	[28]
GO	PIP-TMC/ PSf	Flat sheet	600	87.6	NA	Na_2SO_4: 98%, $MgSO_4$: 97%	FRR (BSA): 95% (HA): 90%	[100]
GO/ZIF-8	PIP-TMC/ PSf	Flat sheet	800	32.5	NA	Na_2SO_4: 100% $MgSO_4$: 70%	FRR (E. coli): 84%	[101]
GO/TiO2	PSf	Flat sheet	150	NA	51.3	NaCl: 99.5%	NA	[102]
rGO/TiO2	PIP-TMC/ PSf	Flat sheet	1000	NA	60.6	Na_2SO_4: 93.5%	NA	[103]
GO	MPD-TMC /PSf	Flat sheet	1000	NA	85	$MgSO_4$: 98%	NA	[25]

Note: APTS, Aminopropyl triethoxy silane; CR, congo red; EDA, ethylene diamine; GOQD, graphene oxide quantum dot; IPDI, isophorone diisocyanate; MB, methylene blue; MO, methyl orange; PIP, piperazine; PSBMA, polysulfobetaine methacrylate); SG, sulfonated graphene; TA, tannic acid.

Table 19.4 Graphene-polymer nanocomposites as reverse osmosis (RO) membranes.

Filler material	Active layer/ polymer support	Membrane configuration	Separation performance			Antifouling/biofouling	References
			Pressure (kPa)	Flux (LMH)	Rejection (NaCl)		
P-Amino phenol modified-GO	MPD-TMC/PSf	Flat sheet	1500	23.6	99.7	*Escherichia coli:* 96.8% *Staphylococcus aureus:* 95.3%	[104]
GOQD/ Ag$_3$PO$_4$	MPD-TMC/PSf	Flat sheet	1600	39.6	98.4%	*E. coli:* 99.9%	[105]
GO/CA	Polyester non-woven fabric	Flat sheet	2500	65	90%	FRR: 77%	[106]
N-GOQD	MPD-TMC/PSf	Flat sheet	1500	27	92.1%	NA	[107]
GO/ZnO	MPD-TMC/PSf	Flat sheet	2000	31.4	96.3%	FRR (HA): 86.3%	[108]
Zwitterion-g-GO	MPD-TMC/ PVDF	Flat sheet	1200	17.5	94.8%	FRR (BSA) 90.5%	[109]
GO/hydrogel	PES	Flat sheet	1000	33.5	98.5%	NA	[110]
GO	MPD-TMC/PSf	Flat sheet	2068	NA	Na$_2$SO$_4$: 97%	NA	[99]
GO	MPD-TMC/PSf	Flat sheet	1500	29.6	98%	FRR (BSA): 85%	[98]
GOQD	MPD-TMC/PSf	Flat sheet	1600	37.5	98.8%	FRR (BSA): 94% (HA): 90%	[111]
GO	MPD-TMC/PSf	Flat sheet	1550	32.8	99.4%	NA	[112]
GO	MPD-TMC/PSf	Flat sheet	1550	84	98.2%	NA	[29]
rGO/TiO$_2$	MPD-TMC/PSf	Flat sheet	1500	51.3	99.5%	NA	[102]

Table 19.5 Graphene-polymer nanocomposites as forward osmosis (FO) membranes.

Filler material	Active layer/ polymer support	Membrane configuration	Draw solution type	FO flux (LMH)	Reverse salt flux (FO) g/m²/h (gMH)	Rejection	Anti-fouling	References
GO	MPD-TMC/PSf	Flat sheet	0.5 M NaCl	24.7	8.4	NA	FRR: 98%	[113]
GO	MPD-TMC/PSf	Flat sheet	2 M NaCl	34.3	1.1	Pb: 99.9%, Cd: 99.7%, Cr: 98.3%	FRR: 96%	[114]
GQDs/MOF (UiO-66-NH$_2$)	MPD-TMC/PES	Flat sheet	1 M NaCl	59.3	19.1	NA	FRR (BSA): 82%	[115]
GQD	MPD-TMC /PVDF	Flat sheet	1 M NaCl	170	NA	MB: 90%, CR: 85% NaCl: 94%, Mg(NO$_3$)$_2$: 99%	FRR: 80%	[116]
GQD	PEI-TMC/ PAN	Flat sheet	0.5 M MgCl$_2$	12.9	1.41	NA	FRR (BSA & HA): 99.8%	[117]
GO	NIPAM-MBA/ Nylon	Flat sheet	0.5 M NaCl	58	2.03	NA	NA	[118]
GO	MPD-TMC/ PDA/ PSf	Flat sheet	1 M NaCl	24.3	3.82	NA	NA	[119]
GO/Fe$_3$O$_4$	MPD-TMC/ PES	Flat sheet	1 M NaCl	55	9.5	NaCl: 94.3%	FRR (BSA): 94.5%	[120]
GO	PAH-m-MXDA -TMC /PES	Flat sheet	0.25 M TSC	13.2	25.8	NA	NA	[121]
GO-CA	MPD-TMC /PSf	Flat sheet	1 M NaCl	14.5	2.6	NaCl: 88%	NA	[122]
GO-chitosan	NA	Flat sheet	1 M NaCl	19	5.8	NA	NA	[123]
GO	SPES/PES	Flat sheet	1 M Na$_2$SO$_4$	25	1.4	NA	FRR (SA): 97.5%	[124]
GO-LDH	MPD-TMC/PSf	Flat sheet	1 M NaCl	33.8	6.4	NA	NA	[125]
GO	MPD-TMC/PSf	Flat sheet	1 M NaCl	13.4	6.2	NA	NA	[31]
GO	MPD-TMC /HPAN	Flat sheet	2 M NaCl	26.5	1.9	94.6%	FRR (SA): 90%	[126]
GO-PDA	PA/PSf	Flat sheet	2 M NaCl	13.6	0.7	NA	FRR (ATP): 98.5%	[127]
GO	MPD-TMC /PSf	Flat sheet	0.5 M NaCl	19.8	3.4	98.7%	NA	[30]
GO/ gC$_3$N$_4$	MPD-TMC /PES	Flat sheet	2 M NaCl	41.4	9.5	NA	NA	[128]

Note: *LDH*, layered double hydroxide; *MXDA*, m-xylylenediamine; *NIPAM*, N-isopropyl acrylamide; *MPD*, m-phenylenediamine; *PAH*, polyallylamine; *PDA*, polydopamine.

agglomeration occurs into the PA layer that lowers the water permeability. Also, solute rejection decreases due to the defect formation by agglomerated nano-filler materials. The surface roughness of the active layer initially decreases with the addition of functionalized GO due to hindrance of IP reaction and then with an increase in filler concentration agglomeration occurs which increases in surface roughness. Thus, the incorporation of graphene and its derivatives into the polymeric membrane matrix have shown some unique features to improve the permeability and selectivity performance of the polymeric membranes.

2. Antifouling property:

Membrane separation performances, mainly the permeation and separation selectivity gradually deteriorate with time during its operation. This fall in performance occurs mainly due to membrane fouling on its surface and inside pores. Different organic, biological, and inorganic foulants are accumulated and cause reversible or irreversible fouling which reduces the membrane lifespan and simultaneously increases the operation cost [32]. Therefore, the development of membranes with low fouling properties is highly desirable. In practice, MF, UF, and the support membrane of TFC which are prepared by phase inversion contain a wide pore size distribution. These membranes go through irreversible fouling by the diffusion of foulants into the pore structure [33,34]. Commonly used polymers such as PSf, PES, PVDF, PAN, etc. are hydrophobic in nature and enhance the adsorption of organic and biological foulants on the membrane surface and pore wall thereby increasing the fouling properties [35]. The active layer of the TFC membrane is also prone to fouling due to the same reason. Surface roughness is another issue of membrane fouling by more interaction of the foulants with the increased surface area.

The incorporation of graphene and its functionalized derivatives into the membrane matrix can reduce the fouling characteristics. This behavior is proved by almost all types of graphene-based composite membranes in the literatures as summarized in Table 19.1—19.5. The flux recovery ratio (FRR) value signifies the potential of the membranes towards organic fouling resistance. Common types of organic foulants are bovine serum albumin (BSA), humic acid (HA), and sodium alginate (SA) that are hydrophobic and negatively charged in normal conditions. They exhibit increased fouling behavior with the unmodified hydrophobic polymeric membranes. The addition of functionalized graphene into the polymer matrix increases the hydrophilicity of the membrane surface which minimizes the interaction with the organic foulants. Also, the hydrophilic groups form strong hydrogen bonds with the water molecules to form a boundary layer and hinder the interaction of organic molecules with the membrane surface. Further, the negative charge of the functionalized graphene derivatives leads to a negative surface charge of the membrane that reveals an electrostatic repulsion with the negatively charged foulants. Thus, the incorporation of graphene nanomaterials into the polymeric membranes increases the antifouling properties of the membrane. In a report, Ayyaru and Ahn [36] studied that sulfonated GO (SGO) was more negatively charged than GO and so SGO incorporated membrane showed higher fouling resistance than GO incorporated membrane. The FRR values of the SGO and GO based membranes after BSA fouling were 88.7% and 75% respectively. Further, several studies described that incorporation of an optimum amount of functionalized GO results smoother surface which exhibited better antifouling property [37—39]. A few research groups explored that GO is a more effective nanofiller than other carbon-based nanomaterials in anti-biofouling or antibacterial membrane application [40,41]. When GO forms a covalent bond to the PA layer surface, the TFC membrane showed very high antibacterial activity [42]. This study exhibited that, direct contact of GO with the *Escherichia coli* cells leads to 65% bacteria inactivation within 1 h.

In addition, to graphene and its functionalized derivatives, GO composite with other nanomaterials when used as fillers in polymer matrix also revealed very high antifouling properties. Particularly, GO with different transition metals or metal oxides and metal organic frameworks (MOFs) are applied to develop antifouling nanocomposite membranes. Sometimes, the release of few metal ions (e.g., Zn, Ag) from the composite showed high fouling resistance property because of their inherent antibacterial activity [43,44]. Also, a few metal oxides such as TiO_2 [45] and Co_3O_4 [37] can generate free radicals under UV light that degrades most of the organic compounds and inactivates the bacteria in the membranes. However, functionalized derivatives of graphene minimize the fouling properties of composite membranes to a greater extent by increasing the hydrophilicity of the membrane surface.

3. Mechanical properties:

The mechanical stress on the membrane surface is generated during its compaction as well as an operational condition through the feed pressure. Under excess pressure, the membrane pore size and shape may change which affects the separation performance and membrane stability. At very high pressure or mechanical stress, the membrane surface may face a permanent plastic deformation or cracks can appear on the active layer which ultimately damaged the membrane.

Graphene exhibits a very high Young's modulus, tensile strength, surface area, and stability [46]. When this nanomaterial is incorporated into the polymeric membrane, those superior properties are obtained in the composite membranes. For example, by functionalization of graphene and its derivatives, uniform dispersion and proper adhesion with the polymer matrix phase are achieved to enable the load dissipation of the nanofillers. Following this mechanism, the mechanical strength of the commonly used polymeric membranes, such as PVDF, PSf, PES, etc., are often reported in the literature [19,29,31]. The incorporation of graphene nanomaterials restricts the polymer chain movement that improves the mechanical stability of the membrane [47]. A few studies also showed that incorporation of organic species functionalized graphene improves the elongation-at-break of the membrane along with the strength and modulus [40,48]. The organic species are mainly long chain organosilanes or polymer chains. These functionalized graphenes merge the stiffness of an inorganic species with the elasticity of the organic materials. Thus, graphene and its derivatives have very high mechanical strength, when incorporated into the polymer matrix withstand excess mechanical stress and impart enhanced stability to the composite membranes.

19.3 Gas separation

Polymer-graphene nanocomposite membranes with their widespread application in water filtration also exhibit some prospective in gas separation application. Graphene and its derivatives such as GO, functionalized graphene/GO, multilayer graphene, porous graphene, etc. have been extensively studied to develop nanocomposite membranes [129]. These nanofiller reinforced polymeric membranes especially have been used as a gas-impermeable layer. A very high aspect ratio of graphene-based nanofillers has been found to reveal gas barrier properties of commercial polymer films. Such nanocomposite membranes possess very high mechanical and thermal properties along with their gas barrier performances [130,131]. Furthermore, any physical and chemical interaction between the nanofiller and polymer matrix has been observed to control the gas transport properties of nanocomposite membranes [132]. The incorporation of graphene-based nanofillers lowers the gas permeability of composite membranes than pure polymer matrix due to the increase in tortuous diffusion paths [133].

Yang et al. [134] reported very low loading of nitrogen-doped graphene (N-G) nanosheets into polyimide matrix and developed thin dense nanocomposite membrane layer for O_2/N_2 and CO_2/N_2 separation. They showed that the presence of nitrogen and oxygen containing functional groups of graphene sheets impart very good compatibility with the polymer matrix that results in higher separation selectivity. Widakdo et al. [135] open up a new research avenue by developing electrically responsive 'smart' PVDF-graphene nanocomposite membranes for gas separation. With increasing graphene content self-piezoelectric property of the PVDF phase was observed by its self-assembly. On application of external voltage to the composite membrane CO_2 single gas permeability and selectivity over N_2 and O_2 increases. Graphene oxide (GO) and poly (dimethyl siloxane) (PDMS) composite membrane was developed by Nigiz et al. for hydrogen purification from its CO_2 mixture [136]. A commercial CO_2-philic co-polymer, polyether-block-amide (Pebax) has been studied widely for CO_2/N_2 separation due to its potential CO_2 permeation and separation performance. In a report, oligomers of poly(ethylene glycol) methyl ether acrylate (PEG-MEA) were added to the Pebax polymer matrix to improve the plasticization of the composite membrane [137]. In this Pebax/PEG mixture GO was incorporated to increase the gas diffusion channel and CO_2 solubility. Polyvinyl chloride (PVC) and GO composite membrane have been synthesized by Raj et al. for CO_2/CH_4 separation [138]. Polyaniline (PANI) and GO composite membranes have been constructed by Kweon et al. for CO_2/N_2 separation [139]. They reported that the incorporation of GO into the PANI matrix enhances CO_2 permeability and thermal stability to a higher extent. A glassy polymer, Poly(2,6-dimethyl-1,4-phenylene oxide) (PPO) has been used as a matrix and few layer graphene as filler to develop PPO/Graphene mixed-matrix membrane (MMM) for CO_2 separation [140]. The study explained that up to a threshold value (<1 wt%) of graphene the permeability and selectivity increases, but at higher loading gas permeability decreases due to the predominance of increased tortuosity. Ge et al. [141] reported in-situ polymerization of poly(amic acid) in presence of aminated-GO filler to form polyimide(PI)/amine-GO MMM. The in-situ polymerization process imparts enhanced interaction and uniform dispersion of the amine-GO filler in the PI matrix. The MMMs showed excellent CO_2 permeability and CO_2/N_2 separation selectivity compared to the pristine polymeric membrane. Polymers of Intrinsic Microporosity (PIM-1) and covalently modified GO MMMs were developed by Aliyev et al. and their different single gas (e.g., O_2, N_2, CO_2, etc.) permeability and selectivity were evaluated [142]. Amine-modified epoxy resin membranes have been developed by Roilo et al. and the polymer chain rigidification was observed upon addition of few-layer graphene nanoplatelets into the polymer matrix [143]. Different single gas such as H_2, N_2, and CO_2 permeation studies, were performed with the developed nanocomposite membranes. With increasing filler content as the polymer matrix becomes more rigid, gas permeability is also found to decrease. Hence, polymer-graphene nanocomposite membranes generally have revealed lower gas solubility and permeability compared to the other conventional polymer nanocomposite membranes [144]. However, graphene-based nanocomposite membranes in gas separation application are still in the early stage of research and more detailed studies in the future will give the fundamental mechanism of gas transport. The future scope of research will include the fabrication of graphene-based nanocomposite membranes for the separation of the actual gas mixture.

19.4   Dialysis/electrodialysis

Polymer graphene composite has emerged as a popular material to be used for dialysis or electrodialysis applications. The last decade has evidenced several advancements in this field that show improved separation ability, selective separation, higher durability, etc. Various types of dialysis process are gaining popularity for many applications like acid recovery from acid waste stream [145], ion dialysis separation [146], dialysis bag for Gd(II) ions [147], hemodialysis [148], blood purification [149], bio-artificial kidney applications [150]. Cseri et al. [151] reported graphene oxide—polybenzimidazolium nanocomposite for use as anion exchange membranes for electrodialysis [151]. According to this study, a small amount of GO is sufficient for achieving good mechanical strength and permselectivity. It is found to exhibit improved tensile modulus (up to 1.2 GPa), Young's modulus (up to 4.3 GPa), and high ion exchange capacity (1.7—2.1 mmol/g). Hosseini et al. [152] used PANI GO composite to fabricate electrodialysis heterogeneous ion exchange membranes. Ion exchange capacity is found to be improved to 1.37 meq/g from 1.2 meq/g of pristine membrane. Transport number and permselectivity are significantly improved due to only 4% GO incorporation in the polymer matrix [152]. Kidambi et al. [153] reported a method to obtain a graphene-based membrane which was found to exhibit higher efficiency for dialysis applications. Polycarbonate track etched supports that are used on graphenes are pressed. IP is performed over it using hexamethylene diamine and adipoyl chloride in a water/hexane system. Size selective separation of small molecules and proteins was evidenced for graphene incorporated membrane. It shows very high selectivity for the separation of potassium chloride from its mixture with Allura and vitamin B12 [153]. Polysulfone/graphene quantum dots composites are reported by Yadav et al. [154] as anion exchange membrane. This membrane shows 35.5%—47.5% acid recovery via diffusion dialysis. Separation factor (S is the ratio of dialysis coefficients of two species) was reported for this membrane in the range of 32—39. In a report, the separation factor is 24.5—34 for the only polysulfone based functionalized membrane [154]. Hence, it can be said, incorporation of graphene enhances the separation efficiency for the membrane used in dialysis. Guo et al. [155] used

the modified Hammer method to obtain modified graphene oxide nanosheet hydrosols. A higher adsorption capacity of 893 mg/g of Pb(II) ion was reported for the
hydrosol incorporated dialysis bags [155]. In another study, Huang et al. pointed
out the enhancement of adsorption of dye (amido black 10B) and Cupric ion from
an aqueous medium with the increase in graphene incorporation in O-(carboxymethyl)-chitosan membranes [156]. The results of adsorption reported by Huang
et al. [156] are shown in Fig. 19.1 (reprinted with the permission taken from
Elsevier) [156].

Reduced graphene oxide-based hydrogels were synthesized by Zeng et al. [157]
using sodium acrylate and N-isopropyl acrylamide as monomers. This study shows
a significant increase in water flux due to the incorporation of 1.2 wt.% rGO in the
hydrogel matrix [157]. Cross-linked chitosan/SiO$_2$-loaded graphene was reported to
have excellent adsorption capacity for bilirubin. Incorporation of graphene results

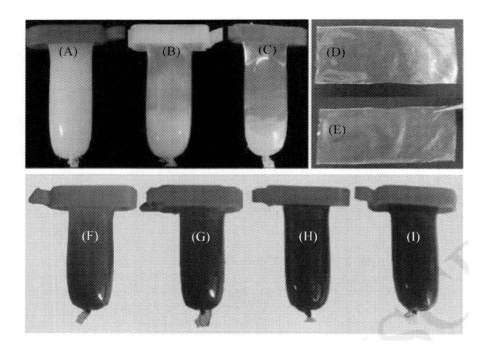

Figure 19.1 Photographs of O-CMC after dialyzed at pH of 4 (A), 3.5 (B), and 4.5 (C),
Dialytic side (D) and depositional side (E) of O-CMC membranes. O-CMC-0.1 GO (F),
O-CMC-0.2GO (G), O-CMC-0.1 rGO (H) and O-CMC-0.2 rGO (I) after being dialyzed at
pH = 4.
Source: Reprinted with permission taken from Elsevier. Q. Huang, G. Li, M. Chen, S. Dong,
Graphene oxide functionalized O-(carboxymethyl)-chitosan membranes: Fabrication using
dialysis and applications in water purification. Colloids Surf. A: Physicochem. Eng. Asp. 554
(2018) 27–33.

in an increase in adsorption capacity from 22.47% to 77.87%. Optical microscopic images of graphene-loaded chitosan microsphere and pristine chitosan microsphere are shown in Fig. 19.2.

Results of the Bilirubin adsorption study for this chitosan/graphene-based hydrogels are shown in Fig. 19.3 (reprinted with permission taken from Elsevier) [158]. The incorporation of graphene in the chitosan matrix was evidenced to improve the adsorption characteristics of the polymer matrix to a large extent. Moreover, it also improves the mechanical strength of the polymer matrix.

Fe_3O_4/Polydopamine/rGO-xerogel was reported by Guo et al. [159] for arsenic and arsenate removal. It was evidenced to have the ability to remove As(III) and As(V) ions in sub-ppm levels from contaminated aqueous medium [159]. A few examples of graphene/polymer composite membranes for application in dialysis are listed in Table 19.6.

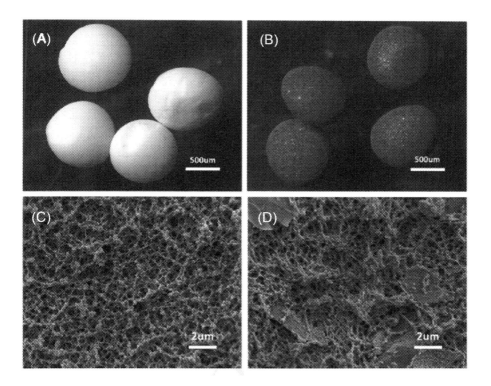

Figure 19.2 Optical microscopic images (A and B) and SEM images (C and D) for CS (A and C) and CS/graphene-SiO_2 beads (B and D).
Source: Reprinted with the permission taken from Elsevier. J. Chen, Y. Ma, L. Wang, W. Han, Y. Chai, T. Wang, et al., Preparation of chitosan/SiO2-loaded graphene composite beads for efficient removal of bilirubin, Carbon 143 (2019) 352−361.

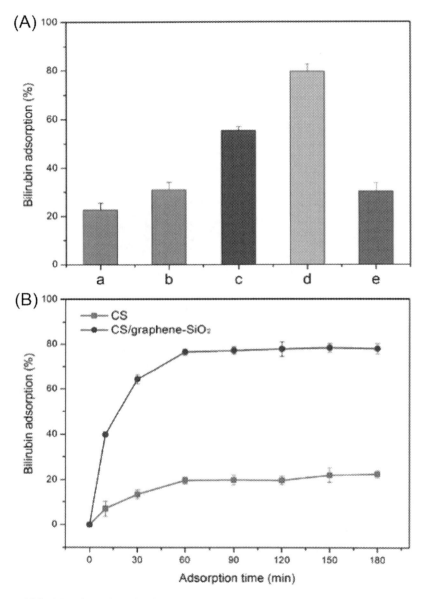

Figure 19.3 (A) Adsorption of bilirubin by using different adsorbents: CBIL = 200 mg/L, adsorbent dosage = 2.3 mg/mL, $T = 37°C$, $t = 2$ h. (a) CS (b) 2.5 wt.% CS/graphene-SiO$_2$ (c) 5 wt.% CS/graphene-SiO$_2$ (d) 10 wt.% CS/graphene-SiO$_2$ (e) 10 wt.% CS/graphene. (B) Adsorption capability of bilirubin by 10 wt.% CS/graphene-SiO$_2$ beads as a function of contact time. CBIL = 200 mg/L, adsorbent dosage = 2.3 mg/mL, $T = 37°C$.
Source: Reprinted with the permission taken from Elsevier. J. Chen, Y. Ma, L. Wang, W. Han, Y. Chai, T. Wang, et al., Preparation of chitosan/SiO$_2$-loaded graphene composite beads for efficient removal of bilirubin, Carbon 143 (2019) 352−361.

Table 19.6 A few examples of graphene incorporated membrane for the application in dialysis.

Materials	Observation	References
Polysulfone/graphene quantum dots composite	Acid recovery via selective diffusion dialysis	[145]
Graphene oxide functionalized O-(carboxymethyl)-chitosan	Amido black 10B dye and Cu(II) ion adsorption was reported	[156]
Reduced GO/ sodium acrylate /N-isopropyl acrylamide	Increase in water flux.	[157]
Graphene-SiO$_2$/Chitosan	>4 fold increase in percentage of bilirubin adsorption	[158]
Fe$_3$O$_4$/Polydopamine/rGO-xerogel	The adsorption capacity is >240 ug/g for As(III) and ~180 for As(V) at pH 7	[159]

19.5 Energy storage devices

Graphene is nothing but a "carbon monolayer packed into a 2D honeycomb lattice" [160]. It is composed of defect-free flat carbon monolayers which are considered as an active material when used in fabricating energy storage devices. Graphene's superior encapsulation ability for entrapping lithium and sodium ions can be used to store electrostatic charge and hence can be used to develop electrochemical double layer capacitor. It is also known to function as a catalyst in metal-air batteries. RGO is used in several research papers for lithium ion storage due to its high capacity (>2000 mAh/g) during the first lithium ion introduction [161]. Olabi et al. [162], mentioned two primary purposes for the use of graphene materials in their review article; (1) decreasing Pt or Pt alloy catalyst loading and (2) replacing the Pt or Pt alloy catalyst. Graphene is very popular as catalyst support for Pt for "oxygen reduction reactions" [162]. The major problem for such materials is associated with defects which results in a decrease in activation for oxygen reduction reaction (ORR) and thus results in deterioration of the stability of HO*. Boron and nitrogen are used to dope the graphene molecules in order to remediate the problem. Such doping process increases the distribution of Platinum and thus improves the ORR activity. Graphene sponges, foam, and aerogels were also developed in the past few years for supercapacitor applications. Production of the superhydrophobic surface using graphene is another achievement by researchers recently. In a recently published review article, Nikolas Natter and coauthors discussed two types of graphenes; (1) graphene nanoplatelets and (2) few-layer graphene flakes which are obtained from plasma-induced exfoliation of graphite. The second one was evidenced to function as a "proper electrode material for ion electrosorption" in a model supercapacitor device [163]. Ultrathin graphene-alumina membrane has a high-power density of 7200 W/kg which has the potential to improve the performance of supercapacitor. Graphene/MnO$_2$ was also evidenced to exhibit excellent specific capacitance (256 F/g at a current density of 500 mA/g) [164].

19.6   Conclusions

Graphene has emerged as a wonder material for developing membranes for several applications. Properties of various polymeric membranes are greatly influenced by the incorporation of graphene in terms of its mechanical strength as well as permeability. Furthermore, ion-specific separation ability is greatly improved due to the incorporation of graphene. As shown in this chapter, graphenes are evidenced to improve the performance of several polymeric membranes significantly and the polymer-graphene composites have excellent potential for use in different applications like filtration, dialysis, gas separation, and energy storage devices.

References

[1] N. Song, X. Gao, Z. Ma, X. Wang, Y. Wei, C. Gao, A review of graphene-based separation membrane: materials, characteristics, preparation and applications, Desalination 437 (2018) 59−72. Available from: https://doi.org/10.1016/j.desal.2018.02.024.

[2] R.K. Joshi, S. Alwarappan, M. Yoshimura, V. Sahajwalla, Y. Nishina, Graphene oxide: the new membrane material, Appl. Mater. Today 1 (2015) 1−12. Available from: https://doi.org/10.1016/j.apmt.2015.06.002.

[3] B.S. Lalia, V. Kochkodan, R. Hashaikeh, N. Hilal, A review on membrane fabrication: structure, properties and performance relationship, Desalination 326 (2013) 77−95. Available from: https://doi.org/10.1016/j.desal.2013.06.016.

[4] X. Wang, Y. Zhao, E. Tian, J. Li, Y. Ren, Graphene oxide-based polymeric membranes for water treatment, Adv. Mater. Interfaces. 5 (2018) 1701427. Available from: https://doi.org/10.1002/admi.201701427.

[5] G.M. Geise, D.R. Paul, B.D. Freeman, Fundamental water and salt transport properties of polymeric materials, Prog. Polym. Sci. 39 (2014) 1−42. Available from: https://doi.org/10.1016/j.progpolymsci.2013.07.001.

[6] L. Badrinezhad, S. Ghasemi, Preparation and characterization of polysulfone/graphene oxide nanocomposite membranes for the separation of methylene blue from water, Polym. Bull. 75 (2018) 469−484. Available from: https://doi.org/10.1007/s00289-017-2046-7.

[7] S. Kong, M. Lim, H. Shin, J.-H. Baik, J.-C. Lee, High-flux and antifouling polyethersulfone nanocomposite membranes incorporated with zwitterion-functionalized graphene oxide for ultrafiltration applications, J. Ind. Eng. Chem. 84 (2020) 131−140. Available from: https://doi.org/10.1016/j.jiec.2019.12.028.

[8] N. Naseeb, A.A. Mohammed, T. Laoui, Z. Khan, A novel PAN-GO-SiO$_2$ hybrid membrane for separating oil and water from emulsified mixture, Materials (2019) 1−13. Available from: https://doi.org/10.3390/ma12020212.

[9] Q. Bao, H. Zhang, J. Yang, S. Wang, D.Y. Tang, R. Jose, et al., Graphene−polymer nanofiber membrane for ultrafast photonics, Adv. Funct. Mater. 20 (2010) 782−791. Available from: https://doi.org/10.1002/adfm.200901658.

[10] S. Khakpour, Y. Jafarzadeh, R. Yegani, Incorporation of graphene oxide/nanodiamond nanocomposite into PVC ultrafiltration membranes, Chem. Eng. Res. Des. 152 (2019) 60−70. Available from: https://doi.org/10.1016/j.cherd.2019.09.029.

[11] S. Nazarenko, P. Meneghetti, P. Julmon, B.G. Olson, S. Qutubuddin, Gas barrier of polystyrene montmorillonite clay nanocomposites: effect of mineral layer aggregation, J. Polym. Sci. Part B Polym. Phys. 45 (2007) 1733−1753. Available from: https://doi.org/10.1002/polb.21181.

[12] H. Yamamura, K. Okimoto, K. Kimura, Y. Watanabe, Hydrophilic fraction of natural organic matter causing irreversible fouling of microfiltration and ultrafiltration membranes, Water Res 54 (2014) 123−136. Available from: https://doi.org/10.1016/j.watres.2014.01.024.

[13] W. Gao, H. Liang, J. Ma, M. Han, Z. Chen, Z. Han, et al., Membrane fouling control in ultrafiltration technology for drinking water production: a review, Desalination 272 (2011) 1−8. Available from: https://doi.org/10.1016/j.desal.2011.01.051.

[14] J. Yin, B. Deng, olymer-matrix nanocomposite membranes for water treatment, J. Memb. Sci. 479 (2015) 256−275. Available from: https://doi.org/10.1016/j.memsci.2014.11.019.

[15] H. Dong, L. Zhao, L. Zhang, H. Chen, C. Gao, W.S. Winston Ho, High-flux reverse osmosis membranes incorporated with NaY zeolite nanoparticles for brackish water desalination, J. Memb. Sci. 476 (2015) 373−383. Available from: https://doi.org/10.1016/j.memsci.2014.11.054.

[16] S.G. Kim, J.H. Chun, B.-H. Chun, S.H. Kim, Preparation, characterization and performance of poly(aylene ether sulfone)/modified silica nanocomposite reverse osmosis membrane for seawater desalination, Desalination 325 (2013) 76−83. Available from: https://doi.org/10.1016/j.desal.2013.06.017.

[17] Z. Xu, T. Wu, J. Shi, W. Wang, K. Teng, X. Qian, et al., Manipulating migration behavior of magnetic graphene oxide via magnetic field induced casting and phase separation toward high-performance hybrid ultrafiltration membranes, ACS Appl. Mater. Interfaces. 8 (2016) 18418−18429. Available from: https://doi.org/10.1021/acsami.6b04083.

[18] E.-S. Kim, G. Hwang, M. Gamal El-Din, Y. Liu, Development of nanosilver and multi-walled carbon nanotubes thin-film nanocomposite membrane for enhanced water treatment, J. Memb. Sci. 394−395 (2012) 37−48. Available from: https://doi.org/10.1016/j.memsci.2011.11.041.

[19] C. Zhao, X. Xu, J. Chen, F. Yang, Optimization of preparation conditions of poly (vinylidene fluoride)/graphene oxide microfiltration membranes by the Taguchi experimental design, Desalination 334 (2014) 17−22. Available from: https://doi.org/10.1016/j.desal.2013.07.011.

[20] D. Cohen-Tanugi, J.C. Grossman, Water desalination across nanoporous graphene, Nano Lett. 12 (2012) 3602−3608. Available from: https://doi.org/10.1021/nl3012853.

[21] Y. Kang, M. Obaid, J. Jang, M.-H. Ham, I.S. Kim, Novel sulfonated graphene oxide incorporated polysulfone nanocomposite membranes for enhanced-performance in ultrafiltration process, Chemosphere 207 (2018) 581−589. Available from: https://doi.org/10.1016/j.chemosphere.2018.05.141.

[22] L. Zhang, B. Chen, A. Ghaffar, X. Zhu, Nanocomposite membrane with polyethylenimine-grafted graphene oxide as a novel additive to enhance pollutant filtration performance, Environ. Sci. Technol. 52 (2018) 5920−5930. Available from: https://doi.org/10.1021/acs.est.8b00524.

[23] G. Zhang, M. Zhou, Z. Xu, C. Jiang, C. Shen, Q. Meng, Guanidyl-functionalized graphene/polysulfone mixed matrix ultrafiltration membrane with superior permselective, antifouling and antibacterial properties for water treatment, J. Colloid Interface Sci. 540 (2019) 295−305. Available from: https://doi.org/10.1016/j.jcis.2019.01.050.

[24] G.M. Geise, H.B. Park, A.C. Sagle, B.D. Freeman, J.E. McGrath, Water permeability and water/salt selectivity tradeoff in polymers for desalination, J. Memb. Sci. 369 (2011) 130−138. Available from: https://doi.org/10.1016/j.memsci.2010.11.054.

[25] S. Bano, A. Mahmood, S.-J. Kim, K.-H. Lee, Graphene oxide modified polyamide nanofiltration membrane with improved flux and antifouling properties, J. Mater. Chem. A. 3 (2015) 2065−2071. Available from: https://doi.org/10.1039/C4TA03607G.

[26] R.K. Joshi, P. Carbone, F.C. Wang, V.G. Kravets, Y. Su, I.V. Grigorieva, et al., Precise and ultrafast molecular sieving through graphene oxide membranes, Science (80−) 343 (2014) 752−754. Available from: https://doi.org/10.1126/science.1245711.

[27] J. Abraham, K.S. Vasu, C.D. Williams, K. Gopinadhan, Y. Su, C.T. Cherian, et al., Tunable sieving of ions using graphene oxide membranes, Nat. Nanotechnol. 12 (2017) 546−550. Available from: https://doi.org/10.1038/nnano.2017.21.

[28] G.S. Lai, W.J. Lau, P.S. Goh, A.F. Ismail, N. Yusof, Y.H. Tan, Graphene oxide incorporated thin film nanocomposite nanofiltration membrane for enhanced salt removal performance, Desalination 387 (2016) 14−24. Available from: https://doi.org/10.1016/j.desal.2016.03.007.

[29] J. Lee, J.H. Jang, H.-R. Chae, S.H. Lee, C.-H. Lee, P.-K. Park, et al., A facile route to enhance the water flux of a thin-film composite reverse osmosis membrane: incorporating thickness-controlled graphene oxide into a highly porous support layer, J. Mater. Chem. A 3 (2015) 22053−22060. Available from: https://doi.org/10.1039/C5TA04042F.

[30] M.J. Park, S. Phuntsho, T. He, G.M. Nisola, L.D. Tijing, X.-M. Li, et al., Graphene oxide incorporated polysulfone substrate for the fabrication of flat-sheet thin-film composite forward osmosis membranes, J. Memb. Sci. 493 (2015) 496−507. Available from: https://doi.org/10.1016/j.memsci.2015.06.053.

[31] P. Lu, S. Liang, T. Zhou, X. Mei, Y. Zhang, C. Zhang, et al., Layered double hydroxide/graphene oxide hybrid incorporated polysulfone substrate for thin-film nanocomposite forward osmosis membranes, RSC Adv. 6 (2016) 56599−56609. Available from: https://doi.org/10.1039/C6RA10080E.

[32] N. Misdan, A.F. Ismail, N. Hilal, Recent advances in the development of (bio)fouling resistant thin film composite membranes for desalination, Desalination 380 (2016) 105−111. Available from: https://doi.org/10.1016/j.desal.2015.06.001.

[33] N. Hilal, O.O. Ogunbiyi, N.J. Miles, R. Nigmatullin, Methods employed for control of fouling in MF and UF Membranes: a comprehensive review, Sep. Sci. Technol. 40 (2005) 1957−2005. Available from: https://doi.org/10.1081/SS-200068409.

[34] S. Jiang, Y. Li, B.P. Ladewig, A review of reverse osmosis membrane fouling and control strategies, Sci. Total Environ. 595 (2017) 567−583. Available from: https://doi.org/10.1016/j.scitotenv.2017.03.235.

[35] E. Matthiasson, The role of macromolecular adsorption in fouling of ultrafiltration membranes, J. Memb. Sci. 16 (1983) 23−36. Available from: https://doi.org/10.1016/S0376-7388(00)81297-1.

[36] S. Ayyaru, Y.-H. Ahn, Application of sulfonic acid group functionalized graphene oxide to improve hydrophilicity, permeability, and antifouling of PVDF nanocomposite ultrafiltration membranes, J. Memb. Sci. 525 (2017) 210−219. Available from: https://doi.org/10.1016/j.memsci.2016.10.048.

[37] G. Ouyang, A. Hussain, J. Li, D. Li, Remarkable permeability enhancement of polyethersulfone (PES) ultrafiltration membrane by blending cobalt oxide/graphene oxide nanocomposites, RSC Adv. 5 (2015) 70448−70460. Available from: https://doi.org/10.1039/C5RA11349K.

[38] H.M. Hegab, A. ElMekawy, T.G. Barclay, A. Michelmore, L. Zou, C.P. Saint, et al., Single-step assembly of multifunctional poly(tannic acid)−graphene oxide coating to reduce biofouling of forward osmosis membranes, ACS Appl. Mater. Interfaces 8 (2016) 17519−17528. Available from: https://doi.org/10.1021/acsami.6b03719.

[39] M. Safarpour, A. Khataee, V. Vatanpour, Effect of reduced graphene oxide/TiO$_2$ nano-composite with different molar ratios on the performance of PVDF ultrafiltration membranes, Sep. Purif. Technol. 140 (2015) 32−42. Available from: https://doi.org/10.1016/j.seppur.2014.11.010.

[40] L. Yu, Y. Zhang, B. Zhang, J. Liu, H. Zhang, C. Song, Preparation and characterization of HPEI-GO/PES ultrafiltration membrane with antifouling and antibacterial properties, J. Memb. Sci. 447 (2013) 452−462. Available from: https://doi.org/10.1016/j.memsci.2013.07.042.

[41] A. Tiraferri, C.D. Vecitis, M. Elimelech, Covalent binding of single-walled carbon nanotubes to polyamide membranes for antimicrobial surface properties, ACS Appl. Mater. Interfaces 3 (2011) 2869−2877. Available from: https://doi.org/10.1021/am200536p.

[42] F. Perreault, M.E. Tousley, M. Elimelech, Thin-film composite polyamide membranes functionalized with biocidal graphene oxide nanosheets, Environ. Sci. Technol. Lett. 1 (2014) 71−76. Available from: https://doi.org/10.1021/ez4001356.

[43] J. Quirós, K. Boltes, S. Aguado, R.G. de Villoria, J.J. Vilatela, R. Rosal, Antimicrobial metal−organic frameworks incorporated into electrospun fibers, Chem. Eng. J. 262 (2015) 189−197. Available from: https://doi.org/10.1016/j.cej.2014.09.104.

[44] S.-F. Cui, L.-P. Peng, H.-Z. Zhang, S. Rasheed, K. Vijaya Kumar, C.-H. Zhou, Novel hybrids of metronidazole and quinolones: synthesis, bioactive evaluation, cytotoxicity, preliminary antimicrobial mechanism and effect of metal ions on their transportation by human serum albumin, Eur. J. Med. Chem. 86 (2014) 318−334. Available from: https://doi.org/10.1016/j.ejmech.2014.08.063.

[45] Z. Xu, T. Wu, J. Shi, K. Teng, W. Wang, M. Ma, et al., Photocatalytic antifouling PVDF ultrafiltration membranes based on synergy of graphene oxide and TiO$_2$ for water treatment, J. Memb. Sci. 520 (2016) 281−293. Available from: https://doi.org/10.1016/j.memsci.2016.07.060.

[46] C. Lee, X. Wei, J.W. Kysar, J. Hone, Measurement of the elastic properties and intrinsic strength of monolayer graphene, Science (80−) 321 (2008) 385−388. Available from: https://doi.org/10.1126/science.1157996.

[47] M. El Achaby, F.Z. Arrakhiz, S. Vaudreuil, E.M. Essassi, A. Qaiss, Piezoelectric β-polymorph formation and properties enhancement in graphene oxide—PVDF nano-composite films, Appl. Surf. Sci. 258 (2012) 7668−7677. Available from: https://doi.org/10.1016/j.apsusc.2012.04.118.

[48] J.A. Prince, S. Bhuvana, V. Anbharasi, N. Ayyanar, K.V.K. Boodhoo, G. Singh, Ultra-wetting graphene-based PES ultrafiltration membrane—a novel approach for successful oil-water separation, Water Res 103 (2016) 311−318. Available from: https://doi.org/10.1016/j.watres.2016.07.042.

[49] J. Zhang, Q. Xue, X. Pan, Y. Jin, W. Lu, D. Ding, et al., Graphene oxide/polyacrylonitrile fiber hierarchical-structured membrane for ultra-fast microfiltration of oil-water emulsion, Chem. Eng. J. 307 (2017) 643−649. Available from: https://doi.org/10.1016/j.cej.2016.08.124.

[50] H. Zhao, S. Chen, X. Quan, H. Yu, H. Zhao, Integration of microfiltration and visible-light-driven photocatalysis on g-C3N4 nanosheet/reduced graphene oxide membrane

for enhanced water treatment, Appl. Catal. B Environ. 194 (2016) 134−140. Available from: https://doi.org/10.1016/j.apcatb.2016.04.042.

[51] C. Ma, J. Hu, W. Sun, Z. Ma, W. Yang, L. Wang, et al., Graphene oxide-polyethylene glycol incorporated PVDF nanocomposite ultrafiltration membrane with enhanced hydrophilicity, permeability, and antifouling performance, Chemosphere 253 (2020) 126649. Available from: https://doi.org/10.1016/j.chemosphere.2020.126649.

[52] X. Liu, H. Yuan, C. Wang, S. Zhang, L. Zhang, X. Liu, et al., A novel PVDF/PFSA-g-GO ultrafiltration membrane with enhanced permeation and antifouling performances, Sep. Purif. Technol. 233 (2020) 116038. Available from: https://doi.org/10.1016/j.seppur.2019.116038.

[53] V.H.T. Nguyen, M.N. Nguyen, T.T. Truong, T.T. Nguyen, H.V. Doan, X.N. Pham, One-pot preparation of alumina-modified polysulfone-graphene oxide nanocomposite membrane for separation of emulsion-oil from wastewater, J. Nanomater. 2020 (2020) 9087595. Available from: https://doi.org/10.1155/2020/9087595.

[54] A. Alammar, S.-H. Park, C.J. Williams, B. Derby, G. Szekely, Oil-in-water separation with graphene-based nanocomposite membranes for produced water treatment, J. Memb. Sci. 603 (2020) 118007. Available from: https://doi.org/10.1016/j.memsci.2020.118007.

[55] R.S. Zambare, K.B. Dhopte, P.R. Nemade, C.Y. Tang, Effect of oxidation degree of GO nanosheets on microstructure and performance of polysulfone-GO mixed matrix membranes, Sep. Purif. Technol. 244 (2020) 116865. Available from: https://doi.org/10.1016/j.seppur.2020.116865.

[56] Y. Jiang, Q. Zeng, P. Biswas, J.D. Fortner, Graphene oxides as nanofillers in polysulfone ultrafiltration membranes: shape matters, J. Memb. Sci. 581 (2019) 453−461. Available from: https://doi.org/10.1016/j.memsci.2019.03.056.

[57] M.S. Algamdi, I.H. Alsohaimi, J. Lawler, H.M. Ali, A.M. Aldawsari, H.M.A. Hassan, Fabrication of graphene oxide incorporated polyethersulfone hybrid ultrafiltration membranes for humic acid removal, Sep. Purif. Technol. 223 (2019) 17−23. Available from: https://doi.org/10.1016/j.seppur.2019.04.057.

[58] S. Yang, Q. Zou, T. Wang, L. Zhang, Effects of GO and MOF@GO on the permeation and antifouling properties of cellulose acetate ultrafiltration membrane, J. Memb. Sci. 569 (2019) 48−59. Available from: https://doi.org/10.1016/j.memsci.2018.09.068.

[59] A. Modi, J. Bellare, Efficient separation of biological macromolecular proteins by polyethersulfone hollow fiber ultrafiltration membranes modified with Fe_3O_4 nanoparticles-decorated carboxylated graphene oxide nanosheets, Int. J. Biol. Macromol. 135 (2019) 798−807. Available from: https://doi.org/10.1016/j.ijbiomac.2019.05.200.

[60] M. Hu, Z. Cui, J. Li, L. Zhang, Y. Mo, D.S. Dlamini, et al., Ultra-low graphene oxide loading for water permeability, antifouling and antibacterial improvement of polyethersulfone/sulfonated polysulfone ultrafiltration membranes, J. Colloid Interface Sci. 552 (2019) 319−331. Available from: https://doi.org/10.1016/j.jcis.2019.05.065.

[61] J.-J. Lu, Y.-H. Gu, Y. Chen, X. Yan, Y.-J. Guo, W.-Z. Lang, Ultrahigh permeability of graphene-based membranes by adjusting D-spacing with poly (ethylene imine) for the separation of dye wastewater, Sep. Purif. Technol. 210 (2019) 737−745. Available from: https://doi.org/10.1016/j.seppur.2018.08.065.

[62] X. Guo, C. Li, C. Li, T. Wei, L. Tong, H. Shao, et al., G-CNTs/PVDF mixed matrix membranes with improved antifouling properties and filtration performance, Front. Environ. Sci. Eng. 13 (2019) 81. Available from: https://doi.org/10.1007/s11783-019-1165-9.

[63] A. Rahimi, H. Mahdavi, Zwitterionic-functionalized GO/PVDF nanocomposite membranes with improved anti-fouling properties, J. Water Process Eng. 32 (2019) 100960. Available from: https://doi.org/10.1016/j.jwpe.2019.100960.

[64] M. Khajouei, M. Najafi, S.A. Jafari, Development of ultrafiltration membrane via in-situ grafting of nano-GO/PSF with anti-biofouling properties, Chem. Eng. Res. Des. 142 (2019) 34—43. Available from: https://doi.org/10.1016/j.cherd.2018.11.033.

[65] F. Abdulraqeb, A. Ali, J. Alam, A. Kumar, M. Alhoshan, M. Azam, et al., Evaluation of antibacterial and antifouling properties of silver-loaded GO polysulfone nanocomposite membrane against Escherichia coli, Staphylococcus aureus, and BSA protein, React. Funct. Polym. 140 (2019) 136—147. Available from: https://doi.org/10.1016/j.reactfunctpolym.2019.04.019.

[66] S. Bala, D. Nithya, M. Doraisamy, Exploring the effects of graphene oxide concentration on properties and antifouling performance of PEES/GO ultrafiltration membranes, High Perform. Polym. 30 (2017) 375—383. Available from: https://doi.org/10.1177/0954008317698547.

[67] A. Abdel-Karim, S. Leaper, M. Alberto, A. Vijayaraghavan, X. Fan, S.M. Holmes, et al., High flux and fouling resistant flat sheet polyethersulfone membranes incorporated with graphene oxide for ultrafiltration applications, Chem. Eng. J. 334 (2018) 789—799. Available from: https://doi.org/10.1016/j.cej.2017.10.069.

[68] M. Sri Abirami Saraswathi, D. Rana, J.S. Beril Melbiah, D. Mohan, A. Nagendran, Effective removal of bovine serum albumin and humic acid contaminants using poly (amide imide) nanocomposite ultrafiltration membranes tailored with GO and MoS2 nanosheets, Mater. Chem. Phys. 216 (2018) 170—176. Available from: https://doi.org/10.1016/j.matchemphys.2018.06.001.

[69] J. Alam, A.K. Shukla, M. Alhoshan, L. Arockiasamy Dass, M.R. Muthumareeswaran, A. Khan, et al., Graphene oxide, an effective nanoadditive for a development of hollow fiber nanocomposite membrane with antifouling properties, Adv. Polym. Technol. 37 (2018) 2597—2608. Available from: https://doi.org/10.1002/adv.21935.

[70] K.H. Chu, Y. Huang, M. Yu, J. Heo, J.R.V. Flora, A. Jang, et al., Evaluation of graphene oxide-coated ultrafiltration membranes for humic acid removal at different pH and conductivity conditions, Sep. Purif. Technol. 181 (2017) 139—147. Available from: https://doi.org/10.1016/j.seppur.2017.03.026.

[71] W. Miao, Z.-K. Li, X. Yan, Y.-J. Guo, W.-Z. Lang, Improved ultrafiltration performance and chlorine resistance of PVDF hollow fiber membranes via doping with sulfonated graphene oxide, Chem. Eng. J. 317 (2017) 901—912. Available from: https://doi.org/10.1016/j.cej.2017.02.121.

[72] T. Hwang, J.-S. Oh, J. Wim, J.-D. Nam, C. Bae, H. Kim, et al., Ultrafiltration using graphene oxide surface-embedded polysulfone membranes, Sep. Purif. Technol. 166 (2016) 41—47. Available from: https://doi.org/10.1016/j.seppur.2016.04.018.

[73] N.J. Kaleekkal, A. Thanigaivelan, D. Rana, D. Mohan, Studies on carboxylated graphene oxide incorporated polyetherimide mixed matrix ultrafiltration membranes, Mater. Chem. Phys. 186 (2017) 146—158. Available from: https://doi.org/10.1016/j.matchemphys.2016.10.040.

[74] Y. Zhao, J. Lu, X. Liu, Y. Wang, J. Lin, N. Peng, et al., Performance enhancement of polyvinyl chloride ultrafiltration membrane modified with graphene oxide, J. Colloid Interface Sci. 480 (2016) 1—8. Available from: https://doi.org/10.1016/j.jcis.2016.06.075.

[75] Y. Jiang, W.-N. Wang, D. Liu, Y. Nie, W. Li, J. Wu, et al., Engineered crumpled graphene oxide nanocomposite membrane assemblies for advanced water treatment

processes, Environ. Sci. Technol. 49 (2015) 6846−6854. Available from: https://doi.org/10.1021/acs.est.5b00904.

[76] Q. Zhao, J. Hou, J. Shen, J. Liu, Y. Zhang, Long-lasting antibacterial behavior of a novel mixed matrix water purification membrane, J. Mater. Chem. A. 3 (2015) 18696−18705. Available from: https://doi.org/10.1039/C5TA06013C.

[77] A. Marjani, A.T. Nakhjiri, M. Adimi, H.F. Jirandehi, S. Shirazian, Effect of graphene oxide on modifying polyethersulfone membrane performance and its application in wastewater treatment, Sci. Rep. 10 (2020) 2049. Available from: https://doi.org/10.1038/s41598-020-58472-y.

[78] M. Pal, M. Malhotra, M.K. Mandal, T.K. Paine, P. Pal, Recycling of wastewater from tannery industry through membrane-integrated hybrid treatment using a novel graphene oxide nanocomposite, J. Water Process Eng 36 (2020) 101324. Available from: https://doi.org/10.1016/j.jwpe.2020.101324.

[79] M.J. Park, S. Lim, R.R. Gonzales, S. Phuntsho, D.S. Han, A. Abdel-Wahab, et al., Thin-film composite hollow fiber membranes incorporated with graphene oxide in polyethersulfone support layers for enhanced osmotic power density, Desalination 464 (2019) 63−75. Available from: https://doi.org/10.1016/j.desal.2019.04.026.

[80] W. Ma, T. Chen, S. Nanni, L. Yang, Z. Ye, M.S. Rahaman, Zwitterion-functionalized graphene oxide incorporated polyamide membranes with improved antifouling properties, Langmuir 35 (2019) 1513−1525. Available from: https://doi.org/10.1021/acs.langmuir.8b02044.

[81] A. Inurria, P. Cay-Durgun, D. Rice, H. Zhang, D.-K. Seo, M.L. Lind, et al., Polyamide thin-film nanocomposite membranes with graphene oxide nanosheets: balancing membrane performance and fouling propensity, Desalination 451 (2019) 139−147. Available from: https://doi.org/10.1016/j.desal.2018.07.004.

[82] W. Shao, C. Liu, H. Ma, Z. Hong, Q. Xie, Y. Lu, Fabrication of pH-sensitive thin-film nanocomposite nanofiltration membranes with enhanced performance by incorporating amine-functionalized graphene oxide, Appl. Surf. Sci. 487 (2019) 1209−1221. Available from: https://doi.org/10.1016/j.apsusc.2019.05.157.

[83] Y. Kang, M. Obaid, J. Jang, I.S. Kim, Sulfonated graphene oxide incorporated thin film nanocomposite nanofiltration membrane to enhance permeation and antifouling properties, Desalination 470 (2019) 114125. Available from: https://doi.org/10.1016/j.desal.2019.114125.

[84] H. Abadikhah, E. Naderi Kalali, S. Khodi, X. Xu, S. Agathopoulos, Multifunctional thin-film nanofiltration membrane incorporated with reduced graphene Oxide@TiO_2@Ag nanocomposites for high desalination performance, dye retention, and antibacterial properties, ACS Appl. Mater. Interfaces. 11 (2019) 23535−23545. Available from: https://doi.org/10.1021/acsami.9b03557.

[85] Q. Xie, S. Zhang, Z. Hong, H. Ma, C. Liu, W. Shao, Effects of casting solvents on the morphologies, properties, and performance of polysulfone supports and the resultant graphene oxide-embedded thin-film nanocomposite nanofiltration membranes, Ind. Eng. Chem. Res. 57 (2018) 16464−16475. Available from: https://doi.org/10.1021/acs.iecr.8b04515.

[86] K. Ko, Y. Yu, M.-J. Kim, J. Kweon, H. Chung, Improvement in fouling resistance of silver-graphene oxide coated polyvinylidene fluoride membrane prepared by pressurized filtration, Sep. Purif. Technol. 194 (2018) 161−169. Available from: https://doi.org/10.1016/j.seppur.2017.11.016.

[87] S. Leaper, A. Abdel-Karim, B. Faki, J.M. Luque-Alled, M. Alberto, A. Vijayaraghavan, et al., Flux-enhanced PVDF mixed matrix membranes incorporating APTS-functionalized graphene oxide for membrane distillation, J. Memb. Sci. 554 (2018) 309−323. Available from: https://doi.org/10.1016/j.memsci.2018.03.013.

[88] H. Abadikhah, E.N. Kalali, S. Khodi, X. Xu, S. Agathopoulos, Multifunctional thin-film nano filtration membrane incorporated with reduced graphene oxide @ TiO_2 @ Ag nanocomposites for high desalination performance, dye retention, and antibacterial properties, ACS Appl. Mater. Interfaces 11 (2019) 23535–23545. Available from: https://doi.org/10.1021/acsami.9b03557.

[89] R. Hu, R. Zhang, Y. He, G. Zhao, H. Zhu, Graphene oxide-in-polymer nanofiltration membranes with enhanced permeability by interfacial polymerization, J. Memb. Sci. 564 (2018) 813–819. Available from: https://doi.org/10.1016/j.memsci.2018.07.087.

[90] S. Fatemeh Seyedpour, A. Rahimpour, H. Mohsenian, M.J. Taherzadeh, Low fouling ultrathin nanocomposite membranes for efficient removal of manganese, J. Memb. Sci. 549 (2018) 205–216. Available from: https://doi.org/10.1016/j.memsci.2017.12.012.

[91] W. Zhao, H. Liu, N. Meng, M. Jian, H. Wang, X. Zhang, Graphene oxide incorporated thin film nanocomposite membrane at low concentration monomers, J. Memb. Sci. 565 (2018) 380–389. Available from: https://doi.org/10.1016/j.memsci.2018.08.047.

[92] A. Karkooti, A.Z. Yazdi, P. Chen, M. McGregor, N. Nazemifard, M. Sadrzadeh, Development of advanced nanocomposite membranes using graphene nanoribbons and nanosheets for water treatment, J. Memb. Sci. 560 (2018) 97–107. Available from: https://doi.org/10.1016/j.memsci.2018.04.034.

[93] G.S. Lai, W.J. Lau, P.S. Goh, A.F. Ismail, Y.H. Tan, C.Y. Chong, et al., Tailor-made thin film nanocomposite membrane incorporated with graphene oxide using novel interfacial polymerization technique for enhanced water separation, Chem. Eng. J. 344 (2018) 524–534. Available from: https://doi.org/10.1016/j.cej.2018.03.116.

[94] A. Al Mayyahi, Thin-film composite (TFC) membrane modified by hybrid ZnO-graphene nanoparticles (ZnO-Gr NPs) for water desalination, J. Environ. Chem. Eng. 6 (2018) 1109–1117. Available from: https://doi.org/10.1016/j.jece.2018.01.035.

[95] C. Zhang, K. Wei, W. Zhang, Y. Bai, Y. Sun, J. Gu, Graphene oxide quantum dots incorporated into a thin film nanocomposite membrane with high flux and antifouling properties for low-pressure nanofiltration, ACS Appl. Mater. Interfaces. 9 (2017) 11082–11094. Available from: https://doi.org/10.1021/acsami.6b12826.

[96] G.S. Lai, W.J. Lau, P.S. Goh, Y.H. Tan, B.C. Ng, A.F. Ismail, A novel interfacial polymerization approach towards synthesis of graphene oxide-incorporated thin film nanocomposite membrane with improved surface properties, Arab. J. Chem. 12 (2019) 75–87. Available from: https://doi.org/10.1016/j.arabjc.2017.12.009.

[97] Q. Xie, S. Zhang, Z. Xiao, X. Hu, Z. Hong, R. Yi, et al., Preparation and characterization of novel alkali-resistant nanofiltration membranes with enhanced permeation and antifouling properties: the effects of functionalized graphene nanosheets, RSC Adv. 7 (2017) 18755–18764. Available from: https://doi.org/10.1039/C7RA00928C.

[98] M.E.A. Ali, L. Wang, X. Wang, X. Feng, Thin film composite membranes embedded with graphene oxide for water desalination, Desalination 386 (2016) 67–76. Available from: https://doi.org/10.1016/j.desal.2016.02.034.

[99] J. Yin, G. Zhu, B. Deng, Graphene oxide (GO) enhanced polyamide (PA) thin-film nanocomposite (TFN) membrane for water purification, Desalination 379 (2016) 93–101. Available from: https://doi.org/10.1016/j.desal.2015.11.001.

[100] J. Wang, C. Zhao, T. Wang, Z. Wu, X. Li, J. Li, Graphene oxide polypiperazine-amide nanofiltration membrane for improving flux and anti-fouling in water purification, RSC Adv 6 (2016) 82174–82185. Available from: https://doi.org/10.1039/C6RA17284A.

[101] J. Wang, Y. Wang, Y. Zhang, A. Uliana, J. Zhu, J. Liu, et al., Zeolitic imidazolate framework/graphene oxide hybrid nanosheets functionalized thin film nanocomposite membrane for enhanced antimicrobial performance, ACS Appl. Mater. Interfaces 8 (2016) 25508—25519. Available from: https://doi.org/10.1021/acsami. 6b06992.

[102] M. Safarpour, A. Khataee, V. Vatanpour, Thin film nanocomposite reverse osmosis membrane modified by reduced graphene oxide/TiO$_2$ with improved desalination performance, J. Memb. Sci. 489 (2015) 43—54. Available from: https://doi.org/10.1016/j. memsci.2015.04.010.

[103] M. Safarpour, V. Vatanpour, A. Khataee, M. Esmaeili, Development of a novel high flux and fouling-resistant thin film composite nanofiltration membrane by embedding reduced graphene oxide/TiO$_2$, Sep. Purif. Technol. 154 (2015) 96—107. Available from: https://doi.org/10.1016/j.seppur.2015.09.039.

[104] Y. Zhang, H. Ruan, C. Guo, J. Liao, J. Shen, C. Gao, Thin-film nanocomposite reverse osmosis membranes with enhanced antibacterial resistance by incorporating p-aminophenol-modified graphene oxide, Sep. Purif. Technol. 234 (2020) 116017. Available from: https://doi.org/10.1016/j.seppur.2019.116017.

[105] S. Li, B. Gao, B. Wang, B. Jin, Q. Yue, Z. Wang, Antibacterial thin film nanocomposite reverse osmosis membrane by doping silver phosphate loaded graphene oxide quantum dots in polyamide layer, Desalination 464 (2019) 94—104. Available from: https://doi.org/10.1016/j.desal.2019.04.029.

[106] S.M. Ghaseminezhad, M. Barikani, M. Salehirad, Development of graphene oxide-cellulose acetate nanocomposite reverse osmosis membrane for seawater desalination, Compos. Part B Eng 161 (2019) 320—327. Available from: https://doi.org/10.1016/j. compositesb.2018.10.079.

[107] M. Fathizadeh, H.N. Tien, K. Khivantsev, Z. Song, F. Zhou, M. Yu, Polyamide/nitrogen-doped graphene oxide quantum dots (N-GOQD) thin film nanocomposite reverse osmosis membranes for high flux desalination, Desalination 451 (2019) 125—132. Available from: https://doi.org/10.1016/j.desal.2017.07.014.

[108] R. Rajakumaran, V. Boddu, M. Kumar, M.S. Shalaby, H. Abdallah, Effect of ZnO morphology on GO-ZnO modified polyamide reverse osmosis membranes for desalination, Desalination 467 (2019) 245—256. Available from: https://doi.org/10.1016/j. desal.2019.06.018.

[109] H. Mahdavi, A. Rahimi, Zwitterion functionalized graphene oxide/polyamide thin film nanocomposite membrane: towards improved anti-fouling performance for reverse osmosis, Desalination 433 (2018) 94—107. Available from: https://doi.org/10.1016/j. desal.2018.01.031.

[110] S. Kim, R. Ou, Y. Hu, X. Li, H. Zhang, G.P. Simon, et al., Non-swelling graphene oxide-polymer nanocomposite membrane for reverse osmosis desalination, J. Memb. Sci. 562 (2018) 47—55. Available from: https://doi.org/10.1016/j. memsci.2018.05.029.

[111] X. Song, Q. Zhou, T. Zhang, H. Xu, Z. Wang, Pressure-assisted preparation of graphene oxide quantum dot-incorporated reverse osmosis membranes: antifouling and chlorine resistance potentials, J. Mater. Chem. A 4 (2016) 16896—16905. Available from: https://doi.org/10.1039/C6TA06636D.

[112] H.-R. Chae, J. Lee, C.-H. Lee, I.-C. Kim, P.-K. Park, Graphene oxide-embedded thin-film composite reverse osmosis membrane with high flux, anti-biofouling, and chlorine resistance, J. Memb. Sci. 483 (2015) 128—135. Available from: https://doi.org/ 10.1016/j.memsci.2015.02.045.

[113] N. Akther, Z. Yuan, Y. Chen, S. Lim, S. Phuntsho, Influence of graphene oxide lateral size on the properties and performances of forward osmosis membrane, Desalination 484 (2020) 114421. Available from: https://doi.org/10.1016/j.desal.2020. 114421.

[114] A. Saeedi-Jurkuyeh, A.J. Jafari, R.R. Kalantary, A. Esrafili, A novel synthetic thin-film nanocomposite forward osmosis membrane modified by graphene oxide and polyethylene glycol for heavy metals removal from aqueous solutions, React. Funct. Polym. 146 (2020) 104397. Available from: https://doi.org/10.1016/j. reactfunctpolym.2019.104397.

[115] M. Bagherzadeh, A. Bayrami, M. Amini, Enhancing forward osmosis (FO) performance of polyethersulfone/polyamide (PES/PA) thin-film composite membrane via the incorporation of GQDs@UiO-66-NH2 particles, J. Water Process Eng 33 (2020) 101107. Available from: https://doi.org/10.1016/j.jwpe.2019.101107.

[116] S. Maiti, P.K. Samantaray, S. Bose, In situ assembly of a graphene oxide quantum dot-based thin-film nanocomposite supported on de-mixed blends for desalination through forward osmosis, Nanoscale Adv 2 (2020) 1993−2003. Available from: https://doi.org/10.1039/C9NA00688E.

[117] S. Xu, F. Li, B. Su, M.Z. Hu, X. Gao, C. Gao, Novel graphene quantum dots (GQDs)-incorporated thin fi lm composite (TFC) membranes for forward osmosis (FO) desalination, Desalination 451 (2019) 219−230. Available from: https://doi.org/10.1016/j. desal.2018.04.004.

[118] A.J. Talar, T. Ebadi, R. Maknoon, A. Rashidi, Investigation on effect of KCl addition on desalination performance of co-polymerized GO/Nylon nanocomposite membrane, Process Saf. Environ. Prot. 125 (2019) 31−38. Available from: https://doi.org/ 10.1016/j.psep.2019.03.008.

[119] H. Choi, A.A. Shah, S.-E. Nam, Y.-I. Park, H. Park, Thin-film composite membranes comprising ultrathin hydrophilic polydopamine interlayer with graphene oxide for forward osmosis, Desalination 449 (2019) 41−49. Available from: https://doi.org/ 10.1016/j.desal.2018.10.012.

[120] M. Rastgar, A. Shakeri, A. Bozorg, H. Salehi, V. Saadattalab, Highly-efficient forward osmosis membrane tailored by magnetically responsive graphene oxide/Fe_3O_4 nanohybrid, Appl. Surf. Sci. 441 (2018) 923−935. Available from: https://doi.org/10.1016/j. apsusc.2018.02.118.

[121] L. Jin, Z. Wang, S. Zheng, B. Mi, Polyamide-crosslinked graphene oxide membrane for forward osmosis, J. Memb. Sci. 545 (2018) 11−18. Available from: https://doi. org/10.1016/j.memsci.2017.09.023.

[122] S.S. Eslah, S. Shokrollahzadeh, O.M. Jazani, Forward osmosis water desalination: fabrication of graphene oxide-polyamide/polysulfone thin-film nanocomposite membrane with high water flux and low reverse salt diffusion, Sep. Sci. Technol. 53 (2018) 573−583. Available from: https://doi.org/10.1080/01496395.2017.1398261.

[123] L. Jin, Z. Wang, S. Zheng, B. Mi, Polyamide-crosslinked graphene oxide membrane for forward osmosis, J. Memb. Sci. 545 (2018) 11−18. Available from: https://doi. org/10.1016/j.memsci.2017.09.023.

[124] H. Salehi, M. Rastgar, A. Shakeri, Anti-fouling and high water permeable forward osmosis membrane fabricated via layer by layer assembly of chitosan/graphene oxide, Appl. Surf. Sci. 413 (2017) 99−108. Available from: https://doi.org/10.1016/j. apsusc.2017.03.271.

[125] S. Lim, M.J. Park, S. Phuntsho, L.D. Tijing, G.M. Nisola, W.-G. Shim, et al., Dual-layered nanocomposite substrate membrane based on polysulfone/graphene oxide for

mitigating internal concentration polarization in forward osmosis, Polymer (Guildf) 110 (2017) 36–48. Available from: https://doi.org/10.1016/j.polymer.2016.12.066.

[126] L. Shen, S. Xiong, Y. Wang, Graphene oxide incorporated thin-film composite membranes for forward osmosis applications, Chem. Eng. Sci. 143 (2016) 194–205. Available from: https://doi.org/10.1016/j.ces.2015.12.029.

[127] H.M. Hegab, A. ElMekawy, T.G. Barclay, A. Michelmore, L. Zou, C.P. Saint, et al., Effective in-situ chemical surface modification of forward osmosis membranes with polydopamine-induced graphene oxide for biofouling mitigation, Desalination 385 (2016) 126–137. Available from: https://doi.org/10.1016/j.desal.2016.02.021.

[128] Y. Wang, R. Ou, H. Wang, T. Xu, Graphene oxide modified graphitic carbon nitride as a modifier for thin film composite forward osmosis membrane, J. Memb. Sci. 475 (2015) 281–289. Available from: https://doi.org/10.1016/j.memsci.2014.10.028.

[129] D.R. Paul, L.M. Robeson, Polymer nanotechnology: nanocomposites, Polymer (Guildf) 49 (2008) 3187–3204. Available from: https://doi.org/10.1016/j. polymer.2008.04.017.

[130] Y. Cui, S.I. Kundalwal, S. Kumar, Gas barrier performance of graphene/polymer nanocomposites, Carbon N. Y. 98 (2016) 313–333. Available from: https://doi.org/ 10.1016/j.carbon.2015.11.018.

[131] H. Kim, A.A. Abdala, C.W. Macosko, Graphene/polymer nanocomposites, Macromolecules 43 (2010) 6515–6530. Available from: https://doi.org/10.1021/ ma100572e.

[132] R. Pal, Permeation models for mixed matrix membranes, J. Colloid Interface Sci. 317 (2008) 191–198. Available from: https://doi.org/10.1016/j.jcis.2007.09.032.

[133] L.E. Nielsen, Models for the permeability of filled polymer systems, J. Macromol. Sci. Part A-Chem. 1 (1967) 929–942. Available from: https://doi.org/10.1080/ 10601326708053745.

[134] E. Yang, K. Goh, C.Y. Chuah, R. Wang, T.-H. Bae, Asymmetric mixed-matrix membranes incorporated with nitrogen-doped graphene nanosheets for highly selective gas separation, J. Memb. Sci. 615 (2020) 118293. Available from: https://doi.org/10.1016/ j.memsci.2020.118293.

[135] J. Widakdo, Y.H. Chiao, Y.L. Lai, A.C. Imawan, F.M. Wang, W.S. Hung, Mechanism of a self-assembling smart and electrically responsive PVDF-graphene membrane for controlled gas separation, ACS Appl. Mater. Interfaces 12 (2020) 30915–30924. Available from: https://doi.org/10.1021/acsami.0c04402.

[136] F.U. Nigiz, N.D. Hilmioglu, Enhanced hydrogen purification by graphene—Poly (Dimethyl siloxane) membrane, Int. J. Hydrogen Energy. 45 (2020) 3549–3557. Available from: https://doi.org/10.1016/j.ijhydene.2018.12.215.

[137] J.E. Shin, S.K. Lee, Y.H. Cho, H.B. Park, Effect of PEG-MEA and graphene oxide additives on the performance of Pebax®1657 mixed matrix membranes for CO2 separation, J. Memb. Sci. 572 (2019) 300–308. Available from: https://doi.org/10.1016/j. memsci.2018.11.025.

[138] K.R. Raj, A.R. Sunarti, Preliminary Fractional Factorial Design (FFD) study using incorporation of Graphene Oxide in PVC in mixed matrix membrane to enhance CO_2/CH_4 separation, IOP Conf. Ser. Mater. Sci. Eng. 702 (2019) 12041. Available from: https://doi.org/10.1088/1757-899x/702/1/012041.

[139] H. Kweon, C.-W. Lin, M.M. Faruque Hasan, R. Kaner, G.N. Sant, Highly permeable polyaniline—graphene oxide nanocomposite membranes for CO_2 separations, ACS Appl. Polym. Mater. 1 (2019) 3233–3241. Available from: https://doi.org/10.1021/ acsapm.9b00426.

[140] R. Rea, S. Ligi, M. Christian, V. Morandi, M. Giacinti Baschetti, M.G. De Angelis, Permeability and selectivity of PPO/graphene composites as mixed matrix membranes for CO_2 capture and gas separation, Polym 10 (2018). Available from: https://doi.org/10.3390/polym10020129.

[141] B.S. Ge, T. Wang, H.X. Sun, W. Gao, H.R. Zhao, Preparation of mixed matrix membranes based on polyimide and aminated graphene oxide for CO_2 separation, Polym. Adv. Technol. 29 (2018) 1334–1343. Available from: https://doi.org/10.1002/pat.4245.

[142] E.M. Aliyev, M.M. Khan, A.M. Nabiyev, R.M. Alosmanov, I.A. Bunyad-zadeh, S. Shishatskiy, et al., Covalently modified graphene oxide and polymer of intrinsic microporosity (PIM-1) in mixed matrix thin-film composite membranes, Nanoscale Res. Lett. 13 (2018) 359. Available from: https://doi.org/10.1186/s11671-018-2771-3.

[143] D. Roilo, P.N. Patil, R.S. Brusa, A. Miotello, S. Aghion, R. Ferragut, et al., Polymer rigidification in graphene based nanocomposites: gas barrier effects and free volume reduction, Polymer (Guildf). 121 (2017) 17–25. Available from: https://doi.org/10.1016/j.polymer.2017.06.010.

[144] F. Moghadam, H.B. Park, Two-dimensional materials: an emerging platform for gas separation membranes, Curr. Opin. Chem. Eng. 20 (2018) 28–38. Available from: https://doi.org/10.1016/j.coche.2018.02.004.

[145] V. Yadav, S.K. Raj, N.H. Rathod, V. Kulshrestha, Polysulfone/graphene quantum dots composite anion exchange membrane for acid recovery by diffusion dialysis, J. Membr. Sci. 118331 (2020).

[146] Z. Jia, Y. Wang, W. Shi, J. Wang, Diamines cross-linked graphene oxide free-standing membranes for ion dialysis separation, J. Membr. Sci. 520 (2016) 139–144.

[147] L. Guo, Y. Xu, M. Zhuo, L. Liu, Q. Xu, L. Wang, et al., Highly efficient removal of Gd(III) using hybrid hydrosols of carbon nanotubes/graphene oxide in dialysis bags and synergistic enhancement effect, Chem. Eng. J. 348 (2018) 535–545.

[148] M.Z. Fahmi, M. Wathoniyyah, M. Khasanah, Y. Rahardjo, S. Wafiroh, A. Abdulloh, Incorporation of graphene oxide in polyethersulfone mixed matrix membranes to enhance hemodialysis membrane performance, RSC Adv. 8 (2) (2018) 931–937.

[149] N.J. Kaleekkal, A. Thanigaivelan, M. Durga, R. Girish, D. Rana, P. Soundararajan, et al., Graphene oxide nanocomposite incorporated poly(ether imide) mixed matrix membranes for in vitro evaluation of its efficacy in blood purification applications, Ind. Eng. Chem. Res. 54 (32) (2015) 7899–7913.

[150] A. Modi, S.K. Verma, J. Bellare, Graphene oxide-doping improves the biocompatibility and separation performance of polyethersulfone hollow fiber membranes for bioartificial kidney application, J. Colloid Interface Sci. 514 (2018) 750–759.

[151] L. Cseri, J. Baugh, A. Alabi, A. AlHajaj, L. Zou, R. Dryfe, et al., Graphene oxide—polybenzimidazolium nanocomposite anion exchange membranes for electrodialysis, J. Mater. Chem. A 6 (48) (2018) 24728–24739.

[152] S.M. Hosseini, E. Jashni, M. Habibi, B. Van der Bruggen, Fabrication of novel electrodialysis heterogeneous ion exchange membranes by incorporating PANI/GO functionalized composite nanoplates, Ionics 24 (6) (2017) 1789–1801.

[153] P.R. Kidambi, D. Jang, J.-C. Idrobo, M.S.H. Boutilier, L. Wang, J. Kong, et al., Nanoporous atomically thin graphene membranes for desalting and dialysis applications, Adv. Mater. 29 (33) (2017) 1700277.

[154] X. Lin, S. Kim, D.M. Zhu, E. Shamsaei, T. Xu, X. Fang, et al., Preparation of porous diffusion dialysis membranes by functionalization of polysulfone for acid recovery, J. Membr. Sci. 524 (2017) 557–564.

[155] L. Guo, L. Liu, M. Zhuo, S. Fu, Y. Xu, W. Zhou, et al., Smaller-lateral-size graphene oxide hydrosols sealed in dialysis bags for enhanced trace Pb(II) removal from water without re-pollution, Appl. Surf. Sci. 445 (2018) 586−595.

[156] Q. Huang, G. Li, M. Chen, S. Dong, Graphene oxide functionalized O-(carboxy-methyl)-chitosan membranes: fabrication using dialysis and applications in water purification, Colloids Surf. A: Physicochem. Eng. Asp. 554 (2018) 27−33.

[157] Y. Zeng, L. Qiu, K. Wang, J. Yao, D. Li, G.P. Simon, et al., Significantly enhanced water flux in forward osmosis desalination with polymer-graphene composite hydrogels as a draw agent, RSC Adv 3 (3) (2013) 887−894.

[158] J. Chen, Y. Ma, L. Wang, W. Han, Y. Chai, T. Wang, et al., Preparation of chitosan/ SiO$_2$-loaded graphene composite beads for efficient removal of bilirubin, Carbon 143 (2019) 352−361.

[159] L. Guo, P. Ye, J. Wang, F. Fu, Z. Wu, Three-dimensional Fe$_3$O$_4$-graphene macroscopic composites for arsenic and arsenate removal, J. Hazard. Mater. 298 (2015) 28−35.

[160] R. Raccichini, A. Varzi, S. Passerini, B. Scrosati, The role of graphene for electrochemical energy storage, Nat. Mater. 14 (3) (2014) 271−279.

[161] C.O.A. Vargas, Á. Caballero, J. Morales, Can the performance of graphenenanosheets for lithium storage in Li-ion batteries be predicted? Nanoscale 4 (2012) 2083−2092.

[162] A.G. Olabi, M.A. Abdelkareem, T. Wilberforce, E.T. Sayed, Application of graphene in energy storage device—a review, Renew. Sustain. Energy Rev. 135 (2021) 110026.

[163] N. Natter, N. Kostoglou, C. Koczwara, C. Tampaxis, T. Steriotis, R. Gupta, et al., Plasma-derived graphene-based materials for water purification and energy storage, C—J. Carbon Res. 5 (2) (2019) 16.

[164] X. Zhao, E. Jiaqiang, G. Wu, Y. Deng, D. Han, B. Zhang, et al., A review of studies using graphenes in energy conversion, energy storage and heat transfer development, Energy Convers. Manag. 184 (2019) 581−599.

Polymer-graphene composites as anticorrosive materials

20

Sheeja Sunil, P. Porkodi, Abhilash J. Kottiyatil and Prosenjit Ghosh
Center for Carbon Fiber and Prepregs, CSIR-National Aerospace Laboratories, Bangalore, India

20.1   Introduction

Corrosion is a serious problem to the growth of economy and the progress of society. It causes a loss of nearly 7%−8% of the nation's Gross Domestic Product (GDP) and is one of the largest sources of drains of the economy because most of the metals and their alloys, for example, magnesium, copper, nickel, carbon steel, etc. suffer from corrosion in industrial productions and applications [1]. Corrosion causes a significant impairment of the material properties such as mechanical strength, chemical reactivity, and electrical conductivity. Corrosion happens by the action of external environments some of which are hostile and accelerate corrosion, for example, fog, humidity, saltwater, alkaline or acidic soils, etc. Corrosion can take place anywhere resulting in structural failures that directly or indirectly affect the surrounding environment including humans. In addition, corrosion reduces the integrity of the material that may lead to severe economic and environmental causalities, especially to the expensive assets employed in adverse conditions like in offshore structures [2].

During industrial and household applications, metals get exposed to a wide range of environments like constant immersion in water, burial in soil, ultraviolet radiation, and atmospheric pollution in industrial areas. Thus, the specific requirements for corrosion protection systems largely depend on the environment, the nature of the corrosive elements, and the time of exposure. Among these, the versatility of the environments plays a decisive role in corrosion protection management because the metals need to protect from chemicals and rain in industrial areas whereas the same metals need protection from bacteria and humidity when buried in the soil. The metals broadly experience three types of environmental exposure depending upon the versatility and corrosiveness of the environment viz. atmospheric, splash zone, and immersion [3]. The corrosive action of the atmosphere is a factor of heat, moisture, ultraviolet radiation, and salt/gas concentration thus changes locally from very low corrosivity to very high corrosivity depending on the climate condition, pollution level, and distance to the sea. Industrial and marine atmospheres have very high corrosivity due to the high content of solid particulates like soot, sand, chloride, and sulphate salts in the atmosphere. In addition, acidic oxides, for example, sulphur dioxide (SO_2) and oxides of nitrogen (NOx) are common in the

Polymer Nanocomposites Containing Graphene. DOI: https://doi.org/10.1016/B978-0-12-821639-2.00007-0

industrial atmosphere and can combine with rain to produce acid rain thereby exposing the metals to an acidic environment [4]. On the other hand, marine atmosphere contains a very high content of chloride ions that play an aggressive role in metal corrosion giving pitting effect [5]. Therefore, the impact of atmospheric exposure has a significant contribution to corrosion protection management. The corrosive actions of splash zone environments are seen in structures situated near the waterline of sea, for example, parts of offshore plants and wind turbine foundations. Splash zone environments provide the unique combination of oxygen-rich atmosphere and continuous splashing of electrolytes from the sea; both have detrimental effects on metal corrosion [6]. Hence, splash zones are extremely aggressive environments toward metal corrosion that get further accelerated due to the exposure of ultraviolet radiation and any forms of mechanical stress. The corrosive actions of the environments on structures immersed in water or buried in soil depend on many factors like temperature, pH, salinity, and the content of dissolved gases, especially oxygen. Therefore, immersion in freshwater, seawater, or buried in soil experiences different corrosiveness of the environment [7].

Corrosion is the physicochemical interaction between the metal and its environment, the fundamentals of which are similar to those of basic electrochemical reactions. Like any other electrochemical reaction, corrosion requires an electrolyte solution and a metallic conductor between two electrodes of different potentials, that is, an anode and a cathode. Therefore, the kinetics of corrosion reaction depend on the two spontaneous half-cells (redox) reactions. Rusting of iron or steel is a well-known example of metal corrosion, so the corrosion of steel can be taken as an example to further elaborate the electrochemical process. In general, different grades of steels are used as primary structural components and thus are readily exposed to various environments. Hence, the redox reactions [shown below Eqs. (20.1)−(20.2)] between iron and oxygen in the presence of water play a distinctive role in the corrosion process of steel. Some areas on the steel surface act as anode, while other areas behave as cathode [8]. The actual corrosion reaction of metal takes place at anode leading to loss of metallic iron and producing ferrous (Fe^{2+}) ions while oxygen gets reduced at cathode on the catalytically active oxidized metal (iron) surface.

Redox reactions:

At cathode,

$$O_2(g) + 2H_2O(l) + 4e^- \rightarrow 4OH^-(aq) \tag{20.1}$$

At anode,

$$Fe(s) \rightarrow Fe^{2+}(aq) + 2e^- \tag{20.2}$$

The transport of ions in metal corrosion takes place through an electrolytic medium to maintain the electroneutrality between the cathodic and anodic corrosion cells on the metal surface. Therefore, the rate of the electrochemical corrosion reaction of metals speeds up in the presence of seawater, and other favorable environments like acid rain. The schematic representation of the corrosion process of steel in the presence of electrolytes, oxygen, and water is given in Fig. 20.1.

The standard electrode potential of the corrosion half-cell could be useful to predict the thermodynamic vulnerability of metals towards corrosion, where the overall potential difference between the cathodic and anodic sites acts as the driving force in the electrochemical corrosion reaction of metals. The specific example of corrosion of steel can elaborate the importance of standard electrode potential on the thermodynamics of a corrosion reaction. The electrochemical equilibrium potential (E^0_{cell}) for the corrosion reaction of steel is 0.848 V, and thereof the calculated free energy change (ΔG^0) is -163.6 kJ/mol [2]. Hence, the corrosion of steel is favorable in ambient conditions from the thermodynamics point of view and will take place spontaneously as the free energy change (ΔG^0) for the reaction has a negative value. However, thermodynamic data can't tell how fast the corrosion happens, that is, the corrosion reaction may be thermodynamically favorable for a given set of environmental conditions but the metal corrodes over a long period due to a slow rate of reaction. In fact, the rate of corrosion depends on several factors; especially temperature and the pressure of the medium as these govern the solubility of the corrosive species in the fluid. In general, the rate of corrosion reaction increases with increasing temperature. Therefore, the thermodynamics, reaction kinetics, and external conditions all must be taken into consideration to predict the corrosion vulnerability of a particular metal.

Metal corrosion has become a major issue in modern industries due to the gradual wearing of metal by the above mentioned electrochemical or oxidation process. In general, atmospheric corrosion requires the involvement of oxygen and water/moisture and does not occur if one of these elements are missing, for example, corrosion of steel does not occur in dry air where the relative humidity is less than

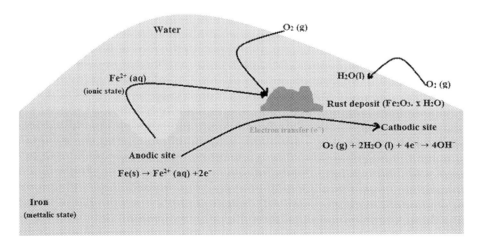

Figure 20.1 Schematic representation of the corrosion process of steel in the presence of electrolytes, oxygen, and water.

30% or when steel is immersed in deoxygenated water [1]. On the other hand, soluble salts and many pollutants significantly alter the course of corrosion. Metal corrosion protection thus plays an important role in the modern industries, and an effective corrosion management practice can save up to 25%−30% of the annual cost of corrosion [9]. Corrosion control is nothing but the prevention of chemical reactions causing the destruction of the metallic state. Several strategies are adopted in industrial practices to control or delay the corrosion of metals, for example, special design of the substrate metal, use of corrosion inhibitor, cathodic protection, and interposing a protective coating between the metal and the surrounding, etc. in which corrosion protective coatings are very popular and effective. These coatings are organic or metallic based and work on the principle of one or more of the following mechanisms [10,11]:

1. Cathodic protection
2. Anodic passivation
3. Electrolytic inhibition
4. Active corrosion inhibition

Basic concepts of these protection mechanisms are discussed in the subsequent section. However, with technological development in the paint and coating industries new corrosion prevention strategies are coming out that aim highly robust protective systems to implement cost-effective methods. Innovation of new classes of corrosion protective coatings such as graphene-based nanocomposites has gained much attention because of scalability and high efficiency compared to the existing methods. These nanocomposites are hydrophobic coatings and are outperforming the conventional coatings in prolonging the life of various metals in corrosive environments [12].

20.2 Anticorrosive coating

Anticorrosive coating protects a metal surface from the attack of external environment by completely inhibiting or slowing down the basic electrochemical reactions responsible for corrosion [13]. Anticorrosive coating provides an effective physical barrier so that the ingress of corrosive particles to the metallic surface is minimized. Hence, the coating must have good adhesion to the metal substrate along with intrinsic durability, flexibility, and toughness to withstand impacts and resist cracking under mechanical stress or weathering but not at the cost of physical appearance [2] (see Fig. 20.2 for more details [14]). Both metallic and organic coatings are used to protect metals from corrosion, and the research related to coating technology has made significant progress in the recent past. However, the problem persists in the long-term protection of metal from aggressive environments and the search for high-performance anticorrosive coatings is continuing. The complexity of the coating-substrate system makes the search rather difficult because the performance and service life of anticorrosive coatings depend on many factors including nature

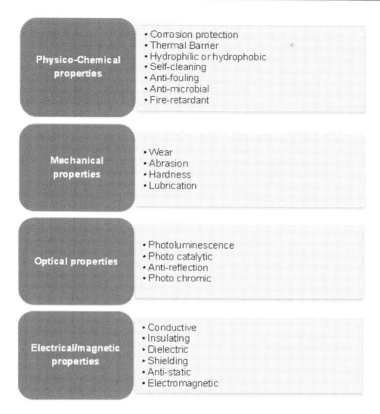

Figure 20.2 Overview of surface properties that can be developed or enhanced through functionalization by organic, inorganic or hybrid coatings.
Source: Reproduced after permission granted from M. Montemor, Functional and smart coatings for corrosion protection: a review of recent advances, Surf. Coat. Technol. 258 (2014) 17–37.

of substrate, pretreatment of substrate, adhesion between the coating and substrate, coating thickness, curing condition, and external environment.

As mentioned earlier, the classification of anticorrosive coating is done based on the basic protective mechanisms against corrosion, that is, barrier protection, cathodic protection, anodic passivation, active corrosion inhibition, and electrolytic inhibition [10,15]. Barrier protection is the primary mechanism for anticorrosive coatings in which the coatings form an electrical double layer in front of a metal surface and prevent the access of corrosive ions as shown in Fig. 20.3 [16]. But, barrier protection does not work in presence of defects like pinholes, air bubble inclusions, inferior adhesion between pigments, and mechanical damages of coating because the ionic impermeability of the coatings gets affected leading to the separation and delamination of the coatings from the metal surface [17]. Delamination creates some discrete low-resistance conductive pathways through which reactants

Defect area: electrolyte covered metal

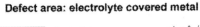

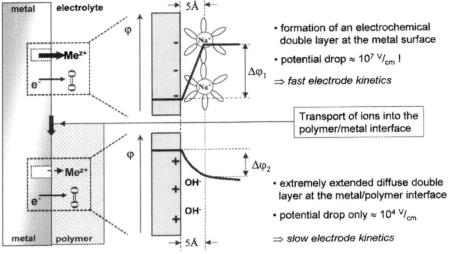

Figure 20.3 Schematic of the shape of the double layer for electrolyte covered metal (top) and polymer-coated metal (bottom).
Source: Reproduced after permission granted from G. Grundmeier, W. Schmidt, M. Stratmann, Corrosion protection by organic coatings: electrochemical mechanism and novel methods of investigation, Electrochim. Acta 45 (15−16) (2000) 2515−2533.

can transport to and from the metal surface and promotes the anodic and cathodic reactions. In cathodic protection, the substrate metal like steel and aluminum alloy is coated with a more electropositive metal, for example, zinc, aluminum, or magnesium alloy which can polarize the substrate and act as sacrificial anode where preferential oxidation takes place because of favorable electrochemical potential thereby inhibiting pitting corrosion of the substrate metal. Galvanization of steel by electroplating or hot dipping is the classical example of cathodic protection of base metal. On the other hand, in anodic passivation method, an ion-impervious or ion-selective barrier is formed by combining a natural oxide layer with a passivation layer or proximate layers of porous and more compact oxides to constitute a bipolar precipitate at the top of the substrate metal to halt ion transport and suppress the redox reaction [18]. Electrolytic inhibition also uses the technique of forming a compact ion-impermeable barrier so that the electrolyte cannot reach the metal surface easily. Here, the coating of a low-ionic-conductivity matrix and diffusion barriers are applied on the metal surface to block the transport of ions through the electrolytic medium. On the other hand, a protective layer at the metal surface reforms upon the damage of original coating in active corrosion inhibition method by the selective release of components like strongly binding unsaturated coordinating ligands or precursors of sparingly soluble oxides with very low solubility product [10,19].

Organic-based anticorrosive coatings consist of ingredients like binder, pigments, solvent, extender, and additives [20]. Earlier chromates were popularly used as pigments in anticorrosive coatings and chromate-rich chemical treatments of metallic structures were extensively used in many industrial applications because of the excellent corrosion resistance properties [21]. However, recent restrictions on the use of hazardous chemicals like chromates which are known for toxicity and potent carcinogenicity have restricted the use of chromates in metal surface finishing. Also, the regulation governing bodies are imposing restrictions on the use of Hazardous Air Pollutants (HAPs) and enforcing manufacturers to reduce Volatile Organic Compounds (VOC) in anticorrosive coatings due to the environmental issues [22]. All these restrictions have triggered the search for more advanced and environment friendly non-toxic coatings, including the exploration of new materials like graphene. For this purpose, a wide range of materials, for example, conductive polymers, engineered polymers, polysiloxanes, encapsulated monomers, metal oxides, self-healing materials, self-assembly monolayers, and sol-gel coatings, etc. have been tested as alternatives to hazardous coatings [23]. However, developing a green and robust coating system from these wide arrays of new materials is challenging because of the lack of knowledge in polymer surface chemistry as well as lack of understanding of the physics of ion transport through the polymer coating medium. In addition, the corrosion process of some metals and their microstructures are not clearly understood which makes the development rather difficult [22]. In the recent past, graphene-polymer composites have emerged as a promising alternative to the conventional anticorrosive coatings, although the commercial viability has to be scrutinized.

20.3   Graphene: a novel material for metal corrosion protection

Graphene is a single atomic monolayer of graphite and is the thinnest two-dimensional carbon material. It is the structural monomer of various carbon allomorphs like graphite, carbon nanotubes (CNTs), and fullerenes (see Fig. 20.4) [24]. Graphene is a versatile material with outstanding properties like lightweight, high strength, large surface area, very good electrical, and heat conductivity equivalent to those of CNTs [25]. Some details of properties of graphene are given in Table 20.1. Graphene and its derivatives have been explored in a good deal of research to produce novel materials for applications including optical components [26], fuel cells [27], biological devices [28], and anticorrosion coating of metals, etc. At present, graphene is a hot scientific topic of industrial importance and a great deal of research is going on the application of graphene and its derivatives in materials, electronics, energy, and biology.

Graphene has the potential to serve as an effective barrier to ionic transport such as ionic barrier for steel [29] and stop charge transfer at the metal-electrolyte

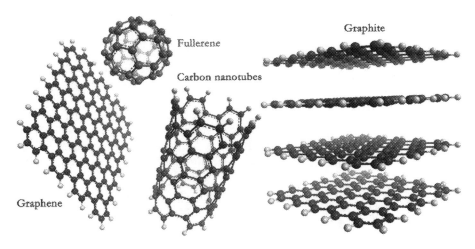

Figure 20.4 Several common carbon materials.
Source: Reproduced after permission granted from R. Ding, W. Li, X. Wang, T. Gui, B. Li,
P. Han, et al., A brief review of corrosion protective films and coatings based on graphene
and graphene oxide, J. Alloy. Compd. 764 (2018) 1039–1055.

Table 20.1 Properties of graphene.

Properties	Unit	Value
Tensile strength	GPa	130
Young's modulus	GPa	1100
Breaking strength	GPa	125
Electrical conductivity	S/m	6×10^5
Carrier mobility	$cm^2/V/s$	2×10^5
Thermal conductivity	W/m/K	3000
Specific surface area	m^2/g	2600
Transparency	%	97.7

interface, thereby minimizing the electrochemical corrosion reaction. Thus, graphene can effectively decouple the metal surface from the environment. Graphene possesses a unique combination of properties specifically required for corrosion inhibiting coating applications. Graphene coating is extremely lightweight and does not significantly alter optical properties and/or decrease bulk properties like electrical and thermal conductivity. Graphene is chemically inert and stable up to 400°C temperature in ambient conditions [30]. In addition, graphene is oxidation resistant even at elevated temperatures in the absence of electrolyte and does not undergo chemical reactions under conditions where other materials would undergo [31]. Graphene has good barrier properties against air and is impermeable to oxygen. More importantly, graphene films can be grown on the meter-scale and easily transferred onto the metal surface by mechanical ways. Both single-layer and multilayer graphene films are in use and exhibit exceptional transparency with more than 90% transmittance, for example, 4-layered graphene films [32].

The excellent electrical conductivity, exceptional high surface area, and other amazing properties of graphene have attracted significant attention to both the scientific and industrial communities from a corrosion protection application point of view [33]. Graphene acts as a protective layer on metal surfaces and recedes gas penetration, oxidation, or corrosion, for example, graphene can protect metallic Cu and Cu/Ni alloys from oxidation up to 200°C in air and by immersion in a 30% hydrogen peroxide bath [31]. It is reported that atom-thick graphene has five times more anticorrosion efficiency than conventional organic coatings because of oxygen inhibition effect of graphene at the metal surface [34]. According to recent research, graphene can change the electrochemical response of underlying metals. Graphene corrosion protection film and graphene-modified anticorrosive coatings are the two main applications of graphene in the field of metal corrosion protection. The following sections focus on the application of graphene polymer composites in anticorrosion coatings. In addition, an overview of pure graphene corrosion protective films is also given.

20.4   Graphene corrosion protective films

Corrosion protective film based on graphene first came into existence around 2011 [31]. It significantly reduces the corrosion rate of metals from common corrosive mediums like water and air because of its barrier effect on these corrosive mediums as shown in Fig. 20.5 [24]. Graphene film acts as a physical barrier between the metal and the corrosive medium and forms the thinnest protective coating as observed through the electrochemical impedance spectroscopy and other experiments [24]. Graphene film thus can protect various metals like copper alloy, aluminum alloy, magnesium alloy, steel, nickel, titanium, etc. [24,35].

20.4.1   Corrosion protective performance of graphene films on various metals

Depending on the properties of the substrate metal, graphene films are prepared on metal surfaces using different techniques such as chemical vapor deposition (CVD), electrodeposition, rapid thermal annealing, mechanical transfer technology, and pyrolysis. Among these, CVD is the most commonly used technique because of its simplicity and more importantly, the prepared film can be directly applied onto the metal surface [36]. The stability of graphene films prepared on different metal surfaces differs largely depending on the bonding strength between graphene and metal crystal faces as well as the distance from metal to carbon atoms. CVD technique for preparing of graphene film is favorable on the metals for which the atomic force with carbon atoms are weaker such as copper, nickel, cobalt but not gold, silver, or palladium [37]. On the other hand, mechanical transfer method allows for covering

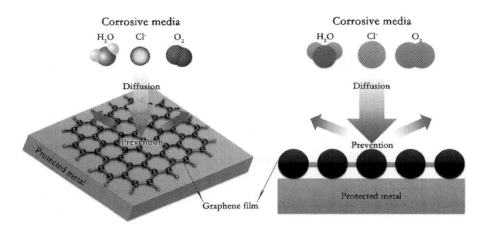

Figure 20.5 Barrier effect on corrosive media of graphene film.
Source: Reproduced after permission granted from reproduced after permission granted from
R. Ding, W. Li, X. Wang, T. Gui, B. Li, P. Han, et al., A brief review of corrosion protective
films and coatings based on graphene and graphene oxide, J. Alloy. Compd. 764 (2018)
1039−1055.

the surface of any metal with high quality graphene films. Special treatments like
electroplating before CVD on the metal surface (e.g. nickel plating of stainless
steel) could improve further the effectiveness of CVD even leading to a 100% coverage of graphene film on the nickel-plated stainless steel as the nickel plating layer
resists the easy formation of metal carbides. Polarization curve experiments show
that graphene coverage on the nickel-plated stainless steel improves the corrosion
resistance many times and also reduces the corrosion current to one-fifth of the bare
nickel-plated stainless steel [38].

Graphene films prepared by CVD method on copper surface reduce the electrochemical corrosion rate by 1.5 orders of magnitude and lowers the corrosion current
by 2 orders of magnitude by increasing the impedance of copper in aggressive solutions containing chloride ions [39]. More importantly, the prepared graphene films
on the copper surface are mechanically transferable to other metal surfaces like
nickel and silver [34,40]. The corrosion prevention ability of graphene film on the
copper surface gets further improved by assembling mercapto derivatives on graphene film. Graphene films also provide a strong corrosion protective effect on
nickel surface and the corrosion resistance increases with an increasing number of
graphene layers, for example, covering nickel surface with four layers of graphene
films reduces the corrosion rate of nickel by four times in sodium sulfate solution
[34]. Monolithic graphene films formed by CVD method and subsequently mechanically transferred onto the surface of silver reduce the corrosion rate of silver by
sixty times at the same time retain the original optical properties of silver which
otherwise is difficult to retain using ordinary protective coating [40].

20.4.2 Practical problems of graphene films

Graphene films show excellent anticorrosion performance because of barrier effects on common corrosive mediums like water and air without affecting the optical properties of the base metal. For this reason, graphene films are called the thinnest protective coating. But graphene films lack many aspects from industrial applications point of view. First of all, scaling up and industrialization of the laboratory-based techniques are very difficult under current technical conditions; hence the production cost is relatively high. Moreover, the electrochemical corrosion of the substrate metal gets accelerated once the protective coating of pure graphene film is damaged as shown in Fig. 20.6 [41]. The excellent conductivity of graphene may also promote the electrochemical corrosion of metals by forming corrosion micro-cells in the damaged region where the metal behaves as anode and further accelerates the corrosion [42]. In addition, graphene film is not able to provide the long-term corrosion protection of metals because of the formation of defects, especially during the preparation of thin graphene films [43]. In fact, wrinkles, cracks, and other defects of the graphene film reduce the long-term protective performance, although do not hamper the short-term protective performance [44]. Repairing the defects by using technology like atomic layer precipitation could be useful to improve the protection ability of the graphene films [45]. However, there are many opposing views on the capability of graphene films in the corrosion protection of metals because the preparation of defect-free graphene films is almost impossible [42]. Hence, the effectiveness of graphene films as anticorrosive material is not yet fully established and there is a big gap between the performance of graphene films and the actual demand, especially in long-term corrosion protection of metals.

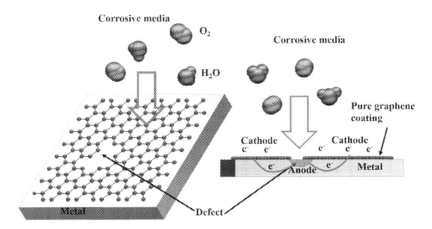

Figure 20.6 Galvanic corrosion occurring in the defects of pure graphene coating.
Source: Reproduced after permission granted from G. Cui, Z. Bi, R. Zhang, J. Liu, X. Yu, Z. Li, et al., A comprehensive review on graphene-based anticorrosive coatings, Chem. Eng. J. 373 (2019) 104–121.

20.5   Polymer-graphene anticorrosive coatings

Graphene filled corrosion protective organic coatings are another important application of graphene in the field of metal corrosion protection where the chemical inertness and barrier properties of graphene are utilized to resist the propagation of the corrosive particles in the coatings. These types of graphene composite coatings first came to knowledge in 2012 [46]. Here, graphene or its derivative materials are added as filler into an organic resin to form the graphene composite corrosion protective coating. However, simple doping or addition and dispersion of graphene in an organic medium by physical mixing does not work because the surface of pure graphene lacks active functional groups and thus the compatibility of graphene and resin is not ideal. In addition, the high surface area and surface energy of graphene pose the problem of particle agglomeration. Hence, it is a common practice to modify the surface of graphene by chemical reaction or other means [47] like oxidation of graphene or reduction of graphene oxide (GO) to make reduced graphene oxide (rGO) thereby decreasing the surface energy of graphene or GO and improving the dispersion and/or compatibility with the resin by repulsion between the functional groups or bonding with the matrix resin, for example, the protective action of epoxy coatings on carbon steel improves many times when doped with GO [48].

According to the research trend, graphene composite anticorrosion coatings could be the future of metal corrosion protection. The strong adhesion properties of graphene and the film-forming properties of the matrix resin are beneficial in improving the overall performance of the graphene composite coating. More importantly, the traditional coating production process could be useful to prepare graphene composite anticorrosive coating, which is of immense importance from industrial synthesis and application point of view. However, as stated earlier, proper dispersion of graphene in the coating matrix is a difficult task that needs serious attention while preparing graphene composite coating. Graphene particles show a strong tendency to agglomerate in the coating matrix because the pure graphene surface does not contain any functional group but has a high aspect ratio and strong van der Waals interactions resulting in a poor corrosion resistance behavior. Improving the dispersion of graphene particles in the coating matrix is thus essential to enhance the corrosion resistance of graphene composite coating. Currently, the coating industry follows three primary methods to improve the dispersion of graphene in the coating matrix viz. physical dispersion, chemical modification, and surface modification of graphene by nanoparticles. These methods are briefly discussed below.

20.5.1   Methods of improving dispersion of graphene in coating matrix

20.5.1.1   Physical dispersion

Physical dispersion is an auxiliary and simple method of dispersing graphene in the coating matrix, although sometimes the dispersion effect is not very prompt.

High-speed magnetic stirring, ultrasonic dispersion, shear emulsification, ball milling, and sanding dispersion, etc. are the common physical dispersion methods in which the latter two methods are found to be the most effective.

20.5.1.2 Chemical modification

Chemical modification of graphene could effectively improve the dispersion of graphene in the coating matrix. Chemical modification requires a chemical coupling agent typically possessing two functional groups, for example, a hydrophilic group that can combine with inorganic graphene particles and an organophilic group that can react with the organic polymer matrix. Thus, chemical modification enables the graphene to graft with the long chain of polymer macromolecules, which improve compatibility with the coating matrix and the overall performance of the graphene composite coating. Silane coupling agents, for example, tetraethoxysilane (TEOS), 3-(triethoxysilyl) propyl isocyanate (TEPI), 3-aminopropyltriethoxysilane (APTES), etc. are the most common and effective modifiers for graphene and its derivatives [49]. Silane coupling modifications improve the compatibility of GO with the silane matrix because the $Si-OH$ group of the silane coupling agent reacts with a reactive functional group of GO, for example, carbonyl, hydroxyl, and epoxy forming $Si-O-C$ covalent bond and thus improving the protective properties of the composite coating significantly compared to the pure silane coating [50]. Dopamine (DA) is another important chemical agent used to modify GO for improving the dispersion of GO in ethanol and the compatibility of GO in waterborne epoxy coatings. DA turns into poly-dopamine (PDA) on the metal surface by self-polymerization and gets adsorbed on the graphene surface by π-π interactions. The functional groups of DA/PDA, for example, catechol, amine, and imine, act as the active sites for the covalent modification of GO and promotes the reduction of GO that improves the dispersion of graphene [51]. The compatibility of graphene in the coating matrix could also be improved through the polar interaction of the nitrile groups of GO. Nitrile group modification of GO results in good dispersion of GO in water and maintain stable dispersion in organic solvents [52].

20.5.1.3 Surface modification of graphene by nanoparticles

Although acid oxidation successfully improves the dispersion of graphene but often ends up destroying the original structure of graphene, thus degrading the inherent performance [53]. Decorating nanoparticles like titanium dioxide, aluminum oxide, calcium carbonate, silicon dioxide, and ferroferric oxide, etc. on the surface of graphene is a relatively new and innovative process that does not disturb the structure and integrity of graphene to a significant extent but can improve the dispersion of graphene in the coating matrix. Nanoparticles decoration can effectively increase the spacing between graphene and its surface, thereby decreasing the agglomeration tendency and increasing the protection performance of the coating [54]. Introduction of one-dimensional multi-walled CNTs can also prevent the accumulation of multi-layer graphene sheets and enlarge the contact area with the polymer matrix and improve the compatibility of graphene [53].

20.5.2 Anticorrosion properties of the graphene composite coating

Graphene and its derivatives enhance the barrier protection of the coatings against the diffusion of corrosive particles like oxygen, moisture, and chloride ions when incorporated in organic coatings. In general, oxygen and moisture are the main reasons for corrosion of steel because there are fine channels inside the commonly applied pure epoxy coatings through which oxygen and moisture can easily penetrate. Graphene-based organic coatings block the permeation channels of oxygen and moisture (see Fig. 20.7 [41]) because of their high surface area and nanometric thickness [55]. However, the anticorrosion properties of graphene composite coatings strongly depend on the dispersion of fillers. Good dispersion of graphene fillers such as proper exfoliation of graphene and its derivatives in the polymeric matrix helps to form an effective physical barrier in the coating and hinders the corrosive particles [56]. Graphene-based composite coatings are commonly prepared via three methods viz. solution mixing, melt mixing, and in situ polymerization. Thus, different exfoliation patterns are expected due to the varying distribution of graphene fillers under different conditions. In addition, the alignment of graphene fillers in the polymer matrix has a significant role in enhancing the protection performance of the composite coating.

As mentioned earlier, the modification of graphene makes strong bonding between the coating and the substrate metal and reduces the gap between them thus decreasing the peeling rate of the coating [57]. Depending on the type of chemical bonds between graphene and other materials, two types of modifications are possible viz. covalent and non-covalent modifications. Covalent modification results when graphene bonds with polymer chains, small molecules, or nanoparticles [58] while non-covalent modification occurs due to ionic bonds, hydrogen bonds, and

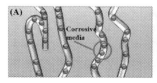 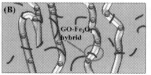 

Figure 20.7 Graphene based organic coatings block the permeation channels of oxygen and moisture (A) Pure epoxy coating, (B) the anticorrosion mechanism of unmodified GO-Fe_3O_4/ EP graphene composite coating, and (C) the modified GO-Fe_3O_4 graphene composite coating.
Source: Reproduced after permission granted from G. Cui, Z. Bi, R. Zhang, J. Liu, X. Yu, Z. Li, et al., A comprehensive review on graphene-based anticorrosive coatings, Chem. Eng. J. 373 (2019) 104−121.

π-π interactions [59]. Covalent modification is done to maintain chemical stability and mechanical strength of graphene, whereas non-covalent modification is aimed to take the advantages of excellent conductivity and ultra-high surface area of graphene [60]. Non-covalent surface modification of graphene is easy to introduce compared to the covalent surface modification because the latter involves complex chemical reactions. However, both covalent and non-covalently modified graphene can be incorporated together in the coatings to get synergistic benefits of different types of modified graphene. Epoxy, polyurethane (PU), polyaniline (PANI), alkyd, polystyrene (PS), polymethylmethacrylate (PMMA), polyvinyl alcohol (PVA), polyvinyl butyral, etc. are the common resins used in making graphene-based organic protective coatings. In the following sections, the applications and protective properties of graphene organic coatings are discussed.

20.5.3   Corrosion protection mechanism of graphene composite coating

The fundamental aspects of corrosion protection mechanism of graphene composite coatings depend on the barrier effect of graphene in the coating and the ability of graphene in enhancing the role of sacrificial anode. The barrier effect of graphene resists the penetration of the corrosive particles and propagation of cracks whereas the high conductivity of graphene enables graphene to act as the channel for electron flow between the anode and substrate metal thus enhancing the protective role of the sacrificial anode.

20.5.3.1   The barrier effect of graphene

20.5.3.1.1   Graphene hinders the penetration of corrosive particles

Pure polymer coatings cannot provide strong barrier effects; hence the corrosive particles easily penetrate along the thickness of the coating and initiate corrosion reactions on the metal surface. Graphene has a specific structure made of sp2-hybridized carbon atoms (see Fig. 20.4). The high electron density of graphene on the aromatic ring enables graphene to block the small molecules making the graphene layer impermeable and thus providing excellent shielding protection against the corrosive particles by prolonging the infiltration pathway [61].

20.5.3.1.2   Graphene hinders crack propagation in the coating

Crack formations during the preparation and application of the coating reduce the corrosion resistance of the coating as the cracks act as the points of defects through which the corrosive particles invade to the metal. Moreover, these cracks continue to expand deeper into the coating and become larger with time, thus promoting severe localized corrosion on the metal surface. Graphene reinforcement increases the toughness of the composite coating hence could resist crack propagation and reduce the penetration of the corrosive particles.

20.5.3.2   Graphene enhances the role of the sacrificial anode

Zinc acts as a sacrificial anode in zinc-rich coatings because of favorable electrode potential and protects the metal surface. However, the polymeric binder in the coatings is in general not electrically conductive and hinders electronic communication between the zinc particles and the metal substrate resulting in isolated and inactive zinc particles. Owing to high conductivity, graphene in zinc-rich coatings conducts electrons to zinc anode and behaves as a cathode for the zinc anodes that are exposed to the external electrolyte. Hence, the electrolyte does not require penetrating through the coating in search of a metal cathode for zinc anode. Graphene being an excellent conductor maintains electron flow between the zinc particles and form a large conductive network in the coating, thus making effective utilization of the sacrificial zinc anodes [62].

20.6   Applications of different polymer-graphene composites in metal corrosion protection

20.6.1   Polyaniline graphene composites

PANI is an electroactive polymer and its functionalized coatings provide excellent corrosion protection to metals compared to other polymers because of high corrosion potential and redox catalytic capability. On the metal surface, PANI forms a passive metal oxide layer responsible for enhanced corrosion protection [63]. PANI/graphene composites with a relatively high aspect ratio of graphene (~ 500) show good potential for advanced gas barrier applications due to the conductive nature of both PANI and graphene [64]. PANI and graphene both possess π-conjugated electrons that help graphene to act as a nucleation center for PANI matrix and between the sheets of graphene. This strategy triggers outstanding sensitivity, improved conductivity, more selectivity, and high capacity. However, developing graphene-based PANI composites is a real challenge, although these materials have already been explored for a range of applications in different fields including metal corrosion applications [65].

PANI/graphene composites show good potential for corrosion protection of metals like steel, copper, etc. [66,67]. Graphene reinforced PANI coatings exhibit better electrochemical corrosion resistance and oxygen and/or water barrier efficiency compared to PANI coatings itself [46]. Further, the two-dimensional graphene layers provide a physical barrier against the migration of corrosive species, which can synergistically couple with the PANI barrier effect. In general, PANI can be found in any of the three oxidation states viz. leucoemaraldine, emaraldine, and pernigraniline. In these, emaraldine base is the most useful form of PANI due to its high stability and becomes highly conducting on doping whereas the other two forms are the poor conductors. Hence, the coating of graphene with emeraldine base PANI can enhance the protective performance of graphene because of the high current barrier and specific anticorrosion property of emeraldine base PANI.

Functionalization of graphene with 4-aminobenzoyl group improves the dispersion of graphene in PANI matrix and lengthens the diffusion path length for active gases, for example, PANI/4-aminobenzoic acid-modified graphene composite with 0.5% (wt.) graphene reduces the diffusion rate of oxygen and water to 24% and 22% compared to that of pure PANI coatings (see Fig. 20.8 [46]) and drops the corrosion current by the orders of two magnitudes [46]. Functionalized graphene/PANI nanocomposites prepared via in situ redox polymerization—dedoping technique increases the corrosion efficiency of the composites from 85.16% to 99.9% compared to base PANI coating with a low corrosion rate of 1.68×10^{-4} mm/year [67]. Graphene/PANI hybrid composite with alternating graphene and PANI layers stack parallel to the metal (copper) surface can act as smart anticorrosive coating layers capable of monitoring the status of metal corrosion. Copper coated with this hybrid composite shows excellent anticorrosion protection efficiencies of 46.6% and 68.4% under electrochemical polarization in 1 (M) H_2SO_4 and 3.5% (wt.%) NaCl solutions, respectively [68]. PANI-grafted graphene nanosheets (0.6% wt.%) can synergistically enhance the barrier properties of the coating and provide cathodic protection on the zinc surface [66].

20.6.2 Epoxy-graphene composites

Epoxy resin is a widely used film-former for high-performance composite coatings because of high tensile strength and modulus, high adhesion and dimensional stability, low shrinkage after curing, good chemical and corrosion resistance, etc. [69].

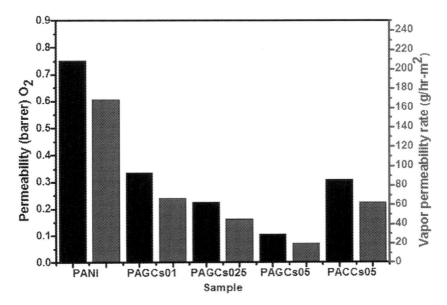

Figure 20.8 Permeability and vapor permeability rates of PANI, PAGCs (PAGCs01, PAGCs025, PAGCs05), and PACCs05.
Source: Reproduced after permission granted from C.H. Chang, T.C. Huang, C.W. Peng, T. C. Yeh, H.I. Lu, W.I. Hung, et al., Novel anticorrosion coatings prepared from polyaniline/ graphene composites, Carbon 50 (14) (2012) 5044–5051.

Graphene-epoxy coatings can protect metal from corrosion by reducing oxygen diffusion and infiltration of salt anions, even at a very low loading of graphene. However, the dispersion of graphene in the epoxy matrix is a tedious job, although could be achieved by polyvinylpyrrolidone (PVP). PVP-rGO disperses evenly in the epoxy matrix and improves the coating adhesion and barrier properties of the coating. The corrosion rate of the epoxy composite coating with 0.7% (wt.) PVP-rGO decreases by 4 times compared to pure epoxy coating [70].

Reinforcing epoxy coatings with graphene nanosheets (GNS) also could enhance the protective performance of the epoxy-based coatings. Here, the size of GNS plays a major role in improving the barrier properties of the coatings. In general, smaller size GNS provides better effectiveness and as the size of graphene continuous to shrink to the level of graphene quantum dots, the protective performance goes on increasing [71]. The same concept also applies to GO nanoplates-polyamine cured epoxy coating system when GO is functionalized by p-phenylenediamine (PPDA). PPDA-GO of less than 1 μm size improves significantly the protective performance and mechanical strength of epoxy coatings because the reduced size GO forms more covalent bonds with PPDA that again bonds with epoxy through hydrogen bonding [72].

Functionalization of graphene by PANI nanofiber also improves the dispersion of graphene in the epoxy matrix and enhances the compatibility of GO with epoxy resin, thereby improves the thermal stability and barrier protection performance [73]. Improving the adhesion of the epoxy coating with the substrate metal through modification of graphene or GO also could improve the protective ability of the coatings, for example, surface modification GO with 3-aminopropyltriethoxysilane through covalent bonding improves the adhesion of epoxy resin with iron metal, thus raises corrosion protection performance and reduces the cathodic delamination rate of the epoxy coating [74].

20.6.3 Polystyrene graphene composites

Vinyl polymer, for example, PS and graphene-based organic coatings are non-conjugated anticorrosive coatings that differ function-wise from conjugated conducting PANI-graphene coatings. PANI is an electroactive conducting polymer that can induce a passive layer on the metal surface, but PANI has limited solubility and its deep blue color is a matter of concern. In addition, PANI is non-plastic, all these limitations of PANI restrict processibility for metal corrosion protection applications [75]. PS-graphene composite coating could be an effective alternative of PANI based systems because of the certain advantages of PS like plasticity, transparency, processability, and low production cost. However, PS-graphene coatings differ widely from PANI-graphene anticorrosive coatings from corrosion protection mechanism point of view and the effectiveness of the latter is far more established, for example, conjugated PANI-graphene coating forms a passive oxide layer on a steel surface, but non-conjugated PS cannot [76]. Improving π-π interaction between the aromatic rings of PS and phenylenediamine and 4-vinylbenzoic acid modified GO (pvGO) through in situ mini-emulsion polymerization method could enhance the dispersion of graphene in non-conjugated PS matrix and thus the gas barrier properties and flexibility of the nanoscale films on the metal surface. The corrosion protection ability of PS coatings increases from 37.90% to 99.53% when doped with 2 wt.% pvGO [48].

20.6.4   Polyurethane graphene composites

Traditional anticorrosive coatings are either based on organic solvents or toxic heavy metals, thus are currently being replaced by environment-friendly water-based coating materials [77,78]. PU-graphene nanocomposites have recently garnered significant interest due to their unique electrical, dielectric, mechanical, and thermal properties [79]. The coatings developed from PU-graphene nanocomposites could act as a diffusion barrier to corroding chemicals, molecules, ions, water, and other solvents [80]. PU coatings are already used extensively in transportation, electronic, textile, and defense sector due to high resistance to biological and radiological warfare. Incorporation of fillers/additives like graphene can give fascinating physiochemical properties [47,81]. PU-graphene nanocomposites are prepared by the following methods: (1) solution method, (2) melt method, (3) in situ polymerization, (4) sol-gel method and (5) solvent-free method. However, recent efforts are focused on developing fabrication technologies for economical and large-scale production of these nanocomposites.

Hydrophobicity and barrier effect of graphene are beneficial in reducing the kinetics of the corrosion process of water-based PU-graphene nanocomposite coatings because enhanced hydrophobic property of the composite could help in preventing the absorption of water through the coatings [47], for example, waterborne PU-GO nanocomposite anticorrosive coating made from modified GNS show inhibition efficiency of 99.8% on mild steel and thus can be considered as a potential candidate for green corrosion protective coating material [78]. Also, the high aspect ratio of graphene and GO favors the preparation of waterborne PU composite coatings for anticorrosive applications [77]. Nanocomposites based on PU-graphene nanoplatelets and PU-GO show a superior anticorrosive property for use in oilsands transportation industry [82]. Again, bilayer coating with electrochemically exfoliated graphene–epoxy PU based nanocomposites is superior compared to GO and rGO [83]. GO reinforced zinc phosphate/PU composites could provide effective anticorrosion property to prevent corrosion of low carbon steel as the composite exhibit low corrosion current density of 4.41×10^{-7} A/cm^2 and coating resistance of 1.04×10^4 Ω cm^2 [84]. PU-amine functionalized (*tert*-butyl amine) GO nanocomposites exhibit good tensile modulus and corrosion resistance due to better π-π interactions between filler and PU [81].

20.6.5   Other polymer-graphene composites

PMMA is an inexpensive and non-conductive polymer but has low thermal stability and mechanical properties. Addition of functionalized graphene or GO could improve the physical properties of PMMA composites. High carboxylic acid group-containing thermally reduced GO-PMMA composite coatings exhibit good corrosion protection of a cold-rolled steel electrode [85]. PMMA-grafted GO nanocomposites prepared via surface-initiated atom transfer radical polymerization (ATRP) show good potential for corrosion protection of copper metal under aggressive saline environments [86]. PMMA nanocomposite coatings embedded with GNS fabricated by nanocasting technique show advanced anticorrosion performance and super hydrophobicity similar to that of a natural lotus leaf [87]. PVA is a non-toxic polymer with good film forming ability and shows strong adhesion with the metal surface. PVA based coatings exhibit excellent corrosion protection on mild steel [88]. Inclusion of graphene into PVA matrix improves further the corrosion resistance of the coating by reducing the pore size. Graphene impregnated PVA coating protects Al alloy efficiently in 3.5% (wt.) NaCl solution [89].

20.7   Summary and conclusions

Graphene-based polymer composite coatings show a huge potential and are of growing interest, although they are at the early stages of development. Polymer-graphene organic coatings provide better corrosion protection to metals than the conventional coatings because of the barrier effect of graphene that resists the ingress of corrosive particles and propagation of cracks. Graphene also enhances the life of sacrificial anode, for example, in the case of zinc-rich coatings. However, the corrosion protection mechanism or theory of graphene-based polymer composite coatings is not fully understood and needs deep insight into it using micro-zone testing techniques, like tow electrodes and microelectrodes. Also, the failure mechanism of graphene composite coating needs to be clarified using more robust accelerated aging experiments.

   Improving the dispersion of graphene in the composite coating can maximize the shielding protection of graphene, but it is another difficult task. Chemical modification of graphene could enhance the stability of graphene dispersion in the polymer matrix, but covalent bond modification is not advisable as that damages the original structure of graphene. Non-covalent modification of graphene structure, for example, ionic bonding, hydrogen bonding, π–π stacking, etc. could be useful. Use of toxic chemical agents in high dosages is also not recommended from the safety of environment and worker's health point of view. Hence, green and environment-friendly modification methods need to explore to improve the dispersion of graphene and reduce environmental pollution and health hazards. Orderly arrangement of graphene is also another important factor in prolonging the diffusion path length of the corrosive particles to delay the corrosion rate of metal. However, the technology of getting ordered distribution of graphene in the coating is not established and will be promising and meaningful research.

   Developing graphene waterborne anticorrosive coatings could reduce the problems associated with graphene solvent-borne anticorrosive coatings such as toxicity of heavy metals and high VOCs. But waterborne coatings lack coherent film-forming abilities due to poor dispersion of graphene and also have low abrasion resistance. Moreover, these coatings exhibit poor shielding protection against water and oxygen due to the presence of residual water-based groups. One important disadvantage of current graphene anticorrosion organic coatings is the lack of self-repairing, thus they fail quickly due to mechanical damage. Galvanic corrosion is another major problem of graphene-based coatings because of the high conductivity of graphene. Hence, it is necessary to develop electrically insulated graphene-like 2D materials like h-BN to reduce the problems associated with graphene and scale up the synthesis of these anticorrosive coatings in an economic and environment friendly way.

References

[1] N.S. Sangaj, V. Malshe, Permeability of polymers in protective organic coatings, Prog. Org. Coat. 50 (1) (2004) 28−39.
[2] P.A. Sørensen, S. Kiil, K. Dam-Johansen, C.E. Weinell, Anticorrosive coatings: a review, J. Coat. Technol. Res. 6 (2) (2009) 135−176.
[3] ISO 12944-1:2017, in Paints and varnishes—corrosion protection of steel structures by protective paint systems-Part 1: General Introduction, DIN Deutsches Institut für Normung e.V., Germany, 2017.
[4] E. Bardal, Corrosion and Protection, Springer Science & Business Media, 2007.
[5] P. Pistorius, G. Burstein, Metastable pitting corrosion of stainless steel and the transition to stability, Philos. Trans. R. Soc. A: Phys. Eng. Sci. 341 (1662) (1992) 531−559.
[6] A. Husain, O. Al-Shamah, A. Abduljaleel, Investigation of marine environmental related deterioration of coal tar epoxy paint on tubular steel pilings, Desalination 166 (2004) 295−304.
[7] K.A. Chandler, Marine and Offshore Corrosion, Butterworths, 1985.
[8] D.A. Jones, Corrosion in selected corrosive environments, second ed., Principles and Prevention of Corrosion, 1996, Prentice Hall, Upper Saddle River, NJ, 1992, pp. 387−390.
[9] R. Bhaskaran, N. Palaniswamy, N.S. Rengaswamy, M. Jayachandran, Global cost of corrosion—a historical review, Corrosion: Mater. 13 (2005) 621−628.
[10] F. Presuel-Moreno, M.A. Jakab, N. Tailleart, M. Goldman, J.R. Scully, Corrosion-resistant metallic coatings, Mater. Today 11 (10) (2008) 14−23.
[11] R. Figueira, C.J. Silva, E. Pereira, Organic−inorganic hybrid sol−gel coatings for metal corrosion protection: a review of recent progress, J. Coat. Technol. Res. 12 (1) (2015) 1−35.
[12] H. Alhumade, A. Abdala, A. Yu, A. Elkamel, L. Simon, Corrosion inhibition of copper in sodium chloride solution using polyetherimide/graphene composites, Can. J. Chem. Eng. 94 (5) (2016) 896−904.
[13] R.V. Dennis, V. Patil, J.L. Andrews, J.P. Aldinger, G.D. Yadav, S. Banerjee, Hybrid nanostructured coatings for corrosion protection of base metals: a sustainability perspective, Mater. Res. Express 2 (3) (2015) 032001.
[14] M. Montemor, Functional and smart coatings for corrosion protection: a review of recent advances, Surf. Coat. Technol. 258 (2014) 17−37.

[15] D.P. Schmidt, B.A. Shaw, E. Sikora, W.W. Shaw, L.H. Laliberte, Corrosion protection assessment of sacrificial coating systems as a function of exposure time in a marine environment, Prog. Org. Coat. 57 (4) (2006) 352−364.

[16] G. Grundmeier, W. Schmidt, M. Stratmann, Corrosion protection by organic coatings: electrochemical mechanism and novel methods of investigation, Electrochim. Acta 45 (15-16) (2000) 2515−2533.

[17] T. Nguyen, J. Hubbard, J. Pommersheim, Unified model for the degradation of organic coatings on steel in a neutral electrolyte, J. Coat. Technol. 68 (855) (1996) 45−56.

[18] B.C. Worley, W.A. Ricks, M.P. Prendergast, B.W. Gregory, R. Collins Jr., J.J. Cassimus, et al., Anodic passivation of tin by alkanethiol self-assembled monolayers examined by cyclic voltammetry and coulometry, Langmuir 29 (42) (2013) 12969−12981.

[19] D. Grigoriev, D. Akcakayiran, M. Schenderlein, D. Shchukin, Protective organic coatings with anticorrosive and other feedback-active features: micro-and nanocontainers-based approach, Corrosion 70 (5) (2014) 446−463.

[20] N. Kouloumbi, L. Ghivalos, P. Pantazopoulou, Determination of the performance of epoxy coatings containing feldspars filler, Pigm. Resin. Technol. 34 (2005) 148−153.

[21] A. Bastos, M. Ferreira, A. Simoes, Comparative electrochemical studies of zinc chromate and zinc phosphate as corrosion inhibitors for zinc, Prog. Org. Coat. 52 (4) (2005) 339−350.

[22] R.B. Figueira, I.R. Fontinha, C.J. Silva, E.V. Pereira, Hybrid sol-gel coatings: smart and green materials for corrosion mitigation, Coatings 6 (1) (2016) 12.

[23] N.H. Othman, M.C. Ismail, M. Mustapha, N. Sallih, K.E. Kee, R.A. Jaal, Graphene-based polymer nanocomposites as barrier coatings for corrosion protection, Prog. Org. Coat. 135 (2019) 82−99.

[24] R. Ding, W. Li, X. Wang, T. Gui, B. Li, P. Han, et al., A brief review of corrosion protective films and coatings based on graphene and graphene oxide, J. Alloy. Compd. 764 (2018) 1039−1055.

[25] B. Aissa, N.K. Memon, A. Ali, M.K. Khraisheh, Recent progress in the growth and applications of graphene as a smart material: a review, Front. Mater. 2 (2015) 58.

[26] J.T. Kim, H. Choi, E. Shin, S. Park, I.G. Kim, Graphene-based optical waveguide tactile sensor for dynamic response, Sci. Rep. 8 (1) (2018) 1−6.

[27] K. Akbar, J.H. Kim, Z. Lee, M. Kim, Y. Yi, S.H. Chun, Superaerophobic graphene nano-hills for direct hydrazine fuel cells, NPG Asia Mater. 9 (5) (2017) e378.

[28] B.R. Goldsmith, L. Locascio, Y. Gao, M. Lerner, A. Walker, J. Lerner, et al., Digital biosensing by foundry-fabricated graphene sensors, Sci. Rep. 9 (1) (2019) 1−10.

[29] S. Sreevatsa, A. Banerjee, G. Haim, Graphene as a permeable ionic barrier, ECS Trans. 19 (5) (2009) 259.

[30] L. Liu, S. Ryu, M.R. Tomasik, E. Stolyarova, N. Jung, M.S. Hybertsen, et al., Graphene oxidation: thickness-dependent etching and strong chemical doping, Nano Lett. 8 (7) (2008) 1965−1970.

[31] S. Chen, L. Brown, M. Levendorf, W. Cai, S.Y. Ju, J. Edgeworth, et al., Oxidation resistance of graphene-coated Cu and Cu/Ni alloy, ACS Nano 5 (2) (2011) 1321−1327.

[32] R.R. Nair, P. Blake, A.N. Grigorenko, K.S. Novoselov, T.J. Booth, T. Stauber, et al., Fine structure constant defines visual transparency of graphene, Science 320 (5881) (2008) 1308.

[33] K.S. Novoselov, A.K. Geim, S.V. Morozov, D. Jiang, Y. Zhang, S.V. Dubonos, et al., Electric field effect in atomically thin carbon films, Science 306 (5696) (2004) 666−669.

[34] D. Prasai, J.C. Tuberquia, R.R. Harl, G.K. Jennings, K.I. Bolotin, Graphene: corrosion-inhibiting coating, ACS Nano 6 (2) (2012) 1102–1108.

[35] J. Liu, L. Hua, S. Li, M. Yu, Graphene dip coatings: an effective anticorrosion barrier on aluminum, Appl. Surf. Sci. 327 (2015) 241–245.

[36] X. Li, Y. Zhu, W. Cai, M. Borysiak, B. Han, D. Chen, et al., Transfer of large-area graphene films for high-performance transparent conductive electrodes, Nano Lett. 9 (12) (2009) 4359–4363.

[37] G.A.K.P.A. Giovannetti, P.A. Khomyakov, G. Brocks, V.V. Karpan, J. van den Brink, P. J. Kelly, Doping graphene with metal contacts, Phys. Rev. Lett. 101 (2) (2008) 026803.

[38] N.W. Pu, G.N. Shi, Y.M. Liu, X. Sun, J.K. Chang, C.L. Sun, et al., Graphene grown on stainless steel as a high-performance and ecofriendly anti-corrosion coating for polymer electrolyte membrane fuel cell bipolar plates, J. Power Sources 282 (2015) 248–256.

[39] R.S. Raman, P.C. Banerjee, D.E. Lobo, H. Gullapalli, M. Sumandasa, A. Kumar, et al., Protecting copper from electrochemical degradation by graphene coating, Carbon 50 (11) (2012) 4040–4045.

[40] Y. Zhao, Y. Xie, Y.Y. Hui, L. Tang, W. Jie, Y. Jiang, et al., Highly impermeable and transparent graphene as an ultra-thin protection barrier for Ag thin films, J. Mater. Chem. C 1 (32) (2013) 4956–4961.

[41] G. Cui, Z. Bi, R. Zhang, J. Liu, X. Yu, Z. Li, et al., A comprehensive review on graphene-based anti-corrosive coatings, Chem. Eng. J. 373 (2019) 104–121.

[42] F. Zhou, Z. Li, G.J. Shenoy, L. Li, H. Liu, Enhanced room-temperature corrosion of copper in the presence of graphene, ACS Nano 7 (8) (2013) 6939–6947.

[43] M. Schriver, W. Regan, W.J. Gannett, A.M. Zaniewski, M.F. Crommie, A. Zettl, Graphene as a long-term metal oxidation barrier: worse than nothing, ACS Nano 7 (7) (2013) 5763–5768.

[44] I. Wlasny, P. Dabrowski, M. Rogala, P.J. Kowalczyk, I. Pasternak, W. Strupinski, et al., Role of graphene defects in corrosion of graphene-coated Cu (111) surface, Appl. Phys. Lett. 102 (11) (2013) 111601.

[45] Y.P. Hsieh, M. Hofmann, K.W. Chang, J.G. Jhu, Y.Y. Li, K.Y. Chen, et al., Complete corrosion inhibition through graphene defect passivation, ACS Nano 8 (1) (2014) 443–448.

[46] C.H. Chang, T.C. Huang, C.W. Peng, T.C. Yeh, H.I. Lu, W.I. Hung, et al., Novel anti-corrosion coatings prepared from polyaniline/graphene composites, Carbon 50 (14) (2012) 5044–5051.

[47] M. Mo, W. Zhao, Z. Chen, E. Liu, Q. Xue, Corrosion inhibition of functional graphene reinforced polyurethane nanocomposite coatings with regular textures, RSC Adv. 6 (10) (2016) 7780–7790.

[48] Y.H. Yu, Y.Y. Lin, C.H. Lin, C.C. Chan, Y.C. Huang, High-performance polystyrene/graphene-based nanocomposites with excellent anti-corrosion properties, Polym. Chem. 5 (2) (2014) 535–550.

[49] J. Li, J. Cui, J. Yang, Y. Ma, H. Qiu, J. Yang, Silanized graphene oxide reinforced organofunctional silane composite coatings for corrosion protection, Prog. Org. Coat. 99 (2016) 443–451.

[50] J. Liang, X.W. Wu, Y. Ling, S. Yu, Z. Zhang, Trilaminar structure hydrophobic graphene oxide decorated organosilane composite coatings for corrosion protection, Surf. Coat. Technol. 339 (2018) 65–77.

[51] L. Wang, D. Wang, Z. Dong, F. Zhang, J. Jin, Interface chemistry engineering for stable cycling of reduced GO/SnO$_2$ nanocomposites for lithium ion battery, Nano Lett. 13 (4) (2013) 1711–1716.

[52] H. Hu, Y. He, Z. Long, Y. Zhan, Synergistic effect of functional carbon nanotubes and graphene oxide on the anti-corrosion performance of epoxy coating, Polym. Adv. Technol. 28 (6) (2017) 754−762.

[53] S.Y. Yang, W.N. Lin, Y.L. Huang, H.W. Tien, J.Y. Wang, C.C.M. Ma, et al., Synergetic effects of graphene platelets and carbon nanotubes on the mechanical and thermal properties of epoxy composites, Carbon 49 (3) (2011) 793−803.

[54] Y. Zhan, J. Zhang, X. Wan, Z. Long, S. He, Y. He, Epoxy composites coating with Fe_3O_4 decorated graphene oxide: Modified bio-inspired surface chemistry, synergistic effect and improved anti-corrosion performance, Appl. Surf. Sci. 436 (2018) 756−767.

[55] B. Ramezanzadeh, E. Ghasemi, M. Mahdavian, E. Changizi, M.M. Moghadam, Covalently-grafted graphene oxide nanosheets to improve barrier and corrosion protection properties of polyurethane coatings, Carbon 93 (2015) 555−573.

[56] Z.S. Pour, M. Ghaemy, Polymer grafted graphene oxide: for improved dispersion in epoxy resin and enhancement of mechanical properties of nanocomposite, Compos. Sci. Technol. 136 (2016) 145−157.

[57] D. Liu, W. Zhao, S. Liu, Q. Cen, Q. Xue, Comparative tribological and corrosion resistance properties of epoxy composite coatings reinforced with functionalized fullerene C60 and graphene, Surf. Coat. Technol. 286 (2016) 354−364.

[58] N. Parhizkar, B. Ramezanzadeh, T. Shahrabi, Corrosion protection and adhesion properties of the epoxy coating applied on the steel substrate pre-treated by a sol-gel based silane coating filled with amino and isocyanate silane functionalized graphene oxide nanosheets, Appl. Surf. Sci. 439 (2018) 45−59.

[59] L. Gu, S. Liu, H. Zhao, H. Yu, Facile preparation of water-dispersible graphene sheets stabilized by carboxylated oligoanilines and their anticorrosion coatings, ACS Appl. Mater. Inter. 7 (32) (2015) 17641−17648.

[60] R. Imani, F. Mohabatpour, F. Mostafavi, Graphene-based nano-carrier modifications for gene delivery applications, Carbon 140 (2018) 569−591.

[61] V. Berry, Impermeability of graphene and its applications, Carbon 62 (2013) 1−10.

[62] R. Ding, Y. Zheng, H. Yu, W. Li, X. Wang, T. Gui, Study of water permeation dynamics and anti-corrosion mechanism of graphene/zinc coatings, J. Alloy. Compd. 748 (2018) 481−495.

[63] B. Wessling, Passivation of metals by coating with polyaniline: corrosion potential shift and morphological changes, Adv. Mater. 6 (3) (1994) 226−228.

[64] S. Stankovich, R.D. Piner, S.T. Nguyen, R.S. Ruoff, Synthesis and exfoliation of isocyanate-treated graphene oxide nanoplatelets, Carbon 44 (15) (2006) 3342−3347.

[65] V. Chabot, D. Higgins, A. Yu, X. Xiao, Z. Chen, J. Zhang, A review of graphene and graphene oxide sponge: Material synthesis and applications to energy and the environment, Energy Environ. Sci. 7 (5) (2014) 1564−1596.

[66] Y. Lei, Z. Qiu, J. Liu, D. Li, N. Tan, T. Liu, et al., Effect of conducting polyaniline/graphene nanosheet content on the corrosion behavior of zinc-rich epoxy primers in 3.5% NaCl solution, Polymers 11 (5) (2019) 850.

[67] X. Sheng, W. Cai, L. Zhong, D. Xie, X. Zhang, Synthesis of functionalized graphene/polyaniline nanocomposites with effective synergistic reinforcement on anticorrosion, Ind. Eng. Chem. Res. 55 (31) (2016) 8576−8585.

[68] S. Kim, T.H. Le, C.S. Park, G. Park, K.H. Kim, S. Kim, et al., A solution-processable, nanostructured, and conductive graphene/polyaniline hybrid coating for metal-corrosion protection and monitoring, Sci. Rep. 7 (1) (2017) 1−9.

[69] T. Jiang, T. Kuila, N.H. Kim, B.C. Ku, J.H. Lee, Enhanced mechanical properties of silanized silica nanoparticle attached graphene oxide/epoxy composites, Compos. Sci. Technol. 79 (2013) 115−125.

[70] Z. Zhang, W. Zhang, D. Li, Y. Sun, Z. Wang, C. Hou, et al., Mechanical and anticorrosive properties of graphene/epoxy resin composites coating prepared by in-situ method, Int. J. Mol. Sci. 16 (1) (2015) 2239−2251.

[71] S. Pourhashem, E. Ghasemy, A. Rashidi, M.R. Vaezi, Corrosion protection properties of novel epoxy nanocomposite coatings containing silane functionalized graphene quantum dots, J. Alloy. Compd. 731 (2018) 1112−1118.

[72] B. Ramezanzadeh, G. Bahlakeh, M.M. Moghadam, R. Miraftab, Impact of size-controlled p-phenylenediamine (PPDA)-functionalized graphene oxide nanosheets on the GO-PPDA/Epoxy anti-corrosion, interfacial interactions and mechanical properties enhancement: experimental and quantum mechanics investigations, Chem. Eng. J. 335 (2018) 737−755.

[73] Y. Hayatgheib, B. Ramezanzadeh, P. Kardar, M. Mahdavian, A comparative study on fabrication of a highly effective corrosion protective system based on graphene oxide-polyaniline nanofibers/epoxy composite, Corros. Sci. 133 (2018) 358−373.

[74] N. Parhizkar, T. Shahrabi, B. Ramezanzadeh, A new approach for enhancement of the corrosion protection properties and interfacial adhesion bonds between the epoxy coating and steel substrate through surface treatment by covalently modified amino functionalized graphene oxide film, Corros. Sci. 123 (2017) 55−75.

[75] L.M. Liu, K. Levon, Undoped polyaniline−surfactant complex for corrosion prevention, J. Appl. Polym. Sci. 73 (14) (1999) 2849−2856.

[76] Y. Wei, J. Wang, X. Jia, J.M. Yeh, P. Spellane, Polyaniline as corrosion protection coatings on cold rolled steel, Polymer 36 (23) (1995) 4535−4537.

[77] J. Li, J. Cui, J. Yang, Y. Li, H. Qiu, J. Yang, Reinforcement of graphene and its derivatives on the anticorrosive properties of waterborne polyurethane coatings, Compos. Sci. Technol. 129 (2016) 30−37.

[78] A. Mohammadi, M. Barikani, A.H. Doctorsafaei, A.P. Isfahani, E. Shams, B. Ghalei, Aqueous dispersion of polyurethane nanocomposites based on calix [4] arenes modified graphene oxide nanosheets: preparation, characterization, and anti-corrosion properties, Chem. Eng. J. 349 (2018) 466−480.

[79] G. Kaur, R. Adhikari, P. Cass, M. Bown, M.D. Evans, A.V. Vashi, et al., Graphene/polyurethane composites: fabrication and evaluation of electrical conductivity, mechanical properties and cell viability, RSC Adv. 5 (120) (2015) 98762−98772.

[80] E. Cunha, M.C. Paiva, Composite films of waterborne polyurethane and few-layer graphene—enhancing barrier, mechanical, and electrical properties, J. Compos. Sci. 3 (2) (2019) 35.

[81] J.G. Um, S. Habibpour, Y.S. Jun, A. Elkamel, A. Yu, Development of π−π interaction-induced functionalized graphene oxide on mechanical and anticorrosive properties of reinforced polyurethane composites, Ind. Eng. Chem. Res. 59 (8) (2020) 3617−3628.

[82] J.G. Um, Y.S. Jun, A. Elkamel, A. Yu, Engineering investigation for the size effect of graphene oxide derived from graphene nanoplatelets in polyurethane composites, Can. J. Chem. Eng. 98 (5) (2020) 1084−1096.

[83] D. Dutta, A.N.F. Ganda, J.K. Chih, C.C. Huang, C.J. Tseng, C.Y. Su, Revisiting graphene−polymer nanocomposite for enhancing anticorrosion performance: a new insight into interface chemistry and diffusion model, Nanoscale 10 (26) (2018) 12612−12624.

[84] Y. Wu, S. Wen, J. Wang, G. Wang, K. Sun, Graphene oxide-loaded zinc phosphate as an anticorrosive reinforcement in waterborne polyurethane resin, Int. J. Electrochem. Sci. 14 (2019) 5271−5286.

[85] K.C. Chang, W.F. Ji, C.W. Li, C.H. Chang, Y.Y. Peng, J.M. Yeh, et al., The effect of varying carboxylic-group content in reduced graphene oxides on the anticorrosive

properties of PMMA/reduced graphene oxide composites, Express Polym. Lett. 8 (12) (2014) 243−255.

[86] K. Qi, Y. Sun, H. Duan, X. Guo, A corrosion-protective coating based on a solution-processable polymer-grafted graphene oxide nanocomposite, Corros. Sci. 98 (2015) 500−506.

[87] K.C. Chang, W.F. Ji, M.C. Lai, Y.R. Hsiao, C.H. Hsu, T.L. Chuang, et al., Synergistic effects of hydrophobicity and gas barrier properties on the anticorrosion property of PMMA nanocomposite coatings embedded with graphene nanosheets, Polym. Chem. 5 (3) (2013) 1049−1056.

[88] S.A. Umoren, E.E. Ebenso, P.C. Okafor, O. Ogbobe, Water-soluble polymers as corrosion inhibitors, Pigm. Resin. Technol. 35 (6) (2006) 346−352.

[89] G.S. Hikku, K. Jeyasubramanian, A. Venugopal, R. Ghosh, Corrosion resistance behaviour of graphene/polyvinyl alcohol nanocomposite coating for aluminium-2219 alloy, J. Alloy. Compd. 716 (2017) 259−269.

Patents on graphene-based polymer composites and their applications

Rajesh Theravalappil[1] and Mostafizur Rahaman[2]
[1]Center for Refining and Petrochemicals, Research Institute, King Fahd University of Petroleum and Minerals, Dhahran, Saudi Arabia, [2]Department of Chemistry, College of Science, King Saud University, Riyadh, Saudi Arabia

21.1   Introduction

The history of graphene starts when K. S. Novoselov and researchers reported the electric field effect of a new material; few-layer graphene (FLG) which are occurring naturally in 2D form. FLG was prepared by repeated peeling (also known as mechanical exfoliation) of highly oriented pyrolytic graphite [1]. Graphene shows excellent thermal conductivity of about 3000 W/m/K and mechanical stiffness of about 1 TPa while having similar fracture strength to that carbon nanotubes and superior electronic transport properties which makes them a suitable candidate for conductivity applications, especially in the field of semiconductor and sensor applications. Polymer composites using graphene as the filler can be usually prepared via solution blending, melt mixing, or in situ polymerization [2,3].

21.2   Polymer/graphene nanocomposites

Here, we are discussing about few patents published on the composites focused on the preparation of graphene nanocomposites and their characterization testing. Graphene nanocomposites of various polymers and methods are mentioned hereafter.

An invention created by Youngkyu Jeong and Jeon-Gil Woo in 2013 (KR20130092234A) was Aramid composites reinforced with mixed carbon nanomaterials of carbon nanotube and graphene in various ratios presented along with their electrical conductivities [4]. In the solution mixture used for the preparation, ratio of solvent: mixed nanoparticles: salt was 10: 88: 2 where DMAc was the solvent and LiCl was the salt, which were stirred at 80°C for 24 h. Mixed nanoparticles were taken in various ratios of MWCNT and graphene as 0:100 to 100:0 with an increment in ratio of 10 wt.%. The dispersion of nanoparticles in the solution mixture was made sure by sonication and stirring. The resultant mixture was then

Polymer Nanocomposites Containing Graphene. DOI: https://doi.org/10.1016/B978-0-12-821639-2.00018-5

poured to a mold and vacuum dried to produce films of thickness of about 0.1 mm. The prepared nanocomposites were tested for their electrical conductivity and have been proven that their conductivities, heat resistance, and mechanical properties were superior to the conventional aramid homopolymers.

Functionalized chemically reduced graphene oxide (CRGO/Pyr-SO$_3$) has been prepared and its suspension has been used as the filler for the nanocomposites with semi aromatic polyamide (SA-PA). CRGO/Py-SO$_3$ was prepared by reacting GO suspension with sodium salt of pyrene sulfonic acid. Adipic acid and *meta*-xylylene diamine required for synthesizing SA-PA were taken in a stoichiometric ratio and was dissolved in the CRGO/Pyr-1 SO$_3$ suspension. This mixture gives rise to a precipitate on addition of cold acetone, later taken to an autoclave reactor maintained at 240°C for 2 h and finally to obtain melt polymer containing 3 wt.% of CRGO/Py-1 SO$_3$ master batch. Pure commercial MXD6 (SA-PA) was used for dilution using a lab scale twin-screw extruder, so that the final sample should have 2 wt.% CRGO/Py-1 SO$_3$. Thermal studies, electrical conductivity, and morphological analysis of these composites were carried out and reported. This invention has been patented by Alexis Bobenrieth and team in 2015 via EP2960205A1 [5].

A process for producing graphene-polymer nanocomposites has been patented in 2016 by Fernand Gauthy and team using perfluoro polymers matrix (EP2820069B1). Graphene oxide which is prepared from graphite flakes was used as the filler in this invention. An aqueous latex of Hyflon PFA M620 was added to a colloidal suspension of GO and mixed well by stirring, which was later cocoagulated using 65% HNO$_3$. After washing and drying, the nanocomposites were obtained in powder form and made into the form of film by compression molding. HCl permeation through this film was studied using a mechanism depicted in the Fig. 21.1. 33% of HCl was filled in a diffusion cell consists of nanocomposite film

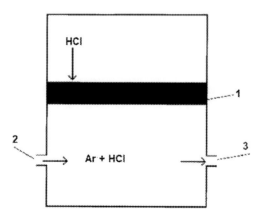

Figure 21.1 Schematic diagram of the diffusion cell used for HCl permeation through nanocomposite film (1—graphene nanocomposite membrane, 2—Argon inlet, 3—Argon outlet). *Source*: F. Gauthy, A. Sanguineti, F. D'aprile, P. D'Orazio, Process for producing graphene-polymer nanocomposites. EP2820069B1, 2016.

of diameter 33 mm and the bottom part was equipped with a facility of inlet and outlet for argon gas to flow through. During permeation, HCl transported through the film was carried along with the argon gas flowing in the bottom part and absorbed by the water outside the membrane holder. Concentration of H^+ and Cl^- ions in the water was calculated by the conductivity measurement of the water containing them which ultimately gave the total amount of HCl passed through the film. A similar film using another polymer with trade name Hyflon MFA P6010 was also prepared and tested for its HCl permeability. It was concluded that the HCl permeability of graphene-filled nanocomposites have been improved considerably along with the extended service life even in the extreme test environments [6].

Allen David Clauss and coworkers has invented a series of polymer/graphene nanocomposites in 2017 (US9790334B2) using PP, PVP, PVP/PA6, LLDPE, HDPE, and Epoxy as matrices [7]. A 5% dispersion slurry of exfoliated graphene nanoplatelets was prepared by using mixture of N-methyl pyrrolidone (NMP) (25%) and dry xylenes (75%). This 5% graphene slurry was added to a solution of PP in xylenes and a 50% (w/w) graphene/PP master batch was obtained after removing the solvents from the viscous suspension in rotary evaporator and followed by vacuum drying. 40% of this master batch was then mixed with 23.7% of PP (MFI = 2), 29.6% of PP (MFI = 12), 5% of Engage 8003 (ethylene octene copolymer) and 1.7% of maleated PP in a corotating twin-screw extruder. A reference composite with 20% of graphite instead of 20% graphene was also prepared by a similar method as mentioned above. Tensile, impact, and electrical properties of graphene/PP composites were analyzed.

In the same way, graphene/PVP master batch also was prepared by dissolving powdered PVP in a 5% slurry of graphene in a solvent system of methanol (95%) and NMP (5%). Graphene/PVP master batch was utilized for preparing graphene/PVP/PA6 composites and tested for its mechanical properties. Analogous of PP composites were prepared by using LLDPE and HDPE also and subsequent composites were subjected for evaluation of mechanical properties. Finally, graphene/epoxy resin master batch was prepared using NMP and acetone solvents and this master batch was used to prepare composites mixing epoxy hardener along with it followed by curing at 120°C for 4 h.

A Chinese patent (CN111253618A) has been awarded to Jay Clark Hanan and Suhir Bandera in 2020 for the invention of graphene nanoplatelets (GNP)/polyethylene terephthalate (PET) nanocomposites [8]. An extensive research has been carried out by these inventors on various aspects of different routes for the production of nanocomposites, factors affecting the final properties and performance of them and their microstructural and morphological evaluation etc.

Two sets of PET-GNP master batches were prepared by twin screw process; one with and another one without ultrasound assistance; with 2, 5, 10 and 15 wt.% GNP. An ultrasound source sonotrode (horn) operating at 40 kHz was placed on the corotating twin-screw extruder in such a way that the tip will touch the polymer melt and the screw speed was adjusted to 200 rpm to facilitate a residence time of 9.2 s in the sonication zone. Various ultrasonic amplitude levels such as 0 (0USM), 3.5 μm (3.5USM), 5 μm (5USM) and 7.5 μm (7.5USM) were employed in this study of PET with and without GNP. Samples for testing were prepared by high-speed injection molding process. Another set of samples were prepared by micro injection molding where the extrudate from twin screw extruder was transferred to the injection molding machine using a special transfer device. Incorporation of GNP has helped in achieving a very high young's modulus of the nanocomposites and high-speed injection molded samples had relatively higher young's modulus values. Sonication helped in better dispersion of GNP particles in the PET matrix and thus reflected in increased toughness. For twin-screw extruded nanocomposites, distance between GNP particles is found to be decreased with increasing concentrations. Addition of GNP has affected the crystallization behavior of PET considerably due to the additional nucleation of graphene which was shown by an increase in crystallization temperature. From the molecular weight studies, it was understood that the reduction happened mainly due to the extrusion process and sonication had a negligible influence.

21.3   Electrically/thermally conducting graphene nanocomposites

Graphene nanocomposites are very well known for their conductive (either electrical or thermal). The majority of them are finding applications in the field of antistatic and electromagnetic shielding materials, electronic equipment and aircraft equipment etc. Very high conductivity of graphene helps to achieve better percolation threshold even at very low loadings and imparts appreciable mechanical properties and thermal stability also. Some of the inventions in this field are mentioned here.

Nanoscale graphene platelets (NGP) nanocomposite supercapacitors were invented by Lulu Song and others in 2009 (US7623340B1) which has exhibited exceptionally high capacitance value. Basically, two forms of mesoporous NGP nanocomposites were produced as depicted in Fig. 21.2; one containing NGPs coated with a thin layer of conducting polymer or functional groups and another with NGPs bonded by a conductive binder coating or a matrix material which is probably a conducting polymer along with four routes of preparation also [9].

A conducting polymer when combined with a NGP type substrate material, can impart pseudo-capacitance to the electrode. Here, an electrochemically p-doped variant of Poly (3-methyl thiophene) also known as pMeT was synthesized in presence of tetra-alkyl-ammonium salts using NGP mat as the electrode. These low-cost composites showed an improved specific capacitance of 93 F/g while that of untreated NGP mat of thickness 5.2 nm was only 38 F/g. Specific capacitance of CNT-based nanocomposites are even higher but at the increased cost of CNT material.

Polypyrrole (PPy) was electrochemically synthesized using NGP mat as electrodes which were initially prepared by slurry molding process. Specific capacitance values of untreated NGP mats of $82-1.9$ nm thickness were in the range of

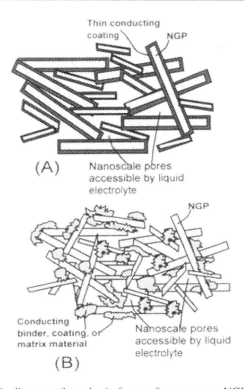

Figure 21.2 Schematic diagram of two basic forms of mesoporous NGP nanocomposites. (A) comprising NGPs coated with a thin layer of conducting polymer and (B) comprising NGPs bonded by a conductive binder, coating, or matrix material.
Source: L. Song, A. Zhamu, J. Guo, B.Z. Jang, Nano-scaled graphene plate nanocomposites for supercapacitor electrodes. US7623340B1, 2009.

5.2−82 F/g which was increased to 41−143 F/g for PPy coated NGP mat, clearly showing the pseudo-capacitance effect.

A dispersion of carbon black (CB) and NGPs were prepared with three different ratios and mixed with a Teflon particle suspension and dried to a dough consistency and coated on a nickel net to form an electrode which was sintered at 320°C under pressure. Specific capacitance of these three samples prepared were in the range of 70−80 F/g. A composition with 5:5 ratio of CB:NGP showed a dramatic rise in electrical conductivity of 100 S/cm compared to a value of 0.1 S/cm corresponding to that of pure CB. Due to high aspect ratio of NGPs, they can easily overlap each other and form continuous conducting paths.

NGP-carbon matrix composite film and NGP-based bucky papers were prepared dispersions of NGP and poly (acrylonitrile-methyl acrylate) copolymer in DMF. Film casting was carried out at 85°C−100°C to obtain films of 5−6 μm thickness. This bucky paper was then treated at 700°C in argon atmosphere for 30 min to obtain NGP-carbon matrix nanocomposites (PAN matrix carbonized). Another set

(PAN matrix carbonized and activated) of film samples were activated in CO_2 at 700°C for 20 min in a tube furnace. A third variant (PAN matrix carbonized and PPy coated) from the first film sample was prepared by electrochemical deposition of PPy on it. Studies on specific capacitance values of these three samples (for various thickness levels) implies that the surface activation and coating by conductive polymer has influenced very much in improving them. Also, it has been noted that specific surface area of modified nanocomposites has been significantly increased.

Composite with high conductivity intercalation structure was prepared using graphene oxide and pyrrole monomer was polymerized simultaneously. Graphene oxide was first dispersed in acidic aqueous solution along with tensio-active agent such as cetyl trimethylammonium bromide under ultrasonication. To this dispersion, pyrrole monomer was added along with ammonium persulfate initiator at 0°C−5°C to carry out emulsion polymerization. Composites with high conductivity (5.44 S/cm) intercalation structure was formed by this method which was patented by Li Chunzhong and team in 2011 via CN101544823A [10].

In 2013, Hu Haiqing has reported electrically conductive graphene nanocomposites using silicone rubber and cis-1,4-polybutadiene rubber as matrices (CN102321379A). Graphite was first oxidized to graphite oxide via Hummer's method using sulfuric acid, nitric acid, and potassium permanganate. This graphite oxide then thermally reduced to graphene at 1000°C and used as filler for the composites. Polylactic acid (PLA)/graphene composites were prepared by solution method by dissolving the matrix in tetrahydrofuran (THF) and mixing graphene/THF dispersion with it. This composite as tested for its electrical conductivity and mechanical properties. PLA/graphene composites showed a volume specific resistance of 4×10^6 Ω.cm and a tensile strength of 60 MPa while their thermal degradation temperature was improved to 340.8°C from 322.7°C which is corresponding to that of pure PLA. The second method of mechanical processing was adopted for the preparation of cis-1,4-polybutadiene/graphene composites by mixing in a banburry mixer, followed by vulcanization. They have exhibited a volume specific resistance of 1.8×10^4 Ω.cm and a tensile strength of 8.22 MPa. Graphene/silicone rubber composites were also produced with a volume specific resistance of 7.0×10^3 Ω.cm [11].

Highly thermal conductive polymer nanocomposites were invented by Ya-Ping Sun and coworkers in 2014 (US20140299811A1). A suspension of carbon nanosheets in DMF was mixed dropwise to epoxy polymer solution in DMF and stirred vigorously from which thin films about 30−80 microns were prepared. Nanocomposites with 33 vol.% of nanosheets exhibited a thermal conductivity of about 80 W/m/K, which is very high compared to composites with similar filler levels reported earlier (less than 80 W/m/K). Even at these high loadings, the nanocomposites showed excellent flexibility. Similarly, composites of polyimide (in DMF) and polyvinyl alcohol (PVA) (in hot water) were also prepared and showed considerably high in-plane thermal diffusivity like its epoxy polymer version. Covalent functionalization of PVA with carbon nanosheets was carried out through carbodiimide-activated esterification reaction between the carboxylic acid moieties present on the nanosheets and hydroxyl part present on PVA where DMSO was used as the solvent. Functionalization has affected the glass transition temperature of the composite which has shown an increase of 20°C. Functionalization was also confirmed using Raman spectra [12].

Graphene-elastomer nanocomposites with high dielectric constant and low dielectric loss were prepared by Tian Ming and his research team and patented in 2015 (CN103183847A). Graphene oxide prepared by Hummer's method was mixed with NIPOL1571H acrylonitrile butadiene rubber latex (with 37 wt.% acrylonitrile content) together with vulcanizing agents. This emulsion was then dried by rotary evaporator at 45°C to remove moisture. Sample preparation was carried out by compression molding with simultaneous vulcanization at 170°C. Dielectric constant and dielectric loss were measured at a test voltage of 1 kV with the frequency range of $10^2 - 10^7$ Hz [13].

In 2016, Tao Chengan and coworkers invented a PDMS/graphene composite which could find application as microfluid chip (CN103436017A). Graphene oxide was modified by an aliphatic amine followed by a hydrophilic group which was then used for preparing composite by mixing with PDMS together with a crosslinking agent. These composites show considerable increase in mechanical properties and thermal (0.15 W/m/K for pure PDMS and 1.3 W/m/K for the composite) and electrical conductivities 8×10^{-13} S/cm for pure PDMS and 1.2×10^{-9} S/cm for the composite). This composite was tested for its application as a microfluid chip. When heated by light sources, fluid temperature of the microfluid chip is increased where the wavelength of heated light source was 650 nm and power was 180 mW. In 1 min, the fluid temperature of the graphene-PDMS composite can achieve a 5°C raise while pure PDMS show change in temperature. As the level of graphene increases in the composite the final temperature also increases. Also, it has been noted that, as the wavelength of the light source applied increases, the temperature also increases [14].

High heat conduction nylon 6/graphene composites were invented by Ding Peng and colleagues in 2016 (CN103450674B). Graphene/Nylon 6 master batch was prepared by in situ polymerization of ε-caprolactam in presence of 6-aminocaproic acid as initiator (in the ratio of 20 weight parts graphene/75 weight parts ε-caprolactam/5 weight parts 6-aminocaproic acid) at 85°C with ultrasonication for 3 h, which was later vacuum dried. Fillers used were a combination of graphene oxide with high radius-thickness ratio and a small amount of nano graphene oxide. This mixture was then mixed with pure nylon 6 in a banburry mixer to produce nanocomposites of various filler levels. Samples for thermal conductivity measurements were prepared by compression molding. Thermal conductivity values of nanocomposites with 0.5 wt.% was 0.279 W/m//K and 10 wt.% composite showed a value of 0.416 W/m//K [15].

21.4   Graphene nanocomposites for barrier applications

Polymers and their nanocomposites especially are well known for their barrier applications. Some limitations in permeability of polymers can be reduced or eliminated by adding nanofillers into it. For example, nano clay is a successful filler in achieving the improved permeability to polymer films. Similarly, exfoliated graphite and graphene are also being used for similar applications.

A composite prepared using polystyrene (PS) and functionalized graphene oxide which was found to be a better candidate for barrier application was invented by Compton and Nguyen in 2011 (US20110223405A1). Initially, graphite powder was transformed to graphite oxide by Hummer's method which was then functionalized by treating with phenyl isocyanate. A dispersion of functionalized graphene in dimethyl formamide (DMF) and a PS solution in DMF prepared at 90°C were mixed. To this, 1,1-dimethylhydrazine as a reducing agent was added and left at 90°C for 18 h. Room-temperature methanol was added to this hot nanocomposite solution to precipitate out the PS/graphene nanocomposite. This precipitate was then filtered and dried at 90°C for 18 h. Polymer-graphene nanocomposite (PGN) films were then compression molded at 130°C. Oxygen permeability of these PGN films were tested in an OX-TRAN 2/21 MH (MOCON) instrument. The improved oxygen barrier property was attributed to the unique crumpled morphology of functionalized graphene nanosheets (see Fig. 21.3). A larger surface area of these nanosheets and better interaction with the polymer chains made it completely "wetted" and resulted in a densified polymer matrix. Ultimately, this has led to the enhancement in barrier property of these PGNs sheets [16].

A series of graphene oxide/polymer composites were prepared using various polymers such as PVA, PS, polymethyl methacrylate (PMMA) and a Chinese patent was issued in 2011 via CN102115566A [17].

Initially, graphene oxide was prepared by treating expandable graphite with potassium permanganate ($KMnO_4$) and conc. Sulfuric acid. PVA composites were prepared by dissolving it in water and adding water suspension of graphene oxide to that followed by ultrasonication. Films were prepared by casting method. A significant drop of about 30 times was observed in the CO_2 permeability of PVA/graphene oxide composite film compared to pure PVA film. This PVA film composite has achieved a CO_2 permeability of 0.022×10^{-15} cm^3.cm/(cm^2.s. Pa) compared to 2.483×10^{-15} cm^3.cm/(cm^2.s.Pa) of the pure PVA film; with more than 100 times

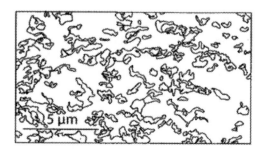

Figure 21.3 Crumpled morphology of functionalized graphene nanosheets.
Source: O.C. Compton, S.T. Nguyen, Composite polymer film with graphene nanosheets as highly effective barrier property enhancers. US20110223405A1, 2011.

improvement. Another composite film was prepared using graphene oxide/PS system in dimethyl formamide as solvent. There was an improvement in CO_2 permeability of about 70 times in CO_2 permeability from 521×10^{-15} cm^3.cm/(cm^2.s. Pa) to 7.5×10^{-15} cm^3.cm/(cm^2.s.Pa) of pure PS cast film. Third composite system was prepared from graphene oxide/PMMA system (with THF as solvent) which has achieved a CO_2 permeability of 0.022×10^{-15} cm^3.cm/(cm^2.s. Pa) compared to 0.689×10^{-15} cm^3.cm/(cm^2.s.Pa) of the pure PMMA film; with more than 30 times improvement. A pure polyimide cast film had a CO_2 permeability of 2.483×10^{-15} cm^3.cm/(cm^2.s.Pa), while the fourth composite system prepared using graphene oxide/polyimide (in DMF) had a CO_2 permeability of 0.018×10^{-15} cm^3.cm/(cm^2.s.Pa) with an increment of about 150 times when compared to that of pure poyimide film. A fifth composite system was prepared using graphene oxide and polyacrylamide using water as solvent system. This polyacrylamide composite system showed an increment of about 110% from 1.436×10^{-15} cm^3.cm/(cm^2.s. Pa) to a CO_2 permeability of 0.013×10^{-15} cm^3.cm/(cm^2.s.Pa). All polymer composite cast films prepared from graphene oxide as the filler has shown an excellent CO_2 barrier property.

One of such inventions has been done by Wang and coinventors in 2012 (CN101812194B) using polypropylene (PP) as the matrix and functionalized graphene as the barrier-improving filler [18]. For this, Vestolen 7052 PP was selected. Functionalization of graphene was carried out by a two-step process as follows. Graphene was first dispersed in deionized water along with vinyl triethoxy silane for about 1 h in ultrasonic bath at 90°C followed by reacting with hydrazine hydrate for 2 h, which was then filtered and dried at 60°C for 12 h in baking oven. Functionalized graphene was the characterized by TEM and Raman spectra. Vestolen PP was dissolved by stirring in carbon tetrachloride (CCl_4) at 90°C for 1 h. Functionalized graphene was then added to this solution and continued stirring for 2 h. The composite was then dried in an oven at 80°C for 24 h. Composites were finally prepared by compression molding and tested for barrier properties. A schematic diagram of obstruct mechanism of graphene-based nanocomposite id shown in Fig. 21.4.

A Korean patent (KR101209293B1) was granted in 2012 for the invention of transparent graphene oxide/polymer films made by Lee Heon-sang. Polymers used here was polyvinyl alcohol, ethylene vinyl alcohol and a blend of the both [19]. Graphene oxide was prepared from graphite flakes by modified Hummer's method and used as the barrier property enhancing filler. Dispersion of the graphene oxide and polymer were casted on a PET surface and films of thickness 20 μm was prepared. These films were used for testing light transmittance and barrier properties. They have shown a light transmittance more than 70%. Oxygen transmittance tested at 25°C and 30% RH was 0.1 cc/m^2.day while moisture permeability was 2.0 cc/m^2. day. These resultant films can find applications in transparent films, food containers, and chemical containers. Same inventor has secured another Korean patent (KR101543260B1) in 2015 for the invention of water vapor barrier transparent polymer complex films based on polyvinylidene fluoride (PVDF). A composition of PVDF and graphene oxide was first prepared in dimethyl fluoride. This was then

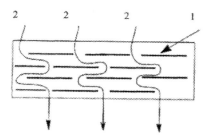

Figure 21.4 Representative diagram of obstruct mechanism of graphene-based nanocomposite material (1—graphene platelets, 2—permeating gas).
Source: W. Xianbao, W. Jingchao, X. Chunhui, W. Li, Graphene-based barrier composite material and preparation method thereof. CN101812194B, 2012.

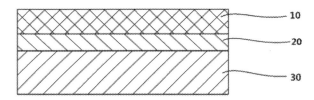

Figure 21.5 Transparent PVDF/graphene oxide composite film (10—composite film layer; 20—inorganic material layer; 30—transparent polymer film (PET) layer).
Source: L. Heon-sang, Water vapour barrier transparent polymer complex films. KR101543260B1, 2015.

layered on a PET film coated with an inorganic layer of silicon oxynitride (SiON) as shown in Fig. 21.5. The product has exhibited a transmittance of 70% and improved moisture resistance when compared to the films without graphene oxide filler [20].

One invention on elastomer-based graphene nanocomposite with improved barrier properties done by Robert Prud'Homme and coinventors which was published via US8110026B2 in 2012 [21]. Japanese patents JP5450069B2 and JP5421110B2 also has been awarded for the same invention in 2014 [22,23]. Barrier, mechanical, thermal, and electrical properties of graphene-filled polymer nanocomposites were investigated through this invention. Through oxidation followed by rapid exfoliation of graphite, graphite oxide (GO) sheet with increased surface area is obtained. This was done by Staudenmaier oxidation-intercalation method, where a combination of oxidant (ratio of sulfuric acid: nitric acid was 5:1 to 1:1) and intercalant (with graphite: potassium chlorate was 1:8 to 1:20 wt./wt.). This process proceeded for 96 h and the reaction mixture was washed with deionized water followed by 5% HCl until absence of sulfate ions was confirmed. pH of the filtrate was controlled around 5. Improved surface area along with a better dispersion even at lower graphene loadings improved all the properties mentioned above. Intercalation of graphene was confirmed by XRD analysis. Exfoliated graphite oxide (FGS) was then prepared from GO using a Lindberg furnace. GO taken in alumina boat was placed in a one-end sealed quartz tube closed the other end using a rubber stopper and flushed with Argon gas. This tube was then inserted into a preheated furnace maintained at 1050°C for a short time span of 30 s. This FGS was then analyzed for C/O ratio.

Graphene nanocomposites were prepared using three different polymers: natural rubber (NR), polydimethyl siloxane (PDMS), and Vector V4111 (a triblock copolymer of NR-isoprene-NR) as matrices. Three various techniques also used for the preparation of composites such as, solution mixing, extrusion and shear mixing. FGS Composites of NR and Vector V4111 were done by solution mixing using THF and toluene as solvents respectively. Vector V4111 and NR composites were prepared by extrusion technique at 170°C/20 min/100 rpm and 90°C/20 min/100 rpm conditions correspondingly. Solution shear mixing technique was adopted for the PDMS/FGS system along with a curing agent once the proper dispersion of the filler in the matrix was ensured. Sheets/films of these composites were then prepared by melt press, mold casting, and rotational molding. Mechanical, electrical conductivity, and permeability tests along with morphological analysis were conducted using these samples. FGS/PDMS composites were found to be performing better in terms of electrical conductivity and improvement in permeability. For NR composites, solution method is found to be showing a better dispersion compared to the extrusion method. A low permeability was shown by the fillers with high aspect ratio in the case of FGS/PDMS composites.

Another Korean patent was secured by Park Min-shim and coworkers in 2013 via KR101308967B1 for the invention of polymer/NGPs nanocomposites with enhanced barrier properties. NGPs solutions were prepared by treating a solution of expanded graphite with N,N-dimethyl acetamide (DMAc) ultrasonically in room temperature for 60 min. This was later mixed with a solution of poly (vinyl alcohol-co-ethylene) in DMAc and films were prepared by casting method. Similarly, another film was prepared by mixing NGP and poly vinyl alcohol in sodium dodecylbenzenesulfonate (SDBS) and casting it on glass substrate. Oxygen permeability of PVA composites were found to be $1.0 \text{ cm}^3/\text{m}^2.\text{day}$ for 0.1% NGP and $0.8 \text{ cm}^3/\text{m}^2.\text{day}$ for 0.5% NGP while that of unfilled PVA film was in the range of $1.8-2.5 \text{ cm}^3/\text{m}^2.\text{day}$, clearly showing the improvement in oxygen permeability properties [24].

Graphene oxide/polymer compositions were discovered and patented for the application of inner liner of tire and inner tube applications through WO2013107229A1 in 2013 by Liu Li and coinventors. This composition has the two following phases; (1) a graphene oxide/rubber composition having three components, namely (a) graphene oxide, (b) a reactive rubber and (c) a solid rubber and the second phase being (2) a dispersion phase of an epoxy natural rubber or a thermoplastic resin. The role of epoxy natural rubber or thermoplastic resin dispersed in the first phase was to form the sea-island structure while graphene oxide in the first phase provided the significant reinforcement and low gas permeability. The reactive rubber in the composition has a reactive group like amine or carboxyl group. The composition was prepared from graphene oxide (1 wt.%), a reactive rubber (butadiene latex) and styrene butadiene rubber as solid rubber. The compositions were showing excellent gas barrier properties [25].

A patent (WO2015133849A1) issued in 2015 which was invented by Park Hobeom and Kim Ho-wyonis dealing with the barrier properties of Graphene oxide nanocomposites with polyethylene glycol diacrylate (PEGDA) matrix. To the dispersion of PEGDA, 1 wt.% of graphene oxide and 0.1% of hydroxycyclohexyl phenyl ketone as photo-initiator was added and sonicated for 2 h followed by stirring for 24 h. The resulting solution was casted on a glass substrate was irradiated by UV radiation of 312 nm for 5 min under nitrogen atmosphere. The oxygen permeability of membranes containing graphene oxide of size of 270 nm and 800 nm was found to be improved significantly than that of pristine PEGDA film. A reduction in oxygen permeability by 90% has been observed in the case of composites with size of 800 nm graphene oxide when compared to that of pristine PEGDA film [26].

Recently, in 2019, a patent has been awarded to Jennifer K. Lynch-Branzoi and team for the invention of graphene reinforced polymer matrix composites (G-PMCs) with improved barrier resistance towards liquid and gas permeants (WO2019143662A1). Highly filled G-PMC master batches such as polyether ether ketone with 50 wt.% graphene (50G-PEEK), PS with 35 wt.% graphene (35G-PS) and 35G-HDPE were prepared and later diluted with same grade polymer to obtain G-PMCs with lower levels of graphene, via melt blending. Film samples of HDPE composites were prepared by extruders fitted with lip die and tested for permeability using small gases such as oxygen, carbon dioxide and water vapor, according to ASTM standards while fuel permeability was tested according to SAE international J2665 cup weight loss procedure at 60°C. Influence of number of processing cycle on mechanical properties such as tensile strength, flexural strength and impact strength were studied and found to have maximum exfoliation was achieved by maximum number of processing cycles (here, 10) and thus reflected in maximum of the mentioned mechanical properties. Effect of mixing time (30, 60, 90 and 120 min) of composites was also reported for 0.5 wt.% HDPE G-PMCs and G-PMC with 0.5 wt.% HDPE was found to have higher barrier property due to increased exfoliation of filler particles. Another factor which was clearly affected the barrier property was the filler level; G-PMCs with 35 wt.% of graphene showed the lowest permeability towards gases tested. A graphical representation of permeability of various gases through the membrane is exhibited in Fig. 21.6 [27].

21.5   Graphene-polymer nanocomposites as strain sensors

Introduction of a filler like graphene helps to serve some polymer composites as sensors by virtue of its high conductivity even at very low loading levels (in other words, low percolation threshold value), especially with polymers with elastomeric nature.

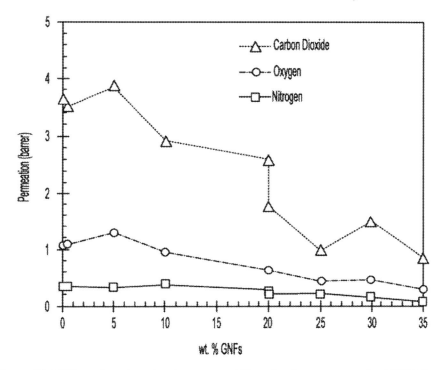

Figure 21.6 Effect of graphene levels on permeability of various test gases for HDPE G-PMCs.
Source: J.K. Lynch-Branzoi, T.J. Nosker, B.H. Kear, C.T. Chang, Use of graphene-polymer composites to improve barrier resistance of polymers to liquid and gas permeants. WO2019143662A1, 2019.

An invention of strain measuring and motion sensing device was reported by Zhou and team in 2011 via CN102506693A. For this, firstly multiple layers of graphene were deposited on copper substrate by chemical vapor deposition or plasma reinforced chemical vapor deposition methods. This graphene layer was then later transferred to polydimethyl siloxane (PDMS) polymer. Ends of this sandwich structure was then coated with conductive silver glue, which was then cured. This electrode was then connected to an ohmmeter which can subjected to bending motions. The surface then receives tensile and compressive strains and can be detected by square wave type resistance variations curve from ohmmeter. Peak-to-valley height difference of this wave form is directly proportional to the strain size. Similar procedure was adopted to prepare electrode using graphene layers and PMMA also. This device was also used to carry out quantitative strain measurements [28].

In 2012, Georg Duesberg and Martin Hegner obtained a patent for the invention (WO2012028748A1) of a sensor which is highly selective for the detection of variety of analytes such as gas, liquids and biomolecules [29]. This sensor is also capable of analyzing cell growth in in vitro conditions and can find applications in in life and food science applications, health care, biopharma, and food production controls.

The sensor works as follows. First layer is the functionalized graphene layer which is providing an interface. The second layer consists of at least one biocompatible component which anchors to the first layer via functionalization. There is a mediation layer between first and second layers which is functionalized by site-specific immobilization of receptors. The mediation layer is generally a biopolymer. This is can be a dielectric polymer, a polymer which can store electrolytes, a polymer suitable for polymer imprinting, a polymer capable of storing a prokaryote and/ or eukaryote cell growth medium or an ion exchange membrane. In operation, the analyte or ligand present in the sample flows through the nanofluidic channel in the direction opposite to the flow of charge which will directly interact with the receptor. This interaction between receptor and ligand reflects as a shift in electrical conductivity of first graphene layer. Functionalization of graphene layer can be carried out by various methods. This functionalization enables the biocompatible buffer layers interface with graphene layer with biological molecules of interest such as receptors, ligands, nucleic acids, proteins or whole cells such as microorganisms and eukaryotic cells. Structure and mechanism of action of the sensor are portrayed in Figs. 21.7 and 21.8. An European patent also was awarded for the same invention in 2012 via EP2426487A1 [30].

Some soft and elastomeric polymers with highly conducting fillers like graphene forms excellent candidates for sensors. It can be strain or pressure sensors. Wang and coinventors reported a pressure sensitive probe with a side electrode through CN102141451B in 2012 [31]. This probe has a sensitive area, a conductive area and an interface area. In the case of sensitive and conductive area, have a three-layer flexible structure. The outermost layers are made by insulated packaging films while the middle consists of a conductive polymer composite material, which was prepared by dispersing conductive powder particles into a soft polymer matrix by solution mixing method. Vulcanized silicone rubber in hexane was used as the matrix while graphene was the conducting filler. Dibutyltin laurate and ethyl

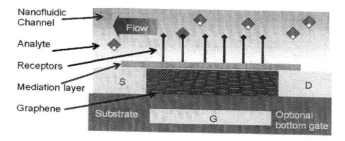

Figure 21.7 Structure of the sensor (S-sink, D-Drain, G-Gate).
Source: G. Duesberg, M. Hegner, Nano-carbon sensor and method of making a sensor. WO2012028748A1, 2012.

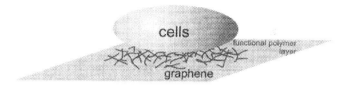

Figure 21.8 Functional polymer layer, overlying the graphene filled with electrolytes. *Source*: G. Duesberg, M. Hegner, Nano-carbon sensor and method of making a sensor. WO2012028748A1, 2012.

orthosilicate were used to vulcanize silicone rubber. This sandwich structure work as a soft pressure sensing probe.

A flexible large deformation flexible strain transducer was prepared using graphene oxide surfactant solution (0.5%−5%) and polymers such as PDMS, PET, or polyimides. The distribution of the electrodes are on the edge or surface of the sensitive porous graphene membrane material. Flexible support layers were distributed on the lower and upper surface and holes of the porous membrane. Porous film material was of 400−1000 μm and pore diameter was in the range of 30−500 μm and was prepared by the scraper plate method. Thickness of the support layers were in the range of 500−1000 μm. The electrode consists of an electrical conductor and a lead. One end of the lead was connected to a conducting body while the other end stretches out flexible support layers to carry out the resistance measurement. The electric conductor used here was silver glue.

As per the invention done by Lee Sang-yeop and colleagues in 2014 (KR101440542B1), biosensors using conductive graphene on polymer substrate (here, PDMS) selectively binds to a target molecule by utilizing the wide surface area and excellent electrical conductivity of graphene. By this, level of detection sensitivity and number of immobilized biomolecules such as DNA can be increased. To detect the biomaterial electrically, it is necessary to have a liquid phase and a space capable of containing a fluid thickness ranging from several millimeters to micrometers between graphene layer and PDMS substrate. Here, any minute potential difference generated in the graphene when the bioreceptor is reacted with the target biomaterial is monitored and detected [32].

Another invention in the field of sensors was made by Simon Park and Kaushik Parmar through EP3058312B1 in 2016. The sensor system comprised of a polymer with a conductive/semiconductive nanoparticles for detecting and monitoring structures for hydrocarbons. The sensor was capable of monitoring leakage, structural change and temperature change in a hydrocarbon transportation/storage structure. A voltage divider connected to sensing elements converted the change in resistance into voltage signal. A multiplexer microswitch which is connected to a voltage divider circuit for detecting and recording the voltage signals and an adaptive neuro-fuzzy inference system (ANFIS) which process the voltage signals and provide information about the sensor network's status [33].

Six compositions were made with polysiloxane as the matrix. Out of these, two (composition 2 and 6) were prepared by using exfoliated graphene nanoplates

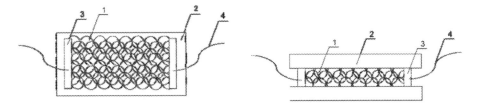

Figure 21.9 Detection of hydrocarbon vapors using sensors made from composition 2 and 6.
Source: S. Park, K. Parmar, Sensing element compositions and sensor system for detecting and monitoring structures for hydrocarbons. EP3058312B1, 2019.

(xGNP) as conductive filler while in rest of the compositions, fillers like MWCNT, ZnO, and TiO2 were used in different levels.

As we can see from Fig. 21.9 Region 1 represents the percentage change in resistance before exposure to crude oil vapors and at Event 1 where the exposure of the sensing element to crude oil vapors takes place, a marakable change in resistance can be observed. Now the sensor is in Region 2, during which the sensor is in contact with the vapor. At Event 2, when the crude oil vapors removed from the sensing element, it enters into region 3 and percentage change in resistance drops. Most of the compositions showed a significantly quick response toward the crude oil vapors and direct contact is also not necessary for detection. This points towards the effectiveness of these compositions as sensors towards hydrocarbons even in very small quantities. It has been also proved that the xGNP composite sensors have higher sensitivity towards crude oil vapors and shows quick response.

A graphene-filled silicone rubber composite piezo resistance sensor was invented by Wang Luheng in 2017 (CN106495085A). For this, graphene, silicone rubber and n-hexane as the solvent were mixed in a ratio of 6:100:900 by volume with the help of ultrasonic mixing instrument, in presence of a catalyst and a crosslinking agent. At 80°C, the solvent was evaporated and the graphene/silicone rubber composite was prepared. This composite layer was sandwiched between two hot-sealed Kapton polyimide layers and the copper electrodes were used to function as the conductor. This piezo resistance sensor was of high sensitivity, high range and appreciable flexibility. This invention can find application in the development of electronic skin and also for curved surface interlayer pressure measurement [34].

Kinloch, Young, and Novoselov claimed the invention of graphene polymer composites and by utilizing the stress sensitivity of the graphene G' band in the Raman spectroscopy, they have studied the stress transfer of their composites in 2018 (US 20180354785A1). Pristine graphene has a stronger Raman signal compared to the functionalized one used in combination with an adhesive when used as a strain sensor. The substrate used was either graphene or functionalized graphene. An optional protective layer also given and in some other cases, graphene dispersed in a liquid also applied to form a film. The role of graphene as reinforcing phase has been demonstrated by stress transfer take place from the PVA polymer matrix to the graphene monolayer. Also, they have monitored the stress transfer efficiency and breakdown of the graphene-polymer interphase. Such graphene/functionalized graphene dispersed in a liquid carrier can find application as wide area strain sensor when they applied in the form of paint on structure such as buildings, ships or aircrafts [35].

Porous graphene layer undergoes cracking by small strain and cause rapid variation in resistance and thus high sensitivity is observed. Flexible support layers provide a covering to the porous graphene layer. Porous layer delivers conductive path and ensure the sensitivity even at high strains. This invention was patented by Zhu and Zhang in 2019 by a Chinese patent with number CN106482628B [36]. A schematic depiction of the sensor can be represented as seen in Fig. 21.10.

A United States patent claimed in 2019 by Jonathan Coleman and coworkers (US 10251604B2) is about graphene-infused rubber bands (G-bands) and its use as strain and motion sensors. This infusion was carried out in various steps. To promote infusion of graphene platelets, it was necessary to open the pores of the rubber band. For this, the normal natural rubber elastic band was soaked in a suitable solvent like toluene along with ultrasonication for about 3.5 h. Based on solubility parameter values, graphene dispersion in a 20:80 mixture of NMP and water was prepared and the swollen rubber band was soaked in it for 1−48 h. Later, these bands were washed with water in a sonicating bath. On drying, it was noticed that the rubber band shrunk than the swollen but not to the initial length. Color of the band also was changed dur to infusion of graphene platelets. After this, the graphene uptake mass and the unstrained electrical resistance were measured for the samples and found that the soaking time has much influential in graphene adsorption to the pores. Strain sensing was studied using these G-bands at various strain levels. Function of G-band as a motion sensor was tested by wearing it around finger, forearm and around throat to sense movements like pulse and breathing [37].

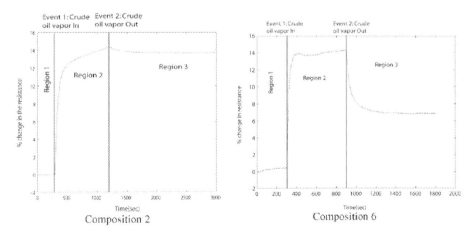

Figure 21.10 Top and side views of flexible strain transducer (1- sensitive material; 2- flexible support layers; 3- electric conductor; 4- lead).
Source: W. Luheng, M. Fangfang, W. Xueting, L. Huanghai, Method for developing flexible pressure sensitive probe with side electrode. CN102141451B, 2012.

21.6   Membranes for water treatment/desalination

Nowadays, polymeric composite membranes are being widely used for the water treatment or desalination for generating drinkable water from polluted or high brine water. Fillers such as graphene adds value to the performance of such polymeric membranes used for this application. Sewage treatment is another important area where polymeric composite membranes are highly being applied for effective utilization of water.

Various types of membranes were prepared and patented by Kalaga Murali Krishna and team in 2014 via WO2014168629A1. Basic format of the membrane was the same with PSf substrate and the following up layers were composed of combination of various polymers/filler types. There are seven models of membranes were presented through this patent. The first membrane consists of a supporting membrane layer and a barrier membrane layer along with embedded graphene particles. Second type was similar to the first type with only difference is that both layers having embedded graphene compound. In another variety of membrane, over the supporting membrane layer, it had a barrier membrane layer and a top polymer layer with graphene compound embedded in it. Fourth category of membrane was comprised of a supporting membrane layer and barrier layer basically having one or more graphene compounds. Another version had primarily two layers. A supporting layer with embedded graphene compound on its surface and a barrier layer on top of that which is made up of one or more graphene compounds. Sixth form of membrane was consisting of a single layer of membrane where embedded graphene was the integral part of it. Seventh and the last variety was the integral membrane with graphene compound embedded on its surface. Schematic representations of the membranes mentioned above are shown in Fig. 21.11 [38].

In 2016, Saira Bano and team has invented a nanocomposite ultrafiltration membrane using polyacrylonitrile (PAN) as the matrix and graphene oxide as filler (US20160303518A1). Graphene oxide dispersions were prepared in DMF (in four concentrations; 0.5, 1.0, 1.5 and 2.0 wt.%) and to this PAN also was dissolved into this. This solution was casted on a glass substrate and dried to obtain nanocomposite membranes. Cross-sectional SEM images show a typical asymmetric porous structure with dense upper layer and a finger-like porous lower layer. For the PAN/GO membranes, finger-like pores have comparatively larger width than pure PAN membrane, which takes place due to the presence of highly hydrophilic GO which and takes place due to rapid exchange between the solvent and nonsolvent during phase transition. This points towards the role of GO in increasing porosity of PAN/GO membranes and its efficiency as effective membrane for water treatment. Introduction of GO to PAN has resulted in many positive ways such that reduction in consolidation coefficient which is related to the mechanical stability of the membrane, reduction in contact angle, increase in porosity and pore size. Functional groups of filler and matrix are connected via hydrogen bonding and is depicted in Fig. 21.12 along with the corresponding FTIR spectra. These membranes exhibited high permeability, high salt rejection ratio, exceptional antifouling property combined with durability and improved mechanical properties [39].

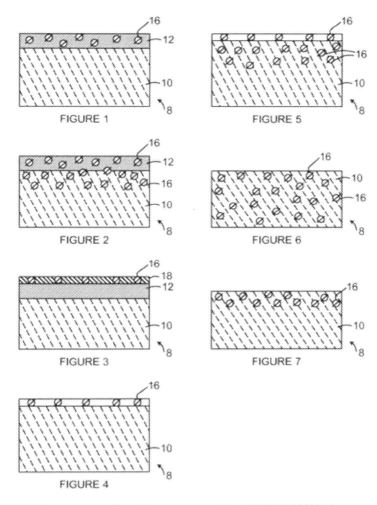

Figure 21.11 Various types of membranes prepared via WO2014168629A1.
Source: K.M. Krishna, A. Bhattacharyya, R.M. Devi, M. Phadke, Membranes comprising graphene. WO2014168629A1, 2014.

Membrane bioreactor (MBR) is a novel kind of wastewater treatment technology where advantages of membrane separation technique combine with biological treating technique. In 2018, Luo Li and researchers have invented a PVDF/rGO composite membrane (CN109081430A). Dispersion of graphene in DMAc was prepared and mixed with a solution of PVDF in DMAc from which a film of 300 μm thickness was prepared by film casting. Membranes prepared were of pure PVDF, composites of PVDF with 0.5 wt.% rGO and 1.0 wt.% rGO and are tested for their antifouling properties where the effective filtration area was 0.01 m². Extracellular polymeric substance (EPS) is considered to be the major element for causing

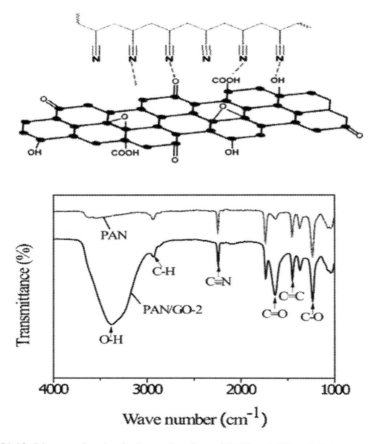

Figure 21.12 Diagram showing hydrogen bonding of PAN and GO and their corresponding FTIR spectra.
Source: S. Bano, A. Mahmood, S.-J. Kim, K.H. Lee, Nanocomposite ultrafiltration membrane containing graphene oxide or reduced graphene oxide and preparation method thereof. US20160303518A1, 2016.

fouling which are classified into two; tightly bound (TB-EPS) and loosely bound (LB-EPS). PVDF/rGO membranes are found to be considerably effective than pure PVDF membranes in their activity [40].

High performance membrane for water reclamation was invented by Thieo Hogen-Esch and coinventor in 2019 (US10456754B2). Polyamide membranes are prepared over polyethersulphone (PES) ultrafiltration membrane. A solution of m-phenylene diamine (MPD) in water and water suspension of graphene oxide were mixed first and poured over the PES membrane. To this, trimesoyl chloride (TMC) solution in hexane was also poured and dried at 65°C to obtain the membrane. PA/GO membrane has shown a threefold increase in permeate flux which represents the efficiency of a membrane, compared to the PA membrane. Both these membranes

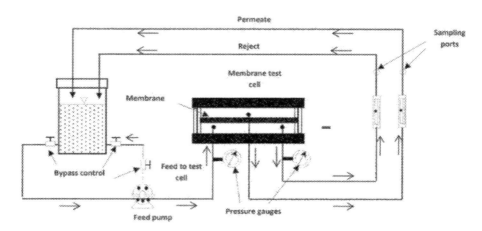

Figure 21.13 Schematic illustration of a plate-and-frame test cell for flat-sheet membranes. *Source*: T. Hogen-Esch, M. Pirbazari, V. Ravindran, H.M. Yurdacan, W. Kim, High performance membranes for water reclamation using polymeric and nanomaterials. US10456754B2, 2019.

had a comparable permeate flux values to that of commercial ultrafiltration PES (UF PES) membrane. A schematic illustration of the test cell is given in Fig. 21.13.

Apart from the type of membranes mentioned above, this team has also invented sulfonated and nonsulfonated Poly (tetrabutyl ammonium styrene sulfonate-co-styrene-co-4-chloromethyl styrene), P(BASS-S-CMS)-PVDF membranes. P(BASS-S-CMS) was dissolved in DMF and to this, sonicated 5 wt.% graphene oxide was added slowly followed by stirring. PVDF solution in DMF also was later added to this. This blend solution was then poured into a Petri dish and the film was prepared by drying in an oven. Sulfonated P(BASS-S-CMS)-PVDF membranes showed a much higher water permeation compared to PA/GO membranes. GO incorporated P(BASS-S-CMS)-PVDF blend membranes had significantly high water permeation [41].

21.7 Conclusions

Research on graphene and its composites using polymer matrices are on high in the current scenario. Since graphene is a versatile material having exceptionally high thermal and mechanical properties combined with superior conductivity, graphene-polymer composites find applications in most of the areas involving human life. This includes electronics where supercapacitors, sensors, OLEDs are involved, applications where high-performance composites are needed like in aerospace, defense and structural applications, membrane applications where desalination and water treatment or packaging films are involved to mention a few. Since pursuit for more efficient and economic materials is in progress, the rank of such composites is also of prime importance. Intense researches on such "hot topics" have helped humanity make more and more gadgets and instruments easily available at a reasonable price, which should ultimately be the aim of scientific research when looking through a common man's perspectives. Researches leading to commercialization of such inventions are going on vigorously all over the world. If we look carefully to the researches leading to patents, one can understand that studies are focused on either to reduce cost without compromising on targeted properties or highly specific applications are leading the way. According to the current pace of research, it is expected that the field of graphene nanocomposites will show exponential growth. The number of research publications and patents in this area are also pointing towards this fact.

References

[1] K.S. Novoselov, A.K. Geim, S.V. Morozov, D. Jiang, Y. Zhang, S.V. Dubonos, et al., Electric field effect in atomically thin carbon films, Science 306 (5696) (2004) 666−669.

[2] S. Stankovich, D.A. Dikin, G.H.B. Dommett, K.M. Kohlhaas, E.J. Zimney, E.A. Stach, et al., Graphene-based composite materials, Nature 442 (7100) (2006) 282−286.

[3] W. Chen, H. Weimin, D. Li, S. Chen, Z. Dai, A critical review on the development and performance of polymer/graphene nanocomposites, Sci. Eng. Compos. Mater. 25 (6) (2018) 1059−1073.

[4] Y. Jeong, J. Gil-woo, Aramid composites reinforced with mixed carbon nanomaterials of graphene and carbon nanotube and process for producing the same. KR20130092234A, 2019.

[5] A. Bobenrieth, P. Dubois, J.-M. Raquez, F. Meyer, C. Frederix, Stable aqueous graphene suspension and its use in producing graphene polymer nanocomposites. EP2960205A1, 2015.

[6] F. Gauthy, A. Sanguineti, F. D'aprile, P. D'Orazio, Process for producing graphene-polymer nanocomposites. EP2820069B1, 2016.

[7] AD Clauss, G. Pan, N.R. Wietfeldt, M.C. Hall, D.D. Taft, Polymer-graphene nanocomposites. United States9790334B2, 2017.

[8] J.C. Hanan, S. Bandera, Graphene reinforced polyethylene terephthalate. CN111253618A, 2020.

[9] L. Song, A. Zhamu, J. Guo, B.Z. Jang, Nano-scaled graphene plate nanocomposites for supercapacitor electrodes. United States7623340B1, 2009.

[10] L. Chunzhong, G. Zheming, W. Gengchao, Z. Ling, Composite material with high conductivity intercalation structure and preparation method thereof. CN101544823A, 2011.

[11] H. Haiqing, Electroconductive graphene/polymer composite material. CN102321379A, 2013.

[12] Y.-P. Sun, J.W. Connell, L.M. Veca, Highly thermal conductive nanocomposites. United States20140299811A1, 2014.

[13] T. Ming, Z. Xiaoqing, N. Nanying, Z. Liqun, F. Yuxing, Graphene elastomer nano composite material with high dielectric constant and low dielectric loss and preparation method thereof. CN103183847A, 2015.

[14] T. Chengan, Z. Xiaorong, X. Hua, W. Jianfang, Z. Hui, S. Liping, et al., Graphene-polysiloxane composite material and preparation method thereof as well as microfluidic chip and application thereof. CN103436017A, 2016.

[15] D. Peng, S. Liyi, S. Shuangshuang, S. Na, T. Shengfu, A kind of high heat conduction nylon 6/graphene nanocomposite material and preparation method thereof. CN103450674B, 2016.

[16] O.C. Compton, S.T. Nguyen, Composite polymer film with graphene nanosheets as highly effective barrier property enhancers. United States20110223405A1, 2011.

[17] R. Penggang, J. Yanling, C. Tao, L. Zhongming, Preparation method for graphene oxide with high barrier property and polymer nanocomposite film. CN102115566A, 2011.

[18] W. Xianbao, W. Jingchao, X. Chunhui, W. Li, Graphene-based barrier composite material and preparation method thereof. CN101812194B, 2012.

[19] L. Heon-sang, Gas barrier transparent polymer composite films. KR101209293B1, 2012.

[20] L. Heon-sang, Water vapour barrier transparent polymer complex films. KR101543260B1, 2015.

[21] R. Prud'Homme, C. O'Neil, B. Ozbas, I. Aksay, R. Register, D. Adamson, Functional graphene-polymer nanocomposites for gas barrier applications. United States8110026B2, 2012.

[22] R. Prud'Homme, C. O'Neil, B. Ozbas, I. Aksay, R. Register, D. Adamson, Functional graphene-rubber nanocomposites. JP5450069B2, 2014.

[23] R. Prud'Homme, C. O'Neil, B. Ozbas, I. Aksay, R. Register, D. Adamson, Functional graphene-polymer nanocomposites for gas barrier applications. JP5421110B2, 2014.

[24] P. Min, S. Hyun-seok, S. Person, K. Tae-an, K. Hee-sook, L. Sang-soo, et al., Process of preparing polymer/NGPs nanocomposite films with enhanced gas barrier properties. KR101308967B1, 2013.

[25] L. Li, M. Yingyan, K. Zhiqiao, C. Junlei, X. Ying, Z. Fazhong, et al., Graphene oxide/polymer composition for inner liner of tyre and inner tube and preparation method thereof. WO2013107229A1, 2013.

[26] H. Park, H. Kim, B. Yoo, S. Jang, Graphene oxide nanocomposite membrane having improved gas barrier characteristics and method for manufacturing same. WO2015133849A1, 2015.

[27] J.K. Lynch-Branzoi, T.J. Nosker, B.H. Kear, C.T. Chang, Use of graphene-polymer composites to improve barrier resistance of polymers to liquid and gas permeants. WO2019143662A1, 2019.

[28] Z. Jianxin, G. Wanlin, Y. Jun, C. Yaqing, Graphene-based strain measuring and motion sensing device and manufacturing method thereof. CN102506693A, 2012.

[29] G. Duesberg, M. Hegner, Nano-carbon sensor and method of making a sensor. WO2012028748A1, 2012.

[30] G. Duesberg, M. Hegner, Nano-carbon sensor and method of making a sensor. EP2426487A1, 2012.

[31] W. Luheng, M. Fangfang, W. Xueting, L. Huanghai, Method for developing flexible pressure sensitive probe with side electrode. CN102141451B, 2012.

[32] L. Sang-yeop, P. Tae-jung, H. Won-hee, J. Hee-tae, P. Ho-seok, C. Bong-gil, Biosensor using the conductive graphene and manufacturing method thereof. KR101440542B1, 2014.

[33] S. Park, K. Parmar, Sensing element compositions and sensor system for detecting and monitoring structures for hydrocarbons. EP3058312B1, 2019.

[34] W. Luheng, Graphene filled silicon rubber composite piezoresistance sensor and its method of production. CN106495085A, 2017.

[35] I.A. Kinloch, R.J. Young, K.S. Novoselov, Graphene polymer composite. United States20180354785A1, 2018.

[36] Z. Hongwei, Z. Rujing, A kind of large deformation flexible strain transducer and preparation method thereof. CN106482628B, 2019.

[37] J. Coleman, C. Boland, U. Khan, Sensitive, high-strain, high-rate, bodily motion sensors based on conductive nano-material-rubber composites. United States10251604B2, 2019.

[38] K.M. Krishna, A. Bhattacharyya, R.M. Devi, M. Phadke, Membranes comprising graphene. WO2014168629A1, 2014.

[39] S. Bano, A. Mahmood, S.-J. Kim, K.H. Lee, Nanocomposite ultrafiltration membrane containing graphene oxide or reduced graphene oxide and preparation method thereof. United States20160303518A1, 2016.

[40] L. Li, C. Hongrui, L. Yanqiu, A method to accelerate the production of membrane from reducing graphite oxide for the water treatment process. CN109081430A, 2018.

[41] T. Hogen-Esch, M. Pirbazari, V. Ravindran, H.M. Yurdacan, W. Kim, High performance membranes for water reclamation using polymeric and nanomaterials. United States10456754B2, 2019.

Polymer-graphene composite in hydrogen production

22

Swarup Krishna Bhattacharyya[1], Susanta Banerjee[1,2] and Narayan Chandra Das[3]
[1]School of Nanoscience and Technology, Indian Institute of Technology, Kharagpur, India, [2]Materials Science Centre, Indian Institute of Technology, Kharagpur, India, [3]Rubber Technology Centre, Indian Institute of Technology Kharagpur, Kharagpur, India

22.1   Introduction

Consumption of energy is inevitable to regulate the wheel of civilization. From the beginning of civilization, different types of fuel are used to produce energy, such as wood, coal, crude oil, nuclear energy, hydrogen fuel, but one thing is shared between all of these sources which is the presence of carbon and hydrogen. We are increasing the ratio of hydrogen: carbon from wood to crude oil to attain better burning energy value [1]. Right now, 95% of consumed energy comes from fossil fuel, while the rest comes from water electrolysis [1]. We cannot deny that consumption of fossil fuels is rapidly increasing day by day to accomplish the ongoing energy demand [2−6]. Simultaneously, we also have to consider that the total amount of fossil fuel is not unlimited, and excessive usage emits a vast amount of greenhouse gases. All these economical, technological, and environmental backgrounds of the energy crisis indicate that there is an urgent requirement fir new, sustainable, clean, low-carbon, and high energy density energy sources [7−9]. Utilization of solar energy to compensate for the uprising energy crisis is an often discussed topic. Solar energy is the most abundant renewable energy resource, and the total amount of absorbed solar energy by the Earth is much more excessive than the requirement. The amount of solar energy received by Earth at any instant is comparable to the power supply by 130 million 500-MW power plants. Therefore the conversion of solar energy into chemical fuels could be an effective way to counter the uprising energy crisis. In this background, the production of hydrogen from water splitting can fulfill all the above criteria and substitute the traditional fossil fuel consumption. Photoelectrochemical (PEC) water splitting by using heat energy from solar, nuclear or other sources [2] and photocatalysis (PC) hydrogen evolution reaction [10−12] (HER) is the most applied way for large-scale hydrogen production. The main difference between the PEC and PC is that the former one involves the reactant in the excited state, whereas the latter one deals with a reactant in the ground state [1].

Different strategies have been employed for utilizing solar radiation in splitting water for hydrogen generation [13,14]; such as solar water splitting cells [15], PEC

Polymer Nanocomposites Containing Graphene. DOI: https://doi.org/10.1016/B978-0-12-821639-2.00022-7

tandem cell [16], Z scheme water splitting device [17], I-S thermochemical process [2]. However, expensive proton exchange membranes [18] and alkaline electrolytes [19] are used for photovoltaic electrolysis of water. Solar to thermochemical water splitting requires a high temperature of $700°C-1000°C$ [20]. Subsequently, PEC and PC have gained much more attention for hydrogen generation because they are inexpensive, convenient, and straightforward, and have massive potential for further development [15].

In the history of hydrogen production, Fujishima and Honda took the pioneering role, and they produced hydrogen through water splitting using TiO_2 as a photoanode under the UV light irradiation [21]. Since then, TiO_2 was vigorously investigated as a photoelectrode/photocatalyst material owing to their high-efficiency, low-cost, non-toxicity, and photostability [22−24]. Lots of research work had been carried out to develop semiconductor-based water-splitting technology to absorb solar energy and convert it into chemical energy and store it for further application. Metal oxides with ABO_3 configuration, such as $SrTiO_3$ [25], $NaTaO_3$ [26,27], effortlessly meeting the thermodynamic requisite for water splitting, were also reported to be efficient photocatalysts. However, all these oxides are only active under UV irradiation, which accounts for about 4% of the total solar spectrum. Therefore, in order to access the visible-light ($\sim$43% of the solar spectrum) of solar spectra, specially designed efficient photoelectrodes or photocatalysts is required.

Now for solar energy conversion water takes a vital role and acts as a reactant. In a typical reaction, water splits into a mixture of hydrogen and oxygen, followed by a separation method in the presence of a suitable catalyst [28]. The overall water splitting reaction is given below:

$$H_2O \rightarrow H_2 + 1/2O_2; \quad \Delta G = 237 \text{ kJ/mol} \tag{22.1}$$

The above two-electron reaction has Gibbs free energy value of 237 kJ/mol. Now, the photon energy "E" can be expressed by the following Eq. (22.2).

$$E = h\nu = hc/\gamma \tag{22.2}$$

where "h" is the Planck's constant, "c" is the speed of light, "ν" is the frequency of photon and "γ" is the wavelength of the photon. Therefore 1.23 eV photon energy or photon equivalent to the wavelength of 1000 nm is required to drive this water-splitting reaction thermodynamically. That means the bandgap of the photocatalyst should be higher than 1.23 eV. Analysis of the solar spectrum is also an essential aspect of solar energy conversion and for the determination of suitable photocatalyst. Now using Gibbs free energy, the STH efficiency can be calculated by using Eq. (22.3).

$$STH = \left(Output\,energy/incidence \text{ solar light energy}\right) = (rH_2 \times \Delta G)/(Ps \times Ag) \tag{22.3}$$

Here, "rH_2" is the rate of hydrogen production, "Ps" is the energy flux of sunlight and "Ag" is the area of the reactor. The solar energy spectrum contains ~ 93 W/m^2 in the ultraviolet region, ~ 543 W/m^2 in the visible region and ~ 363 W/m^2 in the infrared region. At least 10% STH efficiency is required for practical water splitting system [29], which corresponds to the "rH_2" value of ~ 154 μmol H$_2$ cm/h and PEC current of ~ 8.3 mA/cm^2. However, these values are not suitable for large scale industrial application of semiconductor-based photocatalytic (PCt) hydrogen generation. So, it is necessary to develop new types of photocatalyst that can absorb light in the wavelength range of 600−700 nm (~ 1.8−2.0 eV) to achieve this efficiency. That is why scientists are looking for a visible light-responsive catalyst to attain high STH efficiency. Although the main challenge with the visible light responsive catalyst is that most of the organic molecule do not absorb in the visible light wavelength. So, most of the occasion expensive ultraviolet high energy source is used which reduce the industrial-scale production. Here, we will discuss PCt and PEC strategies for hydrogen generation from light-driven water splitting.

A schematic diagram of PEC hydrogen production under light irradiation is demonstrated in Fig. 22.1. A typical PEC cell consists of a working photoanode (made of photoactive materials) for oxygen evolution reaction (OER) and a counter photocathode (made of Pt) for HER. An electrolyte solution and wire complete the current loop between the external circuit and the photoelectrodes. There are very few semiconductor materials that satisfy the requirements for both water OER and HER and simultaneously offer good electrode stability in aqueous electrolyte solutions [30,31]. The reaction mechanism is described in the following half-cell reactions:

$$\text{Photocathode; cathodic reduction:} \quad 2H^+ + 2e^- \rightarrow H_2 \quad (22.4)$$

$$\text{Photoanode; anodic oxidation:} \quad H_2O + 2h^+ \rightarrow 2H^+ + 1/2O_2 \quad (22.5)$$

PCt hydrogen production is generally carried out in a suspended system using a particulate semiconductor photocatalyst excited by photons. In Fig. 22.2, we have

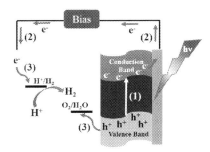

Figure 22.1 Basic principles of photoelectrochemical water splitting for hydrogen production.

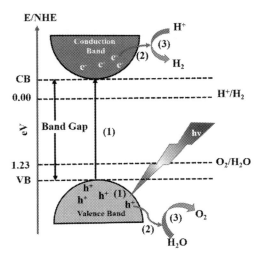

Figure 22.2 Water splitting process on a photocatalyst for hydrogen production. The y-axis E measured against the normal hydrogen electrode (NHE) refers to the potential, in-unit eV.

illustrated the PC hydrogen production schemes. Like PEC, a number of available PC catalysts are also minimal. A typical PC system for solar hydrogen production required a semiconductor with a relative negative conduction band (CB) bottom level compared to the reduction potential of H^+/H_2 (0.0 V vs NHE). In comparison, the location of the valence band (VB) must be more favorable than the oxidation potential of OH^-/O_2 (1.23 V vs NHE) [32,33].

Usually, PCt hydrogen production takes place in an aqueous medium, where alcohol, S^{2-}/SO_3^{2-} or lactic acid serves as sacrificial reagents and donates the electron. This sacrificial reagents consume the photogenerated holes and suppress the recombination of electrons and holes in the semiconductor photocatalyst. As a result, the rate of H_2 generation is enhanced, and the PCt stability is improved [34,35]. The half-reaction of water splitting and the mechanism in the presence of sacrificial reagents is explained as follows:

$$Red + H_2O \xrightarrow[\text{Photocatalyst}]{hv} Ox + H_2O \qquad (6)$$

Here "Red" denotes the electron donor, and "Ox" represents the product by hole oxidation.

This PC water-splitting reaction on a photocatalyst happens in three consecutive steps. These steps are similar to the photovoltaic process and common for both PEC and PCt hydrogen production [35]. In Fig. 22.3, a summarized scheme of PC process is illustrated. At first, a photon is absorbed by the photocatalyst, and electron-hole pairs are generated which get separated and migrated to different sites of catalyst in the next steps. Finally, water is reduced by photogenerated electrons

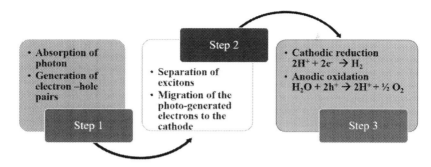

Figure 22.3 Necessary steps of PCt and photoelectrocatalytic water splitting for hydrogen production.

and evolved hydrogen on the surface of the counter electrode of the PEC system, or on the surface of the photocatalyst of the PC system [32,35].

In order to attain higher STH efficiency, all the efficiency determining steps should be accounted for properly. All the light absorption efficiency, charge separation efficiency, transportation efficiency, and surface chemical reaction efficiency should be optimized and considered accurately. To act as a good photocatalyst material, it should have a robust aptitude for efficient harvesting of a large portion of the solar spectrum, supports charge separation and rapid transfer to the semiconductor/aqueous interface, and displays long-term stability [29]. Since the last few decades mostly metal oxides, metal sulfides, metal chalcogenides have been used extensively as a photocatalyst, among them, TiO_2, WO_3, CdS, CdSe, ZnS, $BiVO_4$, ZnO, MoS_2 are the most commonly used semiconductor [36–40]. The bandgap of these semiconductors is good enough to respond in the visible light range. However, at the same time, high recombination of the photogenerated charge carrier, photocorrosion, and aggregation makes its activity and photostability low. Application of cocatalyst, construction of heterojunction structure [29] and coupling of these semiconductors with other semiconductors, nanomaterials or dopant can improve these drawbacks significantly [41].

The rise of nanomaterials for the fabrication of light energy harvesting assemblies has opened up new ways to utilize renewable energy resources. The large surface areas, abundant surface states, and diverse morphologies compared to their corresponding bulk materials make them unique for PEC and PC [15,34,42–50]. Among them, graphene, a single layer sp^2 hybridized honeycomb structure carbon nanomaterial [51], has stimulated tremendous research interest in various energy conversion applications [52–55]. Graphene shows excellent thermal, mechanical, optical properties along with the outstanding electrical property. Its large specific surface area and excellent mobility of the charge carrier at room temperature makes it a unique choice in various field such as supercapacitor, solar cell, PCt degradation, water splitting [56–59]. Further property of graphene can be manipulated by modifying the synthesis route and functional groups [60]. Many researchers already established the fact that incorporation of graphene into the photocatalyst enhanced the water-splitting

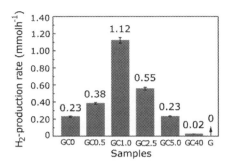

Figure 22.4 Effect of the graphene content in the visible-light catalytic hydrogen production with CdS cluster, 0.5 wt.% Pt as cocatalyst and 10 vol.% lactic acid as a sacrificial agent. *Source*: Reproduced with permission from Q. Li, B. Guo, J. Yu, J. Ran, B. Zhang, H. Yan, et al., Highly efficient visible-light-driven photocatalytic hydrogen production of CdS-cluster-decorated graphene nanosheets. J. Am. Chem. Soc. 133 (28) (2011) 10878−10884. Copyright (2011), with consent from the Journal of the American Chemical Society.

reaction significantly. Probably the graphene's surface and inherent electronic properties which boost the hydrogen production by inhibiting the charge-hole recombination and also by offering an alternative path for electron transportation. Li et al. reported graphene sheet decorated with CdS photocatalyst, Pt cocatalyst and lactic acid as sacrificial agent showed 4.87 times higher hydrogen production rate over CdS nanoparticle alone (Fig. 22.4) [61]. Dubale et al. reported the synergic effect of graphene Cu_2S composite. They claimed the resultant composite system enhanced the visible light absorption, charge carrier separation and also reduced the photocorrosion. Similarly, Xiang et al. and Kong et al. reported better PCt activity of graphene/TiO_2 composite [62] and graphene/Co/CoS_x [63]. Despite this outstanding performance of graphene-based composite, some severe challenges are also there. Literature suggests that graphene loading into the composite matrix cannot be exceeded above a certain weight percentage. After crossing this maximum value, hydrogen production decreases drastically, and this is because of the optical property of graphene, which restricts the light penetration into the reaction mixture. Most of the occasion Pt is used as a cocatalyst in graphene-based photocatalyst, which is expensive and limits its broad scale application. Apart from this, the strong π-π stacking interaction of the reduced graphene oxide (RGO) produces irreversible agglomeration. Moreover, the poor dispersibility of the graphene composite materials reduces the accessible surface area of the catalyst in the catalytic solution [64,65].

The various study suggests that the surface modification of graphene with a functional molecule is an effective way to deal with dispersibility and aggregation related problems [66,67]. Joonsuk et al. proposed that amine-functionalized polymeric microsphere of graphene oxide can successfully handle the issues mentioned above [68]. In this context hybridization of graphene with conjugated and conducting polymers could be an effective way for hydrogen production. Conjugated polymers with versatile chemical structure and facile synthesis route are already used in the optoelectronic field. On the other hand, conducting polymers with the extended π-conjugated electronic system is also recognized for their fast photoinduced charge separation and slow charge recombination properties [69]. Most of the occasion polymers are introduced into this graphene hybridized as a surface modifier and also a source of nitrogen doping [70], Long uniformly dispersed polymeric chain prohibit the catalyst aggregation and inhibits the restacking of graphene sheets [71]. However, to date, polymer-graphene composite for hydrogen production is still not explored widely. In this chapter, we attempt to provide a comprehensive idea about the ongoing research work and progress on polymer-graphene composite in hydrogen production based on published research work since 2015−19. In this chapter, we focus primarily on the basic water splitting techniques, synthesis and property of different polymer-graphene catalysts. Finally, this chapter is concluded with remarks about the remaining challenges and future perspective in this area. Hopefully, this chapter provides a productive insight and motivates the profound exploration of polymer-graphene composite in hydrogen production.

22.2 Polymer-graphene composite for the PCt hydrogen evolution

PC is a light-assisted catalysis where reactive intermediates are formed on the surface due to photoillumination. Many semiconductors behave as a photocatalyst because of their intrinsic electronic structure. Generally, they have filled valance band (VB), and empty CB, when these materials are exposed by the light energy higher than their bandgap electrons are excited and move to CB leaving behind a hole in the VB (as shown in Fig. 22.2). Now this photogenerated electrons and holes drive the oxidation/reduction reaction upon the material surface either by direct charge transfer to the reactant or through the formation of reactive intermediate ($OOH\cdot$, $OH\cdot$, $O_2\cdot$) which accelerates the reaction. Graphene-based photocatalyst acts as a photosensitizer for an extended absorption range and provides an additional electron to the VB. Besides that, low light absorption improves the PC efficiency value. Here we discuss a few of the polymer-graphene composite for PCt hydrogen evolution (Fig. 22.5).

22.2.1 Synthesis strategy of polymer-graphene photocatalyst

Polymer-graphene composite was mostly used in the presence of another photocatalyst. Most of the occasion graphene was synthesized by modified Hummer's method [72], and the following technique prepared the graphene functionalized polymer composite.

22.2.1.1 Suspension polymerization

The suspension polymerization technique synthesized organic polymer microspheres, where all the monomers and initiators were already dissolved in ethanol

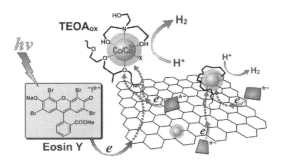

Figure 22.5 Schematic illustration of H_2 evolution from the polymer-graphene composite. *Source*: Reproduced with the permission from J. Wang, K. Feng, B. Chen, Z.J. Li, Q.Y. Meng, L.P. Zhang, et al., Polymer-modified hydrophilic graphene: a promotor to photocatalytic hydrogen evolution for in situ formation of core@ shell cobalt nanocomposites. J. Photochem. Photobio. A. 331 (2016) 247–254, Copyright (2015), with consent from the Elsevier.

under an N_2 atmosphere. Then the N_2 degassed mixture was heated at 343K for 12 h under continuous gentle shaking. Then the product was washed, dried, and dispersed in and allowed to react with ethidenediamine for the amination. Five weight percentage of graphene oxide was dispersed into the freshly prepared amine functionalize microsphere solution and heated under continuous stirring at 343K for 4 h in the water bath to decorate the amine-functionalized microsphere surface with graphene oxide. Finally, the mixture was centrifuged, washed and dried in room temperature.

In the next step incorporation of semiconductor-based photocatalyst and the reduction of graphene oxide take place. The simple solvothermal process was employed for this reaction. In a stainless steel autoclave, all the graphene oxide coated amine-functionalized microsphere, and the precursor of semiconductor catalyst was suspended in DMSO. By varying the reaction time at 453K temperature, different semiconductor decorated polymer-supported graphene structures were produced [73].

22.2.1.2   Ring-opening polymerization

The surface modification of the graphene oxide was carried out with an industrial Jeffamine M-600 polyetheramine by one-pot ring-opening polymerization between epoxy and amide groups [74]. Resultant GJ600 graphene showed excellent water dispersibility and strong affinity toward catalytic metal centers because of the surface modification with tethered polyetheramine chains. In a typical reaction, 80 mg of GO was dispersed in a DI water by prolonging sonication of 2 h, and then 1.2 g J600 and 80 mg of NaOH was supplied to that dispersion. Afterwards, the system was purged with Argon gas and stirred at 80°C for 24 h followed by dialysis and lyophilization process. Similarly, $NaBH_4$ reduced GJ600 graphene was produced by dispersing 0.64 g of $NaBH_4$ in a 160 mL suspension of GJ600 (0.5 mg/mL). The mixture was at 80°C for 1 h, followed by dialysis against water. A possible reaction scheme is presented in Fig. 22.6.

22.2.1.3 Coordination polymerization

Xu et al. proposed this technique for the p-n heterojunction photocatalyst preparation through colloidal blending method. Coordination polymer nanoplates (CPNP) and partially RGO (PRGO) were prepared separately for this purpose [75].

Graphene oxide was first synthesized by the modified Hummers method. In this technique, 1 g of graphite powder was added in 70 mL of 98% H_2SO_4 and kept it in an ice bath under continuous stirring. Then 0.5 g of $NaNO_3$ and 4.5 g of $KMnO_4$ were added and stirred for next 3 h at the same temperature. After that 70 mL of DI water and 20 mL of H_2O_2 were added and with the addition of H_2O_2 the black solution turned in to brown which confirmed the generation of oxidized graphite. Then this slurry was sonicated and exfoliated to produce graphene oxide, which was washed with HCl and DI water and dried in a vacuum oven. Now, for the PRGO preparation, a certain amount of as-prepared GO was dispersed in DI water and sonicated until a clear solution was formed. Then hydrazine hydrate was added and heated at 100°C under stirring for a specific time period. Finally a black precipitated of PRGO was filtered out and washed with water and methanol and dried in a vacuum oven. By varying this reduction time, several PRGO was prepared.

A new coordination polymer (CP) $[Cu(pad)_2(bipy)]_n \cdot n(H_2pad)$ was prepared from $Cu(OAc)_2 \cdot 4H_2O$, H_2pad, and bipy (pad stands for phenylenediacrylatedianion, and bipy represents 4,4′-methyl-2,2′-bipyridine). All these precursors were mixed in DI water, and the pH of the system was adjusted by 1 M NaOH. After

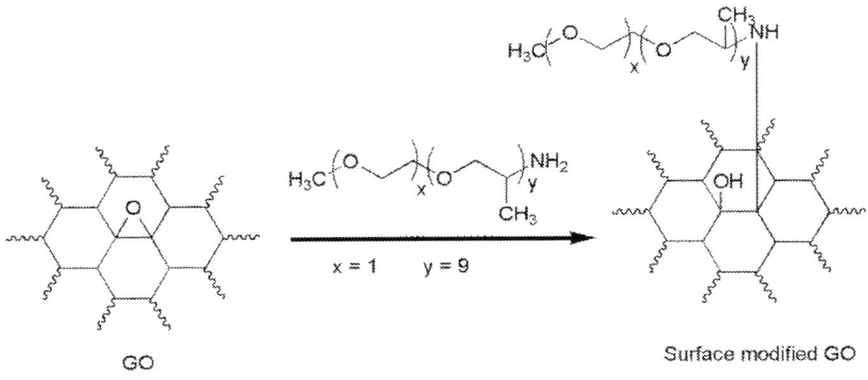

Figure 22.6 One-pot ring-opening polymerization between epoxy and amide groups for the synthesis of GJ600.
Source: Reproduced with the permission from J. Wang, K. Feng, B. Chen, Z.J. Li, Q.Y. Meng, L.P. Zhang, et al., Polymer-modified hydrophilic graphene: a promotor to photocatalytic hydrogen evolution for in situ formation of core@ shell cobalt nanocomposites. J. Photochem. Photobio. A 331 (2016) 247−254, Copyright (2015), with consent from the Elsevier.

stirring the mixture for another 15 min transferred into a Teflon coated stainless steel autoclave, and heated at 150°C under autogenous pressure for 4 days. After 4 days, a large amount of blue crystal of CP was found. This freshly formed crystal was ground in a mortar and pestle. This powder was then dissolved in methanol and placed into the Teflon autoclave, and heated at microwave oven at 250 W for 2 h. The product was separated by centrifugation, washed and dried under vacuum oven at 80°C for 24 h.

Finally, the PRGO/CPNP photocatalysts were prepared by colloidal blending method. For this purpose, at first PRGO was dissolved in DI water and CPNP was also dissolved in DI water separately. Then this CPNP solution was poured into the PRGO solution and sonicated for one hour. A homogeneous solution was obtained and then dried at 70°C.

22.2.2 Experimental technique of PCt hydrogen evolution

In a typical PCt process, a certain amount of photocatalysts were dispersed in a salt solution and sonicated. Then the solution was transferred to a Pyrex tube, and the addition of HCl or NaOH adjusted the pH of the solution accordingly. After that, the solution was degassed by bubbling nitrogen, and when the suspension equilibrated a standard internal gas was injected for the qualitative gas chromatography analysis. Then the Pyrex tube was exposed by 300 W Xenon lamp or some time with visible light LEDs. Evolution of hydrogen was determined by online gas chromatography equipped with a thermal conductivity detector.

22.2.3 Different parameters that influence the PCt performances for hydrogen production

PCt hydrogen evolution from polymer-graphene composite was carried out against sacrificial agents under visible light irradiation. It was observed that the ability of hydrogen production depends upon several factors such as irradiation time, reaction time, loading percentage of graphene, system's pH, Inner Electric Field (IEF), and the photostability of the system.

22.2.3.1 Effects of degree of reduction of graphene oxide

Polymer-graphene composite showed a better hydrogen production rate over the crude product. The hydrogen production rate was greatly influenced by the degree of reduction of the graphene oxide. Xu et al. showed that for the CP PRGO composite rate of hydrogen production increased tediously with the degree of reduction of graphene oxide. According to their theory, prolonged reduction time reduced the graphene oxide such a way which ultimately improved the electron transportation property. At this stage separation of electron and holes became significant enough to enhance the rate of hydrogen production. The conductivity of different PRGO samples with the various degree of reduction value supported this fact [75]. Although, after attaining a particular degree of reduction value, this PCt hydrogen

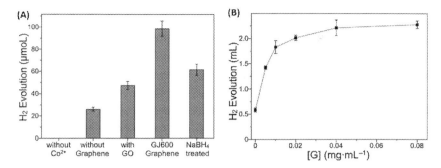

Figure 22.7 PCt H_2 evolution of (A) different graphenes composites at the concentration of 40 mg/mL, (B) as a function reduced graphene contents.
Source: Reproduced with the permission from J. Wang, K. Feng, B. Chen, Z.J. Li, Q.Y. Meng, L.P. Zhang, et al., Polymer-modified hydrophilic graphene: a promotor to photocatalytic hydrogen evolution for in situ formation of core@ shell cobalt nanocomposites. J. Photochem. Photobio. A 331 (2016) 247–254, Copyright (2015), with consent from the Elsevier.

production rate diminished profoundly. The best hydrogen production rate was obtained for the sample corresponding to the 9 h of a degree of reduction. In their view, this was because of complete reduction of graphene oxide surface which decreased the functional groups and promoted the interaction between CP and PRGO, as a results charge separation was hindered and reflected in terms of reduced hydrogen production rate.

Wang et al. also believed that partially reduced features of graphene sheet could effectively facilitate the electron transfer between photosensitizers and catalytic sites through their basal plane. They demonstrated that incorporation of hydrophilic graphene (GJ600) into the Co photocatalyst system increased the hydrogen production amount around 3.78 folds in comparison to graphene free system. They also compared the results against the unmodified graphene oxide and NaBH$_4$ reduced GJ600 system and found that the amount of hydrogen production was 1.82 and 2.36 times higher than the graphene free system. All these results are illustrated in Fig. 22.7A and B. The formation of black aggregate from the photo reduced GO during the process, and the poor hydrophilicity of NaBH$_4$ reduced GJ600 were the reasons for the low hydrogen production value [74].

22.2.3.2 Effect of synthesis method

Cao et al. studied the effects of synthesis methods on different PSGM/RGO/CdS system at a fixed irradiation time. For this study, PSGM/RGO + CdS mechanically mixed the sample, and three different sets of PSGM/RGO/CdS$_X$ system were produced by varying the reaction time 3, 6, and 12 h respectively, and compared their hydrogen production rate with CdS catalyst. All the results are presented here in a tabular form. It was found that the PSGM/RGO/CdS$_3$ system showed the highest hydrogen production rate, and the value was 3.90 times higher than pure CdS catalyst and 19.30 times higher than the physical mixed sample PSGM/RGO + CdS composite (8.62 μmol/h) in 5 h under visible-light irradiation. The formed heterojunction between CdS and graphene took a crucial role in electron promotion and charge transfer from CdS to graphene [73].

22.2.3.3   Effects of surface area and pore diameter

The specific surface area of the catalyst does not control hydrogen production; in fact, photocatalyst having a higher surface area possesses the lowest hydrogen production rate value. Cao and co-worker showed that the photocatalyst sample $PSGM/RGO/CdS_3$ had the lowest specific surface area and still it possessed the maximum hydrogen production rate (as shown in Table 22.1). According to their study, instead of specific surface area, the pore diameter of the catalyst had an impact on the hydrogen production reaction. They showed the sample having a maximum pore diameter also possessed the maximum hydrogen production value. Since the large pore diameter is advantageous for efficient mass transfer (produced hydrogen) from the solution and also provide easy access to contact and react with the photocatalyst's surface [73].

22.2.3.4   Effects of mass ratio

The mass ratio between the polymers: RGO had a significant effect upon the PCt hydrogen production rate. Xu et al. demonstrated this effect by synthesizing CPNP/PRGO photocatalysts with different mass ratios [75]. They noticed that with increasing PRGO amount, the rate of hydrogen production was inconsistent. The maximum hydrogen production rate was achieved for the photocatalyst with a mass ratio of CPNP/PRGO(C) of 100:1. Above this optimized content, the excessive PRGO(C) accumulated on the CPNPs, which boosted the recombination of photogenerated electron-hole pairs and hindered the adsorption of visible light by the CPNPs. This explained the reason behind the poor PCt efficiency for the other compositions ultimately [75].

Table 22.1 Effects of surface area and pore diameters upon hydrogen production rate.

Sample's name	Rate of hydrogen production (μmol/h)	Specific surface area (m²/g)	Average pore diameter (Å)
Pure CdS	35.8	125.23	56.38
$PSGM/RGO/CdS_3$	175	5.67	408.79
$PSGM/RGO/CdS_6$	135	29.21	70.37
$PSGM/RGO/CdS_{12}$	126	36.49	75.15

It is reproduced with the permission from J. Xu, L. Wang, X. Cao, Polymer supported graphene—CdS composite catalyst with enhanced photocatalytic hydrogen production from water splitting under visible light. Chem. Eng. 283 (2016) 816−825, Copyright (2015), with consent from the Elsevier.

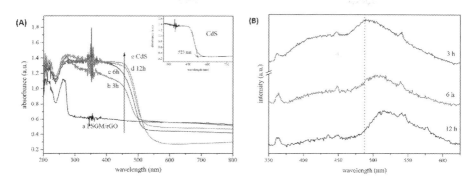

Figure 22.8 (A) UV—Vis spectra of different samples: (a) PSGM/rGO; (b) PSGM/rGO/CdS (3 h); (c) PSGM/rGO/CdS (6 h); (d) PSGM/rGO/CdS (12 h); (e) PSGM/rGO/CdS (24 h); (f) CdS in the wavelength range of 250—800 nm. (B) Photoluminescence spectra of PSGM/rGO/CdS (3 h); PSGM/rGO/CdS (6 h) and PSGM/rGO/CdS (12 h).

Source: Reproduced with the permission from J. Xu, L. Wang, X. Cao, Polymer supported graphene—CdS composite catalyst with enhanced photocatalytic hydrogen production from water splitting under visible light. Chem. Eng. 283 (2016) 816—825, Copyright (2015), with consent from the Elsevier.

22.2.3.5 Effects of reaction time

The reaction time of the solvothermal process also influenced the hydrogen production rate. Generally, the rate of hydrogen production is inversely proportional to the reaction time. This can be explained in terms of quantum size effect which controls the PCt activity of the photocatalyst composite [76,77]. UV—Vis diffuse reflectance spectra and PL spectra suggest when the solvothermal reaction time decreases a widened bandgap composites are formed (as shown in Fig. 22.8A and B). The requirement of the thermodynamic driving force becomes higher for redox reaction as a results hydrogen production and consumption of hole-scavenger becomes less probable [73].

22.2.3.6 Effect of pH

The pH of the solution is an essential factor for hydrogen production. The hydrogen production rate is found optimum at a specific pH value and that value generally matches with the pK_a value of the sacrificial electron donor. In the robust acidic medium photocatalyst system loses its potential because of the protonation of sacrificial electron donor. On the other hand, strong basic pH is also detrimental for hydrogen production [74]. In the strong pH, the redox potential value of H^+/H_2 couple and metal ions shifts toward the more negative position, driving force mediated by photosensitizer is smaller than the same system at low pH to carry out the photoreduction. Wang et al. demonstrated (Fig. 22.9) that the maximum hydrogen production value was obtained at pH 10.65, corresponding to the pK_a value of sacrificial electron donor triethanolamine [74].

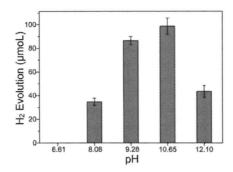

Figure 22.9 The effect of pH on photocatalytic hydrogen evolution from an aqueous system containing $CoSO_4$ 2.0×10^{-4} M, GJ600 40 µg/mL, EY 4.0×10^{-4} M and triethanolamine 0.10 M at pH 10.65 during 4 h with the light intensity of 120 mW/cm^2.
Source: Reproduced with the permission from J. Wang, K. Feng, B. Chen, Z.J. Li, Q.Y. Meng, L.P. Zhang, et al., Polymer-modified hydrophilic graphene: a promotor to photocatalytic hydrogen evolution for in situ formation of core@ shell cobalt nanocomposites. J. Photochem. Photobio. A 331 (2016) 247–254, Copyright (2015), with consent from the Elsevier.

22.2.3.7 Effect of number of cycles

Reusability and the stability of the photocatalyst and the repeatability of the PCt reaction for hydrogen production are essential aspects for the practical application.

Cycling test is generally carried out to evaluate the photostability of the composite photocatalyst by repeating the PCt reaction for four times. That composite catalyst retains their activity after the four times recycling are considered as a photostable one. Before and after of cycling test XPS, XRD and FTIR data are taken off the photocatalyst which confirmed that no structural changes appear and support the photostability statement.

Cao et al. investigated the photostability of their PSGM/rGO/CdS (3 h) catalyst by performing a cycling test. They observed after the 20 h reaction span catalyst retained the hydrogen production capability. However, a slight reduction in hydrogen production was observed due to the consumption of S^{2-}/SO_3^{2-} during the reaction (as shown in Fig. 22.10A). Further XRD measurement was performed upon PSGM/rGO/CdS (3 h) catalyst before and after the cycling test, which revealed the structure of the catalyst was not hampered by the cycling test (as shown Fig. 22.10B) [73]. The repeatability of the PRGO(C)/CPNP catalyst during hydrogen production was also determined by the Xu et al. PRGO(C)/CPNP catalyst showed excellent photostability up to the four cycles and retained the structure, which was further confirmed by the XRD and FTIR analysis [75].

22.2.3.8 Effects of inner electric field

The IEF of the p-n heterojunction significantly enhanced the hydrogen production rates. Xu et al. illustrated this effect for PRGO/CPNP catalysts considering the

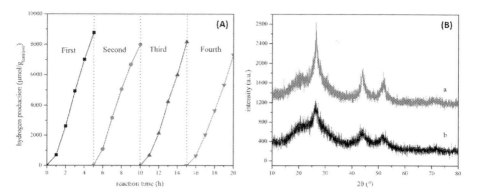

Figure 22.10 (A) Cycling test of catalyst PSGM/rGO/CdS (3 h) under visible light irradiation ($k > 400$ nm) with 0.5 M Na_2SO_3 and 0.5 M Na_2S aqueous solution for 20 h; (B) XRD diffraction patterns of catalyst PSGM/rGO/CdS (3 h) before (a) and after (b) the cycling test.
Source: Reproduced with the permission from J. Xu, L. Wang, & X. Cao, Polymer supported graphene—CdS composite catalyst with enhanced photocatalytic hydrogen production from water splitting under visible light. Chem. Eng. 283 (2016) 816—825, Copyright (2015), with consent from the Elsevier.

PRGO and PRGO/CPNPM (the mechanically blended product of the CPNPs and PRGO) as a reference. They observed that the H_2 production rate of PRGO/CPNP catalysts was superior over the PRGO and PRGO/CPNPM catalysts. This fact can be explained in terms of the IEF of the p—n heterojunction, which facilitates the separation of electrons and holes and enhances the H_2 production rate [75].

22.2.4 Possible mechanism

Polymer-graphene composite shows better performance over the pure semiconductors catalyst for hydrogen production. They play a crucial role in improving PCt activity by receiving photoinduced electron, offering reaction sites [78], and prohibiting the combination of electron and hole. This enhancement in PCt activity can be explained from the energy level diagram of the composite materials by comparing it with the pure photocatalyst. The possible mechanism of PCt hydrogen evolution is as follows. Under the visible light irradiation, photoinduced electrons and holes are produced in CB and VB. Then the photoinduced electrons are transferred to the graphene or semiconductor photocatalyst surface. The photocatalyst surface that absorbs protons earlier, now, transforms into hydrogen by accepting the photoinduced electron. At the same time, the photoinduced holes are consumed by the sacrifice agent. Cao et al. demonstrated the hydrogen evolution mechanism for PSGM/RGO/CdS photocatalyst system by the following reaction steps (Fig. 22.11) [73].

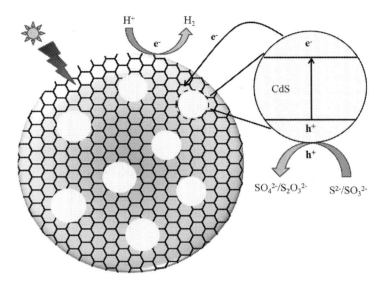

Figure 22.11 Proposed hydrogen evolving mechanism of PSGM/rGO/CdS composite catalyst.
Source: Reproduced with the permission from J. Xu, L. Wang, & X. Cao, Polymer supported graphene—CdS composite catalyst with enhanced photocatalytic hydrogen production from water splitting under visible light. Chem. Eng. 283 (2016) 816—825, Copyright (2015), with consent from the Elsevier.

$$CdS + h\nu \rightarrow CdS + e^- + h^+ \quad (Step\ 1)$$

$$2H_2O + 2e^-_{(CdS)} \rightarrow H_2 + 2OH^- \quad (Step\ 2)$$

$$2H_2O + 2e^-_{(graphene)} \rightarrow H_2 + 2OH^- \quad (Step\ 3)$$

$$SO_2^{3-} + H_2O + 2h^+ \rightarrow SO_4^{2-} + 2H^+ \quad (Step\ 4)$$

$$2S^{2-} + 2h^+ \rightarrow S_2^{2-} (Step\ 5)$$

$$S_2^{2-} + 2SO_3^{2-} + 2h^+ \rightarrow 2S_2O_3^{2-}$$

Xu et al. also described the same mechanism to explain the HER from PRGO/CPNP photocatalyst. They had studied the energy levels of PRGO and CPNP obtained from the slope of the Mott-Schottky (M-S) plot and calculated the bandgap (Eg) from Tauc equation (Fig. 22.12). A negative slope in M-S plot indicates a typical p-type semiconductor whereas the positive slope stands for n-type semiconductor. In Fig. 22.12 scheme of the energy levels of PRGO and CPNP is presented, and a diagram of PCt hydrogen evolution under visible light is also demonstrated [75].

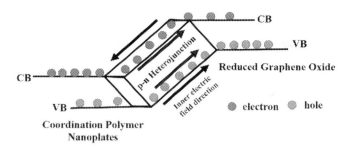

Figure 22.12 Schematic presentation of the photocatalytic mechanism for the PRGO/CPNP photocatalysts under visible-light irradiation [75].

According to their theory, under visible-light irradiation, the CPNPs and the PRGO both were excited in the PRGO/CPN p−n heterojunction photocatalyst and the electrons jumped into the CB of the CPNPs and the PRGO. Afterwards, positive-charged holes formed on the VB of the CPNPs and the PRGO. An IEF was generated with the direction from the n-type CPNPs to the p-type PRGO as the p−n heterojunction existed between them. Then the electrons residing on the CB of the PRGO moved to the CB of the CPNPs under the influence of this IEF. On the other hand, holes existing on the VB of the CPNPs shifted to the VB of the PRGO. This lead to the effective separation of electrons and holes (Fig. 22.12). Moreover, the reductive potential of H^+/H_2 was less negative than the CB energy level of the CPNPs, so the production of H_2 under visible-light irradiation might take place for the PRGO/CPNP p−n heterojunction photocatalyst.

22.3 Polymer-graphene composite for the electrocatalytic hydrogen evolution

Among the all hydrogen evolution techniques electrocatalyst-assisted HER is considered as the most economical, practical, and sustainable method for large-scale hydrogen production. Pt or Pt-based catalyst is generally preferable for their excellent activity toward both HER and Hydrogen oxidation reaction. However, their high cost and low abundance substantially hamper their large-scale utilization [15,79,81]. Thus, the development of low-cost and earth-abundant non-noble-metal catalysts are urgently required for practical requests to replace this Pt-based catalyst. Significant signs of progress have been reported owing to the development of a series of electrocatalysts mostly using transition-metal based carbide, sulfide, nitride, selenide, and phosphide [81−87]. Further, the HER activity of this transition-metal based nanoparticles can be improved by anchoring them onto some conductive supports, such as carbon nanotubes [87−89] or carbon nanosheets [90−92]. All these conductive supports not only inhibit the aggregation of transition-metal based nanoparticles but also improve the dispersion of active sites. Among these conductive supports RGO, especially doped RGO, are worth being mentioned because of their excellent electron-transport property and chemical robustness [93,94]. Nonetheless, transition-metal based nanoparticles tend to aggregate during high-reaction temperature carbonization procedures, thus decreasing the number of exposed active sites and their specific surface area [84,95]. On the other hand, RGO nanosheets usually aggregate because of the strong π-stacking and hydrophobic interactions, which hampers their practical application [96,97]. These issues can be addressed by introducing a conductive polymeric framework during the oxidation polymerization [98]. In general, practice polypyrrole with extended π-conjugated electron structures serve the role of conductive polymer because of their rapid photoinduced charge separation and relatively slow charge recombination [69]. This polymeric framework effectively prevents the RGO from restacking and the transition-metal based nanoparticles from aggregating [70,98−100]. This resultant hybrid shows synergistic electrocatalytic performance for the HER. Generally, a high potential value is preferable to achieve a fast rate of reaction. This over potential value is referred to as the activation energy of the reaction. Electrocatalyst diminishes this over potential values of the reaction and opens an alternative less energetically demanding reactions. Here we discuss a few of this hybrid electrocatalyst along with their synthesis techniques, properties and HER performances.

22.3.1 Synthesis strategy of polymer-graphene-based electrocatalyst

Graphene-based electrocatalyst modified with a various polymeric system can be made by different polymerization strategies. Among them, one-pot redox reaction and carbonization and in situ UV initiated polymerization are mostly used. Here we discuss these polymerization techniques in details for the preparation of polymer-graphene composite.

22.3.1.1 In situ ultraviolet initiated polymerization

Zhuang et al. reported this technique to fabricate dicobalt phosphide (Co_2P) encapsulated nitrogen and phosphorus-doped graphene electrocatalyst. Graphene oxide was synthesized by modified Hummer's method, and that graphene oxide was used for the polyacrylamide, and phytic acid (PA) grafted GO. For this photopolymerization freshly prepared GO, acrylamide, PA and 2-hydroxy-2-methylpropiophenone were mixed in DI water and stirred for half an hour. Then the mixture was deoxygenated by nitrogen gas to remove the trapped oxygen in the solution. Finally, the mixture was exposed under UV light (365 nm, 1.350 W/cm^2) for 10 min. Then the functionalized GO was purified by centrifugation and redispersed in water [101]. Details of the procedure are presented here in Fig. 22.13.

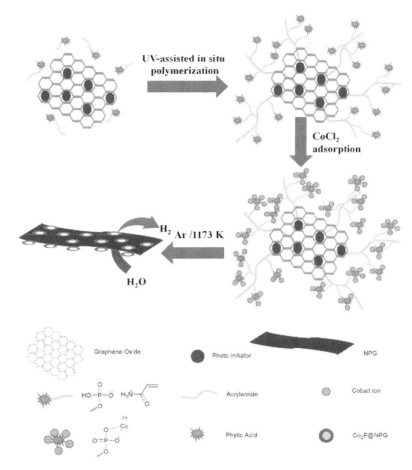

Figure 22.13 Fabrication of Co_2P nanoparticles encapsulated by N, P-doped graphene ($Co_2P@NPG$).
Source: Reproduced with the permission from M. Zhuang, X. Ou, Y. Dou, L. Zhang, Q. Zhang, R. Wu, et al., Polymer-embedded fabrication of Co2P nanoparticles encapsulated in N, P-doped graphene for hydrogen generation. Nano Lett. 16 (7) (2016) 4691–4698, Copyright (2016), with consent from the ACS Nano Letters.

22.3.1.2 One-pot redox reaction and carbonization

In a typical synthesis, route GO was synthesized by the modified Hummers method, and the obtained GO was then dispersed in DI water by ultrasonication to prepare a GO suspension. Transition metal oxide (PW_{12}, PMo_{12}) was dissolved in a DI water by prolonging sonication. Then both of these GO suspension and metal oxide solution was mixed in a three-neck flask under strong ultrasonication. In the next step, the aqueous solution of pyrrole (Py) monomer was added slowly into this mixture,

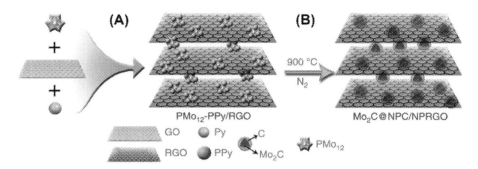

Figure 22.14 Schematic illustration of the synthetic process of Mo₂C@NPC/NPRGO.
(A) Synthesis of PMo₁₂−PPy/RGO via a green one-pot redox relay reaction. (B) Formation
of Mo₂C@NPC/NPRGO after carbonizing at 900°C.
Source: Reproduced with the permission from J.S. Li, Y. Wang, C.H. Liu, S.L. Li, Y.G.
Wang, L.Z. Dong, et al., Coupled molybdenum carbide and reduced graphene oxide
electrocatalysts for efficient hydrogen evolution. Nat. Commun. 7(1) (2016) 1−8, Copyright
(2016), with consent from the Nature Communications.

and the resultant mixture was kept under continuous stirring. Then this flask was
transferred to an oil bath of 60°C under continuous stirring for 1 day. Finally, when
a black precipitate appeared. This black color residue was then collected by vacuum
filtration and centrifugation and washed several times by DI water and alcohol. The
PW₁₂-PPy-RGO or PMo₁₂-PPy-RGO was obtained after 12 h vacuum drying
treatment.

In the second step, PW₁₂-PPy-RGO or PMo₁₂-PPy-RGO was taken in a quartz
crucible and heated at 800°C−900°C in a furnace for less than six hours. During
this carbonization step, a continuous flow of inert gas was maintained inside the
furnace. The obtained sample was then etched by 0.5 M H₂SO₄ for 5 h with continuous agitation at 80°C. During this step, all the unreacted, unstable and inactive
species were removed then the etched sample was washed with DI water until a
neutral pH was accomplished [70,71]. In Fig. 22.14, a schematic of the above synthesis is illustrated.

22.3.1.3 Coordination polymerization

Alex et al. reported the preparation of CP/PRGO electrocatalyst resulting from the
reaction of 1,2,4,5-benzenetetramine (BTA) ligand with palladium(II) chloride
(PdCl₂) in the presence of RGO, and the final material was marked as [Pd(BTA)-
RGO]ᵣₑ𝒹 [102]. In a typical reaction, 25 mL of PdCl₂ aqueous solution received
1 mL of NH₄OH and stirred for 10 min. Then a solution of 1,2,4,5-benzenetetra-
mine tetrahydrochloride was added and stirred for next 12 h at room temperature.
The resulting dark product was then centrifuged, washed, and dispersed in dry acetone. Later on, this dispersed product was refluxed for 24 h and finally collected
after drying in a vacuum oven. On the other hand, GO was prepared by the intercalation − exfoliation method using tetrabutylammonium hydroxide [103]. This GO

was then reduced by the help of hydrazine hydrate, and RGO was formed [104]. In the next step aqueous solution of ammonia (14 mol/L), $PdCl_2$ (1 equiv.) and the freshly prepared RGO (50 wt.%) were mixed and stirred. After 10 min of stirring that solution received 1.5 equivalent of BTA solution, after 12 h of stirring a black precipitate appeared. The resulting black product was dispersed in dry acetone and refluxed for 2 h and dried in a vacuum oven. Now, this [Pd(BTA)-RGO] was suspended in double-distilled water and sonicated for 30 min. Later on, this solution was allowed to cool to until it reached 0°C, and at this temperature, 2 M of $NaBH_4$ aqueous solution was added and stirred until it warmed up to room temperature. This reduction reaction was carried out for 6 h, and finally, the solid product of [Pd(BTA)-RGO]$_{red}$ was formed, which was then filtered out, washed with acetone and dried in the oven. A scheme of the most probable structure of the composite catalyst formed by Pd^{2+}-NH- bridging motifs is illustrated in Fig. 22.15.

22.3.2 Experimental technique of electrochemical hydrogen evolution

An electrochemical workstation with a three-electrode system is used for the electrochemical measurement in a 0.5 M H_2SO_4 aqueous solution. This three-electrode system consists of a working electrode, reference electrode and counter electrode. Generally, Ag/AgCl electrode (saturated KCl) or $Hg/HgSO_4$ are used as a reference electrode, and carbon rod or platinum sheet served the role of the counter electrode. A glassy carbon electrode (GCE) is used as a working electrode which is fabricated with the catalyst powder. For this fabrication process, a certain amount of catalyst powder is dispersed in an isopropanol/DI water mixture (1:4 v/v) along with specific percentages of Nafion by prolonging sonication. Then this mixture is drop-cast onto the GCE surface. Linear sweep voltammetry (LSV) analysis is carried out to record the HER, and the corresponding polarization curves are plotted without IR compensation. Cyclic voltammetry (CV) and electrochemical impedance spectroscopy (EIS) are used after calibrating the electrode by reversible hydrogen electrode (RHE) for the further study of HER.

22.3.3 Influence of different catalytic parameters on the hydrogen evolution reaction performances

22.3.3.1 Overpotential

The electrocatalytic activity of different polymer-graphene composites is assessed by the three-electrode set up in 0.5 M H_2SO_4, as mentioned earlier. The presence of graphene in polymer-graphene composite significantly elevates the conductivity and takes a vital role in determining the current density value (j) [102]. In general, practice, the current density value is kept at 10 mA/cm^2, and the corresponding over potential value is determined. This particular current density value is a significant reference because solar spectrum coupled HER instruments are mostly operated at $10-20$ mA/cm^2 under standard condition (1 Sun, AM 1.5) [15]. The LSV

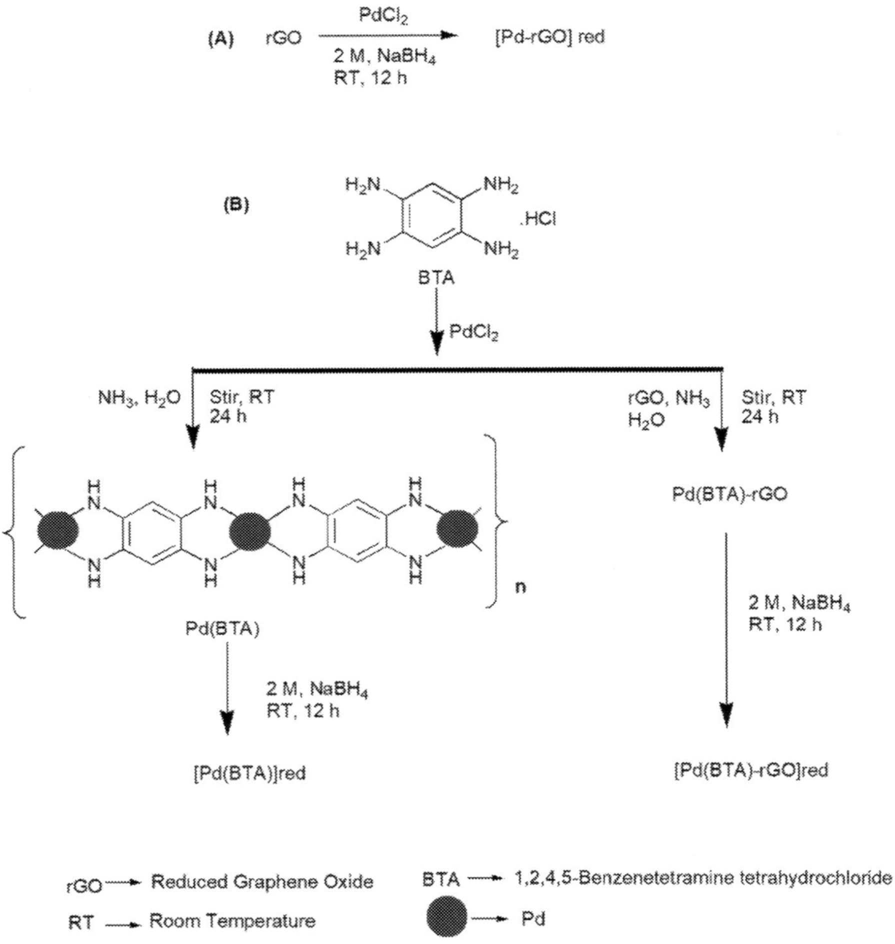

Figure 22.15 A schematic representation of the most probable composite catalyst structure formed by Pd^{2+}-NH- bridging motifs. Several closely related materials are also presented in this scheme for the activity comparison study.
Source: Reproduced with the permission from C. Alex, S.A. Bhat, N.S. John, C.V. Yelamaggad, Highly efficient and sustained electrochemical hydrogen evolution by embedded pd-nanoparticles on a coordination polymer—reduced graphene oxide composite. ACS Appl. Energy Mater. 2 (11) (2019) 8098−8106, Copyright (2019), with consent from the ACS Applied Energy Materials.

profile discloses the required over potential value to attain the 10 mA/cm^2 current density, and the materials with the lowest over potential value at that current density are considered as an appropriate electrocatalyst for HER (as shown in Figs. 22.16 and 22.17).

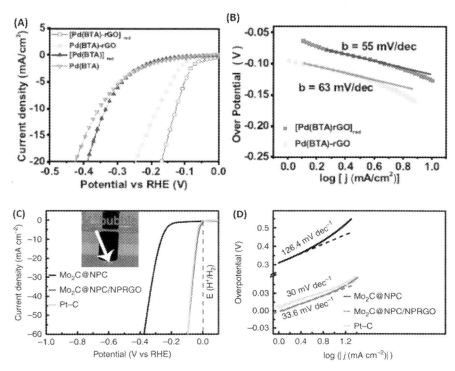

Figure 22.16 HER activity characterization. (A) LSV curves of materials Pd(BTA), Pd [(BTA)]red, Pd(BTA)-rGO, and [Pd(BTA)-rGO]red recorded for HER during the first cycle in 0.5 M H₂SO₄. (B) Tafel plot of Pd(BTA)-rGO and [Pd(BTA)-rGO]red.
Source: (B) Reproduced with the permission from C. Alex, S.A. Bhat, N.S. John, C.V. Yelamaggad, Highly efficient and sustained electrochemical hydrogen evolution by embedded pd-nanoparticles on a coordination polymer—reduced graphene oxide composite. ACS Appl. Energy Mater. 2 (11) (2019) 8098−8106, Copyright (2019), with consent from the ACS Applied Energy Materials. (C and D) Polarization curves and Tafel plots of Mo₂C@NPC, Mo₂C@NPC/NPRGO and Pt−C. (inset: the production of H₂ bubbles on the surface of Mo₂C@NPC/NPRGO). (C) Reproduced with the permission from J.S. Li, Y. Wang, C.H. Liu, S.L. Li, Y.G. Wang, L.Z. Dong, et al., Coupled molybdenum carbide and reduced graphene oxide electrocatalysts for efficient hydrogen evolution. Nat. Commun. 7(1) (2016) 1−8, Copyright (2016), with consent from the Nature Communications.

22.3.3.2 Tafel slope and exchange current density

Like over potential, exchange current density (j_o), and Tafel slope (b) are the two crucial parameters of HER. The explanation of HER mechanism is obtained from the Tafel plot by fitting it in Tafel equation (which is, $\eta = b\log(j) + a$, where "j" is the current density, and "b" is the Tafel slope). The exchange current density denotes the intrinsic catalytic behavior of the material, in ideal practice a catalytic material should have significantly smaller "b" and more immense "j_o" value.

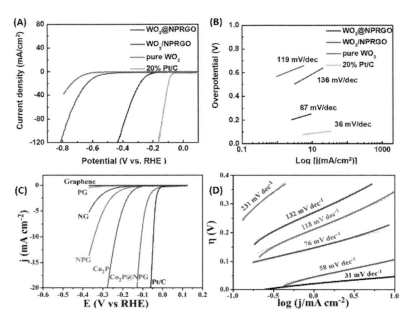

Figure 22.17 HER activity characterization. (A) Polarization curves and (B) Tafel plots of
WO_3@NPRGO WO_3/NPRGO, pure WO_3 and 20% Pt/C. (C) Polarization curves for HER in
0.5 M H_2SO_4 at a glassy carbon electrode modified with G-900, NG-900, PG-900, NPG-900,
Co_2P clusters, Co_2P@NPG-900, and 20 wt.% Pt/C, respectively. (D) Corresponding Tafel
plots derived from a.
Source: (B) Reproduced with the permission from G. Hu, J. Li, P. Liu, X. Zhu, X. Li,
R.N. Ali, et al., B. Enhanced electrocatalytic activity of WO_3@ NPRGO composite in a
hydrogen evolution reaction. Appl. Surf. Sci. 463 (2019) 275−282, Copyright (2019), with
consent from the Elsevier. (D) Reproduced with the permission from M. Zhuang, X. Ou,
Y. Dou, L. Zhang, Q. Zhang, R. Wu, et al., Polymer-embedded fabrication of Co2P
nanoparticles encapsulated in N, P-doped graphene for hydrogen generation. Nano Lett. 16
(7) (2016) 4691−4698, Copyright (2016), with consent from the ACS Nano Letters.

According to the classical theory of hydrogen evolution, the Tafel slope is
related to the rate-limiting step of electrocatalysis. There are three basic steps of
HER procedure in the acid electrolyte which are mainly:

$$H_3O^+ + e^- + M \rightarrow MH_{ads} + H_2O \quad (step\ 1)$$

$$MH_{ads} + H_2O + e^- \rightarrow H_2 + H_2O\uparrow \quad (step\ 2)$$

$$2MH_{ads} \rightarrow 2M + H_2\uparrow \quad (step\ 3)$$

Here, "M," and MH_{ads} denote the catalyst, and the hydrogen adsorbed intermediate.
This three steps of HER procedure are recognized as Volmer step (step 1, 120 mV/dec),

Heyrovsky step (step 2, 40 mV/dec), and Tafel step (step 3, 30 mV/dec). Volmer step is also known as discharge step, where a free proton is consumed in the active site of the photocatalyst surface. This is also the rate-limiting step. Next step is the electrochemical desorption step, where the free proton reacts with the MH_{ads} to produce hydrogen gas. This step is also considered as a rate-limiting step. The final step is the recombination step, where two MH_{ads} combine to release hydrogen gas. A typical HER goes through either reaction (1) and (2) and known as a Volmer- Heyrovsky mechanism or it may follow reaction (1) and (3), known as Volmer-Tafel mechanism. Sometimes these may happen simultaneously [9,70,71,105,106]. In the Table, the "j_o" and "b" are presented for different polymer-graphene composite.

Hu and his co-workers demonstrated that the "j_o" of WO_3@NPRGO (0.34 mA/cm^2) was superior to that WO_3/NPRGO (0.054 mA/cm^2) and pure WO_3 (0.00128 mA/cm^2) and also comparable with the commercial Pt/C electrode (0.39 mA/cm^2). Whereas the Tafel slope of WO_3@NPRGO (87 mV/dec) was relatively higher than Pt/C electrode (30 mV/dec) which suggested that the HER for WO_3@NPRGO composite followed Volmer-Heyrovsky mechanism. Polymer-graphene composite [Pd(BTA)$_{red}$.RGO]$_{red}$ (55 mV/dec) and [Pd(BTA)$_{red}$.RGO] (63 mV/dec) also possessed high "b" value, which indicated that the HER pathway follows the Volmer-Heyrovsky mechanism. This Tafel slope value of [Pd(BTA)$_{red}$.RGO]$_{red}$ and [Pd(BTA)$_{red}$.RGO] was comparatively lower than the other CP composite (presented in Table 22.2), which suggest faster kinetics for HER. Li et al. worked with Mo_2C@NPC/NPRGO composite, which showed higher "j_o" (1.09 mA/cm^2) and minimum "b" (33.6 mV/dec). Here the HER proceeded through Volmer-Tafel mechanism, where recombination was the rate-limiting step. Their results suggested that the Mo_2C@NPC/NPRGO/electrolyte interface was suitable for HER kinetics (as shown in Fig. 22.17A).

Here we tried to summarize all the over potential values at 10 mA/cm^2 current density, exchange current density and Tafel slope of different electrocatalysts.

22.3.3.3   Double-layer capacitance

Electrochemical surface area (ECSA) is also measured from electrochemical double-layer capacitance (Cdl). The CV curves of different composite materials are tested at different scan rates. The Cdl value is directly proportional with the ECSA, which means catalyst composite with higher Cdl value offers higher electrochemically active surface area. Higher ECSA, and exchange current density with the lower over potential, and Tafel slope value are the essential qualities of an ideal catalytic system. All the above polymer-graphene composites show higher Cdl value and thus suitable for the HER procedure (as shown in Fig. 22.18). The Cdl value of different composite materials is presented in Table 22.3.

22.3.3.4   Electrochemical impedance spectroscopy

The interfacial reaction and electrode kinetics in HER are further characterized by EIS. An equivalent circuit model is also used to fit the impedance data. The

Table 22.2 A comparison table of over potential, exchange current density and Tafel slope of different HER electrocatalyst in 0.5 M H_2SO_4 electrolyte.

Catalysts	Over potential at 10 mA/cm^2 (mV vs RHE)	Exchange current density (j_o) (mA/cm^2)	Tafel slope (b) (mV/ dec)	References
Pt/C	40	0.39	30	[71]
Pure WO$_3$	660	0.00128	119	[71]
WO$_3$/NPRGO	589	0.054	169	[71]
WO$_3$@NPRGO	255	0.34	87	[71]
Mo$_2$C@NPC	260	0.00316	126.4	[70]
Mo$_2$C@NPC/NPRGO	34	1.09	33.6	[70]
Pd(BTA)	355	–	–	[102]
Pd(BTA)$_{red}$	334	–	–	[102]
[Pd(BTA)$_{red}$.RGO]	178	–	63	[102]
[Pd(BTA)$_{red}$.RGO]$_{red}$	127	–	55	[102]
Pd-CNx	55		35	[107]
Pt@Pd NFS/RGO	56		39	[108]
PdPSRGO composite	90		46	[109]
Pd$_2$Si	192		131	[110]
Cu[BHT] coordination polymer	450		95	[111]
Zn^{+2}(fcdHP) coordination polymer	340		110	[112]
Co^{+2}(fcdHP) coordination polymer	450		1120	[112]
Co$_2$P@NPG	103	0.21	58	[101]
Co@NG	265	–	98	[113]
Co@NC	200	–	100	[114]
Co@NGF	124.6	–	93.9	[115]
Co@NC/NG	180	–	79.3	[116]
CoP	120	–	46	[117]
CoP/RGO	105	–	50	[118]
CoP/CNT	122	0.13	54	[119]
FeP	250	–	67	[120]
FeP/Ti	155	0.42	38	[121]
Co$_2$P/Ti	100	–	50	[122]

equivalent circuit is constructed with three parts, which are solution resistance (Rs), charge transfer resistance (Rct) and constant phase element. Zhuang et al. stated that "Rct" strongly depended upon the over potential. They showed that with increasing over potential value from 100 to 200 mV, the Rct value reduced to

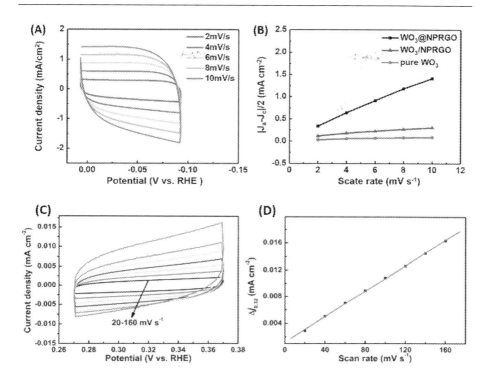

Figure 22.18 (A) The CV curves of WO_3@NPRGO at different scan rates. (B) The plots show the extraction of the double-layer capacitance of WO_3@NPRGO, WO_3/NPRGO, as well as pure WO_3. (C) CVs of Mo_2C@NPC with different rates from 20 to 160 mV/s. (D) The capacitive current at 0.32 V as a function of scan rate for Mo_2C@NPC.
Source: (B) Reproduced with the permission from G. Hu, J. Li, P. Liu, X. Zhu, X. Li, R.N. Ali, et al., B. Enhanced electrocatalytic activity of WO_3@ NPRGO composite in a hydrogen evolution reaction. Appl. Surf. Sci. 463 (2019) 275−282, Copyright (2019), with consent from the Elsevier. (D) Reproduced with the permission from J.S. Li, Y. Wang, C.H. Liu, S.L. Li, Y.G. Wang, L.Z. Dong, et al., Coupled molybdenum carbide and reduced graphene oxide electrocatalysts for efficient hydrogen evolution. Nat. Commun. 7 (1) (2016) 1−8, Copyright (2016), with consent from the Nature Communications.

347 to 28 Ω, which indicates that the faster HER process takes place in higher over potential value. They also observed with increasing dopant percentage (nitrogen, phosphorus) in graphene sheets the Rs value significantly increased. In their opinion, the incorporation of dopant reduced the graphene conductivity and increased the Rs. Hu et al. demonstrated that how robust contact between metal catalyst nanoparticle and RGO sheets influenced the Rct (as shown in Fig. 22.19A) [71]. In Table 22.4, all the Rs and Rct values of pure WO_3, WO_3/NPRGO and WO_3@NPRGO obtained from EIS over the frequency ranging from 1000kHz to 0.1 Hz at the −0.8 V are summarized. From the Table, it is seen that the robust

Table 22.3 A summary of electrochemical double-layer capacitance (Cdl) of various electrocatalyst in different scanning range.

Catalysts	Cdl (F/Cm^2)	Scanning range (V vs RHE)	References
[Pd(BTA)$_{red}$.RGO]$_{red}$	0.097	−0.08 to −0.3	[102]
[Pd(BTA)$_{red}$.RGO]	0.015	−0.08 to −0.3	[102]
Co$_2$P@NPG	0.066	+0.1 to +0.2	[101]
Mo$_2$C@NPC	0.092 × 10^{-3}	+0.27 to +0.37	[70]
Mo$_2$C@NPC/NPRGO	0.017	+0.27 to +0.37	[70]
WO$_3$@NPRGO	0.027	−0.1 to 0	[71]
WO$_3$/NPRGO	0.018	−0.1 to 0	[71]
Pure WO$_3$	0.089 × 10^{-3}	−0.1 to 0	[71]

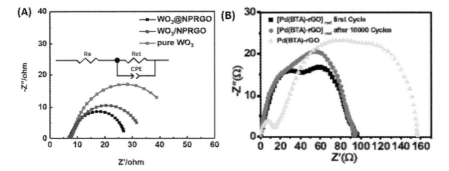

Figure 22.19 (A) Electrochemical impedance spectra (EIS) of pure WO$_3$ and WO$_3$@NPRGO over the frequency ranging from 1000kHz to 0.1 Hz at the −0.8 V. The inset image showing equivalent circuit. (B) Impedance spectra for [Pd(BTA)-rGO]red and Pd (BTA)-rGO after various numbers of cycles.
Source: (A) Reproduced with the permission from G. Hu, J. Li, P. Liu, X. Zhu, X. Li, R.N. Ali, et al., B. Enhanced electrocatalytic activity of WO$_3$@ NPRGO composite in a hydrogen evolution reaction. Appl. Surf. Sci. 463 (2019) 275−282, Copyright (2019), with consent from the Elsevier. (B) Reproduced with the permission from X. Chen, S.S. Mao, Titanium dioxide nanomaterials: synthesis, properties, modifications, and applications. Chem. Rev. 107 (7) (2007) 2891−2959, Copyright (2019), with consent from the ACS Applied Energy Materials.

contact between WO$_3$ and RGO sheets significantly reduces the resistance during the charge transfer process.

22.3.4 Other important parameters that influence the electrocatalytic activity

Apart from the exchange-current density, over-potential, Tafel slope, ECSA, many parameters influence the electrocatalytic activity of the hybrid composite. The

Table 22.4 Solution resistance (Rs) and charge transfer resistance (Rct) values of different WO_3 catalyst.

Catalyst	Rs (Ω)	Rct (Ω)
WO_3	6.98	60.64
WO_3/NPRGO	7.30	40.23
WO_3@NPRGO	7.15	23.72

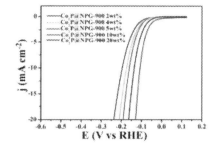

Figure 22.20 Polarization curves for HER in 0.5 M H_2SO_4 for Co_2P@NPG-900 with different Co content to observe the effect of metal content and annealing temperature on the electrocatalytic activity.
Source: Reproduced with the permission from M. Zhuang, X. Ou, Y. Dou, L. Zhang, Q. Zhang, R. Wu, et al., Polymer-embedded fabrication of Co2P nanoparticles encapsulated in N, P-doped graphene for hydrogen generation. Nano Lett. 16 (7) (2016) 4691−4698, Copyright (2016), with consent from the ACS Nano Letters.

extreme condition durability and the types of dopants, weight percentages of dopants, synthesis route [70,71], annealing temperature [101] are also the critical factors that should be taken into account. Here we discussed how these parameters regulate the electrocatalytic activity in HER.

22.3.4.1 Weight percentage of transition metal nanoparticles

Zhuang and co-workers suggested that only a certain weight percentage of metal catalyst could be incorporated into the polymer-graphene matrix during sintering, after crossing that limits aggregation took place [101]. Such aggregation effectively reduced the number of the active site per mass unit; as a result, the integral catalytic activity diminishes. They had observed the electrocatalytic activities of the as-synthesized Co_2P@ NPG composite toward HER with various Co content range from 2 to 20 wt.%. The nonzero cathodic currents were seen at the electrode modified by Co_2P@NPG in 0.5 M H_2SO_4 (as shown in Fig. 22.20). The lowest onset over potential was recorded −45 mV at 1 mA cm^{-2} exchange current density for the Co_2P@NPG with 10 wt.% Co loading [101].

Li et al. also investigated the effects of transition metal nanoparticles contents on the electrocatalytic performances. They used three different catalysts with dissimilar PMo_{12} contents (1.1, 2.2, and 3.3 g) for these purposes. Interestingly they found the optimum HER activity for the catalyst having 2.2 g of PMo_{12} contents among all the three sets under the same experimental conditions. The catalyst corresponding to the 2.2 g of PMo_{12} contents showed the lowest onset over potential and the smallest Tafel slope among the three samples (as shown in Fig. 22.21A and B). They explained these results in terms of the amount and distribution of active sites. The electrocatalyst with 1.1 g of PMo_{12} contents offered more inferior electrocatalytic activity because of the lower amount of transition metal nanoparticles. In contrast, electrocatalyst with 3.3 g of PMo_{12} contents a larger number of transition metal nanoparticles which tended to aggregate together and produce unfavorable HER. These results demonstrate that the amount of PMo_{12} had a noticeable impact on HER performances [70].

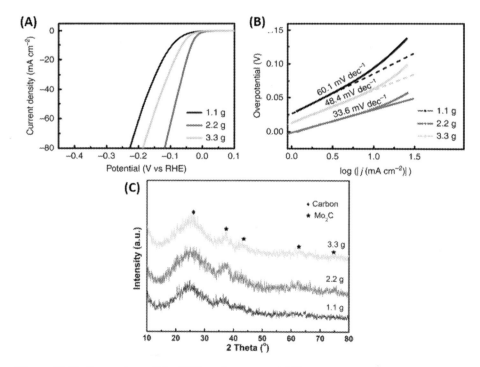

Figure 22.21 Comparison of the HER performance of different electrocatalysts. (A and B) Polarization curves and Tafel plots of $Mo_2C@NPC/NPRGO$ with the different mass of PMo_{12} (1.1, 2.2 and 3.3 g). (C) PXRD patterns of $Mo_2C@NPC/NPRGO$ (1.1, 2.2, and 3.3). *Source*: Reproduced with the permission from J.S. Li, Y. Wang, C.H. Liu, S.L. Li, Y.G. Wang, L.Z. Dong, et al., Coupled molybdenum carbide and reduced graphene oxide electrocatalysts for efficient hydrogen evolution. Nat. Commun. 7 (1) (2016) 1−8, Copyright (2016), with consent from the Nature Communications.

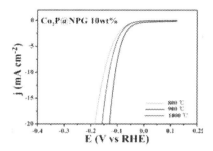

Figure 22.22 Polarization curves for HER in 0.5 M H₂SO₄ for Co₂P@NPG-900 10 wt.% prepared under various annealing temperature to observe the effect of metal content and annealing temperature on the electrocatalytic activity.
Source: Reproduced with the permission from M. Zhuang, X. Ou, Y. Dou, L. Zhang, Q. Zhang, R. Wu, et al., Polymer-embedded fabrication of Co2P nanoparticles encapsulated in N, P-doped graphene for hydrogen generation. Nano Lett. 16 (7) (2016) 4691–4698, Copyright (2016), with consent from the ACS Nano Letters.

22.3.4.2 Annealing temperature of the composite

The annealing temperature of the composite is also essential, it is noticed that a relatively stable crystal structure is formed at higher annealing temperature, but at the same time, it sacrifices the relative content of nitrogen, phosphorus and the metal catalyst. Therefore, optimization of annealing temperature is an essential factor. Zhuang et al. demonstrated that when Co₂P@NPG samples were annealed at 800°C, 900°C and 1000°C, the required over potential value was lowest for the Co₂P@NPG-900 sample to attain current density of 20 mA/cm². They had also presented the polarization curve for HER at a GCE modified with G-900, NG-900, PG-900, NPG-900, Co₂P clusters, Co₂P@NPG-900, and 20 wt.% Pt/C, respectively (Fig. 22.22).

22.3.4.3 Carbonization temperature of the composite

Selection of correct carbonization temperature is also very impactful for the better electrocatalytic activity as the availability of the active sites directly related to the optimization temperature. Li et al. studied the effects of carbonization temperature on HER activity. They prepared three sets of composite catalysts by varying the carbonization temperatures at 700°C, 900°C, and 1100°C. Among these catalysts, the catalyst carbonized at 900°C displayed the optimal HER activity. Possibly the active sites of Mo₂C were not produced for the composite carbonized at 700°C. On the other hand, the high carbonization temperature led to substantial sintering and aggregation of Mo₂C NPs, which further reduced the density of highly active sites. Meanwhile, the N content diminished with increasing carbonization temperature (Table 22.5). All of these results were consistent with the SEM, TEM, and XRD results (Fig. 22.23A−C). Therefore, all these results suggest that the selection

Table 22.5 Comparison of catalytic parameters and atomic percentages (obtained from XPS measurements) of different HER catalysts carbonized at 700°C, 900°C and 1100°C, respectively.

Carbonization temperature (°C)	Over potential[a] (mV vs RHE)	Tafel slope (b) (mV/dec)	C (at. %)	N (at. %)	Mo (at. %)	P (at. %)
700	20	59.7	85.86	2.59	0.64	0.43
900	**34**	**33.6**	**87.76**	**1.82**	**2.5**	**0.39**
1100	27	70.1	88.26	1.31	2.43	0.28

[a]Over potential value is measured at 10 mA/cm^2 current density, at.% = Atomic percent.
The bold values imply the catalyst carbonized at 900°C displayed the optimal HER activity among any other carbonization temperature.

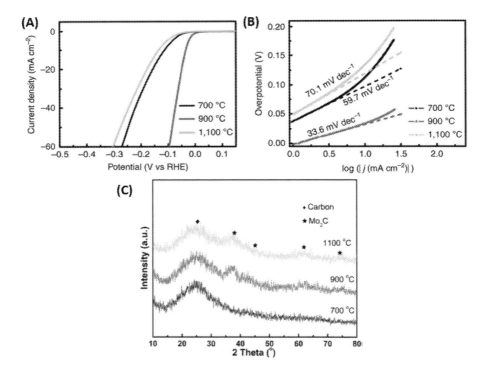

Figure 22.23 (A and B) Polarization curves and Tafel plots of Mo_2C@NPC/NPRGO (2.2 g) at different carbonization temperature. (C) PXRD of PMo_{12}-PPy/RGO-700, Mo_2C@NPC/NPRGO, and Mo_2C@NPC/NPRGO-1100.
Source: Reproduced with the permission from J.S. Li, Y. Wang, C.H. Liu, S.L. Li, Y.G. Wang, L.Z. Dong, et al., Coupled molybdenum carbide and reduced graphene oxide electrocatalysts for efficient hydrogen evolution. Nat. Commun. 7 (1) (2016) 1−8, Copyright (2016), with consent from the Nature Communications.

of the carbonization temperature is a critical factor for the production of high-HER active sites.

22.3.4.4 Types of mixing

Hu et al. reported that the polymer-graphene composite prepared by physical mixing technique was less catalytically active over the chemically modified one. In their view, the dispersion of metal catalyst was non-uniform when they mixed physically, and the number of active sites reduced, which lower down the catalytic performance [71].

22.3.4.5 Stability and durability

Apart from the catalytic activity, the stability of the catalyst is also an essential aspect for HER process. Continuous CV is performed for 1000 or 10,000 continual cycles at a fixed current density in a particular potential range (RHE scale) for the evaluation of the polymer-graphene composite durability. It is worth mentioning that the current density during the durability experiment should not be too high, because the evolution of hydrogen in the form of bubbles mask the HER activity. Alex and co-workers explained how the organic repeating units and the encapsulation of metal catalyst in the RGO influenced the catalytic activity and durability of the polymer-graphene composite ([Pd(BTA)$_{red}$.RGO]$_{red}$). In order to test the contribution of the organic regime, they had studied the LSV profile of the electrocatalyst [Pd(BTA)$_{red}$.RGO]$_{red}$ and Pd-RGO for 10,000 continual cycles at a -85 mA/cm^2 current density in a potential range 0 to -0.36 V (RHE scale). They found that the catalytic activity and the HER efficiency for the [Pd(BTA)$_{red}$.RGO]$_{red}$ catalyst remained unchanged up to 10,000 cycles. However, the over potential observed for Pd-RGO increased from -280 to 350 mV with changing from 5000 to 7000 cycles, which indicates that the organic portion contributes toward the stability of [Pd(BTA)$_{red}$.RGO]$_{red}$ catalyst (as shown in Fig. 22.24A) [102]. Li et al. and Hu et al. showed that the polarization curves for Mo$_2$C@NPC/NPRGO and WO$_3$@NPRGO polymer-graphene composite remained almost the same, after 1000 cycles. According to Hu et al. the long-term stability of the WO$_3$@NPRGO electrocatalyst was because of RGO sheets, which provided substantial support and it protected the structure from bubble corrosion (as shown in Fig. 22.24B) [71]. Li et al. also demonstrated that the electrocatalyst suffers a negligible loss in current density at a static over potential (as shown in Fig. 22.24C) [70].

Zhuang et al. also recorded a minor deviation in the HER activity even after 10,000 cycles for the Co$_2$P@NPG electrocatalyst. Further, they had examined the stability of the electrocatalyst at two fixed over potentials (-100 and -150 mV) over an extended period (30 h). The outcome of the results was entirely satisfactory and suggested the long term sustainability of the Co$_2$P@NPG electrocatalyst in the HER. The durability of the electrocatalyst in the harsh condition was also studied in the media with a pH range from 0 to 14 (shown in Fig. 22.25A−D) [101].

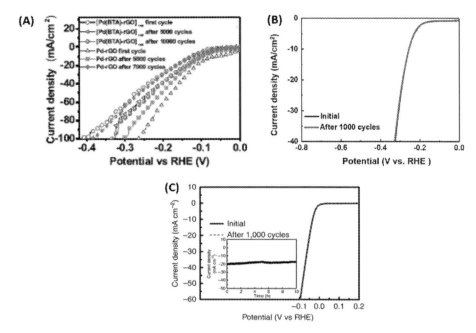

Figure 22.24 (A) Comparison of LSV curves of [Pd(BTA)-rGO]red with those of Pd-rGO obtained for different numbers of cycles. (B) Polarization curves of WO_3@NPRGO at the initial stage and after 1000 CV cycles. (C) Polarization curves of Mo_2C@NPC/NPRGO initially and after 1000 CV cycles. Inset: Time-dependent current density curve of Mo_2C@NPC/NPRGO under a static over potential of 48 mV for 10 h.
Source: (A) Reproduced with the permission from C. Alex, S.A. Bhat, N.S. John, C.V. Yelamaggad, Highly efficient and sustained electrochemical hydrogen evolution by embedded pd-nanoparticles on a coordination polymer—reduced graphene oxide composite. ACS Appl. Energy Mater. 2 (11) (2019) 8098−8106, Copyright (2019), with consent from the ACS Applied Energy Materials. (B)Reproduced with the permission from G. Hu, J. Li, P. Liu, X. Zhu, X. Li, R.N. Ali, et al., B. Enhanced electrocatalytic activity of WO3@ NPRGO composite in a hydrogen evolution reaction. Appl. Surf. Sci. 463 (2019) 275−282, Copyright (2019), with consent from the Elsevier. Reproduced with the permission from G. Hu, J. Li, P. Liu, X. Zhu, X. Li, R.N. Ali, et al., B. Enhanced electrocatalytic activity of WO3@ NPRGO composite in a hydrogen evolution reaction. Appl. Surf. Sci. 463 (2019) 275−282, Copyright (2019), with consent from the Elsevier. (C) Reproduced with the permission from J.S. Li, Y. Wang, C.H. Liu, S.L. Li, Y.G. Wang, L.Z. Dong, et al., Coupled molybdenum carbide and reduced graphene oxide electrocatalysts for efficient hydrogen evolution. Nat. Commun. 7 (1) (2016) 1−8, Copyright (2016), with consent from the Nature Communications.

Co_2P@NPG electrocatalyst possesses lower over potential (61 mV at 1 mA/cm^2) with a very low Tafel slope (96 mV/dec) in the extreme alkaline media (1 M KOH, pH ∼ 14). This result is quite impressive when compared with other HER electrocatalysts. In Table 22.6, a comparative study of HER performance of the different catalyst in 1 M KOH is summarized. Finally, they observed the morphologies of

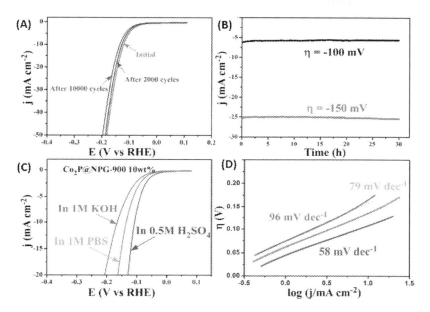

Figure 22.25 (A) The polarization curves of $Co_2P@NPG$-900 of CV cycles. (B) Time dependence of cathodic current density over $Co2P@NPG$-900 during electrolysis at over potentials of 100 and 150 mV. (C) Polarization curves of $Co_2P@NPG$-900 in 0.5 M H_2SO_4 (pH = 0.30 ± 0.05), 1 M PBS (pH = 7.23 ± 0.07), and 1 M KOH (pH = 13.94 ± 0.04) electrolytes, respectively, and (D) the corresponding Tafel plots of C.
Source: Reproduced with the permission from M. Zhuang, X. Ou, Y. Dou, L. Zhang, Q. Zhang, R. Wu, et al., Polymer-embedded fabrication of Co2P nanoparticles encapsulated in N, P-doped graphene for hydrogen generation. Nano Lett. 16 (7) (2016) 4691−4698, Copyright (2016), with consent from the ACS Nano Letters.

Table 22.6 A comparison table of over potential, exchange current density and Tafel slope of different HER electrocatalyst in 1 M KOH electrolyte.

Catalysts	Exchange current density (j_o) (mA/cm²)	Over potential 10 mA/cm² (mV vs RHE)	Tafel slope (*b*) (mV/dec)	References
$Co_2P@NPG$	10	165	96	[101]
Co@BCN	10	183	73.2	[123]
Co@NC	10	210	108	[114]
Co@NG	10	337	—	[113]
CoP/RGO	10	209	129	[118]
FeP	10	218	146	[120]
WP/Carbon	10	170	102	[124]

Co$_2$P@NPG electrocatalyst before and after 1000 cycles testing in the 1 M KOH (Fig. 22.25C). The size and distribution of Co$_2$P nanoparticle along with the graphene layer orientation were found unharmed. These results also support the durability of the Co$_2$P@NPG electrocatalyst in extreme condition

22.4 Summary and future perspectives

In summary, the polymer-graphene structure provided the necessary support to the transition metal photocatalyst. The whole structure could effectively restrict the aggregation of graphene sheets. All these catalysts showed excellent PCt activity under visible light irradiation. Properties of graphene, polymer, and transition metal catalyst strongly influenced the efficiency of the PCt system for HER. The heterojunction formed between the transition metal catalyst and graphene promoted the electrons transfer process and reacted with protons on the active sites on graphene. As a result, the recombination of the charge carriers was inhibited. The quantum size effect caused the widening of the bandgap of transition metal catalyst and the increasing of the thermodynamic driving force. Under the inclusive function of the above factors, the PCt hydrogen production activity was enhanced.

In the case of polymer-graphene electrocatalyst, we can say that polymer-graphene composite was synthesized by a simple and economically feasible synthesis route. The metal catalyst was trapped inside the polymer matrix during polymerization. RGO-supported metal catalyst dispersed in the polymer framework showed better resistance against the aggregation of RGO and catalyst and also prevented the leaching out of the embedded catalyst in the solution during electrolysis. All the polymer-graphene composite showed outstanding performance in HER process. The origin of the vigorous electrocatalytic activity for HER is as follows:

- Small-sized metal catalyst nanoparticles were homogeneously dispersed in the RGO-polymeric framework, which offered the exposure of an abundance of active sites and enhanced the catalytic activity [70,83,90,125].
- The incorporation of heteroatom dopants (N, P) in the RGO results in charge density distribution and asymmetry spin, which improves the interaction with H$^+$, especially, pyridinic N is favorable for HER performances.
- During the carbonization process, heteroatom-doped RGO was dispersed homogeneously in the polymeric network. This resultant system provided outstanding electrical conductivity and accelerated the charge transfer speed and reaction rate.
- The intercalation of metal and polymeric network restricted the RGO aggregation and provided a large surface area.
- The robust conjugation between the metal catalyst and RGO provided a resistance less path for fast electron transfer.
- The unique structure of polymer-graphene composite catalyst was also suitable for the speedy mass transports of reactant and facilitate electron transfer [70,93,126].

Hence the synergistic catalytic effect of polymer-graphene composite along with the factors as mentioned above is the primary origin of excellent catalytic activity for HER process. However, the development of inexpensive and scalable synthesis route for large scale practical applications should be the subject of future research work as the performance of graphene polymer composite cannot match that of noble metal-based counterparts. Significant challenges still need to be overcome to attain practical submissions. Here we have mentioned some challenges that require immediate attention:

First of all, the stability and durability of polymer-graphene catalyst still lies far below the practical requirement. Most of the time polymer-graphene catalyst cannot survive in the harsh electrolytic conditions and shows phase transition, aggregation, and leaching. Further, a study is required for the hybrid polymer-graphene systems to improve the PCt activity under solar light absorption. Apart from all of these understanding, the water-splitting process in the atomic level by the help of catalyst is still not extensively explored. Some research groups tried to provide a meaningful DFT calculation to describe the intermediates' reactions mechanism. However, all these DFT calculations do not introduce realistic reaction environments and complicated surface states of the catalysts. This DFT is calculated by merely considering the adsorption of the proton, formation of surface adsorbed H^* and desorption of hydrogen on the selected surface of the catalysts. Therefore, a computational methodology should be introduced to replicate the whole catalytic system. In situ detection techniques for water, electrolysis is also a critical factor for understanding the mechanistic aspect. It provides more information about the surface chemical and structural evolution of electrocatalysts during electrolysis. The development of such theoretical methods will be beneficial to understand the catalytic mechanism for solar water splitting and hydrogen production. Whereas, the in situ detection techniques will be crucial to fabricate polymer-graphene composite with high spatial distributions and desired structures.

The aforementioned problems suggest the cohabitation of immense prospects and incredible challenges in this area. The combination of experimental techniques and theoretical calculations should be the focus of research in the future to address these issues.

References

[1] M.I. Fadlalla, P.S. Kumar, V. Selvam, S.G. Babu, Emerging energy and environmental application of graphene and their composites: a review, J. Mater. Sci. (2020) 1—28.
[2] A.K. Das, M. Manohar, V.K. Shahi, Acid resistant sulphonated poly (vinylidene fluoride-co-hexafluoropropylene)/graphene oxide composite cation exchange for water splitting by iodine-sulfur bunsen process for hydrogen production, J. Membr. Sci. 552 (2018) 377—386.
[3] Y.S. Sanusi, E.M. Mokheimer, M.R. Shakeel, Z. Abubakar, M.A. Habib, Oxy-combustion of hydrogen-enriched methane: experimental measurements and analysis, Energ. Fuel. 31 (2) (2017) 2007—2016.
[4] A. Goeppert, M. Czaun, G.S. Prakash, G.A. Olah, Air as the renewable carbon source of the future: an overview of CO_2 capture from the atmosphere, Energ. Environ. Sci. 5 (7) (2012) 7833—7853.

[5] P. Markewitz, W. Kuckshinrichs, W. Leitner, J. Linssen, P. Zapp, R. Bongartz, et al., Worldwide innovations in the development of carbon capture technologies and the utilization of CO_2, Energ. Environ. Sci. 5 (6) (2012) 7281−7305.

[6] N. MacDowell, N. Florin, A. Buchard, J. Hallett, A. Galindo, G. Jackson, et al., An overview of CO_2 capture technologies, Energ. Environ. Sci. 3 (11) (2010) 1645−1669.

[7] Z.W. Seh, J. Kibsgaard, C.F. Dickens, I.B. Chorkendorff, J.K. Nørskov, T.F. Jaramillo, Combining theory and experiment in electrocatalysis: insights into materials design, Science 355 (6321) (2017).

[8] C. Tan, X. Cao, X.J. Wu, Q. He, J. Yang, X. Zhang, et al., Recent advances in ultrathin two-dimensional nanomaterials, Chem. Rev. 117 (9) (2017) 6225−6331.

[9] M.S. Faber, S. Jin, Earth-abundant inorganic electrocatalysts and their nanostructures for energy conversion applications, Energ. Environ. Sci. 7 (11) (2014) 3519−3542.

[10] R. Bashyam, P. Zelenay, A class of non-precious metal composite catalysts for fuel cells, Nature 443 (7107) (2006) 63−66.

[11] H. Ashassi-Sorkhabi, B. Rezaei-Moghadam, E. Asghari, R. Bagheri, Z. Hosseinpour, Fabrication of bridge like Pt@ MWCNTs/CoS₂ electrocatalyst on conductive polymer matrix for electrochemical hydrogen evolution, Chem. Eng. 308 (2017) 275−288.

[12] A. Hassan, V.A. Paganin, E.A. Ticianelli, Pt modified tungsten carbide as anode electrocatalyst for hydrogen oxidation in proton exchange membrane fuel cell: CO tolerance and stability, Appl. Catal. 165 (2015) 611−619.

[13] J.A. Turner, A realizable renewable energy future, Science 285 (5428) (1999) 687−689.

[14] O. Akhavan, Graphene nanomesh by ZnO nanorod photocatalysts, ACS Nano. 4 (7) (2010) 4174−4180.

[15] M.G. Walter, E.L. Warren, J.R. McKone, S.W. Boettcher, Q. Mi, E.A. Santori, et al., Solar water splitting cells, Chem. Rev. 110 (11) (2010) 6446−6473.

[16] M.S. Prévot, K. Sivula, Photoelectrochemical tandem cells for solar water splitting, J. Phys. Chem. C 117 (35) (2013) 17879−17893.

[17] K. Maeda, Z-scheme water splitting using two different semiconductor photocatalysts, ACS Catal. 3 (7) (2013) 1486−1503.

[18] P. Millet, F. Andolfatto, R. Durand, Design and performance of a solid polymer electrolyte water electrolyzer, Int. J. Hydrogen Energ. 21 (2) (1996) 87−93.

[19] W. Kreuter, H. Hofmann, Electrolysis: the important energy transformer in a world of sustainable energy, Int. J. Hydrogen Energ. 23 (8) (1998) 661−666.

[20] T. Pregger, D. Graf, W. Krewitt, C. Sattler, M. Roeb, S. Möller, Prospects of solar thermal hydrogen production processes, Int. J. Hydrogen Energ. 34 (10) (2009) 4256−4267.

[21] A. Fujishima, K. Honda, Electrochemical photolysis of water at a semiconductor electrode, Nature 238 (1972) 37.

[22] C.J. Li, G.R. Xu, B. Zhang, J.R. Gong, High selectivity in visible-light-driven partial photocatalytic oxidation of benzyl alcohol into benzaldehyde over single-crystalline rutile TiO_2 nanorods, Appl. Catal. 115 (2012) 201−208.

[23] C.J. Li, J.N. Wang, B. Wang, J.R. Gong, Z. Lin, A novel magnetically separable TiO_2/ $CoFe_2O_4$ nanofiber with high photocatalytic activity under UV−vis light, Mater. Res. Bull. 47 (2) (2012) 333−337.

[24] C.J. Li, J.N. Wang, B. Wang, J.R. Gong, Z. Lin, Direct formation of reusable TiO_2/ $CoFe_2O_4$ heterogeneous photocatalytic fibers via two-spinneret electrospinning, J. Nanosci. Nanotechnol. 12 (3) (2012) 2496−2502.

[25] K. Domen, A. Kudo, T. Onishi, Mechanism of photocatalytic decomposition of water into H_2 and O_2 over NiO.SrTiO₃, J. Catal. 102 (1) (1986) 92−98.

[26] H. Kato, A. Kudo, Highly efficient decomposition of pure water into H_2 and O_2 over $NaTaO_3$ photocatalysts, Catal. Lett. 58 (2−3) (1999) 153−155.
[27] A. Kudo, H. Kato, Effect of lanthanide-doping into $NaTaO_3$ photocatalysts for efficient water splitting, Chem. Phys. Lett. 331 (5−6) (2000) 373−377.
[28] K. Takanabe, K. Domen, Toward visible light response: overall water splitting using heterogeneous photocatalysts, Green 1 (5−6) (2011) 313−322.
[29] J. Chen, D. Yang, D. Song, J. Jiang, A. Ma, M.Z. Hu, et al., Recent progress in enhancing solar-to-hydrogen efficiency, J. Power Sources. 280 (2015) 649−666.
[30] K. Sivula, F. Le Formal, M. Grätzel, Solar water splitting: progress using hematite (α-Fe_2O_3) photoelectrodes, ChemSusChem 4 (4) (2011) 432−449.
[31] I. Oh, J. Kye, S. Hwang, Enhanced photoelectrochemical hydrogen production from silicon nanowire array photocathode, Nano Lett. 12 (1) (2012) 298−302.
[32] D.W. Chang, J.B. Baek, Nitrogen-doped graphene for photocatalytic hydrogen generation, Chem. Asian J. 11 (8) (2016) 1125−1137.
[33] X. Chen, S. Shen, L. Guo, S.S. Mao, Semiconductor-based photocatalytic hydrogen generation, Chem. Rev. 110 (11) (2010) 6503−6570.
[34] A. Kudo, Y. Miseki, Heterogeneous photocatalyst materials for water splitting, Chem. Soc. Rev. 38 (1) (2009) 253−278.
[35] G. Xie, K. Zhang, B. Guo, Q. Liu, L. Fang, J.R. Gong, Graphene-based materials for hydrogen generation from light-driven water splitting, Adv. Mater. 25 (28) (2013) 3820−3839.
[36] M. Ni, M.K. Leung, D.Y. Leung, K. Sumathy, A review and recent developments in photocatalytic water-splitting using TiO_2 for hydrogen production, Renew. Sust. Energ. Rev. 11 (3) (2007) 401−425.
[37] H. Katsumata, Y. Tachi, T. Suzuki, S. Kaneco, Z-scheme photocatalytic hydrogen production over WO_3/g-C_3N_4 composite photocatalysts, RSC Adv. 4 (41) (2014) 21405−21409.
[38] Z. Hu, C.Y. Jimmy, Pt3 Co-loaded CdS and TiO_2 for photocatalytic hydrogen evolution from water, J. Mater. Chem. 1 (39) (2013) 12221−12228.
[39] Z.J. Li, J.J. Wang, X.B. Li, X.B. Fan, Q.Y. Meng, K. Feng, et al., An exceptional artificial photocatalyst, Nih-CdSe/CdS core/shell hybrid, made in situ from cdse quantum dots and nickel salts for efficient hydrogen evolution, Adv. Mater. 25 (45) (2013) 6613−6618.
[40] J. Hou, Z. Wang, W. Kan, S. Jiao, H. Zhu, R.V. Kumar, Efficient visible-light-driven photocatalytic hydrogen production using CdS@ TaON core−shell composites coupled with graphene oxide nanosheets, J. Mater. Chem. 22 (15) (2012) 7291−7299.
[41] Q. Liang, G. Jiang, Z. Zhao, Z. Li, M.J. MacLachlan, CdS-decorated triptycene-based polymer: durable photocatalysts for hydrogen production under visible-light irradiation, Catal. Sci. 5 (6) (2015) 3368−3374.
[42] K. Maeda, K. Domen, Photocatalytic water splitting: recent progress and future challenges, J. Phys. Chem. Lett. 1 (18) (2010) 2655−2661.
[43] M.A. Fox, M.T. Dulay, Heterogeneous photocatalysis, Chem. Rev. 93 (1) (1993) 341−357.
[44] A. Fujishima, X. Zhang, D.A. Tryk, Heterogeneous photocatalysis: from water photolysis to applications in environmental cleanup, Int. J. Hydrogen Energ. 32 (14) (2007) 2664−2672.
[45] P.V. Kamat, Meeting the clean energy demand: nanostructure architectures for solar energy conversion, J. Phys. Chem. C 111 (7) (2007) 2834−2860.
[46] A.J. Esswein, D.G. Nocera, Hydrogen production by molecular photocatalysis, Chem. Rev. 107 (10) (2007) 4022−4047.

[47] H. Tong, S. Ouyang, Y. Bi, N. Umezawa, M. Oshikiri, J. Ye, Nano-photocatalytic materials: possibilities and challenges, Adv. Mater. 24 (2) (2012) 229−251.

[48] X. Chen, S.S. Mao, Titanium dioxide nanomaterials: synthesis, properties, modifications, and applications, Chem. Rev. 107 (7) (2007) 2891−2959.

[49] A.J. Cowan, J.R. Durrant, Long-lived charge separated states in nanostructured semiconductor photoelectrodes for the production of solar fuels, Chem. Soc. Rev. 42 (6) (2013) 2281−2293.

[50] F.E. Osterloh, Inorganic nanostructures for photoelectrochemical and photocatalytic water splitting, Chem. Soc. Rev. 42 (6) (2013) 2294−2320.

[51] K.S. Novoselov, A.K. Geim, S.V. Morozov, D. Jiang, Y. Zhang, S.V. Dubonos, et al., Electric field effect in atomically thin carbon films, Science 306 (5696) (2004) 666−669.

[52] S. Guo, S. Dong, Graphene nanosheet: synthesis, molecular engineering, thin film, hybrids, and energy and analytical applications, Chem. Soc. Rev. 40 (5) (2011) 2644−2672.

[53] F. Bonaccorso, Z. Sun, T.A. Hasan, A.C. Ferrari, Graphene photonics and optoelectronics, Nat. Photonics 4 (9) (2010) 611.

[54] D. Chen, H. Zhang, Y. Liu, J. Li, Graphene and its derivatives for the development of solar cells, photoelectrochemical, and photocatalytic applications, Energ. Environ. Sci. 6 (5) (2013) 1362−1387.

[55] P.V. Kamat, Graphene-based nanoassemblies for energy conversion, J. Phys. Chem. Lett. 2 (3) (2011) 242−251.

[56] Z. Yin, J. Zhu, Q. He, X. Cao, C. Tan, H. Chen, et al., Graphene-based materials for solar cell applications, Adv. Energy Mater. 4 (1) (2014) 1300574.

[57] Y. Sun, Q. Wu, G. Shi, Graphene based new energy materials, Energ. Environ. Sci. 4 (4) (2011) 1113−1132.

[58] T. Xian, H. Yang, L. Di, J. Ma, H. Zhang, J. Dai, Photocatalytic reduction synthesis of $SrTiO_3$-graphene nanocomposites and their enhanced photocatalytic activity, Nanoscale Res. Lett. 9 (1) (2014) 1−9.

[59] Q. Li, H. Meng, J. Yu, W. Xiao, Y. Zheng, J. Wang, Enhanced photocatalytic hydrogen-production performance of graphene−ZnxCd1 − xS composites by using an organic S source, Chem. - Eur J. 20 (4) (2014) 1176−1185.

[60] D. Higgins, P. Zamani, A. Yu, Z. Chen, The application of graphene and its composites in oxygen reduction electrocatalysis: a perspective and review of recent progress, Energ. Environ. Sci. 9 (2) (2016) 357−390.

[61] Q. Li, B. Guo, J. Yu, J. Ran, B. Zhang, H. Yan, et al., Highly efficient visible-light-driven photocatalytic hydrogen production of CdS-cluster-decorated graphene nanosheets, J. Am. Chem. Soc. 133 (28) (2011) 10878−10884.

[62] Q. Xiang, J. Yu, M. Jaroniec, Graphene-based semiconductor photocatalysts, Chem. Soc. Rev. 41 (2) (2012) 782−796.

[63] C. Kong, S. Min, G. Lu, Dye-sensitized NiSx catalyst decorated on graphene for highly efficient reduction of water to hydrogen under visible light irradiation, ACS Catal. 4 (8) (2014) 2763−2769.

[64] J. Liu, S. Xu, L. Liu, D.D. Sun, The size and dispersion effect of modified graphene oxide sheets on the photocatalytic H_2 generation activity of TiO_2 nanorods, Carbon 60 (2013) 445−452.

[65] M. Zhu, Z. Li, B. Xiao, Y. Lu, Y. Du, P. Yang, et al., Surfactant assistance in improvement of photocatalytic hydrogen production with the porphyrin noncovalently functionalized graphene nanocomposite, ACS Appl. Mater. Interfaces 5 (5) (2013) 1732−1740.

[66] J. Luo, J. Kim, J. Huang, Material processing of chemically modified graphene: some challenges and solutions, Acc. Chem. Res. 46 (10) (2013) 2225−2234.

[67] T. Kuila, S. Bose, A.K. Mishra, P. Khanra, N.H. Kim, J.H. Lee, Chemical functionalization of graphene and its applications, Prog. Mater. Sci. 57 (7) (2012) 1061−1105.

[68] J. Oh, J.H. Lee, J.C. Koo, H.R. Choi, Y. Lee, T. Kim, et al., Graphene oxide porous paper from amine-functionalized poly (glycidyl methacrylate)/graphene oxide core-shell microspheres, J. Mater. Chem. 20 (41) (2010) 9200−9204.

[69] H. Ashassi-Sorkhabi, B. Rezaei-moghadam, Ultrasound-assisted synthesis of PPyCuS@ GOPt nanocomposite and investigation of its electrocatalytic behavior towards photo-hydrogen evolution, J. Environ. 5 (3) (2017) 2448−2458.

[70] J.S. Li, Y. Wang, C.H. Liu, S.L. Li, Y.G. Wang, L.Z. Dong, et al., Coupled molybdenum carbide and reduced graphene oxide electrocatalysts for efficient hydrogen evolution, Nat. Commun. 7 (1) (2016) 1−8.

[71] G. Hu, J. Li, P. Liu, X. Zhu, X. Li, R.N. Ali, et al., Enhanced electrocatalytic activity of WO_3@ NPRGO composite in a hydrogen evolution reaction, Appl. Surf. Sci. 463 (2019) 275−282.

[72] H.H. Zhang, K. Feng, B. Chen, Q.Y. Meng, Z.J. Li, C.H. Tung, et al., Water-soluble sulfonated−graphene−platinum nanocomposites: facile photochemical preparation with enhanced catalytic activity for hydrogen photogeneration, Catal. Sci. Technol. 3 (7) (2013) 1815−1821.

[73] J. Xu, L. Wang, X. Cao, Polymer supported graphene−CdS composite catalyst with enhanced photocatalytic hydrogen production from water splitting under visible light, Chem. Eng. 283 (2016) 816−825.

[74] J. Wang, K. Feng, B. Chen, Z.J. Li, Q.Y. Meng, L.P. Zhang, et al., Polymer-modified hydrophilic graphene: a promotor to photocatalytic hydrogen evolution for in situ formation of core@ shell cobalt nanocomposites, J. Photochem. Photobiol. A 331 (2016) 247−254.

[75] X. Xu, T. Lu, X. Liu, X. Wang, An efficient p−n heterojunction photocatalyst constructed from a coordination polymer nanoplate and a partially reduced graphene oxide for visible-light hydrogen production, Chem. - Eur. J. 21 (41) (2015) 14638−14647.

[76] D. Tanaka, Y. Oaki, H. Imai, Enhanced photocatalytic activity of quantum-confined tungsten trioxide nanoparticles in mesoporous silica, ChemComm 46 (29) (2010) 5286−5288.

[77] M.A. Holmes, T.K. Townsend, F.E. Osterloh, Quantum confinement controlled photocatalytic water splitting by suspended CdSe nanocrystals, ChemComm 48 (3) (2012) 371−373.

[78] L. Jia, D.H. Wang, Y.X. Huang, A.W. Xu, H.Q. Yu, Highly durable N-doped graphene/CdS nanocomposites with enhanced photocatalytic hydrogen evolution from water under visible light irradiation, J. Phys. Chem. C 115 (23) (2011) 11466−11473.

[79] J. Durst, C. Simon, F. Hasché, H.A. Gasteiger, Hydrogen oxidation and evolution reaction kinetics on carbon supported Pt, Ir, Rh, and Pd electrocatalysts in acidic media, J. Electrochem. Soc. 162 (1) (2014) F190.

[80] V.R. Stamenkovic, B.S. Mun, M. Arenz, K.J. Mayrhofer, C.A. Lucas, G. Wang, et al., Trends in electrocatalysis on extended and nanoscale Pt-bimetallic alloy surfaces, Nat. Mater. 6 (3) (2007) 241−247.

[81] Y. Zheng, Y. Jiao, Y. Zhu, L.H. Li, Y. Han, Y. Chen, et al., Hydrogen evolution by a metal-free electrocatalyst, Nat. Commun. 5 (1) (2014) 1−8.

[82] R.B. Levy, M. Boudart, Platinum-like behavior of tungsten carbide in surface catalysis, Science 181 (4099) (1973) 547−549.

[83] H.B. Wu, B.Y. Xia, L. Yu, X.Y. Yu, X.W.D. Lou, Porous molybdenum carbide nano-octahedrons synthesized via confined carburization in metal-organic frameworks for efficient hydrogen production, Nat. Commun. 6 (1) (2015) 1−8.

[84] Y. Zhao, K. Kamiya, K. Hashimoto, S. Nakanishi, In situ CO_2-emission assisted synthesis of molybdenum carbonitride nanomaterial as hydrogen evolution electrocatalyst, J. Am. Chem. Soc. 137 (1) (2015) 110−113.

[85] F.X. Ma, H.B. Wu, B.Y. Xia, C.Y. Xu, X.W. Lou, Hierarchical β-Mo_2C nanotubes organized by ultrathin nanosheets as a highly efficient electrocatalyst for hydrogen production, Angew. Chem. 127 (51) (2015) 15615−15619.

[86] L. Liao, S. Wang, J. Xiao, X. Bian, Y. Zhang, M.D. Scanlon, et al., A nanoporous molybdenum carbide nanowire as an electrocatalyst for hydrogen evolution reaction, Energ. Environ. Sci. 7 (1) (2014) 387−392.

[87] D.H. Youn, S. Han, J.Y. Kim, J.Y. Kim, H. Park, S.H. Choi, et al., Highly active and stable hydrogen evolution electrocatalysts based on molybdenum compounds on carbon nanotube−graphene hybrid support, ACS Nano 8 (5) (2014) 5164−5173.

[88] W.F. Chen, C.H. Wang, K. Sasaki, N. Marinkovic, W. Xu, J.T. Muckerman, et al., Highly active and durable nanostructured molybdenum carbide electrocatalysts for hydrogen production, Energ. Environ. Sci. 6 (3) (2013) 943−951.

[89] M. Seol, D.H. Youn, J.Y. Kim, J.W. Jang, M. Choi, J.S. Lee, et al., Mo-compound/CNT-graphene composites as efficient catalytic electrodes for quantum-dot-sensitized solar cells, Adv. Energy Mater. 4 (4) (2014) 1300775.

[90] R. Ma, Y. Zhou, Y. Chen, P. Li, Q. Liu, J. Wang, Ultrafine molybdenum carbide nanoparticles composited with carbon as a highly active hydrogen-evolution electrocatalyst, Angew. Chem. 127 (49) (2015) 14936−14940.

[91] Y. Liu, G. Yu, G.D. Li, Y. Sun, T. Asefa, W. Chen, et al., Coupling Mo_2C with nitrogen-rich nanocarbon leads to efficient hydrogen-evolution electrocatalytic sites, Angew. Chem. 127 (37) (2015) 10902−10907.

[92] W. Cui, N. Cheng, Q. Liu, C. Ge, A.M. Asiri, X. Sun, Mo_2C nanoparticles decorated graphitic carbon sheets: biopolymer-derived solid-state synthesis and application as an efficient electrocatalyst for hydrogen generation, ACS Catal. 4 (8) (2014) 2658−2661.

[93] J. Duan, S. Chen, B.A. Chambers, G.G. Andersson, S.Z. Qiao, 3D WS_2 nanolayers@ heteroatom-doped graphene films as hydrogen evolution catalyst electrodes, Adv. Mater. 27 (28) (2015) 4234−4241.

[94] J. Duan, S. Chen, M. Jaroniec, S.Z. Qiao, Porous C3N4 nanolayers@ N-graphene films as catalyst electrodes for highly efficient hydrogen evolution, ACS Nano 9 (1) (2015) 931−940.

[95] H. Vrubel, X. Hu, Molybdenum boride and carbide catalyze hydrogen evolution in both acidic and basic solutions, Angew. Chem. 5 (2012) 12703−12706.

[96] P.V. Kamat, Graphene-based nanoarchitectures. Anchoring semiconductor and metal nanoparticles on a two-dimensional carbon support, J. Phys. Chem. Lett. 1 (2) (2010) 520−527.

[97] C. Huang, C. Li, G. Shi, Graphene based catalysts, Energ. Environ. Sci. 5 (10) (2012) 8848−8868.

[98] T. Wang, J. Zhuo, K. Du, B. Chen, Z. Zhu, Y. Shao, et al., Electrochemically fabricated polypyrrole and MoSx copolymer films as a highly active hydrogen evolution electrocatalyst, Adv. Mater. 26 (22) (2014) 3761−3766.

[99] J. Yang, X. Wang, B. Li, L. Ma, L. Shi, Y. Xiong, et al., Novel iron/cobalt-containing polypyrrole hydrogel-derived trifunctional electrocatalyst for self-powered overall water splitting, Adv. Funct. Mater. 27 (17) (2017) 1606497.

[100] D. Zhang, Z. Zhang, X. Xu, Q. Zhang, C. Yuan, Flexible MoS_2 nanosheets/polypyr-role nanofibers for highly efficient electrochemical hydrogen evolution, Phys. Lett. A 381 (41) (2017) 3584−3588.

[101] M. Zhuang, X. Ou, Y. Dou, L. Zhang, Q. Zhang, R. Wu, et al., Polymer-embedded fabrication of Co_2P nanoparticles encapsulated in N, P-doped graphene for hydrogen generation, Nano Lett. 16 (7) (2016) 4691−4698.

[102] C. Alex, S.A. Bhat, N.S. John, C.V. Yelamaggad, Highly efficient and sustained electrochemical hydrogen evolution by embedded Pd-nanoparticles on a coordination polymer—reduced graphene oxide composite, ACS Appl. Energy Mater. 2 (11) (2019) 8098−8106.

[103] P.K. Ang, S. Wang, Q. Bao, J.T. Thong, K.P. Loh, High-throughput synthesis of graphene by intercalation−exfoliation of graphite oxide and study of ionic screening in graphene transistor, ACS Nano 3 (11) (2009) 3587−3594.

[104] S. Park, J. An, J.R. Potts, A. Velamakanni, S. Murali, R.S. Ruoff, Hydrazine-reduction of graphite-and graphene oxide, Carbon 49 (9) (2011) 3019−3023.

[105] Y. Lattach, A. Deronzier, J.C. Moutet, Electrocatalytic hydrogen evolution from molybdenum sulfide−polymer composite films on carbon electrodes, ACS Appl. Mater. Interfaces 7 (29) (2015) 15866−15875.

[106] G. Yan, C. Wu, H. Tan, X. Feng, L. Yan, H. Zang, et al., N-carbon coated PW_2C composite as efficient electrocatalyst for hydrogen evolution reactions over the whole pH range, J. Mater. Chem. A 5 (2) (2017) 765−772.

[107] T. Bhowmik, M.K. Kundu, S. Barman, Palladium nanoparticle−graphitic carbon nitride porous synergistic catalyst for hydrogen evolution/oxidation reactions over a broad range of pH and correlation of its catalytic activity with measured hydrogen binding energy, ACS Catal. 6 (3) (2016) 1929−1941.

[108] X.X. Lin, A.J. Wang, K.M. Fang, J. Yuan, J.J. Feng, One-pot seedless aqueous synthesis of reduced graphene oxide (rGO)-supported core−shell Pt@ Pd nanoflowers as advanced catalysts for oxygen reduction and hydrogen evolution, ACS Sustain Chem. Eng. 5 (10) (2017) 8675−8683.

[109] S. Sarkar, S. Sampath, Equiatomic ternary chalcogenide: PdPS and its reduced graphene oxide composite for efficient electrocatalytic hydrogen evolution, ChemComm 50 (55) (2014) 7359−7362.

[110] J.M. McEnaney, R.E. Schaak, Solution synthesis of metal silicide nanoparticles, Inorg. Chem. 54 (3) (2015) 707−709.

[111] X. Huang, H. Yao, Y. Cui, W. Hao, J. Zhu, W. Xu, et al., Conductive copper benzene-hexathiol coordination polymer as a hydrogen evolution catalyst, ACS Appl. Mater. Interfaces 9 (46) (2017) 40752−40759.

[112] R. Shekurov, V. Khrizanforova, L. Gilmanova, M. Khrizanforov, V. Miluykov, O. Kataeva, et al., Zn and Co redox active coordination polymers as efficient electrocatalysts, Dalton Trans. 48 (11) (2019) 3601−3609.

[113] H. Fei, Y. Yang, Z. Peng, G. Ruan, Q. Zhong, L. Li, et al., Cobalt nanoparticles embedded in nitrogen-doped carbon for the hydrogen evolution reaction, ACS Appl. Mater. Interfaces 7 (15) (2015) 8083−8087.

[114] J. Wang, D. Gao, G. Wang, S. Miao, H. Wu, J. Li, et al., Cobalt nanoparticles encapsulated in nitrogen-doped carbon as a bifunctional catalyst for water electrolysis, J. Mater. Chem. 2 (47) (2014) 20067−20074.

[115] X. Dai, Z. Li, Y. Ma, M. Liu, K. Du, H. Su, et al., Metallic cobalt encapsulated in bamboo-like and nitrogen-rich carbonitride nanotubes for hydrogen evolution reaction, ACS Appl. Mater. Interfaces 8 (10) (2016) 6439−6448.

[116] W. Zhou, J. Zhou, Y. Zhou, J. Lu, K. Zhou, L. Yang, et al., N-doped carbon-wrapped cobalt nanoparticles on N-doped graphene nanosheets for high-efficiency hydrogen production, Chem. Mater. 27 (6) (2015) 2026−2032.

[117] H. Yang, Y. Zhang, F. Hu, Q. Wang, Urchin-like CoP nanocrystals as hydrogen evolution reaction and oxygen reduction reaction dual-electrocatalyst with superior stability, Nano Lett. 15 (11) (2015) 7616−7620.

[118] C. Tang, L. Gan, R. Zhang, W. Lu, X. Jiang, A.M. Asiri, et al., Ternary $Fe_x Co_{1-x} P$ nanowire array as a robust hydrogen evolution reaction electrocatalyst with Pt-like activity: experimental and theoretical insight, Nano Lett. 16 (10) (2016) 6617−6621.

[119] Q. Liu, J. Tian, W. Cui, P. Jiang, N. Cheng, A.M. Asiri, et al., Carbon nanotubes decorated with CoP nanocrystals: a highly active non-noble-metal nanohybrid electrocatalyst for hydrogen evolution, Angew. Chem. 53 (26) (2014) 6710−6714.

[120] Y. Xu, R. Wu, J. Zhang, Y. Shi, B. Zhang, Anion-exchange synthesis of nanoporous FeP nanosheets as electrocatalysts for hydrogen evolution reaction, ChemComm 49 (59) (2013) 6656−6658.

[121] P. Jiang, Q. Liu, Y. Liang, J. Tian, A.M. Asiri, X. Sun, A cost-effective 3D hydrogen evolution cathode with high catalytic activity: FeP nanowire array as the active phase, Angew. Chem. 126 (47) (2014) 13069−13073.

[122] J.F. Callejas, C.G. Read, E.J. Popczun, J.M. McEnaney, R.E. Schaak, Nanostructured Co_2P electrocatalyst for the hydrogen evolution reaction and direct comparison with morphologically equivalent CoP, Chem. Mater. 27 (10) (2015) 3769−3774.

[123] H. Zhang, Z. Ma, J. Duan, H. Liu, G. Liu, T. Wang, et al., Active sites implanted carbon cages in core−shell architecture: highly active and durable electrocatalyst for hydrogen evolution reaction, ACS Nano 10 (1) (2016) 684−694.

[124] Z. Pu, Q. Liu, A.M. Asiri, X. Sun, Tungsten phosphide nanorod arrays directly grown on carbon cloth: a highly efficient and stable hydrogen evolution cathode at all pH values, ACS Appl. Mater. Interfaces. 6 (24) (2014) 21874−21879.

[125] H. Yan, C. Tian, L. Wang, A. Wu, M. Meng, L. Zhao, et al., Phosphorus-modified tungsten nitride/reduced graphene oxide as a high-performance, non-noble-metal electrocatalyst for the hydrogen evolution reaction, Angew. Chem. 127 (21) (2015) 6423−6427.

[126] R. Wu, J. Zhang, Y. Shi, D. Liu, B. Zhang, Metallic WO_2−carbon mesoporous nanowires as highly efficient electrocatalysts for hydrogen evolution reaction, J. Am. Chem. Soc. 137 (22) (2015) 6983−6986.

Polymer-graphene composite in aerospace engineering

23

Poushali Das[1,2], Susanta Banerjee[1,3] and Narayan Chandra Das[4]
[1]School of Nanoscience and Technology, Indian Institute of Technology, Kharagpur, India,
[2]Bar-Ilan Institute for Nanotechnology and Advanced Materials and Department of
Chemistry, Bar-Ilan University, Ramat-Gan, Israel, [3]Materials Science Centre, Indian
Institute of Technology, Kharagpur, India, [4]Rubber Technology Centre, Indian Institute of
Technology Kharagpur, Kharagpur, India

23.1 Introduction

The use of polymer nanocomposite in aerospace applications has considerably enhanced owing to their superior unique properties. In aerospace structure, the hunt for improvement of thermal conductivity has gained attention. Aerospace structures were assimilated with electronics that produced substantial heat energy. If this heat energy did not dissipate then it might possibly affect the composite structure integrity. Generally, polymer matrix showed low thermal conductivity for that they were not appropriate for thermal design loads in aerospace applications. To improve the thermal conductivity of the polyme, inclusion of various filler like carbon nanotube (CNT) [1,2], carbon fiber [3], carbon black [4], graphite [5], graphene [6−8], graphene oxide [9], clay [10−13] etc. in the polymer matrix creates an electrical network through the polymer to form a conductive polymer composite. Since a report in 2004 by Geim and coworker [14], graphene has been a great point of attraction because of its unique structure and features. Incorporation of graphene in polymer matrices to make multifunctional nanocomposite is very promising as polymer composites have an outstanding specific modulus and specific strength and applications in defense industries, aerospace, automobile, etc. [15,16]. In comparison to CNTs, potential filler for polymer composites, graphene contain higher surface-to-volume ratio as the inner nanotube surface is not approachable to polymer molecules [17]. Also, graphene is less expensive than CNTs as it can be derived from graphite in large scale. For this reason, graphene is more encouraging for improving the features of polymer matrices including the electrical, thermal, mechanical, and microwave absorption. As air travel continuously growing, multifunctional, easily fabricated, light weight, structural materials like polymer composites are being progressively used in aircrafts. Polymer composites combat high maintenance and fuel expenses, which account for approximately 20% and 50%, respectively, of the operation expenditures, beyond ownership, of a commercial airplane.

Polymer Nanocomposites Containing Graphene. DOI: https://doi.org/10.1016/B978-0-12-821639-2.00001-X

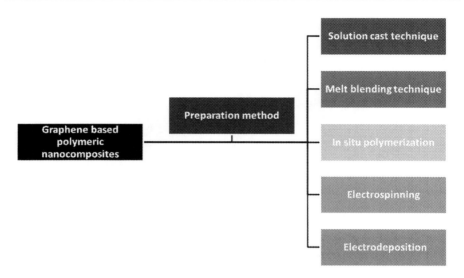

Figure 23.1 Different preparation methods of graphene based polymeric materials.

Composite aerospace structures are different from other structures and often demand challenging properties such as being light weight and bearing high performance. However, there are also strong requirement of multiple demands like processability, risk life cycle, and cost in choosing new structural component. Here, graphene-based polymer composites can act as a game changer in terms of efficiency and performance due to their unique physical and electrical properties [18]. Different preparation methods of graphene-based polymeric materials were shown in Fig. 23.1. The incorporation of graphene in the polymers could afford lightning strike protection (LSP) [19], flame retardancy [20], deicing [21], impact resistance, and others [20,22−26]. To resist degradation [27] through an unlikely incident of a fire in several important applications like skyscrapers, airplanes, or boats, engineered materials is essential [28]. Polymer composites are preferred over metals because of their resistance to fatigue, reduced part count, corrosion suppression, and restrained temperature application (300°C). Therefore graphene-based polymer nanocomposites have widespread uses in aerospace and aeronautical grounds [26]. In this chapter, we will discuss the current state of art and potential of graphene-based polymer composites especially focused on their relevance to aerospace applications.

23.2 Polymers used in aerospace engineering

Substantial challenges and prospects exist in characterization and making of engineering constituents. In the present era, with increasing number of industries, polymers have become progressively more intriguing. Use of polymers is increasing in

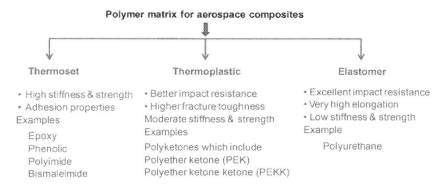

Figure 23.2 Polymer matrix used for aerospace composite.
Source: Reproduced with permission from M. Joshi, U. Chatterjee, Polymer nanocomposite: an advanced material for aerospace applications, in: Advanced Composite Materials for Aerospace Engineering, Elsevier, 2016, pp. 241−264. doi.org/10.1016/B978-0-08-100037-3.00008-0.

the engineering applications because of low cost, good chemical characteristics, lightweight, ease of manufacture/processing [29]. For fabricating aerospace materials, polymers are useful due to their properties like reasonable cost, outstanding corrosion resistance, low density $(1.2-1.4 \text{ g/cm}^3)$, high ductility (except thermosets), etc. [30]. Also transparent and tough polymers are appropriate for aircraft windows and canopies. After the innovation of structural adhesives, many present technological accomplishments have turned out to be feasible. The use of polymers for joining aircraft components as an adhesive is an emerging interest. Adhesives are applied to bond ribs, spars, and stringers to the skins of structural panels used all over the airframe. Polymer adhesive industries are currently concentrating on structural applications including petroleum, protective coating and reparability portions to incorporate it in aeronautics and aerospace manufacture systems [31]. There are different aerospace thermoset and thermoplastic polymers like polyamide, epoxy, polypropylene, polyurethane, and polyaniline. However, polymers can't be used alone as structural materials due to their low stiffness, strength, working temperature, and creep properties. Therefore polymers are employed in aerospace engineering as composite material. A list of some polymers employed in aerospace engineering was shown in Fig. 23.2.

23.3 Nanofillers for aerospace material

Fillers are very vital raw materials applied in material science. They are commonly applied in composites in order to reduce the intake of costly binding material and increased physical features of the composites. Generally, size of the particles, functionalization, geometry, and chemical composition bring about the characteristics of fillers. Introduction of nanofiller in polymer matrix may significantly affect the

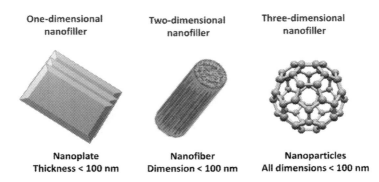

Figure 23.3 Nanoobjects used as nanofillers.

properties of materials including superior thermal, electrical, mechanical, optical, nonflammability features, controlled rheological properties etc. [32]. The ratio of nanofiller and polymer matrix highly influenced the composite property. Nanofiller can be classified as 1D (nanotubes, nanofibers), 2D (shell, laminates, platelets), and 0D or 3D nanofillers (beads and silica). Fig. 23.3 showed the nanoobjects used for different dimensional nanofillers. There are different types of nanofillers such as clay, CNT, graphite, graphene oxide used in aerospace applications. Here, we will discuss about the graphene based nanofillers for aerospace applications.

23.3.1 Graphite as nanofiller

Graphite is one of the well-known allotropes of carbon. Graphite is anisotropic and owing to it's in plane metallic bonding it behaves as brilliant thermal and electrical conductor. Graphite is widely used as electrochemical electrodes and electric brushes due to its electrical conductivity. Owing to its anisotropic nature, graphite can carry out chemical reactions by allowing the reactant molecule termed intercalate between graphene layers. These types of reactions are called intercalation. Compared to graphite, charge transfer between intercalate and graphite takes place to yield electrically conductive material in graphite intercalation compounds. For electromagnetic interference (EMI) shielding applications conductivity bring about great efficiency [33]. On heating most graphite intercalation compounds can be exfoliated. The mechanical interlocking because of compression of exfoliated graphite flakes may produce gasket materials of graphite without binder [34]. Graphene-based different nanofillers were shown in Fig. 23.4 [35].

23.3.2 Graphene, graphene oxide, and reduced graphene oxide as nanofiller

Graphene is considered as a monolayer sheet of graphite. It is a rigid planar nanostructure containing a single layer of carbon atoms organized in a hexagonal crystal lattice [36]. Graphene, 2D macromolecule having repeating structural unit of

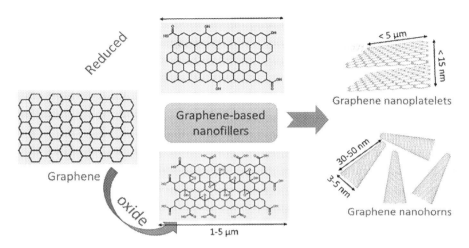

Figure 23.4 Graphene-based different nanofillers.
Source: Reproduced with permission from P. Costa, J. Nunes-Pereira, J. Oliveira, J. Silva, J. A. Moreira, S. Carabineiro, et al., High-performance graphene-based carbon nanofiller/polymer composites for piezoresistive sensor applications, Compos. Sci. Technol. 153 (2017) 241−252.

benzene particularly, comprised of sp^2 hybridized carbon atoms involving both π and σ C-C covalent bonding delocalized into the entire molecular structure. Graphene based nanomaterials have been categorized based on the existence of the layers' number of the sheet, their oxygen content, surface modifications, or orientation as can be seen from Fig. 23.5 [38]. Graphene is more advantageous to use than CNTs, as the former has larger surface area and capability to easily modify by various molecules. Graphene revealed different intriguing physical properties like large surface area ($\sim$ 2,600 m^2/g), high chemical stability for promising applications in the field of super capacitors, lithium secondary batteries, transistors etc. Chemical doping is important to modulate the electrical features of graphene [37]. Owing to enormous carrier mobility and phenomenon arising from linear energy dispersion, graphene is most often highlighted by the researchers. The graphene-based applications have offered significant thermal, chemical, mechanical, and optical properties.

GO or graphite oxide can be acquired from graphite. Graphite oxide, multilayer system, is extremely hydrophilic and generated by layered structure of GO sheets. GO can be prepared through acid/base treatment of graphite oxide and sonication afterward. In GO, different functional groups like carbonyl, epoxy, hydroxyl and phenol groups are anchored the edges of the sheets [39]. Also GO comprises a variety of negatively charged [40] reactive oxygen functionalities at the surface which indicating its water solubility and other polar solvents including dimethylformamide, tetrahydrofuran, ethylene glycol, N-methyl-2-pyrrolidone [39]. GO offers reduced electrical conductivity which can be overcome by the treatment with reducing agent. Although different models have been proposed regarding the atomic

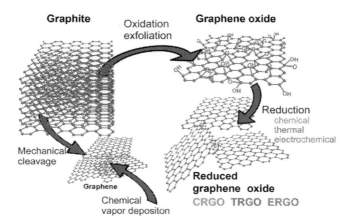

Figure 23.5 Schematic illustration of possible ways for preparation of graphene and rGO.
Left: 3D structure of graphite. Top right: monolayer of GO; GO contains epoxide and
hydroxyl functional groups. Bottom right: GO's reduction leads to production of monolayer
graphene without functional groups containing oxygen. *GO, Graphite oxide.*
Source: Reproduced with permission from S.J. Rowley-Neale, E.P. Randviir, A.S.A. Dena,
C.E. Banks, An overview of recent applications of reduced graphene oxide as a basis of
electroanalytical sensing platforms, Appl. Mater. Today 10 (2018) 218−226.

structure and composition of the GO, among all the most notable is Klinowski-
Lerf's model. This model is based on nuclear magnetic resonance spectroscopy
data which defines GO as an assembly of pristine aromatic "islands" separated
from each other by means of epoxide groups, aliphatic 6-membered rings containing C-COH groups, and double bonds [41]. Although in 2006, some modifications
were made in this model but currently it is the most accepted one [42]. The attractive features of GO is that it can be converted to reduced graphene like sheets partially by eradicating the oxygen functionalities with the retrieval of conjugated
structure. Usually rGO sheets are considered as one type of chemically derived graphene. rGO was prepared through reducing graphite oxide with hydrazine (100°C,
24 h) by Ruoff et al. [43]. Chemical reduction of GO resulted in better hydrophobicity and can cause precipitates formation. This technique of preparing rGO with
hydrazine brings about delamination and pyrolysis of oxygen functionalities to
acquire graphene layers. The chemical reduction can be achieved with other reducing agents like sodium borohydride [44], ascorbic acid [45], and hydroquinone [46].

23.4 Why are graphene and its derivatives suitable for aerospace applications?

Structures materials used in aerospace are different from other structure as they
require some challenging conditions such as light weight and high performance.

Often aerospace structure claim high performance component durable for prolonged time period in a volatile climatically environments that conventional composite material struggle to meet. In this context, graphene based polymer composites can play significant impact based on efficiency of future airframe and performance owing to their unique physical and electrical features. Outstanding structural strength and conductivity of graphene in polymer composites make them appropriate aspirant for aerospace engineering. In comparison to CNTs, potential filler for polymer composites, graphene contain higher surface-to-volume ratio as the inner nanotube surface is not approachable to polymer molecules [17]. Moreover, graphene is less expensive than CNTs as it can be derived from graphite in large scale. Also, graphene is lighter but stronger than carbon fiber. Therefore it could be used to manufacture material which can be used as replacement for steel in the aircraft structure. By this weight can be decreased and fuel efficiency can be improved. Graphene and GO have been extensively used in different applications due to their outstanding electrical conductivity and mechanical properties. These features arise from the 2D crystallographic nature of graphene. Graphene is an one atom thick planar layer of carbon atoms bonded together to form π-conjugated system in a hexagonal arrangement, forming a honeycomb crystal lattice (Fig. 23.6) [57]. The theoretical Young's modulus and tensile strength of graphene was 1.0 TPa and, 130 GPa, respectively [47]. Therefore graphene has higher tensile strength than CNT or nanosized steel. Although individual CNT exhibited high electrical and thermal conductivity (1,000,000 S/m and 2500 W/m/K) than graphene (approximately 7200 S/m and 1800 W/m/K) [40,48,49], but their bulk property of the material highly dependent on the arrangement of the CNTs [50]. The precise control over the position of CNTs is challenging because of intrinsic complications in handling individual nanotubes. Hence, these properties accompanied by large specific area (2630 m^2/g) and the capability to modify the performance of the underlying graphene by the usage of polymer composites, have incorporated this material in aerospace engineering. The exclusive properties of the graphene [51–55] could permit smart incorporation into conditions like flame retardancy [20,28], LSP [19], antiicing/deicing [21,56], and impact resistance etc. [23].

23.5   Polymer nanocomposite in aerospace applications

23.5.1   Polymer/graphite nanocomposite

During the last few decades, the electrical properties of the nanocomposites containing conductive additives dispersed within insulating matrices have been examined. The electrical conductivity of the nanocomposite can be varied in a wide range between the conductivity values of the additive and the matrix, respectively. Polymer graphite nanocomposites are long been extensively used in structural and aerospace application [58]. Generally, electrically conductive polymer nanocomposites are utilized as sensors, temperature-dependent resistors, self-limiting electrical heaters, heating elements, switching devices, and antistatic materials for EMI

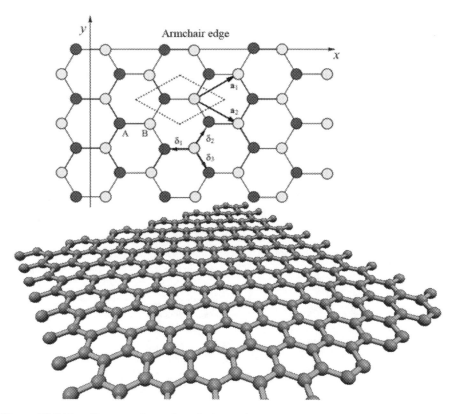

Figure 23.6 Top: Honeycomb graphene lattice and armchair edge. Sublattice An is blue (*dark-gray*); sublattice B is yellow (*light-gray*). Primitive lattice vectors a_1, a_2 have coordinates $a_0/2$ $(3, \pm\sqrt{3})$. Nearest-neighbor vectors: $\delta_1 = a_0 (-1,0)$ and $\delta_{2,3} = a_0/2 (1, \pm \sqrt{3})$. The armchair edge is located at the $y = 0$ line.
Source: Reproduced with permission from P. Maksimov, A. Rozhkov, A. Sboychakov, Localized electron states near the armchair edge of graphene, Phys. Rev. B 88 (2013) 245421. Bottom: Schematic diagram of graphene crystal structure.

shielding of electronic devices [59]. The significance of thermal conductivity in polymer nanocomposites is connected with the requirement of substantial thermal conductivity in circuit boards, heat exchangers, appliances, and machinery [60]. For practical applications, the thermal conductivity of the components is also vital for modeling optimum conditions in the course of material processing along with the analysis of heat transport in materials. The electrical and thermal conductivity of the filled polymers, their mechanical behavior as well as barrier property are greatly influenced by the graphite component. It is often applied for the upgrading of antistatic properties, thermal and electrical conductivity of plastics [61]. The dependence of thermal conductivity of high-density polyethylene (HDPE), and low-density polyethylene (LDPE)/graphite nanocomposites on graphite content was depicted in Fig. 23.7A [62]. With increasing graphite content the thermal conductivity was increased nonlinearly. The higher thermal conductivity of the filled HDPE than filled LDPE was because of high degree of crystallinity of HDPE matrix. The relationship between Young modulus of the LDPE and HDPE nanocomposites with the graphite content was shown in Fig. 23.7B. In both the cases, Young modulus was monotonously elevated with graphite content. Since, the Young modulus of LDPE matrix was lower than HDPE matrix, the Young modulus of the nanocomposites was lower for LDPE/graphite composite. In addition to that, Krupa et al. reported nonlinear increase of thermal conductivities of HDPE, polystyrene (PS)/graphite nanocomposites (Fig. 23.7C,D) with increasing graphite content [61]. Interestingly, the thermal conductivity of the graphite EG composites were lower than the graphite KS-15 composites. At high filler content, this difference was more pronounced for PS matrix. This is probably because of more conductive pathways were created as a consequence of greater agglomeration of KS graphite particles [63]. Also, in all concentration regions the Young modulus was increased with filler content (Fig. 23.7E) [62]. At the same concentration, the increment in Young modulus was more prominent in HDPE/graphite KS than HDPE/graphite EG suggesting more intensive interaction between components due to higher specific surface of the

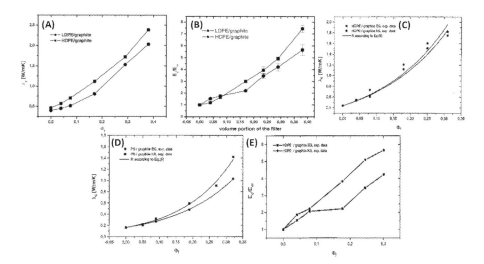

Figure 23.7 (A) Thermal conductivity (λ_c) and (B) Young modulus of the graphite filled LDPE and HDPE as a function of the volume filler content (φ_f). (C) Relationship between the thermal conductivity of the graphite EG and KS filled HDPE composites versus φ_f. (D) Thermal conductivity of the graphite EG and KS filled PS composites as a function of φ_f. (E) Relative Young's modulus of elasticity (E_c/E_m) of the graphite EG and KS filled HDPE graphite composites on the φ_f.
Source: Reproduced with permission from. Krupa, I. Chodak, Physical properties of thermoplastic/graphite composites, Eur. Polym. J. 37 (2001) 2159−2168; Krupa, I. Novák, I. Chodák, Electrically and thermally conductive polyethylene/graphite composites and their mechanical properties, Synth. Met. 145 (2004) 245−252.

graphite KS particles. Therefore due to greatly improved properties and ease of processing, polymer nanocomposites may be a capable alternative for the conventional composites. Certainly, graphene-based polymer nanocomposites are the prospective applicants for the applications in the arena of high potential structural composites involved in aircraft manufacture, spacecraft composites, automotive, marine, and sporting goods.

From the last few decades, graphite/epoxy laminated composites were broadly utilized in principle structures of new generation commercial aircraft due to their potential for reducing weight. In comparison to conventional aluminum alloy, even though these composites showed superior mechanical behaviors, they generally displayed high degradation of strength owing to internal damage for instance matrix cracks and delamination. One of the main reasons of internal damage is impact by fragment hits and tool dropping in the course of maintenance, manufacturing, or operation. Lightning strike is another reason of internal damage to laminated composite structure. Hirano and group studied the damage evolution in graphite/epoxy laminated composites owing to lightning strikes [64]. To elucidate the effect of lightning parameters and specimen size, artificial lightning testing was executed on graphite/epoxy laminated composite. By applying ultrasonic testing, micro X-ray examination, visual inspection, and sectional investigation, damage was evaluated in laminated composites. Three modes of damage were obtained from the results including resin deterioration, internal delamination modes, and fiber damage. The strong electrical orthotropic features of the laminates ruled the damage progression, and the lightning parameters described impulse waveform that displayed a strong connection with definite damage modes, although thickness and specimen size difference hardly disturbed damage size. Using ultrasonic testing, internal damage was investigated. The results of ultrasonic testing for postlightning experiment were showed in Fig. 23.8A–H. The figures illustrated the C-scan image of the typical outcome of the definite test condition. The results of ultrasonic testing were acquired from the opposite side of the lightning attachment surface by a 5.0 MHz transducer.

Delamination is one of the serious issues in the context of damage tolerance and durability evaluation. It is frequently noticed that failure mode in laminated fiber reinforced composite material and originate during manufacture, may be brought by in-service actions like impact, or may be caused by the existence of interlaminar stresses at free-edges or discontinuities. The presence of delamination in a structure can considerably decrease strength and in-plane stiffness of the structure. Donaldson et al. examined brittle graphite/epoxy composite under the condition of tearing mode III to evaluate the delamination growth activities [65]. The author demonstrated that split cantilever beam specimens were experimented under fatigue and static loadings. Growth rates of delamination and associated strain energy release rates were recorded. The static experiments offered the critical strain energy release rate, G_{IIIC}, for mode III and the relation between growth rate of delamination and G_{III} were proven for two displacement ratios from the fatigue experiments. This investigation for the mode III condition was also compared and discussed with previously studied mode I and mode II conditions. For mode III, the critical strain

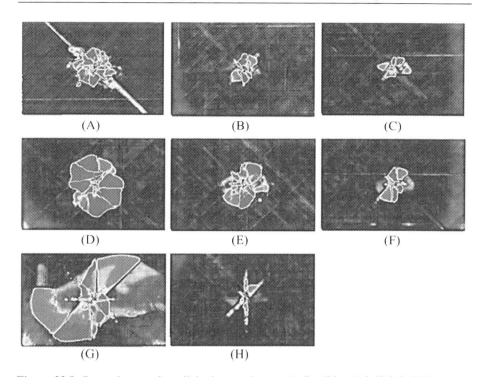

Figure 23.8 C-scan image of postlightning specimen. (A) Condition I (2.6/10.5, 40 kA), (B) Condition II (2.6/10.5, 30 kA), (C) Condition III (2.6/10.5, 20 kA), (D) Condition IV (4/10, 40 kA), (E) Condition V (4/10, 30 kA), (F) Condition VI (4/10, 20 kA), (G) Condition VII (7/150, 20 kA), (H) Condition VIII (7/150, 10 kA).
Source: Reproduced with permission from Y. Hirano, S. Katsumata, Y. Iwahori, A. Todoroki, Artificial lightning testing on graphite/epoxy composite laminate, Compos. A Appl. Sci. Manuf. 41 (2010) 1461−1470.

energy release rate was around one order of magnitude higher than the corresponding value for mode I and twice for mode II for experimented graphite/epoxy composite. Also, the author discussed the outcomes of this study in terms of damage tolerance and design of durability considerations in composite structures. Todoroki and coworkers reported the identification of damage for woven graphite/epoxy composite beams through an electrical resistance change technique [66]. In various practical applications, woven plies were implemented to avoid surface layer peeling. Thus woven graphite/epoxy composite was chosen by the authors as the target material to detect the damage of the size and location of delamination cracks by the electrical resistance change technique using beam type specimens. The delamination crack was produced through an interlamina shear assessment. With the electrical resistance variation, the effect of an in-plane isotropic electrical conductivity of a surface woven layer was examined and also the condition of the electrical contact between specimen and electrode was studied. In a report Kostopoulos et al.

PTFE Film

Impregnated layers by neat epoxy resin

Nanodoped layers

Impregnated layers by neat epoxy resin

Figure 23.9 Schematic representation of the nanodoped lamina with 0.5wt.% graphite oxide and graphene nanoplatelets.
Source: Reproduced with permission from C. Kostagiannakopoulou, T. Loutas, G. Sotiriadis, A. Markou, V. Kostopoulos, On the interlaminar fracture toughness of carbon fiber composites enhanced with graphene nano-species, Compos. Sci. Technol. 118 (2015) 217–225.

illustrated interlaminar fracture toughness of carbon fiber composites improved with graphene nanospecies (Fig. 23.9) [67].

Innovative composites have been extensively utilized in aircraft structures owing to their high elastic moduli and specific strengths. Delamination is one of the primary failure modes in composite laminates because of the shortage of through the thickness reinforcements. Therefore it is of high importance to evaluate interlaminar fracture toughness of composites for emerging fracture criteria. Sun et al. investigated on interlaminar fracture behavior and fracture toughness of a graphite/epoxy multidirectional composite laminate applying end-notched flexure specimens [68]. Both for stable and unstable crack extensions experiments were carried out. Toughness acquired from the experiments of stable crack extension was found to be considerably higher than unstable crack extension tests. Interlaminar fracture toughness was determined and compared with different data reduction techniques including classical laminated plate theory, area method and finite element analysis. The results revealed that the toughness value was dependent on data reduction method. Additionally, the influence of the fiber orientation and specimen geometry on the interlaminar fracture toughness was also estimated. Park et al. reported fabrication (Fig. 23.10A) of basalt fiber reinforced epoxy composites utilizing graphite flakes [69]. Natural graphite flakes considerably improved the mechanical properties of the composite like critical strain energy release rate and critical stress intensity factor. The highest mechanical property was obtained at 20wt.% of graphite flakes loading. The author demonstrated the fracture toughness applying crack theory based on morphology analysis of the fracture surfaces. By the K_{IC} examination, the fracture surface of the specimen was measured by SEM images (Fig. 23.10B–G). The magnified image (Fig. 23.10C) of the Fig. 23.10B showed smooth surface and brittle properties because of regular growth of crack. The magnified SEM image (Fig. 23.10C) of NGB-20 composite revealed irregular distribution and side branch cracks which were accredited to the remarkable energy was essential in the course of crack propagation. From magnified image (Fig. 23.10G) of NGB-40 composite, it can be seen that high amount of NGFs formed agglomeration in the epoxy matrix,

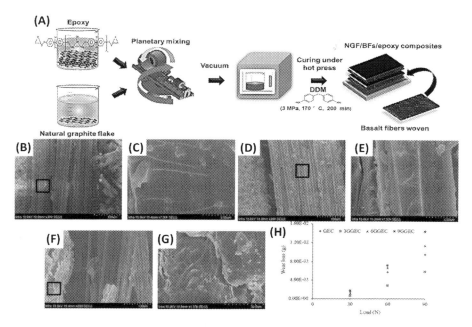

Figure 23.10 (A) Synthesis of NGF/BFs/epoxy composites. SEM images of fracture surfaces: (B) neat sample (D) NGB-20 composite (F) NGB-40 composite. (C), (E), (G) showed the magnified images of enclosed regions in (B), (D), (F) respectively. (H) Wear loss versus normal load of neat and graphite filled glass fabric/epoxy composites.
Source: Reproduced with permission from S.H. Kim, Y.-J. Heo, M. Park, B.-G. Min, K.Y. Rhee, S.-J. Park, Effect of hydrophilic graphite flake on thermal conductivity and fracture toughness of basalt fibers/epoxy composites, Compos. B Eng. 153 (2018) 9−16; B. Shivamurthy, K.U. Bhat, S. Anandhan, Mechanical and sliding wear properties of multi-layered laminates from glass fabric/graphite/epoxy composites, Mater. Des. 44 (2013) 136−143.

leading to tough for dispersion and therefore dropping the mechanical interface characteristics. Additionally, the composites showed high thermal conductivity and stability and a direct linear relationship was found between the thermal conductivity and fracture toughness by certain polar components of surface energy.

Fiber reinforced polymer composites (FRPCs) was widely applied in aerospace and automative applications for manufacturing the component like wheels, rollers, gears, clutches, bearing liner, cams, brakes, seals and bushing [70]. Researchers have been giving efforts to develop wear resistant composites through integrating different forms of reinforcement including abrasive fillers, friction modifiers (solid lubricants), binders and fibers into thermoset matrices [71]. In this context, it is necessary to discuss about the wear behavior of some graphite based polymer composites. Anandhan and his coworkers studied mechanical and sliding wear properties of the multilayered laminates from glass fabric/graphite/epoxy composites [72]. Graphite was applied as filler in the composite preparation which could help in heat dissipation. Using a pin-on-disc wear experiment apparatus, the wear properties of the composites were studied. The effect of loading graphite on mechanical behavior and sliding wear resistance of the composites were investigated.

The hybrid composite with 3wt.% of graphite showed optimum mechanical and wear performances. With the increment in further graphite loading the specific wear rate increased and mechanical property dropped. The authors also described the wear experiments results with morphological examination. At different normal applied loads, the wear loss of the control and the composites for 1.2 km sliding distance was represented in Fig. 23.10H. Increment in wear loss was noticed for all the composites with the increment of normal applied load. At all the loads, 3GGEC displayed low wear loss than the other graphite/glass fabric/epoxy composites. Also, Basavarajappa et al. studied wear behavior of glass/epoxy composites with graphite content loading [73]. Introduction of graphite in glass/epoxy composites showed lower weight loss and the value further falls if the graphite content was increased. This behavior was attributed to a thin coherent and uniform film was transferred on the disc and the interphase comprised of lubricant particles thus dropping the three body abrasion severity. The SEM examination of the composites disclosed fiber breakage, formation of debris, debonding of fiber matrix and exposure of both transverse and longitudinal fibers owing to wear of the matrix. In another report, this author demonstrated the application of Taguchi techniques to examine the impact of applied load, sliding distance and sliding speed on dry sliding wear performance of Al/SiCp and Al/SiCp-Graphite composites [74]. The introduction of graphite particles as secondary reinforcement in the Al matrix increased the wear resistance of the material. The smearing of the graphite and protecting layer formation between the pin and the counter face allowed in decreasing the wear volume loss.

Because of the high specific impulse cryogenic liquid fuels are favored to solid ones as propellants of rockets and launch vehicles in aerospace applications [75]. However, cryogenic liquid fuels make the pressurized tank large and heavy when made of metallic materials due to its low calorific energy to volume ratio. Therefore polymer composites were utilized as substitute materials of the tanks owing to their low thermal expansion and high mechanical performances. Kim et al. reported low temperature tensile behavior of graphite/epoxy composites [76]. Thermo mechanical tensile cyclic loading was applied to T700/epoxy unidirectional laminates at room temperature to $-50°C$, room temperature to $-100°C$, and room temperature to $-150°C$ (CT), respectively. As the temperature decreased, the tensile stiffness considerably increased though the thermomechanical cycling had small influence on it. Nonetheless, the tensile strength declined as temperature reduced to cryogenic temperature, while after cryogenic temperature cycling, the dropping rate of strength was decreased. By applying the double cantilever beam experiment to evaluate the Mode I critical strain energy release rate and the edge delamination tension examination to estimate the mixed Mode I and II critical strain energy release rate of a graphite/epoxy (T300/934) composite Sykes et al. studied the influence of electron radiation on the interlaminar fracture toughness [77]. With radiation exposure, the interlaminar fracture toughness of the composite was boosted

Table 23.1 Graphite/graphene based composites for aerospace applications.

Filler	Applications
Graphene	Provide electrostatic discharge and EMI shielding components as well as weight saving.Transparent electrodes liquid-crystal display screen and integrated circuits.
Graphite polyimides	Used for high speed aircraft for high temperature resistance.
Glass/graphite/ aramid/ boron/epoxy hybrids	Excellent for helicopters and ITL aircraft. Many combinations of fibers may produce a better part than individual fibers alone.
Graphite/epoxy	Structural application for higher stiffness and fatigue resistance; suitable for mosthigher loaded structural parts.

which was ascribed to radiation-induced matrix property changes. Few possible applications of graphite/graphene based polymer nanocomposites were presented in Table 23.1.

23.5.2 Polymer/graphene nanocomposite

Polymers like polyetheretherketone, polyethylenimine, and polyethersulfone (PES) have been used for aerospace applications as plastic matrix because of their excellent friction and wear resistance, chemical resistance, high modulus and tensile strength and higher temperature performance. Graphite, graphene, GO, CNTs, as nanoadditives are ruling enormous market prospective in automobile and aerospace industries. GO/PES composites were fabricated applying high shear melt blending method by K Balasubramanian [78]. Using modified Hummers method, chemical exfoliated GO nanosheets were prepared from graphite flakes and disperse homogeneously in PES matrix with the help of high shear twin screw batch mixer and injection molded to evaluate the mechanical and electromagnetic characteristics. Result showed that introduction of 0.5vol.% of GO improved the flexural modulus and tensile strength by 90% and 40%, respectively. Mechanical characteristics of the GO/PES nanocomposites were discussed thoroughly. The author illustrated that this observation of improved mechanical property with the incorporation of GO in PES matrix was because of good dispersion, formation of continuous network, and strong interfacial interactions. In addition to that, with the addition of GO, noteworthy improvement in the impact resistance was observed for the nanocomposite. Haque et al. investigated the possible improvements to the fracture toughness of the carbon fiber reinforced epoxy composites due to the addition of graphene nanoplatelets [79]. For several engineering fields, fracture toughness is a vital parameter for structural components but it is especially important from a damage tolerance viewpoint in aerospace applications. This authors demonstrated that at a very low loading that is 0.1% weight of graphene considerably increased the fracture toughness

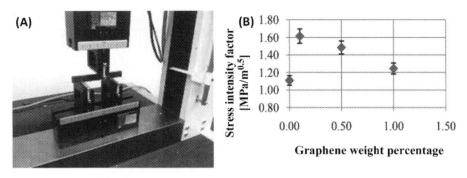

Figure 23.11 (A) Fracture toughness testing setup and (B) fracture toughness for G-Ep fabricated without dispersion agent or high shear mixing.
Source: Reproduced with permission from D.A. Hawkins Jr, A. Haque, Fracture toughness of carbon-graphene/epoxy hybrid nanocomposites, Proc. Eng. 90 (2014) 176−181.

of both carbon epoxy composites and neat epoxy. The testing setup of fracture toughness used in this study was shown in Fig. 23.11A. In comparison to C-Ep composites, the C-G-Ep showed fracture toughness improvements only to be 11.4% due to in the former case carbon fiber played a governing role in the fracture process and it is less affected by G-Ep matrix. At graphene loading higher that 0.1%, the fracture toughness was found to reduce in G-Ep composites (Fig. 23.11B).

The high requirement in aerospace like gas turbine engines creates strong needs on the structural materials. In aerospace industry, to come across the more rigorous necessity, recent reports were mostly focused on toughening and strengthening alloys of TiAl on high Nb incorporation, which certainly leads to density increment. Cui et al. reported carbon fiber coated graphene as toughening fibers and also it acted as supplier of strengthening ceramic particles including TiC and Ti_2AlC [80]. The TiAl alloy composite was prepared through powder metallurgy, melt spun, and vacuum melting. The composites gained significant mechanical features associated with noticeable density reduction. Using continuous technologies such as powder metallurgy, melt spun, and two times' vacuum melting helped the fibers distributed in TiAl matrix homogeneously and enhanced its density and wettability, respectively. In comparison to pure TiAl alloys, the carbon fiber/graphene coated reinforced TiAl alloy composite (CFGRTAC) revealed average fracture strain in the range 16%−26.27%, and an average strength from 1801 up to 2312 MPa. The author demonstrated this study as a new technique to prepare low-density, good mechanical properties TiAl composite for aerospace industry application. The technical process and model of CFGRTAC was shown in Fig. 23.12.

Every aircraft experiences direct lightning strike and must proficient to fly and function after lightning strikes deprived of catastrophic injury according to Federal Aviation Administration regulations Advisory Circular AC 25−21, Section 25.581 "Lightning Protection of Structure" [81]. The current industry standards are the SAE ARP, which comprised guidelines and experiments to evaluate whether the

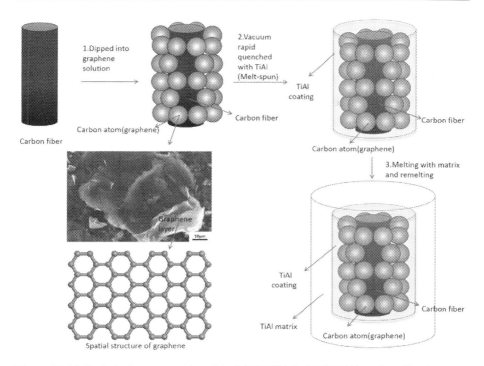

Figure 23.12 Technical process and model of CFGRTAC. *CFGRTAC, carbon fiber/ graphene coated reinforced TiAl alloy composite.*
Source: Reproduced with permission from S. Cui, C. Cui, J. Xie, S. Liu, J. Shi, Carbon fibers coated with graphene reinforced TiAl alloy composite with high strength and toughness, Sci. Rep. 8 (2018) 1−8.

aircraft or aircraft part passes regulations [82]. According to standard SAE ARP 5414, the aircraft is divided into three sections termed lightning strike zones. Every zone displayed the probability to collide by different types of lightning currents, and is one of the way to present the aircraft is protected [81]. Different zones of a commercial aircraft were shown in Fig. 23.13. The severe lightning strikes region was represented by Zone 1. The possible region of a strike hang on was exemplified by Zone 2 and the zones where large currents pass through between attachment points was presented by Zone 3. Both Zones 2 and 3 were requisite to endure lesser currents than in Zone 1 [82,83].

Zone 1 is expected to get initial lightning strikes attaching themselves to the frame which is named attachment points. First return strokes with Zone 1A holding low expectation of hang on and Zone 1B having a high expectation of hang on, and Zone 1C having the first return stroke of decreased amplitude and a low expectation of hang on [83]. Zone 2 is possible to receive succeeding swept strokes or re-strikes along with Zone 2A taking low expectation of hang on while Zone 2B having a high expectation of hang on [9]. Swept strokes take place when an aircraft flies into

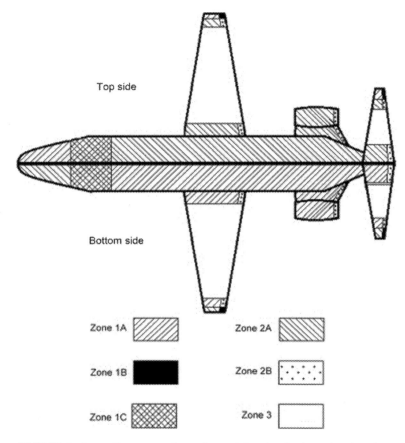

Figure 23.13 Lightning strike zone regions of a straight-wing business jet aircraft. *Source*: Reproduced with permission from M. Gagné, D. Therriault, Lightning strike protection of composites, Prog. Aerosp. Sci. 64 (2014) 1–16.

the lightning channel and making the lightning strike "sweep" across the surface [83,84]. Large lightning current between regions of direct or swept stroke attachment points is supported by Zone 3 [84]. Laboratory examinations of lightning strikes determined the boundaries between zones [84].

The aircraft shell is generally constructed by conductive material like aluminum; therefore the physical damage due to lightning strike often controlled to some burn marks on the skin and trailing edges. The skin of aircraft which is made of metals also acted as Faraday cage during lightning strike, shielding the avionics from EMI [85]. As the fiber incorporated polymer composites are comparatively less conductive in comparison to their metallic counterparts [86,87], LSP against aircrafts has become a significant task. Numerous methods were applied for LSP of composite structures in aerospace industry. LSP are meant to offer continuous conductive path all over the parts of aircraft exterior particularly in the regions which is more vulnerable to lightning strike including wingtips, radomes, nose, nacelles, and the extremities of the empennage. LSP mainly comprised of metallic mesh or foil of Cu or aluminum and low level of titanium, phosphor bronze, and embedded in the outermost laminate ply [88,89]. However, the metallic meshes inserted in carbon fiber enhanced the structure weight and vulnerable to oxidation, pitting, galvanic corrosion and therefore with time their electrical conductivity dropped [90]. For this reason, various aircraft material providers impregnate the metallic meshes with prepregs and adhesive or surfacing films. Currently, nanomaterials like graphene, carbon nanofibers, carbon black, nickel nanostrands have been used in the aerospace industry to increase the mechanical and electrical property of the FRPCs [91−97]. In this context, Asmatulu and coworkers developed 0.5−4wt.% pristine graphene and GO incorporated epoxy resin and the physical properties of the carbon fiber/epoxy and glass fiber/epoxy composites-termed as hierarchical graphene composites were enhanced steadily [96]. Improved mechanical properties were found to be more prominent for the specimens incorporated with GO in comparison to the specimens incorporated with pristine graphene. In presence of 4wt.% GO, the compressive strength of glass fiber composites was improved 84% while the same amount of pristine graphene inclusion the compressive strength was increased by 77% only. For inclusion of both pristine graphene and GO, the thermal conductivity of the both glass fiber and carbon fiber composites increased. With the 4wt.% inclusion of GO and pristine graphene improved the thermal conductivity of the glass fiber composites by 89% and 80%, respectively. More pronounced improved mechanical properties and thermal conductivity with GO inclusion was found in comparison to pristine graphene. This was explained as the functionalities at the graphene edges delivered stronger molecular bonding which leads to more effective heat flux.

The overall results indicated that the thermal, mechanical properties, reliability, and service life of the FRPCs could be considerably improved by nanoinclusions of GO and pristine graphene into the epoxy resin. Hence, the FRPCs offered as a promising replacement for metallic parts in industries like aerospace, automotive, wind energy and marine. Later the author reported flexible thin coating of pristine graphene on the surface of carbon fiber epoxy prepreg laminate coated for LSP [19]. The coating decreased the volume and damage area by 96% and 94%, respectively with comparison to the laminates without coating. The author demonstrated that the coating amended the EMI shielding effectiveness by 49%, 44%, and 22% over the 12−18, 8−12, and 100−2000 MHz range, respectively. It was found that the enhancements were ascribed to the great electrical conductivity of graphene thin film. Therefore the lightweight, flexible thin film (Fig. 23.14A) which was able to take contoured shapes and complex geometries could be a feasible substitute for the metallic meshes presently employed in aerospace industry for the protection of composite structures against EMI shielding and lightning strikes (Fig. 23.14B). Raimondo et al. developed graphene/POSS epoxy resin composite specially to achieve aeronautical structural necessity like electrical conductivity, fire resistance, mechanical performance and thermal stability [98].

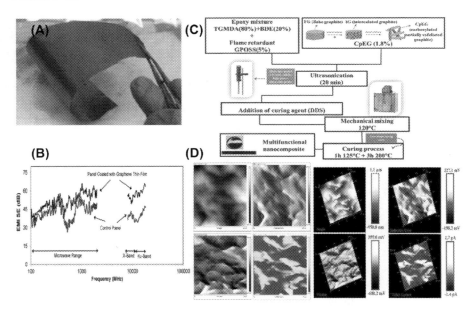

Figure 23.14 (A) Flexibility of the fabricated graphene thin film. (B) EMI shielding
effectiveness as a function of frequency over microwave range, X-band, and Ku-band for
control panel and panel coated with graphene thin film. (C) Scheme for the preparation
technique of the multifunctional epoxy nanocomposite. (D) TUNA-AFM micrographs of the
fracture surface of the T20BD + 5%GPOSS + 1.8% CpEG sample.
Source: Reproduced with permission from B. Zhang, S.A. Soltani, L.N. Le, R. Asmatulu,
Fabrication and assessment of a thin flexible surface coating made of pristine graphene for
lightning strike protection, Mater. Sci. Eng. B 216 (2017) 31–40; M. Raimondo,
L. Guadagno, V. Speranza, L. Bonnaud, P. Dubois, K. Lafdi, Multifunctional graphene/POSS
epoxy resin tailored for aircraft lightning strike protection, Compos. B Eng. 140 (2018)
44–56.

The author discussed their first successful attempt to acquire nanoscale conductivity mapping of graphene/POSS epoxy resin by Tunneling Atomic Force Microscopy.
When GPOSS was used the LOI value was found to be increased while the PHRR
value was decreased and also the time of ignition was enhanced because of the
incorporation of CpEG in epoxy systems. Fig. 23.14C represented the preparation
of technique of the multifunctional epoxy nanocomposite. In the current profile,
different domains brightness represented changes in the current value. By taking a
careful surveillance the authors confirmed that the sample was intrinsically conductive having the current in the range 7.18 fA to 3.5 pA. TUNA-AFM micrographs of the fracture surface of the sample were displayed in Fig. 23.14D. From
left to right on the left side: height, deflection error, friction and tuna current
images and the respective 3D profile was shown in the figure (right side). This
results along with high mechanical and electrical performances and good

thermostability by self-assembly blocks of CpEG nanofiller supported the fabrication of multifunctional composite for aeronautical applications.

One of the most a prevalent problem that degraded the aerospace structure is accumulation of snow/ice on surfaces like aircraft wings and tails, helicopter rotor blades, transmission lines and wind turbines [99,100]. For example helicopter rotor blades were specifically designed and machined to produce airflow to assist buoyancy. However buildup of ice/snow on the blade could leads to modification of shape, rough and uneven surfaces, disrupting smooth air flow and highly dropping the capability of the wings which could compromise aerodynamic performance by means of disturbing airflow around the blade [101,102]. This condition can seriously cause safety risk to the aircraft and therefore to the passengers and pilots. Wing ice management can be categorized into two broad types: one is antiicing; a method for ice prevention and the other one is deicing that is elimination or mitigation of ice. Most technology concentered on deicing, assuming that some ice will form on wing and function to eliminate before it turn into problematic situation (damage the structures or make them unstable). Deicing techniques rely on sprayed chemicals or hot fluid [103], Joule heating [104−106], mechanical force, and infrared radiation [107]. Moreover, recent antiicing systems are insufficient for long-term and complete protection and generally combined with active deicing system which have ability for energy supplies [99,108]. For instance, application of electric power to graphite coating to produce heat, high-voltage capacitors for fast propel ice off from a surface, rubber boots inflated through engine bleed air [109], deicing coating of pristine and functionalized graphene [21,104,110].

Often deicing heating layers were applied in covers of large radio-frequency (RF) apparatus, for instance radar, to eliminate ice/snow. Usually, the deicers were prepared with metal framework and inorganic insulator and frequently nontransparent to RF waves. In the gigahertz frequency, metals generally have skin depth on the order of a micrometer. Therefore it is challenging to get large area subskin depth films for conventional conductive materials like metals. To overcome this problem, in a report Tour et al. developed deicing heating layer composite of thin film graphene nanoribbons (GNRs) with RF transmission ability [104]. Spray-coating method was employed to fabricate GNR films which acted as the conductive layer for deicing coatings of huge RF tools including radome systems and bridge antenna towers. The author demonstrated this ultralight, RF transparent, and metal-free graphene-based conductive coating could considerably decrease the price and size of deicing coatings for RF equipment shields indicating its applications in different marine and aviation prospects. Later the author illustrated preparation of conductive films (Fig. 23.15A) of hexadecylated graphene nanoribbons (HD-GNRs) [21]. The author observed high transparency to radiofrequency (RF) waves of the HD-GNRs at very high incident power density. HD-GNR films that had a thickness in nanoscale and with several square cm area were observed to transmit up to 390 W (2×10^5 W/m^2) of RF power, 99% of RF transmittance. These films also followed electromagnetic skin depth theory. They showed optical transparency for various tinted glass and plastics and were revealed to be capable of voltage-induced deicing of surfaces (Fig. 23.15B,C). In 2016, this author developed GNR filled

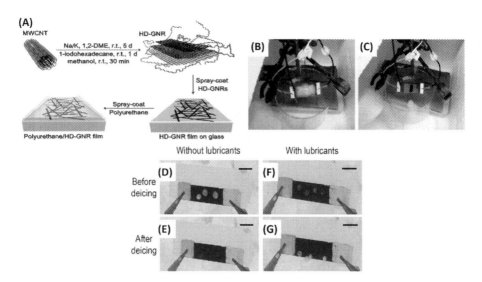

Figure 23.15 (A) Schematic illustration of the HD-GNR synthesis and fabrication of film. Photographs of resistively heated HD-GNR films at −20°C (B) During deicing. (C) After deicing which took 3.7 min. Deicing experiment of the FDO-GNR films. Images of the film without lubricating liquid (D) before and (E) after active deicing by resistive heating, and with lubricating liquid (F) before and (G) after active deicing by resistive heating, respectively. Scale bars 1 cm. *FDO-GNR, perfluorododecylated graphene nanoribbons; HD-GNR, hexadecylated graphene nanoribbons.*
Source: Reproduced with permission from A.-R.O. Raji, S. Salters, E.L. Samuel, Y. Zhu, V. Volman, J.M. Tour, Functionalized graphene nanoribbon films as a radiofrequency and optically transparent material, ACS Appl. Mater. Interfaces 6 (2014) 16661−16668; T. Wang, Y. Zheng, A.-R.O. Raji, Y. Li, W.K. Sikkema, J.M. Tour, Passive anti-icing and active deicing films, ACS Appl. Mater. Interfaces 8 (2016) 14169−14173.

epoxy composites that contained satisfactory electrical conductivity greater than 100 S/m at about ≤5wt.% GNR content [110]. It allowed to execute voltage-induced heating (Joule heating) of the composite for deicing of the surfaces. The author reported that if 0.5 W/cm^2 power density was supplied it could remove about 1 cm thick monolith of ice from a helicopter rotor blade surface in −20°C environment. Later on, this author reported perfluorododecylated graphene nanoribbons (FDO-GNRs- based films that could serve both antiicing and deicing phenomenon [56]). This type of film contained the benefit of low polarizability of perfluorinated carbons and the conductive property of GNRs. The FDO-GNRs films were superhydrophobic and showed antiicing features that could inhibit freezing of ice cold water down to −14°C. Thereafter, application of voltage to the films could resistively heat and deice the surface. To boost the deicing property lubricating liquid was applied to form slippery surface. The author demands these types of films could be promising in extreme environments.

23.6 Conclusion

In this chapter, several prospects of the graphene-based nanofiller such as electrical, mechanical, and thermal properties have been deliberated. The influence of these fillers on various types of polymer matrices has been demonstrated. Notable improvement in mechanical, thermal, electrical, EMI shielding, LSP, antiicing/deicing and flame retardant properties of polymer was discussed with the inclusion of graphene-based nanofiller in the polymer matrix and therefore graphene-based polymer composites reflected as an outstanding material for aerospace engineering. The main focus of this chapter was on the potential of graphene-based filler inclusion in polymer composites for aerospace applications. With the technological innovation, the graphene-based polymer composites have been often exploited in a widespread applications in electronics, automobile, and aerospace parts.

References

[1] S. Mondal, P. Das, S. Ganguly, R. Ravindren, S. Remanan, P. Bhawal, et al., Thermal-air ageing treatment on mechanical, electrical, and electromagnetic interference shielding properties of lightweight carbon nanotube based polymer nanocomposites, Compos. A Appl. Sci. Manuf. 107 (2018) 447–460.

[2] B. Pradhan, K. Setyowati, H. Liu, D.H. Waldeck, J. Chen, Carbon nanotube – polymer nanocomposite infrared sensor, Nano Lett. 8 (2008) 1142–1146.

[3] S. Mondal, L. Nayak, M. Rahaman, A. Aldalbahi, T.K. Chaki, D. Khastgir, et al., An effective strategy to enhance mechanical, electrical, and electromagnetic shielding effectiveness of chlorinated polyethylene-carbon nanofiber nanocomposites, Compos. B Eng. 109 (2017) 155–169.

[4] R. Ravindren, S. Mondal, P. Bhawal, S.M.N. Ali, N.C. Das, Superior electromagnetic interference shielding effectiveness and low percolation threshold through the preferential distribution of carbon black in the highly flexible polymer blend composites, Polym. Compos. 40 (4) (2019) 1404–1418.

[5] S.C. Tjong, Polymer nanocomposite bipolar plates reinforced with carbon nanotubes and graphite nanosheets, Energy Environ. Sci. 4 (2011) 605–626.

[6] S. Ganguly, P. Das, P.P. Maity, S. Mondal, S. Ghosh, S. Dhara, et al., Green reduced graphene oxide toughened semi-IPN monolith hydrogel as dual responsive drug release system: rheological, physicomechanical, and electrical evaluations, J. Phys. Chem. B 122 (2018) 7201–7218.

[7] S. Ganguly, D. Ray, P. Das, P.P. Maity, S. Mondal, V. Aswal, et al., Mechanically robust dual responsive water dispersible-graphene based conductive elastomeric hydrogel for tunable pulsatile drug release, Ultrason. Sonochem. 42 (2018) 212–227.

[8] S. Ganguly, P. Das, M. Bose, T.K. Das, S. Mondal, A.K. Das, et al., Sonochemical green reduction to prepare Ag nanoparticles decorated graphene sheets for catalytic performance and antibacterial application, Ultrason. Sonochem. 39 (2017) 577–588.

[9] P. Bhawal, S. Ganguly, T. Chaki, N. Das, Synthesis and characterization of graphene oxide filled ethylene methyl acrylate hybrid nanocomposites, RSC Adv. 6 (2016) 20781–20790.

[10] K.-t Lau, C. Gu, D. Hui, A critical review on nanotube and nanotube/nanoclay related polymer composite materials, Compos. B Eng. 37 (2006) 425—436.

[11] S. Mondal, T. Das, S. Ganguly, P. Das, R. Ravindren, Oxygen permeability properties of ethylene methyl acrylate/sepiolite clay composites with enhanced mechanical and thermal performance, J. Polym. Sci. Appl. 1 (2017) 2.

[12] S. Ganguly, P. Das, T.K. Das, S. Ghosh, S. Das, M. Bose, et al., Acoustic cavitation assisted destratified clay tactoid reinforced in situ elastomer-mimetic semi-IPN hydrogel for catalytic and bactericidal application, Ultrason. Sonochem. 60 (2020) 104797.

[13] S. Ganguly, N.C. Das, Water uptake kinetics and control release of agrochemical fertilizers from nanoclay-assisted semi-interpenetrating sodium acrylate-based hydrogel, Polym. Technol. Eng. 56 (2017) 744—761.

[14] K.S. Novoselov, A.K. Geim, S.V. Morozov, D. Jiang, Y. Zhang, S.V. Dubonos, et al., Electric field effect in atomically thin carbon films, Science 306 (2004) 666—669.

[15] G. Scarselli, C. Corcione, F. Nicassio, A. Maffezzoli, Adhesive joints with improved mechanical properties for aerospace applications, Int. J. Adhes. Adhes. 75 (2017) 174—180.

[16] J. Njuguna, K. Pielichowski, Polymer nanocomposites for aerospace applications: properties, Adv. Eng. Mater. 5 (2003) 769—778.

[17] S. Stankovich, D.A. Dikin, G.H. Dommett, K.M. Kohlhaas, E.J. Zimney, E.A. Stach, et al., Graphene-based composite materials, Nature 442 (2006) 282—286.

[18] D. Li, R.B. Kaner, Graphene-based materials, Nat. Nanotechnol. 3 (2008) 101.

[19] B. Zhang, S.A. Soltani, L.N. Le, R. Asmatulu, Fabrication and assessment of a thin flexible surface coating made of pristine graphene for lightning strike protection, Mater. Sci. Eng. B 216 (2017) 31—40.

[20] Z. Li, A.J. González, V.B. Heeralal, D.-Y. Wang, Covalent assembly of MCM-41 nanospheres on graphene oxide for improving fire retardancy and mechanical property of epoxy resin, Compos. B Eng. 138 (2018) 101—112.

[21] A.-R.O. Raji, S. Salters, E.L. Samuel, Y. Zhu, V. Volman, J.M. Tour, Functionalized graphene nanoribbon films as a radiofrequency and optically transparent material, ACS Appl. Mater. Interfaces 6 (2014) 16661—16668.

[22] J. Njuguna, K. Pielichowski, Polymer nanocomposites for aerospace applications: characterization, Adv. Eng. Mater. 6 (2004) 204—210.

[23] J. Baur, E. Silverman, Challenges and opportunities in multifunctional nanocomposite structures for aerospace applications, MRS Bull. 32 (2007) 328—334.

[24] P.-y Hung, K.-t Lau, B. Fox, N. Hameed, J.H. Lee, D. Hui, Surface modification of carbon fibre using graphene—related materials for multifunctional composites, Compos. B Eng. 133 (2018) 240—257.

[25] W. Guo, B. Yu, Y. Yuan, L. Song, Y. Hu, In situ preparation of reduced graphene oxide/DOPO-based phosphonamidate hybrids towards high-performance epoxy nanocomposites, Compos. B Eng. 123 (2017) 154—164.

[26] M. Wu, H. He, Z. Zhao, X. Yao, Electromagnetic and microwave absorbing properties of iron fibre-epoxy resin composites, J. Phys. D Appl. Phys. 33 (2000) 2398.

[27] M. Frigione, F. Lionetto, L. Mascia, A. Antonacci, Novel epoxy-silica hybrid adhesives for concrete and structural materials: properties and durability issues, Advanced Materials Research, 687, Trans Tech Publ, 2013, pp. 94—99.

[28] G. Huang, S. Chen, S. Tang, J. Gao, A novel intumescent flame retardant-functionalized graphene: nanocomposite synthesis, characterization, and flammability properties, Mater. Chem. Phys. 135 (2012) 938—947.

[29] M.-G. Périchaud, J.-Y. Delétage, H. Frémont, Y. Danto, C. Faure, Reliability evaluation of adhesive bonded SMT components in industrial applications, Microelectron. Reliab. 40 (2000) 1227−1234.

[30] M. Joshi, U. Chatterjee, Polymer nanocomposite: an advanced material for aerospace applications, Advanced Composite Materials for Aerospace Engineering, Elsevier, 2016, pp. 241−264. Available from: https://doi.org/10.1016/B978−0−08−100037-3.00008−0.

[31] K.L. White, H.J. Sue, Electrical conductivity and fracture behavior of epoxy/polyamide-12/multiwalled carbon nanotube composites, Polym. Eng. Sci. 51 (2011) 2245−2253.

[32] S. Jabeen, A. Kausar, B. Muhammad, S. Gul, M. Farooq, A review on polymeric nanocomposites of nanodiamond, carbon nanotube, and nanobifiller: structure, preparation and properties, Polym. Technol. Eng. 54 (2015) 1379−1409.

[33] S. Ganguly, S. Ghosh, P. Das, T.K. Das, S.K. Ghosh, N.C. Das, Poly (N-vinylpyrrolidone)-stabilized colloidal graphene-reinforced poly (ethylene-co-methyl acrylate) to mitigate electromagnetic radiation pollution, Polym. Bull. 77 (2019) 1−21.

[34] G. Harris, J. Lennhoff, J. Nassif, M. Vinciguerra, P. Rose, D. Jaworski, et al., Lightweight highly conductive composites for EMI shielding, SAMPE J. 36 (2000) 59−63.

[35] P. Costa, J. Nunes-Pereira, J. Oliveira, J. Silva, J.A. Moreira, S. Carabineiro, et al., High-performance graphene-based carbon nanofiller/polymer composites for piezoresistive sensor applications, Compos. Sci. Technol. 153 (2017) 241−252.

[36] A.K. Geim, Graphene: status and prospects, Science 324 (2009) 1530−1534.

[37] X. Wang, X. Li, L. Zhang, Y. Yoon, P.K. Weber, H. Wang, et al., N-doping of graphene through electrothermal reactions with ammonia, Science 324 (2009) 768−771.

[38] S.J. Rowley-Neale, E.P. Randviir, A.S.A. Dena, C.E. Banks, An overview of recent applications of reduced graphene oxide as a basis of electroanalytical sensing platforms, Appl. Mater. Today 10 (2018) 218−226.

[39] S. Park, R.S. Ruoff, Chemical methods for the production of graphenes, Nat. Nanotechnol. 4 (2009) 217−224.

[40] D. Li, M.B. Müller, S. Gilje, R.B. Kaner, G.G. Wallace, Processable aqueous dispersions of graphene nanosheets, Nat. Nanotechnol. 3 (2008) 101−105.

[41] H. He, J. Klinowski, M. Forster, A. Lerf, A new structural model for graphite oxide, Chem. Phys. Lett. 287 (1998) 53−56.

[42] T. Szabó, O. Berkesi, P. Forgó, K. Josepovits, Y. Sanakis, D. Petridis, et al., Evolution of surface functional groups in a series of progressively oxidized graphite oxides, Chem. Mater. 18 (2006) 2740−2749.

[43] S. Stankovich, D.A. Dikin, R.D. Piner, K.A. Kohlhaas, A. Kleinhammes, Y. Jia, et al., Synthesis of graphene-based nanosheets via chemical reduction of exfoliated graphite oxide, Carbon 45 (2007) 1558−1565.

[44] Y. Si, E.T. Samulski, Synthesis of water soluble graphene, Nano Lett. 8 (2008) 1679−1682.

[45] J. Zhang, H. Yang, G. Shen, P. Cheng, J. Zhang, S. Guo, Reduction of graphene oxide via L-ascorbic acid, Chem. Commun. 46 (2010) 1112−1114.

[46] Y. Shao, S. Zhang, M.H. Engelhard, G. Li, G. Shao, Y. Wang, et al., Nitrogen-doped graphene and its electrochemical applications, J. Mater. Chem. 20 (2010) 7491−7496.

[47] C. Lee, X. Wei, J.W. Kysar, J. Hone, Measurement of the elastic properties and intrinsic strength of monolayer graphene, Science 321 (2008) 385−388.

[48] T. Kuilla, S. Bhadra, D. Yao, N.H. Kim, S. Bose, J.H. Lee, Recent advances in graphene based polymer composites, Prog. Polym. Sci. 35 (2010) 1350−1375.

[49] J.-U. Lee, D. Yoon, H. Kim, S.W. Lee, H. Cheong, Thermal conductivity of suspended pristine graphene measured by Raman spectroscopy, Phys. Rev. B 83 (2011) 081419.

[50] A. Saha, C. Jiang, A.A. Martí, Carbon nanotube networks on different platforms, Carbon 79 (2014) 1–18.

[51] S. Ganguly, S. Ghosh, P. Das, T.K. Das, S.K. Ghosh, N.C. Das, Poly (N-vinylpyrrolidone)-stabilized colloidal graphene-reinforced poly (ethylene-co-methyl acrylate) to mitigate electromagnetic radiation pollution, Polym. Bull. 77 (2020) 2923–2943.

[52] M.R. Zakaria, M.H.A. Kudus, H.M. Akil, M.Z.M. Thirmizir, Comparative study of graphene nanoparticle and multiwall carbon nanotube filled epoxy nanocomposites based on mechanical, thermal and dielectric properties, Compos. B Eng. 119 (2017) 57–66.

[53] K.S. Novoselov, V. Fal, L. Colombo, P. Gellert, M. Schwab, K. Kim, A roadmap for graphene, Nature 490 (2012) 192–200.

[54] P. Das, S. Ganguly, S. Banerjee, N.C. Das, Graphene based emergent nanolights: a short review on the synthesis, properties and application, Res. Chem. Intermed. 45 (2019) 3823–3853.

[55] S. Ganguly, S. Mondal, P. Das, P. Bhawal, T.K. Das, S. Ghosh, et al., An insight into the physico-mechanical signatures of silylated graphene oxide in poly (ethylene methyl acrylate) copolymeric thermoplastic matrix, Macromol. Res. 27 (2019) 268–281.

[56] T. Wang, Y. Zheng, A.-R.O. Raji, Y. Li, W.K. Sikkema, J.M. Tour, Passive anti-icing and active deicing films, ACS Appl. Mater. Interfaces 8 (2016) 14169–14173.

[57] P. Maksimov, A. Rozhkov, A. Sboychakov, Localized electron states near the armchair edge of graphene, Phys. Rev. B 88 (2013) 245421.

[58] T. Ezquerra, M. Kulescza, F. Balta-Calleja, Electrical transport in polyethylene-graphite composite materials, Synth. Met. 41 (1991) 915–920.

[59] C. Klason, D.H. Mcqueen, J. Kubát, Electrical properties of filled polymers and some examples of their applications, Macromolecular Symposia, 108, Wiley Online Library, 1996, pp. 247–260.

[60] D. Bigg, Thermal conductivity of heterophase polymer compositions, Thermal and Electrical Conductivity of Polymer Materials, 119, Springer, 1995, pp. 1–30.

[61] I. Krupa, I. Chodak, Physical properties of thermoplastic/graphite composites, Eur. Polym. J. 37 (2001) 2159–2168.

[62] I. Krupa, I. Novák, I. Chodák, Electrically and thermally conductive polyethylene/graphite composites and their mechanical properties, Synth. Met. 145 (2004) 245–252.

[63] D.W. Sundstrom, Y.D. Lee, Thermal conductivity of polymers filled with particulate solids, J. Appl. Polym. Sci. 16 (1972) 3159–3167.

[64] Y. Hirano, S. Katsumata, Y. Iwahori, A. Todoroki, Artificial lightning testing on graphite/epoxy composite laminate, Compos. A Appl. Sci. Manuf. 41 (2010) 1461–1470.

[65] S. Donaldson, S. Mall, Delamination growth in graphite/epoxy composites subjected to cyclic mode III loading, J. Reinf. Plast. Compos. 8 (1989) 91–103.

[66] Y. Hirano, A. Todoroki, Damage identification of woven graphite/epoxy composite beams using the electrical resistance change method, J. Intell. Mater. Syst. Struct. 18 (2007) 253–263.

[67] C. Kostagiannakopoulou, T. Loutas, G. Sotiriadis, A. Markou, V. Kostopoulos, On the interlaminar fracture toughness of carbon fiber composites enhanced with graphene nano-species, Compos. Sci. Technol. 118 (2015) 217–225.

[68] Z. Yang, C. Sun, Interlaminar fracture toughness of a graphite/epoxy multidirectional composite, J. Eng. Mater. Technol. 122 (2000) 428–433.

[69] S.H. Kim, Y.-J. Heo, M. Park, B.-G. Min, K.Y. Rhee, S.-J. Park, Effect of hydrophilic graphite flake on thermal conductivity and fracture toughness of basalt fibers/epoxy composites, Compos. B Eng. 153 (2018) 9–16.

[70] P. Sampathkumaran, S. Seetharamu, S. Vynatheya, A. Murali, R. Kumar, SEM observations of the effects of velocity and load on the sliding wear characteristics of glass fabric–epoxy composites with different fillers, Wear 237 (2000) 20–27.

[71] K.K. Chawla, Composite Materials: Science and Engineering, Springer Science & Business Media, 2012. Available from: https://doi.org/10.1007/978–0–387–74365-6.

[72] B. Shivamurthy, K.U. Bhat, S. Anandhan, Mechanical and sliding wear properties of multi-layered laminates from glass fabric/graphite/epoxy composites, Mater. Des. 44 (2013) 136–143.

[73] S. Basavarajappa, S. Ellangovan, K. Arun, Studies on dry sliding wear behaviour of graphite filled glass–epoxy composites, Mater. Des. 30 (2009) 2670–2675.

[74] S. Basavarajappa, G. Chandramohan, J.P. Davim, Application of Taguchi techniques to study dry sliding wear behaviour of metal matrix composites, Mater. Des. 28 (2007) 1393–1398.

[75] R. Heydenreich, Cryotanks in future vehicles, Cryogenics 38 (1998) 125–130.

[76] M.-G. Kim, S.-G. Kang, C.-G. Kim, C.-W. Kong, Tensile response of graphite/epoxy composites at low temperatures, Compos. Struct. 79 (2007) 84–89.

[77] J.G. Funk, G.F. Sykes, The effects of radiation on the interlaminar fracture toughness of a graphite/epoxy composite, J. Compos. Technol. Res. 8 (1986) 92–97.

[78] K. Balasubramanian, Reinforcement of poly ether sulphones (PES) with exfoliated graphene oxide for aerospace applications, IOP Conference Series: Materials Science and Engineering, 40, IOP Publishing, 2012, p. 012022.

[79] D.A. Hawkins Jr, A. Haque, Fracture toughness of carbon-graphene/epoxy hybrid nanocomposites, Proc. Eng. 90 (2014) 176–181.

[80] S. Cui, C. Cui, J. Xie, S. Liu, J. Shi, Carbon fibers coated with graphene reinforced TiAl alloy composite with high strength and toughness, Sci. Rep. 8 (2018) 1–8.

[81] M. Gagné, D. Therriault, Lightning strike protection of composites, Prog. Aerosp. Sci. 64 (2014) 1–16.

[82] P. Feraboli, M. Miller, Damage resistance and tolerance of carbon/epoxy composite coupons subjected to simulated lightning strike, Compos. A Appl. Sci. Manuf. 40 (2009) 954–967.

[83] J. O'Loughlin, S. Skinner, General Aviation Lightning Strike Report and Protection Level Study, Office of Aviation Research, Federal Aviation Administration, 2004.

[84] F.A. Fisher, J.A. Plumer, Lightning protection of aircraft, National Aeronautics and Space Administration, Sci. Tech. 1008 (1977).

[85] P. Mahapatra, R.J. Doviak, V. Mazur, D.S. Zrnić, Aviation Weather Surveillance Systems: Advanced Radar and Surface Sensors for Flight Safety and Air Traffic Management, 8, Let, 1999.

[86] A.E. Zantout, O.I. Zhupanska, On the electrical resistance of carbon fiber polymer matrix composites, Compos. A Appl. Sci. Manuf. 41 (2010) 1719–1727.

[87] Z. Tianchun, W. Jin, M. Keyi, F. Zhenyu, Simulation of lightning protection for composite civil aircrafts, Proc. Eng. 17 (2011) 328–334.

[88] A.A. Obaid, S. Yarlagadda, Structural performance of the glass fiber–vinyl ester composites with interlaminar copper inserts, Compos. A Appl. Sci. Manuf. 39 (2008) 195–203.

[89] G. Gardiner, Lightning Strike Protection for Composite Structures, 2006. Online Article, 2008.

[90] B. Zhang, V.R. Patlolla, D. Chiao, D.K. Kalla, H. Misak, R. Asmatulu, Galvanic corrosion of Al/Cu meshes with carbon fibers and graphene and ITO-based nanocomposite coatings as alternative approaches for lightning strikes, Int. J. Adv. Manuf. Technol. 67 (2013) 1317–1323.

[91] J. Gou, Y. Tang, F. Liang, Z. Zhao, D. Firsich, J. Fielding, Carbon nanofiber paper for lightning strike protection of composite materials, Compos. B Eng. 41 (2010) 192–198.

[92] D. Zhang, L. Ye, S. Deng, J. Zhang, Y. Tang, Y. Chen, CF/EP composite laminates with carbon black and copper chloride for improved electrical conductivity and interlaminar fracture toughness, Compos. Sci. Technol. 72 (2012) 412–420.

[93] T.-W. Chou, L. Gao, E.T. Thostenson, Z. Zhang, J.-H. Byun, An assessment of the science and technology of carbon nanotube-based fibers and composites, Compos. Sci. Technol. 70 (2010) 1–19.

[94] G. Morales, M. Barrena, J. Gómez de Salazar, C. Merino, Conductive CNF-doped laminates processing and characterization, J. Compos. Mater. 45 (2011) 2113–2118.

[95] D. Domingues, E. Logakis, A. Skordos, The use of an electric field in the preparation of glass fibre/epoxy composites containing carbon nanotubes, Carbon 50 (2012) 2493–2503.

[96] B. Zhang, R. Asmatulu, S.A. Soltani, L.N. Le, S.S. Kumar, Mechanical and thermal properties of hierarchical composites enhanced by pristine graphene and graphene oxide nanoinclusions, J. Appl. Polym. Sci. 131 (2014). Available from: https://doi.org/10.1002/app.40826.

[97] N. Yamamoto, R.G. de Villoria, B.L. Wardle, Electrical and thermal property enhancement of fiber-reinforced polymer laminate composites through controlled implementation of multi-walled carbon nanotubes, Compos. Sci. Technol. 72 (2012) 2009–2015.

[98] M. Raimondo, L. Guadagno, V. Speranza, L. Bonnaud, P. Dubois, K. Lafdi, Multifunctional graphene/POSS epoxy resin tailored for aircraft lightning strike protection, Compos. B Eng. 140 (2018) 44–56.

[99] O. Parent, A. Ilinca, Anti-icing and de-icing techniques for wind turbines: critical review, Cold Reg. Sci. Technol. 65 (2011) 88–96.

[100] K. Al-Khalil, T. Ferguson, D. Phillips, K. Al-Khalil, T. Ferguson, D. Phillips, A hybrid anti-icing ice protection system, in: Proceedings of the Thirty-Fifth Aerospace Sciences Meeting and Exhibit, 1997, pp. 302. Available from: https://doi.org/10.2514/6.1997–302.

[101] W.O. Valarezo, F.T. Lynch, R.J. McGhee, Aerodynamic performance effects due to small leading-edge ice (roughness) on wings and tails, J. Aircr. 30 (1993) 807–812.

[102] F.T. Lynch, A. Khodadoust, Effects of ice accretions on aircraft aerodynamics, Prog. Aerosp. Sci. 37 (2001) 669–767.

[103] J.S. Cornell, D.A. Pillard, M.T. Hernandez, Comparative measures of the toxicity of component chemicals in aircraft deicing fluid, Environ. Toxicol. Chem. Int. J. 19 (2000) 1465–1472.

[104] V. Volman, Y. Zhu, A.-R.O. Raji, B. Genorio, W. Lu, C. Xiang, et al., Radio-frequency-transparent, electrically conductive graphene nanoribbon thin films as deicing heating layers, ACS Appl. Mater. Interfaces 6 (2014) 298–304.

[105] S. Fan, X. Jiang, C. Sun, Z. Zhang, L. Shu, Temperature characteristic of DC ice-melting conductor, Cold Reg. Sci. Technol. 65 (2011) 29–38.

[106] R. Gupta, K. Rao, K. Srivastava, A. Kumar, S. Kiruthika, G.U. Kulkarni, Spray coating of crack templates for the fabrication of transparent conductors and heaters on flat and curved surfaces, ACS Appl. Mater. Interfaces 6 (2014) 13688–13696.

[107] G.G. Koenig, C.C. Ryerson, An investigation of infrared deicing through experimentation, Cold Reg. Sci. Technol. 65 (2011) 79–87.

[108] M. Mohseni, A. Amirfazli, A novel electro-thermal anti-icing system for fiber-reinforced polymer composite airfoils, Cold Reg. Sci. Technol. 87 (2013) 47–58.

[109] Z. Goraj, An overview of the deicing and anti-icing technologies with prospects for the future, in: Proceedings of the Twenty-Fourth International Congress of the Aeronautical Sciences, 29 (2004).

[110] A.-R.O. Raji, T. Varadhachary, K. Nan, T. Wang, J. Lin, Y. Ji, et al., Composites of graphene nanoribbon stacks and epoxy for joule heating and deicing of surfaces, ACS Appl. Mater. Interfaces 8 (2016) 3551–3556.

Packaging applications of polymer-graphene composites

Prashant Gupta[1] and B.G. Toksha[2]

[1]Department of Plastic and Polymer Engineering, Maharashtra Institute of Technology, Aurangabad, India, [2]Basic Sciences and Humanities Department, Maharashtra Institute of Technology, Aurangabad, India

24.1   Introduction

The domain of nanotechnology has already witnessed numerous nanomaterials which have contributed toward a better environmental footprint for use in areas such as biotechnology, pharmaceutics, computation, electronics, medicine, energy, agriculture, textile, defense, food, automotive, etc. One of the wonder materials reported [1] by Professor Sir Andre Geim and Professor Sir Kostya Novoselov in 2004 that is graphene, is a two-dimensional flat sheet of carbon atoms consisting of sp2 hybridized carbon atoms arranged in a honeycomb structure resembling a unit hexagonal lattice. It has been the point of attraction for researchers in both the scientific community and industries [2,3]. A huge extended family of graphene-based nanomaterials includes graphene, graphene oxide (GO), graphite, expanded graphite, and graphite oxide. As one of the thinnest and strongest material, it possesses many outstanding properties such as   high electrical conductivity (around 6000 S/m) [4], high surface area of around 2630 m^2/g [5], quantum hall effect, superior ambient temperature electron mobility (10,000 cm^2/V/S) [1], very high thermal conductivity (3000−5000 W/m/K) [6], optical transmittance (97.7%), and excellent load-bearing characteristics with Young's modulus and fracture strength of about 1 TPa and 125 GPa respectively [7].

The area of nanocomposites came into light since the stronger nature of one material may be beneficial to the other and vice-versa. They quickly became dominant in applications where lightweight and high strength to weight ratios gave the advantage at lower cost in applications where there were net energy savings either through the material processing cost or use of novel materials which were cheaper. GO is the most popular of the lot due to better gas permeation characteristics, mechanical properties, thermal, electrical, and fire-retardant properties to name a few [8,9].

The current market volume and expansions in the future considering the consumer requirements and bringing to fore insights can help stakeholders in identifying the opportunities as well as challenges. According to a new report by Grand View Research Inc., the global food packaging market is going to be valued at around 457 billion USD by 20027, at a Compounded Annual Growth Rate (CAGR)

Polymer Nanocomposites Containing Graphene. DOI: https://doi.org/10.1016/B978-0-12-821639-2.00023-9

of 5% over the period of 5−7 years. The growing demand for packaged food items due to changes in living standards, modernization along with hygienic eating habits of the worldwide population is going to bolster the growth of the food packaging market. The aesthetics of polymers and their versatility on offer in terms of application concerns such as transparency, ease in processing, low cost, chemical and biological inertness, durability, and low gas permeation characteristics further strengthen the cause of an increase in the usage of polymers in the sector of application. Along with the same, the reported statistics for the electronic packaging industry reveal an expansion at a very high rate in the recent past and future at a CAGR of around 18%. The driving force for this market expansion is the rising demand for electronic products, such as televisions, set-top boxes, digital cameras, Wi-Fi chipsets, increasing adoption in automotive (conventional, electric, and hybrid vehicles), memory devices, processors, analog circuits, discrete power devices, and sensors. The history and evolution of the electronics package from the vacuum tube to a multichip "system in a package" is depicted in Fig. 24.1 [10].

Polymers are used in the electronics industry as packaging materials, as insulants, dielectrics, thermal transfer media, and to add protection to fine joints and delicate components against vibration, mechanical and thermal shock, and atmospheric pollutants. Emission of volatiles thermal degradation of a polymeric

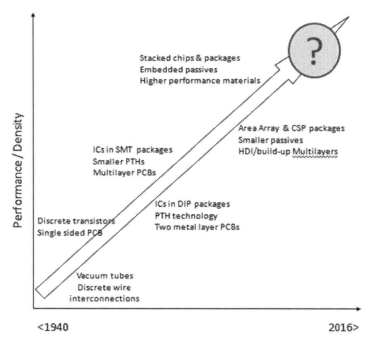

Figure 24.1 History and evolution of the electronics packaging (Ref: An overview of electronic manufacturing technology evolution and trends, Joseph Fjelstad, Verdant Electronics).

Table 24.1 Polymer materials used in electronic packaging and their application [11,12].

Sr. No	Polymer material	Application
1	Cellulose, silk, shellac, gelatin	Substrate; dielectric
2	Poly(vinyl alcohol) (PVA)	Substrate; dielectric
3	Polydimethylsiloxane	Substrate; dielectric
4	Polylactide (PLA) substrate	Substrate; dielectric
5	Polycaprolactone (PCL)	Dielectric
6	Poly(glycerol-co-sebacate) (PGS)	Dielectric
7	Poly(lactic-co-glycolic acid)	Substrate; dielectric
8	(PLGA) substrate	Substrate; dielectric
9	Polyaniline (PANI) conductor (doped)	Conductor
10	Polypyrrole (PPy) conductor (doped)	Conductor
11	Poly(3,4-ethylenedioxythiophene)	Conductor
12	Epoxies	Overmold
13	Filled epoxies	Overmold
14	Silica-filled anhydride resin	Underfills
15	Conductive adhesives	Die bonding, interconnects
16	Laminated epoxy/glass substrates	Substrate
17	Polyimide dielectric	Dielectric
18	Photosensitive polymers	Photomasks

material, excessive wear of brushes and bearings, corrosion of metal parts by PVC and PTFE insulants because of acidic component release, corrosion caused by the breakdown products are some of the challenges in using polymers in the electronic packaging industry. A list of various polymer materials used in electronic packaging and their application is given in Table 24.1.

24.1.1 Polymer graphene nanocomposites in food packaging

The packaging area has seen a lot of technological advancements in the recent past. Despite the versatility on offer, there has been a lot of debate over plastic waste pollution. Polyethylene, a commodity polymer that has enjoyed the top position for decades now has been threatened for some time due to our bad habits. Even though polyethylene has no competition due to the versatility in properties on offer with the price, research today has been focused on the creation of such alternatives which are biodegradable in nature. Biodegradable polymers have been used for some time now in applications such as biomedical, agriculture, and packaging. For applications other than biomedical ones such as wound management, drug delivery, dental, orthopedic devices, tissue engineering, intestinal, and cardiovascular applications, the areas of agriculture and packaging, in particular, have not been able to increase its market enough to replace polyethylene which rules the commodity packaging market. The primary reasons behind the same are the properties and cost of biodegradable polymers. The manufacturing capability of synthetic biodegradable polymers such as polyhydroxy butyrate (PHB), polylactic acid, polycaprolactone, etc. also remains in question since naturally occurring biopoly-mers production is always going to be limited. One can increase the production of corn starch for instance by increasing its agricultural output but to convert the grown corn crops into a biopolymer questions the intention behind it. The questions that arise are:

• Do we really want to grow food on the agricultural land which is decreasing by each min-ute due to modernization?
• Do we want to grow plastic on the food growing soil in a world where people are dying each minute due to the nonavailability of food?

In spite of the high cost of biopolymers, they have tapped into high-end packag-ing markets such as electronics, automobiles, etc. and due to increasing pressures of governments, researchers (both academic and industrial) have been striving to find an alternative to the problem of plastics waste.

The inferior mechanical property problem can be solved by adding graphene into the biodegradable polymer matrix to make it suitable for use and biodegrad-ability. Several researchers have worked on the area to reduce the dependency on petroleum-based plastic products and have published their literature for further research work. The use of graphene (high modulus) has been reported to be effec-tive in the reinforcement of lower modulus polymers exhibiting an increase in the mechanical properties. With GO, the polymer-filler interface interaction along with its distribution and orientation in the polymer matrix decide the outcome of the reinforcing effect [13].

24.1.1.1 Polylactide and graphene-based packaging materials

Narimissa et al. reported the reinforcing effect of nano graphite platelets (NGP) on polylactide polymer. They studied different compositions of 0−10 wt.% of NGP and found that the load-bearing properties deteriorated above 3 wt.% filler loading due to poor intercalation of the filler platelets in the polymer matrix above this level and for-mation of aggregates at higher loadings. (3%−10% filler) [14]. However, the elonga-tional properties that is elongation at break (EAB) of composites were poor right from the addition of even 1% of NGP which has been evidently reported by other authors as well [15,16]. The addition of laminar functionalized graphene sheets on polylactide nanocomposites exhibits the increase in tensile strength (TS) and tensile modulus (TM) at lower levels ($\sim 0.1\%$) of addition. Any further addition of graphene sheets decreased both properties due to agglomeration due to pre-incorporation of filler in oligomers of lactic acid. A vis to vis comparison done by incorporation of filler directly into polylactide polymer by melt mixing wherein lower values were observed at initial loading of filler in the polymer matrix due to agglomeration as confirmed by microscopic analysis. At higher filler levels, TM values increased but were no different to that obtained by pre-incorporation route of the filler. The nature of embrittlement observed in both the filler incorporation routes was quite similar due to filler aggregation in both the incorporation methodologies. The addition via pre-incorporation route inspired enhanced barrier properties evident by the decrease of oxygen and water vapor permeability by 45% and 41% respectively at 2% filler loading. This is largely achieved due to the creation of a tortuous path for the O_2 molecules to permeate through the polymer's free volume [17].

A comparison has been reported over property enhancement of PLA in between physical dispersion of (1) graphene nanoplatelets (GNP) in the polymer (organic) matrix and (2) amino-functionalized silica along with its covalent bonding of polymer (organic) phase. The materials in the form of compression-molded slabs were addressed to ASTM D3363 that is Pencil Hardness testing. The role of fillers can be ascribed by the improved resistance towards indentation or scratch. In comparison with neat PLA where crack onset starts at around 4.8 N PLA/modified silica and PLA/GNP shows the same phenomenon at 7.2 and 7.6 N respectively. Fig. 24.2A show the crack difference between three materials with neat PLA cracks orientation in the same direction whereas the composites show less deformation with smaller fractures seen at higher loads. Due to the highest stiffness of GNP, PLA/GNP composite exhibits increased stiffness values only on physical interaction basis. In nano-silica, amino functionality combines with ester functional groups of PLA via the aminolysis process and thus offers good stiffness characteristics [18].

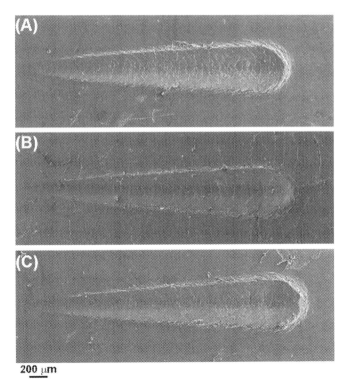

200 μm

Figure 24.2 Scanning Electron micrographs after load (progressive) scratch tests for scratch residue patterns of (A) neat PLA, (B) PLA/modified silica, and (C) PLA/GNP [18]. *PLA*, Polylactide.

Since the dispersion of graphene in the polymer matrix poses problems, a different methodology of application of a uniform coating of a certain masterbatch (prepared from graphene and lactic acid oligomer and polymerization of the same) to improve the dispersion of graphene before melt processing. Table 24.1 shows that there is a slight increment in TS and TM values upon the addition of 0.2% graphene in a simple way. However, upon coating addition, TS values decrease with a slight increase in TM values. The decrease in TS in composites made through masterbatch mode due to the presence of short-chain PLA was used in its oligomeric form to make coat graphene on which polycondensation was carried out to make masterbatch. This masterbatch was then employed to reinforce high molecular weight polymeric lactic acid. The use of short-chain PLA improves the dispersion and compatibility of the composite. The decrease in TS for masterbatch processed composites because short polymer chains align faster than long chains under tensile load and the values further decrease in 0.2% graphene due to increased intensity of short chains in the composite. The increase in TM values in masterbatch processed composites can be attributed to better compatibility (especially in lower graphene content) in comparison to the regular composite. The improvement of EAB (17%) in 0.05% masterbatch processed composites which is the highest among all four tested experimental samples indicates the increase of chain mobility due to short PLA chains of masterbatch in comparison with neat PLA. However, upon the increase of masterbatch percentage, a decrease in EAB is reported which might be due to lesser strength and chain mobility due to which break is faster in comparison with the rest of the samples [19] (Table 24.2).

The addition of GO (5%) in Polylactide-co-glycolide (PLCG-50/50) exhibited an increase in TS and TM by approximately 2.5 and 4.7 times respectively than neat PLCG material is attributed to excellent strength of GO nanosheets. However, the brittleness of the composite was evident from the decrease of EAB values due to physical defects arisen as a virtue of a small percentage of GO resulting in few linkages with the polymer matrix. Also, the DMA measurement of these composites indicated a significant influence of the reinforcement on the thermomechanical properties. The storage TM or the elastic component of PLGA improved around 14

Table 24.2 Mechanical properties of PLA and graphene filled composites processed by various routes [19].

Sample	Ultimate TS (N/mm^2)	TM (N/mm^2)	EAB (%)
Neat PLA	29.5 ± 3.8	4.3 ± 0.56	12.93 ± 3.4
PLA + 0.2% graphene (melt processing mode)	31.3 ± 2.8	4.58 ± 0.61	11.7 ± 1.7
PLA + 0.05% graphene (masterbatch mode)	25.2 ± 3.3	5.47 ± 1.3	15.6 ± 1.7
PLA + 0.2% graphene (masterbatch mode)	24.0 ± 0.3	4.87 ± 0.35	9.3 ± 1.9

times at 2% and 27 times at 5% loading of GO at 25°C. At lower temperatures of around 10°C−15°C, it reduced for polymeric PLGA, and the 5% of DO stabilized the PLGA-GO composite up to 32°C, because of the stiffening of GO nanosheet concentrations at higher temperatures. The glass transition temperature (T_g) of neat PLGA was around 15°C which substantially increased to 32°C with filler addition in the polymeric matrix. This can be attributed to a strong interface build-up with chemical linkages between the polar functional groups of GO nanosheets and polymeric chains in PLGA restricting the mobility of chains near the GO nanosheets resulting in a large volume of polymer matrix thereby increasing the T_g of the system. Also, the Ultra Violet (UV) visible spectra confirmed the dispersion of GO in the PLGA matrix, and the films were physically observed as dark brown. The O_2 transmission rate of the composites was slightly lower than neat PLGA with 5% GO exhibited 47 cc/m^2 day. These values are significantly lower than commercial packaging polymers such as polyethylene by 42−50 times, polypropylene by 50−56 times, and polystyrene by 68 to 80 times [20].

There have been similar trends in mechanical properties for 1% addition of GO into PLA/PEG matrix which improved the TS by 20% and deteriorated the EAB values by 25%. The addition of clove essential oil (CO) decreased the TS by around 38% but its plasticizing effect resulted in a threefold increase in EAB values. The addition of CO and GO had reverse effects on the polymer wherein CO contributed to mobility by plasticization of polymer chains whereas GO restricted the mobility of polymer chain length [21]. Another composite of PLGA/GO where GO sheets are modified by zinc oxide nanoparticles which exhibit more interfacial contact than the usual physical interaction of GO with PLA. This, in turn, improves the tensile properties of the composite (with up to 1% GO modified with ZnO) by improving strength values by around 20% which can be attributed to better dispersion and the planer structural geometry of GO sheets modified by ZnO thereby making it favorable for interfacial interactions with the polymer matrix. However, a similar percentage decrease in the EAB values indicated the restricted mobility of polymer chains thereby increasing the brittleness of the composite. This composite also possesses anti UV abilities which can work as a protection against UV light. It exhibits good efficiency in UV light shielding while having good transmittance in the UV range which is another feature for packaging material. Another food packaging aspect is antimicrobial properties which in this case are boosted due to the presence of GO complemented by the effect of ZnO (a known antimicrobial material) A remarkable effect was observed at lower concentrations of 0.2% GO-ZnO when tested against *Staphylococcus aureus* and *Escherichia coli* [22]. GO was also reported to prepare polyvinyl-N-carbazole-GO nanocomposites with 3 wt.% GO exhibiting excellent antibacterial properties [23].

24.1.1.2 Polyvinyl/Graphene nanofiller (GNF) nanocomposites

The mechanical attributes of PVA/GO composite films vary in concentrations of (0.1%−0.5%) with varying temperatures. The interaction between PVA and GO was evident with the 0.2% curve of storage modulus versus temperature almost four times higher in comparison to the PVA film curve thereby revealing good mechanical strength of the composite. The water uptake analysis exhibits a decrease in the water uptake values with an increase in GO content of the composite. The sample with the highest concentration of GO (0.5%) shows 50% water uptake values in comparison to neat PLA films. Furthermore, the dimensional stability of the composite with 0.5% GO is also 50% to that of neat PLA films [24]. The reinforcement of PVA films by GO nanosheets (GON) up to 2% increases the TS and EAB which attributes to GO nanosheets dispersion in the PVA matrix. The increase of 50% TS is observed in 0.3% exfoliated graphite oxide (XGO) with a corresponding increase of 12% in EAB. The addition of 2% XGO corresponds to an increase in TS and EAB by 50% and 22% respectively which indicates that at lower nanosheet loading, the dispersion achieved was very good. The TM values were reported to be highest for 2% XGO which are indicative of the higher amount of filler that already has a higher modulus. The author proposed the packaging system on the basis of gas transmission rate properties and water vapor barrier properties which are crucial for such applications. For 0.3% and 2% XGO, the O_2 permeation values were reported to be 64% and 24% respectively of permeation in comparison with neat PVA which was around 18.5 cm^3/m^2 per day. For water vapor permeation, the values were 93% and 79% in comparison with neat PVA which was tested to be 709 g/m^2 per day. The system can be employed in delaying the ripening of fruits since ethylene production, a natural plant hormone responsible for initiation in fruit ripening can be reduced by 50% if the O_2 levels in an atmosphere are around 2.5% [25]. The tests were carried out on banana packaging and weight loss along with changes in the physical state were observed over a period of 15 days. It was found that neat PVA bags had 45% weight loss in comparison to 33% in XGO filled films (both 0.2% and 2% had similar results) as shown in Fig. 24.3. It can be seen in the graphical image inset that bananas stored in neat PVA film created holes due to poor mechanical properties [26].

The dependence of tensile properties on relative humidity (20%, 60%, and 80%) for PVA/GO composites has been investigated and reported. The necking behavior is observed in neat PVA and 1% GO filled PVA after the yield point at 20% RH conditions may be due to the semi-crystalline state of PVA. It might also be observed due to the rearrangement of polymer chains in the crystalline portion of the polymer in the direction of application of the tensile forces. Upon increase in GO content to 3%, the strain decreased by almost 10 times 10.1% $\pm$ 2.4%, and necking was not seen at all which can be attributed to the higher GO content interfering with PVA chains rearrangement. The increase in TM with an increase in the GO content is very little since the films were in the glassy state.

At 60% RH, ductility develops in PVA with its TM going down almost 15 times to 101.6 MPa and the EAB increasing around 2.1 times to 386.9%. At 1% GO, TS, and TM increased but EAB decreased slightly to 345.0% $\pm$ 9.8% due to the lack of functional groups which can induce chemical bonding. TM values increased linearly with the increase in the percentage of GO present in the composite. The increase in TM values at 10 wt.% GO incorporation was seven times that of neat PVA films. At 80% RH, TM of neat PVA further went down from 101.6 $\pm$ 11.1 to 5.46 $\pm$ 0.70 MPa and EAB improved from 386.9% $\pm$ 13.4% to 1275.5% $\pm$ 52.6%.

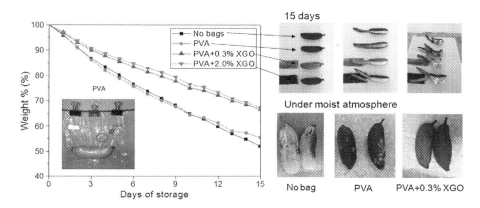

Figure 24.3 Food packaging test carried out with banana and respective weight loss reported over a period of 15 days. The inset image in the graph shows holes in pure PVA film [26]. *PVA*, Poly(vinyl alcohol).

PVA behaves somewhat like rubber as it has adsorbed 4 times more water at 60% RH. The strain values were over 1000%. The TM, in this case, increased almost linearly with the content of GO but the TS and EAB values decreased above 5 wt.% GO attributed to the formation of an interconnected network where GON have no chemical interfacial link with each other leading to worsening of mechanical properties [27].

Quasi static tensile and creep testing was used to investigate the influence of varying PVA molecular weights on GO/PVA composites. The results exhibit that the composites films in comparison to neat PVA films have four times the TS and three times the TM values at around 255.7 and 23.5 GPa respectively. The trends show increasing results in mechanical properties with an increase in molecular weight of PVA due to good connection of PVA with platelets of GO by inter- and intra-linkage with higher molecular weight molecules than low molecular weight ones resulting in better stress transfer through the direction of reinforcement (GO) plane. The introduction of a tender phase (polymer) lowers the fillers volume fraction thereby decreasing the modulus of the system. On the contrary, the hydrogen bonding ensures enhanced interlayer interactions, thereby improving the load-bearing characteristics of GON resulting in an increase of TS and TM. Upon modification, the TS of borate treated GO/PVA films is 360 MPa which is two to three times and the toughness is around four times that of nacre, an organic-inorganic biomaterial [28].

Nanocellulose fiber (CNF) addition with GO into PVA synergistically improves TS. The addition of 8% CNF and 0.6% GO increases the TS from 43 to 80 MPa with an increase of 43 to 69 MPa attributed to the addition of CNF alone [29]. PVA films with an addition of 1% of α-ZrP/GO (5:1) blend exhibits TS of 62.8 and TM of 1643 MPA, which is an increase of 50% and 44% from neat PVA due to increase in crystallinity and intermolecular forces amongst hybrid nanofillers and PVA matrix [30].

24.1.1.3   Chitosan and graphene-based composites

Apart from the lack of mechanical properties, chitosan (CS) a film-forming bio-polymer possesses excellent antimicrobial and antifungal properties with suitability for use as food packaging. The use of GO in CS for mechanical reinforcement (0.5%−2%) increases the toughness with a significant increase in TS and reduction in EAB. This increase observed is due to the dispersion of GO and its surface bonding with CS by covalent bonds and H-bonds [31,32].

GO was reduced using caffeic acid (rGO) by hydrothermal reduction process and incorporated in CS with respect to the weight of the latter with the use of glycerol to make bio-nano composite films. The reduction saved the necessary O_2 containing groups to interact with amine groups of CS and have reinforcing action. The addition of 50% rGO causes the TS and TM of the films to double from 13 MPa and increase five to six times from 0.47 GPa. The decrease in flexibility which is reported through EAB is around 10 times its original value of 20%. The result claims to be the best of tensile values through a green route. With no grafting/other reinforcements, the improvement in tensile properties is due to the effect of rGO. In contradiction to other reported studies which indicate agglomeration of GO at higher concentrations, this study shows the best results as high as 50% rGO-CS composition indicating the dispersion characteristics of rGO in CS polymer matrix enabling an efficient load transfer between matrix-reinforcement phases. The antioxidant activity of CS was tested by the ABTS inhibition method to be just 1%. The introduction of rGO in the CS matrix had a significant antioxidant activity with 50% rGO-CS showing around 82% activity which started at 52% with the initial sample of 25% rGO-CS. The addition of rGO also reinforces the water-resistance of the films with CS films reporting around 34% which is improved by 25% rGO-CS reporting 24% of weight loss (solubility) in an acidic aqueous medium over a period of 7 days [33].

The addition of GO (0.5%−2%) in CS exhibited a linear increase of around 35% in TS and 80% in TM to 44.4 ± 0.5 and 2715 ± 230 respectively with the addition of GO from neat CS films. The addition of glycerol as a plasticizer influences the tensile properties significantly. There is a reduction of TS and TM by 65% and 62% respectively in the same sample processed with the addition of glycerol whereas the EAB has increased by around 111%. This can be attributed to the reduction in intermolecular forces of attraction in CS and the increase in the mobility of polymer chains by the addition of glycerol. The reinforcing effect of GO homogeneously dispersed in the CS matrix further aided by good interfacial adhesion between them enables the efficient load transfer between the filler and the matrix. The functional groups namely, hydroxyl and carboxylic in GO along with amino, primary, and secondary hydroxyl groups in a glucosamine unit of CS enable the strong H-bond between CS and surface of GO may form evident from FTIR results leading to the mechanical property enhancement of CS [34].

The effect of the degree of oxidation of GO on the properties of CS/GO films has been studied. The ratios of graphite:$KMNO_4$ have been varied from 1:2 to 1:8. The increase of concentration of $KMNO_4$ increased the oxygen-containing groups in the composites. Upon comparison with neat CS films, the films (CS/GO4) made using a 1:8 ratio showed a 40% decrease in the moisture content. It is pertinent to note that the addition of only graphite did not have any pronounced effect on the moisture content of the composite films. The water solubility of CS/Graphite films reported a 50% reduction in the values in comparison with neat CS films. This was due to the hydrophobic nature of graphite due to the absence of oxygen-containing functional groups and stacked geometry in graphene layers in which C atoms are covalently bonded which thereby reduced the hydrophilicity of the composite film. The C atoms and H_2O molecular attractions are not strong enough to displace C-C interactions owing to lower water solubility. This also leads to increased wet state mechanical properties which can make it a potential for food packaging material. The water vapor permeability is important to judge the performance of packaging materials as water hampers the shelf life of a packaging product. CS/GO2 films with a 1:4 ratio exhibit a 37% decrease which can be attributed to the rough surface microstructure of the composite. The mechanical properties of neat CS films, TS of 5.94 ± 1.47 MPa, reportedly increased by five times, TM of 1.11 ± 0.12 MPa increased by four times and EAB of 12.92 ± 1.3 decreased by 25% in CS/GO4 samples. This improvement in mechanical properties with an increase in oxidation degree of GO can be attributed to exfoliation and dispersion of GO in the CS matrix. A stronger H-bonding between hydroxyl and amine of CS with oxygenated functional groups in GO is ensured with an increase in oxidation degree [35].

The formation of high quality, smooth and flexible composite films by blending CS with polyvinylpyrrolidone (PVP) and GO was reported wherein the tensile properties of CS are influenced by blending with 50% PVP and GO. (0.75% and 2%) The EAB increased to 10.8% and TM decreased for CS/PVP blended film and it became more ductile than neat CS film. The formation of a compact network leads to the improvement of tensile properties notably TS, TM, and EAB [36].

The increase in antimicrobial properties can be made by incorporating cinnamaldehyde (CA) for functionalization of CS by the formation of a reversible Schiff base which allows the release of active CA. The use of graphene stacks (EGS) to the CA modified CS have reported to increase the mechanical properties and antifungal characteristics. The increase in CA concentration (0.1%−0.5%) in CS increased the TS, TM and EAB of CA-CS system by 25% to 50% ± 7 MPa, 25% to 2.5 ± 0.6 GPA and 33% to 15% ± 3% respectively in 0.5 wt.% CA. In addition to that, an increase in EGS in each of these three concentrations at varying percentages from 1.5−6 wt.% further increased the strength properties identically in all three concentrations of CA. The reduction of EAB was similar too for all three cases. As shown in Fig. 24.4, the antifungal characteristics were checked for neat CS and at different concentrations of CA addition in CS in terms of antifungal growth on packaged white bread slices against controlled samples as PET films. Fig. 24.4 shows an increase in the efficiency against molds (Rhizopus Stolonifer) with an increase in CA concentration. The presence of humidity in an open environment triggered the release of CA which restricted fungal growth in a favorable humid environment. This iterates that CA (in small proportions) is sufficient enough to be effective against the natural action of molds and functionalization of CS [37].

Time (days)	Control (PET)	Chitosan	0.1 CA	0.25 CA	0.5 CA

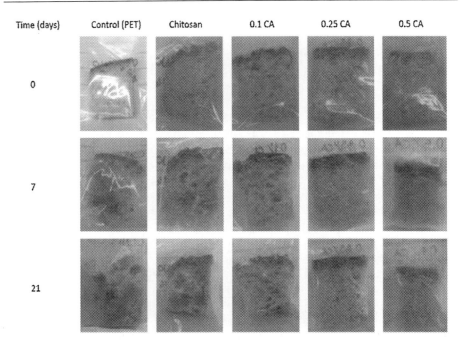

Figure 24.4 Chitosan films with different CA concentrations against *Rhizopus Stolonifer* over a period of 21 days. The control sample is PET and experimental are neat CS, CS-0.1% CA, CS-0.25%CA and CS-0.5%CA [37]. *CA*, cinnamaldehyde; *CS*, chitosan.

24.1.1.4 Miscellaneous biopolymer-graphene based packaging materials

An active packaging has been proposed inspired by natural mussel chemical engineering and based on CNF/gelatin loaded with GO nanohybrids with an increase in functional and antibacterial properties. The nanosheets of GO were reduced and dip coating was carried out to give GO functionalized with polydopamine. (PGO) Ag nanoparticles were further prepared and inserted in the same. The mechanical properties such as TS and toughness of films were reported to be lowest by neat gelatin films (around 7 MPa and 2.5 MJ/m^3 respectively) and highest was for Gelatin/CNF-PGO/Ag films. (35.99 MPa and 13.49 MJ/m^3 respectively) The water vapor permeability of the Gelatin/CNF-PGO/Ag films reduced by around 51.6%, moisture content was lowered by 25%, solubility and water uptake were lowered by 30%. It also exhibited strong antibacterial nature against *S. aureus* and *E. coli* due to the addition of Ag which has strong antimicrobial activity. The film was also tested for UV shielding capacity and thermal stability and exhibited significant improvements in both of them [38].

PHB is a natural microbial polyester with a moderate mechanical, barrier, and thermal properties. The incorporation of GNP at varying concentrations of 0.1%−1.3% has been reported. The TS of the composites (1.3% GNP) increased from 4.5 to 12.2 MPa due to uniform dispersion and EAB reduced from 15% to 8.5% due to brittleness caused by the addition of GNP in the PHB matrix. The oxygen and water vapor permeability values for neat PHB are 1.53 cm^3 mm/m^2 d^2 atm and 9.26 g mm/m^2 d^2 atm respectively. The values in general decreased with the addition of GNP with 0.7% GNP loading as the optimum value with the values going down to 0.4 cm^3 mm/m^2 d^2 atm and 4 g mm/m^2 d^2 atm respectively. At the highest concentration of GNP, the values were still marginally lower than neat PHB. The shelf life of potato chips and milk was found to increase from 60 to 245 days and 6 to 26 days respectively which is roughly four times that of pristine PHB films. An improvement of 10°C was observed in melting temperature (T_m) with an increase in the concentration of GNP from 0.1% to 0.7% which can be attributed to basic core measure reduction necessary for the setting of PHB's stable and thick core. GNP acts as a nucleating agent by providing a core around which the polymer chains reorient themselves thereby changing the Tm of the composite films. The films lacked transparency due to the addition of graphene which I opaque in nature. At 0.7% GNP levels, the reduction was around three times that of neat PHB films to both UV and visible light. However, it can serve as a boon for storing light-sensitive food items such as flavors, pigments, vitamins, etc. The cytotoxicity tests conducted by fluorescence microscopy analysis exhibited a higher mortality rate of the cells with increasing GNP content in the polymer as the cells contained corroded or ruptured cell wall probably due to the slicing effect of sharp corners/edges in GNP. As shown in Fig. 24.5, trypan blue-stained macrophage cells can be only seen in control experimental samples and other experiments performed with neat PHB and PHB/GNP films with a GNP concentration of 0.7 wt.%. The green color indicates live cells. In Fig. 24.5E, the cells stained with red-colored apropidium iodide indicate dead cells. The biodegradation tests carried out in the soil for a

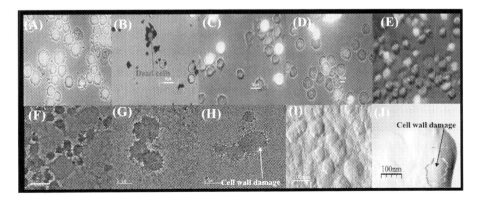

Figure 24.5 FM images of the cells from well containing (A) medium, (B) graphene, (C) neat PHB, (D) PHB-0.7% GNP, (E) PHB-1.3% GNP, TEM analysis of cells from well containing (F) medium, (G) PHB-0.7% GNP, (H) PHB-1.3% GNP, AFM analysis cells from well containing (I) PHB-0.7% GNP (J) PHB-1.3% GNP [39]. *PHB*, Polyhydroxy butyrate; *GNP*, graphene nanoplatelets.

period of 30 days indicate that all the samples biodegraded. However, the addition of graphene increases the resistance of degradation by microorganisms which can be attributed to the antimicrobial activity of graphene [39].

24.1.1.5 Conventional polymer/graphene-derivative nanocomposite films

The effect of screw speed in melt extrusion exhibits a better distribution of GNP in film grade LLDPE (MFI-1 gm/cc) matrix with increased thermal stability and little change in TS of the composite. At different levels of GNP loading, (0%−10%) it was observed that the TS values peak at 2% GNP loading for 100 rpm and 4% loading at 50 and 150 rpm wherein the highest value reported is of 150 rpm indicating almost 47.3% property rise at 4% GNP which can be attributed to good dispersion and distribution in the polymer matrix. Any increase in the filler concentration above that decreases the TS values with lower values reported that neat LLDPE above 8% loading. The ability of the extruder to disperse agglomerates by breaking them up does not work for loadings above 4%. The addition of GNP increases the polymer viscosity thereby leading to higher power requirements while processing. The increase in screw rpm reduced the viscosity of polymer composite due to thermal degradation/chain scission and hence a balance is required to increase the dispersion of filler with minimum effect on viscosity which will thereby decrease the mechanical properties of the composite [40].

Another material that has a film forming ability, that is, Polypropylene was reinforced with exfoliated graphene (EG) via melt compounding. The mechanical properties, that is, TS (yield) and TM values upon comparison with neat PP improved almost linearly with the level of EG up to 10%. However, the TS at break was lower up to 3% of EG than neat PP due to stress hardening for neat PP film after yielding during axial elongation. At 10% loading of EG, the overall increase in TS (yield), TS (break), TM (was around 78%, 43%, and 113%) respectively. On the other hand, EAB went down even with a small addition of EG. The neat PP film exhibited an EAB of 998.5 ± 91.5 which upon addition of just 1% EG decreased to 88.8% ± 18.7% further going down to 5.4% ± 0.4% upon 10% addition of, for example, [41].

24.1.2 Polymer graphene composites in electronics packaging

24.1.2.1 Scope and challenges of electronics packaging

Electronics are used in a wide range of applications including computing, communications, biomedical, automotive, military, and aerospace. Electronic products are fragile and need protection and preservation in transportation from the production houses to installation points. The electronic products working conditions, keeping the packaging as simple as possible, and the package being environmentally friendly are important considerations. The electronic product packaging design must consider protection from extreme temperatures, mechanical damage, electrostatic discharge, and high-frequency noise emission. For example, an IC package protects, powers, and cools the microelectronic device and provides a mechanical connection. The need for packaging material may be thin films, protective gels, rigid encapsulants, and printed wiring board materials and processes.

Electronic packaging is a topic wherein expertise from subjects like Mechanical, Electrical, Industrial Engineering, Physics, Chemistry, and even Marketing is required. Electronic Packaging is broadly housing and interconnection of integrated circuits to form electronic systems. Electronic packaging is one of the most complex levels of packaging as it requires extensive engineering and skill set in order to come up with an optimal overall electronics packaging design compared to other industry packaging designs. The electronic packaging design is a combination of minor electronic packages working together to achieve the desired purpose. They are not only present to protect the electronic components from outside harm but also make sure they do not interrupt each other's functions of built-in components. In recent developments regarding customer satisfaction, unpacking is more and more seen as an experience.

To begin with, let us define the term electronic packaging. At first sight, the nomenclature is a bit confusing. It does not simply mean the plastic cases or goods wrapping covers for transportation. Of course, the design and production of enclosures for electronic devices is a major part of electronic packaging. Protection from mechanical damage, cooling, radio frequency noise emission, and electrostatic discharge are the points to be considered under the packaging of an electronic system. Electronic packaging also comprises the materials and structures used to connect one electronic component with other electronic components. The electronic components have input—output terminals that are used to connect the chip with other chips or with power, resistors, inductors, and capacitors. The spacing between the input-output terminals on the chip is very small compared with the typical spaces on a printed circuit board. In order to "fan out" from the small space on the chip to the larger space on the substrate or circuit board, electronic packaging designs are used. One of the key aspects of polymers utilized in electronic packaging is high temperature stability along with the ease of processing. The polymers in these cases are used as underfills, encapsulating materials, solder masks, hermetic packages, coatings, and mold compounds. The current directions in which electronics packaging researchers exploring graphene-based polymer composites are:

- Thermal dissipation, reduce or remove heat in electronic systems or components, either generated internally during device operation or due to the external environment;
- Provide mechanical support and protect the electronic components from external environment by providing resistance to humidity and moisture and or ionic contaminants;
- Flexible electronics, flexible field emission devices and flexible touch screen made with printed graphene. 3D system integration and Ink—Jet Printed Graphene Electronics
- Miniaturization, high density;
- Pb-free, lower temperature solders and sintering;
- Provide protection against radiation and electromagnetic interference (EMI).

The current market volume and expansions in the future considering the consumer requirements and bringing to fore insights can help stakeholders in identifying the opportunities as well as challenges. As per the reports, the electronic packaging industry has been expanding at a very high rate in the recent past and is going to expand in the future at a CAGR of around 18%. The reason behind this market expansion is the rising demand for products, such as televisions, set-top boxes, digital cameras, wi-fi chipsets, increasing adoption in electric vehicles and hybrid vehicles. As a large number of memory devices, processors, analog circuits, discrete power devices, and sensors are used in electric and hybrid vehicles, the demand in the packaging industry is set to rise at a rapid rate [1−3].

24.1.2.2 Heat dissipation considerations in electronic packaging

Even at low concentrations of graphene are sufficient to effectively improve the mechanical, electrical, and thermal properties of polymers [9]. The polymer-graphene composites conduct electricity and can withstand much higher temperatures than the polymers alone. The addition of a small amount of graphene can modify the polymer properties. They can be converted into tough, lightweight materials. Graphene has special properties, which are superior to those of other materials making graphene believed to be useful in diverse applications in electronics packaging. The mechanical, chemical, electronic, and barrier properties of graphene make it a promising candidate for applications in composite materials by bringing extra functionality to them. The properties such as having the mechanical flexibility and chemical durability very high, its application for touch-screen devices would prove far more long-lasting and would open the door for flexible devices.

The top down approach of exfoliating graphene by mechanical [42], chemical, [12] and thermal reduction [43] and bottom up chemical vapor deposition are the synthesis methods reported in literature. For mass production of graphene, a number of methods which allow a significant choice in terms of sizes, quality and price for any particular application are also reported [44,45]. Among the preparation methods of graphene, the chemical reduction process through GO layers has been the most frequently researched route owing to their cost effectiveness, scalability and production in bulk quantities [46]. The graphene derivative produced by the chemical reduction process termed as reduced GO (rGO). rGO, a derivative of graphene is considered as promising functional nanofiller for electrically conductive adhesives due to their high inherent electrical conductivity [47], excellent mechanical, thermal, optical properties [48] and their ability for large scale production [49].

The ever-increasing demand for higher input/output counts and miniaturization has led to the development of various novel electronics packaging materials [50]. Although the development has been considerably advanced over the past few decades, the vital requirements for the die attach and interconnects like purity, curing rates, stress, package crack resistance, and multichip packaging among all the electronic components remain unchanged. The components should be electrically connected to power, have high purity, fast curing time, lower stress, and high package crack resistance. For such requirements, polymer based electrically conductive adhesives are effectively replacing the tin−lead solders owing to their advantages such as environmental friendliness, lower processing temperature, fewer processing steps, fine pitch capability, long service life, and lower cost, etc. At the same time, more and more functional materials like vapor-grown carbon nanofiber and graphene are being utilized in electronic packaging applications for improving adhesion and electrical conductivity [51,52].

The growth in the electronics industry has followed Moore's law since 1965. In the past decades, chip manufacturers have been increasing the number of transistors with shrunk sizes to achieve higher component density and clock frequency. Moore's law has slowed down in recent years and further miniaturization of transistors will ultimately seize due to the physical limits imposed by quantum effect, multicore design has been proposed and applied to continue the performance evolution [2]. There is a wide scope of research in the development of high-performance graphene-based thermal interface materials. The increasing power density of electronic devices has put forward the urgent demand for the development of thermal interface materials with high through-plane thermal conductivity for handling the issue of thermal management. Heat flux of electronics is dramatically increased with the higher speed, more functionality and feature size decreasing from micron to nanometer, increasing use of three-dimensional chip stack to replace the traditional two-dimensional encapsulation integration [53,54]. Graphene-based heat spreading materials are applied to electronics packaging for thermal management applications. Due to its ultra-high intrinsic thermal conductivity dispersed graphene/polymers, graphene framework/polymers and inorganic graphene-based monoliths exhibited significant potential for the development. The superior thermal conductivity properties exceed the reported results for multiwalled and single-walled CNTs (3000−3500 W/m K) and the best bulk crystalline thermal conductor, that is, diamond (1000−2200 W/m K). On the other hand, today's most electronic packaging applications typically make use of metals Cu, Ag and Al having the thermal conductivity 400, 430, 250 W/m K respectively as heat spreaders or fillers, for solid conduction-based heat transfer. The thermal conductivity values of graphene as reported in the literature vary in the range of 2000−5300 W/m K [55−57].

Thermal conductivity is a crucial factor for the development of thermal interface materials in order to dissipate the large heat flux from high power electronic devices for better performance and longer longevity. The main function of such composites is to fill the gap of solid surfaces between electronic devices and thermal sinks and provide an efficient pathway for heat transfer. Numerous articles have reported that the shrinking size and higher density of transistors and other integrated circuit devices over time have enhanced computing capabilities. However, these technological advances are achieved at the cost of increased power dissipation across the device, die, and system levels [57−59]. The power required for high-performance computing applications on some modern processor modules can reach 200−250 W or more [34], leading to heat loads approaching as high as 1 kW. The high magnitude of the power dissipation at the system and data center levels, and the power density and its spatial distribution at the die level are major concerns and can have important reliability and thermal management implications. The power dissipation with the usage of polymers like fully refined paraffin in chip architectures being nonuniform across the die surface have localized functional areas where the power density is a factor of five to ten higher than the die average [60–62].

These power-dense regions can produce regions where local temperatures are significantly higher than the die average temperature in composites with high density polyethylene and other diverse combinations of organic polymers, metals, and resins [63–65]. Furthermore, nonuniform heat dissipation also leads to the overheating of specific areas in the chip, largely affecting computing performance and reliability of electronics systems [66–68]. Generally, the temperature of the hotspot is such a significant parameter that reduction in the hotspot temperature by 20°C, the transistor's lifetime would be extended by one order of magnitude [66]. Therefore, thermal management of hotspots with localized high heat flux is quite critical in electronics packaging. For electronics, the overall reliability is stated in terms of the hottest region on the die rather than the average die temperature. Consequently, the so-called hot spots can often dictate the required higher-level packaging and thermal management solutions including material selections, heat sink, and cold plate design, and required pumping power at the system and facility levels. Thus, the thermal state at the device and die levels can have a far-reaching influence all the way up to the data center cooling requirements and environmental impact. In addition, if the required cooling cannot be delivered to keep the hottest region of a die under its stated temperature threshold the performance may adversely affect and reduce power and reduction in performance.

As a result, there is a great demand to have enhanced, efficient heat spreading capabilities at the single transistor device level to die and packaging levels in order to minimize the severity and the influence of these hot spots. Thus thermal interface materials are a key to achieving good heat conduction within a package and from a package to a heat sinking device. The graphene polymer composites reportedly have promising results for addressing thermal issues in electronic packaging using selected polymers such as phenolic resin, poly furfural alcohol, polyacrylonitrile, polyimide, polyamide, polyoxadiazole, polybenzoxazole, polybenzobisoxazole, polythiazole, polybenzothiazole, polybenzimidazole, poly(p-phenylene vinylene), polybenzimidazole, polybenzimidazole [69,70]. A considerable reduction in the temperature of the hotspot is reported in a study suggesting graphene-based materials as promising candidates for heat spreading materials for electronics packaging application [71].

24.1.2.3 Flexible, stretchable electronic packaging materials

Various electronics packaging offshoots require flexible, stretchable, and energy efficient transparent conductive films. These applications are still challenging due to the difficulty of preparing large-area crystalline graphene with few defects for transparent conductive films. Many efforts have been made to solve these problems, and several kinds of optoelectronic devices have been successfully fabricated with graphene based transparent conductive films [72–75]. Graphene with superior mechanical flexibility and chemical durability has its application for touch screen devices. Graphene based composites would prove far more long-lasting and would open the door for flexible devices [76]. By the use of graphene, the surrounding polymer matrix of composites can be endowed with gas and moisture barrier properties, electromagnetic shielding, electrical and thermal conductivity, and strain monitoring capability. Graphene-based paints can be used for conductive ink, antistatic, EMI shielding as well as gas barrier applications, and low-cost ink-jet printing could be used for layer deposition and patterning. The resulting flexible transparent electronics would find applications in touch screens, aeronautical fields, solar panels, batteries, actuators, flexible electrodes, sensors, etc. [77].

Flexible electronics and packaging are a rapidly expanding research area. There are various pathways growing such as wearable smart devices like flexible television displays, touch screens, wearable smartwatches and glasses, flexible computers, and flexible electronic substrates with electromagnetic shielding capability. In order to develop these electronic devices into easy-using and low-cost, metal Nanoparticles based conducting inks for direct writing on flexible substrates represent the most promising manufacturing way. More importantly, preparation conductive tracks on heat-sensitive substrates such as paper using carboxylate-functionalized polythiophenes [78], plastic packages using poly(3,4-ethylenedioxythiophene) polystyrene sulfonate (PEDOT:PSS) [79], textile using plastic substrates [80] and silver nanowires polymeric [81] substrates have aroused wide interest as a pathway to fabricate flexible and electronic devices. Polymer nanocomposites packaging materials with functionalization, flexibility and minimal cost are superior alternatives to traditional packaging materials such as glass, paper, and metals. Their inferior mechanical and barrier behaviors create scope for improvement through the inclusion of functionalized reinforcing macro- or nanofillers. Graphene is much lighter, stronger, and harder than steel at similar dimensions has shown to possess an elastic modulus of ~ 1 TPa and high intrinsic strength (~ 30 GPa). These properties pave the way for remarkable mechanical strength combined with geometric flexibility and low density brought in graphene films [82]. In one of such applications, a continuous layer of pure graphene the size of a large television, on a flexible, transparent, 0.63 m-wide polyester foil is produced at Samsung and Sung Kwun Kwan University, in Korea. The team created a flexible touch screen by using polymer-supported graphene to make the screen's transparent electrodes. After building the graphene layer onto a copper foil, it was wrapped around a cylinder and placed in a furnace. With the heat treatment, the graphene sheet was transferred to the polyester base by hot rolling. Silver electrodes were then printed onto the layer.

The material currently used to make transparent electronics is indium tin oxide, which is brittle and expensive. Producing graphene on polyester sheets that bend was the first step to making transparent electronics that are stronger, cheaper, and more flexible. There is a huge potential for application in the production of solar cells, touch sensors, and flat-panel displays. The efficiency, cost, and feasibility of mass production have to be overcome for wide acceptance by most industries.

24.1.2.4 Ink-jet printing in electronic packaging materials

Rubber stamping, embossing and ink-jet printing are the alternatives for preparing networks of field-effect transistors (FETs) by employing plastic substrate and organo silsesquioxane spin-on-glass material [83,84]. Conventionally the networks of FETs are prepared directly on flexible polymer substrates by several coating, curing, and lithographic steps. There is a need to reduce the number of such fabrication steps. Ink-jet printing is one of the most promising techniques for the large area fabrication of flexible plastic electronics. A range of components can be printed, such as transistors, photovoltaic devices, organic light emitting diodes, and displays. Drop on demand ink-jet printing with N-methyl pyrrolidone has been used to produce thin film transistors based on organic semiconducting and conducting inks. However, the performance of such devices is limited, since their mobility is still much lower than that of standard silicon technology [85]. Near-ballistic transport and high mobility make graphene a possible material to solve this problem in nanoelectronics, especially for high frequency applications. One of the concerns in printing conductive graphene stripes which eventually will get the appropriate solution is the environment friendly solvents [86]. The mechanism of making inks viable for printing is their ability to generate droplets, allowing the droplets to pass through the printer nozzle and setting correct fluid dynamic parameters. Ink viscosity, surface tension, density, and nozzle diameter, influence the spreading of the resulting liquid drops while nozzle diameter will cause blocking the nozzle by the dispersed nanoparticles because of clustering of the particles at the nozzle edge [74,87,88]. The final assembly of printed nanoparticle inks depends on the substrate surface energy (SE) as well as ink viscosity and surface tension. Distortion of the drops during solvent drying due to the interplay of ink viscosity and solute transport via solvent motion (arising from surface tension interaction between solvent and substrate). This is one of the most important phenomena affecting the homogeneity of inkjet-printed drops [89,90]. In order to prevent this, it is necessary to seal drop geometry immediately after they form a homogeneous and continuous film on the substrate.

24.1.2.5 Electromagnetic interference shielding in electronic packaging

Severe EMI and radiation pollution poses multifold problems with the increasing usage of electronic devices. It causes interference with electronic devices causing deterioration in the durability and functioning of electronic equipment. It is also harmful to human health causing health concerns such as headaches, sleeping disorders and trepidation [90,91]. Generally, metals such as copper, nickel, silver and aluminum, show good shielding performance and owing to their high reflectivity along with high electrical conductivity. The reflection of energy caused by impedance mismatch of the material and the free space [91—95]. These materials prevent the transmission of EM radiation by reflection and/or absorption of the electromagnetic radiation or by suppressing the EM signals acting as a barrier against the penetration of the radiation passing through the shielding materials. However, heavy weight, high cost, easy corrosion, lack of flexibility and difficulty in processing are the challenges in this approach. The loading level of carbon materials is usually high to achieve the required electrical conductivity for EMI applications, because of the poor dispersion and connectivity of separated carbon. The loading level of carbon and the thickness are the factor that needs to be considered when preparing shielding materials. The appropriate usage of checks wastage of the expensive materials and controls the mechanical properties. While greater thickness will lead to higher EMI, thin and protective layers are required for EMI shielding of sensitive instruments.

The dispersion state, filler nature, and their interaction are the main aspects for the enhanced EMI shielding effectiveness (SE) of hybrid polymeric nanocomposites. Nanocomposites of polymers with GNP are reported to enhance the EMI SE. Improved electrical conductivity along with increase in dielectric constant at lower frequency region enhanced EMI shielding mainly due to the absorption phenomena. The shielding efficiency improved with increasing graphene filler content in PVC/PANI/GNP nanocomposites [96]. Dongyi Ao et al. reported the synthesis of a highly conductive three-dimensional (3D) graphene network employing chemical vapor deposition on a 3D nickel fiber network and subsequent etching process. Reportedly, the synthesized nanocomposite material showed the superior electrical conductivity of 6100 S/m even at a very low loading level of graphene (1.2 wt.%) in polydimethylsiloxane. As a result, EMI SE was achieved. Polymer Composite Films with better EMI shielding and better heat dissipation properties is one of the future scopes of electronic industry in wearable and flexible microwave devices. Polymer graphene nanocomposite films which will demonstrate better response in electrical leakage, sustain impacts of moisture with better EMI shielding and better heat dissipation properties is one of the future scopes of the electronic industry in wearable and flexible microwave devices.

24.1.2.6 Soldering considerations in electronic packaging

In general, soldering is a joining process used to join different types of metals together by melting solder. The solder is an electrical interconnect, a mechanical bond that should serve as a thermal conduit to remove heat from the joined device. Though lead has been traditionally used for soldering purposes, regulations have been enforced to eliminate lead from coatings and solders used in most electronic components due to its toxicity and environmental hazards [97,98]. The new developments in chip size, chip carrier size, and the number of I/Os are demanding a reduction in solder joint size and cost making the interconnects more critical. The extreme condition with heat in the working environment like in automotive electronics and expecting portable devices to withstand severe shock environments such as caused by dropping the phone are additional Work requirements. The use of graphene polymer compositions with different percent of graphene in various polymers (KRATON FG, a high performance elastomeric copolymer that contains polystyrene blocks, rubber blocks, and maleic anhydride blocks, acrylate, methacrylate, vinyl alcohol or vinyl acetate copolymers with polyethylene, polypropylene, polyesters, polyamides, polycarbonates, polytetrafluoroethylene, polyvinyl chloride, polyurethanes, polyacrylonitrile, polysiloxanes, carboxyl terminated butadiene nitrile epoxies) as a lead-free solder material, in which case multiple rework operations could be allowed to take place at a connection and methods for making and using such electrically conductive polymer compositions is patented by Steven Bullock and Vanderwiel [99].

24.2   Conclusion

Various aspects to reach optimal properties for using graphene polymer nanocomposites in food and electronics packaging applications are discussed in this chapter. Appropriate packaging is one of the important parameters in deciding the development in the food and electronics industry. The variations in properties that the graphene polymer composites are bringing are still desired to consistently reach optimal properties but certainly may help fill the gap of the usage of these materials in food and electronic packaging applications.

This chapter describes the potential of graphene-based nanofillers as one of the best candidates for the reinforcement of food packaging materials. The combination of the use of graphene in sensors can do wonders to the category of active and smart packaging materials especially in the domain of food safety, wherein a nanosensor can be put over a food package or a tin can sensor having tinned food which can predict the shelf life, track the package, or its rate of food spoilage and benefit the communication through the supply chain to benefit the world by ensuring that the food does not spoil in the packaged state itself.

The graphene-based nanocomposites show substantial property enhancements at a very low loading concentration compared to traditional composites. The required features like flexibility and better conformal contact brought in by polymers and mechanical strength, electrical, and thermal conductivity by graphene material in combination make these packaging materials promising candidates. The antimicrobial and cytotoxic activity of GO still possesses a concern to go along with all the other properties proving it to be an excellent packaging material. Graphene polymer composite materials need to be explored to meet the ever-changing requirements of electronics packaging applications. These materials must also be tested to determine their compatibility with assembly processes and device performance and meet industry standards. The fundamental characteristics that must be understood and employed include heat dissipation, flexibility, enhancing the reflectivity and electric conductivity for EMI shielding, and usage of these composites to address environmental considerations in food and electronic packaging in a better way along with commercial viabilities.

References

[1] K.S. Novoselov, Electric field effect in atomically thin carbon films, Science 306 (2004) 666−669. Available from: https://doi.org/10.1126/science.1102896.

[2] L. Fu, K. Liao, B. Tang, L. Jiang, W. Huang, Applications of graphene and its derivatives in the upstream oil and gas industry: a systematic review, Nanomaterials 10 (2020) 1013. Available from: https://doi.org/10.3390/nano10061013.

[3] B. Vargas-Quesada, Z. Chinchilla-Rodríguez, N. Rodriguez, Identification and visualization of the intellectual structure in graphene research, Front. Res. Metr. Anal. 2 (2017) 7. Available from: https://doi.org/10.3389/frma.2017.00007.

[4] Z.A. Ghaleb, M. Mariatti, Z.M. Ariff, Properties of graphene nanopowder and multi-walled carbon nanotube-filled epoxy thin-film nanocomposites for electronic applications: the effect of sonication time and filler loading, Compos. Part. Appl. Sci. Manuf. 58 (2014) 77−83. Available from: https://doi.org/10.1016/j.compositesa.2013.12.002.

[5] M.D Stoller, S. Park, Y. Zhu, J. An, R.J. Ruoff, Graphene-based ultracapacitors, Nano Letters 8 (10) (2008) 3498−3502. Available from: https://doi.org/10.1021/nl802558y.

[6] R.R. Nair, P. Blake, A.N. Grigorenko, K.S. Novoselov, T.J. Booth, T. Stauber, et al., Fine structure constant defines visual transparency of graphene, Science 320 (2008) 1308. Available from: https://doi.org/10.1126/science.1156965.

[7] A.A. Balandin, S. Ghosh, W. Bao, I. Calizo, D. Teweldebrhan, F. Miao, et al., Superior thermal conductivity of single-layer graphene, Nano Lett. 8 (2008) 902−907. Available from: https://doi.org/10.1021/nl0731872.

[8] T. Ramanathan, A.A. Abdala, S. Stankovich, D.A. Dikin, M. Herrera-Alonso, R.D. Piner, et al., Functionalized graphene sheets for polymer nanocomposites, Nat. Nanotechnol. 3 (2008) 327−331. Available from: https://doi.org/10.1038/nnano.2008.96.

[9] S. Stankovich, D.A. Dikin, G.H.B. Dommett, K.M. Kohlhaas, E.J. Zimney, E.A. Stach, et al., Graphene-based composite materials, Nature 442 (2006) 282−286. Available from: https://doi.org/10.1038/nature04969.

[10] J. Morris (Ed.), Nanopackaging: Nanotechnologies and Electronics Packaging, second ed., Springer International Publishing, 2018. https://doi.org/10.1007/978-3-319-90362-0.

[11] D. Frear, Packaging materials, in: S. Kasap, P. Capper (Eds.), Springer Handbook of Electronic and Photonic Materials, Springer International Publishing, Cham, 2017, p. 1. Available from: https://doi.org/10.1007/978-3-319-48933-9_53.

[12] H. Liu, R. Jian, H. Chen, X. Tian, C. Sun, J. Zhu, et al., Application of biodegradable and biocompatible nanocomposites in electronics: current status and future directions, Nanomaterials 9 (2019) 950. Available from: https://doi.org/10.3390/nano9070950.

[13] D. Galpaya, M. Wang, M. Liu, N. Motta, E. Waclawik, C. Yan, Recent advances in fabrication and characterization of graphene-polymer nanocomposites, Graphene 01 (2012) 30−49. Available from: https://doi.org/10.4236/graphene.2012.12005.

[14] E. Narimissa, R.K. Gupta, N. Kao, H.J. Choi, M. Jollands, S.N. Bhattacharya, Melt rheological investigation of polylactide-nanographite platelets biopolymer composites, Polym. Eng. Sci. 54 (2014) 175−188. Available from: https://doi.org/10.1002/pen.23550.

[15] E. Narimissa, R.K. Gupta, H.J. Choi, N. Kao, M. Jollands, Morphological, mechanical, and thermal characterization of biopolymer composites based on polylactide and nanographite platelets, Polym. Compos. 33 (2012) 1505−1515. Available from: https://doi.org/10.1002/pc.22280.

[16] C. Thellen, C. Orroth, D. Froio, D. Ziegler, J. Lucciarini, R. Farrell, et al., Influence of montmorillonite layered silicate on plasticized poly(l-lactide) blown films, Polymer 46 (2005) 11716−11727. Available from: https://doi.org/10.1016/j.polymer.2005.09.057.

[17] J. Ambrosio-Martín, A. López-Rubio, M. José Fabra, M. Angel López-Manchado, A. Sorrentino, G. Gorrasi, et al., Synergistic effect of lactic acid oligomers and laminar graphene sheets on the barrier properties of polylactide nanocomposites obtained by the

in-situ polymerization pre-incorporation method, J. Appl. Polym. Sci. 133 (2016). Available from: https://doi.org/10.1002/app.42661.

[18] M. Barletta, M. Puopolo, V. Tagliaferri, S. Vesco, Graphene-modified poly(lactic acid) for packaging: material formulation, processing and performance, J. Appl. Polym. Sci. 133 (2016). Available from: https://doi.org/10.1002/app.42252.

[19] G. Chakraborty, A. Gupta, G. Pugazhenthi, V. Katiyar, Facile dispersion of exfoliated graphene/PLA nanocomposites via in situ polycondensation with a melt extrusion process and its rheological studies, J. Appl. Polym. Sci. 135 (2018) 46476. Available from: https://doi.org/10.1002/app.46476.

[20] J.J. Park, E.J. Yu, W.-K. Lee, C.-S. Ha, Mechanical properties and degradation studies of poly(D,L-lactide-co-glycolide) 50:50/graphene oxide nanocomposite films: poly(D, L-lactide-co-glycolide)/graphene oxide nanocomposite films, Polym. Adv. Technol. 25 (2014) 48−54. Available from: https://doi.org/10.1002/pat.3203.

[21] Y.A. Arfat, J. Ahmed, M. Ejaz, M. Mullah, Polylactide/graphene oxide nanosheets/ clove essential oil composite films for potential food packaging applications, Int. J. Biol. Macromol. 107 (2018) 194−203. Available from: https://doi.org/10.1016/j. ijbiomac.2017.08.156.

[22] Y. Huang, T. Wang, X. Zhao, X. Wang, L. Zhou, Y. Yang, et al., Poly(lactic acid)/graphene oxide-ZnO nanocomposite films with good mechanical, dynamic mechanical, anti-UV and antibacterial properties: the poly(lactic acid)/graphene oxide-ZnO nanocomposites, J. Chem. Technol. Biotechnol. 90 (2015) 1677−1684. Available from: https://doi.org/10.1002/jctb.4476.

[23] I.E. Mejías Carpio, C.M. Santos, X. Wei, D.F. Rodrigues, Toxicity of a polymer-graphene oxide composite against bacterial planktonic cells, biofilms, and mammalian cells, Nanoscale 4 (2012) 4746−4756. Available from: https://doi.org/10.1039/c2nr30774j.

[24] S. Gahlot, V. Kulshrestha, G. Agarwal, P.K. Jha, Synthesis and characterization of PVA/GO nanocomposite films, Macromol. Symp. 357 (2015) 173−177. Available from: https://doi.org/10.1002/masy.201400220.

[25] M. Ščetar, M. Kurek, K. Galić, Trends in fruit and vegetable packaging—a review, Biotechnol. Nutr. (2010) 18.

[26] V. Loryuenyong, C. Saewong, C. Aranchaiya, A. Buasri, The improvement in mechanical and barrier properties of poly(vinyl alcohol)/graphene oxide packaging films: poly (vinyl alcohol)/graphene oxide packaging films, Packag. Technol. Sci. 28 (2015) 939−947. Available from: https://doi.org/10.1002/pts.2149.

[27] K.-J. Lin, S.-C. Lee, K.-F. Lin, One-pot fabrication of poly(vinyl alcohol)/graphene oxide nanocomposite films and their humidity dependence of mechanical properties, J. Polym. Res. 21 (2014) 611. Available from: https://doi.org/10.1007/s10965-014-0611-4.

[28] L. Liu, Y. Gao, Q. Liu, J. Kuang, D. Zhou, S. Ju, et al., High mechanical performance of layered graphene oxide/poly(vinyl alcohol) nanocomposite films, Small 9 (2013) 2466−2472. Available from: https://doi.org/10.1002/smll.201300819.

[29] Y. Xu, W. Hong, H. Bai, C. Li, G. Shi, Strong and ductile poly(vinyl alcohol)/graphene oxide composite films with a layered structure, Carbon 47 (2009) 3538−3543. Available from: https://doi.org/10.1016/j.carbon.2009.08.022.

[30] J. Ding, C. Zhao, L. Zhao, Y. Li, D. Xiang, Synergistic effect of α-ZrP and graphene oxide nanofillers on the gas barrier properties of PVA films: research article, J. Appl. Polym. Sci. 135 (2018) 46455. Available from: https://doi.org/10.1002/app.46455.

[31] J. Ahmed, M. Mulla, Y.A. Arfat, L.A. Thai T, Mechanical, thermal, structural and barrier properties of crab shell chitosan/graphene oxide composite films, Food

Hydrocoll. 71 (2017) 141−148. Available from: https://doi.org/10.1016/j.foodhyd.2017. 05.013.

[32] D. Han, L. Yan, W. Chen, W. Li, Preparation of chitosan/graphene oxide composite film with enhanced mechanical strength in the wet state, Carbohydr. Polym. 83 (2011) 653−658. Available from: https://doi.org/10.1016/j.carbpol.2010.08.038.

[33] A. Barra, N.M. Ferreira, M.A. Martins, O. Lazar, A. Pantazi, A.A. Jderu, et al., Eco-friendly preparation of electrically conductive chitosan—reduced graphene oxide flexible bionanocomposites for food packaging and biological applications, Compos. Sci. Technol. 173 (2019) 53−60. Available from: https://doi.org/10.1016/j.compscitech.2019. 01.027.

[34] M. Cobos, B. González, M.J. Fernández, M.D. Fernández, Chitosan-graphene oxide nanocomposites: effect of graphene oxide nanosheets and glycerol plasticizer on thermal and mechanical properties, J. Appl. Polym. Sci. 134 (2017) 45092. Available from: https://doi.org/10.1002/app.45092.

[35] F. Han Lyn, T. Chin Peng, M.Z. Ruzniza, Z.A. Nur Hanani, Effect of oxidation degrees of graphene oxide (GO) on the structure and physical properties of chitosan/GO composite films, Food Packag. Shelf Life 21 (2019) 100373. Available from: https://doi. org/10.1016/j.fpsl.2019.100373.

[36] M. El Achaby, Y. Essamlali, N. El Miri, A. Snik, K. Abdelouahdi, A. Fihri, et al., Graphene oxide reinforced chitosan/polyvinylpyrrolidone polymer bio-nanocomposites, J. Appl. Polym. Sci. 131 (2014). Available from: https://doi.org/10.1002/app.41042.

[37] C. Demitri, V.M. De Benedictis, M. Madaghiele, C.E. Corcione, A. Maffezzoli, Nanostructured active chitosan-based films for food packaging applications: effect of graphene stacks on mechanical properties, Measurement 90 (2016) 418−423. Available from: https://doi.org/10.1016/j.measurement.2016.05.012.

[38] K. Li, S. Jin, J. Li, H. Chen, Improvement in antibacterial and functional properties of mussel-inspired cellulose nanofibrils/gelatin nanocomposites incorporated with graphene oxide for active packaging, Ind. Crop. Prod. 132 (2019) 197−212. Available from: https://doi.org/10.1016/j.indcrop.2019.02.011.

[39] N.A. Manikandan, K. Pakshirajan, G. Pugazhenthi, Preparation and characterization of environmentally safe and highly biodegradable microbial polyhydroxybutyrate (PHB) based graphene nanocomposites for potential food packaging applications, Int. J. Biol. Macromol. 154 (2020) 866−877. Available from: https://doi.org/10.1016/j.ijbiomac. 2020.03.084.

[40] P. Noorunnisa Khanam, M.A. AlMaadeed, M. Ouederni, E. Harkin-Jones, B. Mayoral, A. Hamilton, et al., Melt processing and properties of linear low density polyethylene-graphene nanoplatelet composites, Vacuum 130 (2016) 63−71. Available from: https:// doi.org/10.1016/j.vacuum.2016.04.022.

[41] J.-E. An, G.W. Jeon, Y.G. Jeong, Preparation and properties of polypropylene nano-composites reinforced with exfoliated graphene, Fibers Polym. 13 (2012) 507−514. Available from: https://doi.org/10.1007/s12221-012-0507-z.

[42] M. Yi, Z. Shen, A review on mechanical exfoliation for the scalable production of graphene, J. Mater. Chem. A 3 (2015) 11700−11715. Available from: https://doi.org/ 10.1039/C5TA00252D.

[43] T.D. Dao, H.M. Jeong, Graphene prepared by thermal reduction−exfoliation of graphite oxide: effect of raw graphite particle size on the properties of graphite oxide and graphene, Mater. Res. Bull. 70 (2015) 651−657. Available from: https://doi.org/ 10.1016/j.materresbull.2015.05.038.

[44] P.T. Yin, T.-H. Kim, J.-W. Choi, K.-B. Lee, Prospects for graphene−nanoparticle-based hybrid sensors, Phys. Chem. Chem. Phys. 15 (2013) 12785. Available from: https://doi.org/10.1039/c3cp51901e.

[45] Y. Zhu, H. Ji, H.-M. Cheng, R.S. Ruoff, Mass production and industrial applications of graphene materials, Natl. Sci. Rev. 5 (2018) 90−101. Available from: https://doi.org/10.1093/nsr/nwx055.

[46] S. Abdolhosseinzadeh, H. Asgharzadeh, H. Seop Kim, Fast and fully-scalable synthesis of reduced graphene oxide, Sci. Rep. 5 (2015) 10160. Available from: https://doi.org/10.1038/srep10160.

[47] Q. Meng, S. Han, S. Araby, Y. Zhao, Z. Liu, S. Lu, Mechanically robust, electrically and thermally conductive graphene-based epoxy adhesives, J. Adhes. Sci. Technol. 33 (2019) 1337−1356. Available from: https://doi.org/10.1080/01694243.2019.1595890.

[48] M. Lundie, Ž. Šljivančanin, S. Tomić, Electronic and optical properties of reduced graphene oxide, J. Mater. Chem. C. 3 (2015) 7632−7641. Available from: https://doi.org/10.1039/C5TC00437C.

[49] J.M. Vazquez-Moreno, V. Yuste-Sanchez, R. Sanchez-Hidalgo, R. Verdejo, M.A. Lopez-Manchado, L. Fernández-García, et al., Customizing thermally-reduced graphene oxides for electrically conductive or mechanical reinforced epoxy nanocomposites, Eur. Polym. J. 93 (2017) 1−7. Available from: https://doi.org/10.1016/j.eurpolymj.2017.05.026.

[50] R. Aradhana, S. Mohanty, S.K. Nayak, High performance electrically conductive epoxy/reduced graphene oxide adhesives for electronics packaging applications, J. Mater. Sci. Mater. Electron. 30 (2019) 4296−4309. Available from: https://doi.org/10.1007/s10854-019-00722-5.

[51] J. Kim, B. Yim, J. Kim, J. Kim, The effects of functionalized graphene nanosheets on the thermal and mechanical properties of epoxy composites for anisotropic conductive adhesives (ACAs), Microelectron. Reliab. 52 (2012) 595−602. Available from: https://doi.org/10.1016/j.microrel.2011.11.002.

[52] M. Li, C. Tang, L. Zhang, B. Shang, S. Zheng, S. Qi, A thermally conductive epoxy polymer composites with hybrid fillers of copper nanowires and reduced graphene oxide, J. Mater. Sci. Mater. Electron. 28 (2017) 15694−15700. Available from: https://doi.org/10.1007/s10854-017-7459-4.

[53] W. Huang, K. Rajamani, M.R. Stan, K. Skadron, Scaling with design constraints: predicting the future of big chips, IEEE Micro. 31 (2011) 16−29. Available from: https://doi.org/10.1109/MM.2011.42.

[54] W. Huang, M.R. Stan, S. Gurumurthi, R.J. Ribando, K. Skadron, Interaction of scaling trends in processor architecture and cooling, 2010 26th Annual IEEE Semiconductor Thermal Measurement and Management Symposium SEMI-THERM, IEEE, Santa Clara, CA, 2010, pp. 198−204. Available from: https://doi.org/10.1109/STHERM.2010.5444290.

[55] S.D. Park, S. Won Lee, S. Kang, I.C. Bang, J.H. Kim, H.S. Shin, et al., Effects of nanofluids containing graphene/graphene-oxide nanosheets on critical heat flux, Appl. Phys. Lett. 97 (2010) 023103. Available from: https://doi.org/10.1063/1.3459971.

[56] E. Pop, V. Varshney, A.K. Roy, Thermal properties of graphene: fundamentals and applications, MRS Bull. 37 (2012) 1273−1281. Available from: https://doi.org/10.1557/mrs.2012.203.

[57] P.K. Schelling, L. Shi, K.E. Goodson, Managing heat for electronics, Mater. Today 8 (2005) 30−35. Available from: https://doi.org/10.1016/S1369-7021(05)70935-4.

[58] W. Haensch, E.J. Nowak, R.H. Dennard, P.M. Solomon, A. Bryant, O.H. Dokumaci, et al., Silicon CMOS devices beyond scaling, IBM J. Res. Dev. 50 (2006) 339−361. Available from: https://doi.org/10.1147/rd.504.0339.

[59] E. Pop, S. Sinha, K.E. Goodson, Heat generation and transport in nanometer-scale transistors, Proc. IEEE 94 (2006) 1587−1601. Available from: https://doi.org/10.1109/JPROC.2006.879794.

[60] H.F. Hamann, A. Weger, J.A. Lacey, Z. Hu, P. Bose, E. Cohen, et al., Hotspot-limited microprocessors: direct temperature and power distribution measurements, IEEE J. Solid-State Circuits 42 (2007) 56−65. Available from: https://doi.org/10.1109/JSSC.2006.885064.

[61] R. Mahajan, Chia-pin Chiu, G. Chrysler, Cooling a microprocessor chip, Proc. IEEE. 94 (2006) 1476−1486. Available from: https://doi.org/10.1109/JPROC.2006.879800.

[62] K.C. Otiaba, N.N. Ekere, R.S. Bhatti, S. Mallik, M.O. Alam, E.H. Amalu, Thermal interface materials for automotive electronic control unit: trends, technology and R&D challenges, Microelectron. Reliab. 51 (2011) 2031−2043. Available from: https://doi.org/10.1016/j.microrel.2011.05.001.

[63] A. Barua, Md.S. Hossain, K.I. Masood, S. Subrina, Thermal management in 3-D integrated circuits with graphene heat spreaders, Phys. Procedia 25 (2012) 311−316. Available from: https://doi.org/10.1016/j.phpro.2012.03.089.

[64] S. Mallik, N. Ekere, C. Best, R. Bhatti, Investigation of thermal management materials for automotive electronic control units, Appl. Therm. Eng. 31 (2011) 355−362. Available from: https://doi.org/10.1016/j.applthermaleng.2010.09.023.

[65] S.M. Sri-Jayantha, G. McVicker, K. Bernstein, J.U. Knickerbocker, Thermomechanical modeling of 3D electronic packages, IBM J. Res. Dev. 52 (2008) 623−634. Available from: https://doi.org/10.1147/JRD.2008.5388568.

[66] K. Azar, Power consumption and generation in the electronics industry: a perspective, Annual IEEE Semiconductor Thermal Measurement and Management Symposium Cat No00CH37068, IEEE, San Jose, CA, USA, 2000, pp. 201−212. https://doi.org/10.1109/STHERM.2000.837085.

[67] R. Skuriat, J.F. Li, P.A. Agyakwa, N. Mattey, P. Evans, C.M. Johnson, Degradation of thermal interface materials for high-temperature power electronics applications, Microelectron. Reliab. 53 (2013) 1933−1942. Available from: https://doi.org/10.1016/j.microrel.2013.05.011.

[68] A. Zhamu, Chemical-free production of 3D (n.d.) 30.

[69] T. Luo, J.R. Lloyd, Enhancement of thermal energy transport across graphene/graphite and polymer interfaces: a molecular dynamics study, Adv. Funct. Mater. 22 (2012) 2495−2502. Available from: https://doi.org/10.1002/adfm.201103048.

[70] M. Shtein, R. Nadiv, M. Buzaglo, O. Regev, Graphene-based hybrid composites for efficient thermal management of electronic devices, ACS Appl. Mater. Interfaces. 7 (2015) 23725−23730. Available from: https://doi.org/10.1021/acsami.5b07866.

[71] G. Yuan, Bo Shan, Y.Y. Shujing Chen, N.W. Jie Bao, S. Peng Su, Y. Huang, et al., Graphene-based heat spreading materials for electronics packaging applications, 2017 IMAPS Nord. Conf. Microelectron. Packag. Nord, IEEE, Gothenburg, Sweden, 2017, pp. 172−174. https://doi.org/10.1109/NORDPAC.2017.7993187.

[72] T. Chen, Y. Xue, A.K. Roy, L. Dai, Transparent and stretchable high-performance supercapacitors based on wrinkled graphene electrodes, ACS Nano 8 (2014) 1039−1046. Available from: https://doi.org/10.1021/nn405939w.

[73] J. Han, J. Lee, J. Lee, J. Yeo, Highly stretchable and reliable, transparent and conductive entangled graphene mesh networks, Adv. Mater. 30 (2018) 1704626. Available from: https://doi.org/10.1002/adma.201704626.

[74] S. Lee, K. Lee, C.-H. Liu, Z. Zhong, Homogeneous bilayer graphene film based flexible transparent conductor, Nanoscale 4 (2012) 639–644. Available from: https://doi.org/10.1039/C1NR11574J.

[75] X. Li, Y. Zhu, W. Cai, M. Borysiak, B. Han, D. Chen, et al., Transfer of large-area graphene films for high-performance transparent conductive electrodes, Nano Lett. 9 (2009) 4359–4363. Available from: https://doi.org/10.1021/nl902623y.

[76] B. Itapu, A. Jayatissa, A review in graphene/polymer composites, Chem. Sci. Int. J. 23 (2018) 1–16. Available from: https://doi.org/10.9734/CSJI/2018/41031.

[77] D. Ponnamma, K.K. Sadasivuni, Graphene/polymer nanocomposites: role in electronics, in: K.K. Sadasivuni, D. Ponnamma, J. Kim, S. Thomas (Eds.), Graphene-Based Polymer Nanocomposites in Electronics, Springer International Publishing, Cham, 2015, pp. 1–24. https://doi.org/10.1007/978-3-319-13875-6_1.

[78] F. Wang, P. Mao, H. He, Dispensing of high concentration Ag nano-particles ink for ultra-low resistivity paper-based writing electronics, Sci. Rep. 6 (2016) 21398. Available from: https://doi.org/10.1038/srep21398.

[79] H. Sirringhaus, T. Kawase, R.H. Friend, T. Shimoda, M. Inbasekaran, W. Wu, et al., High-resolution inkjet printing of all-polymer transistor circuits, Science 290 (2000) 2123–2126. Available from: https://doi.org/10.1126/science.290.5499.2123.

[80] S. Jeong, H.C. Song, W.W. Lee, S.S. Lee, Y. Choi, W. Son, et al., Stable aqueous based Cu nanoparticle ink for printing well-defined highly conductive features on a plastic substrate, Langmuir 27 (2011) 3144–3149. Available from: https://doi.org/10.1021/la104136w.

[81] H.-W. Tien, S.-T. Hsiao, W.-H. Liao, Y.-H. Yu, F.-C. Lin, Y.-S. Wang, et al., Using self-assembly to prepare a graphene-silver nanowire hybrid film that is transparent and electrically conductive, Carbon 58 (2013) 198–207. Available from: https://doi.org/10.1016/j.carbon.2013.02.051.

[82] C. Lee, X. Wei, J.W. Kysar, J. Hone, Measurement of the elastic properties and intrinsic strength of monolayer graphene, Science 321 (2008) 385–388. Available from: https://doi.org/10.1126/science.1157996.

[83] C. Auner, U. Palfinger, H. Gold, J. Kraxner, A. Haase, T. Haber, et al., High-performing submicron organic thin-film transistors fabricated by residue-free embossing, Org. Electron. 11 (2010) 552–557. Available from: https://doi.org/10.1016/j.orgel.2009.12.012.

[84] J.A. Rogers, Z. Bao, K. Baldwin, A. Dodabalapur, B. Crone, V.R. Raju, et al., Paper-like electronic displays: large-area rubber-stamped plastic sheets of electronics and microencapsulated electrophoretic inks, Proc. Natl. Acad. Sci. U. S. A. 98 (2001) 4835–4840. Available from: https://doi.org/10.1073/pnas.091588098.

[85] F. Torrisi, T. Hasan, W. Wu, Z. Sun, A. Lombardo, T.S. Kulmala, et al., Inkjet-printed graphene electronics, ACS Nano 6 (2012) 2992–3006. Available from: https://doi.org/10.1021/nn2044609.

[86] A. Capasso, A.E. Del Rio Castillo, H. Sun, A. Ansaldo, V. Pellegrini, F. Bonaccorso, Ink-jet printing of graphene for flexible electronics: an environmentally-friendly approach, Solid State Commun. 224 (2015) 53–63. Available from: https://doi.org/10.1016/j.ssc.2015.08.011.

[87] B. Derby, N. Reis, Inkjet printing of highly loaded particulate suspensions, MRS Bull. 28 (2003) 815–818. Available from: https://doi.org/10.1557/mrs2003.230.

[88] D. Jang, D. Kim, J. Moon, Influence of fluid physical properties on ink-jet printability, Langmuir 25 (2009) 2629–2635. Available from: https://doi.org/10.1021/la900059m.

[89] R.D. Deegan, O. Bakajin, T.F. Dupont, G. Huber, S.R. Nagel, T.A. Witten, Capillary flow as the cause of ring stains from dried liquid drops, Nature 389 (1997) 827−829. Available from: https://doi.org/10.1038/39827.

[90] M. Singh, H.M. Haverinen, P. Dhagat, G.E. Jabbour, Inkjet printing-process and its applications, Adv. Mater. 22 (2010) 673−685. Available from: https://doi.org/10.1002/adma.200901141.

[91] V. Shukla, Review of electromagnetic interference shielding materials fabricated by iron ingredients, Nanoscale Adv. 1 (2019) 1640−1671. Available from: https://doi.org/10.1039/C9NA00108E.

[92] K. Shi, J. Su, K. Hu, H. Liang, High-performance copper mesh for optically transparent electromagnetic interference shielding, J. Mater. Sci. Mater. Electron. 31 (2020) 11646−11653. Available from: https://doi.org/10.1007/s10854-020-03716-w.

[93] S.K. Vishwanath, D.-G. Kim, J. Kim, Electromagnetic interference shielding effectiveness of invisible metal-mesh prepared by electrohydrodynamic jet printing, Jpn. J. Appl. Phys. 53 (2014) 05HB11. Available from: https://doi.org/10.7567/JJAP.53.05HB11.

[94] Y. Wang, F. Jiang, J. Chen, X. Sun, T. Xian, H. Yang, In situ construction of CNT/CuS hybrids and their application in photodegradation for removing organic dyes, Nanomaterials. 10 (2020) 178. Available from: https://doi.org/10.3390/nano10010178.

[95] C. Zheng, H. Yang, Z. Cui, H. Zhang, X. Wang, A novel Bi4Ti3O12/Ag3PO4 heterojunction photocatalyst with enhanced photocatalytic performance, Nanoscale Res. Lett. 12 (2017) 608. Available from: https://doi.org/10.1186/s11671-017-2377-1.

[96] M.F. Shakir, A.N. Khan, R. Khan, S. Javed, A. Tariq, M. Azeem, et al., EMI shielding properties of polymer blends with inclusion of graphene nano platelets, Results Phys. 14 (2019) 102365. Available from: https://doi.org/10.1016/j.rinp.2019.102365.

[97] A. Ku, O. Oetinscitan, J.-D. Saphores, A. Shapirod, J.M. Schoenunp, Lead-free solders. issues of toxicity, availability and impacts of extraction, 53rd Electronic Components and Technology Conference, 2003. Proceedings, IEEE, New Orleans, Louisiana, USA, 2003, pp. 47−53. https://doi.org/10.1109/ECTC.2003.1216255.

[98] L.J. Turbini, Examining the environmental impact of lead-free soldering alternatives, IEEE Trans. Electron. Packag. Manuf. 24 (2001) 6.

[99] R.W. Vanderwiel, Inventors: Steven Bullock, Tehacahpi, CA (United States) (n.d.) 10.

Polymer-graphene composites as flame and fire retardant materials

Prashant Gupta and Subhendu Bhandari
Department of Plastic and Polymer Engineering, Maharashtra Institute of Technology, Aurangabad, India

25.1 Introduction

There are numerous applications for which synthetic polymers are used in day-to-day life. In spite of innumerable applications due to their versatility, their nature of being derived from petrochemical origin makes them highly flammable. The use of modifying additives (flame retardants in our context) opens up new areas of engineering and industrial applications where the requirement is flame retardancy. The flammable nature of polymer can be turned into self-extinguishing upon removal of flame by means of the addition of conventional flame retardants and converting a polymer into a composite. However, the high dosage of these flame retardants has a detrimental effect on other physicochemical and strength properties of the polymer. This can be addressed by using multifunctional additives such as graphene and its family of additives either alone or in combination with other flame retardants. Also, the nanoform of graphene further improves the strength properties of the composite and opens up newer methods of its modifications and designing a different set of properties by using these modified multifunctional additives.

Fire is a combination of three factors namely: heat, oxygen, and fuel. Heat is responsible for producing flammable gases from burning polymer at a very high temperature (pyrolysis) and ignition occurs via oxygen present in the atmosphere, striking an appropriate ratio between it and the gases formed by pyrolysis as a result of which, heat is exhaled out and inhaled back in the process which continually aids the process of combustion. To break this combustive chain of events, at least one of the three factors needs to be removed from the process and hence there may be several ways to break down the fire triangle [1]. The action of flame retardants inhibits or suppresses the process of combustion by physically or chemically participating in the different stages of the process namely: heating, decomposition, ignition, the spread of flame, and generation of smoke.

25.1.1 Physical action

The physical action of flame retardance can happen through any of the three methods.

Polymer Nanocomposites Containing Graphene. DOI: https://doi.org/10.1016/B978-0-12-821639-2.00010-0

25.1.1.1 Solid phase char formation (protective layer formation)

The addition of fire-retardant additives reacts to form a low thermal conductivity shield/carbonaceous layer of burning matter's surface through an outer heat flux. This layer insulates the polymer by reduction of heat transfer happening inwards that is toward the material and slows down pyrolysis thereby restricting the process for release of gases responsible for fueling combustion. This methodology Is generally exhibited by nonhalogen agents using phosphorus, boron, silicon, nitrogen, and basic base additives. The use of phosphorus additives forms a vitreous protective barrier layer of thermally stable poly or pyrophosphoric compounds generated from pyrolysis of phosphorus containing fire retardant additives.

25.1.1.2 Quench and cool

The additive degradation can greatly influence the energy balances for the process of combustion. These systems are responsible for undergoing an endothermic reaction under the presence of fire which releases H_2O molecules that dilute the combustion process by undergoing cooling to a temperature below the required value for the ignition process to happen. A class of halogen-free flame retardants makes up this category with hydrated minerals and different metal oxides such as alumina trihydrate (ATH).

25.1.1.3 Dilution

There are inert materials that are added to the polymer such as (talc, calcium carbonate, magnesium hydroxide, ash, silica, etc.) and/or additives. (responsible for evolving nonflammable inert gases upon decomposition) Their presence dilutes the fuel in the solid and gaseous phases which ensures that it does not reach the gas mixture lower ignition point. It leads to the decrease in heat conductivity of the filled material ensuring lower thermal degradation of the polymer in the bulk [2].

25.1.2 Chemical action

On the other hand, chemical action is responsible for suppressing the cycle of combustion in the solid and/or gaseous phase.

25.1.2.1 Vapor phase inhibition

The reaction of halogenated flame retardants with the burning polymer in the vapor phase disrupts the production of free radicals at a molecular level thereby interrupting the combustion process. The process of combustion is aided by the generation of free radicals in the gas phase. Its interruption causes the combustion process to stop along with the cooling of the system thereby reducing the supply of fuel gas and suppressing the process completely. The halogenated flame retardants also interfere with chain branching reactions which propagate combustion, taking place due to pyrolysis of polymer reacting with air. This interference avoids hydrogen free radicals and hydroxide to carry out any reactions with O_2 and CO. The exothermic oxidative process of making flames/fire is thus interfered resulting in a shutdown of the combustion process. The reactions that happen in the solid phase during the vapor phase mechanism during the above process are as given below:

$$\bullet H + H - X = H + X \tag{25.1}$$

$$\bullet HO + H - X = HO + X^{\bullet} \tag{25.2}$$

$$\bullet R - H + X^{\bullet} = R^{\bullet} + HX \tag{25.3}$$

In the above reactions, (25.1) and (25.2) exhibit the reaction of radicals HO· and H· with high reactivity, with the halogenated radicals X, generated from the degradation of halogenated flame-retardant additive in the gaseous phase. In reaction (25.3), the reaction exhibits fewer reactive radicals being formed which reduces the kinetics of the combustion. The studies carried out in understanding the effect of halogenated flame retardants reported a decrease in the effectiveness of their function in the hierarchy of Iodine > Bromine > Chlorine > Fluorine. However, the use of bromine and chlorine is predominant in products available due to stability issues and low effectiveness of Iodine and Fluorine respectively. The addition of antimony oxide to halogenated compounds ensures the trapped free radicles reach fire in the gas phase. They form an excellent synergy with antimony trioxide as it allows the formation of volatile antimony halides/antimony oxy halide to interrupt the process of combustion by H· radical inhibition [3,4].

25.1.2.2   Condensed phase inhibition

During processing or application, the breakdown of polymer can be increased by the action of flame retardants. In processing, this may increase the polymer's flow rate and impact the Melt Flow Index (MFI) and other strength properties of the polymer. Alternatively, in use, it decreases the effect of flame which fades away. Also, flame retardants can cause charring on the burning polymer surface through the dehydrating effect leading to the generation of unsaturation in the polymer under attack and these cross-linking process cycle. Using intumescence for flame retardation in polymers is a specific case of condensed phase mechanism and trapping the radicles as in the vapor phase does not appear to the case. The fuel produced during this inhibition mechanism is very less and the formation of char is more evident than gases supporting combustion. This intumescent char consists of a two-way barrier for (1) restricting the passing of gases that can fuel the combustion process and molten polymer to the flame and (2) protecting the attacked polymer from the intensity of the flame. These systems consist of three core ingredients namely a catalyst which works as the source of acid, a charring agent, and a blowing agent. These three basic ingredients, for example, ammonium salts phosphates, polyphosphate as a catalyst, polyhydric compounds as charring agents, and amines/

amides as blowing agents that combine to give an intumescent effect for flame retardance [2,5,6].

25.2 Polymer graphene composites as fire and flame retardants

Graphene and its other forms are capable of exhibiting flame retardance effect due to parameters favoring the reduction of heat and mass transfer such as strong barrier effect, high specific surface adsorption capability, and high thermal stability amongst others. Out of three components required for combustion to continue, graphene works by taking away two components that are heat and fuel [1]. There are three ways through which graphene improves the flame retardancy of the polymers:

1. By promoting the formation of multiple and overlapped char layer via the "labyrinth effect" due to stacked 2-D skeleton structure during decomposition which can potentially work as a barrier to prevent heat transfer from the source of heat and has a delaying effect on the gaseous reaction products generated as a result of pyrolysis of the polymer surface. The improved fire safety, in this case, can be attributed to the creation of a tortuous path as shown in Fig. 25.1. which enables the difficulty in exchange of heat/mass in between the condensed and gaseous phases resulting in improvement of thermal stability.
2. During combustion, adsorption of organic volatiles (flammable) can occur with the virtue of large specific surface area and hindering their diffusion and release, thereby providing a platform for catalysis and carbonization of materials like metallic oxide.
3. Graphene and its oxides contain carboxylic groups at the edges and hydroxyl/epoxy groups at the plane which have abundant in reactive oxygen consisting groups owing to which they can be modified to be used in a wide range of engineering applications. At low temperatures, the decomposition and dehydration of oxygen-containing groups in graphene or (GO) may lead to absorption of heat and thereby cool the polymer substrate during combustion. The gases generated due to dehydration dilute the O_2 concentration around the sidelines of ignition [8].
4. The compatibility of graphene and GO is good with polymeric molecules due to a strong interaction that has the potential to form a 3-D network in the polymeric matrix which is capable of interfering in its rheological behavior by increasing the polymer viscosity upon

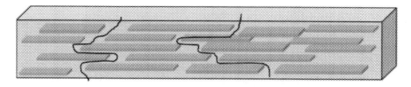

Figure 25.1 Tortuous path creation by graphene nano platelets *(GNP)* modification in polymer composites [7].
Source: X. Wang, E.N. Kalali, J.-T.Wan, D.-Y. Wang, Carbon-family materials for flame retardant polymeric materials. Prog. Polym. Sci. 69(2017) 22−46. https://doi.org/10.1016/j. progpolymsci.2017.02.001.

thermal exposure and prevent dripping of the polymer which is usually the case of a flame attack on the polymer. The labyrinth effect then leads to hindering the release of gaseous volatile decomposition products (fuel for combustion) and affects the flame retardancy of the nanocomposites [9].

Graphene exhibits flame retardance for various polymers including thermoplastic, thermosetting plastics and rubber namely polypropylene (PP) [10,11], polyvinyl chloride (PVC) [12], polyacrylonitrile (PAN) [13,14], polylactic acid (PLA) [15,16], polyvinyl acetate (PVA) [17], polyvinyl formaldehyde [18], polyurethane (PU) [19,20], polystyrene (PS) [9], polymethyl methacrylate (PMMA) [21], polybutylene terephthalate (PBT) [22], PLA and polybutyl succinate (PBS) [23], polyamide-6 (PA-6) [24], cyanate ester [25], epoxy resin [17,25−34], and natural rubber [35] respectively. There are different methodologies of graphene incorporation/use in increasing the flame retardancy of the above polymers namely use of neat graphene, the synergistic action of graphene with other flame retardants, grafting, blending with flame retardants, organic/inorganic graphene-based flame retardants, modified graphite, and its synergism with other flame retardants.

25.2.1   Pure graphene

Graphene and its oxide form with high purity exhibit good stability when combusted in the presence of natural gas flame for a few seconds. As seen in Fig. 25.2A, the burned part turns hot red when reduced GO (rGO) is exposed to natural flame just for a few seconds. However, the reaction as expected does not propagate and is quenched when the flame is removed, exhibiting high intrinsic flame resistance of graphene. After reexposing the part to the flame on multiple occasions, combustion does not occur which is an indication of good thermal stability. Fig. 25.2B shows the effect of leftover potassium salt impurities formed during the synthesis along with GO is capable of undergoing self-propagating thermal deoxygenating reaction which makes GO highly flammable. The flame is triggered with a gentle touch with a hot spot and is capable of undergoing self-propagating combustion of rGO as seen in six images in Fig. 25.2B of time intervals of 0.32, 1.82, 2.5, 2.5, and 2.5 s from left to right. This indicates that KOH is proven to be catastrophic in terms of flame retardance properties for rGO [36].

There are examples to illustrate modifications of graphene with montmorillonite (MMT) and films made via vacuum filtration exhibiting excellent flame-retardant properties to go along with excellent flexibility and electrical conductivity [37]. Their films as shown in Fig. 25.2C and D do not catch fire and become warped under the action of heat. Also, graphene phosphonic acid (made by ball milling operation of red phosphorus with graphite) has also exhibited its nature as an effective flame retardant [38].

The intrinsic flame-retardant nature of neat graphene has been directly used to prepare nanocomposites with flame retardant nature. Graphene, in comparison with other carbon materials such as expanded graphite (EG), nanocarbon black, and multiwalled carbon nanotubes (CNT), graphene uniformly distributes in the

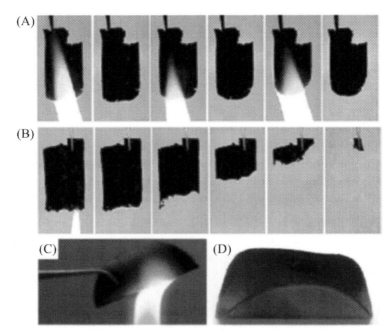

Figure 25.2 Flame treatments on various materials (A) rGO exposed to flame with 3−4 s intervals; (B) Flame treatment of rGO with KOH impurities present in the system with varying intervals of 0.32, 1.82, 2.5, 2.5 and 2.5 s; (C) Flame treatment of MMT-rGO hybrid films and (D) MMT-rGO films after removal of flame [36].
Source: F. Kim, J. Luo, R. Cruz-Silva, L.J. Cote, K. Sohn, J. Huang. Self-propagating domino-like reactions in oxidized graphite. Adv. Funct. Mater. 20 (2010) 2867−2873. https://doi.org/10.1002/adfm.201000736.

system and increases a host of other properties in comparison to PP + flame/fire retardant (FR) such as stiffness and yield stress (increase by 80% and 16% respectively observed at 7.5wt.% thermally reduced graphene) due to ultrahigh aspect ratio and efficient load transfer of graphene nanosheets, thermal property(Tc increased by 10% to 122°C at 7.5wt.% thermally reduced graphene due to better nucleation of graphene in the polymer chains), electrical conductivity (10^{-5} S/cm) and improved flame retardance with smoke suppression (decrease in peak heat release rate or peak heat release rate (PHRR) by 74% to 489 kW/m at 5wt.% thermally total heat evolved/released or THE by 10% to 86 MJ/m [39]). The improvement of flame retardance and smoke evolution of waterborne polyurethane coating (WPC) polymer with the use of solution blending with graphene is carried out. The cone calorimetry studies done to ascertain the flame retardancy characteristics show a reduction in peak heat PHRR and total heat release (THR). In comparison with pure WPC, the PHRR values of 2% rGO filled WPC show a

22% drop in the values with the value reported to be 448 kW/m. Furthermore, a bimodal HRR curve is a clear indication of residue C layer formation which acts as a barrier to prevent heat and mass transfer to and from the polymer. The drop in these values reported with the addition of rGO in WPC can be attributed to the role of rGO sheets that prevent the degradation of polymer inside the C layer thereby suppressing the smoke. The THR curve exhibits the reduction of heat released up to 1.5% rGO concentration which is 72 MJ/m, reportedly 27% lower than neat WPU values. At 2% rGO concentration, the values are higher than 1.5% rGO but the slope of curve is lowest in all concentrations which indicates the smallest flame propagation out of the samples tested. Also, smoke suppression is very important in fire retardance as deaths are also caused due to inhalation of harmful and toxic smoke leading to suffocation and eventual death. Hence, the reduction of total smoke release (TSR) and is important and the TSR values are lower than that of WPC. The values for 1% rGO-WPC were reportedly 475 m^2/m, about 25% lower than neat WPC. The Smoke Factor (SF), a value derived from PHRR and TSR was reported to be 230 kW/m for 1% rGO-WPC, about 38% lower than neat WPC. The values for TSR and SF are higher in 2% rGO-WPC in comparison to 1% rGO-WPC may be because of incomplete combustion of the polymer. The overall flame retardancy mechanism can be summed by a condensed phase flame retardancy by forming a uniform, dense, and compact charred C layer during the combustion process [40].

The intumescent effect in graphene is one of the crucial factors governing the flame-retardant ability in graphene-polymer nanocomposites and it has been reported that the degree of excessive oxidation or higher degree of oxidation in GO is detrimental to the flame retardant effect of GO in the nanocomposites. The fineness in dispersion and disrupting of order in GO due to intercalation of epoxy molecules in GO galleries reduced the intumescent ability of GO. This was evident when the flame retardant effect of GO went down when GO dispersion in the epoxy resin was aided by sonication [41]. The FR effect of different oxidation degrees of GO on PS reveals that the thermal stability decreased and PHRR came down with an increase in the O groups in GO. The optimum loading in terms of flammability properties was observed to be with 5% graphene which exhibited around 50% reduction in PHRR upon comparison with PS [9].

The use of neat GO has also been reported for use in thermosetting resins, for example, benzoxazine [42] and epoxy resin [43]. In graphene/epoxy nanocomposites, the addition of graphene altered the decomposition pattern of epoxy at higher temperatures promoted the formation of char, and increased thermal stability. It led to increasing compactness of both polymer surface and beneath char residues. It also led to effective inhibition of flammable drips of epoxy polymer resin during combustion and decreased flow rate of the polymer. There was an increasing trend in PHRR of graphene/epoxy nanocomposites with respect to neat epoxy samples despite of high values for Limiting Oxygen Index (LOI) and a reduced THR, with virtue of a thermal conductivity balance with barrier characteristics of graphene [43].

25.2.2   Organic flame retardants functionalized graphene (grafting aided molecular modification)

GO, a starting material for graphene, possesses functional groups namely hydroxyl, carboxyl, epoxy, -C=C- groups at the edge and plane, which proves to be an active site for functionalization of graphene [44]. This process turns the hydrophobic surface of GO into hydrophilic thereby facilitating dispersion of sheets of graphene in the polymer matrix [45]. This grafting process is essentially a two-stage process that is put to use by forming a covalent bond between organic flame retardants and GO/graphene. The first stage consists of finding suitable organic flame retardants with functionalities that have the capability to react with either GO and the second deals with reducing functionalized GO to functionalized graphene with the help of reducing agents. Some modifiers may be able to do functionalization and reduction at the same time. Upon comparison to neat graphene and organic flame retardants, they are shown to exhibit superior flame-retardant properties in terms of UL-94 fire rating, LOI, and PHRR. Some of the organic flame retardants grafted to GO along with the change they bring about to the system in terms of flame retardancy is given as below:

1. 9, 10-dihydro-9-oxa-10-phosphaphenanthrene-10-oxide (DOPO) functionalized graphene at 10wt.% improves the yield of char formation both in neat epoxy and GO/epoxy composite (increase of around 50% and 33% respectively) and LOI value (increase of 30% and 13% respectively) [45],
2. 2-(diphenylphosphino) ethyltriethoxy silane (DPPES) functionalized graphene at 10wt.% improves in LOI value by 80% over neat graphene/epoxy nanocomposites and found to pass UL 94 V-0 rating (3.2 mm) which fails otherwise [46],
3. Hyper-branched flame retardant functionalized graphene at 3wt.% shows 29% reduction in PHRR against only 8% reduction observed by neat GO/Crosslinked polyethylene (XLPE) nanocomposite [47],
4. Polyphosphamide (PPA) functionalized graphene at 8wt.% shows 41% and 50% reduction in PHRR and THR respectively in comparison with neat epoxy and 10% and 7% reduction in PHRR and THR respectively when compared to 8% PPA/EP nanocomposites [48]. The effect of PPA in PP reduced the PHRR and THR values by 67% and 24% respectively upon comparison with neat PP and 36% reduction in PHRR and 6% reduction in THR upon comparison with 20% FR filled PP. The time to ignition also went up by 22% in 20% PPA-fGO filled PP; [48,49]. and
5. Poly(piperazine spirocyclic pentaerythritol bisphosphonate)(PPSPB) functionalized graphene at 1wt.% shows 46% and 22% reduction in PHRR and THR respectively in comparison with neat ethylene vinyl acetate (EVA) and 24% reduction in PHRR along with 8.5% reduction in THR respectively when compared to 1wt.% chemically reduced graphene/EVA nanocomposites [50].

The sol-gel chemistry can also be employed for carrying out the functionalization of graphene nanosheets (GNF). The functionalization of graphene was carried out by 3-aminopropyltriethoxsilane (APTS) and the functionalized GNF with the help of reactive functional groups link up with silane-modified PMMA with the help of sol-gel reaction. This approach provides flexibility to graft various SI and P-containing flame retardants for making functionalized graphene which can enhance flame retardancy of polymers and have been employed in various polymers such as PMMA [51], PVA [52], polyurea [53], and epoxy resins [54,55].

25.2.3 Hybrids of inorganic materials/graphene

The modification of graphene through noncovalent bond formation has been reported for inorganic/graphene hybrids such as layered double hydroxide (LDH)/Graphene and metal oxide/graphene. Fig. 25.3 shows a schematic for the formation of LDH/graphene nanocomposites. The stepwise process first involves exfoliation of GO aided by sonication and the subsequent attachment of metal ions over graphene oxide which possesses a negative charge by electrostatic attraction. Then, the reduction of GO to graphene is followed by self-assembly of metal cations possessing graphene sheets into a hybrid LDH/graphene nanostructure owing to the hydrophobic nature of graphene. The attachment of LDH may be able to avoid bundling of graphene sheets during the processing of nanocomposites made using the same. The metal oxide/graphene hybrid is similar in formation to LDH/Graphene hybrid [7].

Table 25.1 exhibits the flame-retardant effects for metal hybrid or LDH/graphene nanocomposites. It is evident to see that around 2% concentration of metal oxide or LDH/graphene reduced the PHRR values by 23%−61% when compared to neat polymer resins. The mechanism of flame retardancy is illustrated for epoxy (EP) resin which was reinforced by EP and Ce−MnO$_2$-f-Graphene nanosheets (GNS)/EP

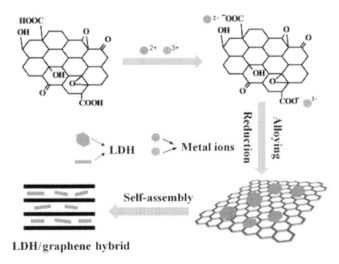

Figure 25.3 Reaction involving the formation of layered double hydroxide /graphene hybrid [7]. *Source*: X. Wang, E.N. Kalali, J.-T.Wan, D.-Y. Wang, Carbon-family materials for flame retardant polymeric materials. Prog. Polym. Sci. 69(2017) 22−46. https://doi.org/10.1016/j. progpolymsci.2017.02.001.

Table 25.1 Flame retardant nature of findings reported in the literature for layered double hydroxide or metal oxide/graphene hybrid containing polymer nanocompo sites.

Polymer	Filler description	Filler (%)	Change in FR properties	Other remarks	Reference
PMMA	NiAl-LDH/r-GO	2	Reduction in PHRR by 25% in comparison to pure graphene (17%) and NiAl-LDH (13%) based composites, increase in time to ignition (TTI) by 10% and 79% reduction in peak carbon monoxide yield (PCOY) upon addition of NiAl-LDH/r-GO	NiAl-LDH-rGO provides physical barrier and NiAl-LDH ensures catalytic carbonization	[21]
PA-6	Co$_3$O$_4$/graphene, NiO/graphene	2	Reduction in PHRR, THR, TSR and PCOY by 23%, 12%, 21% and 9% respectively and increase in TTI by 7% for PA6/GNS-NiO composites and reduction in PHRR, THR, TSR and PCOY by 11%, 5%, 6% and 50% respectively and increase in TTI by 16% for PA6/GNS-Cu$_3$O$_4$ composites when compared to neat PA6 samples	Improvement in thermal stability evidenced by increase in decomposition temperature and reduction in gaseous products with addition of GNS-metal oxide	[24]
Polybutyl succinate (PBS) and Polylactide (PLA)	Co$_3$O$_4$/graphene	1	Reduction in PHRR and PCOY by 31% and 70% respectively Co$_3$O$_4$ f-graphene/PBS composites when compared to neat PBS and reduction in PHRR and PCOY by 40% and 67% respectively for Co$_3$O$_4$ f-graphene/PLA composites when compared to neat PLA samples	Initial degradation temperature increased upon addition of f-graphene by 25°C and 14°C in comparison to pure PBS and PLA	[23]

PBT	MnCo$_2$O$_4$/graphene	1	Reduction in PHRR, THR and PCOY by 37% and 8% respectively Co$_3$O$_4$ f-graphene/PBS composites when compared to neat PBT. Time to ignition is 5% higher in case of composite but overall fire performance index is lower than that of neat PBT samples	Incorporation of MnCo$_2$O$_4$-GNS hybrids decreases the pyrolysis products containing aromatic and carbonyl compounds, CO and CO$_2$	[22]
Acrylonitrile butadiene Styrene (ABS)	Co(OH)$_2$ nanorods/GNS	2	Reduction in PHRR, mMLR and fire growth index (FGI) by 30%, 42%, and 35% respectively with an increase in TTI by 9% upon comparison with neat ABS samples,	Continues charred layers formed upon combustion by catalyzed by Co(OH)$_2$ nanorods	[56]
PU Elastomer	RGO-LDH/Mo	3	Reduction in PHRR and TSR by 63.1% and 50% respectively in comparison to neat PU samples. A significant increase in char yield is also observed	MoO$_3$ generated from the thermal decomposition of RGO-LDH/Mo has the effect of catalyzing carbonization	[57]
Epoxy	Co$_3$O$_4$/graphene, SnO$_2$/graphene	2	Reduction in PHRR and THR by 29%, and 25% respectively and increase in char yield by 52% for SnO$_2$/graphene composites and reduction in PHRR and THR by 27% and 17% respectively and increase in char yield by 23% for Co$_3$O$_4$/graphene composites when compared to neat epoxy resin samples	Decrease in smoke density and CO yield observed with incorporation of f-graphene. Also, decreasing trends in pyrolysis products for hydrocarbons, aromatic and carbonyl compounds and CO.	[58]
Epoxy	MoS$_2$/graphene	2	Reduction in PHRR, THR and TSR by 46%, 25% and 30% respectively for MoS$_2$ f-GNS/Epoxy composites when compared to neat epoxy resin and reduction in PHRR, THR and TSR by 28%, 20% and 23% respectively for GNS/epoxy composites when compared to neat epoxy sample	Char residues increases and MoS$_2$-GNS ensures formation of molybdenum oxide, a highly effective smoke suppressant which can lower the production of smoke	[59]

(Continued)

Table 25.1 (Continued)

Polymer	Filler description	Filler (%)	Change in FR properties	Other remarks	Reference
Epoxy	NiFe-LDH/graphene	2	Reduction in PHRR and THR by 61% each for NiFe-LDH-f-graphene/epoxy composites when compared to neat epoxy resin and time taken for ignition was increased by 31%. The onset thermal degradation temperature was increased by 25°C in comparison to that of pure epoxy	Reduction/Slowdown of release of combustible fuel gases such as hydrocarbons and aromatic compounds	[60]
Epoxy	ZnS/graphene	2	Reduction in PHRR, THR and TSR by 47%, 28% and 63% respectively for ZnS f-graphene/epoxy composites when compared to neat epoxy resin	Micro char formation and reduction in volatiles released during combustion process	[61]
Epoxy	Ce-doped MnO$_2$/graphene	2	Reduction in PHRR, THR and TSR by 54%, 36% and 41% respectively for Ce-doped MnO$_2$-graphene/epoxy composites when compared to neat epoxy resin	Increase in degradation temperatures and char residue aided by formation of pyrolysis products with lower carbon numbers	[62]

LDH, Layered double hydroxide; TSR, total smoke release.

nanocomposites. The layered nanofillers which comprises of LDH and metal oxides increase charring and helps in compaction of the same, thereby working in the condensed phase for flame retardation by standing in as char reinforcers as the efficacy of flame retardant working in condensed phase is dependent upon the structure and composition of char layer during the process of combustion. Fig. 25.4 exhibit the scanning electron micrographs exhibiting the surface morphology of the residues of pure EP and Ce-doped MnO_2/GNS. Fig. 25.4A exhibits loose char surface with a lot of cracks/holes dispersed over the surface at 50 μm magnification. Upon going in to 20 μm magnification as shown in Fig. 25.4B, the surface of the char layer looks porous and very less compact. These images are of course without incorporation of Ce-doped MnO_2-f-GNS after addition of which at 2%, a decrease is observed in the cracks and holes. The surface porosity and its incompactness changes to a more compacted char layer as seen in Fig. 25.4C and D. In the inset of Fig. 25.4D, it can be observed that char contains a nanosphere mass which are interconnected to form a barrier or network (macromolecular chain) structure. The compact barrier char layer alters the heat and mass transfer due to the barrier effect and provides better flame shielding for the epoxy polymer.

The XRD peaks corresponding to MnO can be observed which is due to the in situ formation of its nanoparticles via the reduction of MnO_2 by the action of combustion gases from epoxy resin which in turn can catalyze the carbonization of degradation products. Furthermore, the peak intensity ratio that is I_D/I_G or the intensity

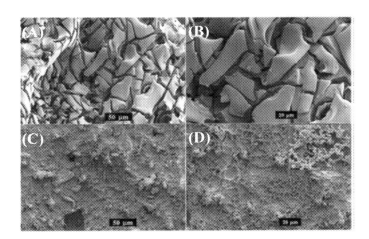

Figure 25.4 Scanning electron micrographs exhibiting the surface morphology of the residues of pure EP (A) and (B) and e−MnO_2−GNS−EP at different magnifications(C) and (D) [62]. *EP, Epoxy.*

Source: Reproduced by permission of S.-D. Jiang, Z.-M. Bai, G. Tang, L. Song, A.A. Stec, T.R. Hull, J. Zhan, Y. Hu. Fabrication of Ce-doped MnO_2 decorated graphene sheets for fire safety applications of epoxy composites: flame retardancy, smoke suppression and mechanism. J. Mater. Chem. A, 2, (2014c)17341−17351. https://doi.org/10.1039/C4TA02882A. The Royal Society of Chemistry.

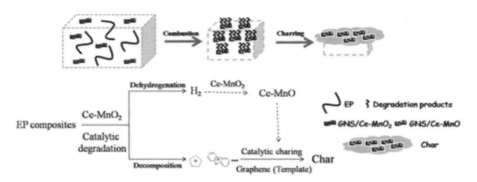

Figure 25.5 Mechanism for flame-retardance by Ce−MnO$_2$−GNS hybrid for epoxy
nanocomposites [62].
Source: Reproduced by permission of S.-D. Jiang, Z.-M. Bai, G. Tang, L. Song, A.A. Stec,
T.R. Hull, J. Zhan, Y. Hu. Fabrication of Ce-doped MnO$_2$ decorated graphene sheets for fire
safety applications of epoxy composites: flame retardancy, smoke suppression and
mechanism. J. Mater. Chem. A, 2, (2014c)17341−17351. https://doi.org/10.1039/
C4TA02882A. The Royal Society of Chemistry.

ratio of D and G bands of Ce-doped MnO$_2$-f-GNS is slightly lower than that of EP
which means that there is improvement in the graphitized carbons in the char residues obtained after combustion indicating a synergy function for char formation
between the dopant and GNS. This higher degree of graphitization is responsible
for the decrease of heat transfer via char compaction and thermal insulation
improvement. As given in Table 25.1, the improvement of flame-retardant properties, in this case, can be attributed to the synergistic effect of Ce−MnO$_2$ and the
barrier effect as described above in GNS.

The mechanism of flame retardance can be explained as shown in Fig. 25.5 with
the help of results of volatile pyrolysis fragment ions obtained by the DP-Ms technique. The catalytic activity of Ce−MnO$_2$ during combustion enables the formation
of lower carbon number pyrolysis products which can be further catalyzed with
ease in the presence of metal oxides for carbonization. At the same time, in situ
reduction reactions of Ce−MnO$_2$ is carried out by redox reactions of hydrocarbons
with reducing ability which are able to catalyze the carbonization of degradation
products. Furthermore, GNS can absorb degraded products by the physical barrier it
creates which in turn extends the flame contact time with metal oxides. The fire
propagates at GNS which provides a micro-char template. The dehydrogenation and
catalytic conversion of degraded products into char is carried out by a combined
effect of GNS's physical barrier and catalysis by Ce−MnO$_2$ [52,61]. The addition
of LDH/metal oxide-f-graphene also improves smoke characteristics such as smoke
production and toxicity in comparison to neat graphene which can be attributed to
the decreased quantity of organic volatiles degraded, a major source of smoke due
to the combustion of polymers [60,63].

25.2.4 Synergy of graphene/other flame retardants

The heat release rate of polymers is effectively reduced by the action of organic FRs and inorganic/graphene hybrids. A majority of these FR systems, however, fail in UL 94 and LOI tests which are important for the polymer to be commercially/industrially established in the market. Thus, the use of graphene has been limited to a synergist with use in other established FR's for providing flame retardancy to the polymer. The materials and their effects on the flame retardancy of the polymer have been reported in Table 25.2. The use of CNT in a synergistic way with graphene (0.5% CNT/0.5% Graphene) is known to improve properties in PP such as PHRR and average mass loss rate (AMLR) above 1% graphene-filled nanocomposites. Furthermore, improvement in mechanical properties (tensile strength and modulus), electrical and thermal conductivity is also observed [56,65].

The preparation of anisotropic foams based on GO and nanocellulose with sepiolite nanorods (SEP) have been studied. The foams were tested for UL 94 vertical flame retardancy setup to optimize the loading of GO and SEP with 10% each. This composition reported very good flame retardancy without self-propagation. The samples without GO or low SEP content don't show the same fire retardancy as the above and shrinks more which helps in understanding that the combination of graphene with SEP works in synergy with each other.

There are reports of excellent antidrip properties during the combustion of composites with 10% IFR and 2% graphene. This also enables a V-0 rating of UL 94, for vertical burning test conducted on the samples. The dependence on increasing the melt viscosity on the antidripping properties has been reported in a UL 94, V-0 formulation having 2wt.% GNS with a combination of 12wt.% ammonium polyphosphate and 6% wt.% melamine in PBS (80%) [67,68].

25.2.5 Pristine graphite, expandable graphite, and modified expanded graphite

The use of graphite, a layered mineral with stacks of graphene sheets where hexagonal cells are formed through covalent bonds in between the nanosheets. The carbon layers are linked by Van der Wall forces of attraction. Expandable graphite is made by chemical treatment and can be easily exfoliated in the polymer matrix which is the major drawback (flakes in graphite stacks) due to which pristine graphene does not find its use in flame retardant applications. It is thereby used either in pristine or modified form for imparting flame retardance to a host of polymers.

EG is synthesized by subjecting the flaky graphite structure to concentrated sulfuric acid or acetic acid along with the use of strong oxidizing agents such as H_2O_2, $NaNo_3$, and $KMnO_4$ [72−75]. The insertion of sulphate ions was accompanied by oxidizing graphite with the above reaction conditions. EG's availability along with low cost, electrical conductivity, mechanical properties, and flame retardance makes it suitable to blend with polymers to make composites as it is a very attractive set of properties that polymers seek. The use of EG in polymers has been evident for

Table 25.2 Flame retardant nature of findings reported in the recent literature for synergistic action of graphene/conventional flame retardants.

Polymer	Filler and its loading (%)	Graphene (%)	Change in FR properties	UL 94 test results/other remarks	Reference
PE	Alumina trihydrate (40wt.%)	0.2	18% reduction in PHRR and 14% reduction in AMLR compared to that of PE/ATH composite	Detrimental effect observed upon addition of higher % of GNP (observed up to 1.5%)	[64]
PP	Carbon nanotubes (CNT) (0.5wt.%)	0.5	Reduction of PHRR and AMLR by 68% and 33% in 1% rGO/PP composites when compared to neat PP. Reduction in PHRR and AMLR by 73% and 37% at 0.5% rGO/0.5% CNT/PP composites in comparison with neat PP samples	Presence of CNT's improve the dispersion of rGO in polymer matrix as evident by 40% reduction in O$_2$ permeability and relative diffusivity	[65]
PP	CNT (1wt.%), Melamine polyphosphate (14.4wt.%) and Pentaerythritol (3.6wt.%)	1	PHRR and AMLR reduced by 83% and 35% respectively for PP/intumescent flame retardants (IFR)/CNTs/RGO composites upon comparison with neat PP. Furthermore, LOI value improved by 44% with respect to neat PP	UL 94 V-0 pass observed in case of PP/IFR/CNTs/RGO which failed in case of PP/IFR, PP/IFR/CNTs and PP/IFR/RGO as well	[66]
ABS	Co(OH)$_2$ (4wt.%)	2	mMLR and Fire Growth index values exhibit reduction of 30%, 42% and 35% respectively in the composites made using the given composition upon comparison with neat ABS	Black color of the compact char obtained after burning indicates promotion of char formation due to inclusion of cobalt compounds	[56]

Polymer	Additives		Property improvement	Flame retardancy	Ref.
Polybutyl succinate (PBS)	Ammonium polyphosphate (12wt.%) and Melamine (6wt.%)	2	LOI value increased to 32% and 33.0% from 23.0% for 1% and 2% graphene when compared with pure PBS	UL 94 V-0 burning test pass and melt dripping restrained with increased MFI with 2% GNS	[67]
Polylactide (PLA)	Phenylphosphinic acid and $(Zn(MeIm)_2)$	2	Reduction in PHRR and THR by 39% and 33% respectively and increase in LOI from 20.5% to 27% when 2% novel ternary hybrid nanoflake filled PLA composite samples are compared with neat PLA samples	UL-94 V-2 rating achieved due to catalytic and cross-linking effects of GO, nano-GIF (zeolitic imidazolate framework) 8 and PPA.	[16]
PU	Microencapsulated ammonium phosphate (12wt.%) and Melamine (6wt.%)	2	LOI value increased from 22.0% to 34.0% and increase in Tg and Tm from −40°C to −29°C and from 235°C to 269°C respectively. Also char residue at 600°C increases to 19% from 6.5% in case of neat PU	UL-94 V-0 rating with excellent antidripping properties and 39% reduction in water swelling percentage and 90% decrease in O_2 permeability	[68]
PMMA	LDH (5wt.%) and 1,2-Bis (5,5-dimethyl-1,3,2-dioxyphospacyclohexane phosphoryl amide) ethane (10wt.%)	1	Reduction in PHRR, THR and AMLR in the given composition in comparison with neat PMMA was 45%, 25% and 45% respectively. Increase in TTI, Char Yield and LOI as per the above is 23%, 15 times and 62% respectively	UL-94 V-1 rating passes on the composite in comparison to PMMA which failed in UL 94 test. Good dispersion of IFR, LDH and rGNS with formation of intercalated/exfoliated composites	[69]

(Continued)

Table 25.2 (Continued)

Polymer	Filler and its loading (%)	Graphene (%)	Change in FR properties	UL 94 test results/other remarks	Reference
Epoxy	Hexagonal Boron nitride sheets 20%	2	Reduction in PHRR, THR and TSP by 33%, 3% and 43% respectively and increase in TTI and FRI by 54% and 2.5 times respectively and LOI from 25 to 29.2 upon comparison with neat natural rubber (NR) samples	UL-94, V-1 rating passes with the given composition with respect to neat samples which fail upon exposed to UL-94 tests	[70]
Natural rubber	Ammonium polyphosphate (APP), pentaerythritol (PER) and melamine (MEL) (3:1:1) 34 phr	16	Reduction in PHRR and THR by 40%, and 23% respectively and increase in TTI and Flame Retardancy Index (FRI) by 36% and 2.5 times respectively and LOI from 18.2 to 26.1 upon comparison with neat NR samples	UL-94, V-0 rating passes with the given composition with respect to neat samples which fail upon exposed to UL-94 tests	[71]

ATH, Alumina trihydrate; LDH, layered double hydroxide; LOI, limiting oxygen index; PHRR, peak heat release rate; PPA, polyphosphamide.

use in PE, high density polyethylene (HDPE), PP, PVC, Nylon 6, PU, epoxy, PMMA, PLA, EVA, ABS, and Polyurea. A table showing the flame-retardant properties is given as Table 25.3.

An effort to understand the mechanism of expandable graphite has been done by loading EG in different particle sizes/loadings (70, 430, and 960 μm/ 0−50 pphp) in water-blown semirigid polyurethane foams (SPFs). The exposure of nanocomposites to fire makes the extrinsic graphite to expand into an intumescent worm-like, low density mass with an instant increase of volume in excess of 200 times. As shown in Fig. 25.6, the formation of thermally insulating layers is evident on the foam's surface exposed to the flame. This char layer/porous carbonaceous layer/expanded EG acts as a physical barrier limiting heat and mass transfer within an underlying layer and the flame of SPF matrix with further retardance action of the flame by the polymer nanocomposite [85]. Overall, the mechanism of EG induced flame retardance in polymers can be summed up as a combined effect of physical barrier provided by its flaky "layered" structure and blowing/intumescent effect by "chemical activity" in the condensed phase of the polymers. The morphology of burnt samples can be observed through TEM analysis. The evidence of layered graphite nanolayers which have an important role in suffocating the flames is visible in the residue after burning can be seen in Fig. 25.7 [86].

The use of sol-gel and microencapsulation methodologies have been put to use to modify EG and improve its compatibility with the PU and epoxy to enhance flame retardant properties [26,81]. The highlight of PU microencapsulated EG particles employed for improvement of flame retardant properties of EVA was the passing of UL-94, V-0 rating even after treatment with water for 7 days at elevated temperatures (70°C) [79].

25.2.6   Synergy of expanded graphite/other flame retardants

The use of synergist to counteract the high percentage of EG for making flame retardant nanocomposites which lead to a decrease in the mechanical properties along with puffy and unstable layers of char formed by combustion is well studied for various polymers. Table 25.4 enlists various polymers in which EG has been put to use with synergistic action of other flame retardants to enable polymers with improved flame retardance. Fig. 25.8 represents the mechanism of flame retardance for PP/EG/PDPFDE. In PP/MA-g-PP/EG nanocomposites, swelling of graphite results in expansion of graphite layers resulting in a popcorn effect forming worm-like structure eventually turning into an incompact charred layer. After the addition of PDPFDE in the above nanocomposite, its thermal decomposition produces 1 actylcyclopentadiene, cyclopentene, benzene ring, and Fe nanoparticles. The carboxyl and hydroxyl groups of PDPFDE and EG oxidize Fe to form Fe_3O_4 which may work as a catalyst to crosslink the polymer thereby ensuring formation of the charred residue. This can happen with the simultaneous reactions by catalyzing effect of Fe nanoparticles via aromatization and rearrangement with the basis of C-C coupling on monomers/dimers of PP, C_5H_5, naphthalene derivatives,

Table 25.3 Flame retardant nature of findings reported in the recent literature for action of expanded graphite on various polymers.

Polymer	Graphite concentration	Change in FR properties	Reference
HDPE	30%	Reduction in PHRR by up to 54% with addition of modified expanded graphite (EG) in HDPE, increase in TTI by 55% and increase in char yield from 8% in neat HDPE to 38% in modified EG/HDPE composites. Hydrothermal process modified EG increases formation of intumescent graphite layers and slow down pyrolysis of HDPE	[76]
PVC	20% EG	Reduction of PHRR and THR by 94% each and TTI increase by 14% upon comparison with neat PVC samples	[77]
Polyamide 6	15% GO	Reduction in PHRR and MLR by 58% and 59% respectively of PA6/GO in comparison with neat Polyamide 6 samples. GO expands voluminously resulting in soft char during burning and is aided by physical barrier mechanism	[78]
EVA	10% EG	LOI % increases to 28% from 18.5% upon comparison of EVA/20% EG composites with neat EVA. Significant reduction (80%) observed for volume resistivity in comparison with neat EVA	[79]
Polylactide (PLA)	9% EG	TTI and total burning time increases by 7.5 times and 3 times respectively. Also, the mean burning rate decreases by 42% for 9% EG filled PLA composites in comparison with neat PLA	[80]
PMMA	30% EG	PMMA/EG20/TEOS30 composite show increase in char yield at 800°C from 2% to 44%, IPDT by 3.5 times and LOI increase from 14% to 47% is exhibited in comparison with neat PMMA samples	[81]
PU	50 php	Increase in char yield at 500°C from 0.12% to 14% and increase in LOI from 22% to 30% in 50 php EG based PU foams in comparison with neat PU foams	[82]
	10%	Rigid polyurethane foam (RPUF)/ pEG@MF = 6:1 sample exhibit an increase in the LOI from 26% to 28% upon comparison to neat RPUF/pEG samples. Vertical burning tests indicate V-0 standard passing in 6:1 sample which is better than V-1 of neat RPUF/pEG	[83]

(Continued)

Table 25.3 (Continued)

Polymer	Graphite concentration	Change in FR properties	Reference
Epoxy	10%	Increase in LOI of 10% EG/Epoxy composite from 24% to 36% when compared to neat epoxy resins	[84]

LOI, Limiting oxygen index.

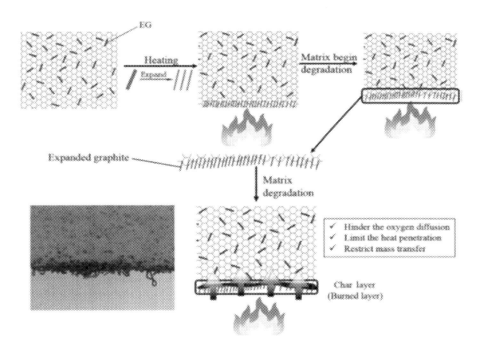

Figure 25.6 Flame retardance of water-blown semirigid polyurethane foams by the action of expanded graphite [85].
Source: Reproduced by permission of W. Luo, Y. Li, H. Zou, M. Liang, Study of different-sized sulfur-free expandable graphite on morphology and properties of water-blown semi-rigid polyurethane foams, RSC Adv. 4 (2014) 37302−37310. https://doi.org/10.1039/C4RA05559D. The Royal Society of Chemistry.

benzene ring, indene derivatives, and biphenyl derivatives turning it into carbon nanoparticles and ultimately generating graphite structure. The chemical cross linking of PP radicals can also be promoted due to Fe_3C presence in Fe-CNT's which are formed at high temperatures. Due to cross linking, the melt viscosity is very high and products formed while melting fill in the gaps in flaky

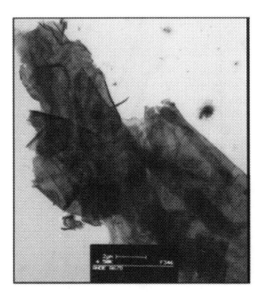

Figure 25.7 Transmission electron micrograph for char formed by exposure of flame to PLA-EG nanocomposite. *PLA-EG, Polylactide- expanded graphite* [86].
Source: M. Murariu, A.L. Dechief, L. Bonnaud, Y. Paint, A. Gallos, G. Fontaine, et al., The production and properties of polylactide composites filled with expanded graphite, Polym. Degrad. Stab. 95 (2010) 889–900. https://doi.org/10.1016/j.polymdegradstab.2009.12.019.

expanded graphite structure during combustion and connected them ensuring the elimination of the popcorn effect. A compact and protective hard char layer possessing Fe-CNT's significantly alter the diffusion of O_2 into PP and migration of decomposed volatiles out toward the flames. PDPFDE doesn't just self-catalyze charring but also works as reactive charring adhesive for EG to suppress further expansion of its layers which is a synergistic effect between PDPFDE and EG to better flame retardance of PP in the condensed phase [88]. A similar effort of creating a halogen-free flame retardant combination has yielded in synergistic effect with 15% each of hydroxyethyl cellulose/15% EG/ 70% PP nanocomposite increasing in LOI from 17% to 31.2% is observed with UL-94, V-0 rating of flame retardance [98].

For PUF, phenyl phosphonic aniline salt (bis(4-aminoanilinium) phenyl phosphonate) has been used with EG in a ratio of 1:12 to achieve LOI value increase of 29.8% from 19.2% with an UL-94 rating of V-0 in comparison with neat PUF. Also, PHRR, THR, and TSR reported a 45%, 22%, and 58% decrease in their values upon comparison with neat PU [99]. The volumetric expansion behavior of EG during combustion makes it a suitable candidate to be synergistically used with phosphorus-containing compounds leading to the formation of an intumescent system which can eventually form a compact intumescent char layer upon

Table 25.4 Flame retardant nature of findings reported in the recent literature for synergistic action of expanded graphite with various flame retardants.

Polymer	Synergist and its concentration	EG (%)	Change in FR properties	Reference
PP	IFR-15%	10	Reduction in PHRR, THR and mass loss by 77%, 63% and 75% respectively and increase in LOI value from 22.3% to 38.8%, TTI by 50% and UL-94 V-0 flame rating	[87]
	Ferrocene-based polymer (PDPFDE)—5%	20	Reduction in PHRR and THR by 83% and 41% respectively. LOI increases from 17% to 28.8% with UL-94 V-0 flame rating	[88]
PS	Aluminum hypophosphate—5%	10	Reduction in PHRR, THR, TSR and TSP by 78%, 39%, 78% and 82% respectively along with LOI value increases from 17% to 25.5% upon comparison with neat PS	[89]
Polyethylene Terephthalate (PET)	Clay-2.5%	2.5	Reduction in PHRR and TSR by 56% and 25% respectively and increase in mass residue by 26% and LOI from 21% to 29% upon comparison with neat PET sample	[90]
ABS	APP-3.75%	11.25	LOI value increases from 19% to 31% with a UL-94 V-0 rating	[91]
PBS	Mg(OH)$_2$-20%	5	Reduction in PHRR and THR by 77% and 41% respectively with TTI increase by 15% and LOI value increase from 21% to 29.4%	[92]
PLA	APP-3.75%	11.25	LOI value of 36.5% in comparison with 22% in neat PLA samples and UL 94 V-0 rating, TTI increased by 18%, residual mass % increased by 8 times and reduction in PHRR and TSR by 38% and 28.5% respectively upon comparison with neat PLA samples	[93]

(Continued)

Table 25.4 (Continued)

Polymer	Synergist and its concentration	EG (%)	Change in FR properties	Reference
EVA	APP-6.5%	23.5	Reduction in PHRR and THR by 63% and 32% respectively along with V-0 flame rating of UL-94 test without dripping upon comparison with neat EVA films	[94]
Polyurethane foam (PUF)	Hexa-phenoxy-cyclotriphosphazene-15%	10	Reduction in PHRR, THR and TSR by 71%, 63% and 45% respectively and increase in LOI and residual mass ratio by 43% and 1.5 times respectively compared to those of neat PU foams	[44]
	Dimethyl methylphosphonate-3.2%	12.8	Reduction in PHRR and THR by 65% and 58% respectively in comparison with neat PU foams	[95]
Epoxy	10-dihydro-9-oxa-10-phosphaphenanthrene-10-oxide (DOPO)-10% hexa-phenoxy-cyclotriphosphazene (HPCP)-10%	10	Increase in LOI from 34.5% to 41.5% in DOPO/EG/EP composites with respect to DOPO/EP samples and 32.5% to 39% in HPCP/EG/EP composites with respect to HPCP/EP samples. Also, both neat samples exhibit V-1 FR rating whereas EG addition gives a synergistic effect to show V-0 FR rating in UL-94	[96]
Unsaturated polyester	Organic Mg(OH)₂–1%	8	Reduction in PHRR, THR and pSPRcd by 67%, 32% and 82% respectively and increase in LOI and from 21.7% to 28.5% along with UL-94 V-0 pass rating when compared to those of neat unsaturated polyester samples	[97]

EG, expanded graphite; LOI, limiting oxygen index; TSR, total smoke release; PBS, polybutyl succinate; PLA, polylactide.

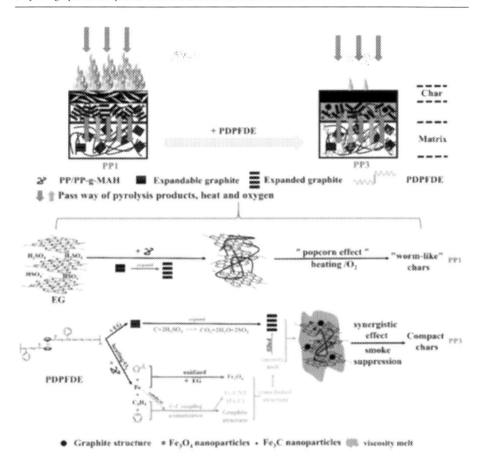

Figure 25.8 Mechanism for flame retardant action of PP systems filled with expanded graphite and PDPFDE [88].
Source: Z. Cheng, D. Liao, X. Hu, W. Li, C. Xie, H. Zhang, et al., Synergistic fire retardant effect between expandable graphite and ferrocene-based non-phosphorus polymer on polypropylene, Polym. Degrad. Stab. 178 (2020) 109201. https://doi.org/10.1016/j. polymdegradstab.2020.109201.

combustion. There are reports for the use of nano magnesium amino-tris-(methylene phosphonate) [100], nano zirconium amino-tris-(methylene phosphonate) [101], novel liquid phosphorus-containing polyol named as PDEO [102], core-shell methodology using aluminum trihydroxide encapsulation over EG [103], phosphorus and nitrogen-containing flame retardant 1,4-bis(Diethyl methylene phosphonate) piperazine [104] in recent literature for successfully improving the flame retardance characteristics of polyurethane polymer.

25.3 Conclusion

There has been considerable work done on the fabrication of polymer nanocomposites with the use of graphene in different methodologies. Graphene poses several advantages over conventional flame retardants in terms of improving fire retardancy of polymer nanocomposites and can be termed as "green flame retardants." At a lower percentage of graphene (up to 5%), there has been considerable improvement in the flame-retardant properties which are reported by reduction of PHRR, THR, and TSR. It also leads to an increase in TTI and LOI. The use of graphene in synergy with conventional flame retardants (nonhalogenated) containing phosphorus and nitrogen containing compounds incorporates flame retardance behavior along with providing UL-94 V-0 pass in vertical flame retardance tests which implies that the flame self-extinguishes upon its removal within 10 s without any dripping is observed. The use of inorganic nanomaterials in synergy with graphene is also found to be satisfactory in improving flame retardance of the polymer but further research needs to be done to study the dispersion of inorganic nanomaterial-graphene in a polymer matrix to ensure better mechanical and flame-retardant properties.

The mechanism of flame retardance of graphene is fairly simple: burning of the nanocomposite leads to the formation of a compact and intumescent char layer which continues, and it functions as a protective barrier to prevent diffusion of oxygen toward the polymer surface and the supply of decomposed volatiles toward the flame. There are compatibility issues of graphene with organic polymers and can be taken care by doing modification/functionalization by means of sol-gel, core-shell methodology, hybridization to name a few. This leads to ease and thereby an increase in dispersion and improved interfacial interaction to be a bridging mechanism. The low cost and huge availability of graphite have enabled its use in industrial production but with graphene, its high cost has still kept it in the research and development stage. Also, the current experimental methods use solvent dispersion technique for graphene which is not commercially feasible. However, with the inception of some processes using low-cost precursors can change this scenario in the coming future.

There has been an effort to have a considerable insight along with the provision of accumulation of knowledge on the roles of graphene family materials, its modified forms, and various flame retardants functioning along with graphene when exposed to the process of combustion which can be substantial if providing further pathways in the development of a newer generation of graphene-based polymer nanocomposites.

References

[1] S.-Y. Lu, I. Hamerton, Recent developments in the chemistry of halogen-free flame retardant polymers, Prog. Polym. Sci. 27 (2002) 1661–1712. Available from: https://doi.org/10.1016/S0079-6700(02)00018-7.

[2] S. Bourbigot, S. Duquesne, Fire retardant polymers: recent developments and opportunities, J. Mater. Chem. 17 (2007) 2283. Available from: https://doi.org/10.1039/b702511d.

[3] M. Lewin, S. Sello, Handbook of Fiber Science and Technology. Vol. II. Chemical Processing of Fibers and Fabrics. Functional Finishes: Part B, Marcel Dekker, New York, 1984.

[4] M. Lewin, E. Weil, Mechanisms and modes of action in flame retardancy of polymers, Fire Retardant Materials, Woodhead Publishing Ltd and CRC Press LLC, Cambridge, 2001, pp. 31−68.

[5] G. Camino, L. Costa, Performance and Mechanisms of Fire Retardants in Polymers: A Review, (n.d.) 24.

[6] J. Green, Mechanisms for flame retardancy and smoke suppression -a review, J. Fire Sci. 14 (1996) 426−442. Available from: https://doi.org/10.1177/073490419601400602.

[7] X. Wang, E.N. Kalali, J.-T. Wan, D.-Y. Wang, Carbon-family materials for flame retardant polymeric materials, Prog. Polym. Sci. 69 (2017) 22−46. Available from: https://doi.org/10.1016/j.progpolymsci.2017.02.001.

[8] B. Sang, Z. Li, X. Li, L. Yu, Z. Zhang, Graphene-based flame retardants: a review, J. Mater. Sci. 51 (2016) 8271−8295. Available from: https://doi.org/10.1007/s10853-016-0124-0.

[9] Y. Han, Y. Wu, M. Shen, X. Huang, J. Zhu, X. Zhang, Preparation and properties of polystyrene nanocomposites with graphite oxide and graphene as flame retardants, J. Mater. Sci. 48 (2013) 4214−4222. Available from: https://doi.org/10.1007/s10853-013-7234-8.

[10] C.I. Idumah, A. Hassan, S. Bourbigot, Influence of exfoliated graphene nanoplatelets on flame retardancy of kenaf flour polypropylene hybrid nanocomposites, J. Anal. Appl. Pyrolysis 123 (2017) 65−72. Available from: https://doi.org/10.1016/j.jaap.2017.01.006.

[11] B. Yuan, A. Fan, M. Yang, X. Chen, Y. Hu, C. Bao, et al., The effects of graphene on the flammability and fire behavior of intumescent flame retardant polypropylene composites at different flame scenarios, Polym. Degrad. Stab. 143 (2017) 42−56. Available from: https://doi.org/10.1016/j.polymdegradstab.2017.06.015.

[12] Z. Wang, Z. Huang, X. Li, J. Zhou, A nano graphene oxide/α-zirconium phosphate hybrid for rigid polyvinyl chloride foams with simultaneously improved mechanical strengths, smoke suppression, flame retardancy and thermal stability, Compos. A Appl. Sci. Manuf. 121 (2019) 180−188. Available from: https://doi.org/10.1016/j.compositesa.2019.03.021.

[13] T. Rahimi-Aghdam, Z. Shariatinia, M. Hakkarainen, V. Haddadi-Asl, Nitrogen and phosphorous doped graphene quantum dots: excellent flame retardants and smoke suppressants for polyacrylonitrile nanocomposites, J. Hazard. Mater. 381 (2020) 121013. Available from: https://doi.org/10.1016/j.jhazmat.2019.121013.

[14] T. Rahimi-Aghdam, Z. Shariatinia, M. Hakkarainen, V. Haddadi-Asl, Polyacrylonitrile/N,P co-doped graphene quantum dots-layered double hydroxide nanocomposite: flame retardant property, thermal stability and fire hazard, Eur. Polym. J. 120 (2019) 109256. Available from: https://doi.org/10.1016/j.eurpolymj.2019.109256.

[15] S. Ran, F. Fang, Z. Guo, P. Song, Y. Cai, Z. Fang, et al., Synthesis of decorated graphene with P, N-containing compounds and its flame retardancy and smoke suppression effects on polylactic acid, Compos. B Eng. 170 (2019) 41−50. Available from: https://doi.org/10.1016/j.compositesb.2019.04.037.

[16] M. Zhang, X. Ding, Y. Zhan, Y. Wang, X. Wang, Improving the flame retardancy of poly(lactic acid) using an efficient ternary hybrid flame retardant by dual modification of graphene oxide with phenylphosphinic acid and nano MOFs, J. Hazard. Mater. 384 (2020) 121260. Available from: https://doi.org/10.1016/j.jhazmat.2019.121260.

[17] O. Zabihi, M. Ahmadi, Q. Li, M.R.G. Ferdowsi, R. Mahmoodi, E.N. Kalali, et al., A sustainable approach to scalable production of a graphene based flame retardant using waste fish deoxyribonucleic acid, J. Clean. Prod. 247 (2020) 119150. Available from: https://doi.org/10.1016/j.jclepro.2019.119150.

[18] S. Araby, J. Li, G. Shi, Z. Ma, J. Ma, Graphene for flame-retarding elastomeric composite foams having strong interface, Compos. A Appl. Sci. Manuf. 101 (2017) 254–264. Available from: https://doi.org/10.1016/j.compositesa.2017.06.022.

[19] W. Du, Y. Jin, S. Lai, L. Shi, Y. Shen, H. Yang, Multifunctional light-responsive graphene-based polyurethane composites with shape memory, self-healing, and flame retardancy properties, Compos. A Appl. Sci. Manuf. 128 (2020) 105686. Available from: https://doi.org/10.1016/j.compositesa.2019.105686.

[20] L. Ye, X.-Y. Meng, X. Ji, Z.-M. Li, J.-H. Tang, Synthesis and characterization of expandable graphite–poly(methyl methacrylate) composite particles and their application to flame retardation of rigid polyurethane foams, Polym. Degrad. Stab. 94 (2009) 971–979. Available from: https://doi.org/10.1016/j.polymdegradstab.2009.03.016.

[21] N. Hong, L. Song, B. Wang, A.A. Stec, T.R. Hull, J. Zhan, et al., Co-precipitation synthesis of reduced graphene oxide/NiAl-layered double hydroxide hybrid and its application in flame retarding poly(methyl methacrylate), Mater. Res. Bull. 49 (2014) 657–664. Available from: https://doi.org/10.1016/j.materresbull.2013.09.051.

[22] D. Wang, Q. Zhang, K. Zhou, W. Yang, Y. Hu, X. Gong, The influence of manganese–cobalt oxide/graphene on reducing fire hazards of poly(butylene terephthalate), J. Hazard. Mater. 278 (2014) 391–400. Available from: https://doi.org/10.1016/j.jhazmat.2014.05.072.

[23] X. Wang, L. Song, H. Yang, W. Xing, H. Lu, Y. Hu, Cobalt oxide/graphene composite for highly efficient CO oxidation and its application in reducing the fire hazards of aliphatic polyesters, J. Mater. Chem. 22 (2012) 3426. Available from: https://doi.org/10.1039/c2jm15637g.

[24] N. Hong, L. Song, T.R. Hull, A.A. Stec, B. Wang, Y. Pan, et al., Facile preparation of graphene supported Co_3O_4 and NiO for reducing fire hazards of polyamide 6 composites, Mater. Chem. Phys. 142 (2013) 531–538. Available from: https://doi.org/10.1016/j.matchemphys.2013.07.048.

[25] Z. Zhang, L. Yuan, Q. Guan, G. Liang, A. Gu, Synergistically building flame retarding thermosetting composites with high toughness and thermal stability through unique phosphorus and silicone hybridized graphene oxide, Compos. A Appl. Sci. Manuf. 98 (2017) 174–183. Available from: https://doi.org/10.1016/j.compositesa.2017.03.025.

[26] C.L. Chiang, J.M. Yang, Flame retardance and thermal stability of polymer/graphene nanosheet oxide composites, Novel Fire Retardant Polymer Composite Materials, Elsevier, 2017, pp. 295–312. Available from: https://doi.org/10.1016/B978–0–08–100136-3.00011-X.

[27] F. Fang, P. Song, S. Ran, Z. Guo, H. Wang, Z. Fang, A facile way to prepare phosphorus-nitrogen-functionalized graphene oxide for enhancing the flame retardancy of epoxy resin, Compos. Commun. 10 (2018) 97–102. Available from: https://doi.org/10.1016/j.coco.2018.08.001.

[28] Y. Feng, C. He, Y. Wen, Y. Ye, X. Zhou, X. Xie, et al., Superior flame retardancy and smoke suppression of epoxy-based composites with phosphorus/nitrogen co-doped

graphene, J. Hazard. Mater. 346 (2018) 140−151. Available from: https://doi.org/ 10.1016/j.jhazmat.2017.12.019.

[29] F.-L. Guan, C.-X. Gui, H.-B. Zhang, Z.-G. Jiang, Y. Jiang, Z.-Z. Yu, Enhanced thermal conductivity and satisfactory flame retardancy of epoxy/alumina composites by combination with graphene nanoplatelets and magnesium hydroxide, Compos. B Eng. 98 (2016) 134−140. Available from: https://doi.org/10.1016/j.compositesb.2016.04.062.

[30] N.P. Singh, V.K. Gupta, A.P. Singh, Graphene and carbon nanotube reinforced epoxy nanocomposites: a review, Polymer. 180 (2019) 121724. Available from: https://doi. org/10.1016/j.polymer.2019.121724.

[31] F. Sun, T. Yu, C. Hu, Y. Li, Influence of functionalized graphene by grafted phosphorus containing flame retardant on the flammability of carbon fiber/epoxy resin (CF/ER) composite, Compos. Sci. Technol. 136 (2016) 76−84. Available from: https://doi.org/ 10.1016/j.compscitech.2016.10.002.

[32] Y. Xiao, Z. Jin, L. He, S. Ma, C. Wang, X. Mu, et al., Synthesis of a novel graphene conjugated covalent organic framework nanohybrid for enhancing the flame retardancy and mechanical properties of epoxy resins through synergistic effect, Compos. B Eng. 182 (2020) 107616. Available from: https://doi.org/10.1016/j. compositesb.2019.107616.

[33] W. Xu, B. Zhang, X. Wang, G. Wang, D. Ding, The flame retardancy and smoke suppression effect of a hybrid containing CuMoO$_4$ modified reduced graphene oxide/layered double hydroxide on epoxy resin, J. Hazard. Mater. 343 (2018) 364−375. Available from: https://doi.org/10.1016/j.jhazmat.2017.09.057.

[34] J. Zhang, Z. Li, L. Zhang, J. García Molleja, D.-Y. Wang, Bimetallic metal-organic frameworks and graphene oxide nano-hybrids for enhanced fire retardant epoxy composites: a novel carbonization mechanism, Carbon 153 (2019) 407−416. Available from: https://doi.org/10.1016/j.carbon.2019.07.003.

[35] L. Li, X. Liu, X. Shao, L. Jiang, K. Huang, S. Zhao, Synergistic effects of a highly effective intumescent flame retardant based on tannic acid functionalized graphene on the flame retardancy and smoke suppression properties of natural rubber, Compos. A Appl. Sci. Manuf. 129 (2020) 105715. Available from: https://doi.org/10.1016/j. compositesa.2019.105715.

[36] F. Kim, J. Luo, R. Cruz-Silva, L.J. Cote, K. Sohn, J. Huang, Self-propagating domino-like reactions in oxidized graphite, Adv. Funct. Mater. 20 (2010) 2867−2873. Available from: https://doi.org/10.1002/adfm.201000736.

[37] C. Zhang, W.W. Tjiu, W. Fan, Z. Yang, S. Huang, T. Liu, Aqueous stabilization of graphene sheets using exfoliated montmorillonite nanoplatelets for multifunctional free-standing hybrid films via vacuum-assisted self-assembly, J. Mater. Chem. 21 (2011) 18011. Available from: https://doi.org/10.1039/c1jm13236a.

[38] M.-J. Kim, I.-Y. Jeon, J.-M. Seo, L. Dai, J.-B. Baek, Graphene phosphonic acid as an efficient flame retardant, ACS Nano 8 (2014) 2820−2825. Available from: https://doi. org/10.1021/nn4066395.

[39] D. Hofmann, K.-A. Wartig, R. Thomann, B. Dittrich, B. Schartel, R. Mülhaupt, Functionalized graphene and carbon materials as additives for melt-extruded flame retardant polypropylene: functionalized graphene and carbon materials as additives . . ., Macromol. Mater. Eng. 298 (2013) 1322−1334. Available from: https://doi.org/ 10.1002/mame.201200433.

[40] J. Hu, F. Zhang, Self-assembled fabrication and flame-retardant properties of reduced graphene oxide/waterborne polyurethane nanocomposites, J. Therm. Anal. Calorim. 118 (2014) 1561−1568. Available from: https://doi.org/10.1007/s10973-014-4078-7.

[41] Y.R. Lee, S.C. Kim, H. Lee, H.M. Jeong, A.V. Raghu, K.R. Reddy, et al., Graphite oxides as effective fire retardants of epoxy resin, Macromol. Res. 19 (2011) 66−71. Available from: https://doi.org/10.1007/s13233-011-0106-7.

[42] G. Xu, T. Shi, Q. Wang, J. Liu, Y. Yi, A facile way to prepare two novel dopo-containing liquid benzoxazines and their graphene oxide composites, J. Appl. Polym. Sci. 132 (2015). Available from: https://doi.org/10.1002/app.41634. n/a-n/a.

[43] S. Liu, H. Yan, Z. Fang, H. Wang, Effect of graphene nanosheets on morphology, thermal stability and flame retardancy of epoxy resin, Compos. Sci. Technol. 90 (2014) 40−47. Available from: https://doi.org/10.1016/j.compscitech.2013.10.012.

[44] L. Qian, F. Feng, S. Tang, Bi-phase flame-retardant effect of hexa-phenoxy-cyclotriphosphazene on rigid polyurethane foams containing expandable graphite, Polymer 55 (2014) 95−101. Available from: https://doi.org/10.1016/j.polymer.2013.12.015.

[45] S.-H. Liao, P.-L. Liu, M.-C. Hsiao, C.-C. Teng, C.-A. Wang, M.-D. Ger, et al., One-step reduction and functionalization of graphene oxide with phosphorus-based compound to produce flame-retardant epoxy nanocomposite, Ind. Eng. Chem. Res. 51 (2012) 4573−4581. Available from: https://doi.org/10.1021/ie2026647.

[46] K.-Y. Li, C.-F. Kuan, H.-C. Kuan, C.-H. Chen, M.-Y. Shen, J.-M. Yang, et al., Preparation and properties of novel epoxy/graphene oxide nanosheets (GON) composites functionalized with flame retardant containing phosphorus and silicon, Mater. Chem. Phys. 146 (2014) 354−362. Available from: https://doi.org/10.1016/j.matchemphys.2014.03.037.

[47] W. Hu, J. Zhan, X. Wang, N. Hong, B. Wang, L. Song, et al., Effect of functionalized graphene oxide with hyper-branched flame retardant on flammability and thermal stability of cross-linked polyethylene, Ind. Eng. Chem. Res. 53 (2014) 3073−3083. Available from: https://doi.org/10.1021/ie4026743.

[48] X. Wang, W. Xing, X. Feng, B. Yu, L. Song, Y. Hu, Functionalization of graphene with grafted polyphosphamide for flame retardant epoxy composites: synthesis, flammability and mechanism, Polym. Chem. 5 (2014) 1145−1154. Available from: https://doi.org/10.1039/C3PY00963G.

[49] B. Yu, X. Wang, X. Qian, W. Xing, H. Yang, L. Ma, et al., Functionalized graphene oxide/phosphoramide oligomer hybrids flame retardant prepared via in situ polymerization for improving the fire safety of polypropylene, RSC Adv. 4 (2014) 31782. Available from: https://doi.org/10.1039/C3RA45945D.

[50] G. Huang, S. Chen, S. Tang, J. Gao, A novel intumescent flame retardant-functionalized graphene: nanocomposite synthesis, characterization, and flammability properties, Mater. Chem. Phys. 135 (2012) 938−947. Available from: https://doi.org/10.1016/j.matchemphys.2012.05.082.

[51] Y.-L. Li, C.-F. Kuan, C.-H. Chen, H.-C. Kuan, M.-C. Yip, S.-L. Chiu, et al., Preparation, thermal stability and electrical properties of PMMA/functionalized graphene oxide nanosheets composites, Mater. Chem. Phys. 134 (2012) 677−685. Available from: https://doi.org/10.1016/j.matchemphys.2012.03.050.

[52] S.-D. Jiang, Z.-M. Bai, G. Tang, Y. Hu, L. Song, Fabrication and characterization of graphene oxide-reinforced poly(vinyl alcohol)-based hybrid composites by the sol−gel method, Compos. Sci. Technol. 102 (2014) 51−58. Available from: https://doi.org/10.1016/j.compscitech.2014.06.029.

[53] X. Qian, B. Yu, C. Bao, L. Song, B. Wang, W. Xing, et al., Silicon nanoparticle decorated graphene composites: preparation and their reinforcement on the fire safety and mechanical properties of polyurea, J. Mater. Chem. A 1 (2013) 9827. Available from: https://doi.org/10.1039/c3ta11730h.

[54] X. Qian, L. Song, B. Yu, B. Wang, B. Yuan, Y. Shi, et al., Novel organic−inorganic flame retardants containing exfoliated graphene: preparation and their performance on the flame retardancy of epoxy resins, J. Mater. Chem. A 1 (2013) 6822. Available from: https://doi.org/10.1039/c3ta10416h.

[55] Z. Wang, P. Wei, Y. Qian, J. Liu, The synthesis of a novel graphene-based inorganic−organic hybrid flame retardant and its application in epoxy resin, Compos. B Eng. 60 (2014) 341−349. Available from: https://doi.org/10.1016/j.compositesb.2013.12.033.

[56] N. Hong, J. Zhan, X. Wang, A.A. Stec, T. Richard Hull, H. Ge, et al., Enhanced mechanical, thermal and flame retardant properties by combining graphene nanosheets and metal hydroxide nanorods for Acrylonitrile−Butadiene−Styrene copolymer composite, Compos. A Appl. Sci. Manuf. 64 (2014) 203−210. Available from: https://doi.org/10.1016/j.compositesa.2014.04.015.

[57] W. Xu, B. Zhang, X. Wang, G. Wang, D. Ding, The flame retardancy and smoke suppression effect of a hybrid containing $CuMoO_4$ modified reduced graphene oxide/layered double hydroxide on epoxy resin, J. Hazard. Mater. 343 (2018) 364−375. Available from: https://doi.org/10.1016/j.jhazmat.2017.09.057.

[58] X. Wang, W. Xing, X. Feng, B. Yu, H. Lu, L. Song, et al., The effect of metal oxide decorated graphene hybrids on the improved thermal stability and the reduced smoke toxicity in epoxy resins, Chem. Eng. J. 250 (2014) 214−221. Available from: https://doi.org/10.1016/j.cej.2014.01.106.

[59] D. Wang, K. Zhou, W. Yang, W. Xing, Y. Hu, X. Gong, Surface modification of graphene with layered molybdenum disulfide and their synergistic reinforcement on reducing fire hazards of epoxy resins, Ind. Eng. Chem. Res. 52 (2013) 17882−17890. Available from: https://doi.org/10.1021/ie402441g.

[60] X. Wang, S. Zhou, W. Xing, B. Yu, X. Feng, L. Song, et al., Self-assembly of Ni−Fe layered double hydroxide/graphene hybrids for reducing fire hazard in epoxy composites, J. Mater. Chem. A 1 (2013) 4383. Available from: https://doi.org/10.1039/c3ta00035d.

[61] S.-D. Jiang, Z.-M. Bai, G. Tang, Y. Hu, L. Song, Synthesis of ZnS decorated graphene sheets for reducing fire hazards of epoxy composites, Ind. Eng. Chem. Res. 53 (2014) 6708−6717. Available from: https://doi.org/10.1021/ie500023w.

[62] S.-D. Jiang, Z.-M. Bai, G. Tang, L. Song, A.A. Stec, T.R. Hull, et al., Fabrication of Ce-doped MnO_2 decorated graphene sheets for fire safety applications of epoxy composites: flame retardancy, smoke suppression and mechanism, J. MCater Chem. A 2 (2014) 17341−17351. Available from: https://doi.org/10.1039/C4TA02882A.

[63] Y. Dong, Z. Gui, Y. Hu, Y. Wu, S. Jiang, The influence of titanate nanotube on the improved thermal properties and the smoke suppression in poly(methyl methacrylate), J. Hazard. Mater. 209−210 (2012) 34−39. Available from: https://doi.org/10.1016/j.jhazmat.2011.12.048.

[64] Z. Han, Y. Wang, W. Dong, P. Wang, Enhanced fire retardancy of polyethylene/alumina trihydrate composites by graphene nanoplatelets, Mater. Lett. 128 (2014) 275−278. Available from: https://doi.org/10.1016/j.matlet.2014.04.148.

[65] P. Song, L. Liu, S. Fu, Y. Yu, C. Jin, Q. Wu, et al., Striking multiple synergies created by combining reduced graphene oxides and carbon nanotubes for polymer nanocomposites, Nanotechnology 24 (2013) 125704. Available from: https://doi.org/10.1088/0957-4484/24/12/125704.

[66] G. Huang, S. Wang, P. Song, C. Wu, S. Chen, X. Wang, Combination effect of carbon nanotubes with graphene on intumescent flame-retardant polypropylene nanocomposites, Compos. A Appl. Sci. Manuf. 59 (2014) 18−25. Available from: https://doi.org/10.1016/j.compositesa.2013.12.010.

[67] X. Wang, L. Song, H. Yang, H. Lu, Y. Hu, Synergistic effect of graphene on antidripping and fire resistance of intumescent flame retardant poly(butylene succinate) composites, Ind. Eng. Chem. Res. 50 (2011) 5376−5383. Available from: https://doi.org/10.1021/ie102566y.

[68] J.N. Gavgani, H. Adelnia, M.M. Gudarzi, Intumescent flame retardant polyurethane/reduced graphene oxide composites with improved mechanical, thermal, and barrier properties, J. Mater. Sci. 49 (2014) 243−254. Available from: https://doi.org/10.1007/s10853-013-7698-6.

[69] G. Huang, S. Chen, P. Song, P. Lu, C. Wu, H. Liang, Combination effects of graphene and layered double hydroxides on intumescent flame-retardant poly(methyl methacrylate) nanocomposites, Appl. Clay Sci. 88−89 (2014) 78−85. Available from: https://doi.org/10.1016/j.clay.2013.11.002.

[70] Y. Feng, G. Han, B. Wang, X. Zhou, J. Ma, Y. Ye, et al., Multiple synergistic effects of graphene-based hybrid and hexagonal born nitride in enhancing thermal conductivity and flame retardancy of epoxy, Chem. Eng. J. 379 (2020) 122402. Available from: https://doi.org/10.1016/j.cej.2019.122402.

[71] N. Wang, H. Liu, J. Zhang, M. Zhang, Q. Fang, D. Wang, Synergistic effect of graphene oxide and boron-nitrogen structure on flame retardancy of natural rubber/IFR composites, Arab. J. Chem. (2020). Available from: https://doi.org/10.1016/j.arabjc.2020.05.016.

[72] W. Hummers, R. Offeman, Preparation of graphitic oxide, J. Am. Chem. Soc. 80 (1958) 1339.

[73] F. Kang, T.-Y. Zhang, Y. Leng, Electrochemical behavior of graphite in electrolyte of sulfuric and acetic acid, Carbon 35 (1997) 1167−1173. Available from: https://doi.org/10.1016/S0008-6223(97)00097-3.

[74] F. Kang, T. Zhang, Y. Leng, Influences of H_2O_2 on synthesis of H_2SO_4-GICs, J. Phys. Chem. Solids 57 (1996) 889−892.

[75] F. Kang, T. Zhang, Y. Leng, Electrochemical synthesis of sulfate graphite intercalation compounds with different electrolyte concentrations, J. PhysChem Solids (1996) 883−888.

[76] K.-C. Tsai, H.-C. Kuan, H.-W. Chou, C.-F. Kuan, C.-H. Chen, C.-L. Chiang, Preparation of expandable graphite using a hydrothermal method and flame-retardant properties of its halogen-free flame-retardant HDPE composites, J. Polym. Res. 18 (2011) 483−488. Available from: https://doi.org/10.1007/s10965-010-9440-2.

[77] W.W. Focke, H. Muiambo, W. Mhike, H.J. Kruger, O. Ofosu, Flexible PVC flame retarded with expandable graphite, Polym. Degrad. Stab. 100 (2014) 63−69. Available from: https://doi.org/10.1016/j.polymdegradstab.2013.12.024.

[78] A. Dasari, Z.-Z. Yu, Y.-W. Mai, G. Cai, H. Song, Roles of graphite oxide, clay and POSS during the combustion of polyamide 6, Polymer 50 (2009) 1577−1587. Available from: https://doi.org/10.1016/j.polymer.2009.01.050.

[79] B. Wang, S. Hu, K. Zhao, H. Lu, L. Song, Y. Hu, Preparation of polyurethane microencapsulated expandable graphite, and its application in ethylene vinyl acetate copolymer containing silica-gel microencapsulated ammonium polyphosphate, Ind. Eng. Chem. Res. 50 (2011) 11476−11484. Available from: https://doi.org/10.1021/ie200886e.

[80] K. Fukushima, M. Murariu, G. Camino, P. Dubois, Effect of expanded graphite/layered-silicate clay on thermal, mechanical and fire retardant properties of poly(lactic acid), Polym. Degrad. Stab. 95 (2010) 1063−1076. Available from: https://doi.org/10.1016/j.polymdegradstab.2010.02.029.

[81] C.-F. Kuan, W.-H. Yen, C.-H. Chen, S.-M. Yuen, H.-C. Kuan, C.-L. Chiang, Synthesis, characterization, flame retardance and thermal properties of halogen-free expandable graphite/PMMA composites prepared from sol—gel method, Polym. Degrad. Stab. 93 (2008) 1357—1363. Available from: https://doi.org/10.1016/j.polymdegradstab.2008.03.030.

[82] M. Thirumal, D. Khastgir, N.K. Singha, B.S. Manjunath, Y.P. Naik, Effect of expandable graphite on the properties of intumescent flame-retardant polyurethane foam, J. Appl. Polym. Sci. 110 (2008) 2586—2594. Available from: https://doi.org/10.1002/app.28763.

[83] H.-J. Duan, H.-Q. Kang, W.-Q. Zhang, X. Ji, Z.-M. Li, J.-H. Tang, Core-shell structure design of pulverized expandable graphite particles and their application in flame-retardant rigid polyurethane foams: core-shell structure design of expandable graphite, Polym. Int. 63 (2014) 72—83. Available from: https://doi.org/10.1002/pi.4489.

[84] C.-L. Chiang, S.-W. Hsu, Novel epoxy/expandable graphite halogen-free flame retardant composites — preparation, characterization, and properties, J. Polym. Res. 17 (2010) 315—323. Available from: https://doi.org/10.1007/s10965-009-9318-3.

[85] W. Luo, Y. Li, H. Zou, M. Liang, Study of different-sized sulfur-free expandable graphite on morphology and properties of water-blown semi-rigid polyurethane foams, RSC Adv. 4 (2014) 37302—37310. Available from: https://doi.org/10.1039/C4RA05559D.

[86] M. Murariu, A.L. Dechief, L. Bonnaud, Y. Paint, A. Gallos, G. Fontaine, et al., The production and properties of polylactide composites filled with expanded graphite, Polym. Degrad. Stab. 95 (2010) 889—900. Available from: https://doi.org/10.1016/j.polymdegradstab.2009.12.019.

[87] G. Bai, C. Guo, L. Li, Synergistic effect of intumescent flame retardant and expandable graphite on mechanical and flame-retardant properties of wood flour-polypropylene composites, Constr. Build. Mater. 50 (2014) 148—153. Available from: https://doi.org/10.1016/j.conbuildmat.2013.09.028.

[88] Z. Cheng, D. Liao, X. Hu, W. Li, C. Xie, H. Zhang, et al., Synergistic fire retardant effect between expandable graphite and ferrocene-based non-phosphorus polymer on polypropylene, Polym. Degrad. Stab. 178 (2020) 109201. Available from: https://doi.org/10.1016/j.polymdegradstab.2020.109201.

[89] Z.-M. Zhu, W.-H. Rao, A.-H. Kang, W. Liao, Y.-Z. Wang, Highly effective flame retarded polystyrene by synergistic effects between expandable graphite and aluminum hypophosphite, Polym. Degrad. Stab. 154 (2018) 1—9. Available from: https://doi.org/10.1016/j.polymdegradstab.2018.05.015.

[90] J. Alongi, A. Frache, E. Gioffredi, Fire-retardant poly(ethylene terephthalate) by combination of expandable graphite and layered clays for plastics and textiles, Fire Mater. 35 (2011) 383—396. Available from: https://doi.org/10.1002/fam.1060.

[91] L.-L. Ge, H.-J. Duan, X.-G. Zhang, C. Chen, J.-H. Tang, Z.-M. Li, Synergistic effect of ammonium polyphosphate and expandable graphite on flame-retardant properties of acrylonitrile-butadiene-styrene, J. Appl. Polym. Sci. 126 (2012) 1337—1343. Available from: https://doi.org/10.1002/app.36997.

[92] H. Chen, T. Wang, Y. Wen, X. Wen, D. Gao, R. Yu, et al., Expanded graphite assistant construction of gradient-structured char layer in PBS/Mg(OH)$_2$ composites for improving flame retardancy, thermal stability and mechanical properties, Compos. B Eng. 177 (2019) 107402. Available from: https://doi.org/10.1016/j.compositesb.2019.107402.

[93] H. Zhu, Q. Zhu, J. Li, K. Tao, L. Xue, Q. Yan, Synergistic effect between expandable graphite and ammonium polyphosphate on flame retarded polylactide, Polym. Degrad. Stab. 96 (2011) 183—189. Available from: https://doi.org/10.1016/j.polymdegradstab.2010.11.017.

[94] G. Moradkhani, M. Fasihi, T. Parpaite, L. Brison, F. Laoutid, H. Vahabi, et al., Phosphorization of exfoliated graphite for developing flame retardant ethylene vinyl acetate composites, J. Mater. Res. Technol. 9 (2020) 7341−7353. Available from: https://doi.org/10.1016/j.jmrt.2020.04.085.

[95] F. Feng, L. Qian, The flame retardant behaviors and synergistic effect of expandable graphite and dimethyl methylphosphonate in rigid polyurethane foams, Polym. Compos. 35 (2014) 301−309. Available from: https://doi.org/10.1002/pc.22662.

[96] S. Yang, J. Wang, S. Huo, M. Wang, J. Wang, B. Zhang, Synergistic flame-retardant effect of expandable graphite and phosphorus-containing compounds for epoxy resin: strong bonding of different carbon residues, Polym. Degrad. Stab. 128 (2016) 89−98. Available from: https://doi.org/10.1016/j.polymdegradstab.2016.03.017.

[97] J. He, W. Zeng, M. Shi, X. Lv, H. Fan, Z. Lei, Influence of expandable graphite on flame retardancy and thermal stability property of unsaturated polyester resins/organic magnesium hydroxide composites, J. Appl. Polym. Sci. 137 (2020) 47881. Available from: https://doi.org/10.1002/app.47881.

[98] Z. Zheng, Y. Liu, B. Dai, C. Meng, Z. Guo, Fabrication of cellulose-based halogen-free flame retardant and its synergistic effect with expandable graphite in polypropylene, Carbohydr. Polym. 213 (2019) 257−265. Available from: https://doi.org/10.1016/j.carbpol.2019.02.088.

[99] P. Acuña, X. Lin, M.S. Calvo, Z. Shao, N. Pérez, F. Villafañe, et al., Synergistic effect of expandable graphite and phenylphosphonic-aniline salt on flame retardancy of rigid polyurethane foam, Polym. Degrad. Stab. 179 (2020) 109274. Available from: https://doi.org/10.1016/j.polymdegradstab.2020.109274.

[100] L. Liu, Z. Wang, Synergistic effect of nano magnesium amino-tris-(methylenephosphonate) and expandable graphite on improving flame retardant, mechanical and thermal insulating properties of rigid polyurethane foam, Mater. Chem. Phys. 219 (2018) 318−327. Available from: https://doi.org/10.1016/j.matchemphys.2018.08.010.

[101] L. Liu, Z. Wang, M. Zhu, Flame retardant, mechanical and thermal insulating properties of rigid polyurethane foam modified by nano zirconium amino-tris-(methylenephosphonate) and expandable graphite, Polym. Degrad. Stab. 170 (2019) 108997. Available from: https://doi.org/10.1016/j.polymdegradstab.2019.108997.

[102] W.-H. Rao, W. Liao, H. Wang, H.-B. Zhao, Y.-Z. Wang, Flame-retardant and smoke-suppressant flexible polyurethane foams based on reactive phosphorus-containing polyol and expandable graphite, J. Hazard. Mater. 360 (2018) 651−660. Available from: https://doi.org/10.1016/j.jhazmat.2018.08.053.

[103] Y. Wang, F. Wang, Q. Dong, M. Xie, P. Liu, Y. Ding, et al., Core-shell expandable graphite @ aluminum hydroxide as a flame-retardant for rigid polyurethane foams, Polym. Degrad. Stab. 146 (2017) 267−276. Available from: https://doi.org/10.1016/j.polymdegradstab.2017.10.017.

[104] Z. Zhang, D. Li, M. Xu, B. Li, Synthesis of a novel phosphorus and nitrogen-containing flame retardant and its application in rigid polyurethane foam with expandable graphite, Polym. Degrad. Stab. 173 (2020) 109077. Available from: https://doi.org/10.1016/j.polymdegradstab.2020.109077.

Index

Printed in the United States
by Baker & Taylor Publisher Services